21st Ethylene Producers Conference 2009

Topical Conference at the 2009 AIChE Spring National Meeting

Tampa, Florida
26 – 30 April 2009

Volume 1 of 2

ISBN: 978-1-61567-256-1

Printed from e-media with permission by:

Curran Associates, Inc.
57 Morehouse Lane
Red Hook, NY 12571

Printed by Curran Associates, Inc. (2009)

For permission requests, please contact AIChE
at the address below.

AIChE
3 Park Avenue
New York, NY 10016-5991

Phone: (203) 702-7660
Fax: (203) 775-5177

www.aiche.org

Additional copies of this publication are available from:

Curran Associates, Inc.
57 Morehouse Lane
Red Hook, NY 12571 USA
Phone: 845-758-0400
Fax: 845-758-2634
Email: curran@proceedings.com
Web: www.proceedings.com

TABLE OF CONTENTS

VOLUME 1

VOLUME 2

Ethylene Producers' Conference Introduction

The Ethylene Producers' Committee (EPC) of the AIChE is proud to present these Proceedings of the 21st Ethylene Producers' Conference, Volume 18.

The EPC was organized in 1988 by a group of six gentlemen who were passionate about the safety and effectiveness of ethylene-producing facilities in North America. These men organized the first Ethylene Producers' Conference in 1988, consisting of one session of a few papers at an AIChE meeting that year. From that humble beginning, the committee has grown to include representatives from ninety percent of all companies operating ethylene-producing facilities in North America and from all the major engineering firms associated with the design and construction of ethylene plants.

Besides the conference, another great advancement that has grown out of the EPC's work is the creation of www.ethyleneweb.com, a free and valuable resource for anyone involved with ethylene production or usage. Ethyleneweb.com contains many links to other resources including http://epc.omnibooksonline.com, where, for a membership fee, users can search an online database of the proceedings from all the past conferences since 1992. This website also contains a database of all ethylene cracking furnace-related papers presented at past conferences and a searchable database of short summaries of all ethylene plant safety incidents, their causes, and corrective actions presented at the conference since 1992. Information on subscribing to this service is available on the website.

The EPC would like to thank all the authors, presenters, session chairs, and co-chairs for their hard work planning, organizing, and conducting the sessions in this year's conference. We also thank all the companies and educational institutions which have recognized the value the conference provides to their employees and students.
Their financial support has enabled the conference to grow year-after-year for twenty-one years.

We welcome you to the 21st Ethylene Producers' Conference. We hope you find this Conference is valuable in your efforts to improve your ethylene production business.

2009 Ethylene Producers' Committee Members

Doug Badon	Williams Olefins L.L.C.
Brian Bahr	Ineos Olefins & Polymers USA
Colin Bowen	Shaw Stone and Webster Inc.
Jon Bowen	Dow Chemical
Ricardo Callejas	Independent Contractor
Toby Darbonne	Westlake Petrochemicals, LP
David Gent	Nova Chemicals LTD
Jim Hackney	LyondellBasell
Sanjeev Kapur	Fluor Corporation
Michael Kennedy, Treasurer	LyondellBasell
Aivars Krumins	CB&I
Ravi Lal	Technip USA, Inc.
David Mullenix, Chairman	Sasol North America, Inc.
Bill O'Brien	Chevron Phillips Chemical Company LP
Rick Prickett	Chevron Phillips Chemical Company LP
John Robinson	BASF Fina Petrochemicals
Richard Rodriguez	Shell Chemical Co.
Rashmi Shah	CB&I
Bob Simon	Huntsman
Stephen Stinnette	Flint Hills Resources
Bob Strack	ExxonMobil Chemical
Ray Stroud, Vice Chairman	Eastman Chemical
Michael Tallman	KBR
Jack Tindel	E.I. DuPont de Nemours & Co.
Paul Van Helmond	Selas Fluid Processing Corporation
Greg Yeo	ExxonMobil Chemical

2009 Ethylene Producers' Conference Session Chairs and Co-Chairs

Jon Bowen, Conference Chairman
Greg Yeo, Conference Co-Chairman

David Mullenix	Ethylene Producers' Conference Keynote Chair
Ray Stroud	Ethylene Producers' Conference Keynote Co-Chair
Marilyn Blaschke	Ethylene Plant Contaminants: Myths and War Stories Chair
Michael Kennedy	Ethylene Plant Environmental Issues Chair
Doug Badon	Ethylene Plant Environmental Issues Chair Co-Chair
John Ellis	Ethylene Plant Maintenance and Reliability Chair
Vance Grieshop	Ethylene Plant Operations Chair
Bill O'Brien	Ethylene Plant Operations Co-Chair
Vance Ham	Ethylene Plant Operations - Acetylene Hydrogenation Tutorial Chair
Dwain Raduenz	Ethylene Plant Operations - Acetylene Hydrogenation Tutorial Co-Chair
Jim Hackney	Ethylene Plant Process Control Chair
Satish Baliga	Ethylene Plant Process Control Co-Chair
Kevin Kisor	Ethylene Plant Rotating Equipment Chair
Robert Parrott	Ethylene Plant Rotating Equipment Co-Chair
Kevin Kisor	Ethylene Plant Rotating Equipment Roundtable Chair
Robert Parrott	Ethylene Plant Rotating Equipment Roundtable Co-Chair
Jack Tindel	Ethylene Plant Safety Chair
Richard Rodriguez	Ethylene Plant Safety Co-Chair
David Brown	Ethylene Plant Technology - Fundamentals and Innovation Chair
Jack Buehler	Ethylene Plant Technology - Fundamentals and Innovation Co-Chair
Michael Tallman	Ethylene Plant Technology – Energy Cons. and Opt. Chair
Greg Dunnells	Ethylene Plant Technology – Energy Cons. and Opt. Co-Chair
Sanjeev Kapur	Ethylene Plant Technology – Trends in Olefin Feedstocks Chair
Aivars Krumins	Ethylene Plant Technology – Trends in Olefin Feedstocks Co-Chair

pH MYTHs: Simple to Measure...Easy to Control

Paper 13a

Susan Brunner – GE Water & Process Technologies
Mark Brayden – Dow
Dwight Hines - BASF

presented at the
American Institute of Chemical Engineers
21st Ethylene Producers Conference
Tampa, Florida
April 27, 2009

imagination at work

Presentation Outline

- Refresher on pH
- Factors impacting pH measurement accuracy
- Temperature effects on pH
- Practical applications

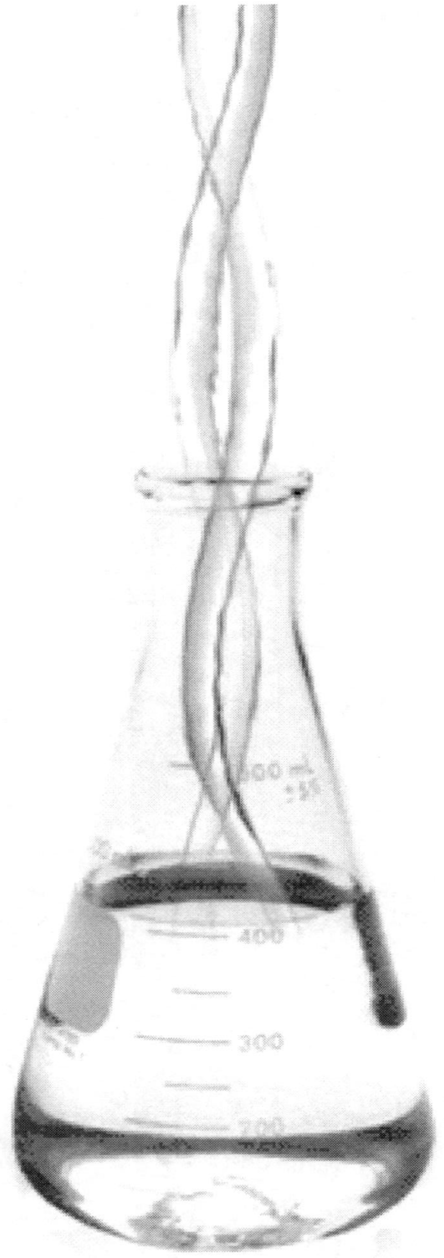

GE imagination at work

pH Refresher

$pH = -log(H+)$

- Measure of ion activity related to the concentration of hydrogen ions in a solution
- pH Measurement System
  - Measuring electrode
  - Reference electrode
  - pH meter
  - Temperature sensor

GE imagination at work

Response of pH Probe

Nernst Equation:

$E = E^0 + 2.3(RT/F)\log aH^+$

E = mV response of the pH sensor

E^0 = a calibration constant

2.3 (RT/F) = slope in mV/pH units

R = gas constant

T = absolute temperature

F = Faraday constant

aH^+ = activity of the hydrogen ion

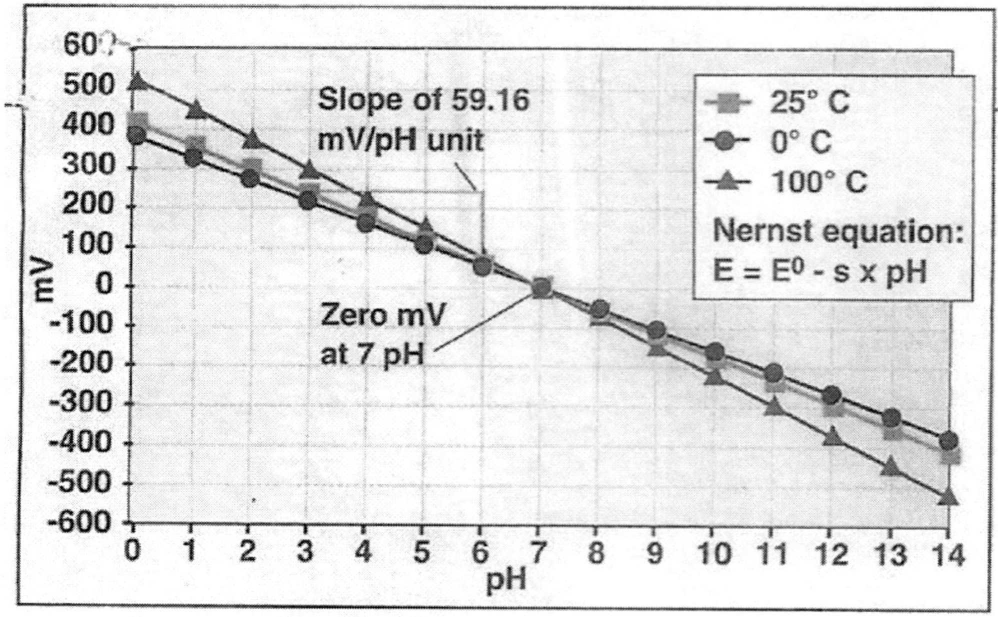

(Reference 2)

GE imagination at work

pH Probe Calibration using Buffers

- Calibrate using two buffers that bracket expected sample pH
- Calibrate with buffer that is at the same temperature as the sample
- Do not use expired buffers
- Discard buffer after use
- Rinse pH probe with DI water
- Do not wipe probe
- Store the pH electrode in pH 4 buffer

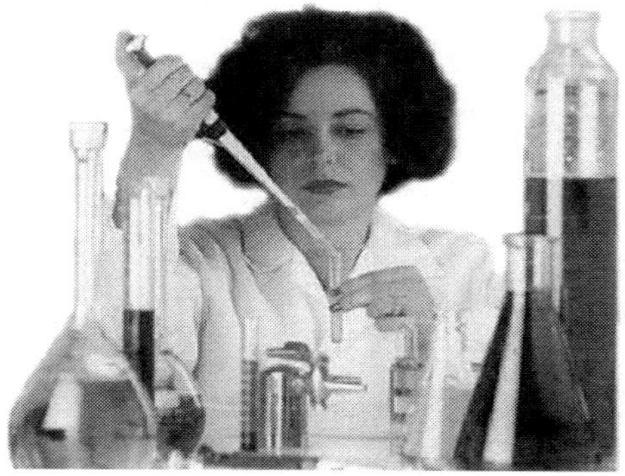

(Reference 3)

GE imagination at work

Factors Impacting pH Measurement Accuracy

- pH probe not temperature compensated
- pH probe is not properly calibrated
- Samples tested at different temperatures
- Low ionic strength water
- High purity sample poured into open cup
- Stirring sample for extended period of time
- Storing pH probe in DI water
- Lab discipline for offline analysis
- Hydrocarbon contamination
- Contamination of sample from pH probe electrolyte

GE imagination at work

Temperature Effects

- Higher temperatures increases dissociation of hydrogen and hydroxide ions
- A neutral pH varies from 7.45 at 0°C to 6.2 at 100°C

(Reference 2)

Table 1
Comparison of pH of the Same Sample at Different Sample Temperatures

Day	pH at 46 C	pH at 29 C
1	7.9	8.2
2	8	8.2
3	7.9	8.3
4	8.1	8.5
5	8.4	8.7
6	8.5	8.8
7	8.5	8.9
8	8.4	8.9
9	9	9.4
10	9	9.5

Ref: GE Technical Paper
Improved Water Treatment for High Pressure Boilers

Temperature Effect On pH
(Ammonia in otherwise pure water)

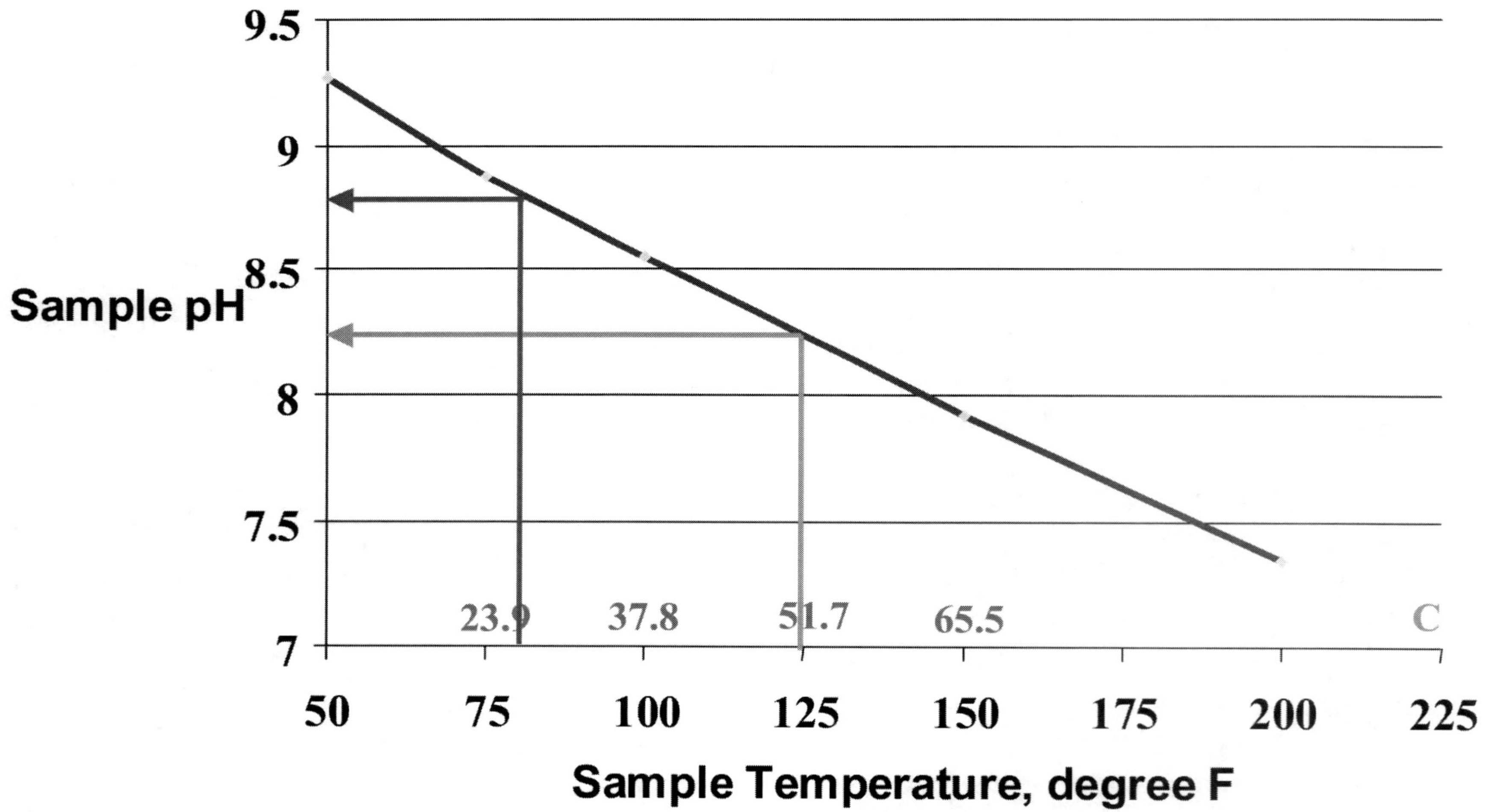

GE imagination at work

Practical Application in Ethylene Plants

Control pH to mitigate process corrosion and fouling

GE imagination at work

TLE and Boiler Systems
pH Measurement in High Purity Samples

- Susceptible to contamination during sampling and testing
- Low buffering capacity/low conductivity
- Proper sampling
  - Cool sample to <90 F
  - Stainless steel sample stations
  - Thoroughly purge sample station
  - Control sampling rate
  - Seal sample container
  - Minimize lapse time to sample analysis
  - Addition of ionic strength adjuster
  - Use pH meter dedicated to high purity samples

GE imagination at work

pH Testing Accuracy Implication

- A 0.5 pH point measurement error can impact program control
- Example:
  - > To increase pH from 8.8 to 9.2 in BFW/condensate requires 3 times increase in amine product feed rate
  - > Congruent pH/PO_4 – in or out of the box
    - – Resulted taking the wrong corrective action

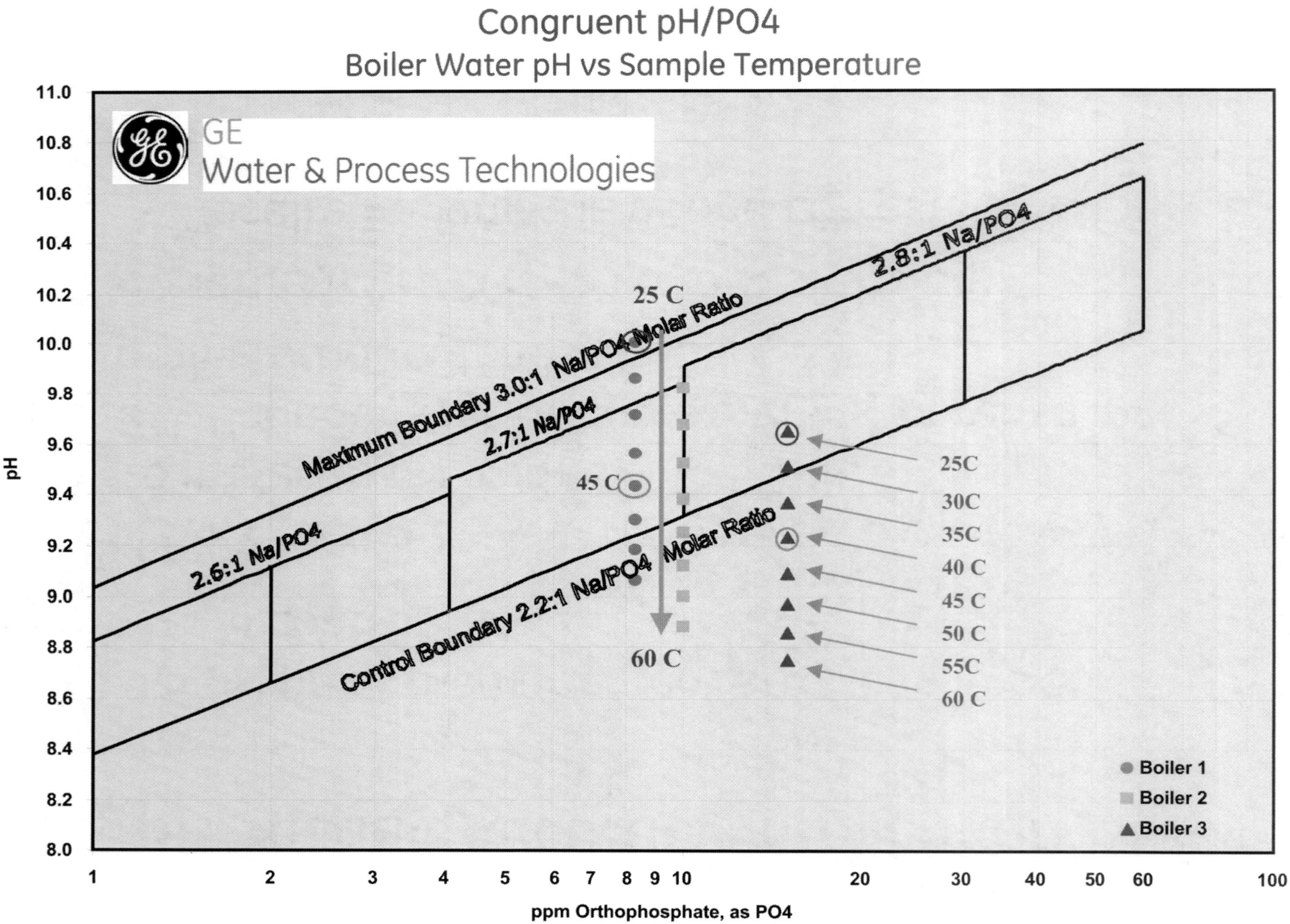
Congruent pH/PO4
Boiler Water pH vs Sample Temperature
GE
Water & Process Technologies
Maximum Boundary 3.0:1 Na/PO4 Molar Ratio
Control Boundary 2.2:1 Na/PO4 Molar Ratio
2.6:1 Na/PO4
2.7:1 Na/PO4
2.8:1 Na/PO4
25 C
45 C
60 C
25C
30C
35C
40 C
45 C
50 C
55C
60 C
Boiler 1
Boiler 2
Boiler 3
pH
11.0
10.8
10.6
10.4
10.2
10.0
9.8
9.6
9.4
9.2
9.0
8.8
8.6
8.4
8.2
8.0
1
2
3
4
5
6
7
8
9
10
20
30
40
50
60
100
ppm Orthophosphate, as PO4

Quench Cycle Water

- pH control is necessary to prevent corrosion
  - Indicated by iron levels (e.g., < 1 ppm)
  - Minimize emulsion tendencies
    - Some emulsion breakers work best in a "neutral pH range"
- Typical control range: 6-7

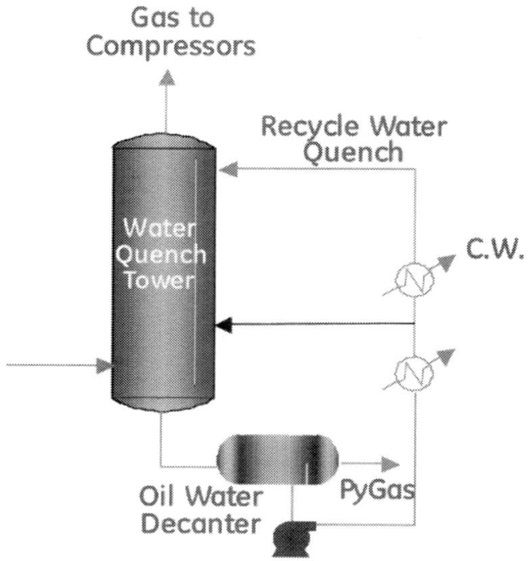

GE imagination at work

Quench water cycle
pH measurement

- Obtain a representative sample
- Off-line (laboratory)
  - Infrequent sampling frequency
  - Improper methods
- On-line
  - Reliability is an issue
    - Improper installation design
    - Hydrocarbon fouling
      - » Frequent maintenance required
    - Proposed solution
      - » Self-cleaning probe (Mettler-Toledo)

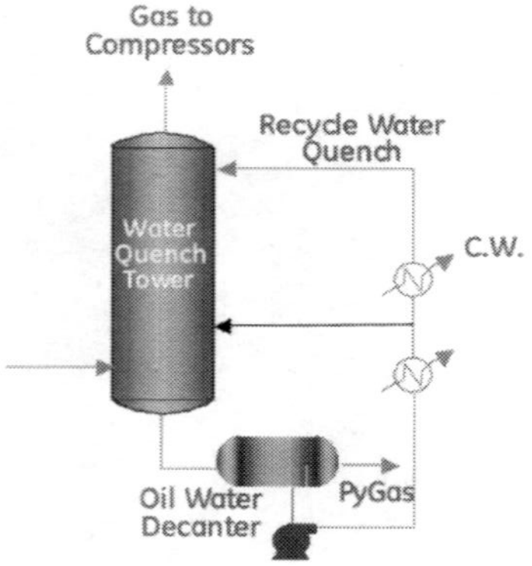

GE imagination at work

Process Water Stripper and Dilution Steam Generators

- Higher temperature = higher corrosion potential
  - Higher pH than quench cycle (e.g., 7-9)
- DSG system similar to low pressure boiler applications
  - pH control can be similar (8-10), but
  - Oligomers/polymers can form in steam stripper and foul pH probes.
  - pH measurement typically done off-line

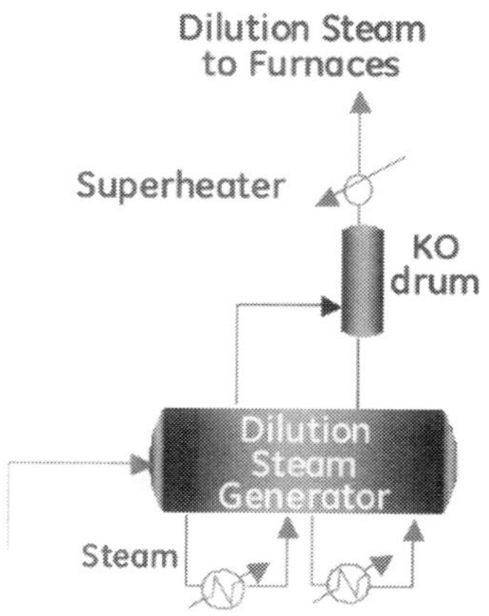

GE imagination at work

References

1. Istre, Richard, Improved Water Treatment for High Pressure Boilers, Betz Laboratories, AIChE Ammonia Producers Conference, November, 1991
2. Joseph, David, What's the Real pH of that Stream?, Chemical Engineering, March, 2005, pp 60-62
3. http://www.emersonprocess.com/raihome/documents/Liq_Handbook_44-6033_Theory_And_Practice_Of_pH_Measurement_199909.pdf

GE imagination at work

13b Polythionic Acid Stress Corrosion Cracking

Bala Devakottai
LyondellBasell Industries
Channelview, TX 77530
bala.devakottai@lyondellbasell.com

Pinakin Mistry
LyondellBasell Industries
Channelview, TX 77530
pinakin.mistry@lyondellbasell.com

Abstract:

The most common fluid that causes Inter Granular Stress Corrosion Cracking (IGSCC) in petrochemical plants and refineries is polythionic acid. Polythionic acid is formed in the presence of sulfur, moisture and oxygen. Sulfur can come from feed stock, additives or fuels. This failure mechanism is possible from both inside and outside of a tube, depending on the sulfur source. IGSCC is most common in sensitized 300 series stainless steel and higher nickel base austenitic alloys.

This paper will describe the details of polythionic acid formation, case histories of failures and its mitigation.

13c Polythionic Acid Stress Cracking in Furnace Tubes

Tom Kopecky
Flint Hill Resources
Port Arthur, TX 77640
tom.kopecky@fhr.com

Jim McCarter
Flint Hill Resources
Port Arthur, TX 77640
jim.mccarter@fhr.com

Bruce Scott
Flint Hill Resources
Port Arthur, TX 77640
bruce.scott@fhr.com

Abstract:

This paper will provide details of an olefin unit's actual experience with Polythionic Acid Stress Corrosion Cracking (PASCC) in a furnace convection section. Investigation into the suspected cause of the tube failures along with the supporting metallurgical data will be presented.

Naphtha Contamination and Furnace Coil Plugging

AIChE 2009 Spring National Meeting
Ethylene Producers Conference
Session T4A13, Paper No. 13d

John Robinson

BASF The Chemical Company

TOTAL

War Story
Naphtha Contamination and Furnace Coil Plugging

- BASF Fina Petrochemicals Overview
- Naphtha Feed System Overview
- Coil Plugging History
- Naphtha Contamination
- Plugging Mitigation Efforts and Results
- Future Issues and Concerns

BASF The Chemical Company

TOTAL

2009 EPC Conference – Tampa, FL

John Robinson

BASF FINA Petrochemicals

BASF Fina Petrochemicals Overview

- Joint Venture Between BASF Corporation and TOTAL PETROCHEMICALS USA, INC.
- Started Up in 4 QTR 2001 – Lummus Design
- 2 Billion Pounds per Year Ethylene Capacity
- Naphtha Base Cracker – Some LPGs
- Located in Pt. Arthur, TX
- Next to TOTAL's Refinery and Sabina JV
- 1 S&W, 1 Lummus SRT III, and 7 Lummus SRT VI Furnaces

BASF Fina
Petrochemicals

Naphtha Feed System Overview

- **Naphtha Feed System – Original Design**

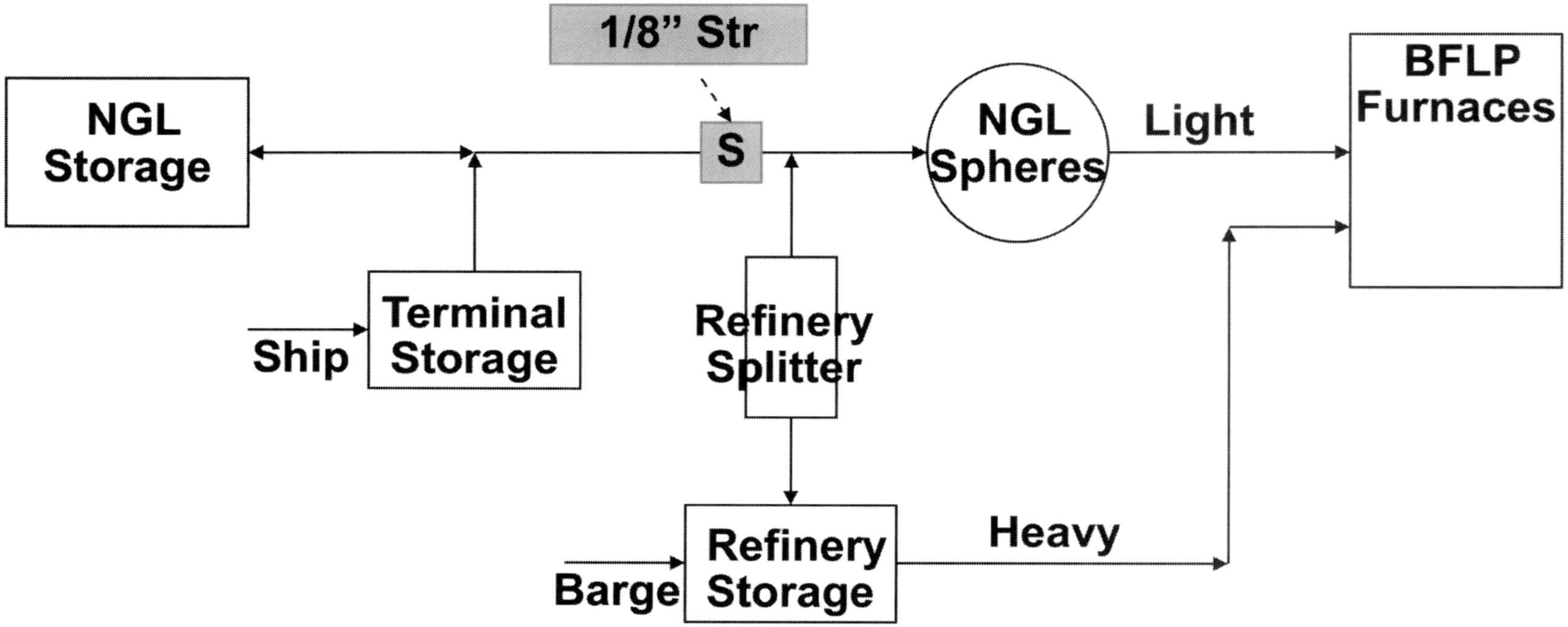

Coil Plugging History

- 2002-2005 – 93 Plugged Coils – Averaged 2 per Month
- 83 Inlets and 10 Outlets – Entire Furnace Worth
- Filter Installed on Original Project Were By-Passed
- Furnace Average Run Length Decreased to 62% of Design
- A Lot of Lost Production and Maintenance Expense

Coil Plugging History

- Began Work with both Lummus and Pipeline Company

Needed a Solution!

Naphtha Contamination

Naphtha Sample Pictures

R	Graphic Images for Furnace Operators

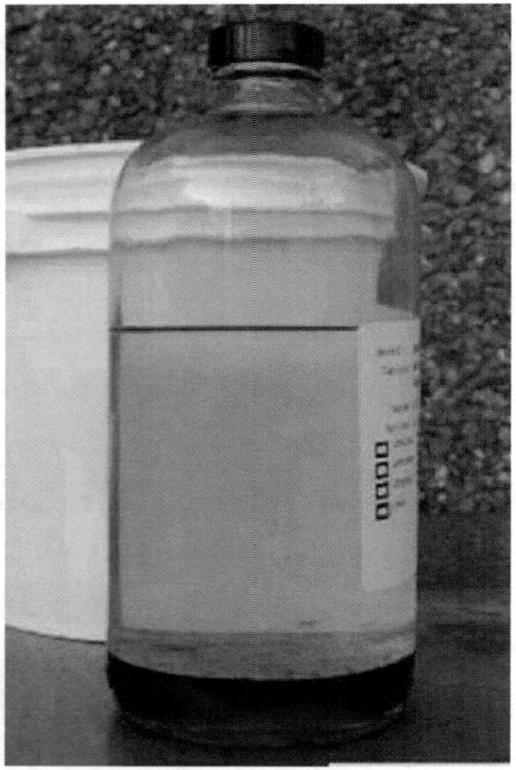

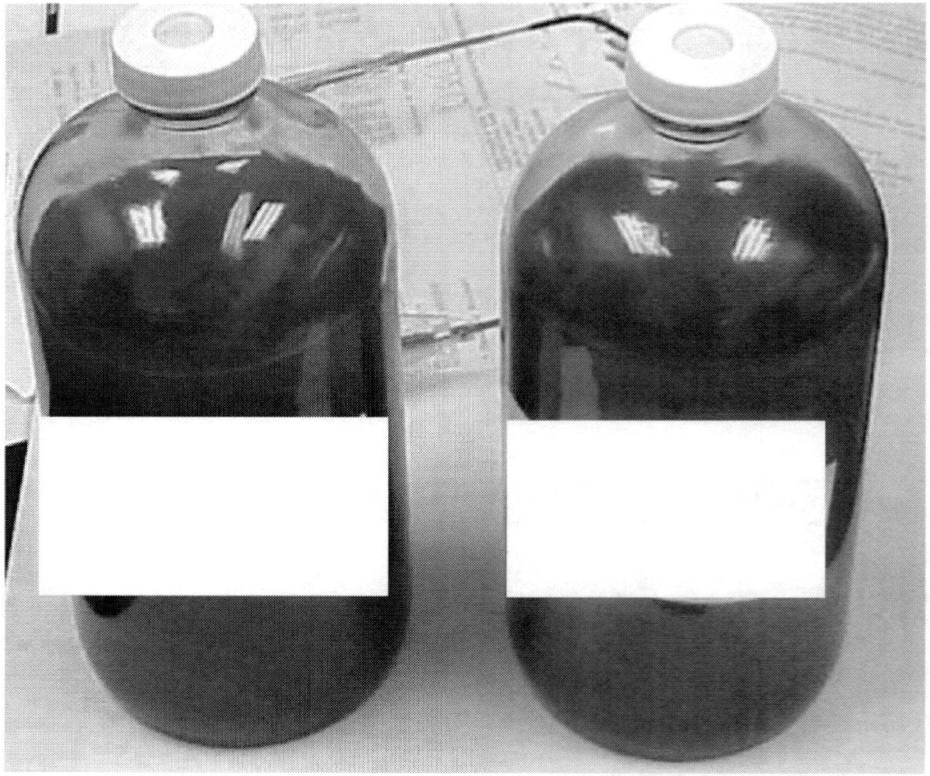

Naphtha Contamination

- Water and Salt – Brine and/or Seawater
- A Lot of Iron/Rust – Pipeline and/or Brine Corrosion
- Debris – Fractionator Bottoms and/or Brine Rag Layer
- Accelerated Coking Rates – Na and Fe
- Affect Last Heater on Naphtha Header First
- Accumulated In Outer Radiant Inlet Coils in Bends
- Fouling of Radiant Coil Header Strainers

Naphtha Contamination

- Learned How to Operate with Plugged Coils

Naphtha Contamination

- Incidents Increase When Rate is Increased
- Plant Rate Was Constrained by Custody Meter Strainer

Naphtha Contamination

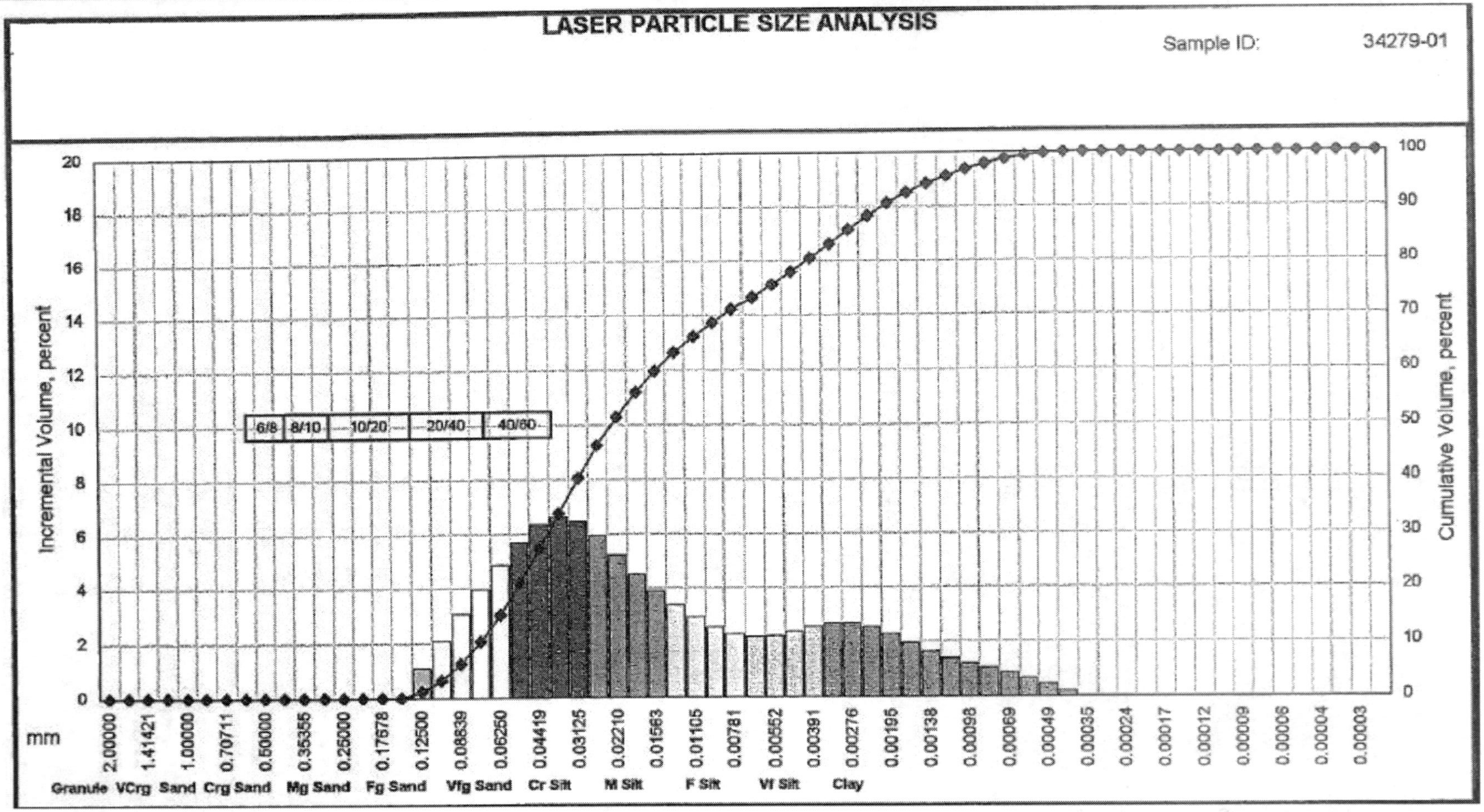

2009 EPC Conference – Tampa, FL

John Robinson

BASF FINA Petrochemicals

Naphtha Contamination

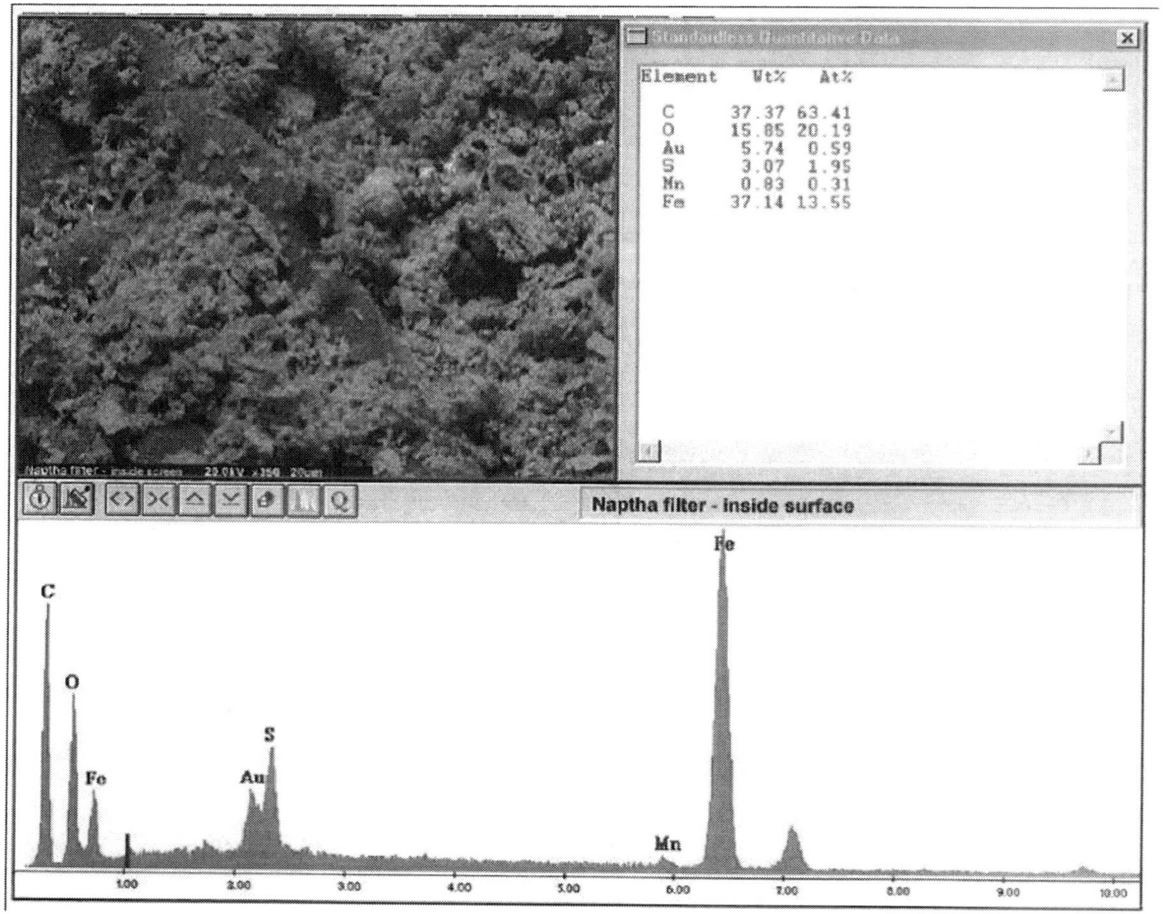

Plugging Mitigation Efforts and Results

- Initially Installed Temporary Filter on Last Furnace

Plugging Mitigation Efforts and Results

- Installed Permanent Filters and Coalescers in Plant
- Cleaned Pipeline During 2006 and 2007 Outages
- Installed 5 Filters Throughout Pipeline System

2009 EPC Conference – Tampa, FL
John Robinson

BASF FINA Petrochemicals

Plugging Mitigation Efforts and Results

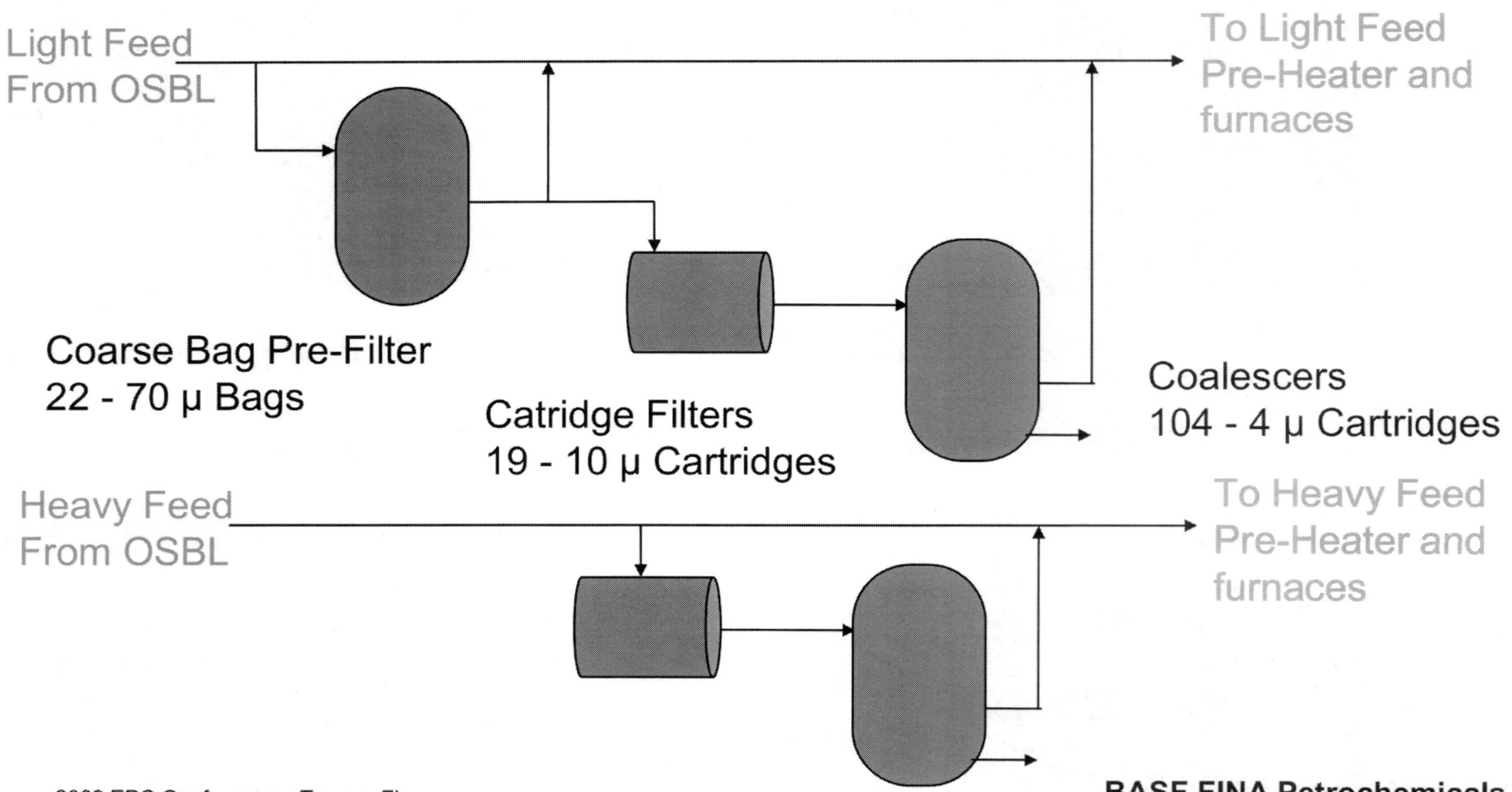

2009 EPC Conference – Tampa, FL

John Robinson

BASF FINA Petrochemicals

Naphtha Feed System Overview

- **Pipeline Filtering**

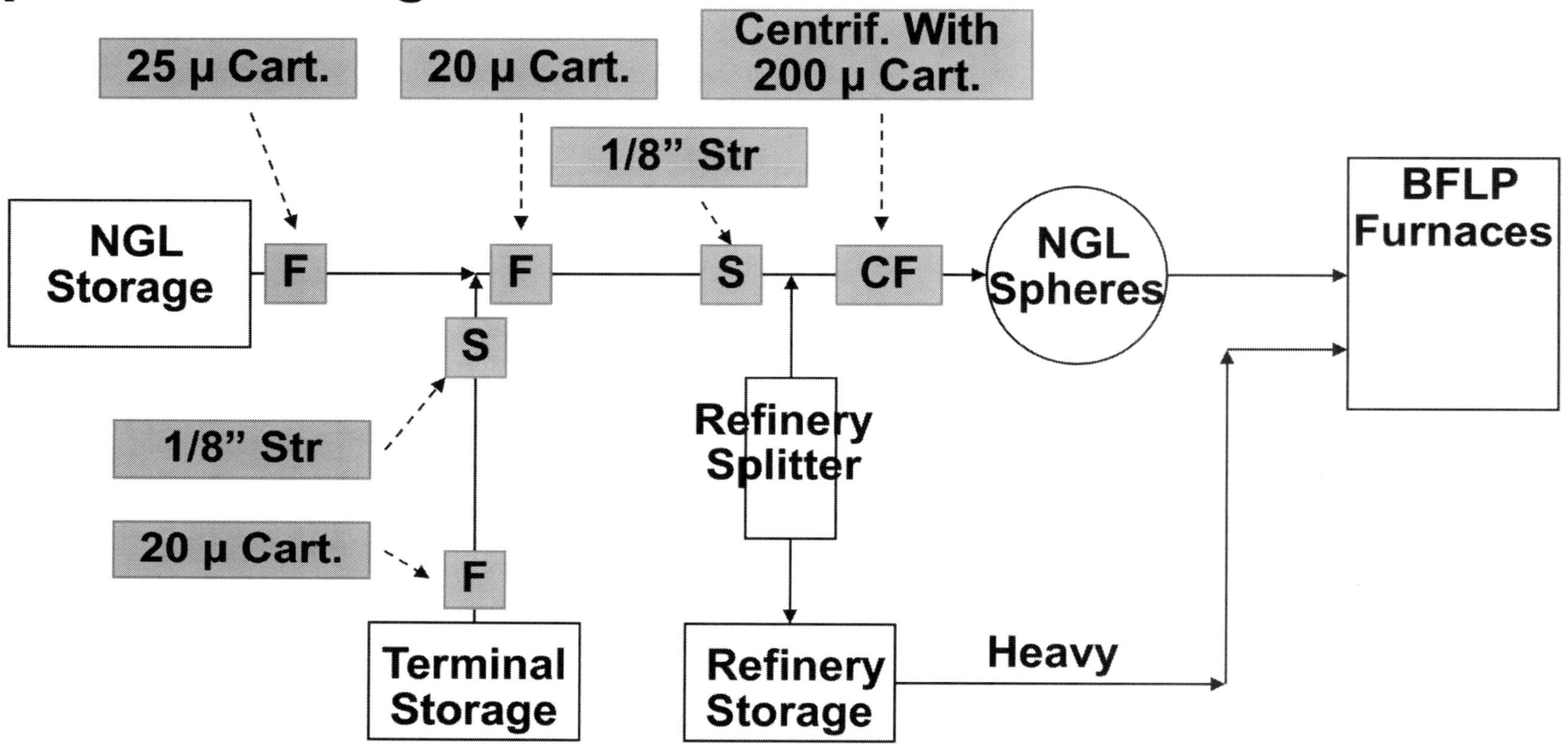

2009 EPC Conference – Tampa, FL

John Robinson

BASF FINA Petrochemicals

Plugging Mitigation Efforts and Results

- With Filters On-Line – Plugged Coils Eliminated
- Change Elements About Every 5 Days
- Difficult to Keep Pre-Sphere Filter On-Line

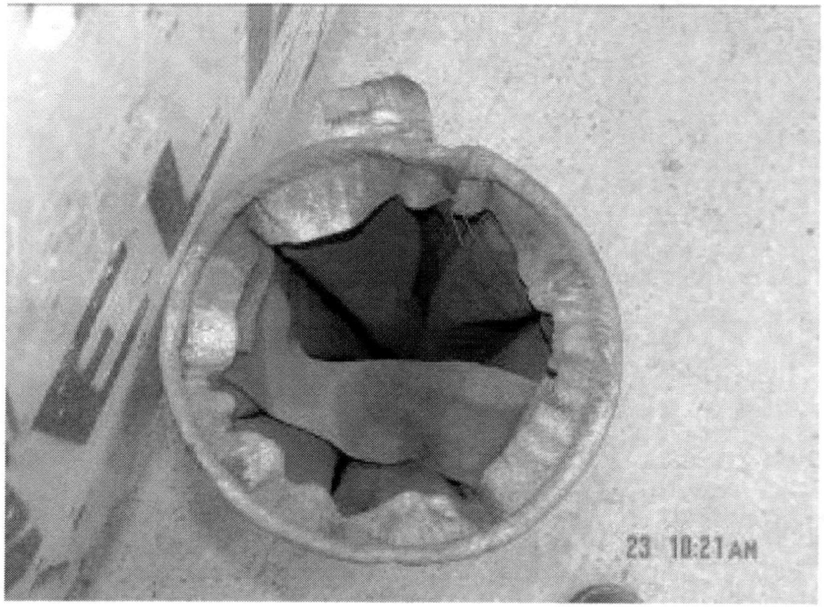

Success!

SRT VI 12 Month Rolling Average Run Length

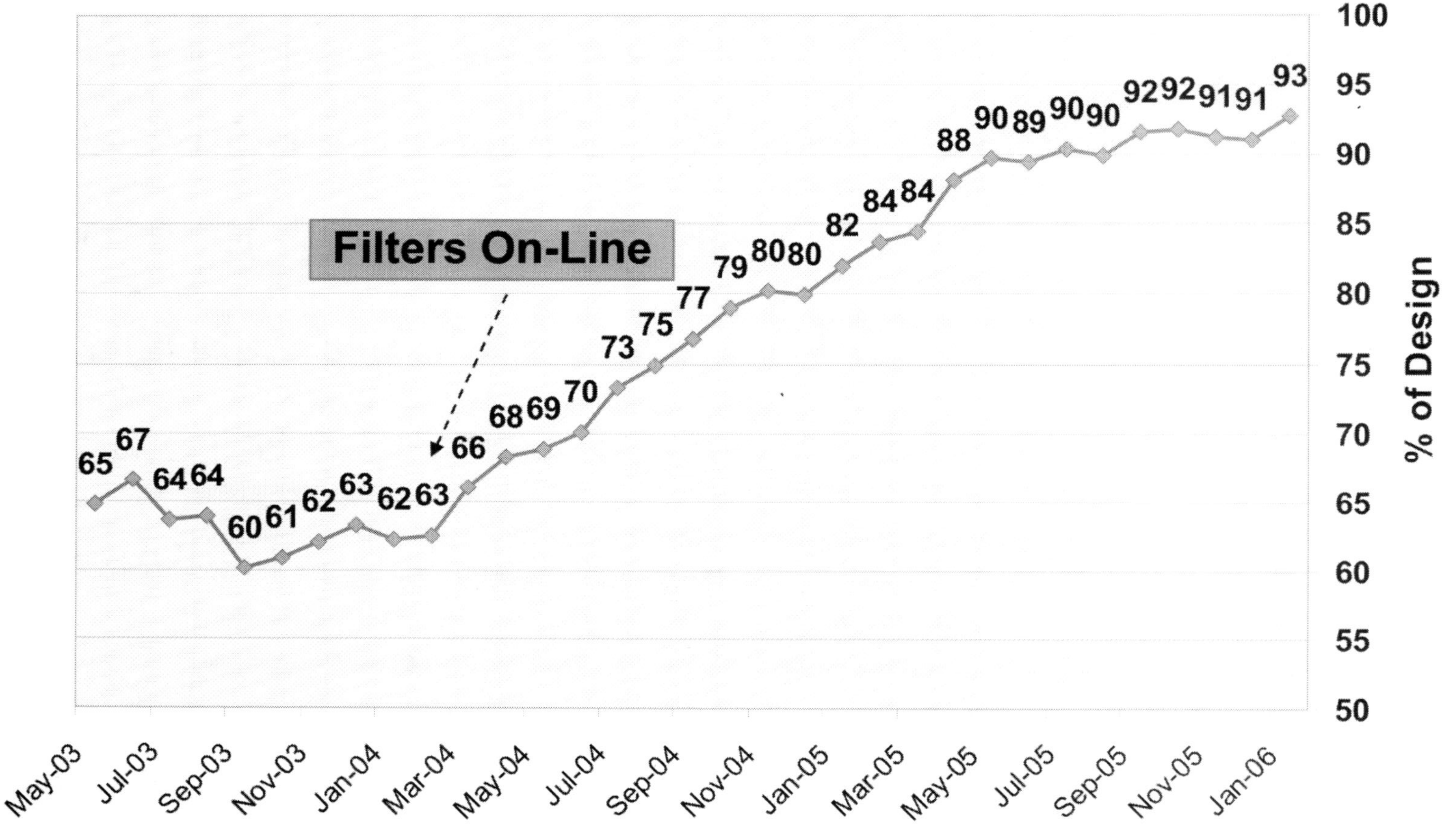

2009 EPC Conference – Tampa, FL

John Robinson

BASF FINA Petrochemicals

Future Issues and Concerns

- Pyrophoric Iron Sulfide – H2S and Rust In Naphtha
  - Handling of Pigs and Filter Elements
- Silica Contamination of Light Feed – Catalyst Deactivation
- Disposal of Coalescer Water
  - Deactivation of Pygas 1st Stage in Heavy Gasoline
- Filters Still By-Passed Occasionally
- Optimization of Filter Sizes

Future Issues and Concerns

- Coalescer Water Sample Results

Chloride	127.8	ppm (mg / L)
Nitrite	794.4	ppm (mg / L)
Bromide	270.6	ppm (mg / L)
Silicon	25	ppm (mg / kg)
Sulfur	1170	ppm (µg / g)
Iron	64.0	ppm (mg / kg)
Sodium	104.2	ppm (mg / kg)
Nitrogen	1456	ppm (mg / kg)

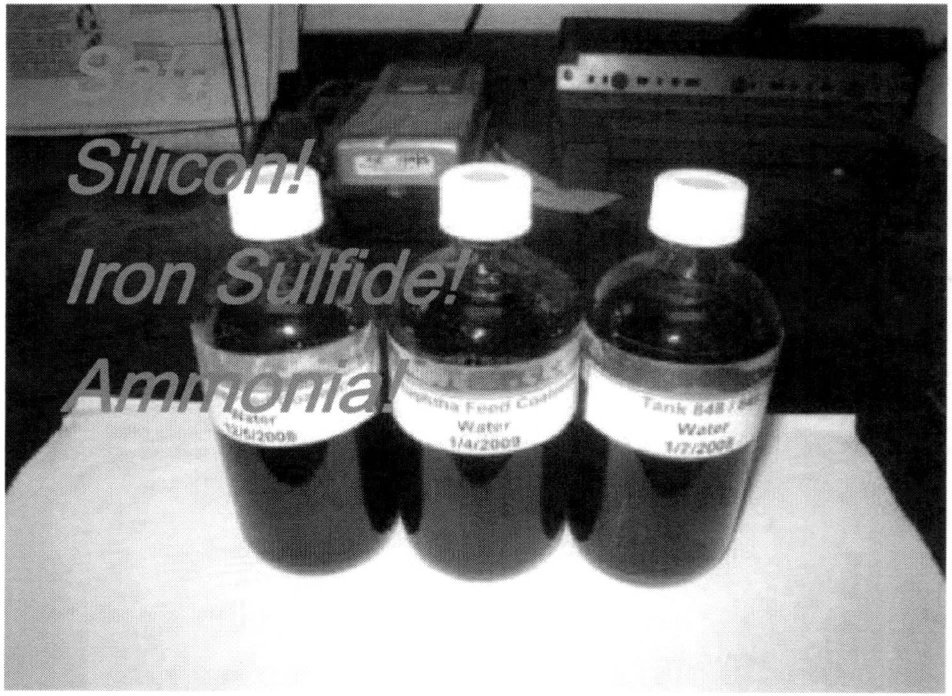

Cracker Poison!

2009 EPC Conference – Tampa, FL

John Robinson

BASF FINA Petrochemicals

Questions?

High Performance Catalysts
For Chemicals

BASF
The Chemical Company

Copper-Based Catalysts, Copper Acetylides, and Safety in Olefins Plants

AIChE 2009 Spring National Meeting
Ethylene Producers Conference
Session T4A13, Paper No. 13e

David J. Artrip

Acknowledgements

BASF
The Chemical Company

- Professor Michael D. Amiridis, University of South Carolina
- Dr. Michael Bender, BASF AG
- Dr. Peter Cheung, ChevronPhillips Chemical
- Dr. Dane Grenoble, Consultant
- Dr. Daniel Heineke, BASF Corporation
- Mr. Bruce Kennedy (ret.), BASF Corporation
- Mssrs. Klausmeyer, Wilson and Rohrback, OxyChem
- Dr. Sabah Kurukchi, Stone & Webster
- Dr. Zheng Liu, Alstom (former post-doc at USC)
- Professor Jimmie Oxley, University of Rhode Island
- Dr. Heiko Urtel, BASF AG

Copper-Catalyst Safety 101

BASF
The Chemical Company

- Copper-based catalysts are used in a wide range of Processes in the CPI in which they come into contact with Acetylene.
- In some of these Processes, explosive copper acetylides (ECA) can be formed.
- In using catalysts in which the copper is highly dispersed, absent any aqueous phase, no ECA can be formed. However…
- …the Reactivity of Cu must be Respected !

Agenda

BASF
The Chemical Company

1) Facts about Copper Acetylides

2) BASF's Research on Cu-Acetylides

3) 'Do's and Don'ts' w/ Cu-Catalysts

4) Case Histories & Lessons Learned

Copper Acetylides: *Facts vs. Myths*

BASF
The Chemical Company

- Quote: 'The great enemy of the truth is very often not the lie – deliberate, contrived, and dishonest – but the Myth – persistent, persuasive and unrealistic.'

  - John F. Kennedy

- Let us try to separate the Facts from the Myths, about Copper Acetylides !

Copper Acetylides

BASF
The Chemical Company

- The interaction of copper with acetylene can result in the formation of copper acetylides with a wide range of chemical compositions.
- The environment in which they are formed will determine the composition, the appearance, and the degree of explosivity of copper acetylides.
- Copper acetylides can be
  - Highly explosive
  - Mildly explosive
  - Non-explosive

Explosive Copper Acetylides

BASF
The Chemical Company

- Explosive copper acetylides (ECA's) may result when copper acetylides are formed in solution and the resulting precipitate allowed to dry out.
- Copper acetylides can only be explosive when dry.
- ECA's may also form from any copper metal surface or hi-Cu alloy surface in a material of construction.
- Most industrial incidents involving ECA's result from the use of copper metal or Cu-alloy as a material of construction where acetylene is present.
- The greatest danger is from ECA as ignition source.

Reppe Chemistry

BASF
The Chemical Company

- In the Reppe process, a Cu-based catalyst reacts Acetylene with formaldehyde.
- The activated catalyst is copper acetylide!
- The catalyst bed is immersed in formaldehyde solution, keeping the catalyst wetted.
- Reppe Rule: ‘ Never, never, never let the catalyst dry out. Always keep it wet!

Non-explosive Copper Acetylides

BASF The Chemical Company

- Precipitated copper acetylides may be explosive (ECA), or non-explosive (NECA).
- ECA are at least 1 mm3 in size.
- Absent an aqueous phase, only NECA can form when the copper surfaces are ‘small’ ($< 1\ mm^2$).
- PuriStar® catalysts: Cu-crystallites are < 20 nm !
- PuriStar catalysts in non-aqueous systems can only form NECA’s !

Agenda

BASF
The Chemical Company

1) Facts about Copper Acetylides

2) BASF Research on Cu-Acetylides

3) 'Do's and Don'ts' w/ Cu-Catalysts

4) Case Histories & Lessons Learned

BASF's Research on Cu-Acetylides

BASF
The Chemical Company

- Work was performed in recent past (6-8 yrs).
- Motivated by customer concerns about ECA.
- BASF has the consent of all parties involved, to publish the findings of this work.

Phases of Research Project

BASF
The Chemical Company

1) Spectroscopic method for acetylide detection on catalyst surfaces

2) Quantitative measurements of copper acetylides on CuZn catalysts

3) Explosivity Testing on acetylated CuZn catalysts

Phase 1

BASF The Chemical Company

- **Spectroscopic method**
  - Developed in BASF's Central Research Division in Ludwigshafen, Germany, by Dr. Heiko Urtel
  - Combination of DRIFTS and XPS (ESCA)
  - Confirms the presence of Surface Acetylides
    - √ Samples acetylated in the lab (USC)
    - √ Samples from a commercial C3-treater

Phase 2

BASF
The Chemical Company

- Method for quantifying copper acetylide on catalyst
- Dr. Zheng Liu & Professor Amiridis (USC)
- Collaboration with Dr. Peter Cheung (CPChem)
- CuZn samples are exposed to acetylene or MA.
- Acidolysis w/ HCl releases the acetylene or MA.
- Acetylene/MA in vapor space is measured by GC.
- Amount of copper acetylides is calculated from GC results with volume and pressure correction.

Equipment Used at USC (1 of 2)

BASF
The Chemical Company

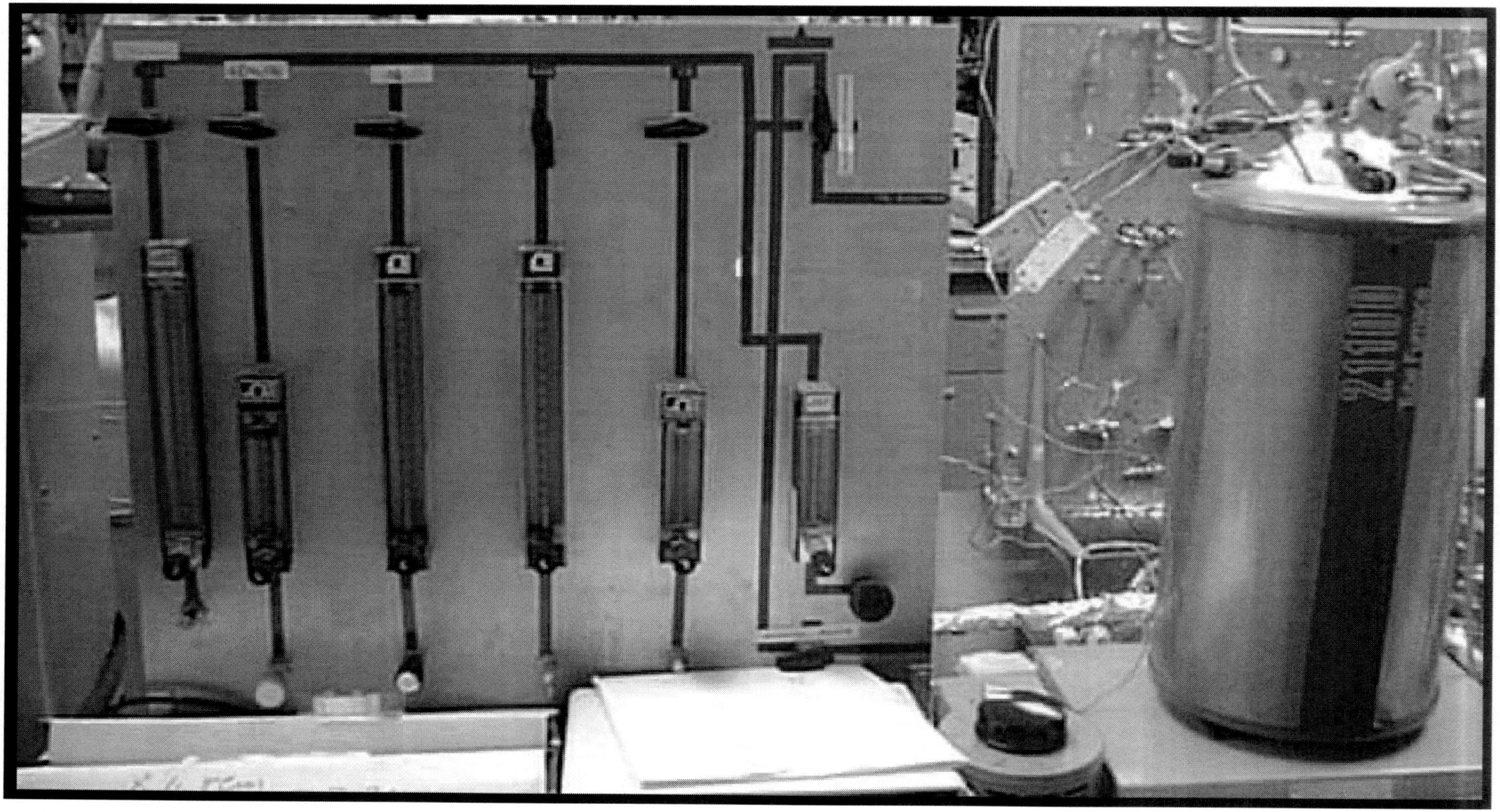

Equipment Used at USC (2 of 2)

BASF
The Chemical Company

Apparatus for acidolysis of acetylated CuZn-samples w/ dilute-HCL, and determination of moles of acetylene and MA recovered in the enclosed vapor space.

Results of Phase 2

BASF
The Chemical Company

Acetylide Formation Measurements via HCl Acidolysis

Sample	Pretreatment			MA/AC Released (mol x 10^4)	wt. % Acetylide on Sample
	Gas	T (°C)	Days		
Fresh CuOZnO	1%MA/He	100	7	9.54 (MA) [1]	0.9
Fresh CuOZnO	3%MA/P-P	55	14	5.63 (MA) [1]	0.6
Fresh CuOZnO	3%MA/P-P	100	14	14.10 (MA) [1]	1.4
Reduced CuZn	3%MA/P-P	100	7	0.00 [1]	0.0
Fresh CuOZnO	1%AC/He	25	7	3.07 (AC) [2]	0.8
Fresh CuOZnO	1%AC/He	50	7	7.90 (AC) [3]	1.7
Commcl-Spent	Liquid C3	15-20	~2 yrs	14.60 (MA) [1]	1.5

1) Sample amount 4 g

2) Sample amount 1 g

3) Sample amount 1.2 g

Phase 3 - Explosivity

BASF
The Chemical Company

- **Explosivity testing by Prof. Jimmie Oxley (URI)**
  - Acetylated CuZn samples shipped to URI in water
  - Samples were dried, then tested for Explosivity
- **Special apparatus using 'RP-3 detonator'**
  - Detonator comprises 29 mg of PETN (penta-erythritol tetranitrate), a sensitive high explosive used for triggering secondary explosions.
  - Catalyst sample is packed into a brass casing (like a rifle shell) together with the detonator.

Explosivity Post-Mortem Photo

BASF The Chemical Company

Each test '...produced no more shattering of the brass cartridge than would have been caused by the RP-3 detonator alone.'

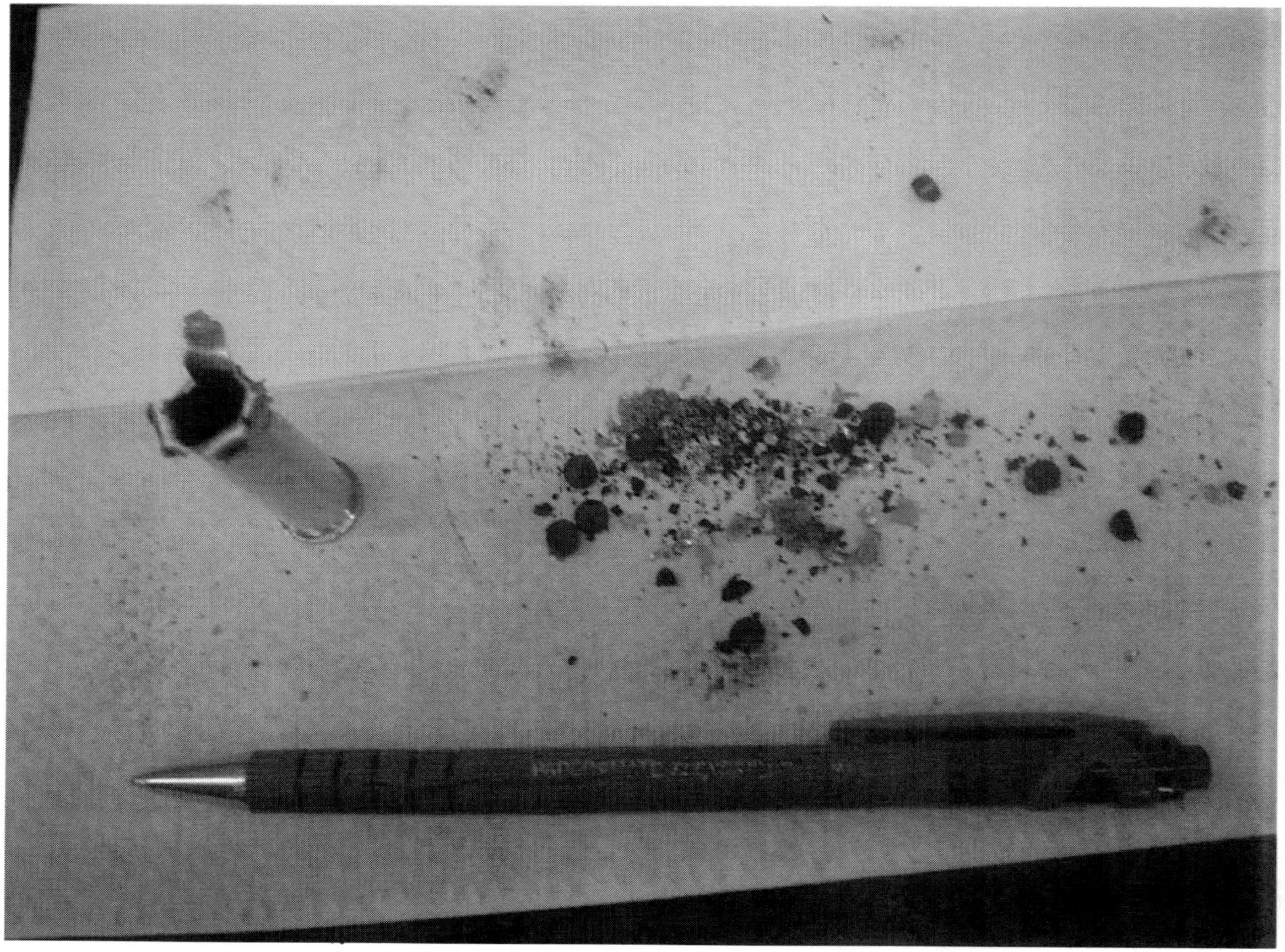

Results of Phase 3

BASF
The Chemical Company

Conclusions of Explosivity testing:

'… this material is not an explosive hazard.'

Agenda

BASF
The Chemical Company

1) Facts about Copper Acetylides
2) BASF's Research on Cu-Acetylides
3) 'Do's and Don'ts' w/ Cu-Catalysts
4) Case Histories & Lessons Learned

'Do's and Don'ts' #1

BASF The Chemical Company

- *Do Not Use Unreduced, Unsulfided Copper Oxide Catalyst on Streams Containing >3% H2.*
- **REASON: Copper oxide can be reduced via reaction with H_2 in a highly exothermic reaction.**
  - Limiting the H_2 concentration to 3% will limit the temperature rise in case of sudden reduction.
  - With syngas or 90+% H2, reduced-Cu can be used.
  - In Refinery Offgas, however - rich in both H2 and olefins – only <u>sulfided</u> copper can be Safely used.

‘Do’s and Don’ts’ #2

BASF
The Chemical Company

- *Do Not Allow Air into Reaction Vessel containing Cu/CuO bed, except*
    - when the catalyst is in the fully oxidized state such as when loading a fresh CuO type; or
    - at the end of an oxidation procedure, after the catalyst has been fully oxidized.
- **REASON: Copper oxide after operation in olefin service is generally in a partially reduced state and therefore capable of “self-heating” in air.**

'Do's and Don'ts' #3

BASF
The Chemical Company

- ***Do*** ***Passivate or Inert-Blanket Spent Copper Catalysts before Dumping, and use Steel for Containment of the Spent Catalyst.***
  - Passivate preferably by wetting; or alternatively by an oxidation procedure.
  - However, if doing oxidation procedure, be on the look-out for secondary exotherms, from oxidation of any carbon deposits on the catalyst.
- **REASON: Otherwise … self-heating may occur after exposure to air.**

‘Do’s and Don’ts’ #4

BASF
The Chemical Company

- ***Do Not use a Hydrogen-containing Gas for Hot-purging of a Copper Catalyst Bed.***
  - **Low levels of H_2 (<< 1%) are acceptable.**
- **REASON: If the bed is heated to > 100°C, the H_2 may suddenly start reducing CuO, causing a significant temperature spike. (See D-D#1.)**

‘Do’s and Don’ts’ #5

BASF
The Chemical Company

- ***Don’t be Surprised by Sudden Exotherms During the Oxidizing or Hot-Purging of a Cu/CuO Bed.***
  - “Oxidation procedure” by definition is exothermic.
  - At elevated temperatures, CuO can do Catalytic Oxidation on hydrocarbons or carbon deposits, even if no free Oxygen is present in the gas phase.

Agenda

BASF
The Chemical Company

1) Facts about Copper Acetylides
2) BASF's Research on Cu-Acetylides
3) Some 'Do's & Don'ts' w/ Cu-Cat's
4) Case Histories & Lessons Learned

The Self-Heating Dumpster

BASF
The Chemical Company

- **ISSUE: Spent arsine catalyst was removed into a dumpster without passivation.**
- **CONSEQUENCE: A plastic tarp placed over the dumpster melted.**
- **LESSONS LEARNED: (1) If a spent copper catalyst is removed into air without passivation, be prepared for some self-heating upon exposure to air. (2) Leave it to oxidize in the open, away from process areas (uncovered), for a few weeks.**

The Melting Supersacks

BASF
The Chemical Company

- ISSUE: Spent arsine treater catalyst was removed from vessel into Supersacks without passivation.
- CONSEQUENCE: The supersacks melted.
- LESSONS LEARNED:

1) Same as in Case of the Self-Heating Dumpster, and additionally,

2) use STEEL for containment of the spent material.

The Fused Bed

BASF
The Chemical Company

- **ISSUE: Spent arsine treater catalyst was steamed in place and then "cooled" with air.**
- **CONSEQUENCE: Reaction with air was not detected in time, causing bed to fuse. Jack-hammers were needed to remove the catalyst.**
- **LESSONS LEARNED ("Perfect Storm" scenario)**
  1) never use air for "cooling" on a spent Cu/CuO bed
  2) know the elevations of any thermocouples in the bed
  3) gas flow must be sufficient to avoid channeling
  4) properly monitor potentially exothermic procedures

The Unexpected Exotherm

BASF The Chemical Company

- **ISSUE: CuZn-Oxide was used as arsine guard on a Refinery Offgas (>10 vol% H_2 & > 10% C_2H_4).**
- **FACILITATOR: Upstream adsorbent beds, when switched, put temperature spike into the CuO bed.**
- **CONSEQUENCE: Reduction of the CuO in the CuZn bed was initiated, even though the 'Design' operating temperature was < 50°C.**
- **LESSONS LEARNED:**
  - **Use PbO for arsine removal from ROG.**

When Considering a New Catalyst from Process Safety angle, know:

BASF
The Chemical Company

1) its reactivity during normal operation;
2) its reactivity during loading, activation (if any), regeneration (if any), start-up, shut-down, decommissioning and removal from the plant;
3) procedures for each of the steps in (1) and (2);
4) amounts needed (if any) of steam, nitrogen, air, hydrogen, sulfiding agents etc. for loading, activation, regeneration, decommissioning and removal;
5) specialized equipment or contractors needed for catalyst loading, activation or unloading;

When Considering a New Catalyst from Process Safety angle, know:

BASF The Chemical Company

6) level of assistance needed or expected from the catalyst vendor;

7) the expected change-out frequency;

8) sampling requirements, procedures, and any special equipment/tools needed, for sampling the catalyst as well as the process;

9) any training needed for engineering, operations or maintenance; and

10) personal protective equipment needed for operators or workers doing the sampling or catalyst handling.

The Final Key to Safety

BASF
The Chemical Company

Even with the best in safety concepts, technology and engineering, an excellent safety performance in the plant can only be achieved by a keen safety awareness and commensurate conduct, in everyday operations and maintenance.

The final key to Safety is ... YOU !

For More Information

BASF
The Chemical Company

For questions or additional information,

please email david.artrip@basf.com

or give me a call at: 713-759-3017

Use of Molecular Sieves and other Adsorbents for Contaminant Removal in Ethylene and Propylene Production

Jay Gorawara

Senior Development Associate
Catalysts, Adsorbents and Specialties
UOP LLC

Abstract: Ethylene and propylene producers have faced the challenge of contaminants for decades. Adsorbents, both molecular sieve and alumina based, have been successfully used to help olefin producers manage their operations as contaminated feeds and increasingly sensitive polymerization catalysts continue to offer them increasing challenges. The most widely used adsorbents are the 3A based molecular sieve products utilized to remove water from the cracked gas, liquids, ethylene-ethane and hydrogen. Elsewhere in the production of ethylene, adsorbents are used to remove not only water but also other contaminants from product ethylene and propylene.

As we move downstream to the polymer plant, contaminants are a problem for sensitive polymerization catalysts and polymer quality. Catalyst life can be greatly diminished by the presence of oxygen compounds, sulfur compounds, nitrogen compounds and even arsine and phosphine. In this part of the plant, adsorbent beds of various combinations can be utilized to remove a variety of compounds. Key to trace contaminant removal from olefinic streams is the use of adsorbents with low reactivity towards the process stream. Although adsorbents are a cost effective option for olefin stream dehydration and purification, it is critical to select the right product.

Selection criteria for matching adsorbent products with specific streams and contaminants as well as tips for optimal operation will be covered.

Use of Molecular Sieves and other Adsorbents for Contaminant Removal in Ethylene and Propylene Production

Introduction

Adsorbents such as molecular sieves and activated aluminas are widely used in the processing industries to remove water and other contaminants. The use of adsorbents in ethylene and propylene production and downstream processing, however, is typically more demanding because of the reactive nature of the streams processed. Type 3A based molecular sieves, promoted activated aluminas, and combinations of larger pore molecular sieve and promoted aluminas have the advantage of being able to operate in reactive environments without special processing steps. They are capable of removing detrimental impurities to extremely low levels to avoid endangering plant equipment, achieving design throughput, or poisoning sensitive downstream catalysts. The process typically involves passing the contaminated stream through one or more beds at ambient temperature to remove (adsorb) the contaminants, while simultaneously, heating a previously contaminant loaded bed with hot gas at a high temperature (>400 F) to desorb the contaminant. The hot bed is then cooled and is ready for another adsorption step. Although dehydration of reactive process streams is still one of the largest applications for adsorbents in ethylene and propylene production, in recent years there has been a growing need for the removal of other contaminants in ethylene and propylene production. This paper will discuss the capability of these adsorbents to remove many contaminants in reactive streams using both regenerative and non-regenerative adsorbents, show where current applications exist, and help processors identify potential applications to improve their operation or upgrade the quality of their products.

Zeolite Molecular Sieves

Zeolite adsorbents are synthetically produced molecular sieves that are microporous, crystalline, metal aluminosilicates. Metal cations contained in the crystalline structure of molecular sieve adsorbents balance the negative charge of the framework. These metal cations create an electrical field resulting in a strong attraction for polar molecules. Higher polarity generally results in a higher attraction for a molecular sieve. Table 1 shows the relative polarity of common impurities and carrier streams found in ethylene and propylene production and processing. Because of its high polarity, water can be removed by molecular sieves from most process streams.

Table 1

Order of Selectivity

Water
Ammonia
Methanol
Hydrogen Sulfide
Carbon Dioxide
Carbonyl Sulfide
Propylene
Ethylene

The size of the pore opening on each face into the internal structure where the contaminants pass through is determined during manufacturing by the type of crystal and occupying cation (Figure 1). Thus some molecules can be readily adsorbed or completely excluded (even highly polar molecules) according to their relative molecular size.

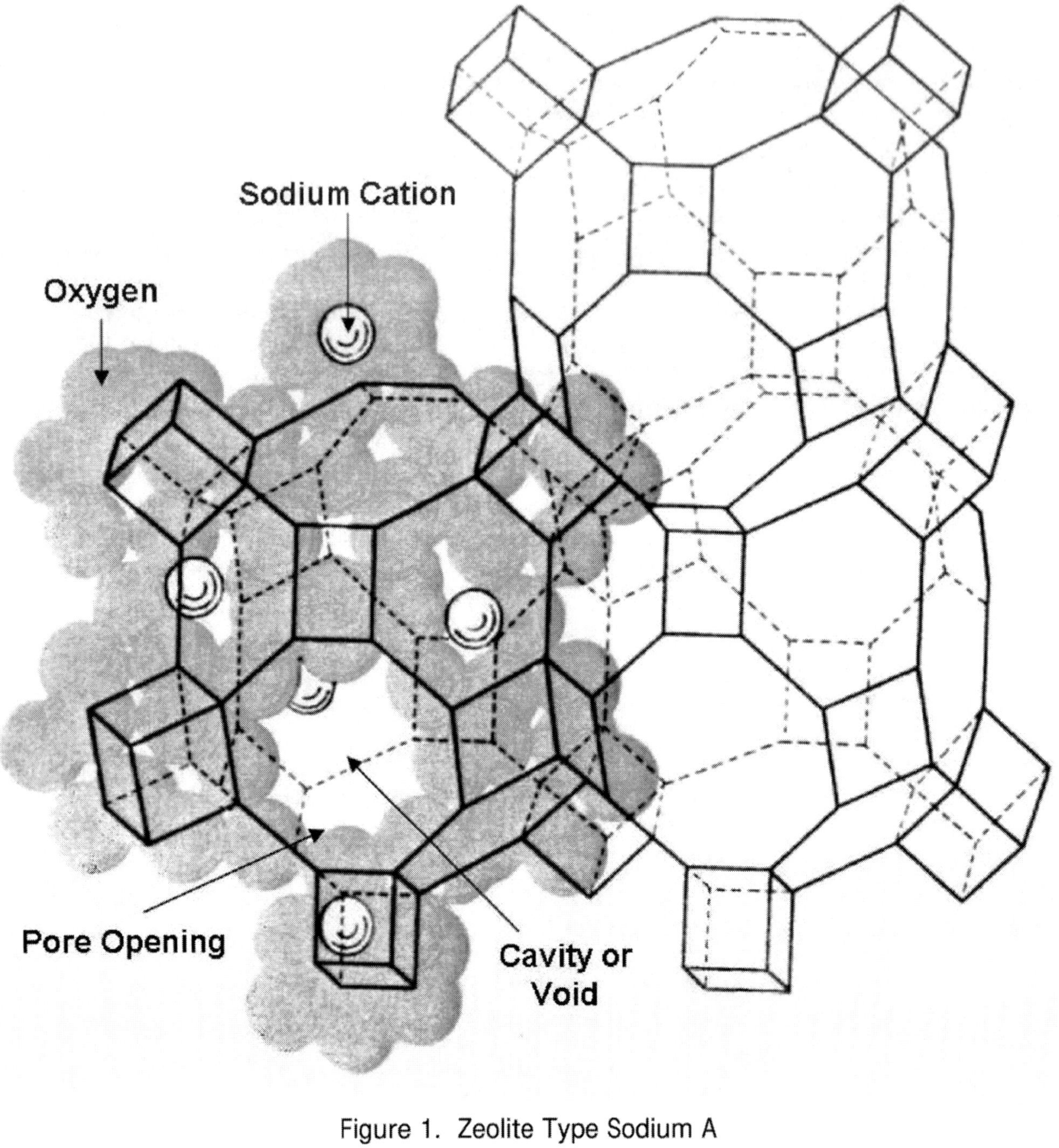

Figure 1. Zeolite Type Sodium A

Table 2 lists the two types of molecular sieve adsorbents most commonly used for olefin processing and the molecules that they will adsorb and exclude.

Table 2

*General Type	Molecules Adsorbed	Molecules Excluded
3A	<3 angstroms effective pore diameter H_2O , NH_3,	>3 angstroms effective diameter acetylene, ethylene, propylene and larger hydrocarbons
13X	<8 angstroms effective diameter H_2O,NH_3, CH_3OH, H_2S, CO_2, C_2H_4, C_3H_6 **Larger Oxygenate and Sulfur compounds	>8 angstroms effective diameter – All common contaminants and hydrocarbons in ethylene and propylene processing are smaller than 8 angstroms

* Chart depicts basic molecular sieve types only. In all applications, these basic forms are customized for specific use.

** Typically found in C4+ olefins present in refinery streams

Type 3A molecular sieve is highly effective for removal of water and ammonia to extremely low levels from reactive streams in ethylene plants because of the polarity of these contaminants and the fact that the reactive carrier stream components are excluded. As long as a non-reactive binder with high breakup resistance is used to form the finished molecular sieve product, long life is expected.

Because Type 13X zeolite molecular sieve has one of the highest capacities of any molecular sieve for most polar compounds and will adsorb larger contaminants such as branched chain oxygenates and sulfur compounds, it is frequently used for contaminant removal from olefinic streams. Type 13X and other large pore molecular sieves require special operating procedures when used in an olefin stream where they readily adsorb the carrier stream. When a molecular sieve adsorbs a compound, heat is released. Most contaminants are at low enough levels that the heat release is minimal. But when the adsorbed compound is at high level such as in a stream containing significant olefins, a large amount of heat can be released triggering possible exothermic reactions that endanger equipment and personnel. To prevent this situation, a preload step is used before the initial exposure of a fresh or regenerated adsorbent to the olefin containing stream. This procedure involves using a low (<5 wt%) concentration of the olefin in a non-adsorbed carrier stream to slowly saturate the adsorbent with the carrier stream olefin and safely remove the heat of adsorption. The operating sequence is illustrated in Figure 2.

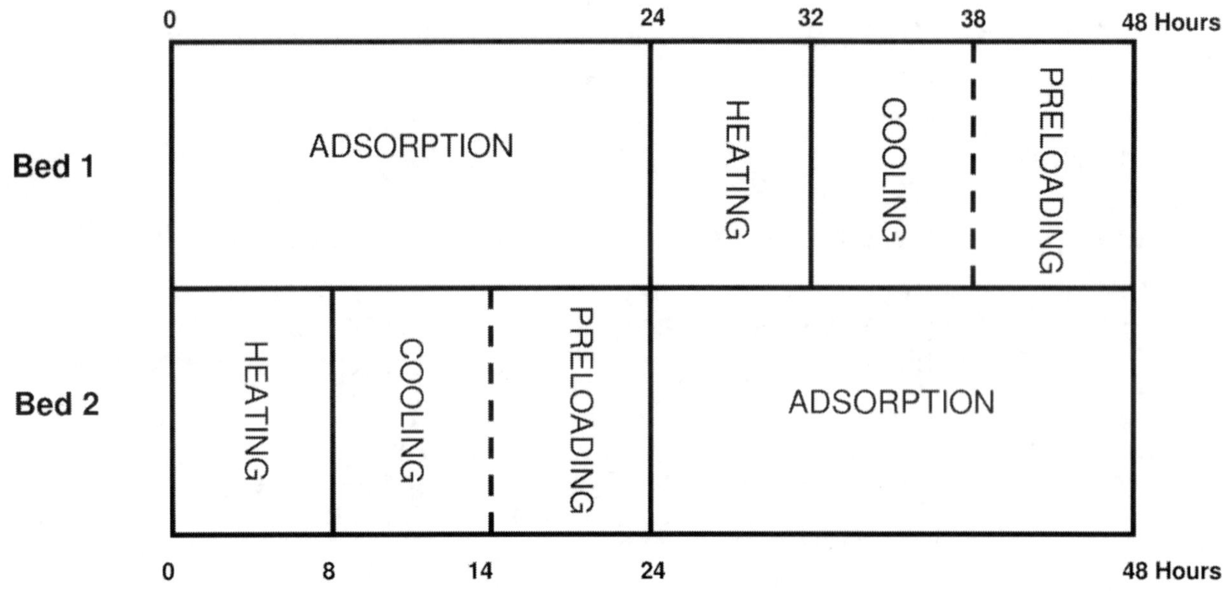

Figure 2. Typical Adsorber Operating Sequence

It is also possible to use type 4A and 5A molecular sieves to dry and purify olefin streams. However, there is typically no advantage over 13X since they adsorb the reactive components such as ethylene and propylene, have lower overall capacity, and cannot adsorb larger oxygenate and sulfur compounds.

Promoted Aluminas

Modified activated alumina adsorbents are synthetically produced, transitional phase aluminas. Unlike zeolites, activated aluminas have a pore system that is accessible to practically all compounds present in ethylene, propylene and polymer plants. Hence, the size exclusion phenomenon cannot be used in alumina based adsorption processes. Instead, the complex surface chemistry of aluminas offers tremendous opportunities for designing adsorbents for particular applications. The modification of surface functionality by inorganic additives, or promoters, during manufacturing is an especially useful method for enhancing adsorbent performance. The promoted alumimas have excellent performance for removal of specific contaminants to very low levels.

In olefinic streams, light acid gases such as CO_2, H_2S and COS are common contaminants that can be most effectively removed with this type of modified activated alumina. As these aluminas rely on chemisorption to remove CO_2, H_2S and COS, they exhibit high capacities for these contaminants even at very low concentrations.

These promoted alumina adsorbents also have the additional advantage in that unlike molecular sieves, they do not have a strong affinity for reactive, polar compounds such as olefins. With little or no selective capacity for olefins, there is a very low heat of adsorption when olefins are exposed to promoted aluminas, hence a preload step is not required. In addition, these adsorbents exhibit very low reactivity towards the main carrier stream. The formation of additional contaminants, such as olefin oligomers, is practically non-existent at typical process conditions.

Hybrid Adsorbent

By combining high selectivity and capacity for light acid gases with low reactivity and heat of adsorption of hydrocarbons, a properly modified activated alumina is very suitable for the purification of olefinic streams. Correspondingly, molecular sieves have a high capacity for polar molecules. Ideally, one adsorbent could exhibit the properties of both zeolitic and promoted alumina adsorbents. A hybrid adsorbent, a homogenous combination of a promoted alumina and a large pore molecular sieve adsorbent can take advantage of the complementary performance characteristics of both materials. A properly designed high quality hybrid adsorbent can simultaneously have good capacity for light acid gases as well as a broad range of polar molecules.

Though a hybrid adsorbent contains zeolite, when properly produced it typically may not require a preload step when processing olefinic hydrocarbons. The elimination of the preload step, while retaining the effective removal of polar compounds and light acid gases, provides the olefins producer and polymer manufacturer with significant process advantages. The unique properties of a hybrid adsorbent may enable the processor to use a single product for the adsorptive removal of the broadest range of contaminants without a preload step. The result is an adsorbent that provides additional protection against unanticipated contaminants with a simplified and safer process.

There are various hybrid adsorbents available in the market today and the characteristics discussed above may vary. Consult your adsorbent supplier to clarify operating procedures for your operations.

Metal Oxide Containing Adsorbents

This class of adsorbents is a "must have" for propylene purification, where the removal of contaminants such as AsH_3, PH_3, COS to low part per billion (ppb) levels is critical for the economic production of polypropylene. These trace contaminants adversely effect performance of the very sensitive modern polymerization catalysts. The metal oxide component in these adsorbents irreversibly binds the hydride contaminants via an oxidation-reduction mechanism; hence water is a by-product of the reaction.

It is critical for these advanced metal oxide adsorbents to have high contaminant capacity combined with low reactivity towards the carrier stream. An optimized pore structure is also essential for this liquid phase application. A high binder surface area can help to retain the water produced in the arsine and phosphine scavenging reactions.

Note: There is a potential for the metal oxide products to form hazardous (potentially explosive compounds) in the presence of acetylene under certain conditions. Consult your adsorbent supplier for recommendation of the proper adsorbent, as well as operating procedures for your operations.

Comparative Performance of Various Adsorbents in Olefin Service

Table 3 shows the comparative performance and typical processing recommendations of various adsorbents for specific contaminants found in ethylene plants, propylene production, downstream storage systems and polymer plants. Depending on the contaminant type, either a molecular sieve, a promoted alumina or a metal oxide containing adsorbent has the highest capacity. Type 13X requires a preload step; however, hybrid adsorbents offer the broadest range of contaminant removal typically without requiring a preload step, although it may be at the expense of somewhat lower capacity.

Ideally, adsorbents should not exhibit any appreciable catalytic activity towards the stream to be purified. Most adsorbents fulfill this requirement when used in non-reactive streams, but potential problems arise in the case of reactive feeds. This topic was discussed in a previous AICHE paper written by Vlado Kanazirev and Dave Smith, "Adsorbent Reactivity: An Important Factor in Contaminant Removal" presented in 2004. One of the important potential reactions is the oligomerization of olefins which can occur over strongly acidic sites on the adsorbent. These oligomerization reactions are exothermic, can cause uncontrolled temperature excursions, and are hence a safety concern. In less severe cases, "green oil" can be produced which results in deposition of heavy hydrocarbons and coke precursors on adsorbents, thereby reducing their efficiency and operational life. In addition, it is possible to form different contaminants than those present in the carrier stream by reactions catalyzed between the carrier stream and the contaminants, or between the contaminants themselves. The challenge in such cases is that some of the newly created contaminants may be more difficult to remove than the original ones. Some metal oxide containing adsorbents, can create by-products while removing the targeted contaminant.

Figure 3 is a plot measuring the temperature of initiation of polymerization of isobutylene over various adsorbents. This type of testing is useful to rank the relative reactivity of various types of adsorbents. The temperature at which the pressure of isobutylene decreases is an indication of the onset of polymerization activity. Hence, more reactive adsorbents will polymerize isobutylene at lower temperatures. For example, both of the 13X materials tested are more reactive than the AZ-300 adsorbent, and the 3A shows no measurable reactivity. Note from the plot that the reactivity of similar materials can also vary between adsorbent manufacturers.

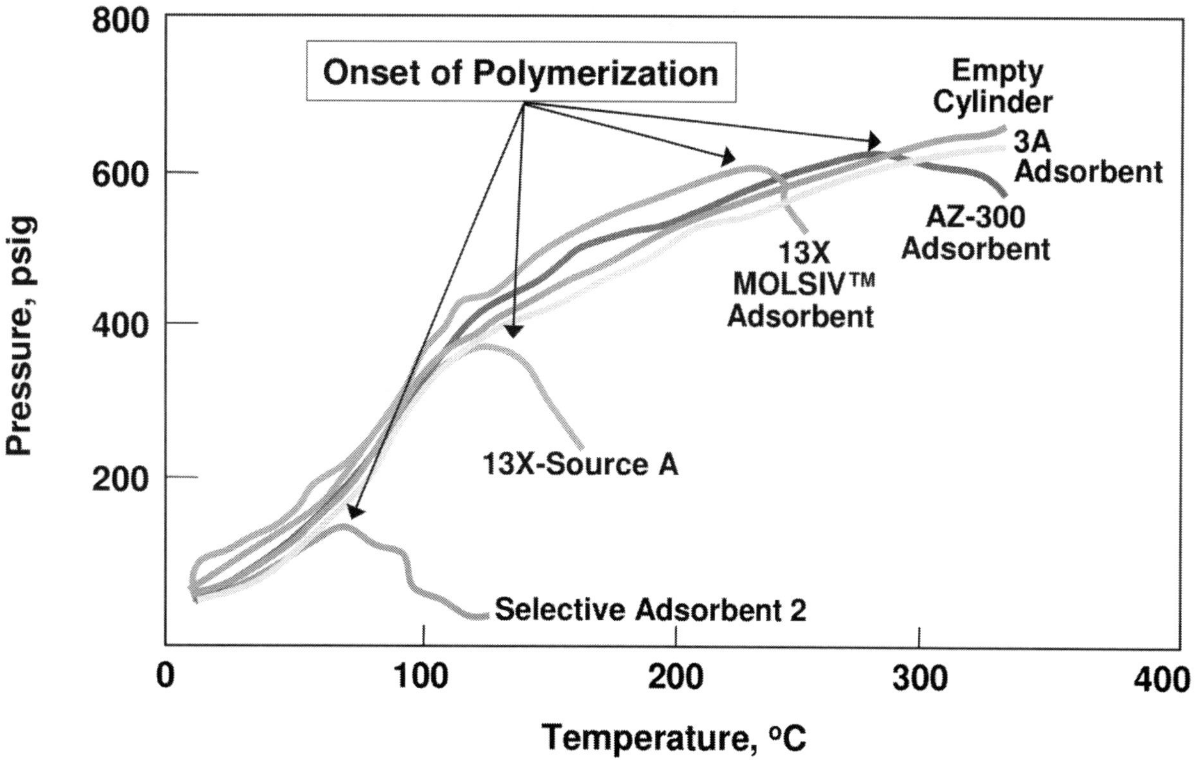

Figure 3: Isobutylene Conversion at Process Relevant Conditions – Adsorbent Reactivity Testing

Table 3

Comparative Product Capabilities for Contaminant Removal from Olefins*

Contaminant Family	Contaminant	Product Type	Adsorbent Capacity	Olefin Preload?
Oxygenates	Water	Type 3A	High	No
		Hybrid Adsorbent	Moderate	No
	CO_2	Promoted Alumina	High	No
		Hybrid Adsorbent	Moderate	No
	Methanol	Modified 3A	High	No
		Hybrid Adsorbent	Moderate	No
	**other Alcohols, Ketones, Aldehydes, Ethers and Peroxides	Type 13X	High	Yes
		Hybrid Adsorbent	Moderate	No

Sulfur compounds	H_2S	Promoted Alumina	High	No
		Hybrid Adsorbent	Moderate	No
	COS	Promoted Alumina	High	No
		Hybrid Adsorbent	Moderate	No
	**Mercaptans, Sulfides and Disulfides	Type 13X	High	Yes
		Hybrid Adsorbent	Moderate	No
Nitrogen compounds	**Amines and Nitriles	Type 13X	High	Yes
		Hybrid Adsorbent	Moderate	No
	Ammonia	Modified 3A	High	No
		Hybrid Adsorbent	Moderate	No
Other	Mercury	HgSIV-3	High	No
	Arsine	***Metal Oxide Product	High	No
	Phosphine	***Metal Oxide Product	Low	No

* Based on UOP products and experience
** These compounds are typically found in C4 olefins sourced from crude oil refineries
*** Non-regenerative

The many locations where adsorbents are currently removing contaminants from ethylene and propylene production streams and downstream processing are shown in Figures 4A, 4B and 4C. Table 4 shows the contaminants being removed, the products employed, and the typical reactive components in the carrier stream for those locations.

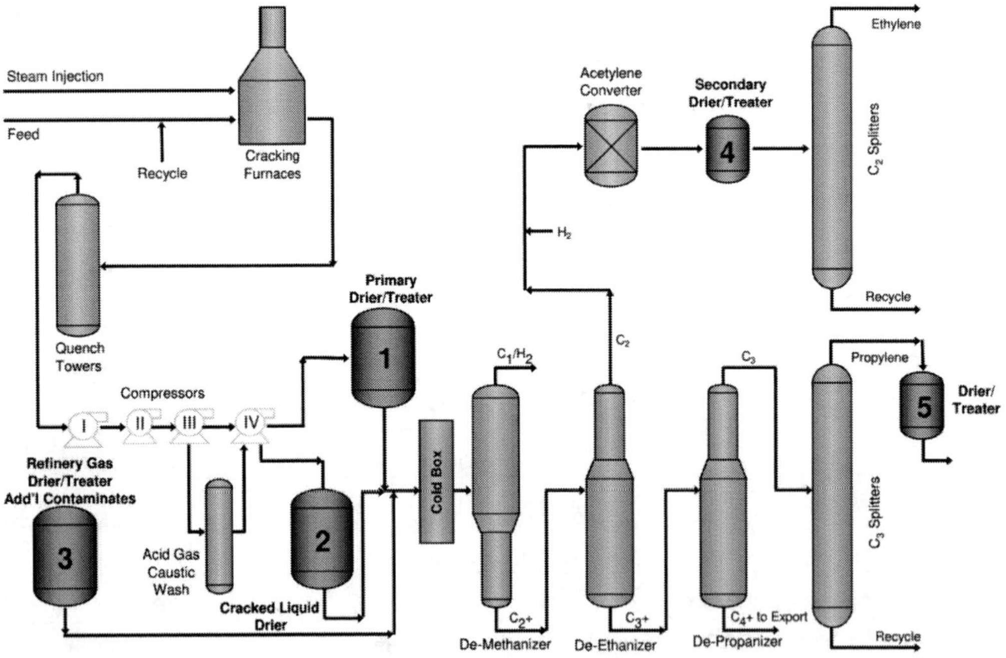

Figure 4A: Treating Reactive Feeds with Adsorbents in Ethylene Plants

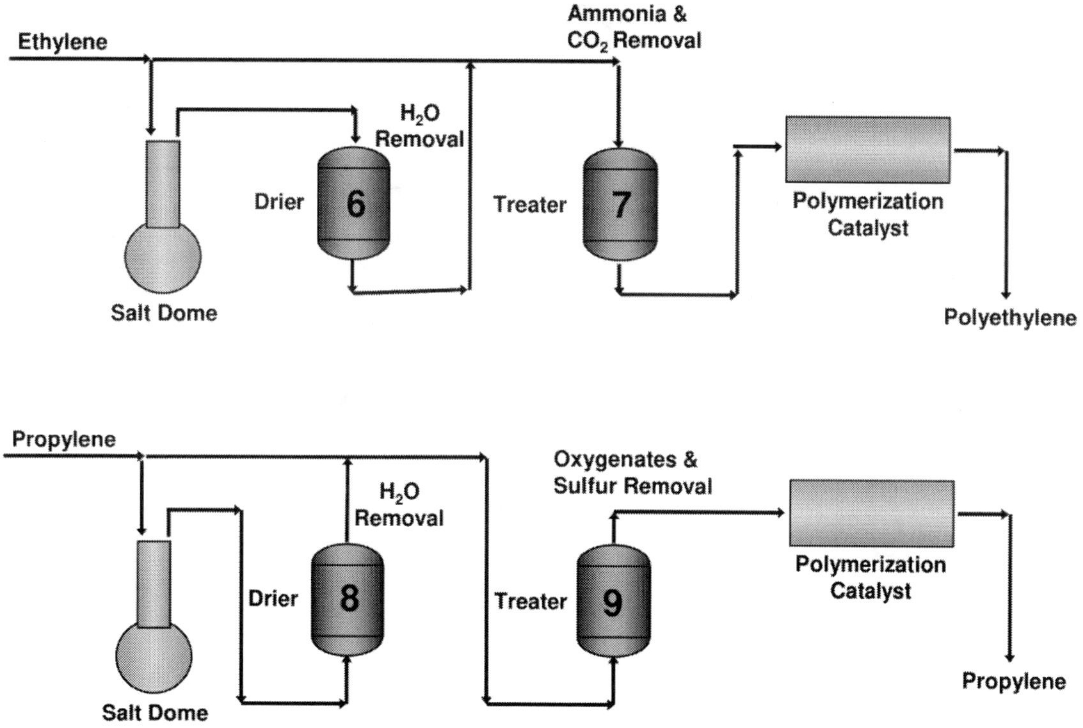

Figure 4B: Olefin Plant Adsorbent Applications

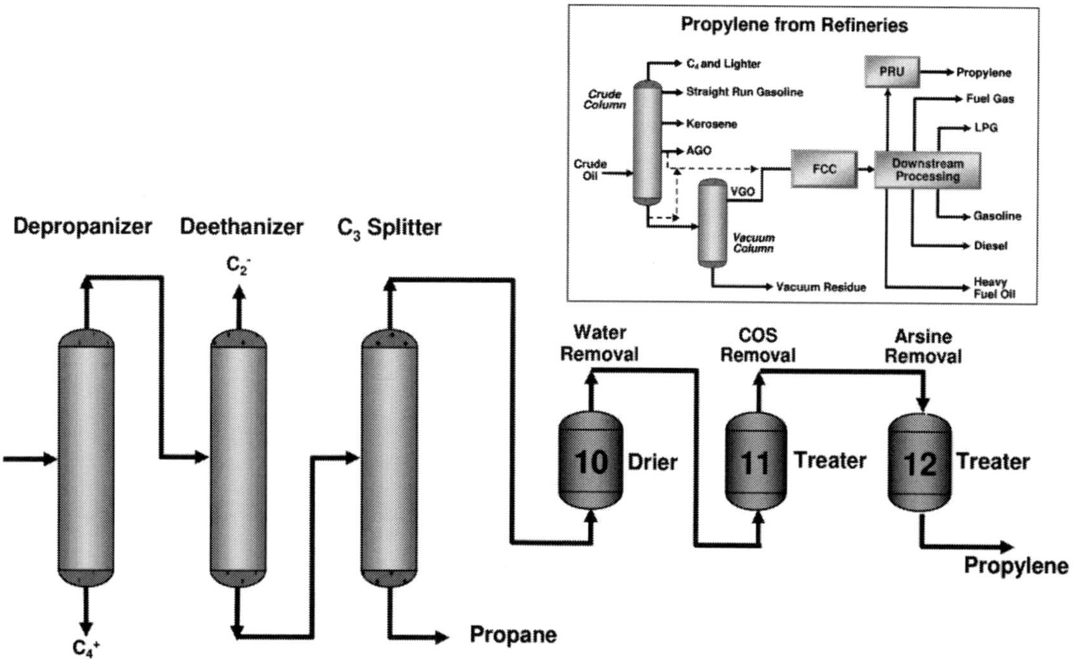

Figure 4C: Propylene Recovery Unit

Table 4

Current Adsorbent Contaminant Removal Locations in Reactive Streams for Ethylene and Propylene Processing

Location	Adsorber Designation	Reactive Components	Contaminants	Type Adsorbents Used
1	Cracked Gas Drier/Treater	Diolefins, Olefins	H_2O	Type 3A
			Ammonia	Modified 3A
			Mercury	HgSIV 3
2	Cracked Liquids Drier/Treater	Diolefins, Olefins	H_2O	Type 3A
			Mercury	HgSIV 3
3	Refinery Gas Drier/Treater	Diolefins, Olefins	H_2O	Type 3A
			Ammonia	Modified 3A
4	Secondary Drier /Treater	Ethylene	H_2O	Type 3A
			CO_2	Modified Activated Alumina
5	Propylene Drier/Treater	Propylene	H_2O	Type 3A
			Methanol	Hybrid Adsorbent
				Modified 3A
			COS, H_2S	Modified Activated Alumina
				Metal Oxide Product
			Arsine / Phosphine	Metal Oxide Product
6	Dome Storage Ethylene Drier	*Ethylene	H_2O	Type 3A
7	Dome Storage	*Propylene	H_2O	Type 3A
8	Polyethylene Plant Feed Drier/Treater	Ethylene	H_2O, Ammonia, CO_2	Hybrid Adsorbent
9	Polypropylene Plant Feed Drier/Treater	Propylene	H_2O, MEOH, COS, RSH	Hybrid Adsorbent
Not Shown	Polybutylene Plant Feed Drier/Treater	**Butenes	H_2O, Alcohols, Aldehydes, Carbonyls, Ethers, Ketones, Peroxides Mercaptans, Sulfides and Disulfides	Hybrid Adsorbent
				Type 13X

* Regeneration at high pressure with drier product or feed recycled to drier feed
** Contaminants typically found in crude refinery sourced butenes

Tables 3 and 4 can be useful for processors to help identify potential opportunities to use adsorbents to remove contaminants from olefin streams to help protect plant equipment, throughput, and expensive catalyst downstream.

Commercial Experience

Ethylene Plants

In a typical cracked gas drier where dehydration is the only duty required, one would look for adsorbents with good attrition resistance, low reactivity, good mass transfer efficiency and predictable performance. UOP recommended the following configuration to a large ethylene producer in the Middle East, but this is very similar to numerous other installations around the globe.

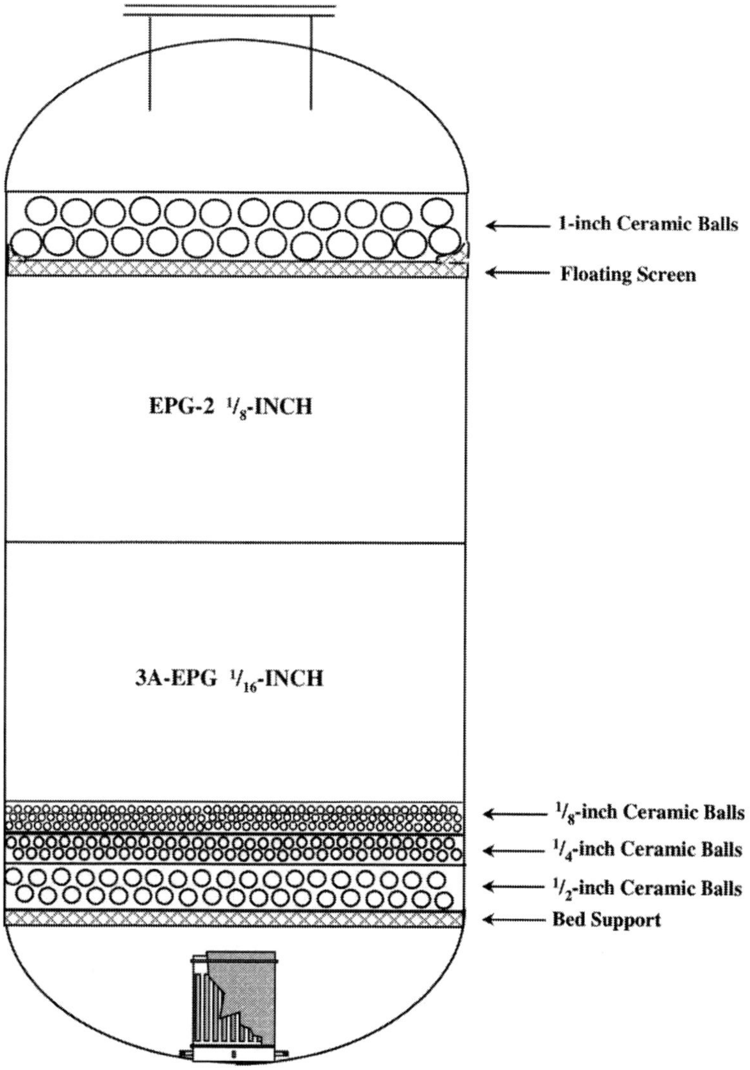

Figure 5: Typical Cracked Gas Drier for Dehydration

If additional contaminants, such as mercury and ammonia were required to be removed, UOP would recommend adding HgSIV-3 and MOLSIV™ EPG-N adsorbents as layers in the cracked gas driers such as seen in Figure 6.

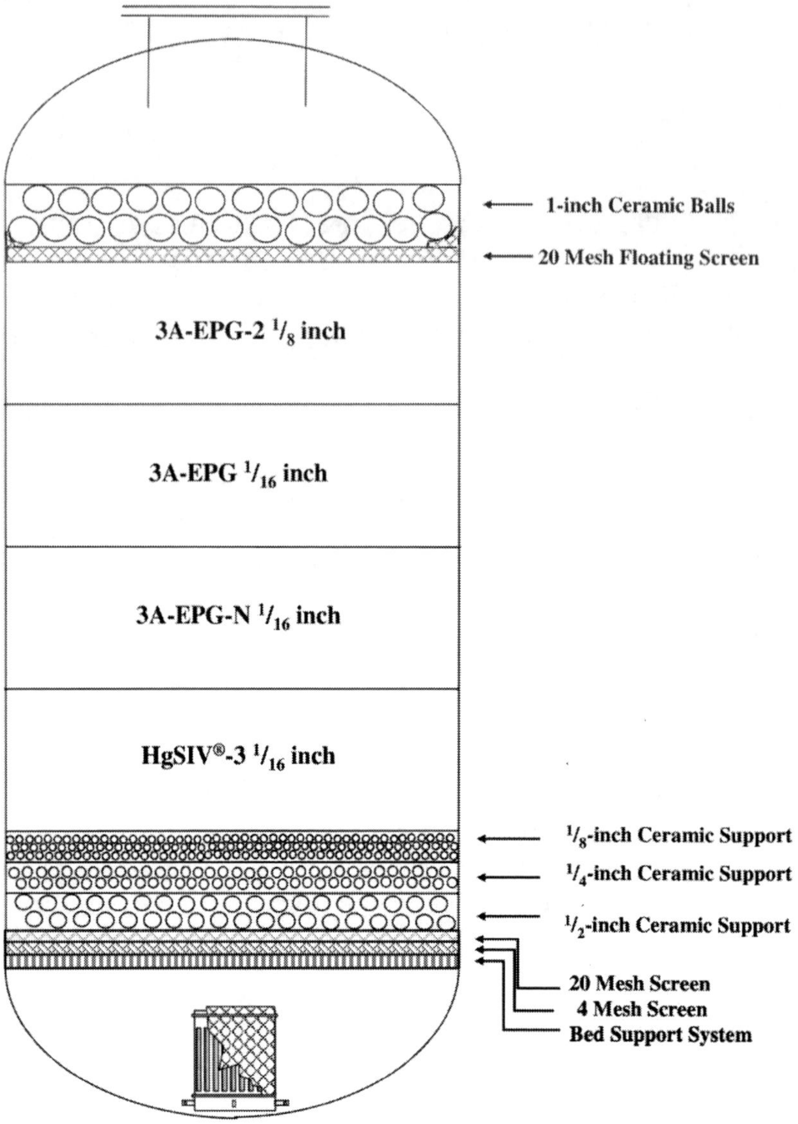

Figure 6: Cracked Gas Drier with Hg and NH_3 Removal

Propylene Production/Polymer Plants

There are three main sources of propylene; co-product from an ethylene plant, on purpose production by dehydrogenation of propane, and propylene from FCCU and coker units. Each of these sources contains varying contaminants. The propylene from an ethylene plant has fairly low levels of contaminants. A full bed of AZ-300 is an excellent choice as it will capture a wide range of unidentified contaminants. The treatment of propylene from a refinery is much more complex, as there are many more contaminants to be handled.

Figure 7 shows the treating system designed for a Propylene Recovery Unit (with the mixed C_3 feed supplied from both FCCUs and coker units) to assure very tight product specifications.

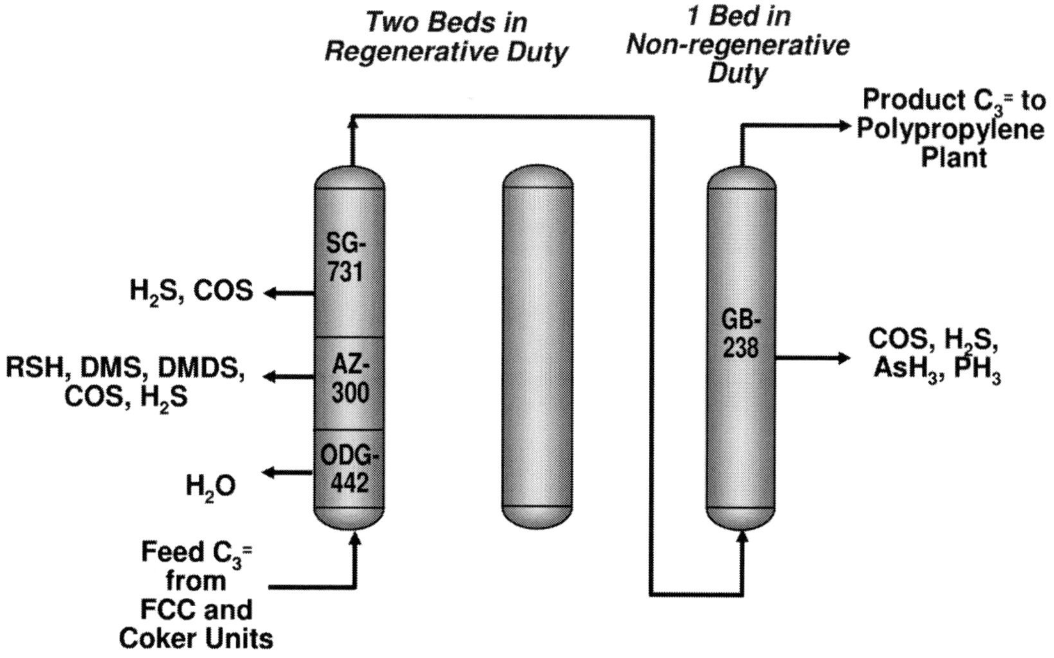

Figure 7: An Asian Propylene Recovery Unit's Purification System

Conclusion

Removal of trace contaminants using adsorbents is one of the most economical methods of purification of a process stream. However, purification of very reactive hydrocarbon streams such as those present in ethylene and propylene processing is not straight forward. Selection of proper adsorbent products is critical in economically achieving product specifications, and operating the processes reliably and safely.

References

Brayden, M. et al., “Top 5 Contaminants in Ethylene Production Unit Feedstocks”, AIChE 2008 Ethylene Producer's Conference.

Coleman, S. et al., “Mercury Removal from Cracked Gas and Liquid Streams”, AIChE 2004 Ethylene Producer's Conference.

Kanazirev, V., and Smith, D. L., “Adsorbent Reactivity: An Important Factor in Contaminant Removal”, AIChE 2004 Ethylene Producer's Conference.

Reid, J. A., and McPaul, D. R., “Control Contaminants in Olefin Feedstock and Products”, *Hydrocarbon Processing*, (September, 1996).

Reid, J. A., “An Overview of Hydrogenation Catalyst Poisons in Olefin Units”, AIChE 2001 Ethylene Producer's Conference.

Sicard, J., and Richman, P., “Product and Process Improvements in Cracked Gas Drying”, UOP Brochure F-4343.

CENTRIFUGAL PUMPS 101 FOR PROCESS ENGINEERS

J. Howard Kelly

Rotating Equipment Engineer

Shaw Energy & Chemicals Group

Abstract

This paper will present an overview of important concepts for the process engineer regarding selection of centrifugal pumps, from the perspective of the rotating equipment engineer. These concepts were chosen based on the process engineer's relative control of these items and their impact on pump selection.

The paper will begin with an overview of the theory behind how a centrifugal pump adds energy to a liquid stream to provide the discharge pressure required by the process. In this section, the effect of head and flow rate on the appropriate speed, size and type of pump will be discussed.

The paper will then review items that affect the pump's performance, namely net positive suction head (NPSH) and minimum flow. NPSH required by the pump and its association with flow rate and pump speed will be discussed. An explanation of the importance of respecting pump minimum flow requirements is provided because of the corresponding effect on pump longevity.

Switching gears slightly, the paper will then consider how the pump selection affects downstream piping. The pump discharge pressure should increase as the flow (head rise to shut-off) decreases to better control the pump with reasonable precision. Downstream piping must be designed to accommodate this increase in pressure.

Finally and with these concepts in mind, the paper will consider the selection of an ethylene product pump. End-user preferences on pump orientation for this service and the importance of considering this factor in setting the parameters of upstream process conditions also are reviewed.

Introduction

Pumps comprise a significant portion of the machinery used in an ethylene process plant, and as such, increasing their efficiency and reliability can significantly impact the bottom line of an ethylene producer. Since the process engineer is often focused on overall plant economic and operating issues involved in efficient production of ethylene and other by-products, s/he often does not consider other important details that can lead to improved reliability and efficiency. S/He relies on the rotating equipment engineer to specify, purchase, and maintain the pumps that meet the process requirements. In areas critical to process design, the client or the process engineer might request spare pumps so that the process flow is not interrupted. This paper highlights areas of concern within the rotating equipment discipline that can be considered in promoting more dialogue during the early stages of a project to increase efficiency in the requisitioning process as well as long-term reliability. Although there are many different types of pumps that could be discussed, this paper focuses on the radial flow centrifugal pump which is the most common type used in the process industry.

Theory

Head

A centrifugal pump increases the absolute pressure of a fluid by adding velocity energy to the fluid ($\frac{1}{2}mv^2$) and then converting that to pressure energy (mgh) in the volute. Consider Figure 1 below. At point 1, the fluid is sucked into the impeller eye at a velocity v_1, essentially the volumetric flow rate divided by the cross-sectional area of the impeller eye.

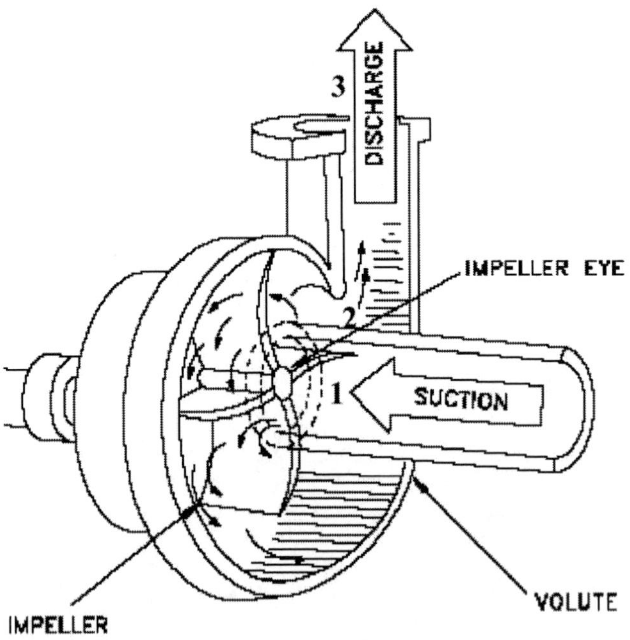

Figure 1. Fluid Flow of Pump

At point 2, the velocity and pressure of the fluid have been increased by the rotation of the impeller. At point 3, the velocity is the volumetric flow rate divided by the area of the exit, the fluid being slowed down through the volute leading to the discharge nozzle. The energy added by the impeller has been converted to pressure head as discussed below.

Figure 2 below shows the velocity diagram of a radial flow impeller. The variable v is the velocity of the fluid relative to the rotating impeller and is approximately equal to the flow rate divided by the cross sectional area through which that portion passes. Euler's pump equation relates the head to the change in velocity of the fluid through the impeller shroud as:

$$\Delta H = \eta_{HY}\Delta(UV_u)/g \qquad (1)$$

Where:

η_{HY} = hydraulic efficiency (does not include mechanical losses),
U = centrifugal velocity component Ωr (angular velocity times impeller radius),
V_u = circumferential velocity component of the absolute velocity vector V.

The absolute velocity **V** at any point along the fluid's route up the impeller vane is found by adding the vectors **U** and **v**.

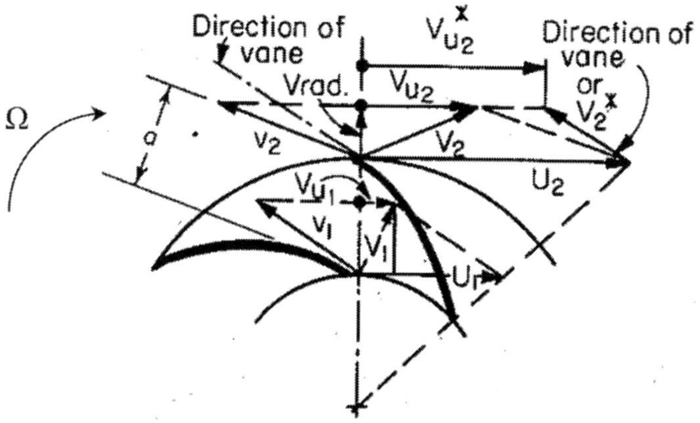

Figure 2. Velocity Diagram of a Radial Flow Impeller [Source: Ref. 8]

The pressure head H is defined as the increase in height of a column of fluid that the pump would create if the velocity head were converted without loss into elevation head. Equation 1 above shows that head is not dependent on the fluid being moved. However, to get the actual change in pressure, ΔP, resulting from this head, the following formula is used:

$$\Delta P = \rho g H \qquad (2)$$

Where:

ρ = the density of the fluid
g = acceleration of gravity

The above formula shows that a low density fluid such as ethylene requires more head to produce the same differential pressure as a higher density fluid such as water. This means that more and/or larger impeller stages are required for a low density fluid to produce the same differential pressure as a higher density fluid.

To summarize, the differential pressure required by the process flow is created by adding energy to the fluid through the spinning impeller. The amount of pressure increase is directly related to the density of the fluid (ρ) and the product of the impeller radius and the shaft rotation speed squared $(r\Omega)^2$. The initial cost of the pump is related to the radius of the impeller which determines the pump size. Reliability of the pump is somewhat related to the speed of the shaft, with slower speeds generally more reliable and requiring less NPSH as will be discussed later. Keep in mind that a slow speed pump needs a larger impeller to produce the same head as a fast speed pump, but the increase in reliability may offset the initial cost of the larger pump.

Specific Speed

Specific speed is a convenient variable used to optimize the geometry of a pump. It is based on the scaling and similitude characteristics of centrifugal pumps which state that the head coefficient, Ψ, and the flow coefficient Q_s, are constant for similar geometries, where:

$$\Psi = (g\ \Delta H)/(r\Omega)^2 = \text{constant} \tag{3}$$

$$Q_s = Q/(r^3\Omega) = \text{constant} \tag{4}$$

and Q is the volumetric flow rate of the fluid. The specific speed is defined by the dimensionless quantity:

$$\Omega_s = (Q_s)^{1/2}/\Psi^{3/4} \tag{5}$$

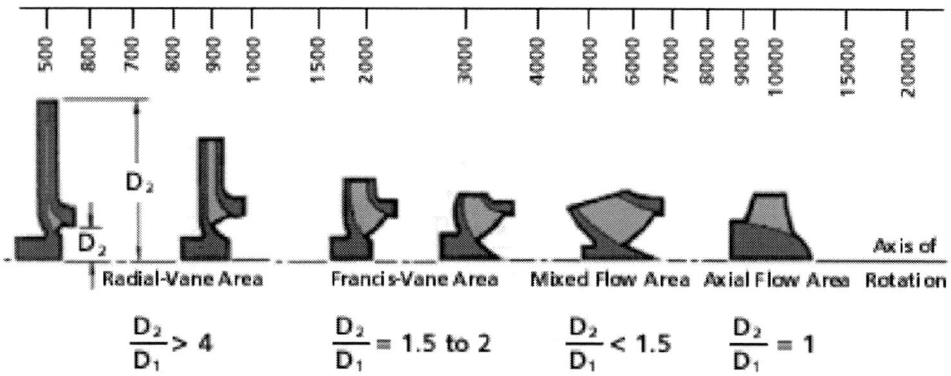

Figure 3. Impeller Design vs. Specific Speed N_s (Source: Ref. 6)

For convenience, specific speed is generally represented in terms of N (rpm), Q, and ΔH in U.S. or metric units as:

$$N_s(\text{U.S.}) = N(\text{rpm})[Q(\text{USgpm})]^{1/2}/[\Delta H(\text{ft})]^{3/4} \tag{6}$$

$$n_q(\text{metric}) = N(\text{rpm})[Q(\text{m}^3/\text{s})]^{1/2}/[\Delta H(\text{m})]^{3/4} \tag{7}$$

$$n_q = N_s/51.64 \quad (8)$$

Figure 3 above demonstrates how impeller shape varies with specific speed. High flow, low head pumps at the right end of the spectrum require axial flow impellers. Radial flow impellers are to the far left. Notice that the further left you go, the larger the diameter of the impeller and the narrower the passage. There comes a point where the friction in the passage decreases efficiency and produces heat such that centrifugal-style impellers are not practical. Left of the scale shown requires positive displacement-type pumps that can handle extremely low flow and high head applications.

NPSH and Minimum Flow

Net Positive Suction Head (NPSH)

NPSH is the difference between the suction pressure and the vapor pressure of the fluid. It is defined by the equation:

$$NPSH = (P_{in} - P_v)/(\rho g) \quad (9)$$

Where:

P_{in} = inlet or suction pressure
P_v = vapor pressure of the fluid

For convenience, we define net positive suction head available (NPSHA) as the NPSH available to the pump based on the configuration of the upstream piping and anticipated fluid levels of upstream vessels. Net positive suction head required (NSPHR) is the amount of NPSH required by the pump to operate properly.

As the fluid enters the pump inlet, losses occur, resulting in a decrease of static pressure. Referring to Figure 4 below, these drops are due to inlet passage losses (A to B), frictional losses (B to C), and blade loading and loss within the impeller (C to D).

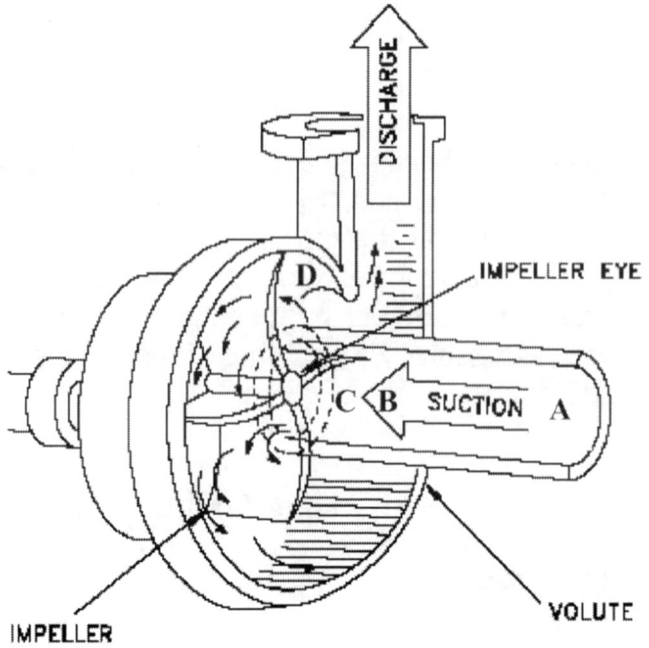

Figure 4. Pump Suction Head Losses

If the static pressure drops below the vapor pressure, the fluid begins to boil, creating vapor bubbles and decreasing the density of the fluid. When this occurs, the differential pressure created by the dynamic head of the impeller decreases (see equation 2). NPSHR is defined by API 610 (ISO 13709) as the amount of suction head resulting in a 3 percent loss of head at the first stage of the pump using water as the test fluid. Reduction factors for hydrocarbons are not allowed.

In addition to pressure loss, insufficient NPSH can also result in pump cavitations. Cavitations occur when vapor bubbles that have formed in low static pressure areas rapidly collapse. This occurs as the bubbles move along the impeller vanes into higher pressure areas. As these bubbles implode, the subsequent forces begins to erode the impeller vane surfaces, causing progressive pitting damage such as that shown in Figure 5 below. There is a distinct noise associated with cavitation described as a crackling sound like the noise when a fluid begins to boil. It should be noted that at 3 percent head loss, cavitation is already occurring. Acceptable margin between NPSHR and NPSHA is therefore required to promote pump reliability, particularly for high suction energy pumps. A minimum NPSH margin of 1 meter is an industry rule of thumb. However, since NPSH required increases with increased flow (see Figure 8 below), thought should be given to the maximum expected flow as well as the rated and normal flow when deciding on the required margin.

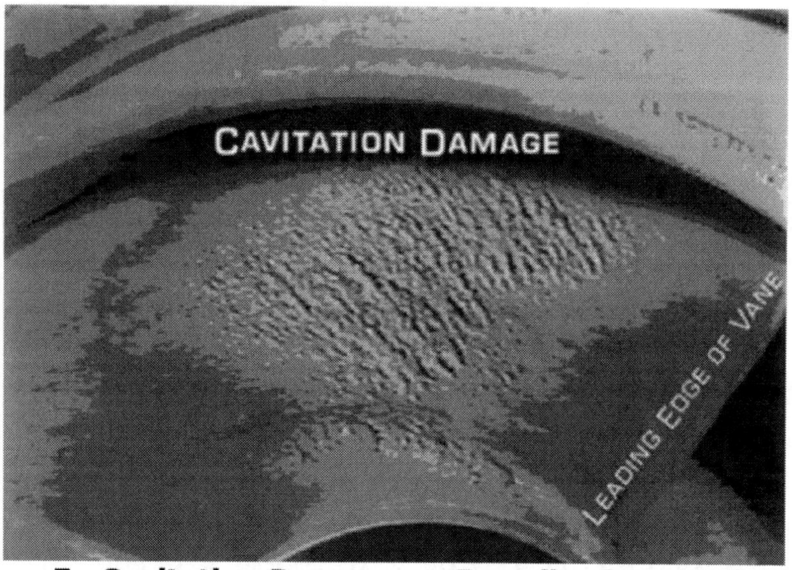

Figure 5. Cavitation Damage on Impeller (Source: Ref. 5)

NPSHA is the suction head available to the pump. As established in the above discussion about NPSHR, there is a need for sufficient margin between the NPSH available and that required by the pump. NPSHA can be easily calculated based on the piping configuration and process fluid levels in upstream vessels. Consider Figure 6 below. NPSHA is calculated as:

$$NPSHA = (P - P_v)/\rho g + L_h - f_L \qquad (10)$$

Where:

P = absolute pressure on the surface of the fluid
L_h = static suction head/lift in feet/meters from the impeller centerline
f_L = suction pipe friction loss in feet/meters

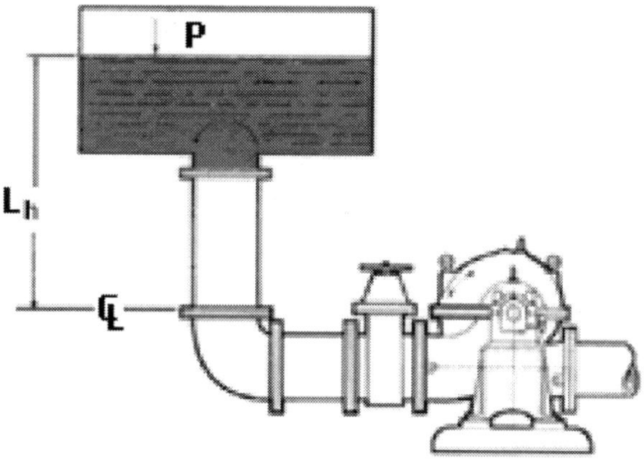

Figure 6. Pump NPSHA (Source: Ref. 4)

If the fluid is in a tank open to the atmosphere, P is the absolute barometric pressure. The sign of L_h depends on whether the fluid is located above or below the centerline of the impeller. Fluid above the centerline produces a positive head while fluid below the centerline imposes a negative lift on the value of NPSHA. As stated above, there should be a safe margin of at least 1 meter between NPSHA and NPSHR to account for all possible process conditions as well as increased losses due to aging.

Suction Specific Speed

As described above, NPSHR is the point where the difference in inlet pressure and vapor pressure of the fluid begins to cause a 3 percent drop in performance head. This NPSHR is equal to the inlet velocity head at this point or:

$$NPSHR = V_1^2/2g \tag{11}$$

At the inlet, the velocity is a function of the volumetric flow rate and inlet area:

$$V_1 = Q/A_1 \tag{12}$$

Suction specific speed is a term used by pump designers to quantify the suction capability of the pump as it relates to NPSHR:

$$S = NQ^{1/2}/NPSHR^{3/4} \tag{13}$$

As can be seen, a high S corresponds to a low NPSHR. It would, therefore, be reasonable to assume that higher S is better. However, comparing equations 11 and 12 above demonstrates that decreasing the NPSHR for a particular flow is most easily accomplished by increasing the inlet area. This is not always a good practice as it increases the minimum flow value for the pump as will be discussed below. Industry practice is to try to keep S between 7,000 and 12,000 depending on the fluid and the experience of the engineer. When a pump cannot be found that meets the suction specific speed criteria set for a particular project, a recirculation line can be added to meet the minimum flow requirements.

Minimum Flow

Although the NPSH required by the pump decreases with decreased flow, there are other problems encountered that need to be considered. As flow decreases, the tendency of the fluid to fall back on itself or re-circulate increases. Every pump has a best efficiency point (BEP) flow rate. At this flow rate, fluid moves smoothly through the impeller passage. As the flow decreases, there is not enough fluid to fill in the volume between the impeller vanes. The fluid gets pressed against the pressure side of the vane and recirculation occurs in the suction side of the vane. This is illustrated in Figure 7 below. The horseshoe-shaped vortices erode the impeller both at the discharge and inlet. Pumps with high suction specific speed will have correspondingly high minimum flow requirements that will likely require a recirculation line to provide satisfactory operation.

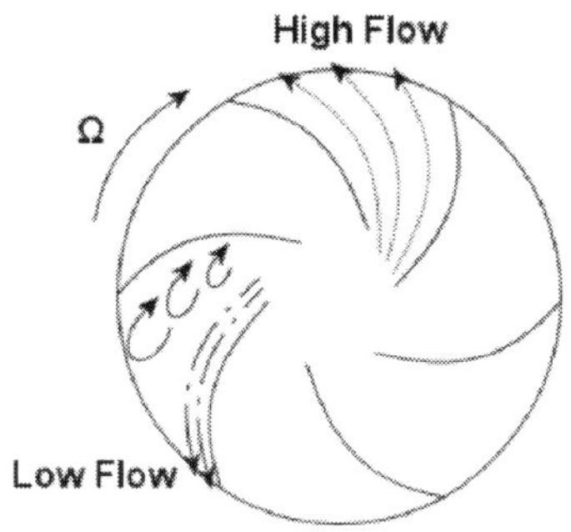

Figure 7. Flow Patterns

One needs to distinguish between minimum stable flow described above and minimum thermal flow. API 610 (ISO 13709) defines each as follows:

> minimum continuous stable flow (MCSF) - lowest flow at which the pump can operate without exceeding the vibration limits imposed by this International Standard
>
> minimum continuous thermal flow (MCTF) - lowest flow at which the pump can operate without its operation being impaired by the temperature rise of the pumped liquid

MCSF as discussed above relates to recirculation of the fluid that can result in cavitation and vibration. MCTF is concerned with temperature rise. A pump efficiency curve (see Figure 8 below) shows that efficiency drops as flow decreases below the best efficiency point. This decrease in efficiency is characterized by an increase in temperature. MCTF is the point at which this rate of temperature increase impairs the operation of the pump. For the purpose of defining the minimum operating flow, one should take the highest of MCTF and MCSF. Generally, MCSF occurs at a higher flow rate than MCTF and becomes the defining variable.

Shut-Off Head and Downstream Piping

There are times in a new plant construction project that one has to execute the system design, layout, piping design, and procurement of equipment concurrently. As a result, careful coordination is required with regard to the shut-off head and downstream piping design pressures. Shut-off head refers to the pressure head at zero flow, or blocked-in condition. Figure 8 below shows a typical performance curve for a radial flow impeller.

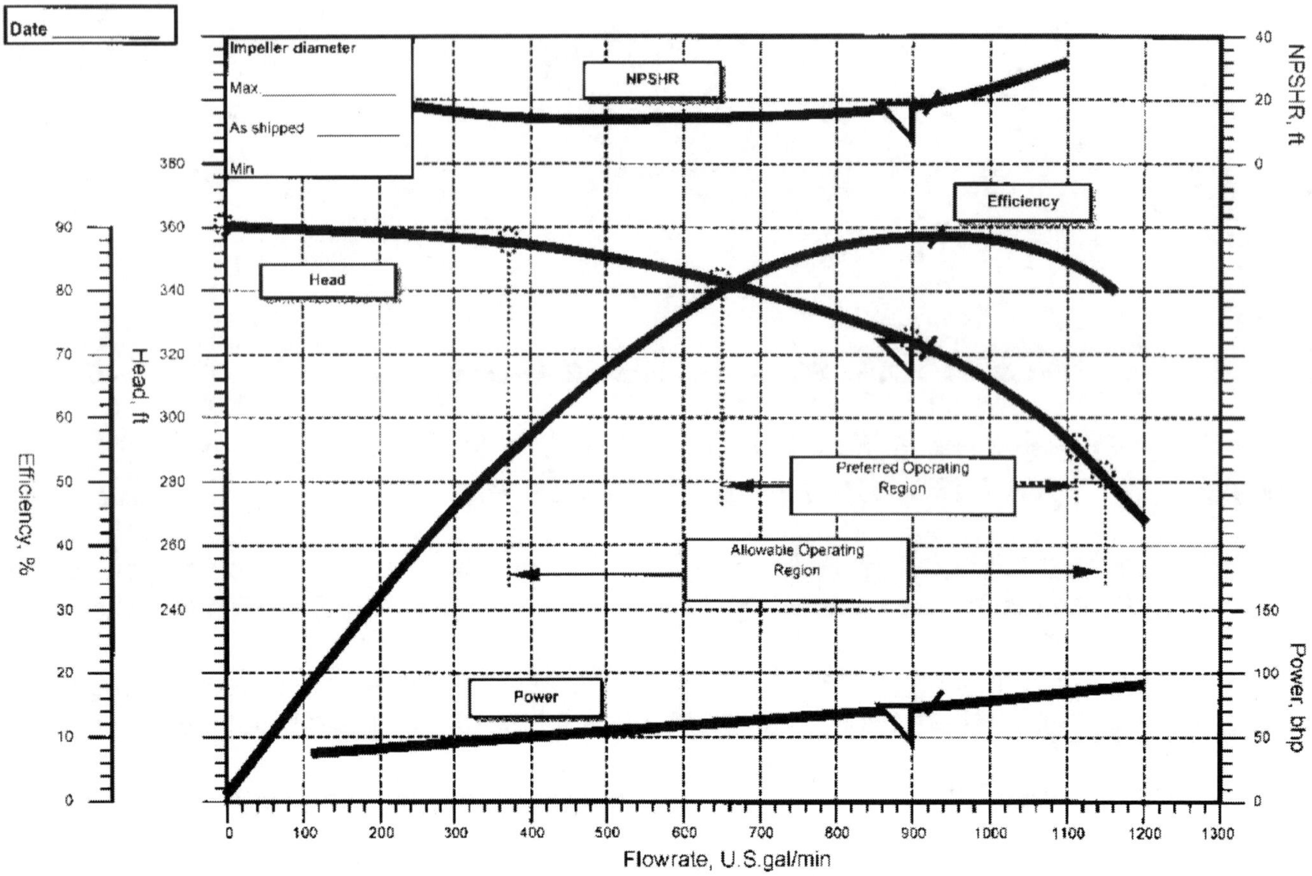

Figure 8. Pump Performance Curve (Source: Ref. 1)

The rise from the rated point to the shut-off point should always be positive and should be from 10 percent to 20 percent of the rated head. However, what is quoted by the pump vendor and what is finally shipped to the site might be different due to the margins allowed by API 610 (ISO 13709). Figure 9 below displays the API tolerances allowed for the pump performance test.

Condition	Rated point %	Shutoff %
Rated differential head:		
— 0 m to 150 m (0 ft to 500 ft)	− 2 + 5	+ 10 − 10 [a]
— 151 m to 300 m (501 ft to 1 000 ft)	− 2 + 3	+ 8 − 8 [a]
— > 300 m (1 000 ft)	− 2 + 2	+ 5 − 5 [a]
Rated power	+ 4 [b]	—
Rated NPSH	0	—
NOTE Efficiency is not a rating value.		
[a] If a rising head flow curve is specified (see 5.1.13), the negative tolerance specified here shall be allowed only if the test curve still shows a rising characteristic.		
[b] Under any combination of the above (cumulative tolerances are not acceptable)		

Figure 9. Performance Tolerances from API 610 (ISO 13709)

Consider the following scenario where the systems engineer has specified the downstream piping and equipment design pressure based on a maximum head rise of 120 percent.

Rated Suction Pressure = 320 kPag
Maximum Suction Pressure = 1450 kPag
Differential Pressure = 525 kPa
Downstream Design Pressure = 2080 kPag
Specific Gravity = 0.8
Head = (525 kPa)/(800 kg/m^3)/(9.81 m/s^2)*1000 = 67 m
Quoted curve gives a rise to shut-off of 17%
Shut-off Head = (67 m)*1.17 = 78.4 m

At the given rated head, the maximum shut-off head can vary by as much as 10 percent.

Shut-off Head at Performance Test = (78.4 m)*1.1 = 86.2 m
$P_{d,mx}$ = (86.2 m)*(800 kg/m3)*(9.81 m/s^2)/1000 + 1450 kPag = 2130 kPag

As shown above, this gives a maximum discharge pressure of 2130 kPag, which is greater than the downstream design pressure. Although this may not be a problem for piping, it could be an issue for heat exchangers or other equipment already purchased based on the rise to shut-off pressure of 120 percent. If the downstream design pressure was never communicated to the rotating equipment engineer who purchased the equipment, the pump could end up with a maximum shut-off pressure greater than the design of the downstream system design. If it was never communicated to the pump vendor, there could be a problem at the performance test if the pump meets API margins on the high end.

The scenario above assumes that an induction motor is used to drive the pump. If a turbine is used, the problem is amplified. NEMA SM23 allows the maximum continuous speed of the turbine to be 5 percent greater than the normal running speed. The pump head changes in proportion to the change in speed, N (rpm), as in the equation below:

$$H_2 = H_1*(N_2^2/N_1^2) \qquad (14)$$

Assuming a shaft speed of 3600 rpm for the pump above and using the maximum API tolerance for shut-off head, the maximum continuous speed shut-off head and pressure could be as high as:

$$H_2 = 86.2 * (3600*1.05)^2/3600^2 = 95 \text{ m}$$

$$P_{d,mx} = (95 \text{ m})*(800 \text{ kg/m3})*(9.81 \text{ m/s}^2)/1000 + 1450 \text{ kPag} = 2196 \text{ kPag}$$

This can create greater problems for downstream equipment. When equipment is purchased in parallel to systems design, the systems engineer should consider the API 610 (ISO 13709) and NEMA 23 tolerances in setting downstream design pressures. Assuming a maximum allowable shut-off head of 120 percent, the design head tolerances are shown in Table 1 below.

Table 1. Design Head Tolerances

Rated Differential Head	Design Head Tolerance (Motor Driven)	Design Head Tolerance (Turbine Driven)
0 m to 150 m (0 ft to 500 ft)	132%	145.5%
151 m to 300 m (501 ft to 1000 ft)	129.6%	142.9%
> 300 m (1000 ft)	126%	139%

Ethylene Product Pump Example

There are good reasons why some customers prefer horizontal pumps over vertical pumps. This preference must be considered from the beginning for the ethylene product pump since NPSHA for horizontal pumps is difficult to change once the system design, including suction vessel elevation and the operating pressure of the main compressor, has been set. Horizontal pumps have the advantage over vertical pumps of being easier to maintain and less susceptible to excessive vibration problems. They have the disadvantage of being less flexible with regard to NPSH requirements since the first stage of a vertical can pump can be lowered to compensate for low available NPSH.

Regarding the purchase of an ethylene product pump, it is particularly important to specify the critical exposure temperature (CET) and the NPSH available. The CET greatly affects material selection for the metal parts as well as o-rings. The metallurgy selection changes at -46°C and again at -73°C. O-ring selection changes at -40°C. Knowing the NPSHA will allow the rotating equipment engineer to determine if a vertical pump is required or if a horizontal pump is an option.

As an example of how process conditions affect pump selection, consider an ethylene product pump with the following process conditions.

Rated Flow = 400 m^3/hr
Rated Suction Pressure = 1600 kPag
Discharge Pressure = 5700 kPag
Differential Pressure = 4100 kPa
Head = (4100 kPa)/(460 kg/m^3)/(9.81 m/s^2)*1000 = 909 m
NPSHA = 6.2 m
Specific Gravity = .46

First determine the number of stages required with the assumption that each stage produces the same head and that an N_s value (see Figure 3 above) of about 1100 or an N_q value of about 21.3 is desired. Using a modified version of equations 7 and 8 and assuming a shaft speed of 3000 rpm:

$$H_{stg} = (3000*(400/3600)^{1/2}/(1100/51.64))^{4/3} = 169 \text{ m}$$

Dividing total head required, 909, by 169 and rounding up to the next highest integer gives six stages producing 151.5 m head per stage. This gives an acceptable specific speed value of:

$$N_s = 51.64*3000*(400/3600)^{1/2}/(151.5)^{3/4} = 1195$$

The size of the impellers can be approximated by assuming that the head is mainly produced by impeller OD tip speed head:

$$H \approx (r*\Omega)^2/g \quad \text{or}$$
$$d \approx 2*(H*g)^{1/2}/\Omega$$
$$= 2*((151.5 \text{ m})*(9.81 \text{m/s}^2))^{1/2}/(3000*2\pi/60 \text{ rad/sec}) = 245 \text{ mm}$$

Since at least 1 meter margin for NPSH is desired, an NPSHR of 5.2m is assured. Using equation 13 above:

$$S = 51.64*3000*(400/3600)^{1/2}/(5.2)^{3/4} = 15{,}000$$

This value is too high, so if a double suction style pump is used, the suction specific speed becomes more acceptable. Note that for a double suction style pump, the flow rate is halved in the suction specific speed calculation.

$$S = 51.64*3000*(200/3600)^{1/2}/(5.2)^{3/4} = 10{,}604$$

Using the on-line program for a major API pump supplier, the selection is shown below, which correlates fairly well with the assessment.

Number of Stages = 6
Rated Impeller Diameter = 352 mm
NPSHR = 4.3 m
S = 10,965
Efficiency = 77.8%
Minimum Flow = 121.8 m^3/hr (30% of rated flow)
Shut-Off Head = 1110 m (122% rise)

Conclusions

Close coordination between the process engineer and the rotating equipment engineer is essential throughout the design and specification phases of a project. This paper presented several important concepts for pump selection that the process engineer should know. Items highlighted were flow rate, head, suction pressure, NPSH available, and maximum discharge pressure. These parameters affect the impeller selection (specific speed), minimum flow requirements, NPSH required, and shut-off head tolerances. Finally, an example showed how these parameters affect the selection of a pump.

References

1. API Standard 610 (ISO 13709), 2004, "Centrifugal Pumps for Petroleum, Petrochemical and Natural Gas Industries," Tenth Edition, American Petroleum Institute, Washington, D.C.
2. Karassik, Igor J., Joseph P. Messina, Paul Cooper, Charles C. Heald, *Pump Handbook*, 3rd Ed., McGraw-Hill Professional, 2001.
3. Elliott, Hurlel G., "Ensuring Centrifugal Pump Reliability Through Proper Selection/Procurement & Design Audits", written for *Rotate '96*, IRR, Dubai, UAE, 1996.
4. *Net Positive Suction Head*, http://www.gouldspumps.com/cpf_0007.html, Goulds Pumps, accessed 18-May-2006.
5. *About Pumps (2004-01:3)*, http://www.pumps.org/content_detail.aspx?id=2180, Hydraulic Institute, accessed 23-January-2009.
6 *Specific Speed and Pump Type*, http://www.gouldspumps.com/cpf_0005.html, Goulds Pumps, accessed 31-January-2009.
7. Tuzson, John, *Centrifugal Pump Design*, Wiley-IEEE, 2000.
8. "Centrifugal and Axial Pumps" by Igor J. Karassik and William C. Krutzsch in "Marks Standard Handbook for Mechanical Engineers", 8th Ed., ed., Theodore Baumeister, Eugene Avallone, Theodore Baumeister III,McGraw-Hill Book Company, NY, 1978.
9. Bloch, Heinz P., Allan R. Budris, *Pump User's Handbook Life Extension*, 2nd Ed., CRC Press, 2006.
10. NEMA SM 23, 1991 (R1997, R2002), "Steam Turbines for Mechanical Drive Service," National Electrical Manufacturers Association, Rosslyn, Virginia.
11. API Standard 682 (IS0 21049), 2004, "Pumps—Shaft Sealing Systems for Centrifugal and Rotary Pumps," Third Edition, American Petroleum Institute, Washington, D.C.
12. Andrews, Dale, "Minimum Continuous Thermal Flow", http://www.lawrencepumps.com/documents/news_vol3_i1_jan.pdf, Lawrence Pumps, accessed 14-February-2009.

14b Gas Turbine Generator and Cracking Furnace Integration

Daniel J. Barnett
Shaw Energy & Chemicals Group
Houston, TX 77077
daniel.barnett@shawgrp.com

Abstract:

This paper includes a review of the benefits of integrating a Gas Turbine Generator with the Cracking Furnaces in an ethylene unit, review of critical design considerations for proper integration, and discussion of associated emissions issues.

Review of benefits will include:

· Effective GTG cycle efficiency

· Impact of integration on Cracking Furnace

o Fuel Firing

o Steam Production

o Thermal Efficiency

· Overall site benefits

Review of critical design considerations will include:

· Full range of GT operations

· Full range of Cracking Furnace operations

· Emergency/upset conditions

· Provisions for maintenance

Discussion of associated emissions issues will include:

· NOx emissions impact of non-integrated and integrated

· CO2 emissions impact of non-integrated and integrated

Design Audits of Machinery – Rotor Dynamics

C. Hunter Cloud, Ph.D.

President

BRG Machinery Consulting

Abstract: Turbomachinery's overall reliability depends heavily on its underlying rotordynamic characteristics. This tutorial briefly reviews some of the important rotordynamic principles and phenomena. Case history examples are presented illustrating the influence of operating conditions on machinery vibration performance.

Introduction

The costs associated with critical machinery downtime in ethylene and other petrochemical plants continue to increase. While machinery OEMs may guarantee their hardware, they do not guarantee these downtime losses in the event the machine does not work reliably over the expected life. Therefore, the end-user bears substantially greater costs and is the one most concerned with minimizing risks. To mitigate the risks of such machinery downtime, there are many types of machinery design audits that can be performed: aerodynamic and thermodynamic performance, blade and impeller stresses, auxiliary system sizing, coupling sizing and rotordynamic performance [1].

The fundamental objective of all such audits is to identify problems before manufacturing to avoid delays during shop testing and downtime in the field. If a rotordynamic problem is encountered after manufacturing begins, the end-user may have to deal with delays in machine delivery. If the problem is not identified until after field commissioning, the implications include reduced production due to restricted operating range (speed, pressure, power, flow, etc.), poor long term reliability/performance reducing time between outages, or even failure threatening life and environment.

Given these implications, it is important to perform rotordynamics audits for not only new machines but also for any modification of existing machines in critical applications. Debottlenecking projects that may involve gas composition or speed changes can adversely affect a machine's rotordynamics just as much as a bearing, seal or coupling design upgrade.

Fluid dynamics determine how a particular fluid will move through and over surfaces. A machine's rotordynamics determine how the rotating shaft will move or vibrate as it spins and works with the process fluid. For petrochemical turbomachinery, the vibrations of primary concern are the shaft's lateral and torsional motions. As shown in Figure 1, lateral vibrations involve the radial or whirling movement of the shaft. Large lateral vibrations can threaten stationary components with tight clearances such as the bearings and seals. Torsional vibration results in twisting of the shaft, where repeated stresses can cause fatigue failure of the shaft.

This paper will focus primarily on rotordynamics involving lateral vibrations. After discussing a few of the key vibration principles and phenomena, the aspects of rotordynamics analysis will be reviewed with specific considerations involving some ethylene plant machinery. A more comprehensive and detailed tutorial on rotordynamics is provided by the American Petroleum Institute (API) [2].

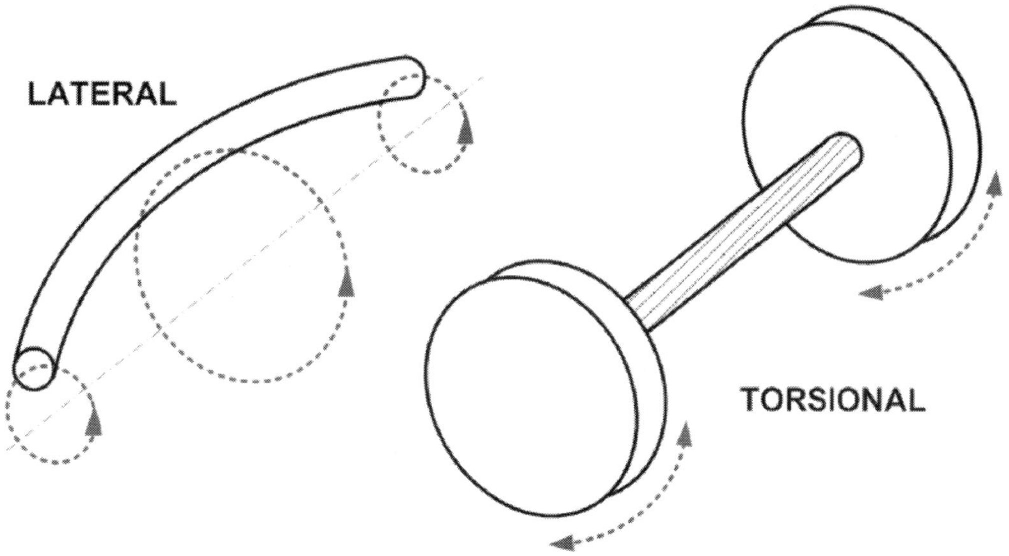

Figure 1: Lateral versus torsional shaft vibrations

Vibration Principles and Phenomena

Throughout a machine's operating range and life (between turnarounds), maintaining small rotor vibration amplitudes is our basic desire. Small rotor vibrations result in low stresses on the shaft, bearings and surrounding equipment. Furthermore, when lateral vibrations are kept at low levels, machine designers are able to implement tighter seal clearances yielding better thermodynamic efficiency of the machine. Though numerous vibration phenomena threaten these benefits, two deserve the most attention, resonance and instability.

Resonance

Resonances are a concern because small excitations can be amplified to yield very large vibrations. Strictly speaking, a resonance condition occurs when an excitation's frequency nearly coincides with a system natural frequency. An opera singer breaking a wine glass is a classic example of resonance, where the singer's pitch nearly equals the natural frequency of the glass. Even though the force of the sound waves is very small, their frequency causes large vibrations in the glass that eventually cause it to crack.

Just like wine glasses, guitar strings and tuning forks, shafts and machine structures have frequencies where they naturally prefer to oscillate when excited. These natural frequencies are dependent on various properties of the system. For a simple weight hung from a spring, its natural frequency, ω_n, equals:

$$\omega_n = \sqrt{\frac{k}{m}}$$

where k is the spring's stiffness and m is the weight's mass. A turbomachine will have many natural frequencies, not just one, that are also dependent on its stiffness and mass properties.

As mentioned earlier, the recipe for resonance demands not only a natural frequency, but also the presence of an excitation source with a similar frequency. The biggest excitation source of concern within turbomachinery is shaft unbalance. Just like an unbalanced washing machine load or an unbalanced tire, shaft unbalance occurs when the center of mass is not coincident with the center of rotation. Caused by material imperfections within the shaft, manufacturing tolerances and anything that affects the rotational mass distribution (such as fouling), unbalance can never be totally eliminated.

The excitation force produced by unbalance has the same frequency as the rotor speed. Therefore, any natural frequency from zero to trip speed can be excited by shaft unbalance. As shown in Figure 2, whenever the rotor speed coincides with a natural frequency, the lateral vibration amplitudes can increase dramatically. These resonance situations are called "critical speeds."

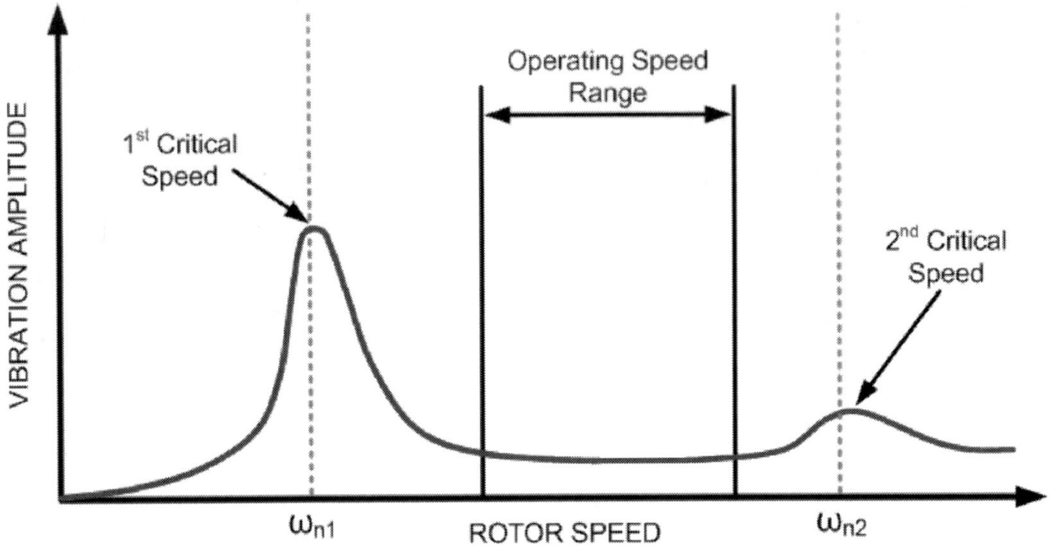

Figure 2: Bode plot of vibrations versus rotor speed

Whether or not a critical speed will cause large vibrations is highly dependent on damping in the system. Adding wine to the glass makes it more difficult, if not impossible, for the singer to crack the glass. Relative to the glass material by itself, the wine has very large damping helping to decay any vibrations quickly. The oil film within the journal bearings provides the main source of damping for many turbomachines. However, for pumps, the main source of damping comes from the process fluid within the close clearance wear rings.

Figure 3 shows how increasing the bearings' damping can reduce vibrations when going through a critical speed. However, bearings have a limit to how much damping they can provide and this may not be enough to restrict critical speed vibration levels. When this is the case, the critical speed must be avoided through startup procedures (quickly accelerating through it) as well as ensuring adequate separation margin with the operating speed range. As shown in Figure 2, the first critical speed often has less damping than the second critical speed. Most large compressors and turbines are designed to operate between their first and second critical speeds.

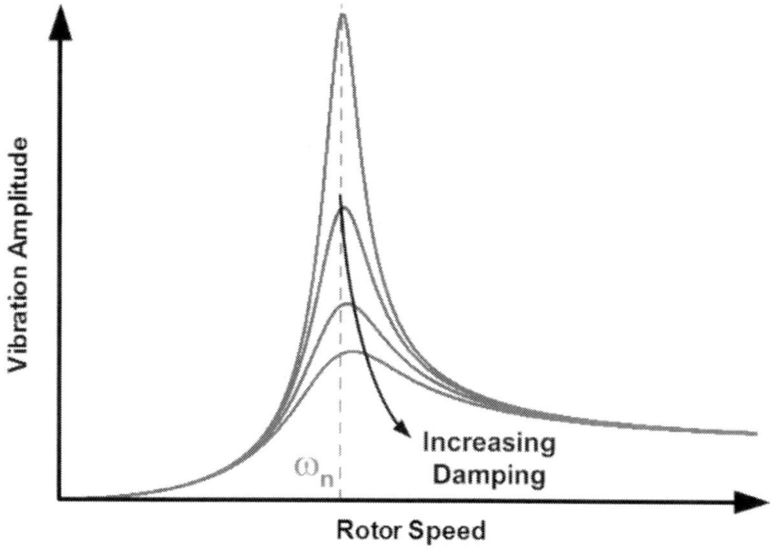

Figure 3: Bearing damping's influence on critical speed amplification

Instability

Resonances and critical speeds are phenomena falling into the class of forced vibration problems. Without the presence of an excitation source like unbalance, the vibration would not occur. Another class of vibration problems exists, however, which are self-induced or self-excited in nature. Self-excited implies that the forces causing the vibration are controlled by the vibration itself. Turbomachines can experience these self-excited vibrations through a phenomenon called rotor instability [3].

A common example of self-excited instability is the unpleasant squeal sometimes heard to due feedback within a microphone-amplifier-speaker system. The frequency of the squeal is actually the natural frequency of the sound system, and it is so loud because the system has no damping allowing the squeal to grown uncontrollably. Eliminating the squeal is accomplished by reducing the amount of feedback by either turning down the volume, moving the speakers, or using a directional microphone.

Acting just like feedback mechanisms, various components within a turbomachine produce forces whose magnitude and direction are functions of the rotor's motion. Their strength is often dependent on rotor speed and the process fluid's pressure and density. Figure 4 shows how these destabilizing mechanisms encourage vibration by acting tangential to the rotor's whirling motion. Battling against these destabilizing forces is damping. When the damping forces are larger than the destabilizing forces, the vibrations remain stable.

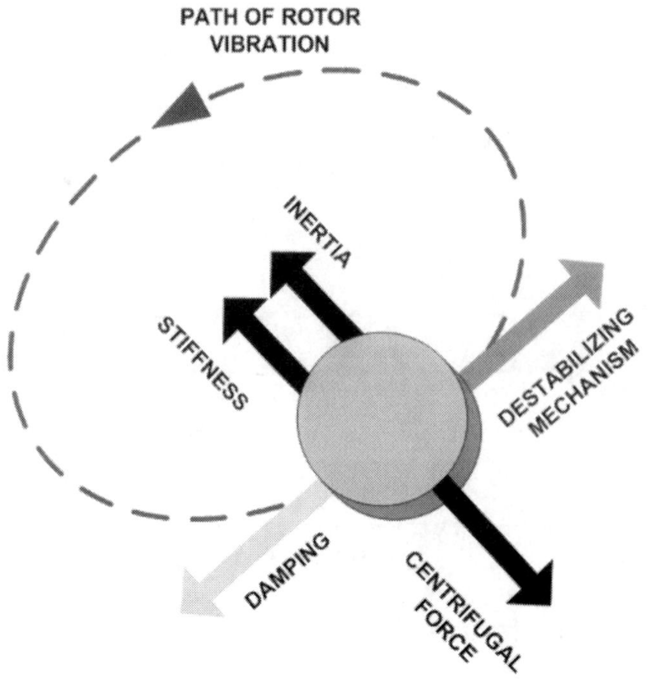

Figure 4: Forces battling over rotor stability

Figure 5 shows how rotor vibrations can grow when destabilizing forces win the battle. For this experiment on an actual rotor, the destabilizing mechanism could be manually turned on or off. When turned on, the rotor vibrations grew uncontrollably to very large levels until 3.2 seconds when the mechanism was turned off and stable vibration returned. Unfortunately, it's difficult to simply turn off these destabilizing mechanisms on industrial turbomachines.

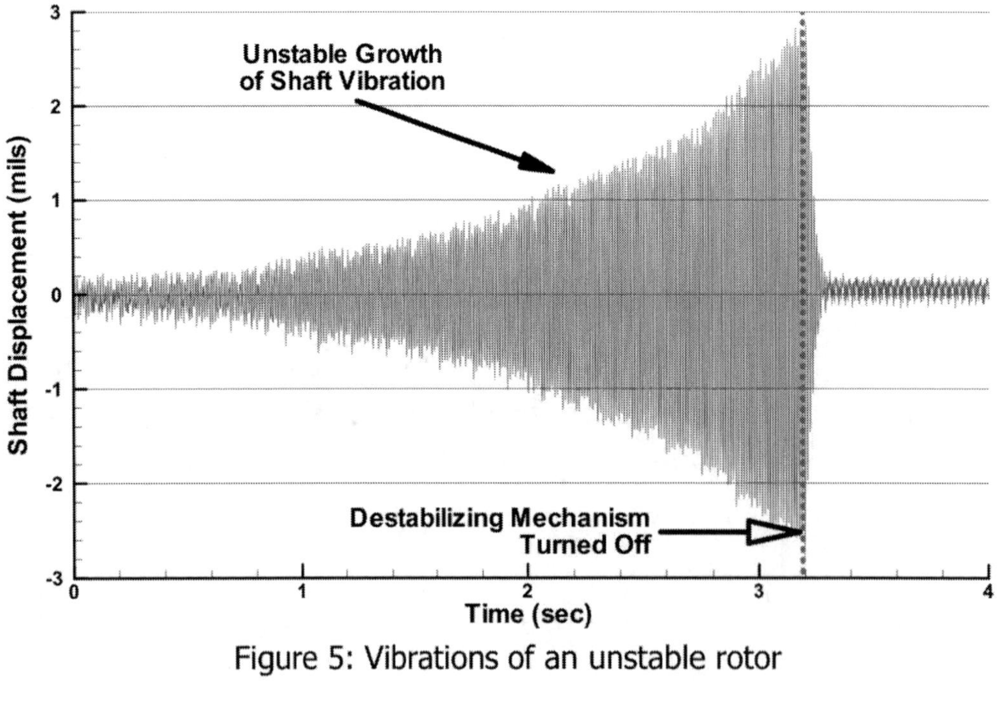

Figure 5: Vibrations of an unstable rotor

Unstable vibrations are able to reach such large levels because of the immense energy available within the spinning rotor. Like a flywheel which stores energy, the spinning motion of a turbomachine's rotor creates a large kinetic energy reservoir. The whirling vibration of the shaft is a separate reservoir that is hopefully small. Destabilizing mechanisms provide a pipeline between these two energy sources, stealing rotational energy and pumping it into whirling vibration energy [4]. When the energy injected into the whirling vibration exceeds the energy dissipated by damping, the rotor vibrations become unstable and large because their source, rotational energy, is so large.

Another reason rotor instability can lead to large vibrations is because the rotor's natural frequency is excited. Figure 6 shows the typical situation when a compressor experienced instability. Each line represents the frequency spectrum at a particular rotor speed. At 6,500 rpm, the only vibration present was due to unbalance with a frequency component at 108 Hz (equal to 6500 cpm). As speed increased, however, destabilizing forces began winning the battle over damping causing the rotor to also vibrate at its natural frequency near 70 Hz. When the rotor speed reached 9,500 rpm, the rotor instability generated very large vibrations at the natural frequency. Unable to reach its maximum continuous speed of 10,000 rpm, operation of the compressor was limited to 9,000 rpm to avoid the damaging nature of this instability.

Restriction of the operating speed range is just one of the costly implications possible with resonance and instability vibration problems. Other restrictions may be imposed such as limiting the machine's inlet flow or the discharge pressure. The vibrations, in some cases, may be such that the machine is essentially inoperable. In other extreme cases, the machine can be completely destroyed before there is even time to shut it down. Developing an accurate model of the rotor's dynamics allows the design to be audited, helping to avoid these collateral problems.

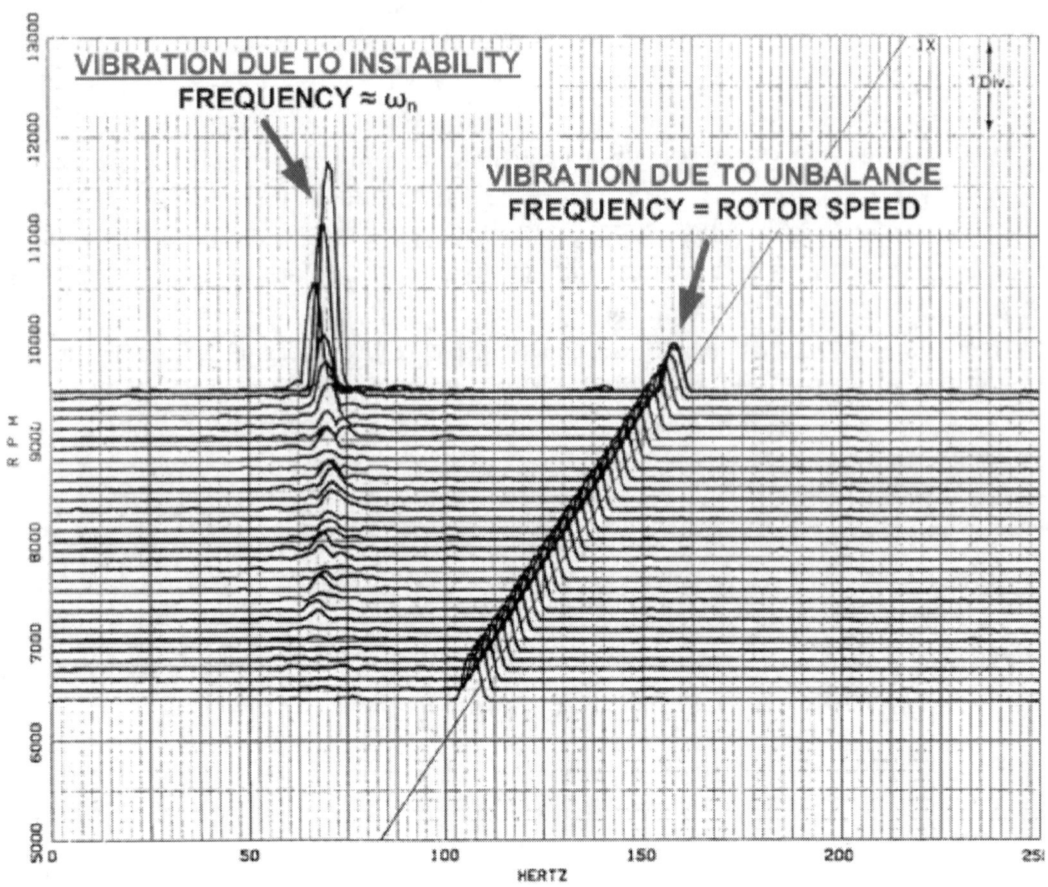

Figure 6: Vibrations due to rotor instability versus unbalance

Rotordynamic Modeling and Auditing Aspects

Predicting or simulating the vibrations of a turbomachine involves transforming the physical system into a mathematical representation. This is accomplished by modeling all the various components within the machine (rotor, bearings, seals, coupling, etc.) as a combination of masses, springs or dampers. Various engineering disciplines are necessary to develop these component models. Solid mechanics principles are applied to transform the shaft into a series of masses interconnected by springs. Modeling the stiffness and damping properties of the fluids within the bearings and seals relies heavily on fluid dynamic and heat transfer principles [5]. Once all the components are modeled, their masses, springs and dampers are assembled together into a rotordynamic system like that shown in Figure 7.

To develop this rotordynamic model requires extensive design information from the machinery manufacturer. Maintaining the highly proprietary and confidential nature of this information is essential for the end-user or independent party conducting the audit. For the rotor, this information includes detailed dimensions across the entire length as well as the weights and locations of all mounted components such as impellers and couplings. Journal bearing design information needs include dimensions, clearances, as well as material and lubricant properties. API provides a detailed listing of the necessary information.

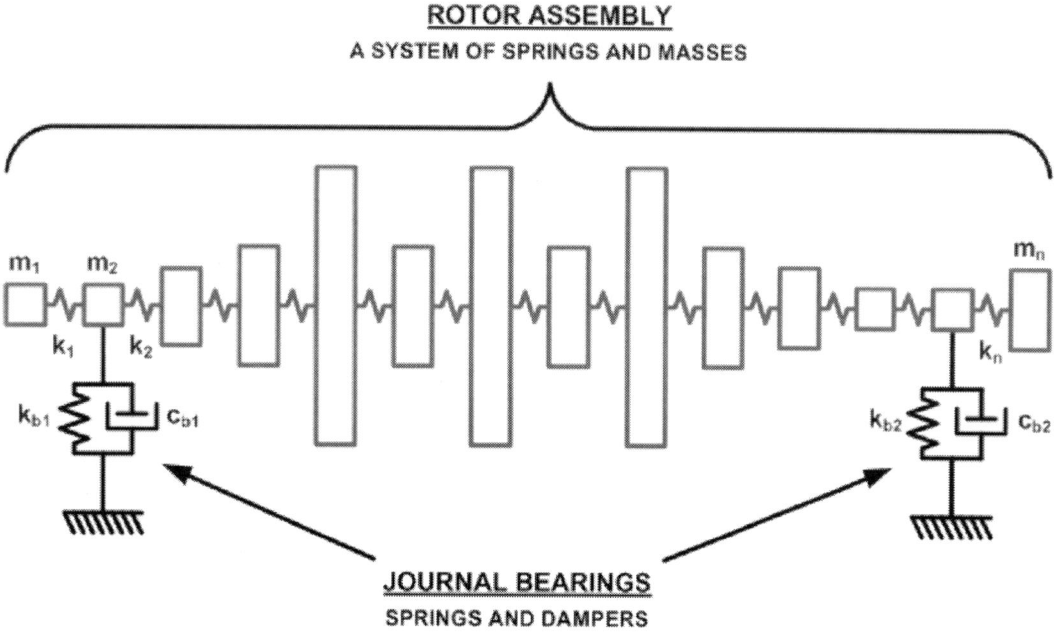

Figure 7: Dynamic model of rotor system

Once the rotordynamic model is completed, it is used in various analyses to predict how the machine will vibrate during operation. The main lateral rotordynamic analyses required by API are unbalance response and stability. These two analyses address the two largest vibration threats that were highlighted earlier, resonance and instability.

Unbalance Response Analysis

Unbalance response analysis involves applying known unbalance(s) to the rotordynamic model and simulating startup/shutdown of the machine. This analysis predicts whether or not the following design objectives will be achieved:

- Avoid resonance situations with respect to the operating speed range (maintain an adequate separation margin away from lightly damped critical speeds)
- Maintain low vibrations (much smaller than the smallest clearance) in the operating range that are robust to balance degradation over time (such as due to fouling, up to a certain practical limit)
- Ensure vibrations do not cause damage during startup and trip (no reduction in efficiency/performance due to rubbing/wear of seals)

Figure 8 presents the results of an unbalance response analysis of a 40,000 hp (30 MW) propylene refrigeration compressor. The first critical speed occurs at 1,545 rpm with a large vibration peak. Within the operating speed range, the vibration is much lower than the API limit of 0.001 inch (25.4 μm) p-p which is very desirable for long term operation. No resonance problems are present, since both the first and second critical speeds are well separated from

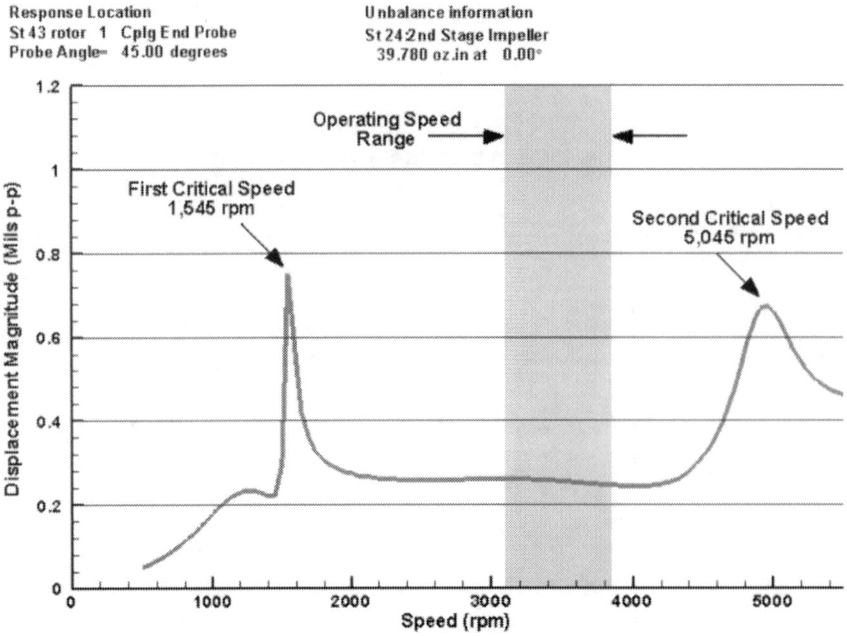

Figure 8: Propylene refrigeration compressor unbalance response

the operating speed range. API requires a certain amount of separation margin depending on the critical speed's level of amplification (the peak's sharpness).

Manufacturing tolerances exist within the journal bearings, and oil supply conditions may change. These potential variations must be also examined to ensure the vibration design objectives are maintained.

Even though the propylene refrigeration compressor's vibrations are low in the operating speed range of Figure 8, they become much larger when going through the first critical speed. During startup, these vibrations at the first critical speed will be avoided by quickly accelerating through this speed. However, during a coastdown, the deceleration rate may be very slow allowing these vibration levels to threaten the tight clearances within the machine.

A rub check analysis is conducted to assess whether or not rubbing will occur. Figure 9 presents the deflected shape of the rotor at the first critical speed. It is important to notice that vibrations at the five impellers within the machine are much larger than those near the bearings, where vibration measurements are typically acquired. These vibrations at the impellers must be compared against the impellers' seal clearances to make sure that rotor-seal contact will not occur, damaging the seals and reducing compressor performance. Many compressors are designed with abradable seals that allow for some limited contact/wear without loss of performance.

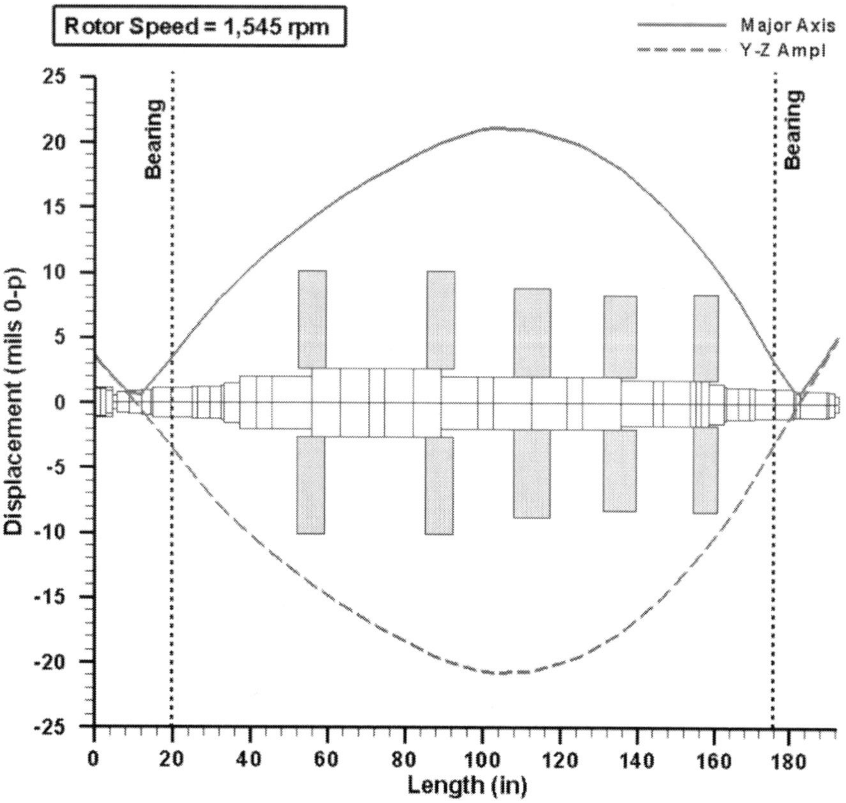

Figure 9: Vibration levels along propylene refrigeration compressor's rotor at first critical speed

Stability Analysis

To avoid the rotor instability phenomenon discussed earlier, stability analysis becomes another vital part of the rotordynamic audit process. Unlike the unbalance response analysis which is conducted across the entire speed range, stability analysis is typically only conducted at maximum continuous speed. Two levels of stability analysis are conducted, where the first consists of a screening analysis to assess:

- Stability level with an empirical estimate of the destabilizing forces present,
- Stability robustness, and
- Stability relative to industrial experience (centrifugal compressors only).

If the screening analysis reveals a marginal machine, a second analysis level is required involving state-of-the-art, detailed modeling of seals and other potentially destabilizing components. API provides specific screening and acceptance criteria for both of these levels of analysis [2].

The stability sensitivity curve is one of the main products of the stability screening analysis. Figure 10 presents this curve for the propylene refrigeration compressor discussed earlier. As the level of destabilizing force is increased, stability decreases until it reaches the threshold and unstable vibrations would occur. Because of the uncertainty in the actual bearing condition, this sensitivity analysis is conducted for the extremes of the bearing

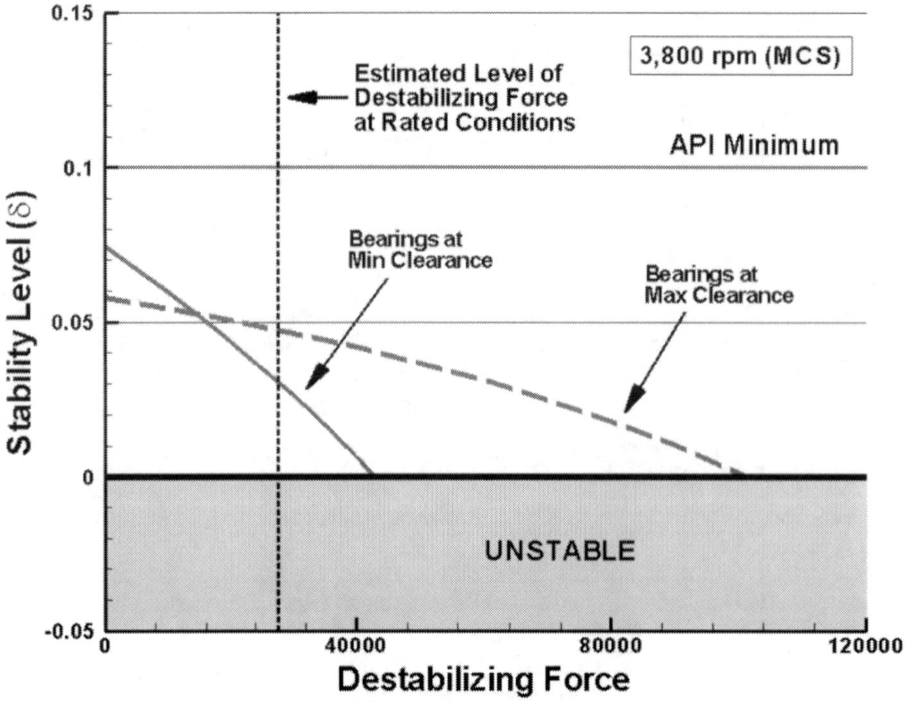

Figure 10: Stability sensitivity curve for the propylene refrigeration compressor

tolerances. Figure 10 shows that, at minimum clearance bearing conditions, the compressor is less robust to destabilizing forces than under maximum clearance conditions.

The estimated level of destabilizing force in Figure 10 is determined based on the speed as well as suction and discharge gas densities at rated compressor operating conditions. Because operating conditions can change, it is important that the machine be robust enough to resist the resulting variations in destabilizing forces. Figure 10 indicates this propylene refrigeration compressor will not be able to handle a large increase in destabilizing forces if the bearings are at their minimum clearance.

Specific Concerns with Ethylene Plant Machinery

During the audit of critical ethylene plant machinery, there are several rotordynamic concerns that typically arise. The rotors within modern cracked gas and refrigeration trains tend to be very large and heavy. This causes their first critical speeds to be very amplified with large vibrations, threatening bearing and seal clearances during startup and shutdown. These heavy rotors also often lead to the machines' supporting pedestals becoming important dynamically. Therefore, rotordynamic analyses need to consider these pedestal dynamics in addition to those of the shaft and bearings [6].

Today, many cryogenic turboexpanders are being designed with active magnetic bearings in place of fluid film bearings. Active magnetic bearings have many attractive mechanical and operational features, but also add significant complexity to the rotordynamic auditing process. The rotor, magnetic bearings, as well as their electronic controls must all be modeled and combined into a single system for the dynamic analyses [7].

Shop Testing versus Analytical Auditing

Some may argue that rotordynamic auditing is not necessary since their machine will be tested in the manufacturer's shop before shipment. Shop testing is certainly an important measure for avoiding field vibration problems. Such testing usually consists of the following:

- a mechanical run test where the machine achieves full speed, but under essentially no load conditions,
- a performance test using an inert gas mixture at reduced speeds and pressures.

Mechanical run tests are useful for identifying critical speed or resonance problems. Unfortunately, neither type of test does a good job of uncovering stability problems because the level of destabilizing forces is not present. An alternative performance test at full speed, power and gas density can be conducted exposing the machine to realistic levels destabilizing forces. However, the high costs associated with this alternative type of performance testing often preclude it from many projects.

Regardless of the shop testing conducted, field conditions can never be totally replicated in the manufacturer's shop. This reality, combined with the fact that any problem identified during shop testing will likely cause a delay in the machine's delivery, makes auditing during the design process a worthwhile initiative. Furthermore, rotordynamic auditing provides the flexibility to examine the machine's robustness to various situations that are likely in field operation, but may not be practical for shop testing. These include variations in gas molecular weight, process fluid density, bearing oil supply temperature, as well as additional unbalance due to fouling.

Case Histories

To demonstrate some of these rotordynamic principles, several case histories will now examined. Two of these cases illustrate how rotordynamic auditing during the design phase could have easily avoided subsequent vibration problems.

Cryogenic Turboexpander

In 1939, Kapitza [8,9] introduced a radial inflow, expansion turbine for cryogenic air separation, a design that would become the father of modern turboexpanders used in ethylene plants. Figure 11 presents a cross section of the Nobel laureate's turboexpander. During the development of this turboexpander, Kapitza encountered what is likely the first documented case of a turbomachine's process fluid (in this case, air) acting like a destabilizing mechanism and causing rotor instability.

To achieve the high thermal efficiency desired, the turboexpander's seal clearances had to be very small (approximately 0.006 inch (0.15 mm)) between the stator and rotor spinning at 40,000-50,000 rpm. Therefore, the rotor's vibrations had to be kept sufficiently small in order to avoid mechanical contact with these tight seal clearances.

With room temperature air supplied to the turboexpander, no vibration problems were encountered. However, at the low air temperatures required for air liquefaction, the

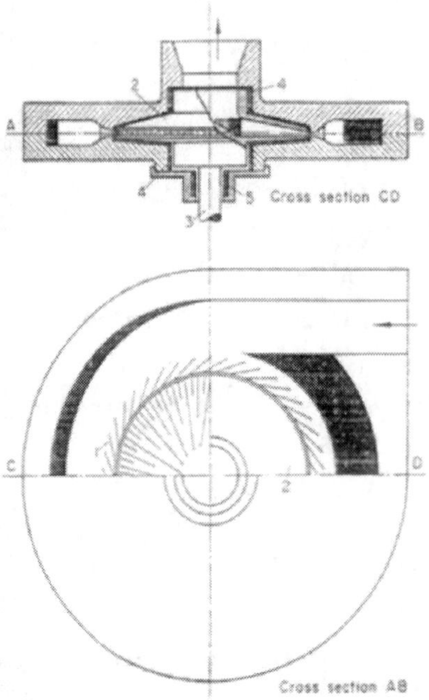

Figure 11: Kapitza's cryogenic expansion turbine [8]

turboexpander could not operate at full speed because of large rotor vibrations damaging the seals. Figure 12 illustrates the type of shaft vibrations observed by Kapitza. The shaft's orbital motion grew with each cycle eventually making contact with the stationary seals. The small, higher frequency oscillations present in Figure 12 are due to unbalance.

Kapitza correctly identified the cause of this instability to be the increased air density, which was three to five times higher than that at room temperature. Recognizing the need for damping to win the battle against air's destabilizing force (see Figure 4), Kapitza implemented a squeeze film damper that eliminated the instability. The turboexpander was then able to achieve normal operating speed and efficiency because its seals were not being damaged. Such squeeze film dampers are commonly employed in modern jet aircraft engines.

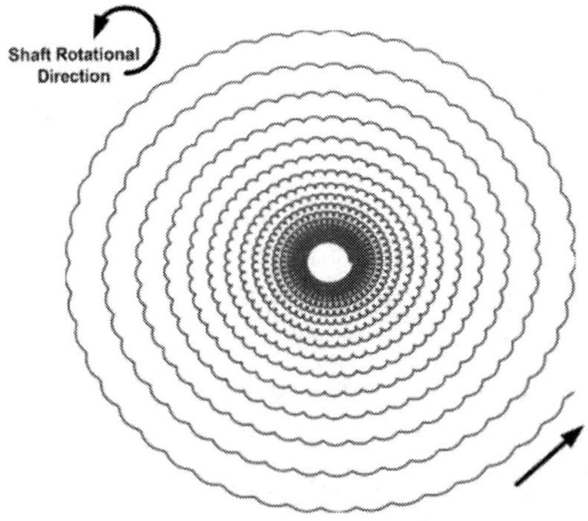

Figure 12: Unstable orbital shaft motion observed by Kapitza

Kapitza's turboexpander provides a good example of the implications when a rotordynamic vibration problem is encountered. Until a time-intensive design solution could be implemented, the machine's operation was limited and his development project delayed. The stakes are raised, however, when a vibration problem occurs on a machine critical to a modern petrochemical plant. Such was the case on a rerated cracked gas compressor.

Cracked Gas Compressor

During commissioning of the rerated 11-stage cracked gas compressor, a surge event occurred causing large shaft vibrations that wrecked the machine before it could be shutdown. Figure 13 presents the orbital motion at one of the compressor's bearings an instant before failure. The amplitude of vibration was over twelve times the normal level.

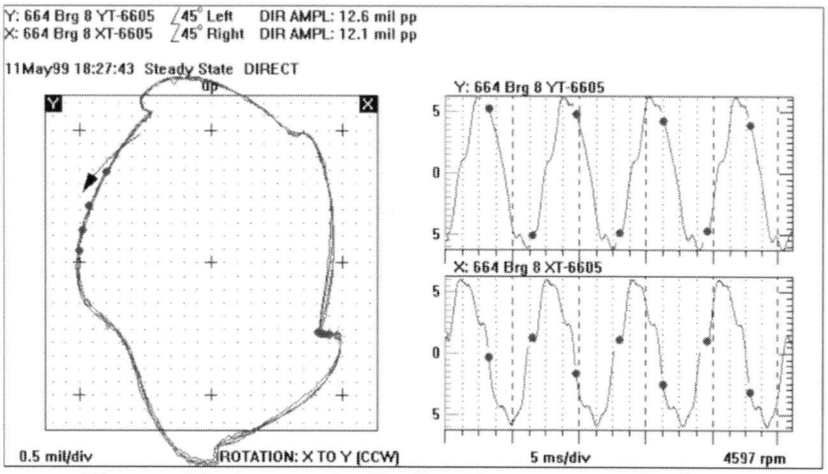

Figure 13: Large shaft vibrations on cracked gas compressor

Using spare parts, the compressor was rebuilt and improvements were made to the surge control system. However, upon restart, the compressor surged wrecking the entire compressor once again. While not desirable in centrifugal compressors, surge does not typically lead to large radial or lateral vibrations of the shaft like those experienced here. Surge, in this case, acted as a catalyst or trigger for the underlying cause of the high vibrations, rotor instability.

Post-mortem rotordynamic analyses by the manufacturer and end-user confirmed that the machine was prone to instability problems. Like Kapitza, more damping in the bearing system was required and a squeeze film damper design was implemented. Figure 14 presents the resulting improvement in the machine's stability characteristics.

Until the machine's design could be modified, the entire ethylene plant's startup was delayed by one month due to this vibration problem. The root cause of this problem was eventually attributed to a simple human error in the manufacturer's original stability analysis. Furthermore, no design audit had been performed by the end-user. If such an audit had been performed, it would have likely avoided the millions of dollars in production losses. A simple

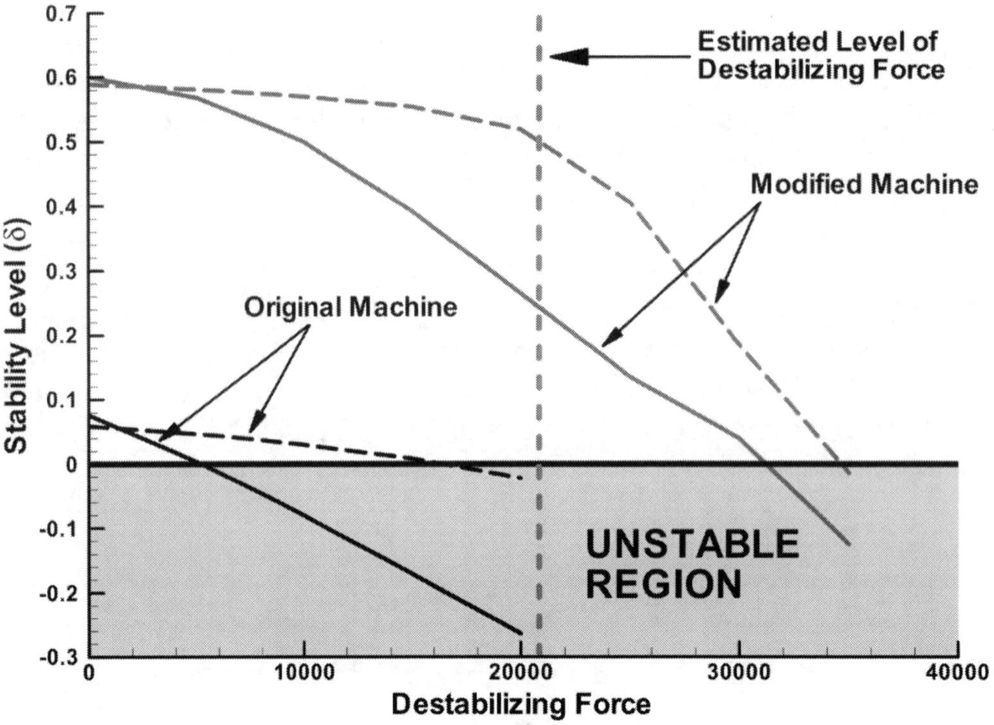

Figure 14: Stability sensitivity curves for original and modified cracked gas compressor at minimum (solid line) and maximum (dashed) bearing clearance conditions

benefit-cost ratio calculation for this audit would be on the order 1000 to 1. In other words, every dollar invested in the rotordynamic audit would have avoided $1000 in production losses.

Ethylene Product Pump Motor

During shop no-load testing, high vibrations were measured on a vertical 1,200 hp (895 kW) induction motor spinning at 3,600 rpm. This new motor was being purchased to replace the pump's existing motor with a higher power model. No rotordynamic design audit had been conducted by the end-user.

Figure 15 presents the motor's vibrations measured during coastdown. At running speed, vibration levels are shown to exceed 0.2 in/sec (5 mm/sec) 0-peak, well above the specified limit. The manufacturer attributed the peak at approximately 1,200 rpm to be the motor's first critical speed, but could not explain the large increase in vibration as speed was increased.

The manufacturer had conducted no rotordynamic analysis on this motor design. Therefore, models of the motor's stator casing and rotor were developed and coupled into a single dynamic system. Figure 16 presents a schematic of this model. Unbalance response analysis identified the vibration problem to be caused by a critical speed separation margin problem. The motor's running speed and a critical speed at 3,700 rpm were in a state of resonance.

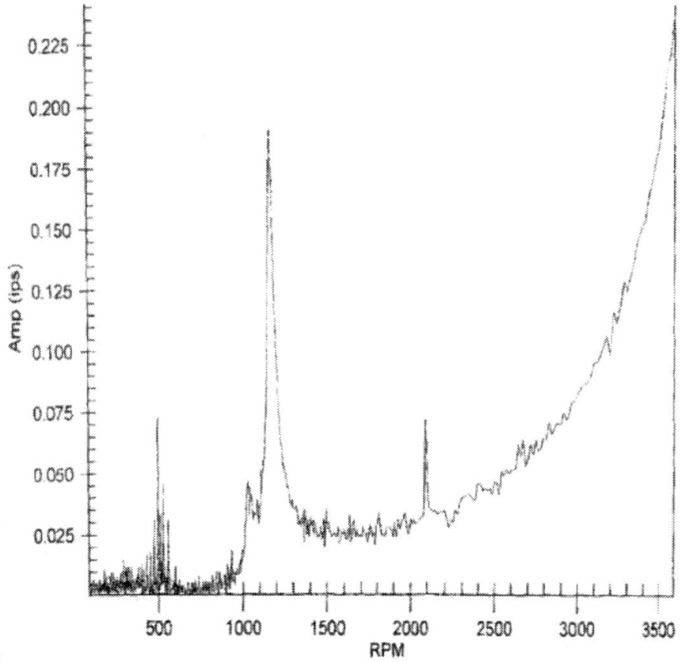

Figure 15: Motor vibration level versus speed

The analysis also identified the 1,200 rpm peak to be associated with a natural frequency where the entire motor housing is pivoting about its bottom mounting flange. Commonly referred to as a "reed critical frequency" [10], this natural frequency is not associated with the rotor's critical speed as thought by the motor manufacturer.

Major shaft modifications were required to correct the resonance problem with the rotor's critical speed. These modifications delayed motor shipment by three months.

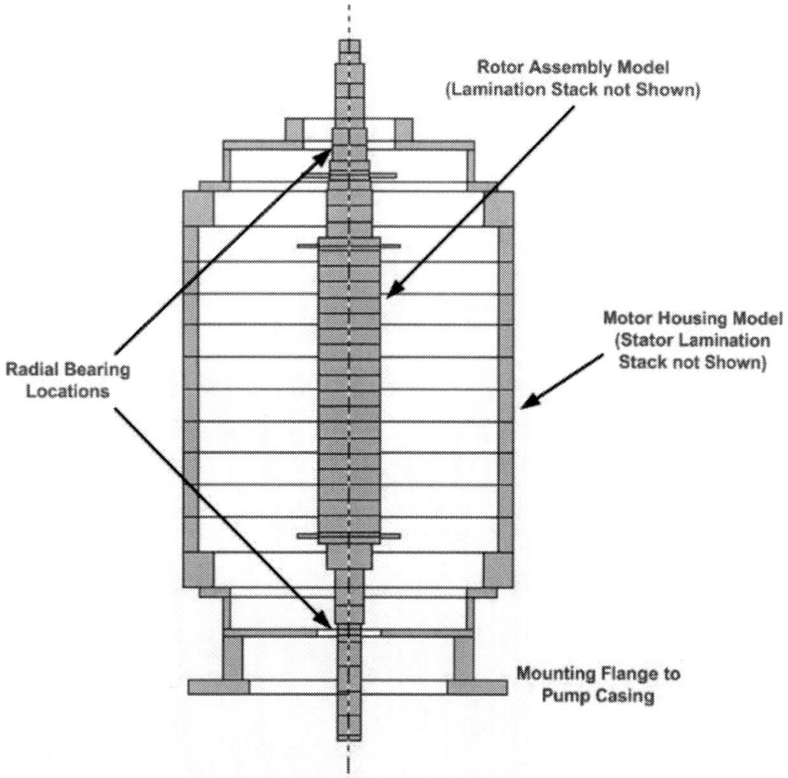

Figure 16: Schematic of motor's casing and rotor models

Summary and Recommendations

A turbomachine's underlying rotordynamic characteristics determine how its shaft will vibrate. Two vibration phenomena, resonance and instability, are the greatest threats to the turbomachine being delivered on time, achieving the process's performance requirements, and running reliably between planned outages. Independent rotordynamic audits during the design phase can help identify these problematic vibration phenomena before they cause shipment delays or downtime in the field.

One of many types of design audits that can be performed, rotordynamic auditing should especially be considered for critical, non-spared machinery as well as for new machine designs without significant experience. Auditing is also highly recommended for machine rerates and any modifications to the rotor assembly, bearings, seals and coupling(s).

Even as modeling and design tools improve, the threat of poor rotordynamic performance will always be a concern. Several realities drive this continued threat. First, continuing improvements and increasing demands of industrial processes require machinery to operate at higher speeds, pressures and power levels. Second, competitive pressures force OEMs to push existing designs beyond their proven envelopes.

References

1. Bloch, H. P. and Geitner, F. K. (1994) *An Introduction to Machinery Reliability Assessment*, 2nd ed., Houston, Texas: Gulf Publishing Co..
2. API (2005) *API Standard Paragraphs Rotordynamic Tutorial: Lateral Critical Speeds, Unbalance Response, Stability, Train Torsionals, and Rotor Balancing*, 2nd ed., American Petroleum Institute, no. RP 684.
3. Ehrich, F. F. and Childs, D. W. (1984) "Self-excited Vibration in High-performance Turbomachinery," *Mechanical Engineering*, vol. 106, no. 5, pp. 66–79.
4. Crandall, S. H. (1983) "The Physical Nature of Rotor Instability Mechanisms," in *Rotor Dynamical Instability*, vol. 55, ASME, Applied Mechanics Division, pp. 1–18.
5. He, M., Cloud, C. H., and Byrne, J. M. (2005) "Fundamentals of Fluid Film Journal Bearing Operation and Modeling," in *Proceedings of the Thirty-Fourth Turbomachinery Symposium*, Texas A&M University, pp. 155–175.
6. Vázquez, J. A., Barrett, L. E. and Flack, R. D. (2001) "A Flexible Rotor on Flexible Bearing Supports. Stability and Unbalance Response," ASME, *Journal of Vibrations and Acoustics*, vol. 123, pp. 137–144.
7. Swanson, E. E., Maslen, E. H., Li, G., and Cloud, C. H. (2008) "Rotordynamic Design Audits of AMB Supported Machinery," in *Proceedings of the Thirty-Seventh Turbomachinery Symposium*, Texas A&M University, pp. 133–158.
8. Kapitza, P. L. (1939) "Expansion Turbine Producing Low Temperatures Applied to Air Liquefaction," *Journal of Physics USSR*, vol. 1, p. 7.
9. Kapitza, P. L. (1939) "Influence of Friction Forces on the Stability of High-speed Rotors," *Journal of Physics USSR*, vol. 1, p. 29.

10. Finley, W. R., Hodowanec, M. M., and Holter, W. G. (2000) "An Analytical Approach to Solving Motor Vibration Problems," *IEEE Transactions on Industry Applications*, vol. 36, no. 5, pp. 1467–1480.

RECOMMENDED PRACTICES FOR SUCCESSFUL DRY GAS SEAL OPERATION IN ETHYLENE PLANTS

Joe Delrahim

Director - Gas Seal Sales for Latin America

John Crane Inc.

Abstract: Never before has there has been so much pressure among ethylene producing plants to achieve maximum compressor reliability and product quality to comply with today's strict local and global emission standards. Compressor shaft seals and sealing systems are critical to plant reliability. This paper will cover shaft seal types and arrangements installed on compressors serving the ethylene industry, advantages and disadvantages of the different seal types, and a description of control and auxiliary system required for each seals type. Critical parameters and milestones necessary to insure long-term continuous operations will be discussed.

"Recommended Practices for Successful Compressor Seal Selection and Operation of Dry Gas Seals in Ethylene Plants"

Never before has there been so much pressure among ethylene-producing plants to comply with today's strict local and global environment standard. Compressors are critical to ethylene plant reliability and in turn, compressor shaft seals are the single most important part of any compressor. Plant reliability, compressor reliability and product quality are all dependent on properly functioning seals.

Compressor reliability and uninterrupted operation is essential for successful ethylene production, especially in situations where a spare compressor is not available. As compressor sizes have increased and compressor-operating requirements push the limit, sealing technology has also continued to evolve. Advancements in sealing technology include reducing emissions, eliminating the high costs of discarding and disposing contaminated oil, and preventing oil from contaminating the process. A 10-year rate of continued compressor operation is now achievable.

Compressor Seal Evolution

Labyrinth/Carbon Bushings

In the early days of compressors, labyrinth/carbon bushings were used to seal process fluid and prevent it from entering the bearing cavity and the outside environment. The labyrinth or bushing clearance was closely controlled to limit the amount of process leakage or emission. Leakage flow rate across the labyrinth, which is a function of the clearance and diameter, can measure as high as 100 scfm.

Circumferential/Contacting Seals

Circumferential/contacting seals require an oil buffer at a higher pressure than the sealing process pressure. Oil leakage from the seal face is drained to a reservoir and is separated from the gas. Speed and pressure is limited to the seal size. Oil leakage may be as high as eight-gallons or more per day. Seal oil support system reliability is vital to the operation of these seals. Process fluid contaminated from oil during the operation or massive seal failure has been a major reliability concern. The high cost of degassing or disposing oil mixed with process gas has forced plants to seek alternative solutions to circumferential/contacting seals.

Non-Contacting Spiral Groove Gas Seals

Demand from regulatory agencies to reduce emissions and a global concern on fugitive emissions is forcing industries to use all available seal and system technology to meet current and expected emission regulations.

The development of non-contacting spiral groove technology is one of the most outstanding achievements in the science of sealing technology. Understanding the related environmental and tribological problems has resulted in a significant decrease in sealing problems along with a significant increase in rotating equipment Mean Time Between Failures (MTBF).

In the early 1980s, non-contacting spiral-groove gas seals were introduced to the compressor industry in response to demand for increased operating design and to comply with Volatile Organic Compound (VOC) regulations. As a result, discarding or degassing contaminated oil became expensive and process fluid contaminated with buffered oil was no longer acceptable.

The operating principal for non-contacting dry-running gas seals is based on face separation due to hydrodynamic lift. Seal face separation is the only way to eliminate the effects of friction due to contact. Spiral-groove design is the most efficient method for separating the seal faces as the shaft rotates.

Pressure distribution at the seal face is illustrated in Fig.1.

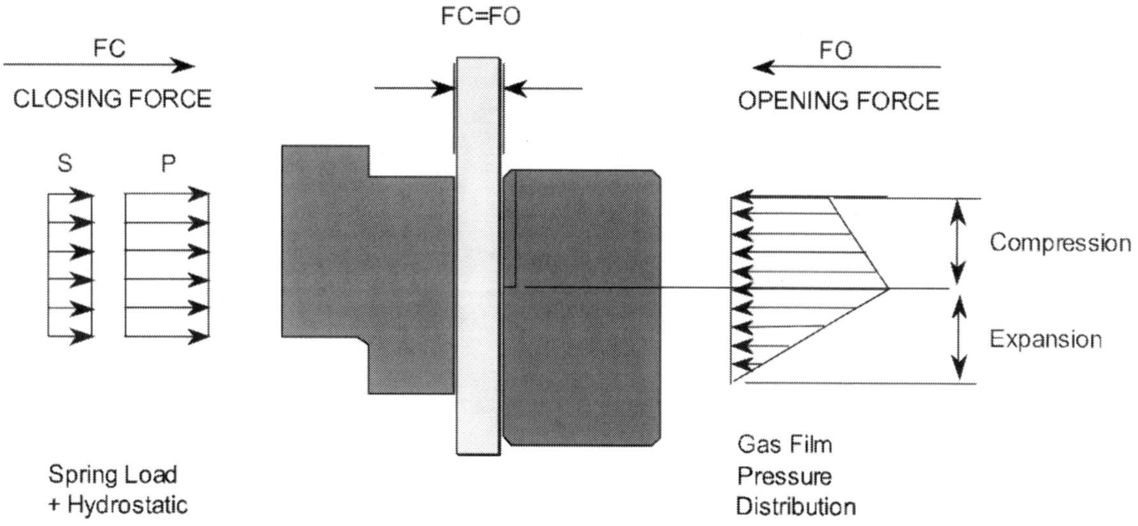

The sealing dam at the inside of the seal face allows the gas to compress and provides the necessary force to separate the seal faces (Fig. 2).

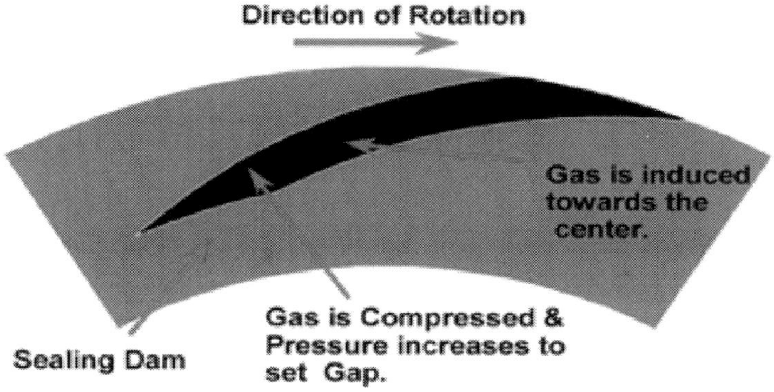

This face separation results in slight leakage across the faces. The leakage flow rate is a function of sealing pressure and speed (Fig. 3).

Dynamic Test Leakage For a 4.937" Dia. Seal

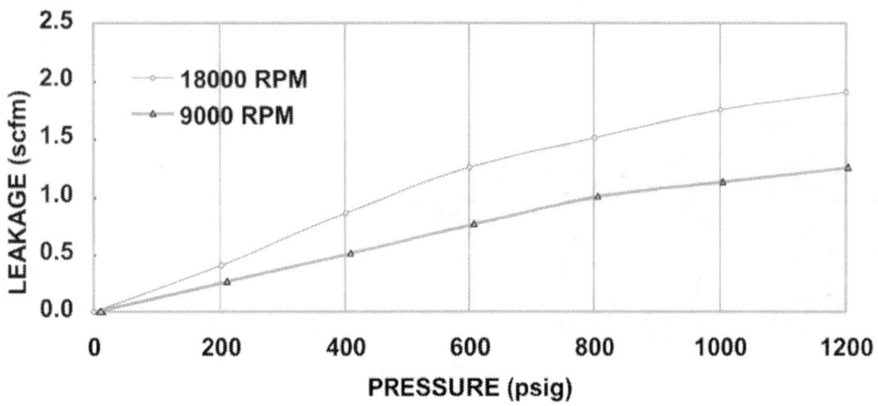

A tandem arrangement gas seal is commonly used for ethylene and propylene applications (Fig. 4).

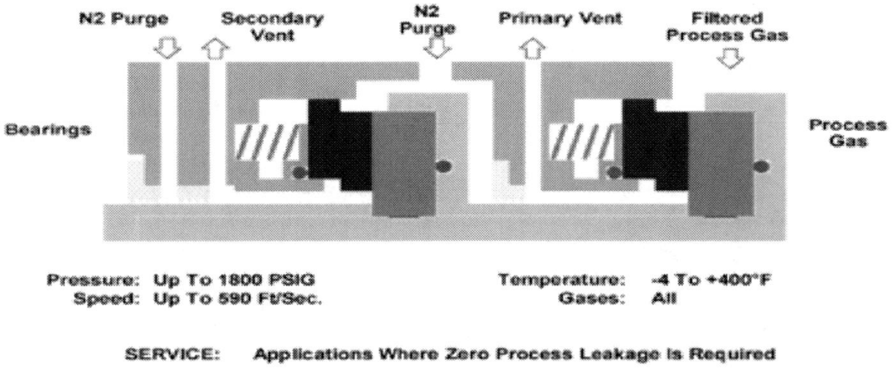

The first seal operates as a primary seal while the second seal functions as a back-up seal. Process gas from the discharge line of the unit is piped to the control system supply line. The control system regulates and filters the buffer flow before it is injected into the primary seal. Leakage across the primary seal is piped to the flare. The pressure and or leakage flow rate are monitored and recorded to insure that the seal functions properly.

To comply with emission requirements, an intermediate labyrinth is installed between the primary and secondary seals. A controlled nitrogen flow rate of one acfm at two-to-three psi above flare pressure is injected between the two seals. As a result, leakage flow across the secondary seals remains as inert nitrogen gas. This sealing arrangement meets and exceeds the fugitive emission regulation, issued in 1991 by the EPA and the California South Coast Air Quality Regulation Management District, limiting fugitive VOC emissions to 1000 PPM.

Retrofitting from Wet Seals to Gas Seals

More and more ethylene plants are retrofitting their older compressors wet seals to dry gas seals. The reasons for converting wet seals to gas seals include unsafe oil sealing systems, the unreliability of oil seals, the high maintenance/operating costs for oil seals, oil usage, process gas loss, the elimination of vent casing during shutdown, environmental/emission requirements, and the operating simplicity of gas seals. When considering a gas seal retrofit, an economic justification/evaluation should be undertaken. As part of such an evaluation, maintenance and operating costs for the actual oil sealing system and the gas seal system should be gathered, if available. Experience shows that the majority of these retrofits are accomplished with little or no modification to the compressor sealing cavity or concerns about the change in unit rotodynamics in almost all cases, the modifications to the wet sealing cavity are reversible if it is necessary to revert to the wet seal system.

Control Systems

The control system insures that clean, dry buffer gas is available at the sealing faces at all times; regulates and maintains buffer gas pressure at above sealing pressure; and monitors seal performance. Clean, dry buffer gas is achieved by careful analysis of process gas composition. This analysis shows where liquid condensate develops due to changes in fluid pressure and temperature. The results of gas composition analysis determines the type of filtration suitable for the condition, such as a simple coalescent, liquid knock-out or, in extreme cases, a cooler to drop the fluid temperature and collect as much liquid as possible.

Normally, the source of the buffer gas supply is from the discharge of the same compressor and/or other reliable fluids such as fuel gas, natural gas or fluid that is compatible with the process gas. It is critical to maintain a supply of clean, dry buffer gas particularly during the first phase of plant commissioning. The clean, dry buffer gas protects the seals from any fluid, objects or contaminants left behind in suction lines during the constriction to reach the sealing environment.

Typically, buffer gas regulation is achieved by a flow control and or a pressure control device. Either method is acceptable as long as steps are taken to assure adequate flow of clean fluid velocity is maintained across the process labyrinth. The seal performance is monitored by measuring seal leakage flow across the sealing faces. Leakage flow is measured locally by flow meter and/or pressure drop across a fixed orifice plate located at the primary vent. In addition, a pressure and/or flow transmitter is installed to validate the leakage rate remotely in a control room. The transmitter's preset alarm and shutdown can be installed to initiate a corrective action.

Training is Key

Another common source of compressor problems is lack of planning and training. For example, many maintenance technicians in plants with dry running gas seals have never received training in seal operation and maintenance. Technicians familiar with wet seals are used to flooding the seal cavity with no consequences; however, in contrast, dry running seals require no lubrication, which means the faces have to be isolated from bearing oil at all times.

Training must be an integral part of the maintenance and operations process prior to the commissioning period and any short-cuts will likely result in a costly plant start-up. Contrary to common practice by many in the industry, mechanical failure or delivery of a new pump or seal is the wrong time to begin a training program, let alone acquaint oneself with the compressor-, pump- or seal-operating manual.

Rather, training programs should be considered as part of supplier selection. Compressor, pump and mechanical seal manufacturers, as well as the engineering firm, should offer training as part of their service. In addition, it is essential that a seal manufacturer provide on-site technicians to help you improve equipment reliability – Mean Time Between Repair (MTBR) – and overall plant productivity. Be wary of any supplier that promises the best technology, yet simply ships you a box containing a product without delivering the proper support.

Finally, communication among all players is vital. Too often, finger-pointing occurs among the engineering firm, compressor technician and the seal manufacturer. Without a clear understanding and effort to bring all parties to the table at the design onset, there are sure to be problems.

Successfully Commissioning the Compressor System

After performing the necessary planning and training, it is time for the all-important commissioning process. However, once again, without a thorough review of the environment and all factors, even the best plans can result in seal failure. In fact, 99-percent of all dry gas seal failures that occur during commissioning are not caused by seal failure. Rather, failure results from a variety of factors including: a compressor flooded with oil or liquid, foreign contaminates in the seal's cavity, and problems with alignment, piping and/or instrumentation.

With so many factors necessary for a successful compressor start-up, it is important to develop a team of critical component experts for the commissioning. This team should include, but not be limited to, the engineering company that selected the critical components, the compressor manufacturer, the seal manufacturer, and the plant owner/operator. Clearly defined areas of responsibility for each party are essential, especially for the many steps that must occur in conjunction with another function or team member. More often than not, unproductive and lengthy start-up periods are a result of parties not fully understanding the responsibilities and roles of each of the critical partners.

Ironically, the time when everything must come together -- the start-up -- is often the only time the team physically comes together. As such, it is crucial to change this model by opening up the lines of communication at the beginning of the project -- not at its culmination -- when it is too late to effect change.

The following guidelines have proven to aid sustainability and increase productivity:

- Focus on creating a shared document – **a strategic agenda** -- that leads participants to action, creates a forum for comments, and outlines ideas, concerns and possible bottlenecks. Goals and objectives must be commonly understood and agreed upon by all parties involved with the commissioning.

- Create a clear, agreed upon **process for reaching these goals**. This process must include reviewing the compressor start-up checklist and assigning definitive deadlines and responsibilities. To be effective, requests should be recorded, tracked and reviewed at each compressor start-up.

- Recognize that other critical partners have data that is crucial to successful implementation, so **share information**. This step includes asking questions and communicating before the compressor start-up, as well as gathering any necessary materials. Make sure someone captures the agenda, discussion points and action items.

- Encourage each critical partner in the commissioning process to develop a **sense of involvement and empowerment**. Such an environment creates ownership, as well as fosters honesty and teamwork.

- **Review the compressor start-up.** There is always room for improvement, so schedule time near the completion of the start-up to review what happened and acknowledge the effort. Such an effort will give you the opportunity to troubleshoot future issues that may occur based on the compressor's history and documentation.

Final Thoughts for Success

Although it once seemed unthinkable, many members of the ethylene industry are now benefiting from 10-plus years of continuous operation from their compressor and seal arrangements. By selecting the proper dry gas seal and control system, as well as engaging in adequate training and a thorough commissioning process, reaching beyond the old criteria of the 40,000-operating-hours-milestone is a realistic goal – as is the resulting reliability and improved uptime.

# # #

14e How Cracked Gas Compressors Must Be Monitored for Machinery 3 "P" - Protection, Prediction, and Performance

Bill Broussard
Emerson Process Management
Knoxville, TN 37932
Bill.Broussard@Emerson.com

Abstract:

The Cracked Gas Compressor is the bottle neck equipment in the ethylene process. So monitoring this key asset for machinery health is essential. This oral discussion will provide a fundamental understanding behind the common methods utilized in monitoring machinery health that API 670 refers to as the 3 "P"s.

The paper will cover the essential terms and aspects communicated behind API 670 Protection systems, including vibration sensors, locations, installation methods, monitoring practices, hardware requirements, integration to existing process control, and user graphics standards.

Machinery health prediction systems provide advanced warning of common issues (imbalance, alignment, coupling issues, sleeve bearing behavior, etc.). A fundamental tutorial behind the extra processing techniques required to provide predictive information are addressed, including a brief vibration primer for the non-expert listeners.

Machinery health performance is an application that has strategic alignment with cracked gas compressors. The discussion covers methods and challenges of calculating equipment performance efficiencies and how this information is communicated to operators and reliability resources.

The paper will cover bottom line financial impact to monitoring the 3 "P"s on the cracked gas compressor as well as the refrigeration compressors.

The listener will have a good fundamental understanding of 3P for Turbomachinery.

22a Syngas - Chemical Feedstock of the Future?

Chris Turner
Sasol Synfuels International
Houston, TX 77079
chris.turner@us.sasol.com

Abstract:

Traditional feedstock for light olefins and other basic chemicals are subject to the volatility of oil price. Syngas routes allow the producer to use alternative carbon sources such as coal and stranded natural gas to hedge oil price volatility.

This keynote is a high level overview of syngas as a basic chemical feedstock, the routes available to produce syngas and the challenges and opportunities.

Basic Care of Machinery

For *unspared* compressor trains:

1. Outside operator periodic activities: each shift or day, week, month, quarter, etc.
2. Inside or board operator tools and checks: operating window parameters defined and monitored, performance or efficiency trending.
3. Shutdown systems testing: who and how often.
4. PM programs.

For *spared* equipment such as pumps:

1. Outside operator periodic activities: each shift or day, week, month, quarter, etc.
2. Inside or board operator tools and checks: operating window parameters defined and monitored, performance or efficiency trending.
3. PM programs.

40a Future Feedstock for Olefin Plants

Mark Eramo
CMAI
Houston, TX 77079
meramo@cmaiglobal.com

Abstract:

This paper will review some of the historic data for olefin plant feedstock in different regions of the world and the changes in feed mix. This in combination with future projections of olefins demand and availability and relative pricing of feedstock will dictate the future trends in olefin plant feedstock. This paper will also review the developments in alternative feed sources (such as methane, coal and bio) and how the economic viability of these feeds relative to conventional feed sources will impact the industry.

Alternatives to Steam Cracking: Will the Steam Cracker Go the Way of the Dodo?

Mark Jones
Technology Strategy Development Scientist
Dow Chemical

Abstract: Market forces and technology development are converging to portend the unthinkable: olefin production without a steam cracker. The Alternative Feedstock Program at Dow Chemical is reviewing and implementing routes to olefin derivatives that would have been unthinkable a decade ago. This talk will describe these efforts and, in particular, highlight the emergence of bio-based polyethylene made by catalytic dehydration of ethanol to form ethylene. Other routes from coal and natural gas will be reviewed. The portfolio of options will be described. The talk concludes that the extinction of steam crackers is not imminent, but new technologies will find their place. These new technologies will not compete in all niches, but will replace steam cracking in some. This will bring new vibrancy and uncertainty to the production of olefins

Alternatives to Steam Cracking: Will the Steam Cracker Go the Way of the Dodo?

No consumer currently purchases ethylene. Consumers purchase products derived from ethylene, but have no use for raw, unconverted ethylene. The products industry makes from ethylene are placed in consumers hands in a variety of ways. They are dominated by, but not limited to, polymers that form such an integral part of modern life. Everything from cell phones, computers, automobiles, plumbing, soaps and detergents, packaging, textiles,….the list is virtually endless.

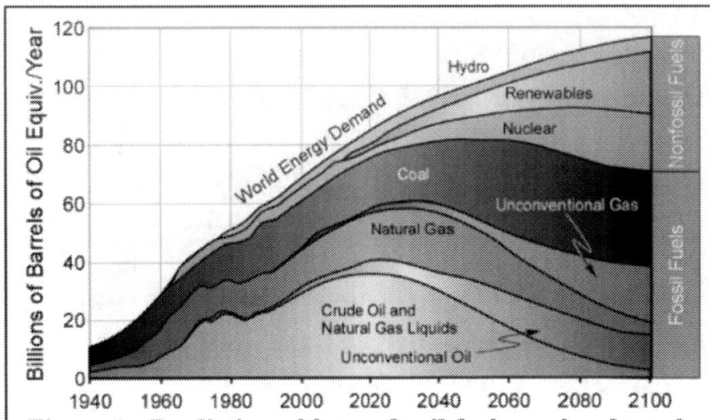

Figure 1: Prediction of future fossil fuel supply adapted from David Wood, Oil and Gas Journal, 9 February 2009.

The steam cracker is currently the foundation upon which all these products are based. The current industry takes refined products from natural gas processing or from petroleum processing and passes them through a steam cracker to produce a variety of products.

Consumers are beginning to ask for green products. Frequently this means materials derived from biological raw materials. Fortunately, this doesn't mean changing the polymers used to make goods. It can mean changing the methods by which the same polymers are made.

The feedstocks used in steam cracking are products of the energy infrastructure. No one has a perfect crystal ball, but virtually all views of the future indicate a shift from the current patterns of energy use. Recently, the data from which Figure 1 is adapted appeared in Oil and Gas Journal [1]. Typical of many predictions, it shows petroleum and natural gas diminishing in importance, and increasing use of renewables and coal. Renewables, in this plot, include biomass, wind, solar, geothermal wave and biofuels. Unconventional gas includes coal-bed

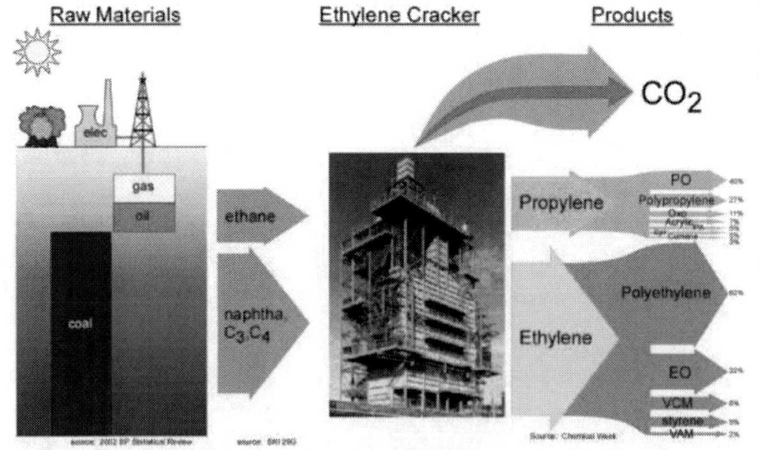

Figure 2: Schematic showing the modern chemical industry methods for getting from raw feedstocks to chemical products.

methane, shale gas and gas hydrates. Unconventional oil includes bitumen, tar sands, oil shale and ultra-heavy crude.

As stated, the steam cracker is the foundation upon which the modern chemical industry is based. Refined products from natural gas processing and petroleum processing are converted to olefins in the steam cracker. The two simplest olefins are also the most important. Ethylene and propylene form the majority of the high value products from the steam cracking of natural gas liquids and petroleum naphtha. These are shown in Figure 2. Ethane and propane are the largest of the natural gas liquids. An ethane-only cracker is currently the lowest capital, most selective means of producing ethylene from fossil feedstocks. As the molecular weight of the feed is increased, co-products appear. Propylene, butadiene and BTX all are produced by naphtha cracking.

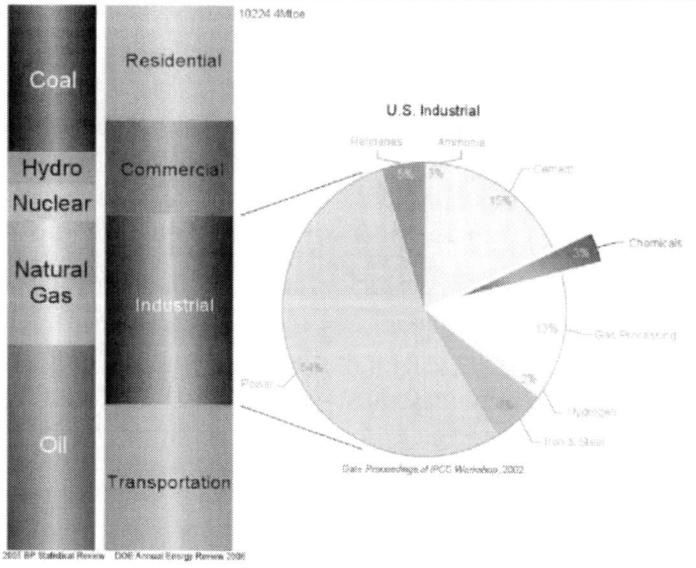

Figure 3: Chemical industry position in the global energy flow.

The endothermic reaction that occurs in the furnace tubes produces both hydrogen and methane. These materials are commonly used as fuels for the cracker. Carbon emissions from the cracker depend on the feed, the cracking severity, and the amount of off-gas consumed. Depending on how these emissions are allocated among the cracker's products, the amount of CO_2 can vary widely.

As we look to the future of our industry, several forces are marshalling.

- carbon management seems all but inevitable.
- consumers are asking for renewable materials in the products they purchase.
- prices for our conventional feedstocks have recently experienced unprecedented volatility.
- the U.S. and Europe, traditional centers of our industry, now command less of the industry's capacity.
- new markets are growing in areas where our conventional feedstocks simply don't exist.

Our industry extracts refined products from the fuels industry. The chemical industry, for all its heft, consumes only a small fraction of the world's energy. We consume as feedstocks materials that most of the world burns as fuel. As a result, we are buffeted by markets as opposed to controlling feedstock markets.

The steam cracker, for all its positive attributes, does have some shortcoming. Paramount among these is the capital. These are formidable structures. They are large, complex devices. Trends to go ever larger continue. Currently, the industry awaits an inrush of new capacity currently under construction in the Middle East. Many pushing scale to new levels. As this is happening, there is a growing question of feedstock availability. Even in the resource-rich Middle East, feedstocks present a challenge. Much of the Middle Eastern capacity is base upon ethane. Ethane is a dilute component of natural gas. Outlets must be available for the natural gas in order to extract the ethane. Resources of sufficient size to support the multiples of gargantuan scale are themselves uncommon and limited. Naphtha is an easily transported, world commodity and is a global feedstock. Ethane is a local commodity, difficult to economically ship.

An alternative feedstock is anything that isn't the natural gas liquids and petroleum naphtha used today. For a feedstock to be considered:

- it must exist in sufficient quantities in the geography where product is needed
- technology must exist that converts the alternative feedstock to either an olefin intermediate or an olefin derivative
- the feedstock economics must be favorable when the total cost of production is considered

Bio-renewable feedstocks are constantly in the news. Figure 4 shows the relative scales of conventional production and the available biological feedstocks [2, 3, 4]. Several points warrant mention. The first is that the current industry scale is close in scale to ethanol production and considerable larger than global biodiesel production. Dow controls approximately 10% of the global ethylene capacity spread over more than 10 crackers. A single cracker, at about one million tones per annum, is an imperceptibly small perturbation on the distillate market, but a 5-10% perturbation on the ethanol market. It is reasonable that some ethylene could be produced from ethanol, but inconceivable that ethanol could meet the entire ethylene demand. Raw materials, such as corn and cane, are clearly not the limiting factor.

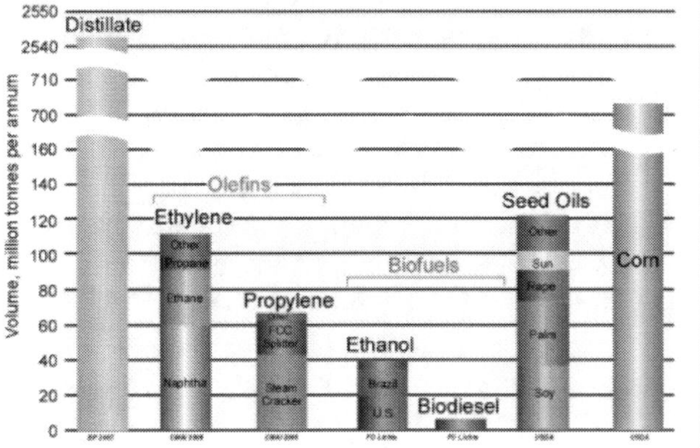

Figure 4: Comparison of conventional feedstocks and biological feedstocks.

The plot also shows that while the hype around biofuels continues, they actually are quite small when compared with the total gasoline and diesel production. Biofeedstocks have clear pluses and minuses.

Biological ethylene holds a place of importance. In 1901, ethylene was shown to ripen fruit. In the 1930s, ethylene was recognized as a plant hormone, produced by plants through normal metabolism. Plants produce ethylene in huge amounts, but not in an industrially useful way.

A different flavor of biological ethylene holds an important place in the history of chemistry. The first report of an ethylene derivative was the production of 1,2-dichloroethane, also called ethylene dichloride or EDC, in 1795. At that time, it was called Dutch oil due to the Dutch chemists that discovered it. Today, it is an important precursor to vinyl chloride and, ultimately, polyvinyl chloride.

Figure 5: Production of ethylene from alcohol.

Since the gas formed an oil, albeit an oil that sinks rather than floating on water, it was named olefiant, or oil-forming, gas. This history persists today in the word *olefin*.

There were no steam crackers in 1795, so it is reasonable to ponder where the ethylene used in the experiments that led to making Dutch oil. It is likely that ethylene was produced in the coal processing of the day, but no methods were available for the separations necessary to make pure ethylene. To get pure ethylene, the chemists of the day turned to ethanol. Production of ethylene was accomplished by the reaction of "one part of alcohol with six or seven parts, by weight, of oil of vitriol."[5]. The apparatus for ethylene production, as described in 1872 [6], is shown in Figure 5. Ethanol is dehydrated by sulfuric acid, also know as oil of vitriol, to produce relatively pure ethylene, with only a slight contamination of diethyl ether.

Ethanol conversion to ethylene took a great step forward with the discovery of solid acid catalysts [7]. Alumina was reported as an active and selective catalyst for the gas-phase decomposition of ethanol to ethylene. In fact, the first chemical reaction ever to be put through the rigors of kinetic analysis was the dehydration of ethanol over acidic alumina. Sabatier, in 1909, studied ethylene production from ethanol. Subsequent studies have shown this to be a good, industrially viable catalyst.

Commercial use did not begin until the early part of the 20th century. Prior to World War I, ethane for use as a refrigerant was produced by hydrogenating ethylene produced from ethanol dehydration. Large amounts of ethylene were produced during the war for use in the manufacture of mustard gas. Ethanol dehydration continued to play an important role in supplying ethylene for chemicals and polymers through the 1930s and 1940s. The

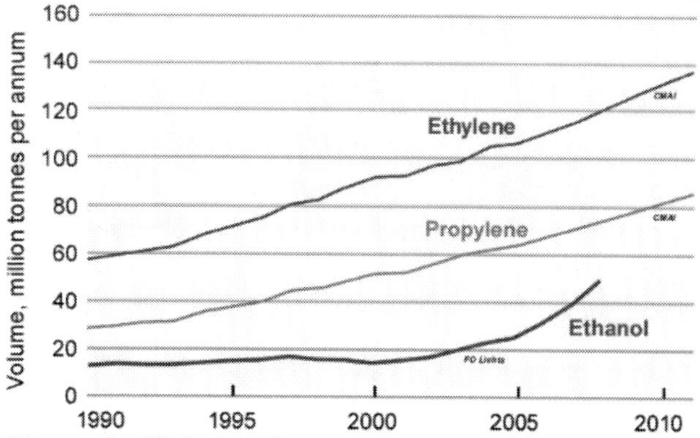

Figure 6: Global olefin production compared with global ethanol. Data as reported by CMAI and F.O. Lichts.

technology was largely superseded by steam cracking after World War II [8]. During this time, commercial production of ethanol was based upon hydration of ethylene, the opposite of ethanol dehydration.

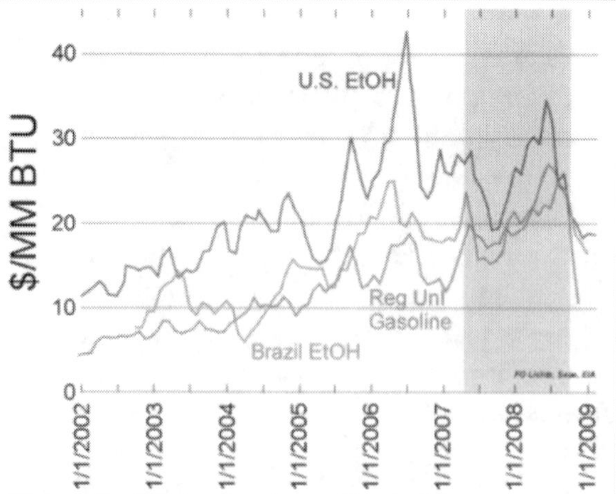

Figure 7: Ethanol and gasoline pricing compared.

Two countries dominate the production of ethanol with approximately equal production capacity. Brazil's production is based on sugarcane, while corn forms the basis for the large production capacity in the United States. The rest of the world contributes only a small amount to total.

The rapid growth in ethanol production needs to be viewed in the context of ethylene production. Year-over-year growth in ethylene is approximately equal to the current rate of ethanol growth. Figure 6 plots these data.

In reviewing the prerequisites for a successful alternative feedstock, it appears that biologically derived ethanol has both sufficient supply and available technology. All that remains is favorable economics.

The economics of ethanol as a feedstock depend largely on the cost of ethanol and the capital required for the conversion. Addressing the ethanol cost requires several taking several views. The first is a comparison of ethanol price versus the price of more conventional feedstocks. For this comparison, Figure 7 shows the price history of ethanol, in both the U.S. and Brazil, against regular unleaded gasoline [9]. Unleaded gasoline is a relatively accurate, readily available surrogate for naphtha. The important thing to note is that in Brazil, ethanol achieved BTU parity with fossil-based fuels. This has yet to occur in the U.S. where ethanol trades at a premium for reasons of both incentives and oxygenate regulations. Cellulosic ethanol promises to reduce the price for ethanol in essentially all markets. The BTU parity is important because the work required to convert ethanol to ethylene is considerably less than the effort required to convert naphtha to ethylene.

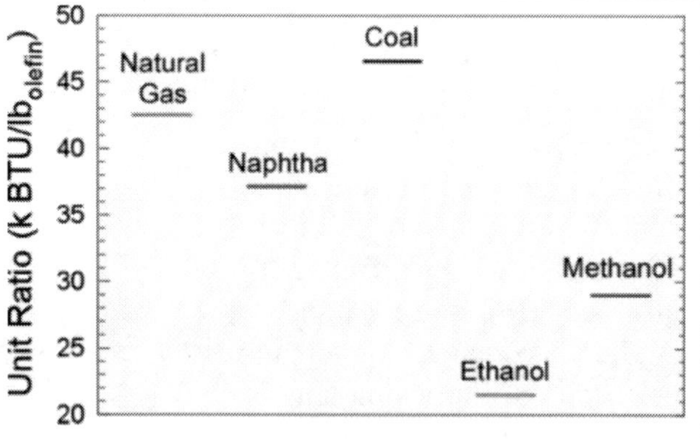

Figure 8: Unit ratio for olefin production expressed in BTUs. Olefins are the combined ethylene and propylene production and the energy represents the higher heating value (HHV).

Unit ratios are the amount of feed required, usually in pounds, per pound of product. For petrochemical feedstocks, the amount of feed can also be expressed as an energy content. Taking the higher heating value, HHV, for the energy content, the unit ratio for a conventional naphtha cracker [10] and ethanol dehydration [11] can be compared. This comparison is shown in Figure 8, along with the unit ratios for natural gas, coal and methanol assuming use of zeolite-based methanol-to-olefins (MTO) technology [12]. The data show that the unit ratio for ethanol is approximately half that for naphtha, when expressed in energy. Therefore, when naphtha and ethanol trade at BTU parity, ethanol yields a lower cash cost.

Cash cost is only part of the story since it ignores capital. In order for ethanol to be truly advantaged, its cost for production with a return on capital must be considered. To make this comparison, data from several published sources were collected to allow the comparison of ethanol dehydration with steam cracking of naphtha [8, 13, 14, 15, 16]. These reports are not in complete agreement, presenting a range of capital estimates resulting in the bands shown in Figure 9. The minimum and maximum scales are those reported in the literature. All capital estimates have been escalated to fourth quarter 2008 dollars using the SRI PEP index.

The naphtha capital is considerably higher than the ethanol dehydration capital. It is generally assumed that the scale of ethanol is limited relative to naphtha due to supply constraints. This is due to an implicit assumption that the dehydration will be linked to the fermentation asset. Ethanol plants are limited in size due to difficulty in transporting biomass over long distances. In the case of sugarcane, the timing between cut and crush must be kept short to limit sugar degradation. This limits the amount of land that any single mill can utilize, limiting the ultimate plant scale.

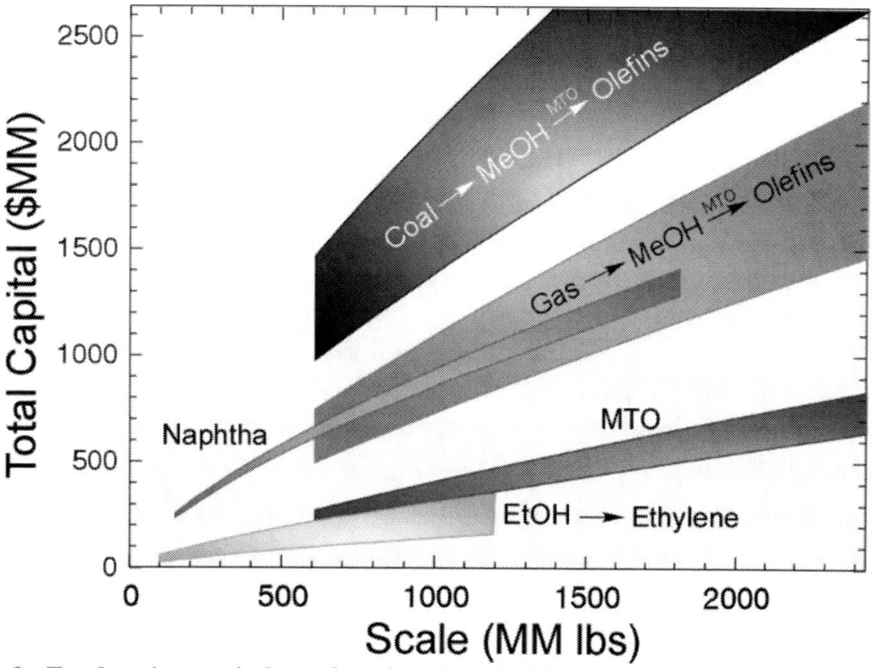

Figure 9: Total project capital as a function of total olefin production. Total olefins are the sum of ethylene and propylene production. Uncertainty bands represent the range of literature reported values.

The last piece of the economic puzzle is in place. Favorable unit ratios and capital favor ethanol dehydration over when the feedstock prices are near BTU parity. Reduction of carbon emissions provides yet another driver favoring dehydration of bioethanol for chemical production [17].

Dow Chemical and a partner announced plans for an ethanol to polyethylene plant in Brazil. The facility will have a polyethylene capacity of 350,000 metric tons per year of DOWLEX* polyethylene [18]. Polyethylene derived completely from biological materials is also produced by other manufacturers. Ethylene and butene copolymers were launched in 2008 using raw materials completely derived from sugarcane [19], with an expansion of 200,000 metric tons per year of polyethylene [20] planned. In China, plans have been unveiled for a 300,000 metric tons per year ethanol-based ethylene plant in the Jilin Province. [8]

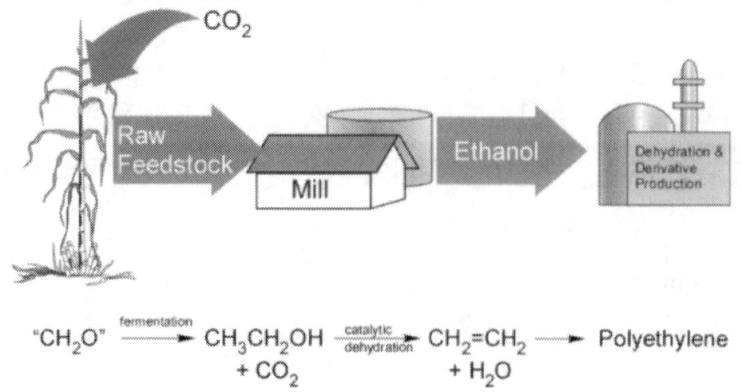

Figure 10: Simplified schematic of the conversion of sugarcane to polyethylene.

This is a bit of back to the future. From the 1950s to the early 1980s, Union Carbide, now part of Dow Chemical, and other companies produced ethylene from ethanol. In addition to UCC, Halcon/Scientific Design, ICI, Lummus, Petrobras, Compagnia Tecnica Internazionale Progetti (CTIP)/Solvay, and Centro de Tecnologia Promon (CTP)/Haldor Topsøe built and operated small (3,000 to 60,000 metric tonnes per year) ethanol-to-ethylene plants in India, Pakistan, Peru, and Brazil [8].

The products derived from ethanol are "green" since it is made with carbon that came from atmospheric CO_2 metabolized by sugarcane. These carbons are contained in the polymer that will end up in durable goods rather than released to the atmosphere. This also rewrites the paradigm for polyethylene production. Typically, polyethylene is one of several products produced on the periphery of a large, maybe even gargantuan, steam cracker. The economics of polyethylene production hopelessly entwined with the cracker operation. Multiple plants are served by the cracker, making scale and reliability key metrics. Ethanol dehydration allows much smaller plants to be competitively built serving only a single derivative.

Ethanol is not the only alternative for chemical production. Referring again to Figure 1, coal is also predicted to play a more important role in energy and, therefore, as a potential chemical feedstock. Another alcohol, in this case methanol, will be a integral part of using coal for chemical production. Unlike ethanol, coal is likely to be capital intensive [21].

---

* DOWLEX is a trademark of The Dow Chemical Company

Coal does not play a role in the current production of olefin derivatives. Production for methanol continues to grow, led by new capacity being added in China [22]. Conversion of coal to olefins is assumed by most by methanol-to-olefins [23, 24]. Natural gas is a common feedstock for methanol today, so will be used as comparison case for the discussion that follows.

Coals are divided according to rank, a measure of the fuel quality of the coal. Plotted in Figure 11 are the ranges of coal heating values, with three liquid fuels for comparison [21, 25].

Conversion of coal to methanol first requires synthesis gas production. This is a capital intensive process, and does not scale particularly well [26]. Methanol-to-Olefins, or MTO, has been well described elsewhere [27, 28]. The process produces a mixture of light olefins with approximately equal amounts of propylene and ethylene [29]. Starting from methanol, the capital requirement is actually similar to ethanol dehydration and well below that of naphtha, as shown in Figure 9.

The issue is a bit different when conversion from the gas or coal feedstock is considered. Conversion of natural gas to synthesis gas is capital intensive, though not as intensive as for coal. The result is that the capital for a natural gas to methanol to olefins process being similar to a naphtha cracker [30, 31].

Use of coal requires still higher capital. Coal to olefins must appropriately trade off coals very low cost against the higher capital required. Figure 8 illustrates that coal has the highest unit ratio when expressed in energy units [32]. For most of 2008, world coal prices were in the area of $1/MM BTU, making it the least expensive world-traded fossil fuels.

There are currently no MTO plants in operation based on either gas or coal. Companies have announced plans for both. Recent economic turmoil has delayed virtually all construction in the chemical industry. Only time will tell whether this nascent technology will prove to be a worthy competitor for the steam cracker.

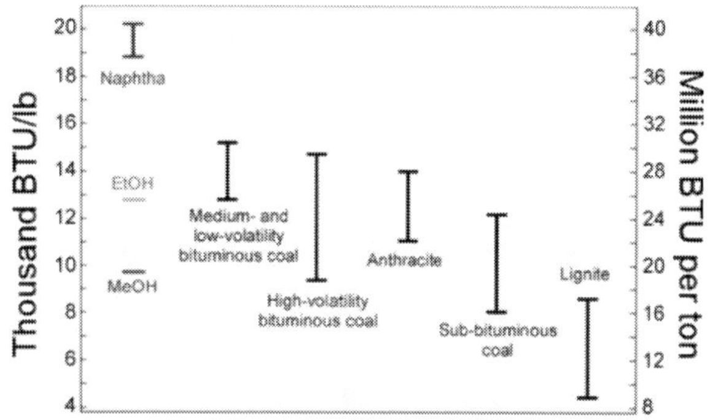

Figure 11: Energy content of coals, naphtha, ethanol and methanol.

In conclusion, options that are competitive with steam cracking are here today. Other wait in the wings. These technologies will likely coexist with steam cracking for the foreseeable future. They are not applicable everywhere and steam cracking will dominate certain geographies. Ethanol dehydration is compelling in certain geographies. Of most importance is the notion that cost effective derivative production can now occur without a cracker complex in some situations. For certain products, this may prove to be an important advance.

References

1 Wood, David; "Global perspectives required for risk, opportunity analyses", Oil and Gas Journal, 107(6), 9 February 2009, page 24.
2 Eramo, Mark; "Ethylene, propylene demand will experience increased growth in 2005-10", Oil and Gas Journal,
3 "BP statistical review of world energy", June 2007. Download from http://www.bp.com/ on 2 August 2007.
4 Foreign Agricultural Service, "Oilseeds: World Markets and Trade", USDA FOP 07-07, July 2007.
5 Parnell, Edward Andrew; "Applied Chemistry" (Appleton, New York), 1844.
6 Bloxam, Charles Loudon; "Chemistry Inorganic and Organic", 2nd Edition (J. & A. Churchill, London), 1872.
7 Sabatier, P.; Mailhe, A., Ann. Chim. Phys. 20 (1910) 289.
8 Noguiera Cesar, Marcos A. ; "Chemicals from Ethanol", SRI Process Economics Program Report 235, September 2007.
9 Data sources include CARD.isu.edu, Seae, EIA, F.O. Lichts.
10 Naphtha cracking is assumed to produce 1 lb of ethylene and 0.5 lb of propylene for every 2.78 lb of feed.
11 Ethanol dehydration is assumed to yield 1 lb of ethylene for every 1.68 lb of feed.
12 Cunningham, Richard; "GTO: a breakthrough in the petrochemical industry", The Lamp, volume 86(3), Fall 2004, pages 11-14.
13 T. K. Ng, R. M. Busche, C. C. McDonald, and R. W. F. Hardy ; "Production of Feedstock Chemicals"(11 February 1983) Science 219 (4585), 733.
14 Ren, Tao; "Petrochemicals from Oil, Natural gas, Coal and Biomass: Energy Use, Economics and Innovation", Utrecht University Thesis (ISBN 978-90-393-5019-5), 4 March 2009.
15 Tsao, Utah; Reilly, James; "Dehydrate ethanol to ethylene", Hydrocarbon Processing, February 1978, pages 133-136.
16 Nexant ChemSystems PERP Report 06/07S11 Green Polyethylene, March 2008.
17 Banholzer, William F.; Watson, Keith J.; Jones, Mark E.; "How Might Biofuels Impact the Chemical Industry", Chemical Engineering Progress, March 2008, pages S7-S14.
18 Dow Chemical press release on July 19, 2007, available on www.dow.com
19 "World's first 'green' linear polyethylene launched", Chemistry World, a Royal Society of Chemistry publication available at www.rsc.org, 31 March 2008.
20 Voegele, Erin; "Sugarcane plays role in development of green polyethylene", Ethanol Producer, March 2009.
21 Bartis, James T.; Camm, Frank; Ortiz, David S.; "Producing Liquid Fuels from Coal: Prospects and Policy Issues" a RAND Corporation report, ISBN/EAN: 9780833045119, 10 December 2008.
22 Jordan, J.; "World methanol overview: 2007 and beyond", Hydrocarbon Processing, April 2008, pages 61-68.
23 Morgan, Mark; "Recent Converts", European Chemical News, 14/20 July 2003, pg 30-31.
24 Sutton, Mark; Roberts, Peter; "Consider coal for olefins production" Hydrocarbon Processing, July 2007.
25 Natural gas has a heating value of 23,600 BTU/lb.
26 Wan, Victor; Smith, Ronald; SRI Process Economics Program Report 43E Methanol From Coal, December 2006.
27 Nexant ChemSystems PERP Report 00/01S9 Methanol to Olefins, July 2002.
28 Wan, Victor; SRI Process Economics Program Report 261 Methanol To Olefins, November 2007.
29 5.9 lb of methanol are assumed to produce 1 lb each of ethylene and propylene, for a unit ratio of 2.95 lb methanol per pound of olefin.
30 Pavone, Anthony; SRI Process Economics Program Report 43D Mega Methanol Plants, December 2003.
31 Nexant ChemSystems PERP Report 03/04-4 Methanol, December 2004.
32 Unit ratios for natural gas and coal to methanol are assumed to be 14,300 and 15,700 BTU/lb of methanol, respectively.

TRENDS IN ETHYLENE PLANT FEEDSTOCKS

Alan J. Nizamoff
Senior Consultant, Nexant, Inc.

ABSTRACT

If one examines the regional capacity split over the last 30 years a number of clear patterns can be seen: North America and Europe have lost their dominance of ethylene markets; both North American and European capacities stagnate during "Bad Times"; Asia Pacific has developed to become a key producer and grows almost irrespective of industry profitability; and although the Middle East capacity is lower it has become globally significant in the last decade. Naphtha cracking remains the dominant source of ethylene globally; however gas cracking has been gathering interest. Naphtha crackers took the brunt of the pressure after overbuilding in the late 1970s, with significant closures in the United States. Ethane cracking has come from nowhere to be a significant source of ethylene, principally as a result of capacity developments in the Middle East. The LPG cracker segment (albeit chiefly comprising ethane/propane crackers) has increased steadily since the late 1980s.

This paper will review the historical, current and forecast feeds for steam crackers with an eye towards the impact that feeds from Natural Gas and Refinery sources will play. In addition, this presentation will also look at alternative feed sources such as coal, methane and bio sources and how the economics of these feeds relative to conventional feed sources will impact the industry. Finally we will review under what crude/natural gas pricing scenarios these technologies will make economic sense.

INTRODUCTION

This paper depicts the capacity development for steam cracker ethylene production over the last 30 years and a preview of what's to come. This paper will focus on the following questions:

- How will future olefin demand impact the conventional/opportunity feeds from Natural Gas sources (ethane/propane/butane, NGLs) and Refinery sources?
- How do the economics of alternate feed sources - coal, methane, biomass – stack up relative to conventional feed sources and at what crude oil/natural gas pricing do these technologies will make sense?
- What would be the impact of these on our industry?

Both global and regional information is presented, and each time the capacity profile is overlaid by a "profitability profile" depicting when the industry was experiencing "good times" with strong profitability and "bad times" when laggard producers were operating at or below breakeven. This profitability profile (cash margin index) is explained further on in this paper.

The regional section then also looks at the feedstock supply resources available in each region and considers how market characteristics for these feedstocks affected their suitability as cracker feedstock. All three feedstock types are "by-products" from crude oil production, refinery operations, or gas processing. All availability considerations have therefore been analyzed by considering these sources and the potential each offers for exploitation. Actual production may have differed from this theoretical availability.

Segmenting crackers by feedstock type is necessarily an arbitrary process as, in reality, many producers can operate individual crackers on a range of feedstocks. This paper segments crackers as follows:

Designation	Segmentation
Ethane crackers	Pure ethane cracking
LPG crackers	Any cracker including propane / butane cracking but no large naphtha component. E/P crackers are included in this category
Naphtha/LPG crackers	Crackers usually processing both naphtha and a significant proportion of propane and/or butane
Flexible feed crackers	Crackers able to process ethane through naphtha and heavier as required
Naphtha crackers	Crackers focused on cracking a naphtha or condensate feedstock
Heavy crackers	Crackers able to process heavy feedstocks such as vacuum gas oil and waxy residue

CASH MARGIN INDICES

Nexant's Cash Margin Indices demonstrate the functioning of the petrochemical cycle as the industry swings from good times to bad. The Cash Margin Indices are calculated as the capacity weighted average cash margin for around 30 commodity petrochemicals in each region. They are presented as indices and commonly have 1984 Q1 set at a default value of 100. Each regional index consists of basic chemicals, intermediates and polymers and provides a good guide to the overall profitability of the petrochemical industry.

West European index has data available from the mid 1970s, with our analysis of the United States starting in 1982 and Asia in 1986. Figure 1 includes much generalized time periods during which the industry either experienced strong profitability "good times", or poor profitability "bad times". There is a definite tendency for these latter "bad times" to last for some considerable time as the industry struggles with either weak demand or excessive capacity.

This paper will use these profitability indices to try to understand regional cracker investments. Correlations to this industry measure should be more meaningful than a simple analysis against cracker profitability as new cracker projects needs to be sure of successful derivative operation as well as profitability at the cracker itself.

Figure 1
Petrochemical Cash Margin Index
(U.S. dollars, 1984Q1 = 100)

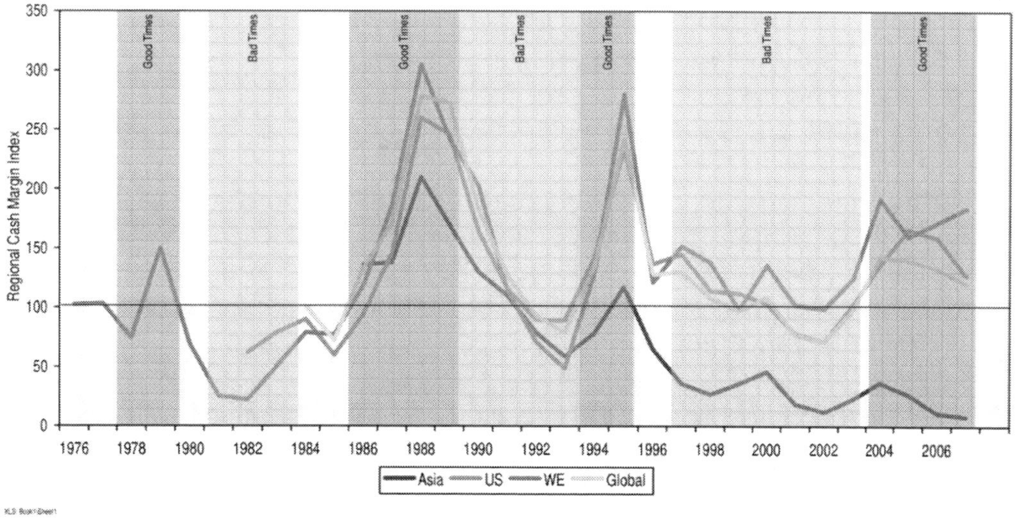

GLOBAL ETHYLENE CAPACITY

While North America and Western Europe used to dominate ethylene production, new project developments are focused in Asia Pacific and Middle East regions.

The global cracker portfolio is dominated by naphtha crackers, although the proportion identified as designed for this hydrocarbon feed has decreased over the last 30 years.

The United States has seen the largest range of feedstock types selected, with crackers designed for feeds ranging from pure ethane to ones designed for heavy liquids present in the market. The United States also has a number of crackers that were either built or modified to be able to process a wide range of feedstock.

The Middle East has meanwhile seen a strong focus on gas cracking, with ethane crackers dominating the market. More recent projects however are tending to be ethane/propane crackers since incremental ethane supplies are limited.

Asia's focus is on heavier liquid feeds, although many crackers are built to accept a wide variety of feeds owing to the cut throat competition in the region.

Western Europe continues to expand existing units although no new builds have been started or are expected in the near term. Naphtha designs dominate this region.

Nexant's capacity database allows us to analyze ethylene capacity information all the way back to the mid 1970s. Figure 2 presents the regional capacity split over the last 30 years, overlaid with a broad description of petrochemical industry profitability (as portrayed by *ChemSystems* Cash Margin Index).

- "Good Times" – periods of strong profitability

- "Bad Times" – periods of weak margin

A number of clear patterns can be seen from the graphic:

- That North America and Europe have lost their dominance of ethylene markets

- Both North American and European capacity stagnate during "Bad Times"

- Asia Pacific has developed to become a key producer and, as a region has grown almost irrespective of industry profitability (although capacity build did slow in the early 2000s after the Asian financial crisis in the late 1990s)

- Middle East capacity is lower but has become globally significant in the last decade

The largest proportion of crackers are still naphtha-based, however gas cracking has been gathering interest. The global breakdown of capacity by feedstock shown in Figure 3 provides further insight:

- Naphtha cracking remains the dominant source of ethylene globally.

- Naphtha crackers took the brunt of the pressure after overbuilding in the late 1970s, with significant closures in the United States.

- Ethane cracking has come from nowhere to be a significant source of ethylene, principally as a result of capacity developments in the Middle East (total ethane cracking however remains significantly below total liquid cracker volume).

- The LPG cracker segment (albeit chiefly comprising ethane/propane crackers) has increased steadily since the late 1980s.

Figure 2
Global Ethylene Capacity by Region
(thousand tons per annum)

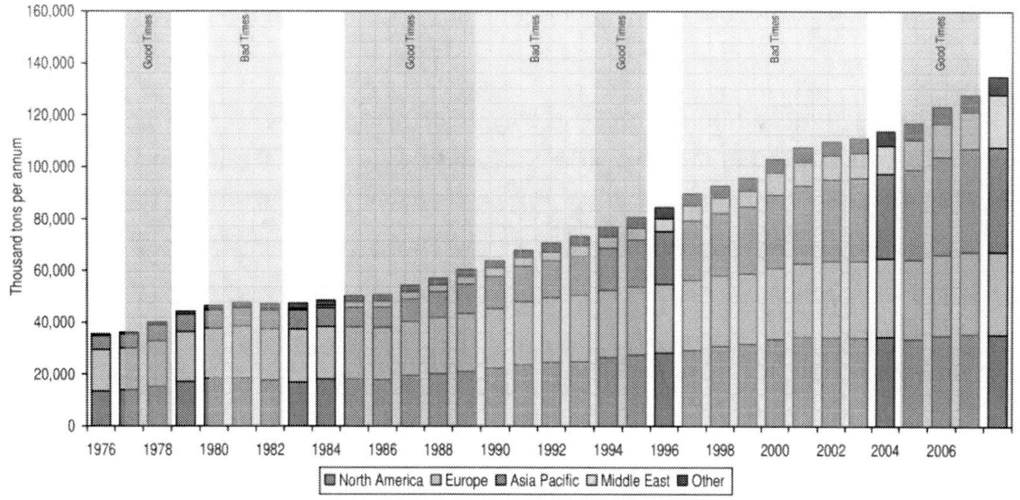

Figure 3
Global Ethylene Capacity by Source
(thousand tons per annum)

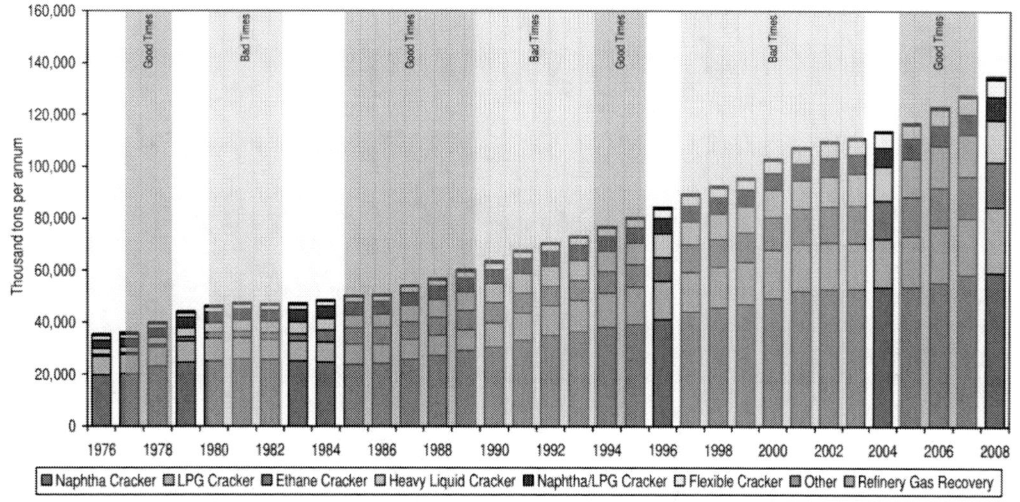

NORTH AMERICAN DEVELOPMENT

North America is a mature market with high flexibility to alter feedstock slates to crack the most competitively priced feedstocks. The North American market was the birthplace of the petrochemical industry, with liquid crackers associated with new refineries allowing the development of downstream olefin and aromatic derivatives. The total installed capacity had been growing rapidly in the 1970s supported by an enthusiasm for technological development (re. Figure 4). The LPG cracker segment (ethane/propane) was the dominant configuration in the 1970's, however developments favored both naphtha and ethane only cracking, with LPG capacity stagnating until the late 1980s. LPG cracking had a further surge in development in the late 1980s (again largely ethane/propane cracking) on the back of the high profitability seen in the mid to late 1980s and propane's positioning as a competitive feedstock at that time.

Naphtha crackers meanwhile suffered badly from the effects of the second oil crisis and associated high crude oil and hence naphtha costs relative to lighter gas based feeds in the early 1980s. Total installed naphtha capacity fell back as old uncompetitive plants were forced to close.

Ethane crackers started to find a presence in the market in the early 1980s, taking advantage of the disconnect between natural gas and escalating crude oil prices, and positioning the U.S. Gulf Coast as the low cost exporter of ethylene derivatives to Asia.

North American capacity was severely impacted by the recession in the early 1980s. Capacity has since almost doubled, although momentum has been lost again this decade.

Figure 4
North American Ethylene Capacity by Source
(thousand tons per annum)

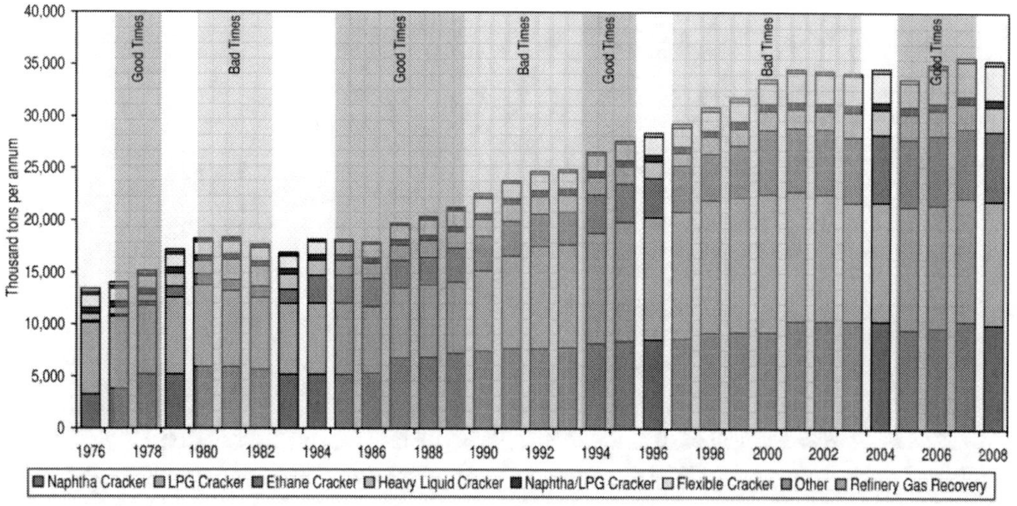

North America crackers make heavy use of LPG components as a feedstock. LPG availability has shown very little growth during the past decade as production associated with crude oil fell. North American LPG is sourced on an almost equal basis from gas associated with crude oil, separation from dry natural gas and from processing operations in refinery centers in the United States (re. Figure 5). Availability from all sources has been fairly stable across the last 30 years, with total production now back above the levels seen in 1980, after falling due to reduced refinery activity in the early 1980s.

Figure 5
North American LPG Availability by Source
(million tons per annum)

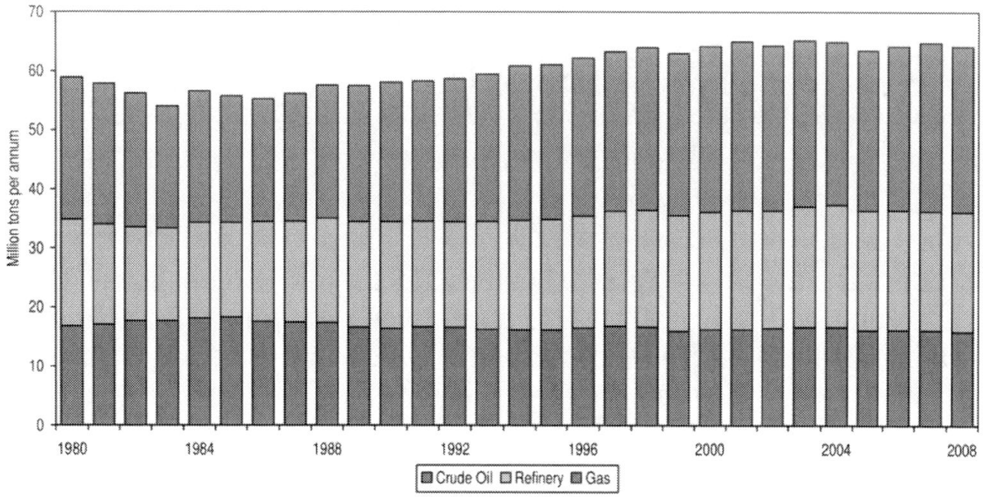

This total LPG make serves a wide range of markets, the most significant being domestic heating and cooking, industrial heating and petrochemical feedstock applications. Significantly more LPG is also available at the refinery as large volumes are consumed as blend stock in gasoline (subject to vapor pressure constraints). Current U.S. crackers consume approximately 14 million tons of LPG per year – equivalent to around one quarter of the total LPG available.

Only around one quarter of North American LPG is consumed by the petrochemical industry, leaving plenty of opportunity for increased use despite static LPG availability.

North America naphtha availability has increased slowly with incremental oil processing. The vast majority of naphtha-like material in North America is consumed in the automotive gasoline market. The North American naphtha stream is sourced from refineries and as condensate from the processing of natural gas (natural gasoline). Refinery operations are related to gasoline and fuel demand, which has a tendency to track below GDP growth rates as a result of energy efficiency improvements.

Total naphtha availability (re. Figure 6) has therefore tended to grow more slowly than ethylene demand (which has grown at GDP+ rates until recently). Total naphtha consumption in steam cracking is estimated at 26 million tons in North America in 2007.

Figure 6
North American Naphtha Availability
(million tons per annum)

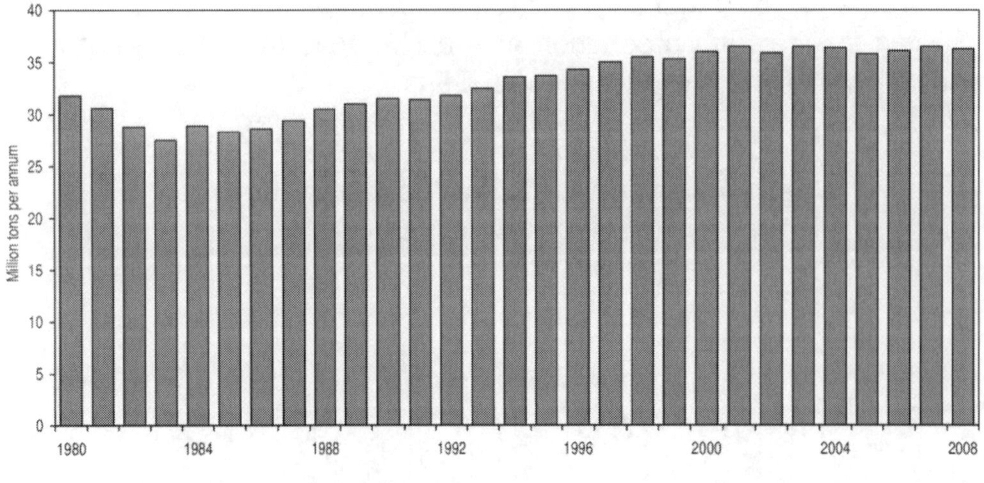

Ethane is dominant as an ethylene feedstock (largely in conjunction with propane in ethane/propane crackers) and is available in significant volumes in the North American market (re. Figure 7). Ethane is available from both associated and non-associated gas streams, and is separated from rich gas at dedicated separation units (some more specifically "straddle plants"). Ethane is more readily available in associated gas streams, although some non-associated fields such as those in Canada have sufficient ethane present to warrant extraction. Removing ethane from gas is an expensive process, involving high pressures and cryogenic separation.

Figure 7
North American Ethane Availability
(million tons per annum)

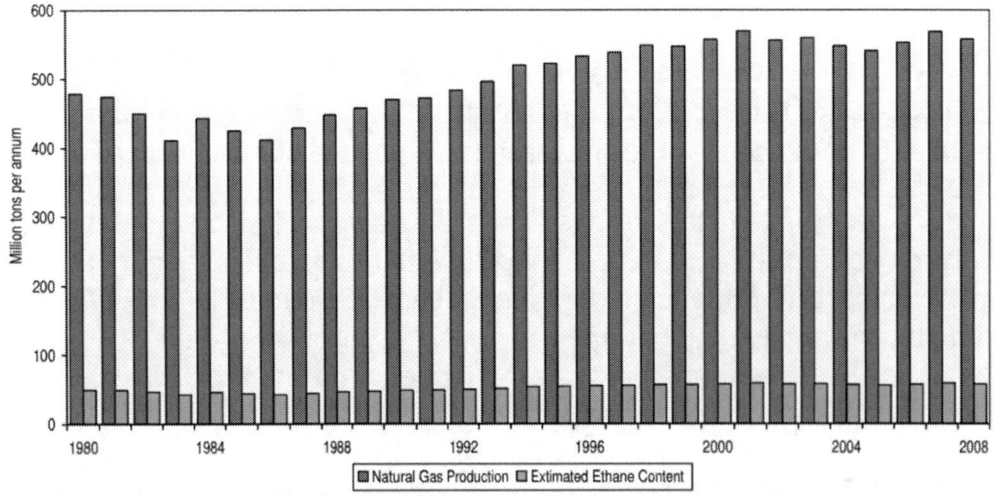

The "shrinkage" impact on the gas stream is also significant as the gas stream loses both from the volume of gas extracted and also from the fact that the ethane had a higher calorific value than the remaining methane. It is only economic to extract ethane in those places where there is sufficient gas available to justify investment in a separation plant and where the demand for gas is easily met by the remaining methane stream. New sources of natural gas including shale and coal beds yield lean natural gas with no ethane. These sources provide nearly 20 percent of current U.S. natural gas supplies.

Total ethane consumed by the crackers in North America is estimated at just over 16 million tons in 2007. The petrochemical industry is estimated to be using approximately one quarter the potentially available ethane in North American natural gas.

New developments in steam cracking within North America will be mainly through the use of naphtha, although there will be some development of oil sands for petrochemical use in Canada.

EUROPEAN DEVELOPMENT

European ethylene capacity continues to increase, with growth coming from the debottlenecking of existing capacity. Naphtha is the dominant steam cracker feedstock in Europe. Naphtha is also the most available feedstock in the region and is the feedstock of choice (re. Figure 8). Europe also has a well developed petrochemical industry. The lower bias towards gasoline in the region's refined product demand profile has allowed the development of a heavily liquid based cracker industry (the United States has a far higher relative gasoline requirement).

Figure 8
European Ethylene Capacity by Source
(thousand tons per annum)

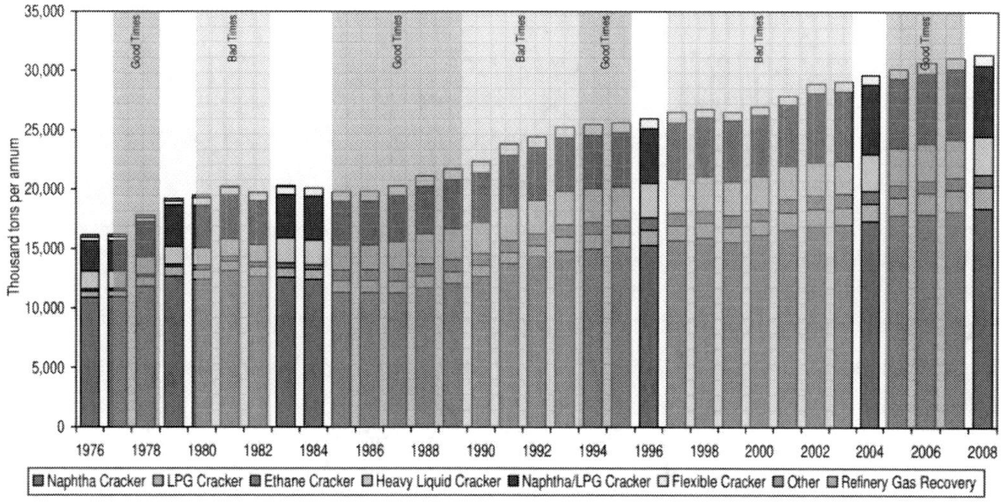

Total installed capacity had been growing rapidly in the 1970s. In contrast to the United States however capacity only stagnated in the 1980s recession as the region struggled to close uncompetitive units. As seen for North America, current capacity is now double what it was in 1976. Naphtha cracking is the dominant configuration in the region, although many crackers are able to process some LPG as well as heavier feedstocks such as vacuum gas oil and waxes. Ethane cracking has very little presence in the region, with only one major unit at Mossmorran able to take advantage of this feedstock.

Europe LPG supply relies heavily on refinery throughput (re. Figure 9). Falling supply of North Sea crude is balanced by increased availability from Russian sources. European LPG is heavily dependant on refinery sources as the region is relatively lacking in oil and gas production. Total LPG availability is reasonably stable at around 40 million tons per year.

Figure 9
European LPG Availability by Source
(million tons per annum)

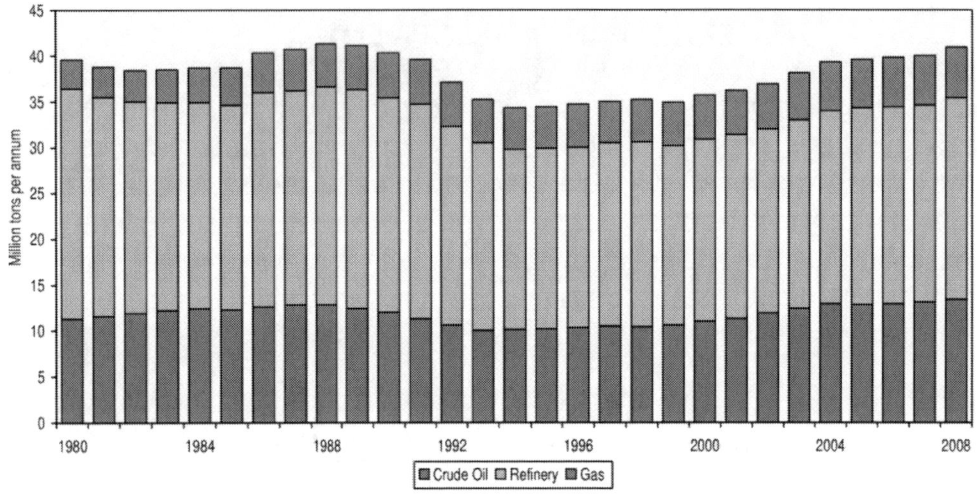

The petrochemical industry is a relatively small consumer, with demand accounting for around one fifth of total availability, with the majority of LPG consumed in heating markets (consumer and industrial). Significantly more LPG is also available at the refinery as large volumes are consumed as blend stock in gasoline (subject to vapor pressure constraints). Even though European crackers consume only one fifth of the total available LPG, more is readily available, however economics have not always proved advantageous.

European naphtha supply has decreased over the last thirty years as the region shut uncompetitive refineries (re. Figure 10). Condensates and vacuum gas oil have allowed the region to continue to operate significant liquid cracking capacity. Naphtha production in Western Europe has been falling over the last twenty years as uncompetitive refineries shut and refineries gradually increased their gasoline and middle distillate (diesel) yields. Total naphtha availability from refineries in the region has stabilized at around 60 million tons –

sufficient to produce around 18 million tons of ethylene. Almost all of this naphtha is consumed at crackers in the region, with additional imports of condensates from Algeria and vacuum gas oil from refineries needed to fully utilize liquid cracking capacity in the region.

Figure 10
European Naphtha Availability
(million tons per annum)

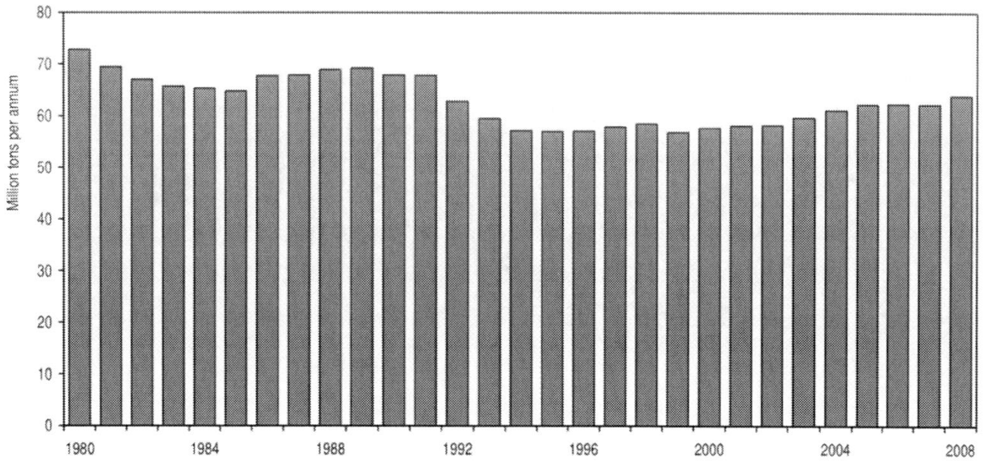

Europe consumes almost all its spare naphtha as cracker feedstock. As shown in Figure 11, ethane is hardly used in the European market despite the high volumes of natural gas available in the market – heating demand has denied access to the petrochemical industry.

Natural gas is supplied as a convenient piped fuel and clean burns without soot formation – as such it has achieved a dominant presence in the North West European domestic market. European obtains its natural gas supplies from four main areas – dry gas fields off the Dutch coast, associated and dry gas from the North Sea, pipelines from North Africa and now of growing importance, dry gas supply via transit pipelines from Russia. The preponderance of dry gas and long distance pipeline supply means that the average gas composition in the region is depleted of ethane and LPG fractions. This low ethane content, and the high value placed on gas for heating use have prevented the region from developing significant ethane extraction capability, and hence little is available as feedstock for petrochemicals.

The bulk of capacity additions are forecast to come through the expansion of existing facilities as the only planned cracker project remains highly tentative. Any new developments in steam cracking within Western Europe will be mainly through the use of naphtha. Future capacity additions in Eastern and Central Europe are expected to be based on liquid feedstock, primarily naphtha and gas oil.

Figure 11
European Natural Gas and Ethane Availability
(million tons per annum)

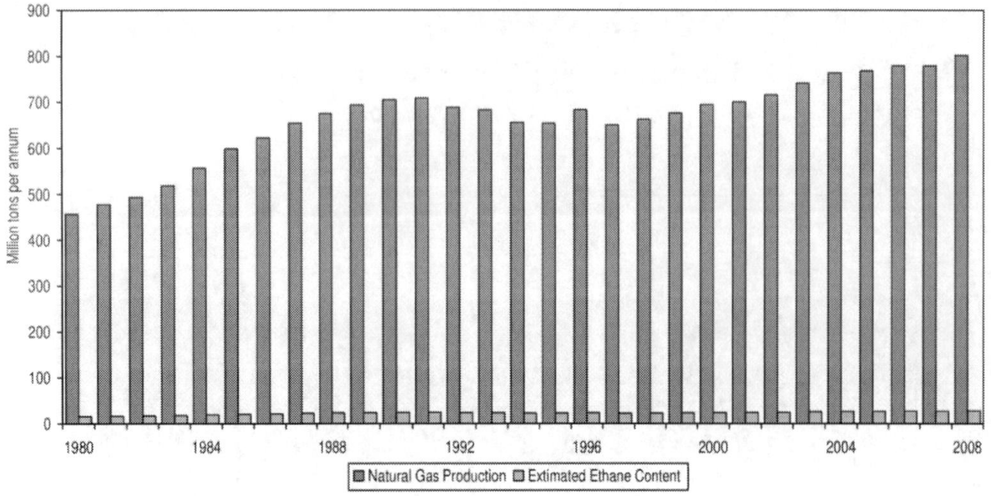

ASIA PACIFIC DEVELOPMENT

Asia Pacific ethylene capacity has grown dramatically over the last thirty years as Korea, South East Asia and now China developed advanced petrochemical sectors (re. Figure 12). The Asian petrochemical industry developed rapidly in the 1950s as Japan recovered from the devastation of the Second World War and built a modern petrochemical industry. Developments in the region took off when South Korea started building an industry in the 1980s, followed by South East Asia and China in the 1990s. Since overtaking North America in 2005, Asia Pacific is set to remain the largest ethylene producing region globally. The Asia Pacific region now accounts for just over 40 million tons installed ethylene capacity, with a heavy focus on naphtha and heavier feedstocks. Despite the rapid growth, the region is still impacted by industry profitability, with a definite slowdown in capacity development apparent at the start of this decade.

Asia Pacific LPG availability has been increasing due to refinery expansion, extraction from gas production, and, in China, the substitution by natural gas of regasified LPG for pipeline distribution (re. Figure 13). Despite this strong reliance on naphtha and heavier feeds there are significant quantities of LPG available in the region, primarily sourced from refinery processing of crude oil. The Asian LPG market also receives significant imports from the Middle East, further boosting the supply base. LPG demand however is firmly focused on domestic and industrial heating applications with only around 7 percent being consumed by the petrochemical industry.

Figure 12
Asia Pacific Ethylene Capacity by Source
(thousand tons per annum)

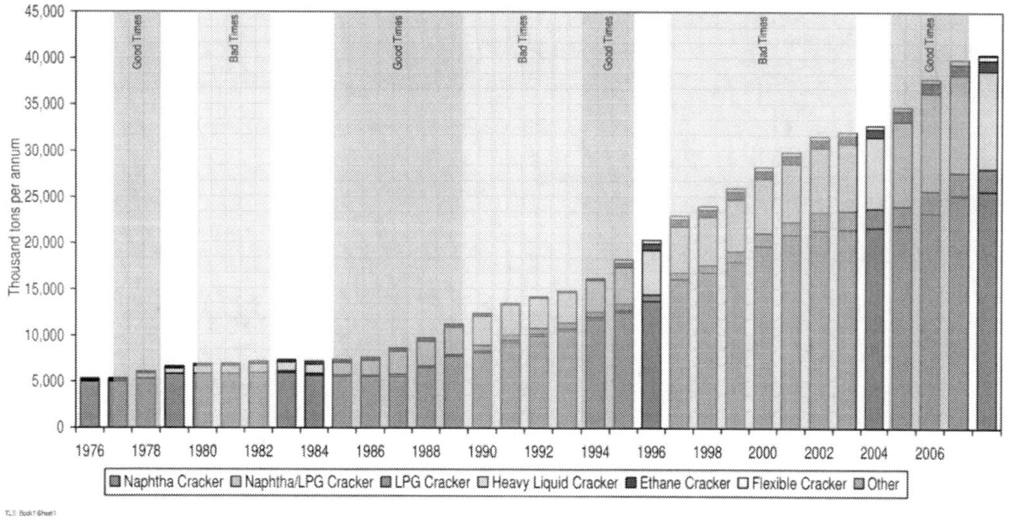

Figure 13
Asia Pacific LPG Availability by Source
(million tons per annum)

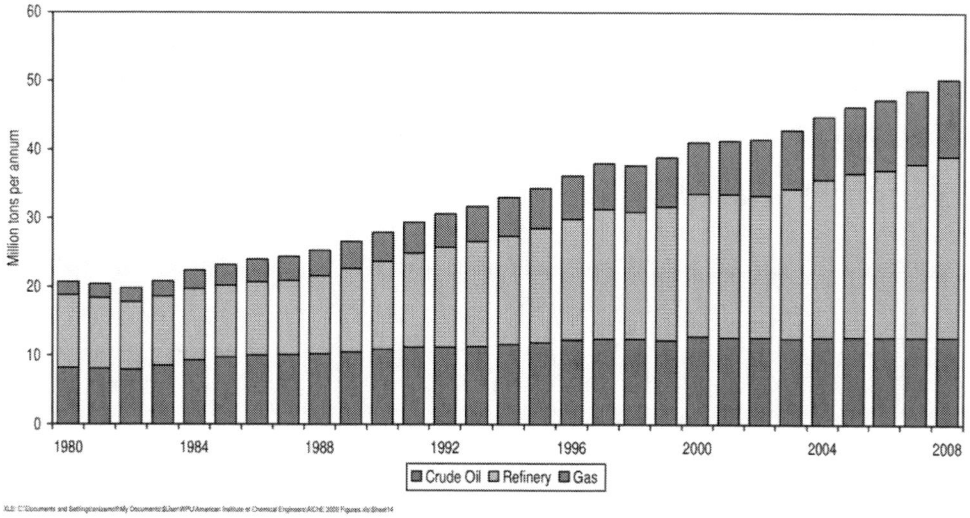

Asia Pacific naphtha supply has also grown rapidly, although at a lesser rate than cracker investments (re. Figure 14). The region has moved from having a naphtha surplus to a significant deficit. Ongoing expansions to refinery capacity have added significantly to naphtha availability in the Asia Pacific region. This availability is further boosted by the fuel focus of many Asian refineries allowing more molecules to be diverted into naphtha streams (which in North America or Europe would have ended up in gasoline or diesel). Despite this rapid increase in production, the region has become short of naphtha, with significant volumes of naphtha and condensate moving from the Middle East, and Reliance even building a world scale refinery based on exports to the Asian market. Thus, Asian crackers have a major impact on the global naphtha markets and have a strong influence on naphtha values.

Figure 14
Asia Pacific Naphtha Availability
(million tons per annum)

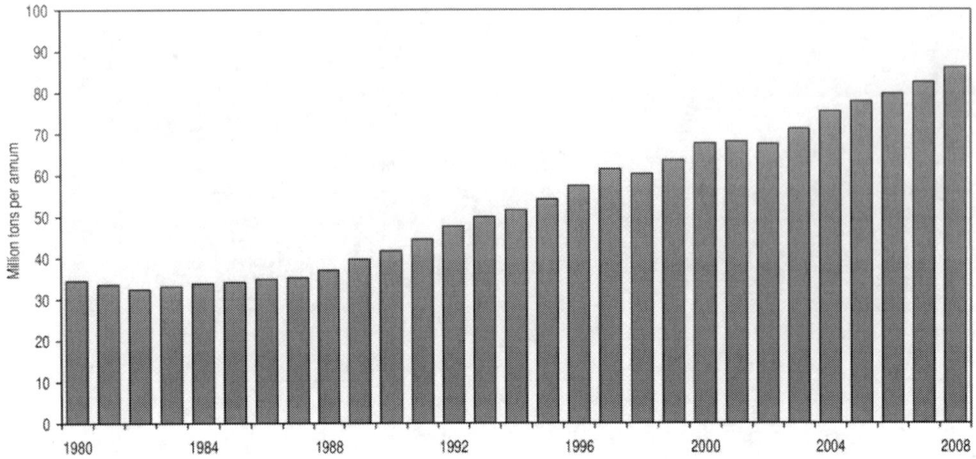

Consumption of ethane in steam crackers is limited to South East Asia and Australia. The Asian gas market is relatively undeveloped, especially considering the population in the region. Production in the region is largely associated with crude oil production and as such tends to have a higher ethane component. Little of this ethane component is currently extracted from natural gas. Exploitation has been limited to date to crackers in South East Asia and Australia. The relative small size and dispersed nature of the Asian market has precluded the development of an ethane extraction centre (re. Figure 15).

Total ethylene capacity in the Asia Pacific regions is expected to increase by almost 20 percent over the next three years. Most ethylene in Asia Pacific is produced by naphtha cracking and the majority of future developments in the region will continue to be employing liquid feedstocks. Asia will continue to rely on imports in order to satisfy the growing demand of naphtha.

Figure 15
Asia Pacific Ethane Availability
(million tons per annum)

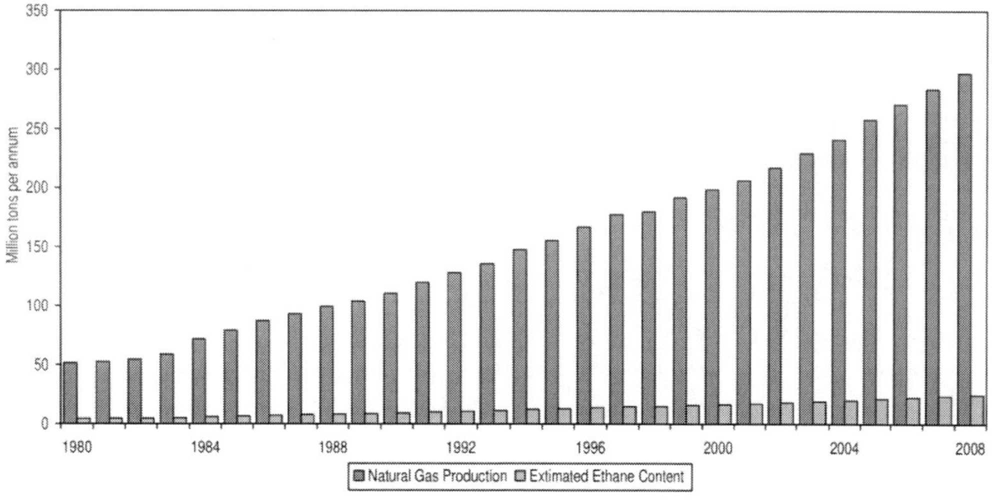

MIDDLE EAST DEVELOPMENT

Middle East ethylene capacity has shown rapid growth since the start up of the first gas crackers in Saudi Arabia in the mid-1980s. Massive hydrocarbon reserves have given the region considerable feedstock advantage, which is now being fully exploited. Although there had been an established petrochemical industry in the Middle East even back in the 1970s, its remit was to focus on satisfying local demand. It was not until SABIC took a lead in the early 1980s that the region took off as a major petrochemical player, using access to advantageously priced feedstocks to target global export markets.

The first wave of crackers proved the viability of using fixed price ethane at world scale petrochemical units in the harsh Gulf climatic environment. Subsequent waves have been a mixture of ethane and ethane/propane, ethane/butane and even pentane crackers (re. Figure 16). The region (and Saudi Arabia in particular) is now struggling to identify sufficient additional ethane feedstock to support continued expansion, with numerous plans on hold awaiting an "ethane allocation." LPG cracking, although still attractive, is more capital intensive, while at the same time offering lower returns. Middle East LPG prices have a much closer linkage with international energy prices. The Middle East has proved that low cost ethane from stranded gas streams is the most competitive feedstock for commodity petrochemical production.

Figure 16
Middle East Ethylene Capacity by Source
(thousand tons per annum)

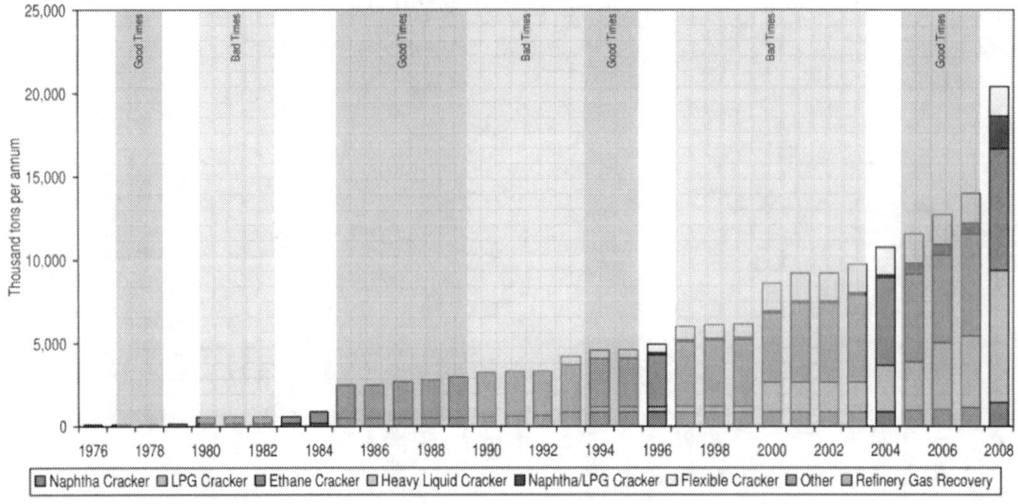

Middle East LPG availability (re. Figure 17) is dominated by associated gas production, although a growing portion is now coming from non-associated gas. Associated gas is produced in conjunction with crude oil in Saudi Arabia's major oil fields. The massive "Master Gas" project undertaken in the 1970s and 80s eliminated the flaring of this huge gas stream, by building a huge pipeline and separation system to process the hydrocarbon resources. Associated gas is far richer in heavier components than dry "non-associated" gas as it is produced directly in conjunction with liquid crude oil. These components include both ethane and LPG as well as smaller quantities of heavier components in equilibrium in the gas. The bulk of the stream however is methane, and is used in the power and desalination sectors. Saudi Arabia's petrochemical industry has taken full advantage of the ethane available from associated gas production. LPG production meanwhile is split between consumption for petrochemicals and export as a heating fuel. The Middle East is the world's largest LPG exporting region of which Saudi Arabia, Kuwait and the USE account for 30 percent of global trade. Gas developments in the Qatari North Field and Iranian South Pars field are expected to significantly boost LPG exports from the Middle East.

Middle East naphtha is sourced from local refineries and as condensate from gas processing units in the region (re. Figure 18). The Middle East has historically exported its liquid hydrocarbons (i.e., naphtha) for processing in refineries located in the major demand centers in Asia and the West. Local refining operations have to date been modest, although this will change going forwards as new world scale refineries are built in the region. The majority of available naphtha has been exported out of the Middle East with significant volumes heading to Asia. Around 7 million tons is fed to liquid crackers in the region. The Middle East is a major source of traded naphtha, with significant flows heading to Asia Pacific.

Figure 17
Middle East LPG Availability by Source
(million tons per annum)

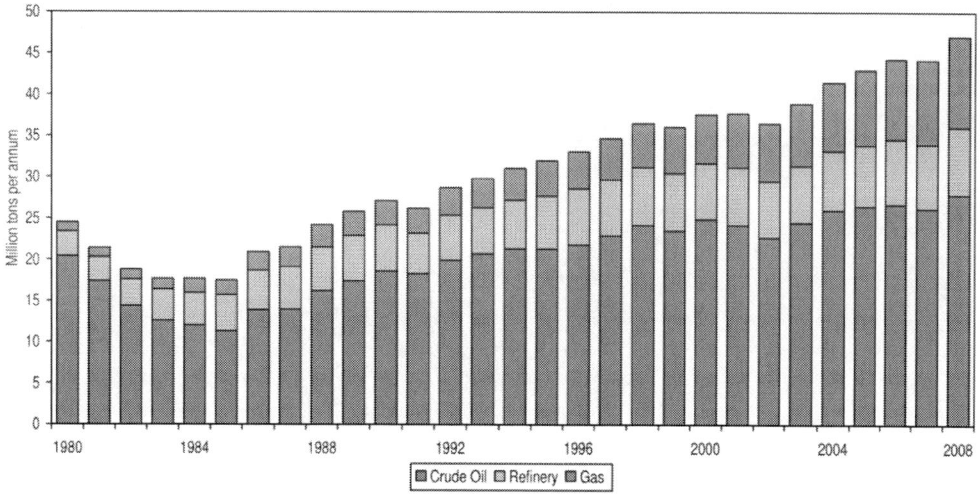

Figure 18
Middle East Naphtha Availability
(million tons per annum)

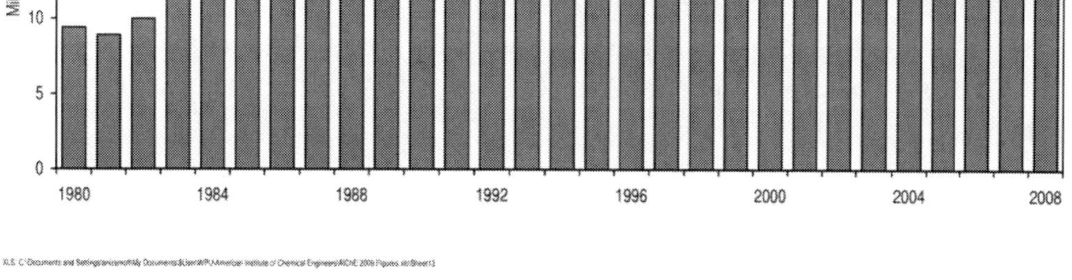

The Middle East ethane cracking industry has been built on finding an economic use for natural gas components that were previously flared. Unlike the West, where demand for gas has driven the search for gas supply, the Middle East has for a long time had access to abundant gas produced in conjunction with crude oil (re. Figure 19). This supply however is fundamentally tied to crude oil production – limiting the ability to easily produce additional volume. The low population in the region prevented the underlying lean natural gas demand from catching up with supply, although that situation has become tighter as industry, desalination and power sectors increase demand for gas.

Figure 19
Middle East Ethane Availability
(million tons per annum)

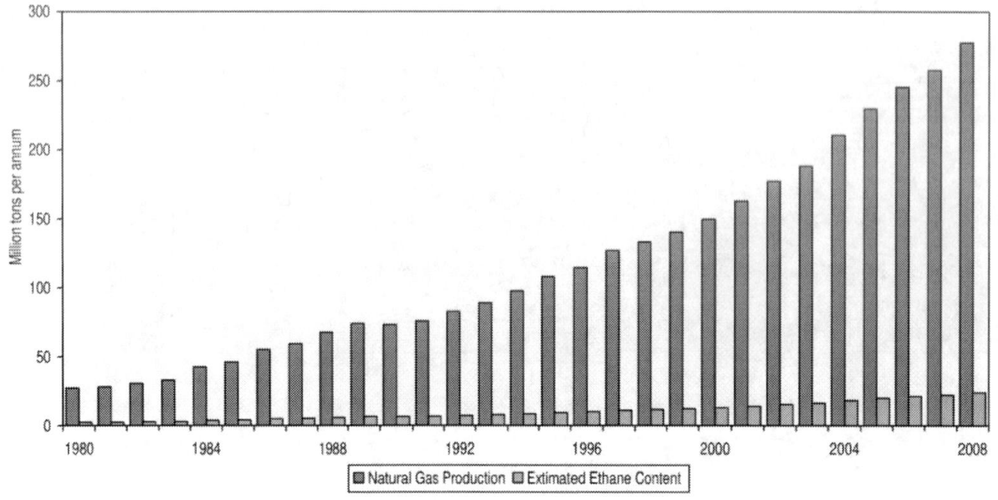

Low demand allowed Saudi Arabia to argue that ethane and natural gas could initially be charged at a nominal extraction fee of US$0.50-0.75 per million BTU (for ethane) – significantly below market prices for natural gas in developed markets (US$8 – US$12 per million BTU in global markets). This low pricing has allowed the rapid development of a very profitable commodity petrochemical industry – however this high profitability has meant that all available gas is rapidly absorbed into the petrochemical sector as soon as it becomes available. The Middle East has exploited ready access to associated gas to develop a highly profitable ethylene industry. Ethane supply however remains tied to associated gas production and future volumes will be limited by crude production rates.

Pure ethane based crackers which have previously been the norm are expected to be less common in the future, as the region seeks to diversify and exploit other feedstocks such as propane and liquids.

OLEFINS SUPPLY & DEMAND

Ethylene

As ethylene feeds into all major polymers, with derivatives spanning all industrial sectors, global demand growth closely follows economic activity. As ethylene itself is very expensive to ship, trade is mainly in ethylene derivatives such as the polyethylenes and ethylene glycol. Aside from areas such as the United States Gulf Coast and North-West Europe, which have ethylene pipeline networks, most ethylene is consumed at plants adjacent to the steam crackers.

Global consumption of ethylene is forecast to grow at a higher rate over the next ten years than has been seen over the previous decade. The fundamental driver for this is growing domestic consumption in developing economies. Asia Pacific has become the largest consumer of ethylene since 2005, now accounting for approximately 30 percent of global consumption. Ethylene demand in the region is expected to grow at above global rates, but slower than in the previous decade. Ethylene consumption growth in the Middle East will be highest at an average of 13 percent per year until 2015 because of the numerous world-scale cracker projects in the region which will be integrated with derivatives. Similarly, consumption in South America and Africa is forecast to rise sharply in the future driven by an increase in polyethylene capacity in both regions. On the other hand, the demand in North America and Western Europe, collectively accounting for about 47 percent of global consumption, is forecast to show only modest growth owing to the lack of domestic demand growth, the import of ethylene derivative containing goods, and declining derivative export competitiveness.

Current ethylene capacity additions are located to exploit two key aspects of competitive advantage: low cost feedstock and proximity to market. Gaseous feedstocks, like olefins themselves, are very costly to transport, and it is therefore preferable to locate plants close to the feedstock sources, and transport the derivatives to market. The attractiveness of liquids cracker projects depends on a high level of refinery integration, low fixed/labor costs and favored access to derivative markets. Most new Chinese steam cracker projects enjoy all of these benefits.

The availability of advantaged feedstock in the Middle East has driven a wave of major new steam cracker developments, particularly in Saudi Arabia and Iran. Middle Eastern producers can serve both the European and Chinese markets. Projects in China lack the feedstock price advantage of those in the Middle East, but their economics are greatly enhanced by refinery integration and lower construction and labor costs, as noted previously. Seasonal fluctuation in the value of LPG relative to naphtha offers a potential competitive advantage, and many producers in Asia Pacific are modifying their plants to be able to crack LPG to exploit this.

North America and Western Europe are quite mature markets with minimal capacity additions expected and some closures likely. Capacity development in South America will accelerate, to take advantage of relatively high derivative demand growth and advantaged feedstock in countries such as Venezuela and Trinidad. There is some bio-ethylene production planned in Brazil but it is very limited at this time. Central and Eastern Europe will develop new world-scale capacities in the future. There is little advantaged feedstock in Central Europe, but new capacities will be encouraged by the desire to develop national industries, the relatively high economic growth rates and the access to the West European market.

Propylene

Propylene demand growth has historically been greater than global GDP growth but is expected to align closer to GDP levels as the major derivative markets mature. Demand growth of propylene has required supply in addition to that available from steam cracking and refinery operations. On-purpose propylene production has been developed to fill this gap and allow polypropylene and other derivatives to increase market penetration.

Regional development is heavily focused in the Middle East and Asia. Development in the Middle East is almost entirely for export-oriented polypropylene complexes, while Asian demand is expanding to meet local demand for a broader base of derivatives. Forecast demand growth in Western Europe is particularly low, and the region is expected to move from being a net exporter of polypropylene to being a net importer.

Propylene capacity is quite evenly divided between the three major regions of North America, Western Europe and Asia Pacific. The share held by Middle Eastern producers is comparatively small due to the prevalence of ethane-based steam crackers, but growing rapidly as a number of world-scale projects come on stream with on-purpose propylene production processes. Western Europe and Asia Pacific derive most propylene from naphtha crackers, while supply in the United States is based to a greater extent on refinery production.

Propylene availability in the Middle East is increasing due to a move towards heavier cracker feed slates, investment in refinery FCC units and the development of new on-purpose propylene production processes namely, propane dehydrogenation and metathesis. Various companies are currently evaluating methanol-to-olefins projects in the Middle East, although projects proposed in Africa are progressing more quickly. The expansion of demand in Asia is supported by the large scale naphtha crackers under development in China and South-East Asia. Propylene availability from refinery sources will however be limited by slow capacity addition rates of the refineries and an increasing preference for hydrocrackers instead of fluid catalytic crackers in new refineries.

OLEFINS FROM OTHER SOURCES

Coal

Higher prices for conventional petrochemical feedstocks have driven technology development to exploit alternative feedstocks such as coal and methane. Coal-based olefins production was developed in South Africa as a means of exploiting freely available coal at a time when oil-based hydrocarbon feedstocks were difficult to obtain due to trade restrictions. The economics of coal-based production are not attractive in many regions, although coal-based projects are under development in China, which has abundant coal reserves and a shortfall in domestic availability of liquid cracker feedstocks. The first two plants producing propylene from coal employing Lurgi's methanol-to-propylene (MTP) process are already under development in China. China has developed its own MTO (methanol-to-olefin) and MTP technology, but commercialization has not yet been achieved.

Nexant has considered the economics for a coal-to-olefins (CTO) plant employing a coal gasification plant followed by methanol synthesis and a methanol-to-olefins plant (producing 500 kta of ethylene) and compared this to a world-scale, conventional steam cracker employing light naphtha as a feed (producing 1,000 kta of ethylene). The results are shown in Figure 21. The base case coal price is $33.60 per short ton (Illinois coal, average 2007). Crude oil was averaging $72 per barrel during this period.

Figure 21
CTO Sensitivity to Economy of Scale and Cost of Coal
(500 kta ethylene, cost plus return, USGC, 2007)

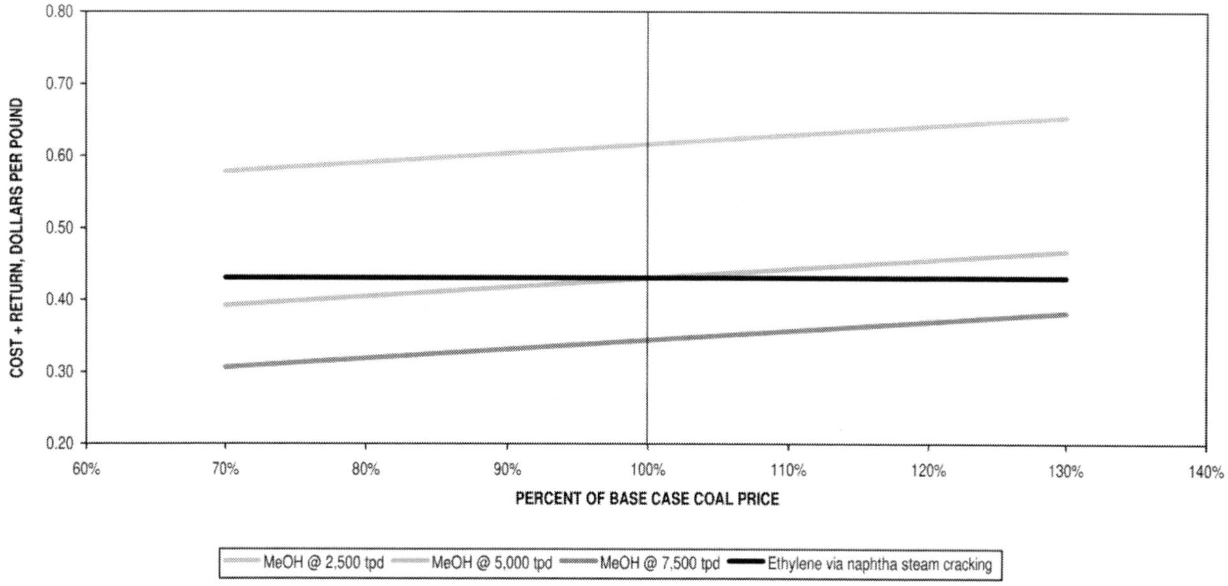

One of the critical aspects of this route to olefins is the size of the methanol synthesis unit. The maximum capacity for methanol synthesis units has increased significantly over the past ten years. Units as large as 5,000 tons per day have been built and are operating. Licensors are now stating that the maximum single train capacity is approaching 7,000-7,500 tons per day, although none has yet been built at this larger capacity. Therefore, the sensitivity of the CTO economics has been examined for smaller and larger methanol capacities of 2,500 and 7,500 tons per day. At the larger capacity, 2 MTO trains will be required and this will have a noticeable negative affect on the cost of production. The results for the USGC are depicted in Figure 21. For the U.S. the cost of production (including depreciation and return) increase significantly at the smaller capacity such that the CTO route is simply not competitive with conventional technology. At the larger capacity, the CTO route is competitive and, in fact shows a small advantage over the conventional naphtha-based technology. The higher investment for 2 MTO trains producing 500 thousand tons per annum of ethylene (total) results in higher capital charges when compared to the smaller plants. If the single train capacity of the MTO unit would be increased to better match the methanol facility, even lower costs for the ethylene could be expected.

At the 2007 pricing level, the coal-to-olefins technology is only competitive at very large methanol plant capacities and marginally competitive at smaller capacities with petroleum-based sources in the U.S. However, in China where coal has historically been less expensive and by-products have had a higher value than in the U.S, the coal-to-olefins technology could be very competitive with steam cracking.

It should be noted that although the analysis provide herein shows some economic advantages to coal-to-olefins, there are other considerations besides economics. Countries looking at developing coal to liquids projects (transportation fuels as well as petrochemicals) should be:

- Net energy importers with state backing
- Have deregulated fuel prices
- Not have any carbon constraints
- Have sufficient resources (i.e., both coal and water resources)

While the conclusion is that some technologies allow/may allow production of petrochemicals from coal at competitive conditions in today's/ future environment and since coal is abundant, cheap to produce, and may be forced to be an energy of the future if nuclear is not relaunched fast enough, coal-based petrochemicals produce approximately 3 times (!) more CO_2 than oil & gas-based petrochemicals. Hence, coal should not be seen as an acceptable alternative until a viable solution can be found for CO_2 storage.

Bio Sources

Brazil is the world's largest producer of ethanol from sugarcane, and hence has the potential for ethanol-based ethylene. One such project, together with polyethylene capacity is being developed between Dow and Crystalsev and is expected to be onstream in 2011. Brazil's ETH Bioenergy, an Odebrecht Company, also plans to invest in a 300 thousand ton per year ethanol-based ethylene and polyethylene complex around 2010-2011.

Nexant has developed the economics for the production of 350 thousand metric tons per annum of ethylene from bio-ethanol and compared it to the cost plus return for a world-scale, conventional steam cracker employing a light naphtha feed (1,000 thousand metric tons per annum of ethylene produced). We have considered fermentation of sugarcane ($16/ton), corn ($3.46/bushel) and wheat ($182/ton) as well as a speculative process employing switchgrass ($46.70/ton). The latter is integrated with a power co-generation unit. The results are shown in Figure 20. While switchgrass is well suited for the North American climate, sugarcane is ideal for more tropical climate zones (e.g., Brazil).

Figure 20
Sensitivity of Ethylene Cost of Production to Bio-Mass Price
(350 kta, USGC, 2007)

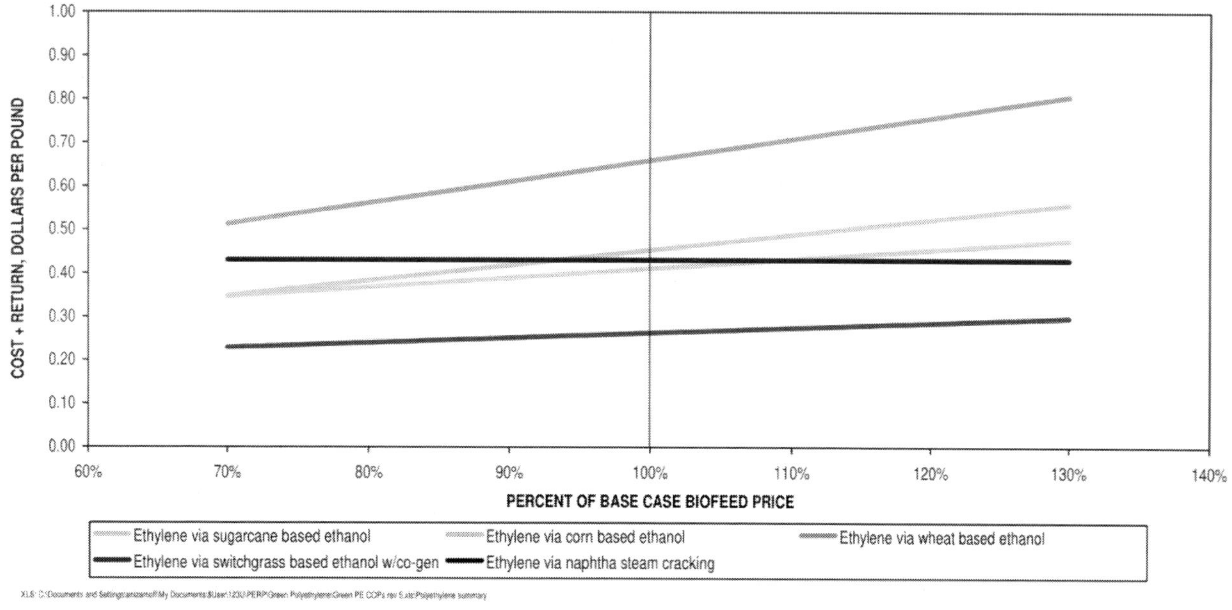

At present, "green" ethylene produced via ethanol produced from corn (100 million gallons per year) in the United States is not competitive with ethylene produced using world-scale (i.e., 1,500 thousand metric tons per annum) naphtha-based ethylene. However, in Brazil where ethanol from sugarcane is significantly less expensive than from corn in the United States, the ethanol to ethylene to green polyethylene route may be attractive.

Economics not withstanding, there is certain cachet in being able to offer "green" polyethylene to the marketplace. Braskem has received some amount of positive attention by being able to advertise that their "green" polyethylene has been certified by a well known carbon analysis laboratory to be made with 100 percent renewable materials. Both Braskem and Dow Chemical expect their "green" polyethylene to be essentially identical to conventional petroleum and gas based polyethylene and to be able to be used by converters without them having to invest in any new equipment.

Certain end-use applications may find value in advertising that their product is 100 percent "green" and may be able to charge a premium to their customers.

WHAT'S NEXT?

Nexant has undertaken some modeling work on LPG consumption in steam crackers and have had discussions with a number of producers about the potential increase in flexibility. As cracker operating rates decrease in the next few years the existing flexibility automatically increases: operators can decide to cut back their naphtha or liquids crackers and run their gas crackers flat out. This increases the proportion of ethylene from LPG and decreases the co-product make. In addition, as more natural gas comes from unconventional sources such as coal bed, shale, and tar sands without associated natural gas liquids (NGLs), the premium for NGL feed to steam crackers will rise and more naphtha will be cracked.

In addition Nexant just completed some work on the technical requirements to retrofit gas cracking flexibility to naphtha crackers and are considering producing a multiclient report on the subject as part of ChemSystems' Process Evaluation/Research Planning (PERP) program. Nexant also has prepared an analysis of the feedstocks markets for a feedstocks report as part of ChemSystems' Petroleum and Petrochemical Economics (PPE) program (*Petrochemical Market Dynamics: Feedstocks*).

40d Panel Discussions

Sanjeev Kapur
Fluor Corporation
Sugar Land, TX 77478-3899
Sanjeev.kapur@fluor.com

Aivars Krumins
CB&I
Houston, TX 77252-2833
akrumins@cbi.com

Abstract:

The panel discussion will address the issues and challenges our industry will face in future to compete with other demands for the limited hydrocarbon resources that are essential for meeting the increased demand of petrochemicals. Today's competitive industry environment requires special attention to the availability and access to cheaper feedstock, as the cost of production of olefins is highly dependent on feed costs. The shared knowledge of petrochemical producers, consultants, refiners, process licensors and engineering firms, catalysts and chemicals suppliers, and other industry experts as our panelists and attendees will keep the focus on future trends and help our industry to be ready for the opportunities.

The panelists include keynote speaker of Ethylene Producer Conference, speakers from this session, industry experts from different regions of the world.

If you have a question (or questions) that you would like to ask these experts, related to trends in olefin plant feedstock, please e-mail those to sanjeev.kapur@fluor.com or AKrumins@cbi.com prior to March 1, 2009. In e-mail, please make sure to include your full name, company name and address. We will also take questions from the floor during this session.

EUROPEAN ETHYLENE PRODUCERS COMMITTEE (EEPC) CORROSION UNDER INSULATION (CUI) STUDY GROUP REPORT "HOW BIG IS THE IMPACT OF CUI?"

Hendrik J de Bruyn

Borealis Group Services AS, Stathelle Norway

Antoine Baecke

DOW Benelux B.V., Terneuzen, the Netherlands

Kevin W McQuillan

SABIC UK Petrochemicals, Wilton, United Kingdom

Abstract: The European Ethylene Producers Committee (EEPC) has an active Corrosion Under Insulation (CUI) Issue Group. The Issue Group has participation from 12 European ethylene producers and 4 engineering and construction companies with global orientation. The Issue Group was formed early 2007 because of serious safety and equipment CUI incidents reported at annual EEPC conferences.

The Issue Group is working on its plan to provide information that will help EEPC cracker operators to measure, analyze, improve and secure the situation of CUI in their crackers. The Issue Group wants to be pro-active and has developed and distributed a survey on the impact of CUI.

The purpose of this paper is to share the results of the EEPC CUI survey to a wider audience. The Issue Group plans to work further with these results in order to establish best practices and processes and implementation experiences for CUI inspection and remediation.

Introduction

The Corrosion under Insulation (CUI) Issue Group of the European Ethylene Producers Committee (EEPC) was chartered during April 2007. In the original terms of reference the following was stated about the intended activities of this group:

The Corrosion Under Insulation (CUI) Issue Group will present and document experiences good and bad in the management of CUI on carbon and austenitic steel as well as other materials if applicable for pressure equipment in Petrochemical plants. The presentations and documents will cover issues related to the life-time of equipment as listed below:

- *CUI basics*
- *Design phase*
- *Construction phase*
- *Inspection*
- *Operational phase*
- *Maintenance phase*
- *Legal aspects*

The CUI Issue Group meets twice per year, usually at a cracker site of one of the member companies. These meetings focus on the above and also include the following:

- A plant tour, where appropriate, focused on the discussion topic for the meeting.
- Presentations on the chosen topic giving members' experiences (good or bad!) in the targeted area.
- Presentations, where appropriate, from external organizations with relevant expertise.
- Discussion of incidents related to the chosen topic.
- Discussion and formation of guidance on presentations and documents.

As with all EEPC Issue Groups, the CUI Issue Group has a limited scope of work and will be terminated after approximately 4 years. The purpose of this paper is to share the results of the EEPC CUI Survey to a wider audience. The Issue Group plans to work further with these results in order to establish best practices and processes and implementation experiences for CUI inspection and remediation.

Corrosion under Insulation (CUI) Survey

The CUI Issue Group has developed and distributed a survey on the impact of CUI during 2007. This survey attempted to obtain the status quo of CUI in European cracker plants in terms of:

- Lost production, health, safety and environmental impact.

- CUI Cost for inspection, maintenance, remediation during normal operation and at Turn Arounds.
- Design, construction, operation and maintenance practices and modes related to CUI.
- Inspection techniques, painting and insulation specifications applied.

The detailed list of questions in the survey is presented in Table 1.

General Survey Results

Age, Capacity and Location

All EEPC cracker operating companies have completed the survey and the CUI Issue Group has data from 19 European crackers, 2 crackers in the Middle-East, and 1 in Australia.

The average age of the cracker plants is 29 years, with the oldest plant aged 47 and the newest plant 8 years old. These cracker plants processes a wide variety of feeds, averaging 545 kT per year.

The location of the crackers in the survey is shown in Figure 1. 50% (11/22) of the cracker plants in the survey are located in a coastal industrial area (Marine), 41% (9/22) in an inland industrial area, and 9% (2/22) in a dry climate.

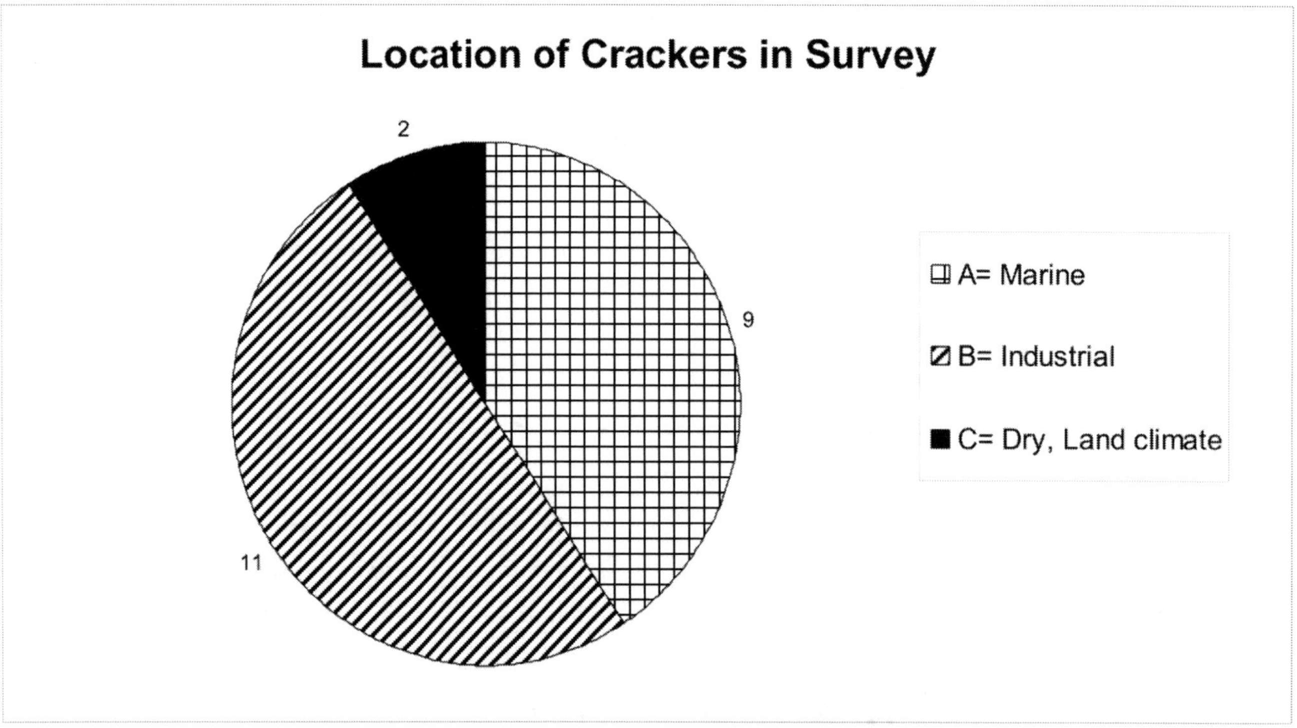

Figure 1: Location of Crackers in the CUI Survey

Outages and Volume Losses due to CUI

Nine (40%) crackers in the survey have reported outages due to CUI during the last 5 years. This resulted in a total loss of ethylene production of 63 kT, an average of 7 kT per outage.

CUI Leaks and Repair Strategies

A surprisingly high number of plants (18/22 = 81.8%) reported leaks due to CUI. This amounted to a total of 292 reported CUI leaks during the last 5 years. The type of leaks was reported as follows:

- 72 product leaks
- 220 utility leaks

It should however be noted that more than half of the respondents could not provide data on utility leaks. The reported numbers for utilities may therefore be significantly higher.

The repair methods employed appeared to be dependent upon the type of equipment involved. Most repairs were fixed by clamping, followed by weld repairs, and in a few rare cases total replacement of the equipment involved.

Preventative maintenance to avoid leaks has been carried out by the majority (20/22) of the cracker plants in the survey. A total of 643 preventative repairs/replacements were carried out during the last 5 years. Of these, 106 had a credible leak scenario within the next 5 years. Preventative maintenance has avoided 4.8 potential leaks per plant over the last 5 years. This is an average of about 1 leak per plant per year.

Best in Class

The best cracker plants (4/22), in terms of CUI damage avoidance, had no leaks during the last 5 years. The ages of these plants are 6, 26, and 35 years. Two of these plants are located in a marine industrial environment, and two in an inland industrial environment.

While it might be argued that one of the plants that reported no CUI leaks is very young, the others definitely seem to have a strategy in place to avoid CUI becoming a serious operational problem.

Causes of CUI

The perceived causes of CUI are presented graphically in Figure 2. It is interesting to note that operations are listed as the highest cause of CUI. This related to the care with which insulated equipment is generally treated. Own actions such as walking on equipment and pipelines, insulation removed and left open, and draining operations over insulated equipment are often causing more damage than anything else. The lack of draining water from wetted insulation systems can also contribute significantly to CUI issues.

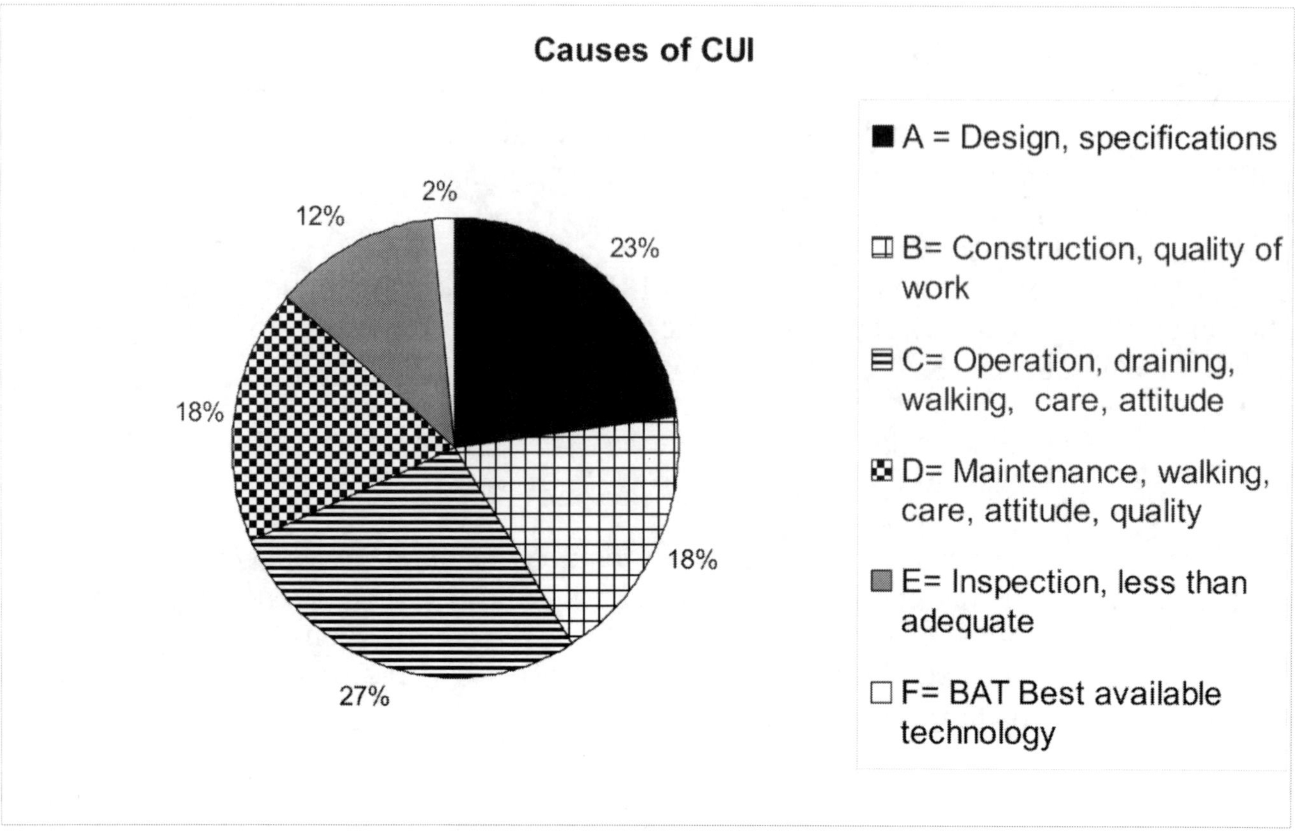

Figure 2: Causes of CUI

Design and specifications are rated second highest as a cause of CUI. This is probably due to the fact that many of the European cracker plants were constructed at a time when insufficient attention was paid to CUI. For example:

- Until the late 1990's, painting systems under insulation often consisted of a single primer layer.
- Sheeting (insulation jacketing) details with numerous horizontal surfaces allowing water entrapment.
- Sheeting details with openings instead of roof tile constructions.
- Horizontal insulation penetrations e.g. at supports and horizontal lines without rainwater deflectors.
- Supports and hangers without rain water deflectors.

It is not known whether most European cracker operating companies now have better specifications in place to minimize CUI.

Construction and the quality of installation of insulation systems contribute significantly to CUI in cracker plants. Significant improvements can be made by ensuring that insulation systems are installed as intended, and that inferior construction details are avoided. Inadequate constructions at plant start-up expose operators and maintenance personnel to wrong examples. Optimum insulation systems necessary for CUI prevention are therefore never given a chance.

Damage due to maintenance activities and the lack of attention to details when reinstalling insulation systems, together with inadequate inspection, are also stated as major causes of CUI. It is well known that some companies opted for short-term savings by adopting inadequate painting and insulation practices during weak economic times in the early 1990's. Today, due to the impact of CUI leaks, costs and outages and operating older plants than originally planned, more companies start to realize that the CUI problem is a culture and a management issue requiring long-term financial planning. CUI mitigation should be seen as an investment to ensure lowest cost of ownership.

Parameters Influencing CUI

Insulation Type

The survey suggests that there is not clear influence of insulation type, although 6 plants reported that polyurethane insulation appears to be more problematic in terms of CUI. CUI seems to occur regardless of insulation type.

Cyclic Conditions

Five Plants reported that cyclic conditions have a strong influence on CUI. This is in agreement with well accepted guidelines on CUI[1,2]. Cyclic conditions are likely to cause severe general corrosion over a wide area instead of localized damage. This might increase the risk for sudden failure/rupture instead of leaks.

Sheeting and Sealing

The survey suggests that the importance of sheeting (insulation jacketing) details is under estimated. Sheeting is often not considered an important maintenance element. This is especially true as the average plant life is well beyond the original design life. Too little attention is paid to the design, application and maintenance of insulation systems. At the end of the original plant life, few companies are actually refurbishing insulation systems, and often expect such systems to continue performing as new.

Temperature

Two crackers with more than 36 years of history made an extensive CUI lifetime investigation. This showed that equipment operating in the temperatures range 100 - 150°C (210 - 300°F) is most prone to CUI. However, CUI can also occur at temperatures as low as 0°C (sweating areas).

Equipment operating at higher temperatures is not immune to CUI. The survey respondents reported CUI damage to equipment operating up to 200°C at dead ends and remote parts that are lower in temperature.

Equipment Type

The survey results relating equipment type to the occurrence of CUI can be summarized as follows:

- Piping: 50%
- Columns: 15%
- Heat Exchangers: 10%
- Pressure Vessels: 6%
- Other: 19%

The fact that most CUI damage occurs on piping is worrying. Traditionally most inspection resources are used on pressure vessels. The survey also pointed out that instruments and instrument connections are also prone to CUI. These types of equipment are often not inspected together with other pressure bearing equipment.

Process Areas

In cracker plants, process areas that are most prone to CUI are:

- Depropaniser & debutaniser
- Dryers (cyclic operation)
- Propylene fractionator
- Acetylene conversion/ removal
- Oil Quench
- Furnace feed piping

All other areas that are within the critical temperature ranges, are in cyclic operation, and in sweating/condensation service are also susceptible to CUI.

Cracker Age

The number of CUI failures increase with plant age. Several plants reported complete refurbishment/replacement of equipment and piping due to CUI. CUI refurbishment/ replacements are carried out while in operation and during turn arounds.

Approaches to CUI Mitigation

Approximately half of the plants in the survey reported using a systematic approach, such as risk-based inspection (RBI), to CUI. There is however almost no correlation that these plants are doing better.

The reason for failure is often cited as a lack of follow-up. It seems that most companies are not good at closing the loop, and need to be better at planning, checking, acting and following up.

Painting Practices

The survey revealed that for the protection of carbon steel under insulation, mostly conventional painting systems are used. These often consist of zinc silicate primers followed by zinc epoxy or epoxy top coats. Thermal sprayed aluminum (TSA) is not widely used yet.

Many plants do not repaint at all, while a few repaint every 15 – 25 years. The majority of the respondent plants seem to only repaint when damage is detected.

Paint systems have improved vastly during the last 20 years. However, many plants are not benefiting from these improvements.

Inspection Methods and Inspection Frequency

Removal of insulation followed by visual inspection is employed by most of the crackers in the survey as the preferred approach to inspection of insulated equipment. This is performed on 80 – 100% of the insulated surfaces.

Radiography is used by 7 plants for 20 – 50% of the insulated surfaces. Only 2 plants have reported the use of digital radiography to find CUI damage. The use of other inspection methods is even lower:

- Guided wave UT: 3%
- Pulse eddy current (PEC/Incotest): 1%

The average inspection frequency in the survey is 6 years for insulation and fire proofing systems.

Manpower Employed

The cracker plants in the survey have an average of 10 – 15 persons per plant occupied with CUI. These numbers exclude TA activities or manpower levels when problems are detected. The majority are insulators and scaffold builders. Some painting personnel and a few inspectors are also included in the personnel numbers. One Inspector is employed for every 7 - 9 other crafts.

The Cost of CUI in European Crackers

The survey revealed that it is very difficult to get the real cost data. This might be because CUI costs are often hidden in the following budgets:

- Painting program
- Refurbishment projects
- Maintenance budget
- Equipment replacements

Five plants have between 15 - 20 people working on CUI inspection and remediation. On average, all plants have about 10 people working on CUI.

The average CUI spending is between 1 - 3 million € per plant per year, excluding turn arounds and special refurbishment projects. Inclusive of turn arounds, the range is 1 - 4 million € per year.

Most money is spent during turn arounds. Six Plants have reported spending 0.6 - 4 million € during a turn around on CUI. Eight Plants reported 50 - 250 locations where replacements or welding was required for CUI remediation during a turn around.

Conservation Cost over Lifetime

Following the CUI survey, the CUI Issue Group has attempted to compare CUI costs as a fraction of the total investment cost of a plant for the different CUI conservation methods available. These are:

- Thermal sprayed aluminum (TSA) applied on shop fabricated equipment during new construction.
- Thermal sprayed aluminum (TSA) applied in a running plant after 25 years.
- Top class painting, insulation and sheeting (P+I+S), followed by top class maintenance.
- Average painting, insulation and sheeting (P+I+S), followed by average maintenance.
- Average painting, insulation and sheeting (P+I+S), followed by average maintenance with a CUI outage in year 25 and year 35.

Figure 3 shows a graphical presentation of this cost comparison, and clearly illustrates how dramatic cost increases due to inaction can be. Over the lifetime of a plant continuing to do little to no maintenance on insulations systems, and allowing CUI leaks and outages to occur, will result in very high costs.

Top class maintenance of insulation and weather barriers is a very good investment compared to average/no maintenance. Such investments will avoid outages.

The application of thermal sprayed aluminum (TSA) can potentially result in large savings after 15 to 20 years of plant life, and is likely to outperform top class maintenance. The above analysis showed that TSA is not a measurable extra cost when applied to new plants. TSA applied after 25 years is 40% more expensive than repainting, but is still considered a good economic alternative considering lower future maintenance and turn-around costs.

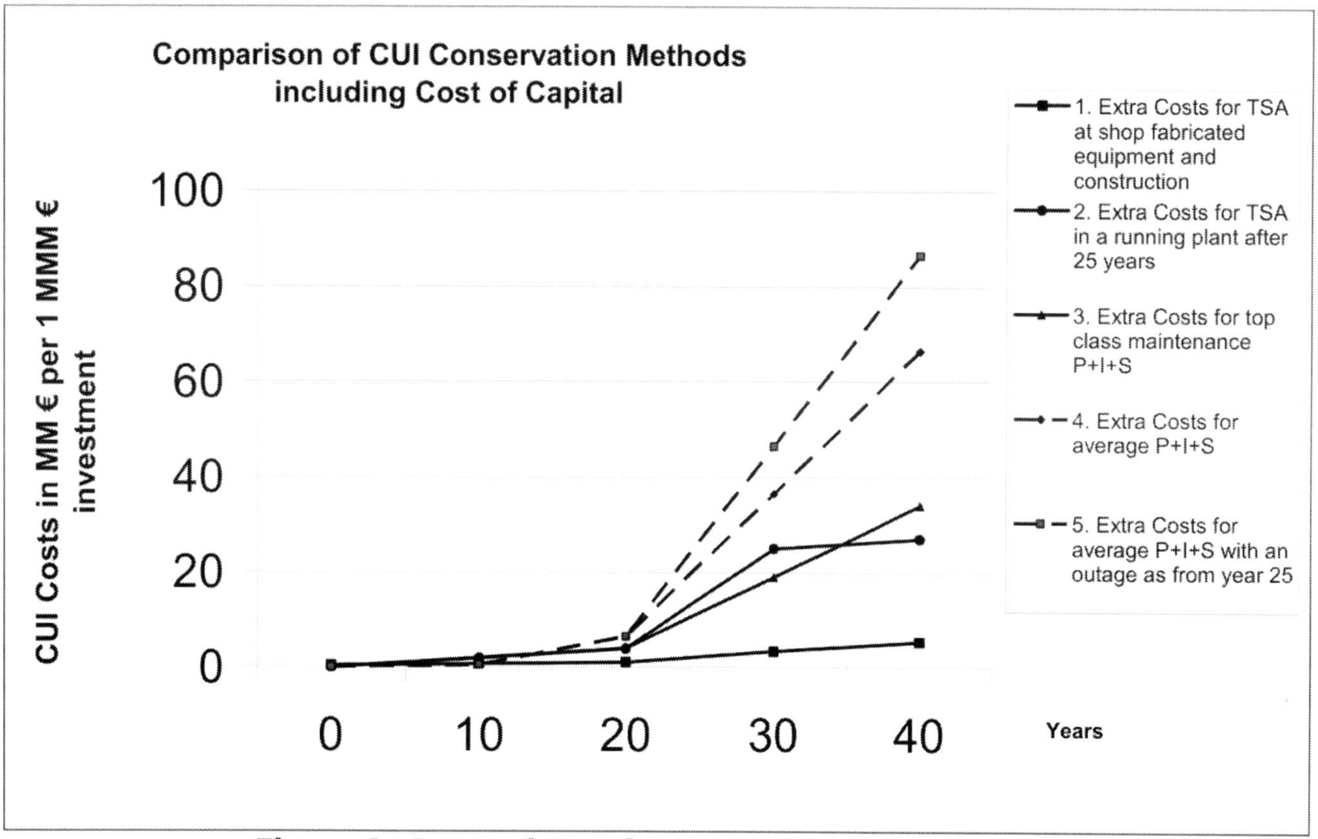

Figure 3: Comparison of CUI Conservation Methods

Conclusions

The activities of the CUI Issue Group in the European Ethylene Producers Committee (EEPC) have provided member companies with useful insights into maintenances practices and costs associated with combating CUI. The CUI Survey was an excellent tool to show the size of the problem to this industry in Europe.

The work of the CUI Issue Group has also indicated that for plants older than 20 years, the lowest total cost of ownership can be achieved by the application of TSA in the CUI susceptible temperature range. Intangible benefits in the area of safety, loss of primary containment, environment, reliability and license to operate have not been quantified.

Second best is to ensure top class maintenance of painting, insulation and sheeting. This means no compromise on painting preparation, quality of application and insulation application and continuous care by operation and maintenance.

Both of the above strategies are capital intensive investments that will save costs over the lifetime of a plant, and will improve safety and reliability with the lowest environmental impact.

Further Activities

There are excellent publications on CUI available. Yet, most plant inspectors and maintenance engineers are struggling to utilize and implement the experiences of others, and are frequently unable to convince their own management teams to take action. The EEPC CUI Issue Group will therefore during 2009 and 2010 work on a "roadmap" to help guide inspectors and maintenance engineers through the numerous publications and standards related to CUI. At the moment a pictorial database is being established to show both good and bad insulation practices. Contributions from the EPC are most welcome.

References

1. Winnik, Stefan (2008), *Corrosion-under-insulation (CUI) guidelines*, European Federation of Corrosion Publication No. 55.
2. NACE Standard RP0198-2004, *The Control of Corrosion Under Thermal Insulation and Fireproofing Materials - A Systems Approach*, National Association of Corrosion Engineers.

Table 1: CUI Survey Questions

Number	Question
1	**Age in years of cracker**
1.1	Type of cracker. *Naphtha, Condensate LPG, Ethane, etc*
1.2	Size of cracker: in tons ethylene/year
2	**Location of cracker** *A= Marine environment* *B= Industrial environment* *C= Dry, land climate*
3	**Plant outages last 5 years due to CUI**
3.1	Duration of outage in days
3.2	Volume loss of ethylene in tons
3.5	Safety impact
3.6	Environmental impact
3.7	Health impact
3.8	Property loss impact
4	**Number of leaks due to CUI last 5 years**
4.1	Number of product leaks
4.1.1	Repair methods product leaks
4.1.2	What are the yearly costs involved to repair product leaks (in €)?
4.2	Number of utility leaks
4.2.1.	Repair methods utility leaks
4.2.2	What are the yearly costs involved to repair the utility leaks (in €)?
5	**Number of preventive repairs (clamps, remediation, renewal of lines etc) for CUI last 5 years.** *Ongoing preventive actions in place?*
5.1	Average number of preventive repairs and replacements per year
5.1.2	What are the average costs per year for preventive repairs (in €)?
5.2	Average credible leak scenario's prevented per year

Number	Question
6	**Related causes for CUI in %** *A = Design, specifications.* *B= Construction, quality of work.* *C= Operation, draining, walking, care, attitude.* *D= Maintenance, walking, care, attitude, quality.* *E= Inspection, less than adequate. F= BAT Best available technology*
7	**Average number of crafts for CUI prevention and remediation per year last 5 years**
7.1	Crafts by skill, numbers: *A= Inspectors* *B= Supervisors/project manager* *C= Planners* *D= Job rep's / Mechanics* *E= Scaffolders* *F= Insulators* *G=Painters / blasting* *H= NDT technicians* *I= Others*
8	**Topics on CUI** *Systematic CUI approach RBI based*
9	**Painting** *A= Typical type of painting* *B= Repainting frequency* *C= Thermal spray aluminum (TSA)* *D= Yearly painting cost for CUI in €*
10	**Type of inspection/ detection**
10.1	*A= Visual and stripping* *B= PEC: Pulsed Eddy Current (incotest)* *C= RT: Radiography digital and conventional* *D= UT: Guided Wave Ultrasonics* *E = Other* *F= Yearly NDT cost in €*
10.2	Average Yearly total cost of CUI inspection and remediation in €
11	**Inspection frequency for CUI in years**
12	**Inspection for Fireproofing in years**

Number	Question
13	**CUI related to Insulation type**
13.1	A= Cold insulation 1. *Cellular glass (foam glass)* 2. *Cellular foam (polyurethane)* 3. *Other*
13.2	B= Hot insulation 1. *Calcium silicate* 2. *Mineral Fiber* 3. *Other*
14	**Upcoming / Next TA.** *In case your upcoming TA is not scoped yet, fill out for your last TA.*
14.1	How many locations are inspected for CUI before your upcoming/last TA?
14.2	How many locations need cold work in your upcoming/last TA? (cleaning/blasting and repainting)
14.3	How many locations need hot work in your upcoming/last TA?
14.4	How many locations need replacements due to CUI in your upcoming/last TA?
15	**Totaling of costs**
15.1	Average total (yearly) cost for CUI repair, inspection coating and other remediation efforts in €
15.2	Typical cost for CUI inspection and remediation at a Turn Around in €
15.3	Typical TA frequency in years

Hot Tapping and Line Stopping

Edward J Buckley

HTS Service Line Manager

TEAM Industrial Services

Abstract: Demand for continuous production at high rates requires minimum process down time. This demand creates challenging situations for engineers, planners and maintenance personnel required to perform scheduled or preventative maintenance on existing equipment. It also limits the time and opportunity available for modifications or upgrades of existing equipment or tie-ins for new equipment.

Hot Tapping and Line Stopping provides options to achieve the required connections or isolation to allow for safe maintenance, repair and modifications of piping systems while minimizing impact on plant production. This paper will review the basic functions, designs, and uses of hot tapping and line stopping equipment and some of the issues to be considered for their use.

Hot Tapping and Line Stopping

Introduction

Piping System Modifications without Interruption of Production

Today's demand for continuous production reduces turnaround frequency and duration. This creates a challenge for engineering, planning and maintenance to be able to perform scheduled and preventative maintenance on existing equipment, and limits any modifications or upgrades. Hot Tapping and Line Stopping provides a means to achieve the required connections or isolation to allow for safe maintenance, repair and modifications of piping systems.

Hot Tapping is the use of an under pressure drilling machine to cut a hole in an operating pipeline for a new tie in from an original pipe. Line Stopping is a means of temporarily stopping flow and holding line pressure in an operating pipeline. Understanding the application and capabilities of these services allows you to provide alternatives to expensive and unplanned shutdowns.

Hot Tapping and Line Stopping are commonly referred to in the same context and by many are perceived as one in the same. This is not true. Hot Tapping is the access into an operating piping system for a tie-in or some other type of connection. It has very detailed condition of service requirements for the operation to be done safely and correctly. Line Stopping, although it requires a hot tap to be performed, is temporarily stopping the product in the line. Therefore a separate explanation of what each process is and how it is achieved will be discussed.

Hot Taps and Line Stops require the use of specialized equipment and highly trained technicians in order to cut a hole in a pipe without taking the system out of service, or reducing the pressure or temperature. These applications are performed in chemical plants, refineries, power plants, manufacturing plants, water/wastewater systems or any other location where loss of service is not acceptable. This paper has been written to provide a basic introduction about why hot tapping and line stopping applications are performed and their advantages. This work is highly technical and should not be attempted without the proper equipment and technical certifications to perform the service.

Hot Tapping

Hot Tapping as stated above is the process of drilling or cutting a hole in an existing pipe, at operating pressure and temperature. This can be done for various reasons such as a new tie-in, flow meter installation, product sampling, chemical injection and many more. Although this sounds relatively simple it requires equipment specifically designed for this application, proper fittings for the piping system and qualified technicians to perform the operation. To safely and successfully perform this operation there is critical information required. This information can be obtained by field inspection of the system, and/or from operations, maintenance and engineering departments. To understand why this information is required we have to understand the tapping process.

The first step is to identify the piping system:

Pipe size, material, O.D. and I.D. or wall thickness
Product in the line
Design pressure and temperature, code and/or ANSI rating
Operating pressure and temperature during the tapping process
Flow rate during tapping process
Size of tap required
Orientation of the tap to the pipe
Clearance for the equipment
Scaffolding and pipe support requirements

The basic understanding of a Hot Tap, how it is achieved and the importance of the proper fitting selection will allow engineering, planning and maintenance to work with the service company to execute this process effectively and safely.

The three primary components of a Hot Tap consists of a pressure retaining tapping machine (including the housing, cutter and pilot drill), full port valve and fitting. The fitting and valve remain permanently on the line therefore must meet the design parameters of the system to which they are attached. In most plant applications the fitting is welded onto the pipe. The fitting in many cases is a full encirclement tee with a flanged outlet which is welded onto the pipe. The valve is attached to the fitting and the Hot Tap machine is attached to the valve. The fitting and valve are normally installed and pressure tested before the hot tap service provider arrives on site. Once the Hot Tap machine is attached to the valve with the proper studs, nuts and gaskets, an integrity test (pressure test) is performed on the assembly to make sure there are no leaks before the tapping process begins. API 2201 guidelines recommend the test pressure should be at least equal to the operating pressure of the line being tapped, but should not exceed the internal pressure by more than 10 percent in order to avoid internal collapse of the pipe.

During this setup process the Hot Tap technician will measure and record critical dimensions of the fitting on the pipe, valve flange to flange and pilot drill and cutter location on the tapping machine. He will also verify that there are no foreign materials in the fitting and valve, components are properly aligned and the bore of the assembly is correct for the size cutter being used. With this information he will calculate the Total Travel Distance (TTD) for his pilot drill and cutter travel to complete the hot tap operation. The machine diagram and measurement identification chart below is an example of where these measurements are taken and the relation to the overall tapping process. Once the TTD is calculated the technician then verifies that the tapping machine has the required travel to reach the Maximum Cut Out (MCO) distance.

Although there are various manufacturers and types of Hot Tapping machines in the market place, each has a design pressure and temperature rating, maximum travel distance and recommended maximum tap diameter. During the final job review this information should be reviewed, verified and recorded.

During the tapping process the coupon, cut out piece of pipe, will break free before reaching the MCO position. Stopping before MCO location will leave a ridge in the cut hole, which for a Hot Tap will not affect the flow or operation of the system. However, for a Line Stop this ridge would not allow the Line Stop Head to properly enter the pipe.

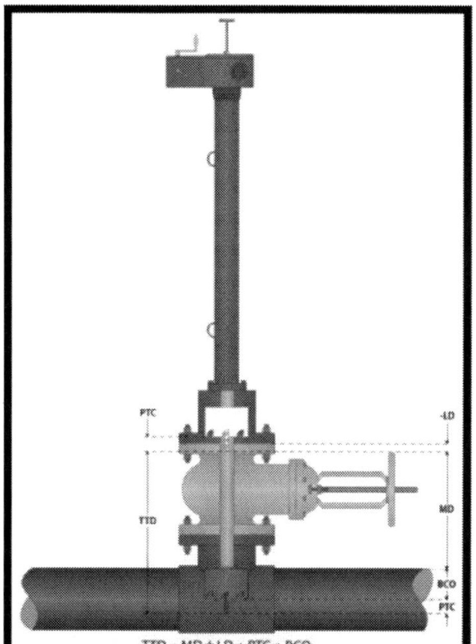

TTD = MD ± LD + PTC + BCO
Total Travel Distance

TTD - Total Travel Distance
MD - Measured Distance
LD - Lost Distance
PTC - Pilot to Cutter
LI - Lower In
BCO - Book Cut Out
MCO - Max Cut Out
MT - Machine Travel

Note: The BCO or Book Cut Out dimension is obtained from the Hot Tap, Line Stop Field Pocket Reference.

Hot Tap Operation

The Hot Tap is made in one continuous process one it begins. With the dimensional data recorded above, the technician advances the machine until the pilot touches the top of the pipe. His dimension verified, he engages the machine and the pilot drill penetrates the pipe. Once the pilot drill penetrates the pipe, the tapping machine is stopped and the product inside the pipe fills the fitting, valve and housing assembly. This assembly is pressure retaining and is bled off through a bleed off valve on the tapping adapter. This bleed off is

controlled and is properly disposed of into a drain or container designed for the product. When the fitting, valve and housing are full of the line product the bleeder valve is closed and the pressure is equalized inside this assembly. The tapping machine is then advanced until the cutter is on the pipe.
This is verified and then the tapping machine is engaged and the cutter begins the cut.

The tapping machine is advanced either manually or automatically depending on the type of machine to the Book Cut Out (BCO) dimension. Upon completion of the tap the machine is turned off and the technician retracts the cutter and pilot assembly to the initial starting dimension. The tapping valve is closed and the pressure bled off the tapping machine. After draining the product from the machine, the machine is removed from the line and new service is established. The coupon is removed from the pilot drill and normally given to the inspection department for evaluation.

Cutters

For smaller holes in pipes, drill bits that are attached to special holders for the respective tapping machine will be used (commonly this is up to two inch diameter). For sizes above one inch Hot Taps and Line Stops the cutters are essentially a specialized hole saw. The hole saw is attached to a holder which also has a pilot drill. The pilot drill has two functions, first to guide or hold the cutter in line during the cut and second to provide a means to retain the coupon upon completion of the hot tap. The industry standard pilot drill retains the coupon with "U" wires that are set into the pilot which move into a slotted area when entering the pipe. Once past the coupon the "U" wires rotate out of the slot to retain the coupon. This type of coupon retention is limited by the flow of the product in the pipe. In most cases the limitations for this type of pilot drill is 10 Ft/Sec for liquid and 30 Ft/Sec for gas.

As with the dimensional information discussed above, confirmation of the flow rate during the tapping process is critical to coupon retention. In most instances, if the flow rate is above these limits, plant operations can lower the flow during the tapping process to be within these requirements. If the flow is above the recommended rates there are specialized pilot drills with multi wires and other methods for high flow applications that could be used. These are special components which may require additional preparation time and cost. Although every effort is made to retain the coupon, unforeseen conditions of the pipe, service and flow in the line make it impossible to guarantee. Accurately knowing the conditions of service discussed above greatly increases the probability of coupon retention.

Below is a diagram of a cutter and pilot drill penetrating the pipe during a Hot Tap. The adjacent chart shows the standard cutter sizes for Hot Taps and Line Stops. The cutter size for a standard Hot Tap is smaller than for a Line Stop. These are industry standard dimensions. The reason for a Hot Tap being smaller is that standard valves have slight variances in the bore dimensions. Having a smaller cutter diameter insures it will pass through the valve without causing any internal damage. The valve used for Line Stops are temporary and are manufactured or bored oversize to allow for the cutter, Line Stop head and completion plug to pass through.

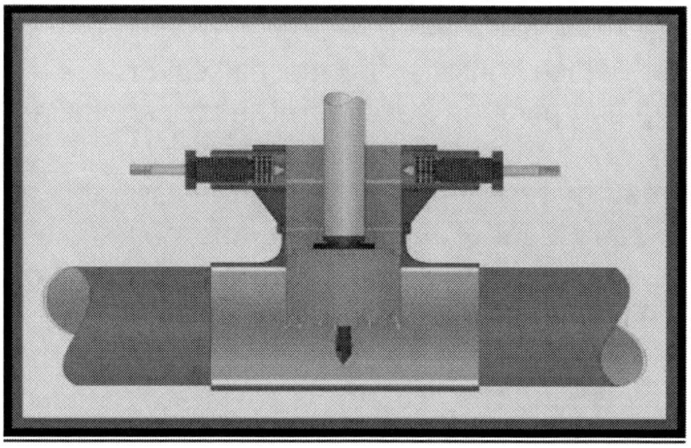

Common Sizes of Cutters		
Nominal Pipe Size OD	HT Cutter OD	LS Cutter OD
3"	2.483"	2.938"
4"	3.438"	3.938"
6"	5.469"	5.938"
8"	7.313"	7.875"
10"	9.500"	9.875"
12"	11.500"	11.813"
14"	12.750"	13.063"
16"	14.688"	15.063"
18"	15.063"	17.000"
20"	17.000"	19.000"
24"	21.000"	23.000"
30"	27.000"	29.000"
36"	33.000"	35.000"

Line Stop

Line Stopping is a means of temporarily stopping the flow and holding the pressure in an operating piping system to allow for upgrades, repairs or modifications. As mentioned previously the Line Stop procedure starts with a Hot Tap. Similarly as for a Hot Tap the system operating and design conditions determine the type of Line Stop method to be used. The pivoting head is the most common method used in the industry.

As with a Hot Tap, the Line Stop fitting, normally utilizes a full encirclement tee is welded to the pipe. This fitting has a specially modified flange that will accept a completion plug which is manufactured to meet the design conditions of the system, and when secured in place will allow for removal of the valve after the Line Stop operation is complete. There are various designs of this flange-completion plug arrangement. The function it provides is to allow for the insertion of a completion plug through the valve into the flange for a permanent seal. Once in place the plug with a metal to metal seat, "O" ring seal or combination of both, allows for removal of the valve and a blind flange, nuts, studs and gasket to be installed.

Line Stop Process

After the Hot Tap is completed, the temporary valve, sometimes called a sandwich type valve, is closed, pressure bled off and the tapping machine is removed. Then a Line Stop assembly which consists of a mechanical or hydraulic actuator, Line Stop housing and pivoting head with the proper sealing element are bolted to the temporary valve. The recommended procedure is before the system is equalized an integrity test is performed on the assembly to make sure there are no leaks in the equipment or flange make-ups. Once the integrity test is complete and before the valve can be opened the pressure above the valve must be equalized with the line pressure. This is done through the temporary valve by-pass or externally though the Line Stop housing. When the pressure is equalized the valve is opened and the actuator is advanced to insert the Line Stop head into the line. This advancement of the head is done by measurements that have been previously calculated and reviewed by the technician. Each actuator has a means of measuring the travel distance for verification of reaching the proper set point.

As seen in the picture below the elastomeric seal is ground to the proper O.D. to match the pipe I.D. This seal is compatible with the product in the line and can be one of the following common materials, Buna, EPDM, Viton, or Aflas. MSDS sheets of the line product are required when specifying the sealing element material to insure compatibility. These seals also have pressure and temperature ratings and are selected based on the system operating pressures, temperatures and length of time expected to be in the line. Depending on these conditions of service and the work being performed a tandem or back up stop may be required to provide a safe environment for the workers.

Completion Plug

The completion Plug when installed is manufactured to meet the system design pressure. The completion plug is installed into the fitting by the tapping machine or a plug setter. The completion plug is bolted to a plug holder that attaches to the tapping machine. The holder and plug assembly under pressure is advanced to the proper measurement for locking into place. Once locked into place the plug and holder can be released from the tapping machine boring bar. This allows for the tapping machine to be retracted, the pressure bled off and the machine and temporary valve removed.

Shown below, the Sealtite® flange has jack bolts that when the plug is at the determined measurement are advanced to lock it into position. There is a tapered shoulder below the plug which is a sealing surface and mechanical stop for the plug. The "O" ring is a secondary seal. Other types of completion plugs have segments that advance into a groove in the top of the plug with an "O" ring that seals against the fitting bore. This type does not have a mechanical stop and the "O" ring is the primary seal. The limitations on this type of completion plug are dictated by the compatibility and temperature rating of the "O" ring material. The "O" ring material like the sealing element is determined in accordance with the MSDS sheets and material compatibility charts.

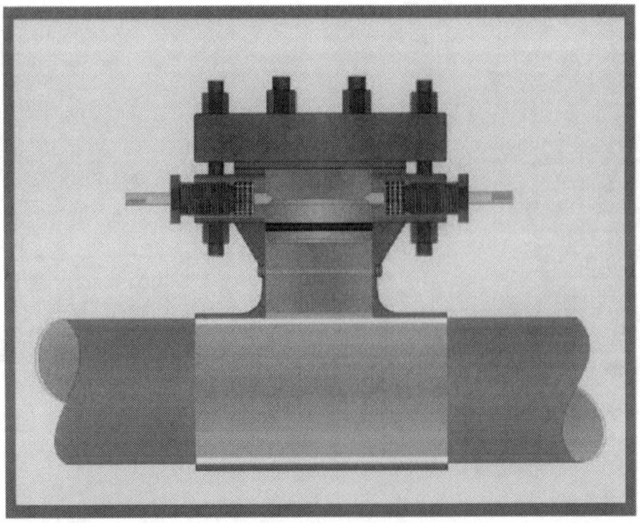

Fittings

Integral with the Hot Tap and Line Stop is the fitting through which the service is performed. The fitting becomes a permanent part of the plant piping system and must meet the system design parameters. The selection of the fitting type is very important. The simplest being the flanged nozzle which is welded to the pipe. These are usually applications that are not size on size. An example is a 12" X 6", which designates a 12" pipe run with a 6" Hot Tap outlet. Included in the fitting selection process engineering calculations for the area of replacement of the material removed during the Hot Tap are reviewed. This is to insure the fitting will meet the design conditions of the system. The fitting manufacturer provides this information with the material certification package. Nozzles are most commonly used for low pressure tie-ins other than size on size and some low pressure Line Stops (less that ANSI 150). Some plants may require full encirclement tees for all applications.

For Line Stops, size on size, or high pressure Hot Taps full encirclement tees are usually required. They are designed to meet the ANSI requirements of the system and provide support for the weight and stress the Line Stop operation may place on the pipe. This type of fitting will also help with out of round conditions that may exist.

Listed below are the Hot Tap and Line Stop design codes that fittings must meet for the respective application.

ASME B 31.1 for power piping, primarily for power plant design.

ASME B 31.3 for process piping, primarily for chemical plants, refineries etc.

ASME B 31.4 for liquid pipelines, for transportation of fluids.

ASME B31.8 for gas transmission and distribution systems.

Fitting Installation

Included with each fitting is a recommended installation procedure that details how to install the fitting. The first section of this procedure will reference all applicable codes and ordinances. It will then outline the precautions and recommended steps for proper and safe welding of the fitting. Note: Minimum flow should be maintained while welding. See API 2201 regarding flow and welding recommendations.

The pipe material, wall thickness and NDE examination data must be considered for the proper welding procedure. The welding procedure is the responsibility of the plant or welding contractor installing the fitting. The pipe surface must be clean and the fitting properly secured in place before welding. JSA, hot work permit, and all plant safety requirements must be completed and approved before any work begins. The installation procedure provides a step by step guide on welding the longitudinal seams first then the radial welds on each end one at a time. When the welding is complete the fitting is pressure tested. To prevent collapsing of the pipe, the recommended maximum test pressure is 1.10 times operating pressure (See API code 2201).

Summary

The information above is an overview of the requirements for Hot Taps and Line Stops. These applications are often times a small part of a major outage, expansion or modification but are integral to the success of the larger task. Therefore during the planning stages of the operation it is important to review the requirements with your service provider to make sure the required operation can be done safely and effectively within the design criteria, time and space allowed. The information provided discusses the most common applications and design requirements. This is an overview for informational reference only. All applications must comply with any and all applicable design codes.

In addition to the conventional Hot Tap and Line Stop there are various different types of applications that have been done to meet specialized requirements. These listed below, have required engineering, specialized procedures and planning to complete the job safely and successfully

Hot Stop: Pivoting head type Line Stop for temperatures of 400F to 700F to 700 PSIG.

Hi-Stop®: Across the line metal to metal stop for high temperature and pressure, ½" to 16", temperatures to 1100F, pressures to 2600 PSIG

Across the Header Stop: A method of isolating a smaller stub outlet branch connection by performing a Hot Tap opposite the smaller outlet, then inserting a radius plug across the line to seal on the inside of the pipe at the stub outlet.

Angle Taps: Hot Taps that are performed at an angle other than 90 degrees to the pipe. These taps require a special fitting for the angle desired. This also may require a target for the pilot drill in addition to extended length pilot drills and cutters to meet the extended travel requirements.

Elbow Taps and Stops: Hot Taps and Line Stops are sometimes done through an elbow of a piping system. This is done only where a straight run of pipe in the area is not available. Special fittings, cutters and pilot drills are required for this application and may require a test mock up for a Line Stop.

Permanent Stop: Some applications require the line top head be left in the line permanently. Depending on the conditions of service this can be done with a pivoting head or Hi-Stop®. Utilizing specifically designed equipment the head is inserted into the line and locked into place with the Sealtite® flange and completion plug assembly.

Bag Stops: A process where a smaller tap is made into the system and an inflatable bag is inserted into the system to stop the flow in the line. This is used in low pressure applications usually less than 50 PSIG.

Line Freezing: Line Freezing can be done on lines that have freezable products in the line, the line is full and the flow can be Isolated (no flow) during the freeze process.

Hi-Stop®, and Sealtite® are registered trademark of Team Industrial Services.

American Petroleum Institute, Procedures for Welding or Hot tapping on Equipment Containing Flammables. API Publication 2201

TRP-5537 Full Encirclement Split Tee Fitting Installation. Engineering Repair Procedure Manual – Rev.4. Technical Repair Procedure. Team Industrial Services, Alvin TX

GENERAL PERFORMANCE COMPARISON AND PREVENTATIVE MAINTENANCE OF ETHYLENE FURNACE ISOLATION VALVES

Todd Peltier

Engineered Valve Repair Specialist

Crane Energy Flow Solutions®, Houston, Texas

Craig Kempken

Valve Repair Specialist

Crane Energy Flow Solutions®, Houston, Texas

Abstract: The consideration of furnace isolation valves and respective preventative maintenance (P/M) programs follows the course as with most valve type applications. Standard in the decision of valve selection are the normal and variable process operating parameters. In addition, consideration should be afforded to the length of process runs between scheduled maintenance outages, severity of the process, and availability of the isolation valves for maintenance time during outages. These factors should help answer the valve selection questions below:

- Select the lower upfront cost of general service valves with a more intensive maintenance and replacement cost?
- Select the lower upfront cost of general service valves and replace at determined intervals?
- Select the higher upfront cost of specialized or "Engineered Service Valves," anticipating longer service life, enhanced reliability, and lower total cost of ownership?

Introduction

In this presentation, we hope to utilize what we have learned from years of working with end users, valve manufacturing, and valve repairs to convey a condensed perspective aimed at increasing reliability of furnace isolation valves. We are not endorsing one brand or type of valve design over another. We strive only to bring out what we have found as positive and negative attributes from a point of failure and refurbishment. With this information, we hope you will come away with a clear understanding of the different valve types to enable you to make good purchasing decisions. We also hope you will realize the value of a well-defined maintenance program and the impact the process has upon the various valve types.

Findings

In the Ethylene processing industry, furnace isolation valve failure is commonly attributed to valve quality issues. Other times, failures are perceived as simply the cost of operating in a harsh service and "just has to be lived with." In working with various end users, we have found this to sometimes be the case, although very rare. More often than not, the root cause of furnace isolation valve failure is attributed to a poor understanding of the valve type and design, consequently, the requirements to properly maintain the performance and service life of the valves chosen. We find that even the most robust valve design can and will fail in service when subjected to the following:

- Incorrect start-up precautions,
- Incorrect control of process conditions,
- Incorrect maintenance plan.

On the other side of this coin, we occasionally find valves that were not designed for furnace isolation service working successfully for years in this service simply due to correctly addressing the above.

Defined Valve Failures and Causes

Considering the multiple designs of valves available for use in furnace isolation, the types of failures encountered are relatively few and consistent. The various designs of valves most commonly used for furnace isolation are:

- Rising Stem Gate
- Rising Stem Slide (Through Conduit)
- Quarter Turn (Butterfly or Ball)
- Plug Valve

The failures associated with these valves in furnace isolation service are mostly confined to:

- Failure to operate (Seized)
- Leak by when closed
- Faulty position indication

Inevitably, root cause of valve failure can and is usually brought about by the harsh operating conditions of the process and the valve design's compatibility (or incompatibility) with the process conditions.

The design and manufacture of most industrial block valves fall under the guide lines of API and ASME. These institutions have done a great job standardizing manufacturing of industrial valve designs for service in *general* process conditions. Consideration by these institutions is generally aligned around ensuring materials of construction are compatible with process temperatures, pressures, flow rates and interchangeability between brands of industrial valves. Valves manufactured under these respected guidelines typically will exhibit the same performance characteristics.

Take for example, a raised face flanged, ASME Class 300 Gate Valve in C-12 material with hard-faced seats. No matter the manufacturer, specified are:

- Minimum wall thickness of the casting
- Minimum stem diameter
- Face-to-face dimension
- Composition of the metal forming the casting

All these things and many more are specified so that performance between brands will be the same. When process conditions are stable, the performance of these valves is stable.

Problems arise when process conditions are variable, such as with furnace isolation valves. Valves used in this service have to contend with high temperature swings during decoke cycles. They must be able to control back pressure to the firebox in a fairly precise manner during the "Switching" phase on and off line. Once lined up and exposed to the process, these valves must be able to withstand the constant bombardment of erosive particulates entrained in the cracked gas. Finally, these valves must be able to sit for extended periods of time in an opened or closed position without seizing. Also, considering that valves used in this service are usually very large in size, the performance expectations border upon extreme. The respected institutions governing valve design and manufacture fall somewhat short when performance expectations of this process are required.

Process Specific Engineered Valves come into play at this point. As the name implies, these engineered valves adhere to the requirements of the governing manufacturing standards as to pressure, temperature, and shut-off ratings. They are also engineered to perform in specific services with additional performance requirements such as Ethylene furnace isolation. When manufacturers are requested to provide "Process Engineered Valves," they start by examining details of the process parameters. The obvious is looked at first, such as proposed size, process pressure and temperature, CV, and flow characteristics. After all, the very nature and specific purpose of a valve is to open and allow flow, and close to prevent flow. Then details of system control are examined. Will the valve be required to control flow via pressure, volume, or temperature and what is the desired control range? From that point, design turns to how the valve will be controlled, actuated (electric, pneumatic, hydraulic), or manually operated. Virtually everything in an engineered valve package is designed to work in these specific conditions. Many times, failures of furnace isolation valves are due to process conditions exceeding the design criteria of the valve system provided. For example, a manufacturer is provided with a maximum operating pressure of 40 psig with a normal process pressure of 22 psig on a Transfer line valve. The actuator and valve package are set up to open and close at this maximum pressure. If during a furnace switch the back pressure on the Transfer line valve is maintained at 50 psig, the valve may fail to operate correctly because the valve package was not designed for this condition. If the valve does operate, chances are that abnormal wear of the actuator, seats, and seals will be experienced, thus, leading to premature failure of the valve package.

Specific Valve Type Comparison

The valve designs most commonly found in furnace isolation service are produced by numerous manufacturers. These designs in terms of initial procurement cost range from in-expensive to extreme in comparison to each type. The total cost of ownership varies greatly and depends on the furnace design and maintenance regimen. The following is a listing of the most commonly used valve designs in order lowest to highest upfront cost. In addition are performance characteristics that should be considered when using these type valves.

Standard Gate

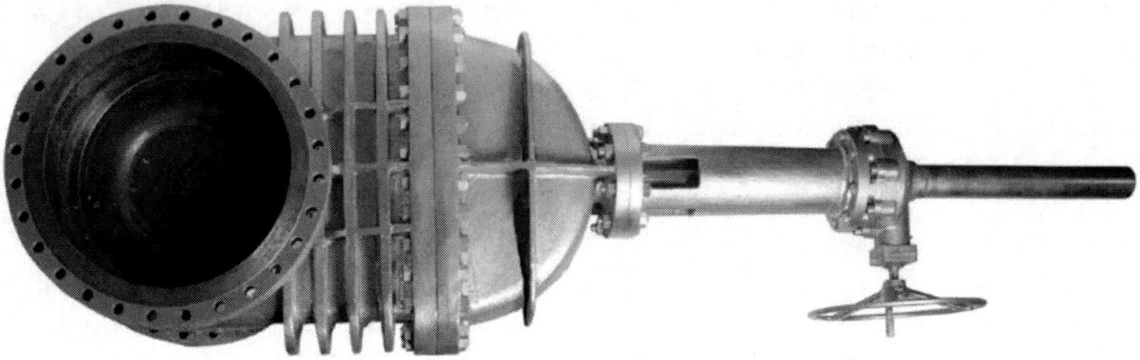

Standard gate is the least costly of the valve types. Standard gate valves typically use Stellite® hard-facing on Seat rings and Wedge face to off set abrasion caused by coke fines. These valves are easily actuated via electric, pneumatic, or piston style units and tend to have adequate flow characteristics as well as seats that feature bi-directional seal. Draw backs of this design are:

- Entry angle of the Wedge / Seat Rings is sharp, leading to Thermal Sticking
- Stellite® hard-facing of Wedge and Seat Rings tends to crack due to thermal cycles
- Packing must constantly be adjusted due to long rising stem
- Stem nuts are subjected to great wear. They must be kept clean and well lubricated
- Seat ring faces are in direct path of process and particulates, leading to damage and erosion
- Valve design is light as compared to other types

The standard gate valve lends itself to the highest variability or range of pressure variances. Of great advantage is the number of repair facilities around the world capable of performing quality repairs on this valve type. Replacement parts are also easily obtainable.

Process Engineered, High Performance Butterfly

High performance Butterfly valves typically have triple off set seats exhibiting tight sealing capabilities with above average control characteristics. The design is the most compact of the valves considered, which is good for weight considerations in large sizes. The draw backs to this compact design are:

- Valve is subject to line stress causing Seizing while in service
- Seats are Uni-directional
- Disc is exposed to direct flow of service media, resulting in damage to disc face and sealing area of disc
- Stem bushing areas are exposed to and are subject to collection of coke fines, causing seizure, sticking, and/or excessive friction

High performance butterfly valves lend themselves to a wide variety of actuator applications and, due to the rotary action of this type of valve, packing leaks are rare. This type of valve requires some specialized equipment and training to make proper repairs. This is especially true for valves with Integral body seats. Quality repair facilities are numerous although consideration should be given to factory authorization. Major replacement parts are generally long lead time items.

Process Engineered Ball (Metal Seated)

This is a very ruggedly designed valve and will give years of reliable use in service. It is uni-directionally seated and features a very wide point of contact between the ball and seat ring. The ball and seat rings are typically hard-faced via proprietary coatings or solution carburized, making them very hard and wear resistant leading to tight shutoff. This valve type works with purge and flush steam introduced at a higher-than-process pressure to keep the body cavity and packing area clean. In some cases, the purge steam is also used to load the seat rings further enhancing the seat seal. Draw backs of this design are:

- Although rotary operated, actuator selection is limited due to the high amount of pre-load applied to the seats by the springs and/or steam
- Valves require increased force to operate, thus requiring large actuators
- Seats / Ball are subject to galling due to the amount of pre-load required
- Even though protected from process media when fully opened or closed, seat damage can occur due to abrasion, although minimal

C_V ratings are at the top of the spectrum, featuring very low pressure drops. Control characteristics are reasonable, as well. Repairs are generally precise in nature and should only be performed by factory trained personnel in shops possessing specialized equipment. Replacement parts are long lead time items and are generally required for intensive repairs. Engineered Ball (Metal Seated) Valves are typically some of the costliest valves featured, but these valves are consistent and reliable over time. Authorized repair facilities are few, and manufacturer involvement should be enlisted when planning maintenance projects.

Process Engineered Slide Valve (Through Conduit Gate)

Engineered slide valves used in Ethylene furnace isolation are linearly operated, rising stem, through conduit type valves. These valve types are competitively priced with Ball and Wedgeplug® valves. In the larger sizes, these valves are some of the heaviest, and so, consideration has to be given to line support. These valves do not utilize force pressing down on angled seats to facilitate sealing. To provide sealing force on the seat rings and/or disc, springs, purge steam, process pressure or a combination of all is used to press the upstream seat ring to the valve disc. Purge steam is usually provided to prevent coke build-up in the cavities. As these valves are "Through Conduit" design, the seat rings and disc are protected from direct abrasion from process flow. Downfalls of this valve type are:

- Seats and disc are in constant contact under high load, subjecting them to "galling"
- Packing must constantly be adjusted due to long rising stem
- Stem nuts are subjected to great wear. They must be kept clean and well lubricated
- Large actuators or reduction gears are required due to size and weight of disc as well as the high loads applied by the seats
- Size and overall weight of the valves must be taken into consideration
- Valves are subject to seizure due to line stresses

Flow control characteristics are fair in comparison to other types of valves noted. Generally, replacement parts are long lead time items, and repairs require factory trained personnel and specialized equipment to grind the large disc. Time has shown this type of valve to be reliable when maintained correctly.

Wedgeplug® Valve

The Wedgeplug® valve is a true integral, metal-seated valve. This valve is competitively priced with Metal Seated ball and Slide Valves. It utilizes features from both the ball design and standard gate valve. It comes in three port sizes, but is most often full port for furnace isolation service. As with the Ball and Slide valves, the water way is through conduit. Overall weight of the larger valves is considerable but usually between that of Ball valves and Slide valves. The feature that really sets this valve apart from the other frequently used valves is operational. When the plug is seated and the valve is operated, the plug rises up and away from the body seats to a preset point approximately ½". As the gear head shaft continues to turn, the plug rotates 90 degrees. Once the plug position indicator arm touches the opposing yoke leg, the plug then is forced back down onto the body seats. Because the plug loses contact with the body seats before turning, this type valve has the lowest force requirement to operate. This translates into smaller actuation packages. In addition, this greatly reduces the wear of the seats.

Wedgeplug® valves are bi-directionally seated and, because the sealing force to the seats is applied mechanically when opened or closed, the body cavity is completely closed off to the process. In particulate laden services, Wedgeplug® uses purge steam introduced into the body cavity to displace solids. Unlike uni-directionally seated valves, the purge steam is captured in the body cavity and does not continuously flow out into the process line. Wedgeplug® valves are very rugged in design to provide the end user an above average life expectancy. This is due to the simple design with relatively few moving parts in contact with the process. The draw back to these type valves are as follows:

- Subject to fail when exposed to high pressure drops
- Valve can "stall" during operation when pressure parameters are exceeded

- Process pressure and purge steam pressure must be closely monitored
- Weight of larger valves is considerable

Repairs are typically minor in comparison to other engineered valves and are mostly confined to replacement of yoke and gear head parts. Repairs and parts replacement of gear head and yoke assemblies can be performed online. Plug and Body seats have to be refurbished by grinding or welding occasionally, typically averaging 12-15 year intervals. To perform this type of refurbishment, the complete valve must be pulled and sent to an off-site facility. Very specialized equipment is required, and only specific repair shops with factory trained personnel should be commissioned for this work. Due to the volume of Wedgeplug® valves in the market, common replacement parts are readily available with short lead times. However, plugs and bodies have very long delivery times.

Conclusion

Based on the evidence provided by repairs of the various furnace isolation valve designs, the success of application and reliability is directly attributable to the following when selecting the valve design:

Operating parameters – How precise pressure, temperature and switches are controlled and, in some cases, purge steam

Size of the process lines – Weight, Thermal growth and shrinkage, actuation limitations

Anticipated maintenance opportunities – Intervals between individual furnace outages for minor maintenance and adjustment; intervals between entire unit outages for major maintenance opportunities; level of maintenance required during opportunities

Quality outside support – Access to trained, qualified and equipped outside repair facilities; OEM support for both the end user and repair facility.

Level of training – Experienced, factory trained personnel exposure to end user operations and maintenance personnel.

As noted here, the five most commonly used furnace isolation valves have very different design characteristics. Along with these differences come distinct considerations to the above, how well the valves initially will perform, and how well the valves will perform long term.

References (pictures)

(Ball valve) www.valv.com/catalog/products/?prod=4

(Slide valve) www.Friatec-rheinhute.com/pdfs/valve%20brochures/valves%20chemical%20and%20petrochemical%20industry.pdf

Acknowledgements

API

ASME

Crane Energy Flow Solutions® Valve Services Division

Crane Co. is a diversified manufacturer of engineered industrial products. Crane Co. is traded on the New York Stock Exchange (NYSE: CR)

RADIANT CRACKING TUBES – QUALITY FROM THE FOUNDRY

G. Matthew Webb
Manager Customer Service
Kubota Metal Corporation
Michael Györffy
Research and Development Metallurgist
Kubota Metal Corporation

Abstract: Heat resistant nickel chromium alloy tubes have refractory properties which make them an ideal material for ethylene pyrolysis coils. These alloys are typically cast, without a universally accepted standard to cover the manufacturing and testing. As a result individual users have developed their own specifications for inspection and testing. The applied testing has varying degrees of relevance and effectiveness.

A batch casting process results in individual centrifugally cast tubes and therefore it is important that the manufacturing methods are fully understood to evaluate tube quality, without adding unnecessary costs to the product. Tube chemistry is a strong factor in quality, since each cast tube is poured individually, control of variation in chemistry is vital. Subsequently, physical inspection methods in the as-cast, machined and welded states are important to ensure the quality of the castings and the final assembly.

The tube manufacturing process, including casting, machining and fabrication, is presented with the typical inspection operations.

Material requirements for steam cracking furnaces are extremely demanding, requiring high creep strength, hot corrosion resistance balanced with retained metallic characteristics associated with good heat transfer and weldability. Material processing and fabrication are therefore critical quality components to successfully utilize the full range of available properties and maximize furnace performance.

Metallurgical Background

Metal alloys used to form wrought parts have to be formable at high temperatures. To manufacture parts by forming, metal is softened by heating, and then worked so the metal flows under stress to a near net shape. The microstructure formed by the thermo-mechanical processing makes these alloys very strong at lower temperature due to the grain refinement of the component. High temperatures will rapidly coarsen the refined grains making them unsuitable for the extreme temperatures of cracking furnaces.

Casting achieves near net shape in the liquid state. Figure 1 shows the microstructure of a wrought and cast alloy, both austenitic and with approximately 20% chromium and 32% nickel composition. The wrought microstructure has been subject to strain, and this is apparent due to twinned grain boundaries. The cast microstructure is dendritic and the distribution of carbides is determined by the heat transfer during solidification. There are two processes for petrochemical castings where the alloys are poured cast into spinning moulds or dies, or into stationary or static moulds.

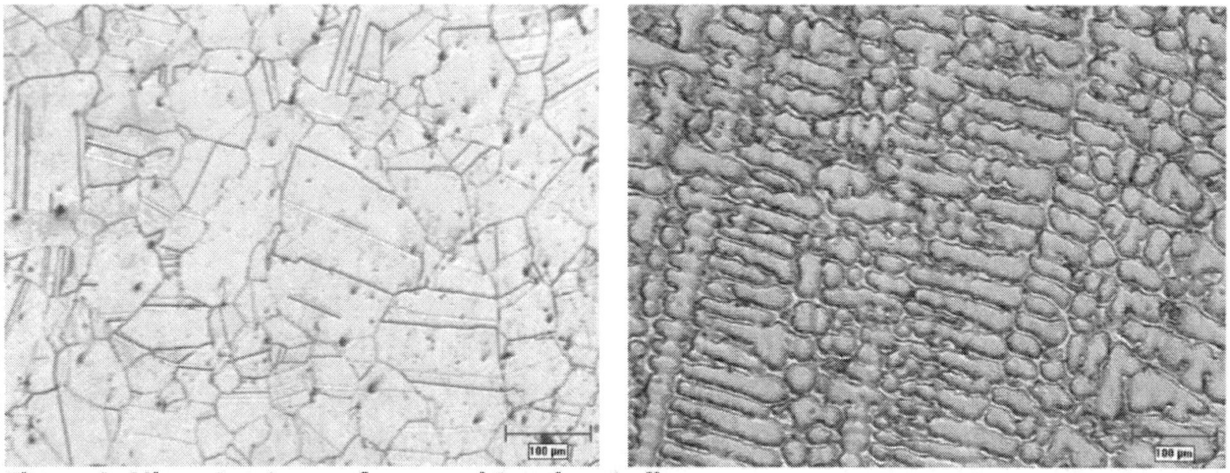

Figure 1: Microstructures of a wrought and cast alloy.

Static moulding allows the production of near finished parts such as return bends, elbows or collectors. Shapes are formed by static disposable sand moulds, permanent metal or graphite dies. The shapes are made hollow by sand or ceramic cores held together by a polymer binder. The heat from the solidifying metal burns off the binder holding the refractory sand in place, which allows the parts to be removed easily.

Parts with radial symmetry can be formed by pouring molten alloy into a rapidly spinning die. The centrifugal forces of the spinning die promote the separation of lighter phases, such as glass phase or gases, from the sound metal. These move to the center of rotation and freeze onto an inner layer of unsoundness on the casting where they can be later removed by machining. The material of the die is usually steel, providing a large heat sink that causes rapid, directional solidification of the alloy from the outside diameter toward the inner diameter. The rapid solidification ensures fine grain size and a uniform dispersion of carbides for high temperature strength by a dispersion strengthening mechanism.

The alloys used are modifications of the ASTM cast heat resistant steels, with evolution from HK40 to HP microalloys to 35-45 grades being use in the history of steam cracking furnaces. These alloys have a high fraction of nickel which has exceptional high temperature strength and resists creep. They also have large amounts of chromium which provides the hot corrosion resistance needed for an oxidizing and reducing environment. Also, compared to wrought grades, cast heat resistant steels have higher carbon content, leading to strong carbide network formation within the alloy that strengthens it in service. The carbide type and distribution will determine the properties of the alloy and this will be a result of thermal history from the foundry practices of the producer.

Since carburization is the principal mode of failure for cracking tubes, the alloys are designed to resist this form of hot corrosion.[1] High levels of silicon are desirable;[2] however, silicon acts to increase the creep rate and reduces the usable mechanical life of the tubes. To counter the effect of high silicon, strengthening additions of strong carbide forming elements are added that tend to partition from the primary carbide network formed during casting. These are referred to as microalloys and are elements such as small amounts of titanium or zirconium in a carbide form. In the proper form and dispersion these carbides will resist dislocation movement even at high temperatures. If the microalloys are not present in the proper form and sufficient concentration the tubes will creep rapidly in operations, possibly resulting in the coils extending to the bottom of the furnace box or bowing severely towards the burners.

Upon casting the cooling rate from liquid to solid is quite fast. This results in a super saturated alloy that has not reached equilibrium. Much of the microalloying additions are dissolved in the austenite, but at high temperatures where diffusion is possible the microstructure ages; a process where the microalloy precipitates in the austenite. This is illustrated in Figure 2, and the microalloy precipitates act to increase the amount of strain energy the material can absorb, effectively increasing its strength and creep resistance. Therefore it is critical during casting to have the correct concentration of microalloying elements in solution in the melt. If the kinetics are inconsistent from tube to tube, there will be differing distributions in the microalloy carbides.

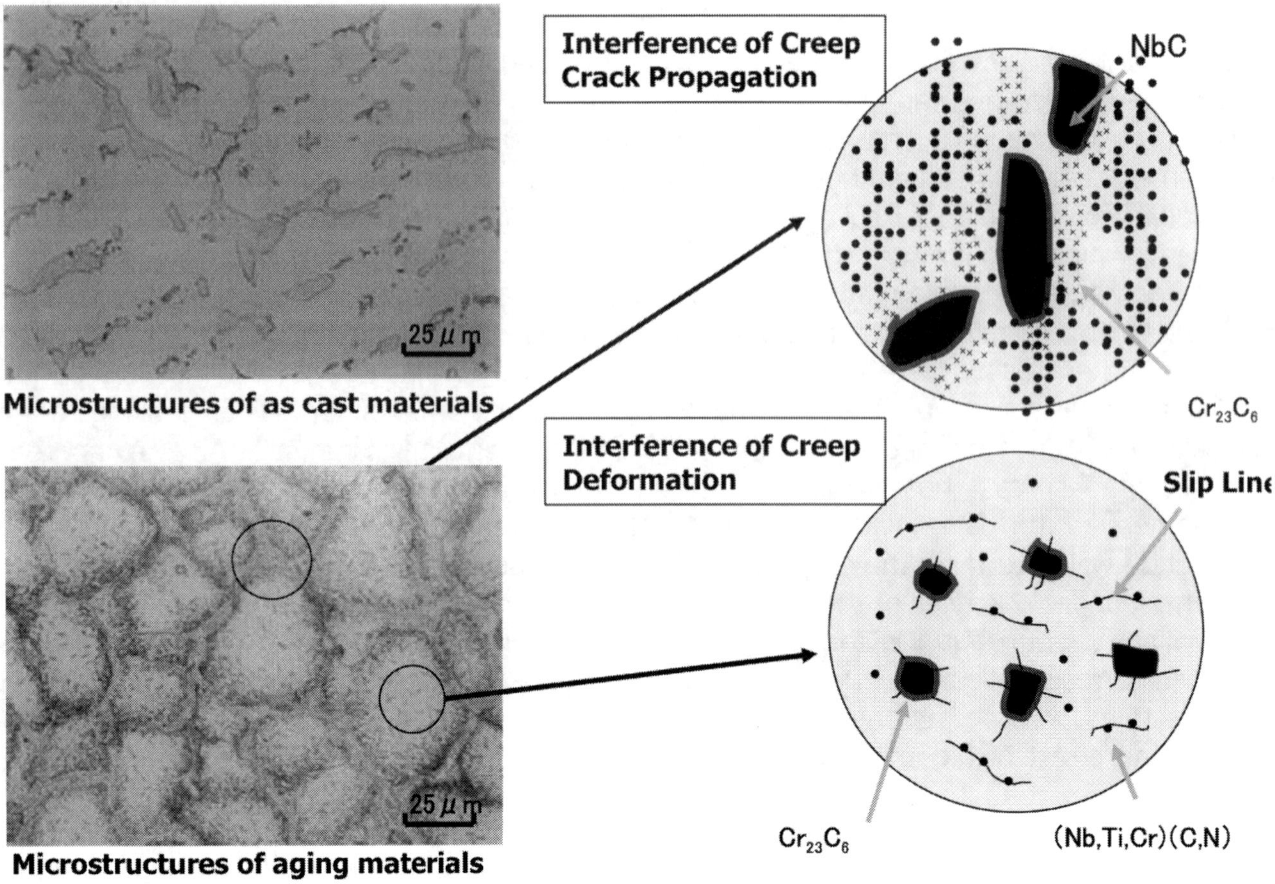

Figure 2: Microalloy precipitation and strengthening schematic.

Melting Process

A wide range of compositions have been developed for specific furnace needs. Usually the alloy is mixed from virgin raw materials in the furnace. The batch method allows for a large range of specific chemical compositions to be available. To ensure that the final product is correct, it is important to have stringent controls on the charge going into the furnace and the chemistry of the melt.

Raw materials are controlled through certified assays and positive material indications (PMI) upon receipt at the foundry. A furnace charge is based on the composition of raw materials such as nickel, ferrochromium, ferroniobium and steel plate. Samples from the melt are analyzed on optical emission or X-ray fluorescence spectrometers. These methods have rapid response that allow for high production rates. Spectrometers use computers to scan multiple channels simultaneously. This gives very fast, precise measurements for the metallic elements used. Since the spectrometer programs use comparative methods, it is key to have known standards. The accuracy of the results depends on the accuracy of the standards.

Spectrometers can detect trace, undesirable elements, but only if the standards also contain the target elements. The level of detection of deleterious elements like lead, antimony or arsenic in the unknown sample has to be comparable to the standard. Since the alloys used in the furnace are all proprietary to the manufacturers, each one is required to develop standards for testing.

The composition of the melt can be adjusted prior to pouring to meet the required composition target limits. When the chemistry in the furnace is in range it can be tapped into a ladle and moved to the waiting mould. Certain alloying elements are reactive or volatile at the temperatures required to melt the alloy. This necessitates controls on the timing of these additions.

Microalloy additions, in particular, are highly reactive in the melt. They oxidize and migrate to the glass phase (slag) in the melt. If they are added to the ladle too early they will burn off, and if they are added too late they will not properly disperse in the heat. Since there is such strong time dependence to the retention of microalloy additions pouring multiple tubes from one ladle will result in insufficient microalloying in subsequent tubes. Checking the analysis from the product verifies the manufacturing control processes. If only one tube is poured per heat, and the pouring process is control controlled, the product analysis only needs to be performed on a random basis.

The final melting parameters to be controlled is the die preheat and the superheat, temperature above the melting temperature, of the melt. These parameters will determine the cooling rate and consistency of the product. Also the superheat will determine how soluble gasses and microalloys will be in the heat and have a direct impact on cast quality.

Quality Control Process

Physical testing can be performed on a periodic basis when casting repeated heats of an alloy to validate material properties. This testing can include mechanical tests performed on representative samples from the cast heat, as well as examination of the micro and macrostructure. Croppings from tubes or keel blocks (bars cast concurrently with the static mould) are machined into bars conforming to tensile testing standards. Room temperature, high temperature tensile and stress rupture testing can be performed. Additionally, etch rings from tubes can be taken to visually review the macrostructure of the casting and dendrite secondary arm spacing (DSAS) measurements made from examination of the microstructure of the tubes.

Both stress rupture and high temperature tensile testing validate the material at cracking furnace operating conditions. Room temperature tensile testing does not qualify the material for high temperature service. High strength, low temperature alloys frequently need specific heat treatments to produce desired room temperature properties; however, austenitic alloys are not affected by heat treatments. Therefore, for austenitic heat resistant alloys room temperature tensile testing does not provide a value-added validation of high temperature performance.

A low tensile testing frequency in addition to etch rings and DSAS measurement in addition to a statistical process control (SPC) program dictating effective and consistent pouring conditions can replace full tensile testing of every heat in an order. For austenitic materials, the cast cooling rate will determine the secondary dendrite arm spacing and this can be correlated to the mechanical properties of the tube. If the SPC program can consistently maintain the casting parameters then tensile testing of every tube is not necessary to verify bulk material properties.

A centrifugal casting defect can occur due to the large thermal stresses on the tube while still in the die. Circumferential tears may occur as the tube contracts axially on cooling and encounters localized friction on the surface. When a solidifying tube is not free to contract, the hot soft metal can tear along grain boundaries originating on the OD. Methods to detect tears include visual, liquid penetrant testing (LPI) or eddy current inspection of the OD surface. Tears that do not penetrate through the wall will not be detected by pressure testing. Tears are common around the ends of the tubes where there is interaction with the end plates that act as a mechanical lock against the tubes thermal contraction. The die end plates are disks that prevent the molten metal from running out of the ends and while necessary they produce a high possibility of mechanical locking of the newly cast tube due to the high heat transfer rates at the die ends. LPI is effective at detecting tears in the outer surface, although it relies on skilled inspectors. Eddy current inspection has proven effective at detecting both surface tears and internal defects. Eddy current systems can be automated to inspect tubes and mark indications without an operator but act as a secondary check system that cannot replace visual inspection.

Tapers on the inside diameter result from insufficient metal reaching the cold end of the die before freezing. Most small diameter tubes have tapers to some extent. The taper is there as a result of the chilling effect during pouring and the length of the die. The fluid poured does not solidify into a uniformly thick layer, or more precisely there is more sound thickness at one end than the tapered end. If the tapers are severe, there will be insufficient material on the ID on the cold end to machine to sound metal. Unsoundness is detected by liquid penetrant inspection of the bore of the cold end. As the inside diameter is tapered the machining stock increased toward the hot end, therefore cropping the tube and re-inspecting will usually result in finding the sound metal at the ID.

Lamination type defects can occur mid-wall on the cold end of heavier walled tubes when pouring flow is inconsistent. The first metal reaches the end and freezes; subsequent metal fills in the wall of the tube leaving a lamination type defect between the two layers. Since these defects extend from the cold end, they are detected at the machined end of the tube with liquid penetrant inspection. Again, tubes can be cropped back to sound metal.

Radiography is not an effective control method for evaluating the cast quality of tubes since defects are found faster by cheaper inspection methods. Ultrasonic inspection of castings is not effective since the coarse grain structure of the casting gives high background noise and masks flaws. Ultrasonic probes can be used to measure tube wall thickness as the reflection from the opposite wall is stronger than the noise.

Finally, these defects highlight the need for a statistical process control program which monitors production variables and correlates rejections. By controlling production variables product quality can be maintained within consistent limits.[3]

ID Machined to Remove Unsoundness

A unique property of centrifugal cast tube allows all of the unsound metal to collect on the inside surface where it can efficiently be removed by machining. The centrifugal forces from the die rotation promote separation of the lighter phases to the ID of the liquid metal. Metal solidification starts at the OD from the chilling effect of the steel die and progresses to the ID. The movement of this solidification front also pushes melt soluble impurities to the ID.

The unsound metal is removed to reduce the weight of the tubes and reduce the wall thickness for improved heat transfer. The tubes can be machined by piloting the cutting head on the machined surface (push boring) or on the as cast surface (pull boring). Push boring produces a straight cut, which if the OD is also machined produces concentric walls. Pull boring follows any curvature in the tubes, which allows a concentric cut. (See Figure 3).

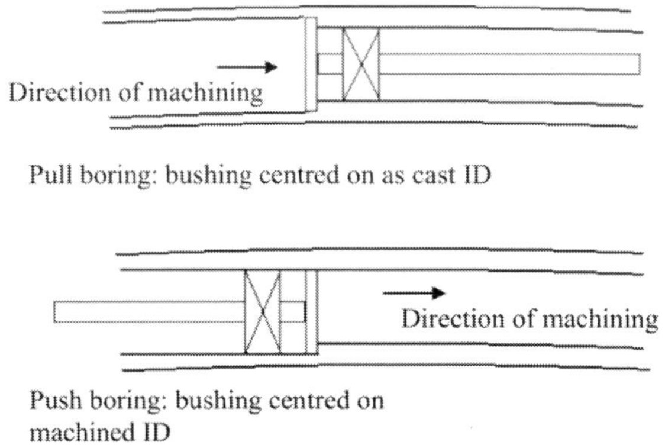

Figure 3: Schematic showing pull boring versus push boring through a bent tube.

In an example of poor machining a premature failure of a reformer tube was documented (See Figures 4 and 5). The failure analysis reported that there was a 25% variation in the machined wall thickness. The tube failed as a result of creep rupture accelerated by the eccentric geometry of the tube.[4]

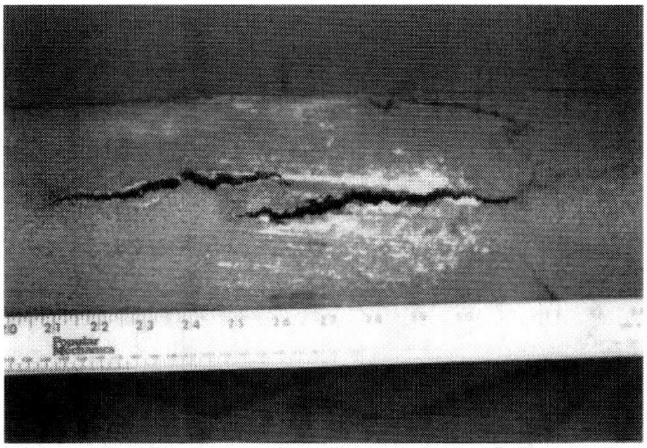

Figure 4: Close-up of reformer tube failure site

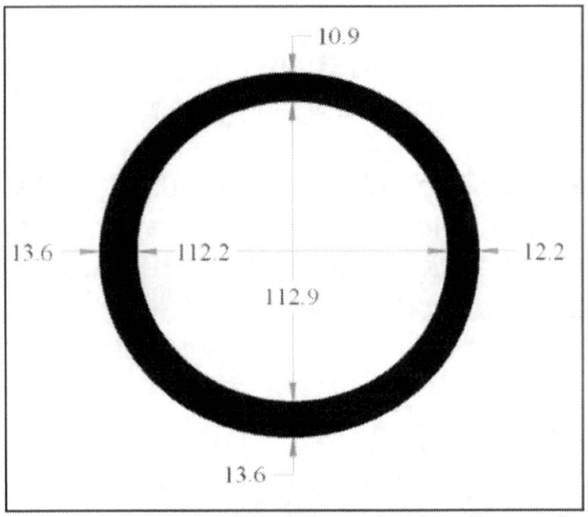

Figure 5: Critical measured tube dimensions (mm) of failed tube at crack tip

This demonstrates the importance of concentricity to furnace tube applications. Tubes should be verified after machining to ensure that machining is consistent and concentric. Tube ends can be verified manually.

Once the tube bores are machined, the weld preparations are machined. The dimensions of the weld preparation are important to ensure proper fit-up at welding. The ends of the tubes are examined by liquid penetrant inspection. This will detect porosity or delamination type defects.

Fabrication

Welding seams between tubes are normally rotated under a fixed torch. Gas shielded arc welding (GTAW or TIG) is the preferred method for these welds because of relatively low heat input. TIG uses a non-consumable electrode with separately fed filler metal. It is desirable that when welding two components with the same base metal the filler metal will closely match the base metal. Completed welds should meet the allowable stresses of the base metal to avoid having the welds become weak points in the system. To test welds they are inspected by radiography and LPI. Radiography from gamma radiography from a source such as iridium is effective at detecting injurious, internal defects. LPI detects cracks in the weld, heat affected base metal and components. Fabrication control can be achieved with techniques such as machine or orbital welding.

Although high carbon cast alloys are susceptible to aging from the heat of welding, centrifugal castings are more forgiving than static castings because of the smaller grain size resulting from the chill effect of the die.

Welds, where accessible, should be ground smooth on the bore before radiography. The bore should not be machined after welding, as weld contraction reduces the diameter and subsequent machining could lead to a thin wall.

Pressure Testing

Pressure testing is the final inspection operation on finished assemblies, and usually involves hydrostatic testing at elevated pressures to compensate for the low temperature of the test. This testing is often a statutory requirement of jurisdictions governing pressure vessels. Elevated pressure for testing due to the temperature correction factor can pose risks to the product and safety. Pressure should be limited for safety.

If exposure to water is undesirable, such as internal refractory, pneumatic testing is the alternative. Pneumatic testing has elevated risks from the contained energy and therefore is used only at lower pressure for worker safety.

Conclusion

The quality of the materials from a foundry is not dependent on the inspection that takes place but on the production control. Inspection methods are effective at identifying problems after parts are produced, but do not necessarily produce good quality. The chemistry in the product is of principal importance. Not only does the chemistry have to be accurately measured, methods must be in place to ensure consistency throughout production. Specifically, this is true for microalloy additions as they are important for tube longevity and can only be monitored on individual tubes.

Room temperature tensile testing is not representative of high temperature properties and does not accurately reflect the quality of a tube. There is no value in room temperature testing for high alloy tubes.

Pull boring is effective at removing unsound material from the ID of cast tubes. This process should always be specified for cast tubes where the OD remains as cast.

Automated and machine welding techniques provide improved consistency for fabrication.

---

[1] Parks S.B. and Schilmoller, C.M, Use alloys to improve ethylene production, Hydrocarbon Processing, March 1996.

[2] Smith G.M, Young D.J. and Trimm D.L., Carburization Kinetics of Heat-Resistant Steels, Oxidation of Metals, 18, 5/6, 1982.

[3] C. Steel, D. Fraser, "Statistical Process Control of the Microstructure and Properties of Cast Heat Resistant Alloy Tubing," NACE Corrosion 92

[4] Reformer Tubes: Not a Commodity – G. Matthew Webb and W.K. Taylor, Ammonia Plant Safety, Vol. 47, Vancouver 2006

RADIANT COIL SPRING HANGER MAINTENANCE CONSIDERATIONS

Mark S. Karrs (Author)
Director of Engineering
Lummus Technology Heat Transfer, Bloomfield, NJ
Ken Webb (Co-author)
Principal Mechanical Engineer
Eastman Chemical Company, Longview, TX

Abstract: In order to meet design flux rates and target run-lengths, cracking heater lower convection coils, radiant coils and crossovers must be designed to operate at elevated temperatures. However, they are normally constructed and maintained at ambient conditions. As the pipes and tubes are heated from ambient to operating temperatures (in the range of 2000°F+ for the radiant coils), significant thermal expansion of the crossover and radiant tubes will occur and must be accommodated in the design of the supports.

At operating temperatures, the radiant coil materials are near the mechanical limits for the metallurgies, exhibit limited mechanical strength, and are easily deformed. After several heat up and cool down cycles, the tubes will not return completely to their original ambient or "cold" positions because axial tube stresses during heat-up/cool-down and operation will "stretch" the tubes such that the overall length increases slightly. This irreversible process occurs over each operating cycle, with the net result that the coil's overall dimensions will progressively change over the coil life. This progressive length increase is known as "creep elongation" and must also be accommodated by the support system.

The radiant coil support system is normally designed to nullify unbalanced forces and moments, and also to allow the crossover and radiant coil to expand and move as needed over any single warm-up (expansion) and cool-down (contraction) cycle. However, since the rates and extent of creep elongation will vary with operating parameters, it may be required to make adjustments to the support systems over the life of the coil, in order to maintain the range of free motion needed to avoid generating moments and stress on the coils. Such unbalanced forces and stresses will usually be evident as deformation (e.g. bending and bowing) of the affected vertical radiant tubes.

In order to properly adjust and maintain the coil support systems, it is necessary to understand the requirements for coil expansion with respect to the amounts and directions of the expected movements. The direction and extent of coil expansion is tied to the coil configuration and support concept. Measurements must be taken and data collected in order to allow the coil creep behavior to be tracked and projected. With the coil expansion defined and the expected range-of-motion limits established, techniques and procedures to adjust and monitor the radiant coil support spring hangers can be developed.

Coil Expansion

The economic forces associated with higher cracking yields have gradually and progressively pushed pyrolysis coil designs toward lower residence times, higher flux rates and higher coil metal temperatures. Pyrolysis coil metallurgy has pushed the limits of alloy development beyond commercially common wrought and cast tube materials to proprietary high Chrome-Nickel alloys that can maintain strength and carburization resistance at temperatures exceeding 2000°F.

Although HK40 (nominal 25% Chrome-20% Nickel covered by ASTM A297) is still in limited use, most contemporary pyrolysis coil designs are based on HP40 (nominal 25% Chrome-35% Nickel) or even higher (35% Chrome-45% Nickel) proprietary alloys. The coefficient of thermal expansion for these alloys is similar, as shown in Table 1, and the small differences do not require any special consideration in the case of bi-metallic coil designs or the substitution of one alloy for another in any specific component. However, welding of dissimilar parts typically will require special procedures and/or filler to be used.

Table 1
Coefficient of Thermal Expansion for Select Heat Resistant Alloys

Alloy	Expansion Coefficient 10E-6 in./in. °F	Temperature Range °F
HK-40	10.2	70-1800
HP-40 mod.	10.3	70-1800
35Cr/45Ni	9.1	68-1832

Source: www.kubotametal.com/alloys

Using the defined expansion coefficients, it is possible to estimate coil expansion from the as-built condition (ambient) to the normal operating Start-of-Run (SOR) and maximum operating End-of-Run (EOR) conditions. For the case of a typical vertical coil of tube length 40 feet between return bends, constructed from HP40 alloy at 70°F, and designed for a SOR Tube Metal Temperature (TMT) of 1920°F, the expected linear expansion is:

40 feet x 0.00001 ft/ft °F x (1920 - 70°F) = 0.76 feet, or roughly 2% of the cold length

Assuming a design EOR TMT of 2012°F, the additional expansion from the SOR to EOR condition, associated with the increase in TMT due to coke buildup inside the tubes, is:

40 feet x 0.00001 ft/ft °F x (2012 -1920°F) = 0.04 feet, or an additional 0.1% of the cold length

Pyrolysis tubes are normally designed to be straight in order to limit pressure drop. If the necessary level of thermal expansion cannot be accommodated by the support design and available clearances, a mechanical stress may be imposed on the hot tubes through the restraint of the needed expansion. This stress will typically be relieved by "bowing" of the restrained tube, out of the coil plane, to allow for relaxation of the compressive stress.

Coil Creep Elongation

Creep is the permanent deformation of materials over time when placed under stress. Typically, the rate of creep damage becomes significant when the materials are exposed to temperatures that are 40% to 50% or more of their absolute melting point (e.g., normal operations with ethylene furnace radiant coils). Ethylene furnace operators commonly observe creep damage in the form of changes in the tube ovality and stretching (or elongation) of the radiant tubes. In addition to hoop stress from internal pressure, the tubes in a radiant coil are also subject to axial stress from the weight of the tubes, differential thermal expansion of adjacent tubes, flow-induced thrust loading at bends, and the longitudinal stress component from internal pressure. By design, the radiant outlet tubes and those tubes immediately adjacent typically operate at the highest temperatures and axial loads. Because the rate of creep damage increases with higher temperature and/or stress levels, these tubes usually develop the most significant and earliest signs of creep damage.

On a microscopic scale, creep damage occurs from structural and metallurgical changes. Dislocation slip, grain boundary sliding, and diffusional flow are theoretical mechanisms responsible for the motion and deformation at the grain boundaries. Tube manufacturers of high temperature cast alloys, such as HP-40, use sophisticated alloying and casting techniques to optimize creep strength and reduce the rate of creep damage by controlling the shape, size, and orientation of the grains. Foundries also tightly control the carbon content (~0.4% for most HP-40 grade alloys) and other rare-earth elements (i.e., Nb) that form stable carbides at the grain boundaries. These carbides help stabilize the matrix by resisting the movement of dislocations that will eventually result in deformation of the grains and development of voids or cavities.

A theoretical curve of creep behavior and resultant damage is presented in Figure 1. (Figure 1 should not be interpreted as absolute and is presented only to aid in understanding typical creep behavior from as-cast to failure with the possible progression of microscopic changes.) The creep curve represents the total strain over time for a constant temperature and stress. The most notable feature of a traditional creep curve is the three distinct regions that represent the three stages of creep life.

First stage creep occurs relatively quickly and begins with application of the load where there is an instantaneous deformation; however, this initial strain is so small (and mostly elastic), it is not always represented on the creep curve (not included in Figure 1). The continuously decreasing creep rate in the first stage after the initial deformation, indicates the material is hardening and becoming more difficult to deform.

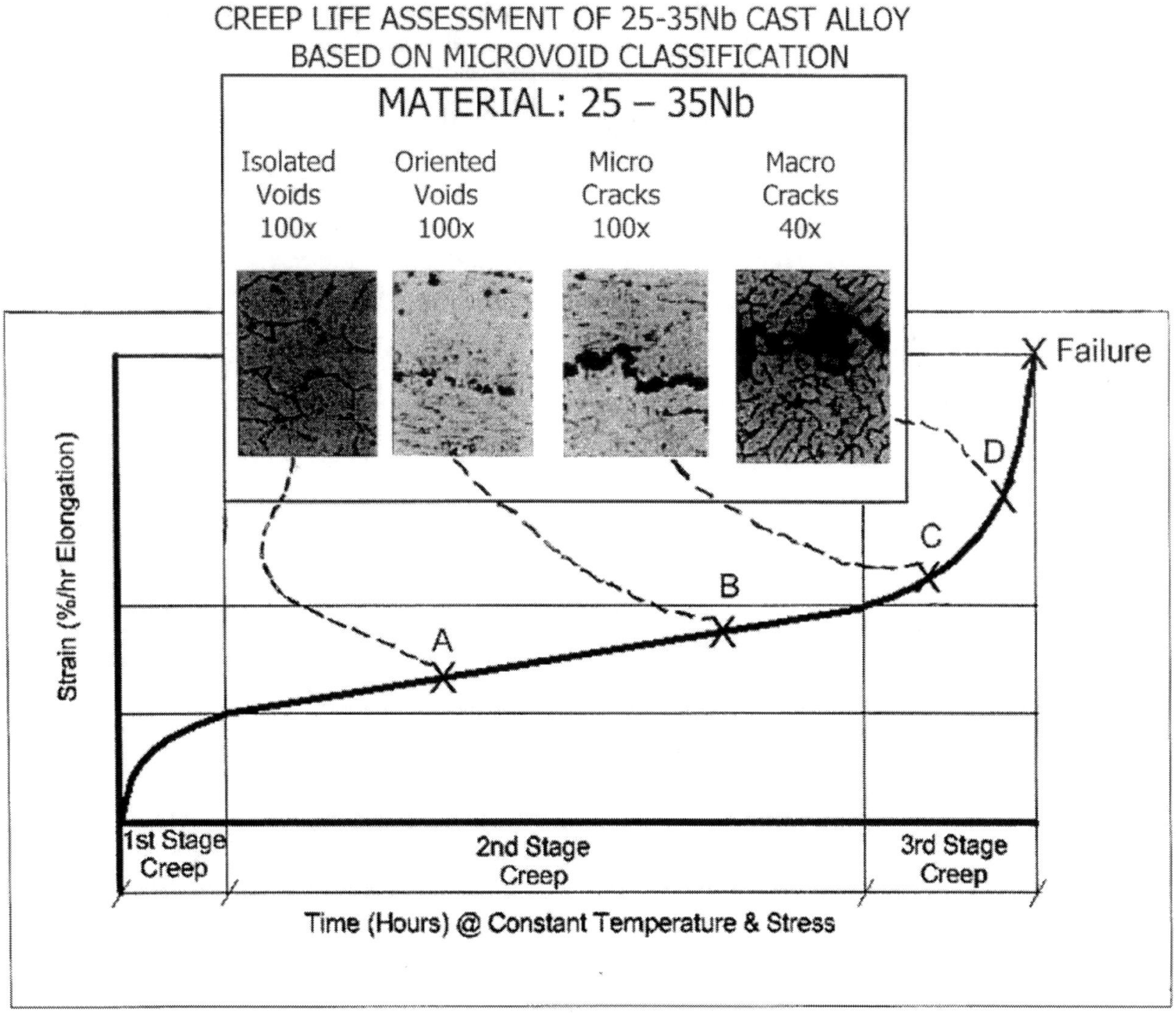

Figure 1 – Theoretical Creep Curve with Microvoid Damage Progression
Courtesy of MetalTek International [1]

The second stage has the longest duration with a creep rate that closely resembles a linear function. This essentially constant creep rate is indicative of an additive softening process that balances with the hardening process, allowing further deformation. The average creep rates that tube manufacturers publish reference the creep behavior in this linear section of the creep curve. In Figure 1, the isolated microvoids or cavities might begin to develop near Point "A" and theoretically progress to oriented microvoids (Point "B"). These microvoids are very tiny and usually not immediately detrimental by themselves.

The third (and last) stage has a slope that increases dramatically to represent the accelerated creep rate, ultimately leading to failure or rupture. Referring to the theoretical progression of damage in Figure 1, by the third stage of creep life, the directionally oriented microvoids have evolved into very small micro-cracks (only visible with magnification) at Point "C." These micro-cracks could progress even further to macro-cracks (cracks visible without magnification) at Point "D," just ahead of the ultimate failure. In addition to voids and cracks, the tubes may also experience a narrowing of the cross-sectional area at failure.

Because radiant tubes are not considered long-life applications and tubes are expected to eventually fail, creep rate is not the limiting design factor. Ethylene furnace radiant tubes are traditionally designed using "100,000 hour stress rupture" values for the allowable stress rather than the stress value for a minimum creep rate. The reality is that ethylene furnace radiant tubes almost always fail long before the 100,000 hour life from a number of failure mechanisms ranging from bulging to rupture caused by cracking from carburization damage.[2,3]

No discussion on coil creep life would be complete without some words on the effects of thermal cycles, decoking, and carburization damage. Hendrix and Clark[2] published a very good paper on the impact that decoking practices and carburization damage have on the increase of creep rates over as-cast tubes. For example, Hendrix and Clark observed creep rates of 2.5%/yr in carburized HP-40+Nb tubes, which is dramatically higher than the published average creep rates of as-cast tubes.

Coil Support Concept

The amount of movement of the radiant coil components resulting from thermal expansion will depend on the temperature, and the direction of movement will be determined by the location of fixed points in the support concept. The Transferline Exchanger (TLE) is a relatively massive piece of equipment, located outside the firebox, that must be supported independently of the radiant coil. The TLE support then becomes a fixed point at the radiant coil exit, and expansion must occur away from this fixed point. The coil configuration and support concept determine the direction and amount of movement.

Bottom-Outlet Serpentine Coil

Early designs of the vertical serpentine coil arrangement included horizontal TLEs located under the fireboxes and supported on independent foundations. Many plants incorporating this design remain in operation. The outlet tubes of the multi-pass serpentine coils direct flow vertically down and out the bottom of the fireboxes; thermal expansion of these bottom outlet coils must be vertically up from the TLEs.

These coils are normally supported from above via spring hangers attached by cast hangers to the upper return bends between adjacent vertical radiant tubes (Figure 2). Because the outlet tube spring hanger support is at the opposite end from the fixed point, it must absorb the entire movement from cold condition to the ultimate EOR temperature.

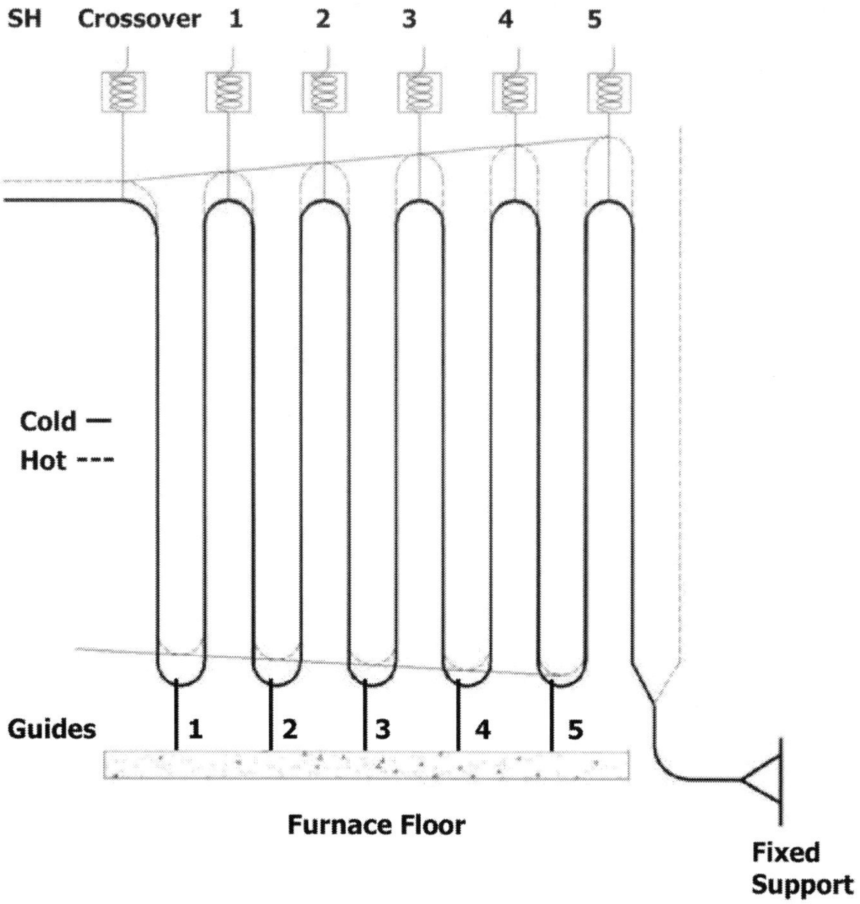

Figure 2 – Expansion of Bottom Outlet Serpentine Coil

All tubes are normally designed to be the same length in the cold condition, but the expansion of individual tubes will vary as a result of TMT differentials. Due to the process temperature profile and the tendency for coke formation to be heaviest near the coil outlet, the resulting TMTs are expected to be highest for the outlet tubes at both SOR and EOR conditions.

The TMT of any individual tube also varies over the length of the tube as a result of process temperature profile, internal coke layer variation, and especially, burner flame pattern. However, computing the expected tube expansion as a function of an average TMT (averaged over the tube length and circumference) can give an indication of the expected movements of individual tubes and supports.

A typical EOR TMT profile for an 11-pass, vertical, bottom-outlet serpentine coil is shown in Table 2, together with the calculated hot lengths for the individual tubes and the associated movements of the support points. The constant load spring hangers must maintain the required support loads over this range of movement in order to avoid shifting the dead loads onto adjacent supports.

Table 2 – Thermal Movement of Bottom Outlet Serpentine Coil

	Avg TMT deg F	Length Cold ft.	Length Hot ft.	Movement ft.	Movement inches
Fixed point				0	0.0
Outlet tube	1950	40	40.752		
lug @ SH-5				0.752	9.0
Tube 10	1920	40	40.740		
Guide 5				0.012	0.1
Tube 9	1900	40	40.732		
lug @ SH-4				0.744	8.9
Tube 8	1880	40	40.724		
Guide 4				0.020	0.2
Tube 7	1860	40	40.716		
lug @ SH-3				0.736	8.8
Tube 6	1830	40	40.704		
Guide 3				0.032	0.4
Tube 5	1800	40	40.692		
lug @ SH-2				0.724	8.7
Tube 4	1770	40	40.680		
Guide 2				0.044	0.5
Tube 3	1740	40	40.668		
lug @ SH-1				0.712	8.5
Tube 2	1700	40	40.652		
Guide 1				0.060	0.7
Tube 1	1650	40	40.632		
lug @ Crossover SH				0.692	8.3

It should be noted that this illustration is based on as-built coil dimensions (without consideration for bending or creep as a result of operation) and a theoretical TMT profile. Actual conditions may result in significant differences in observed movements. It should also be noted that the figures represent only the calculated movements due to thermal expansion. Additional displacements due to creep elongation will be additive to the length increases, and will occur over time.

Methodology to compute the coil weight (dead loads) and properly allocate the associated support loads for a bottom outlet serpentine coil has been covered previously[4].

Top-Outlet Serpentine / Split Coil

Moving the TLE to a vertical location above the radiant firebox allowed for the elimination of flow direction changes at the outlet and for direct support of the outlet pass tubes from the TLE (Figure 3). In this configuration, there is no spring hanger supporting the outlet pass. Spring hangers attached to the intermediate return bends need only absorb the movement associated with the differential expansion between adjacent radiant tubes.

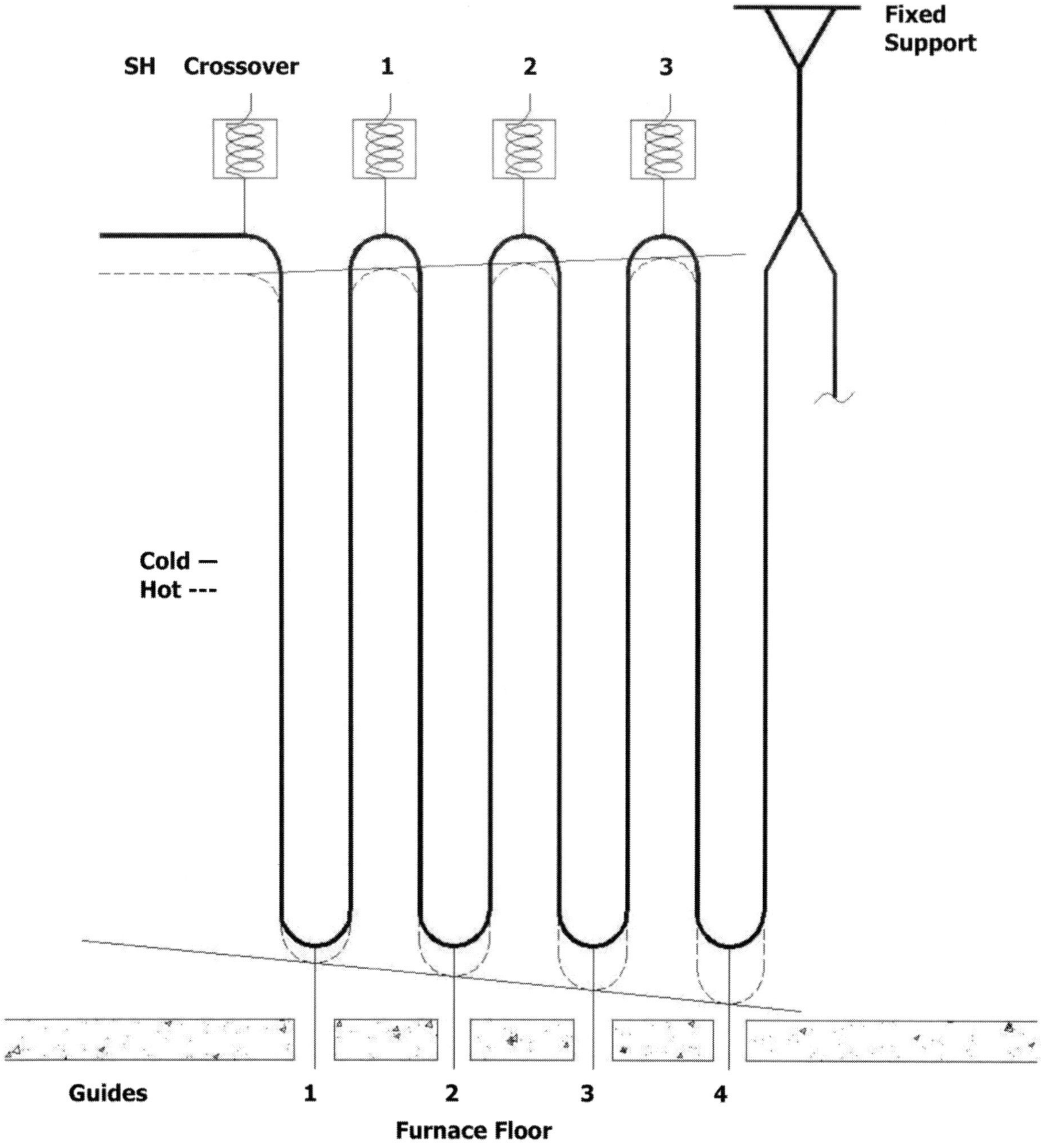

Figure 3 – Expansion of Top Outlet Serpentine Coil

Table 3 presents a typical EOR TMT profile for an 8–pass, vertical, top-outlet serpentine coil. It should be noted that the spring hanger's required travel (cold to hot movement) is in the downward direction and is significantly less than the travel for the bottom-outlet design.

Table 3 – Thermal Movement of Top Outlet Serpentine Coil

	Avg TMT deg F	Length Cold ft.	Length Hot ft.	Movement ft.	Movement inches
Fixed point				0	0.0
Outlet tube	1950	40	40.752		
Guide 4				-0.752	-9.0
Tube 7	1920	40	40.740		
lug @ SH-3				-0.012	-0.1
Tube 6	1870	40	40.720		
Guide 3				-0.732	-8.8
Tube 5	1820	40	40.700		
lug @ SH-2				-0.032	-0.4
Tube 4	1770	40	40.680		
Guide 2				-0.712	-8.5
Tube 3	1740	40	40.668		
lug @ SH-1				-0.044	-0.5
Tube 2	1690	40	40.648		
Guide 1				-0.692	-8.3
Tube 1	1650	40	40.632		
lug @ Crossover SH				-0.060	-0.7

The support concept and TMT profile of vertical split coils (multiple inlet pass/single outlet pass coils) is similar to the serpentine top outlet, as shown in Figure 4.

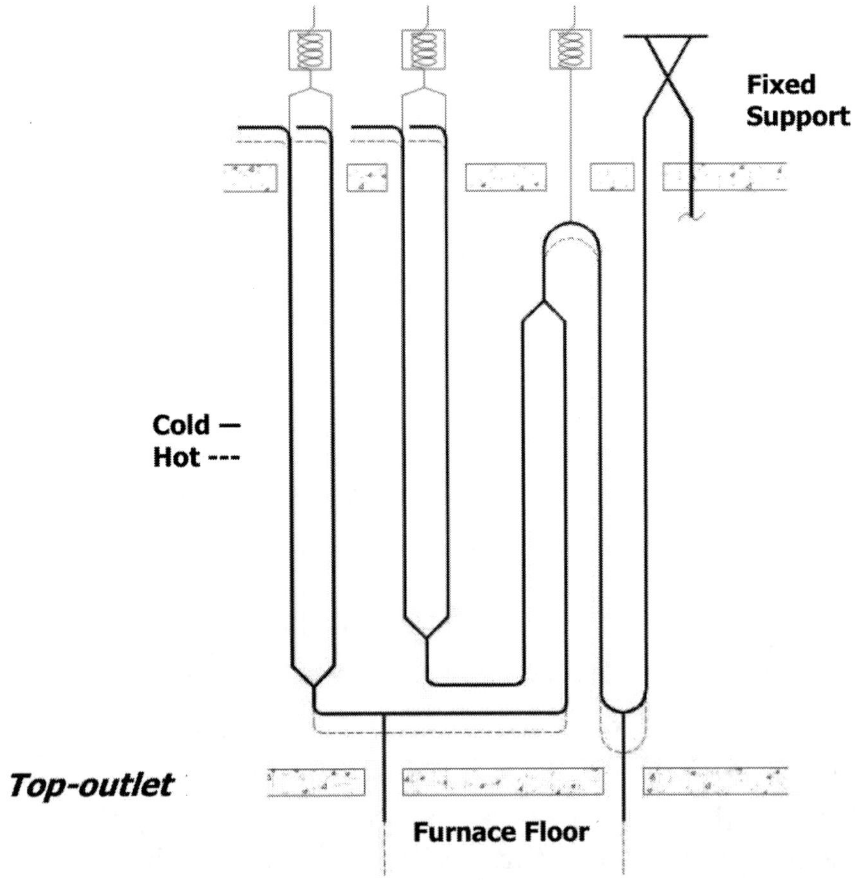

Figure 4 – Expansion of Vertical Split Coil

Modern two-pass, low-residence-time coils are typically top outlet, with the outlet pass hanging vertically down and supported from the TLEs (Figure 5). The inlet pass tubes move downward by the amount of differential thermal expansion between inlet- and outlet-pass tubes.

These coils usually have multiple inlet pass tubes for each outlet tube, and the inlet pass tubes are normally smaller in diameter, operating at significantly lower TMTs. This means that the relatively complex inlet pass supports (complex because of the number of inlets) must accommodate the differential thermal expansion and the differential creep behavior between the inlet and outlet pass tubes.

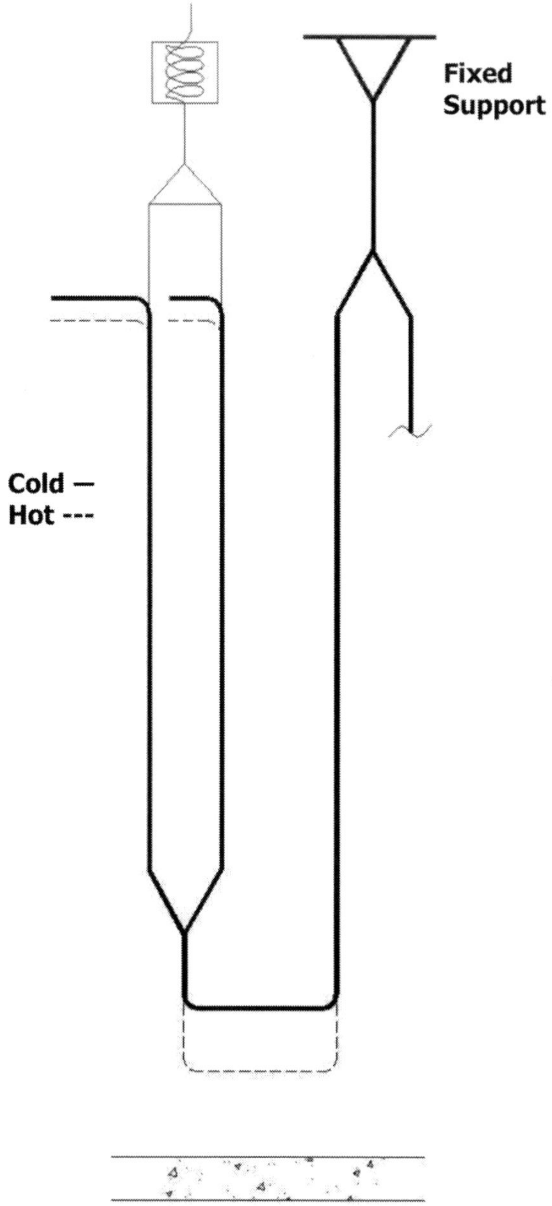

Figure 5 – Expansion of Top Outlet Two Pass Coil

Table 4 presents a typical EOR TMT profile and expected movements for a 2-pass, vertical, top-outlet split coil.

Table 4 – Thermal Movement of Top Outlet Split Coil

	Avg TMT deg F	Length Cold ft.	Length Hot ft.	Movement ft.	Movement inches
Fixed point				0	0.0
Outlet tube	1980	40	40.764		
Interm. Guide				-0.764	-9.2
Inlet Tubes	1750	40	40.672		
lug @ Inlet SH				-0.092	-1.1

Support Adjustments

Each spring hanger is typically furnished with a position indicator, with both the cold and expected hot positions clearly marked (Figure 6). For the initial construction, the position of each spring hanger in relation to the supported pipe should be adjusted so that the position indicator falls at the expected cold position marker.

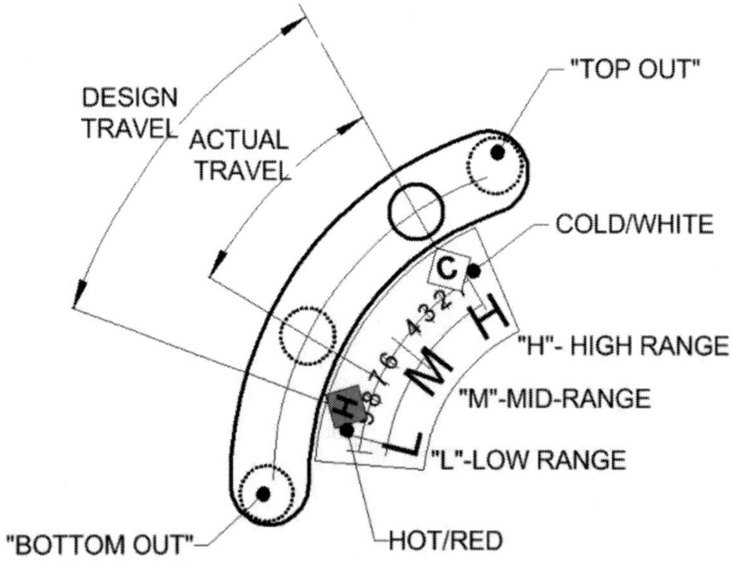

Figure 6 - Typical Spring Hanger Position Indicator

It is highly recommended that each spring hanger be attached to the support hanger with rods that include a turnbuckle to allow for position adjustment (Figure 7).

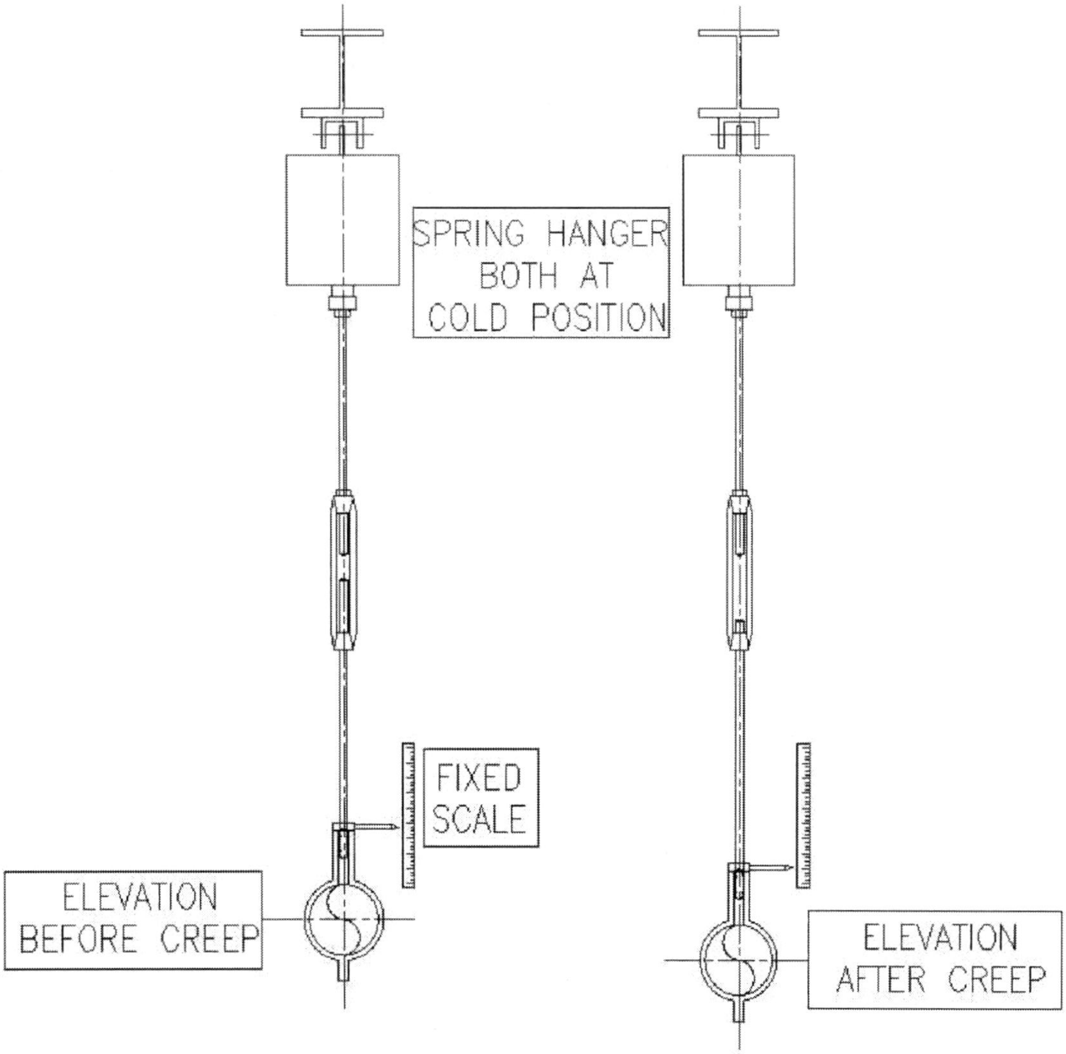

Figure 7 – Adjustment of Spring Hanger Position with Turnbuckle

Upon start up and feed introduction, the spring hanger position should move from the cold position in the direction of the hot position. Because of variations in the process and TMT profiles, the normal operating hot position may or may not match the expected hot position indicated on the spring hanger. However, the direction of expansion should be from cold toward hot, and the travel should be less than the maximum travel possible for the hanger. If any hanger reaches its limit of travel ("top out" or "bottom out"), it is no longer functional as a constant load spring hanger. A spring hanger that has topped out is providing zero support; one that has bottomed out is providing unlimited support, up to the point of component failure, and is effectively a fixed support.

During the cool down period, the radiant coil should move back toward the starting cold position, but due to creep, will likely end somewhere between the last hot and initial cold positions. Therefore, it is important to record and log the coil positions between equivalent conditions in order to determine and project the rate and amount of creep elongation.

At each cold shutdown, the turnbuckles on spring hanger supports should be used to bring the position indicators back to the indicated cold position. Afterward, the movement should be confirmed to be toward the expected hot position.

Radiant support spring hangers operate in a difficult environment, and deterioration of the spring hangers over several years of service is expected. However, peeled paint, rust and bent components are not necessarily indicators that any spring hanger has reached the end of its useful life and needs to be retired.

The basic function of a constant load spring hanger is to furnish a constant support level over a defined travel length. As long as it can be confirmed that a given spring hanger can provide the required functionality, replacement is not necessary. It is relatively simple to remove a spring hanger during a maintenance outage and take it into the shop where tests can confirm the ability to meet the required load and travel. If testing at the manufacturer was specified in the original purchase of the hanger, the test results can be directly compared to the capability at original conditions and a judgment can be made with respect to deterioration.

Monitoring, Recording and Analysis

Over the life of the radiant coil, creep elongation will result in a gradual shift in the position of most coil components away from the fixed points. At some point, physical limitations to free expansion of the radiant coils will require length correction. This is sometimes referred to as "tube trimming", "creep cutting" or "creep adjustment."

Creep adjustment requires at minimum, an additional one week downtime for the affected furnace since it needs to be de-inventoried, decoked, and brought down to a cold shutdown before the radiant coils can be temporarily anchored, trimmed, and re-welded. Re-welding of coils that have been in service may be complicated by carburization of the tube wall; significant carburization makes welding of "aged" tubes very difficult and is frequently the factor that limits the tube life and requires that aged coils be replaced. These factors make it very important to record and be able to project the creep rate of radiant tubes so that the need for creep cutting can be projected forward, planned, and completed before the coil becomes significantly carburized.

Table 5 presents a hypothetical calculation based on coil position measurements and Figure 8 presents a plot of coil and crossover position for a typical 2-pass split coil over the first five years of coil life. Such plots can be helpful to visualize and analyze the coil movements during transitions from cold-hot-cold and over the life of the coils. It is important to note and record the coil and crossover positions – not the spring hanger positions – as the spring hangers should be adjusted back to the cold position at every cold shutdown.

Table 5 – Position History

Date		@SH-A	@SH-B	@SH-C	@SH-D	@SH-1	@SH-2	@ Guide
14-Mar-03 cold	position	800	800	850	900	1200	1200	1400
14-Mar-03	displacement	0	0	0	0	0	0	0 (as built)
16-Mar-03	movement	-2	-10	-10	-12	-15	-15	-225
16-Mar-03 hot	position	798	790	840	888	1185	1185	1175
16-Mar-03	displacement	-2	-10	-10	-12	-15	-15	-225
23-Jun-04	movement	2	9	8	7	10	10	175
23-Jun-04 cold	position	800	799	848	895	1195	1195	1350
23-Jun-04	displacement	0	-1	-2	-5	-5	-5	-50 creep
28-Jun-04	movement	-2	-10	-10	-13	-15	-15	-225
28-Jun-04 hot	position	798	789	838	882	1180	1180	1125
28-Jun-04	displacement	-2	-11	-12	-18	-20	-20	-275
17-Oct-05	movement	1	9	6	7	12	12	200
17-Oct-05 cold	position	799	798	844	889	1192	1192	1325
17-Oct-05	displacement	-1	-2	-6	-11	-8	-8	-75 creep
20-Oct-05	movement	-2	-9	-9	-12	-15	-15	-225
20-Oct-05 hot	position	797	789	835	877	1177	1177	1100
20-Oct-05	displacement	-3	-11	-15	-23	-23	-23	-300
12-Nov-06	movement	2	8	5	7	12	12	185
12-Nov-06 cold	position	799	797	840	884	1189	1189	1285
12-Nov-06	displacement	-1	-3	-10	-16	-11	-11	-115 creep
15-Dec-06	movement	-2	-10	-10	-12	-15	-15	-225
15-Dec-06 hot	position	797	787	830	872	1174	1174	1060
15-Dec-06	displacement	-3	-13	-20	-28	-26	-26	-340
6-Oct-07	movement	1	9	8	7	11	11	140
6-Oct-07 cold	position	798	796	838	879	1185	1185	1200
6-Oct-07	displacement	-2	-4	-12	-21	-15	-15	-200 creep
16-Oct-07	movement	-2	-11	-11	-12	-15	-15	-225
16-Oct-07 hot	position	796	785	827	867	1170	1170	975
16-Oct-07	displacement	-4	-15	-23	-33	-30	-30	-425
25-Dec-08	movement	2	10	9	8	10	10	100
25-Dec-08 cold	position	798	795	836	875	1180	1180	1075
25-Dec-08	displacement	-2	-5	-14	-25	-20	-20	-325 creep

In this example, the "position" of the coil and crossover are recorded over time as a distance from some stationery reference point, either on the structure or the heater casing. In this case, the measurements are in millimeters and the movements are calculated as positive when moving vertically up and negative when moving vertically down.

Table 5 and Figure 8 illustrate how the measurements can be used to analyze the progressive movements of a single crossover and determine the necessary adjustments to the spring hangers supporting that crossover. The upstream end of the crossover is normally welded to the convection outlet where it exits the header box. This connection is generally fixed (in terms of vertical movement) and therefore, the crossover piping in the vicinity of this connection is not expected to move either up or down to any significant extent, either as a result of radiant coil expansion or progressive creep.

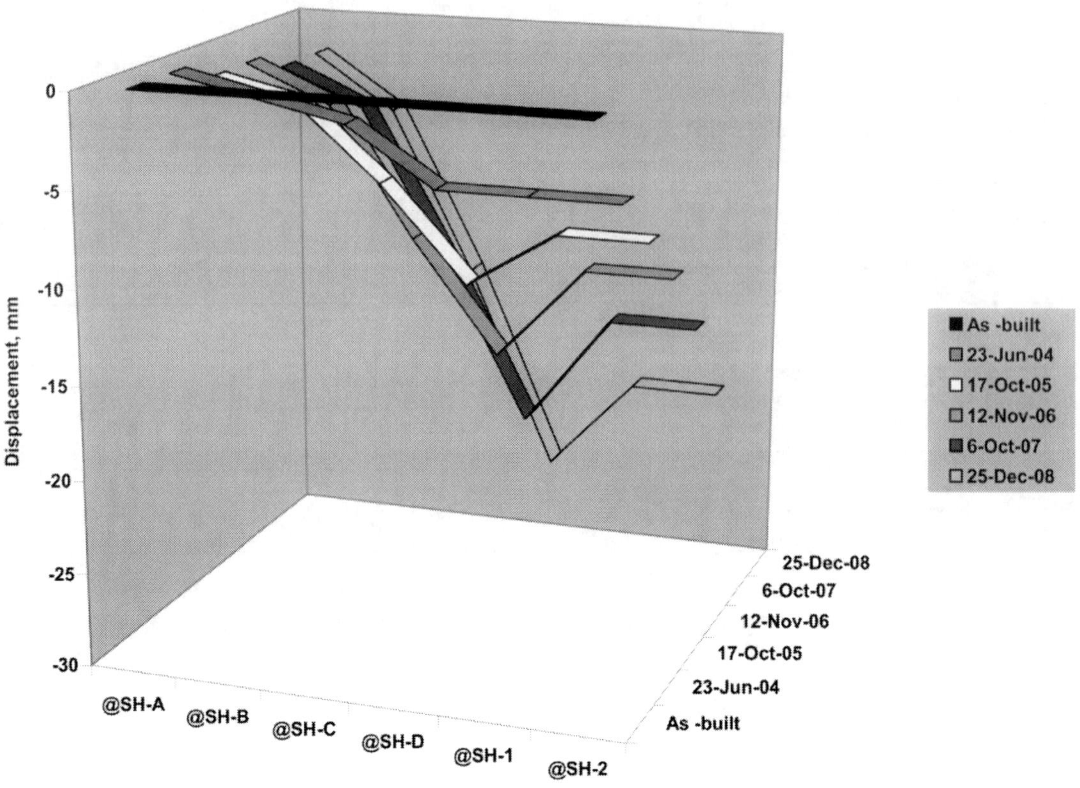

Figure 8 – Crossover Position Plot (Cold)

In the design associated with this example, the radiant inlet connections (or manifold) will move down to a hot position. Consequently, the downstream end of the crossover attached to the radiant inlet must also move down to a hot position. Differential creep rates between the outlet and inlet pass radiant tubes will cause the radiant inlet manifolds to move progressively lower as the outlet tube exhibits progressively more creep elongation over the coil life. The net effect over the long term is a gradual downward slope of the crossover, with the slope increasing over time. If the spring hangers supporting the crossover are re-adjusted back to the cold position at every cold shutdown, the spring hanger travel absorbs only the thermal expansion plus creep differential between cold shutdowns: the permanent displacement due to creep elongation differentials of the radiant coil will be absorbed by the turnbuckles.

In this particular example, the downward displacement of the crossover in question is actually greater than the associated inlet manifold to which it is connected (crossover position at SH-D displacement greater than inlet manifold displacement at SH-1 and -2). This indicates "sagging" of the crossover at SH-D, which is typically the result of inadequate support. The level of support required in this area many be higher than anticipated during the design, if the installation varies in some way from the basis used for support calculations. Additional instrumentation or transmitters mounted on the pipe, instrument purge piping or tubing supported from the pipe, or rainwater accumulation under the insulation sheeting are all factors that would increase the support required. The root cause of sagging should be investigated and if it cannot be corrected, the level of support should be increased to compensate. Most spring hangers have a load adjustment screw to allow the load set point to be adjusted, within certain limits.

Table 5 and Figure 9 illustrate how the measurements of radiant coil guide position can be used to track outlet tube creep elongation, project coil expansion limits, and plan creep adjustments (tube trimming). Figure 9 plots the bottom guide displacement from the hypothetical example at both hot and cold conditions on a time scale. In theory, the outlet tube length increase between any cold condition and the subsequent hot condition will be essentially constant – equal to the thermal expansion of the outlet tube to operating temperatures.

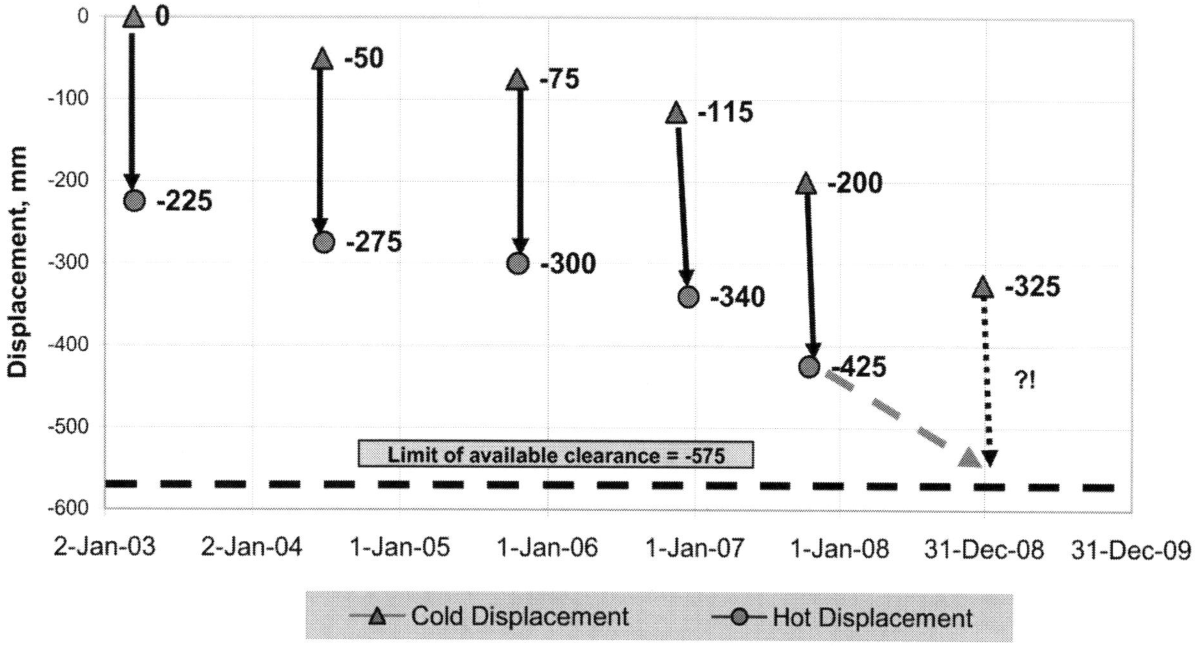

Figure 9 – Bottom Guide Displacement

However, displacement between hot conditions will be affected by progressive creep and the rate of creep can be determined if the data are gathered over time in a systematic way, recorded, and plotted. In the hypothetical example, the coil experiences five cold shutdown/heat-up cycles at roughly one year intervals, starting with the initial installation. The limit of acceptable displacement is fixed by the firebox dimensions at -575 mm from the initial cold position. It is clear from Figure 9 that the heat-up from the current cold condition would result in a displacement to roughly -550 mm, and another one-year cycle between cold shutdowns will not be possible before the coil "bottoms out" (bottom of the intermediate manifold contacts the top of the floor or trough refractory).

Creep adjustment during this shutdown is clearly needed, but since this coil has been in service for five and a half years, the outlet tube is likely significantly carburized. If the outlet tube is cut in a carburized region to remove a tube section, the two ends intended to be re-welded may be too carburized for a satisfactory closing weld to be made, and the entire carburized section of the coil may need to be scrapped and replaced.

Faced with this situation, the engineer should consider removal of a tube section outside the firebox, in the transition area between the firebox and TLE. The tube material in this area has not been exposed to temperatures exceeding the process temperature and can be expected to be significantly less carburized than tube material that has been exposed to radiant heat input. However, this option will make the creep adjustment work slightly more complex, since the outlet pass will also need to be temporarily supported before it can be cut loose.

Conclusion

The physical changes in high temperature piping and furnace coils can be quantitatively examined, tracked, and projected by analyzing data associated with movements of these components over time. In order to be useful for this purpose, such data must be gathered, and recorded over long time periods in a systematic, consistent and reproducible way. In consideration of the amount of data, the associated time period, and likely changes in staff during that time, a written data collection procedure seems warranted.

References

1. Eynon, G and Gapinski, J. with MetalTek International, Private Correspondence, "Creep Curve with Typical Microvoid Damage", 2002
2. Hendrix, D. E. and M. W. Clark (1985), "Contributing Factors to the Unusual Creep Growth of Furnace Tubing in Ethylene Pyrolysis Service", *Corrosion 85*, Paper No. 21, March, Houston, TX, NACE International.
3. Jakobi, D. Dr., Gommans, R. (2002) "Typical Pyrolysis Coil Failures and How to Combat Them", in *14th Ethylene Producers Conference*, Paper No. 98e, New Orleans, LA, AIChE 2002 Spring National Meeting.
4. Karrs, M. and Cannon, T.C. (2002), "Cracking Heater Radiant Coil Balancing", in *14th Ethylene Producer's Conference,* Paper No. 98f, New Orleans, LA, AIChE 2002 Spring National Meeting.

ETHYLENE FURNACE MAINTENANCE SURVEY

Paul van Helmond
Manager Process Engineering Department
Selas Fluid Processing Corporation

David Oulton
Metallurgist
Nova Chemicals

David Eisenhawer
Furnace Mechanical Engineer
Nova Chemicals

Steve Guinn
Reliability Engineer
DOW Chemicals

Abstract: The Ethylene Producer's Committee sponsored a worldwide furnace maintenance survey among ethylene producers in the winter of 2008. The survey focuses on the reliability and maintenance aspects of specific furnace components including radiant coils, burners, quench exchangers, and refractory among others.

The survey was conducted by the Maintenance and Reliability Subcommittee. Data from the survey was reviewed and interpreted by a team consisting of a metallurgist, a reliability engineer, and furnace specialists. Each of these team members has many years of experience with ethylene production.

This paper is a review of the results from the survey along with correlations and trends identified by the survey team.

Introduction

A survey team consisting of four members of the Maintenance and Reliability sub-committee sent a furnace maintenance survey to operating companies in North America, South America, Europe, Middle East & Africa and South East Asia. The survey was conducted in accordance with the guidelines that the EPC has EPC sponsored surveys. These rules ensure that there are provisions in place to ensure that participating company's data is not trace-able. The rules also require that all survey questionnaires are handled confidentially and are destroyed after analysis. Previous surveys of the EPC were conducted on paper; this survey was conducted as an anonymous web based survey through a secure website supplied by Cvent Inc.

This survey report contains all responses to the survey questionnaire. Due to space limitations in the printed conference proceedings we have not included cross tabulation analysis in this paper. These cross tabulations will be available at the conference and on the website of the Ethylene Producers Committee, ethyleneweb.com.

Question 1: Where is your plant located?

	Choice	Response Percent	Response Total
1	North America (Canada - USA - Mexico)	26.9%	18
2	South or Central America	10.4%	7
3	Europe	49.3%	33
4	Middle East & Africa	7.5%	5
5	South East Asia (Thailand - Malaysia - Singapore - Indonesia - Australia)	6.0%	4

Total Respondents:	**67**

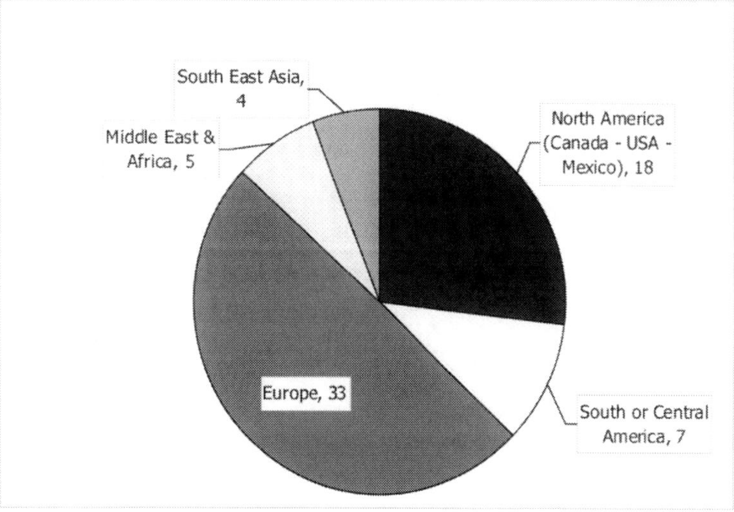

Question 2: How many furnaces of this type do you have in operation?

Total Respondents	**67**
Total No. of Furnaces	**423**

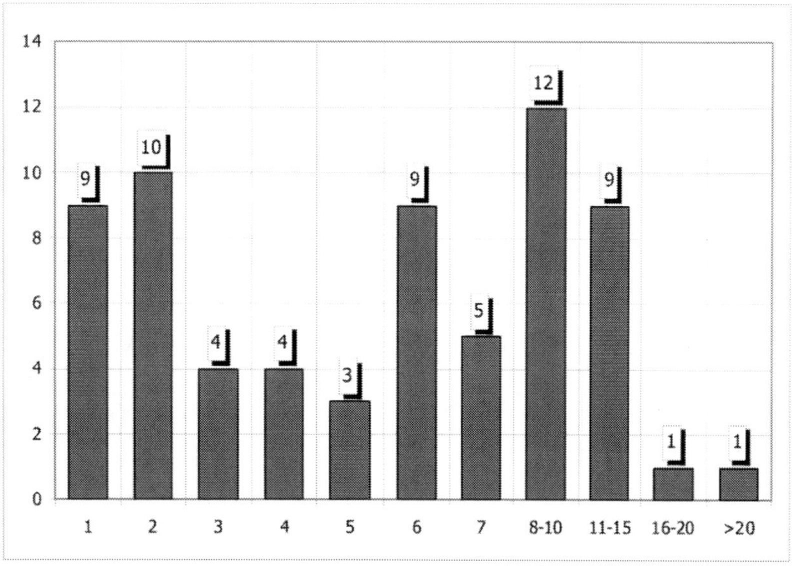

Question 3: What is the feed stock for this type of furnace?

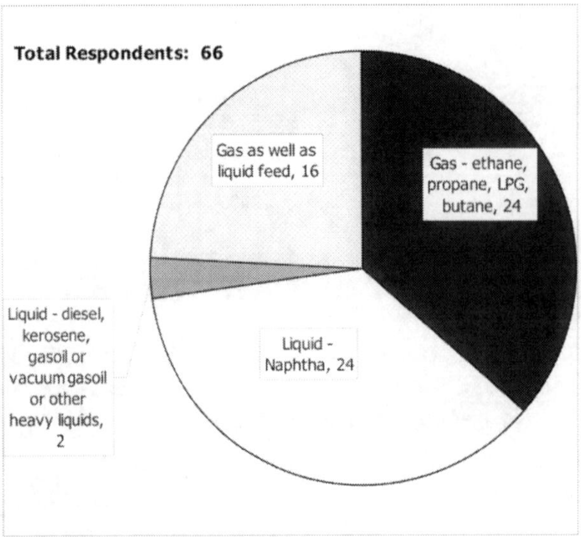

Comment: Few responses only for heavy liquid feedstocks. Some heavy feedstock responses may be covered in 'Gas as well as liquid feed' response.

Question 4: When were your furnaces started up? If your furnaces underwent a significant revamp please enter the date of the start-up after this revamp.

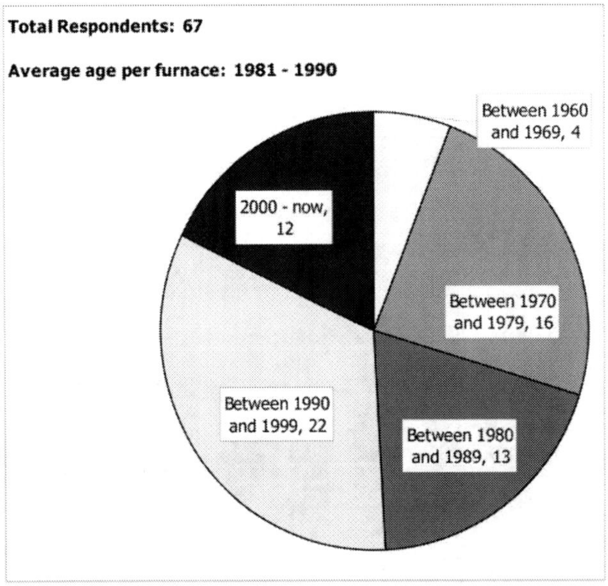

Question 5: How would you describe your radiant coil?

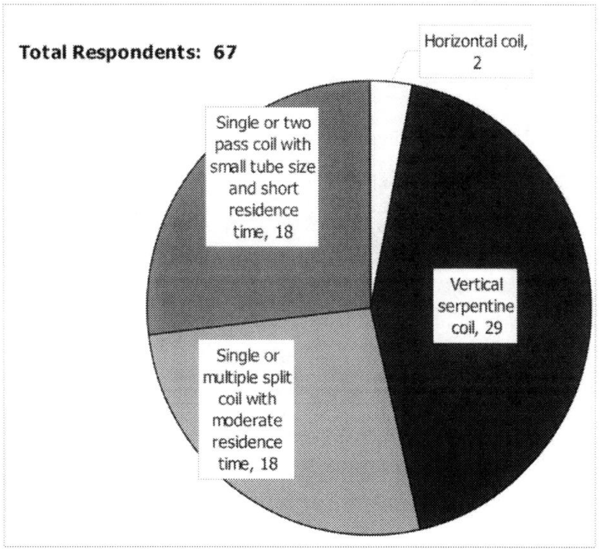

Comment: Four pass coils without split are sorted into 'Vertical serpentine coil'.

Question 6: How would you describe furnace criticality of the furnaces for your plant production?

	Choice	Response Percent	Response Total
1	Our plant is completely furnace limited, decoke of a furnace and/or furnace outage for maintenance leads to equivalent loss in plant production	9.2%	6
2	Our plant is furnace limited, decoke of a furnace and/or furnace outage for maintenance leads to some loss in plant production	24.6%	16
3	We can maintain plant capacity for a scheduled outage by modifying our decoke schedule and/or increase capacity on the other furnaces	38.5%	25
4	A single furnace outage does not affect plant capacity or decoke schedule	27.7%	18
5	Comment	0.0%	0
	Total Respondents:		**65**

Comment: There is no clear correlation between number of furnaces or even age of furnaces and furnace criticality. This is most likely due to different plant operation practices as severity of cracking, sparing philosophy etc.

Question 7: What metallurgy do you use in your radiant coil?

	Choice	Response Percent	Response Total
1	Cast HK 40 (25 Cr - 20 Ni)	3.0%	2
2	Cast HP 40 (25 Cr - 35 Ni)	4.5%	3
3	Cast HP 40 modified (25 Cr - 35 Ni + others + micro-alloys)	19.4%	13
4	Cast 35 Cr - 45 Ni + others + micro-alloys	23.9%	16
5	Combination of 25 Cr - 35 Ni + micro-alloys and 35 Cr - 45 Ni + micro-alloys	49.3%	33

Total Respondents:	**67**

1 comment has been sorted into HP 40 modified representing majority of the furnaces

Comment: Almost 75% of the respondents utilize 35 Cr- 45 Ni material at least partially in their radiant coil.

Question 8: What is the frequency of unscheduled furnace outages due to radiant coil failure? This should be answered on a per furnace basis, i.e. if you have 10 furnaces and in an average year you have 5 individual furnace outages, your reply shall be once per 2 years.

	Choice	Response Percent	Response Total
1	More than once per year	24.6%	16
2	Once per 1 - 3 years	33.8%	22
3	Once per 4 - 5 years	18.5%	12
4	Once per 6 - 10 years	10.8%	7
5	once per 11 - 20 years	7.7%	5
6	never	4.6%	3

Total Respondents:	**65**

Question 9: How would you characterize your radiant coil maintenance program?

	Choice	Response Percent	Response Total
1	Repair after failure - a purely reactive program	11.2%	7.5
2	Inspection based - You inspect your coil to determine life and use this to schedule replacement	36.6%	24.5
3	Fixed schedule - maintenance takes place after a certain fixed number of years of operation	52.2%	35

Total Respondents:	**67**

1 comment has been split into answers 1 & 2

Question 10: After how may years of operation do you replace your complete radiant coil? Note that there will be a separate question for those of you who only replace partial radiant coils.

	Choice	Response Percent	Response Total
1	0 - 2 years	1.5%	1
2	3 - 5 years	32.8%	22
3	6 - 8 years	49.3%	33
4	more than 8 years	9.0%	6
5	We do not replace complete radiant coils	4.5%	3
6	Comment	3.0%	2

Total Respondents:	**67**

2 comments: Complete radiant coils were not yet replaced

Question 11: What is the normal number of furnace down days for a complete radiant coil replacement? Please read "down" days as days from feed-out to feed-in. If it is standard to use radiant coil replacement as part of a furnace outage in which you do more than just a coil replacement, please enter the number of days you would use in case you had a coil replacement only.

	Choice	Response Percent	Response Total
1	0 - 5 days	3.2%	2
2	6 - 10 days	9.7%	6
3	11 - 20 days	58.1%	36
4	More than 20 days	29.0%	18

Total Respondents:	**62**

Question 12: After how may years of operation will you execute a partial coil replacement, for example outlet sections?

	Choice	Response Percent	Response Total
1	0 - 2 years	3.0%	2
2	2 - 5 years	26.9%	18
3	5 - 8 years	11.9%	8
4	more than 8 years	0.0%	0
5	We do not replace partial radiant coils (only complete coils)	53.7%	36
6	Comment	4.5%	3

Total Respondents:	**67**

3 comments: Partial replacement upon failure

Comment: No correlation has been found between furnace criticality and any of questions 10 thru 12.

Question 13: What is the normal number of furnace down days for partial radiant coil replacement? Please read "down" days as days from feed-out to feed-in. If it is standard to use radiant coil replacement as part of a furnace outage in which you do more than just a coil replacement, please enter the number of days you would use in case you had a coil replacement only.

	Choice	Response Percent	Response Total
1	0 - 5 days	17.6%	6
2	6 - 10 days	38.2%	13
3	11 - 20 days	38.2%	13
4	More than 20 days	5.9%	2

Total Respondents:	**34**

Question 14: Which of the following techniques do you use to measure remaining radiant tube life? Mark all that apply.

	Choice	Response Percent	Response Total
1	Carburization testing	46.3%	31
2	Visual inspection	91.0%	61
3	Recording of creep	37.3%	25
4	Statistical methods (Weibull, Crow-AMSAA)	13.4%	9

Total Respondents:	**67**

Comments: 5 respondents do not measure remaining life time
2 of these run to failure
1 of these uses historical data

Question 15: Do you perform end-of-life coil inspection to determine the reason of coil failure, if so which types of inspection do you do? Please mark all that apply.

	Choice	Response Percent	Response Total	Corrected *) Response Percent	Corrected *) Response Total
1	Carburization testing	52.7%	29	43.3%	29
2	Metallurgical lab analysis	72.7%	40	59.7%	40
3	Measurement of creep elongation	23.6%	13	19.4%	13
4	We do not inspect after failure	25.5%	14	38.8%	26

Total Respondents:	**55**
Total Non-Respondents:	**12**

*) Note: It is assumed that those who did not respond, do not perform end-of-life inspection

Question 16: Please rank the three (3) most common mechanisms of radiant coil failures, 1 being the most common.

Carburization: "Carbon enrichment due to carbon migration"
Stress Rupture: "Fracture resulting from advanced creep damage"
Brittle Fracture: "Fracture that exhibits very little or no plastic deformation"
Oxidation/Scaling: "An oxygen reaction that commonly results in scaling with the potential for excessive metal loss"
Metal Dusting: "Catastrophic carburization resulting in severe loss of material"
Sulfidation: "Corrosion of materials in contact with sulfur-rich compounds at high temperatures"
Fuel Ash Corrosion: "Accelerated loss of material under deposits of furnace combustion residues at high temperatures"

Total Respondents: 65

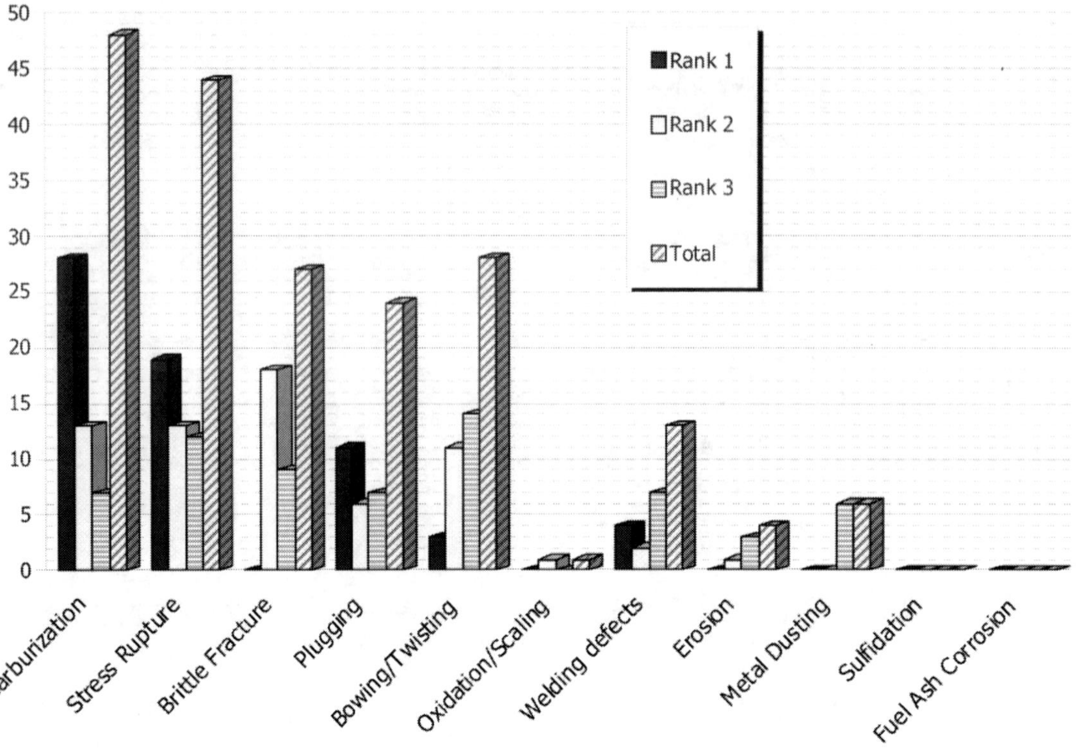

Comment: Carburization and Stress Rupture, induced by advanced creep, are clearly the most common radiant coil failures, followed by Brittle Fractures, Plugging, and Bowing/Twisting. Welding defects are also contributing to coil failures.

Question 17: What are the most common causes of radiant coil failure in your furnaces?

	Choice	Response Percent	Response Total
1	Upset operating condition	63.5%	40
2	Flame impingement / burner control	39.7%	25
3	Casting defects	17.5%	11
4	Welding defects	22.2%	14
5	External moment (unbalanced coil)	17.5%	11
6	Comment	6.3%	4

Total Respondents:	**63**

Comments: 2 x End of service life
1 x Carburization
1 x Thermoweld cover

Question 18: Who carries out the following inspections in your plant?

	Maintenance	Operations	Others	Comment		Response Total
Recording of metal temperature	0.0% (0)	95.3% (61)	3.1% (2)	1.6% (1)	1 comment: IR TMT control	**64**
Burner/flame shape visual inspection	4.8% (3)	95.2% (60)	0.0% (0)	0.0% (0)		**63**
Inspection of coil balancing system	75.0% (46.5)	23.4% (14.5)	1.6% (1)	0.0% (0)		**62**
Recording of creep elongation of the radiant coil	58.8% (30)	5.9% (3)	35.3% (18)	0.0% (0)		**51**

Note 1: It is assumed that those who did not respond, do not perform the particular inspection; comments are not listed if stating that a particular inspection is not performed.

Note 2: If both maintenance as well as operations carry out inspections, this is listed as 0.5 for each

Question 19: What is the frequency of metal temperature (pyrometer) measurements?

	Choice	Response Percent	Response Total
1	Every shift	17.5%	11
2	Daily	38.1%	24
3	Weekly	33.3%	21
4	Only at end of run conditions	3.2%	2
5	Never	6.3%	4
6	Comment	1.6%	1

Total Respondents:	**63**

Comment: As required

Question 20: What is the frequency of burner/flame shape visual inspection?

	Choice	Response Percent	Response Total
1	Every shift	53.1%	34
2	Daily	31.3%	20
3	Weekly	9.4%	6
4	Only when problems are encoutered	4.7%	3
5	Never	0.0%	0
6	Comment	1.6%	1

Total Respondents:	64

Comment: Quarterly

Question 21: What is the frequency of inspection on the coil suspension system (spring hangers and/or counterweights)?

	Choice	Response Percent	Response Total
1	Daily	11.5%	7
2	Weekly	1.6%	1
3	After each outage	41.0%	25
4	After coil repairs and replacements only	26.2%	16
5	Only when we encounter problems	11.5%	7
6	Never	3.3%	2
7	Comment	4.9%	3

Total Respondents:	61

Comment: 3 x After coil repairs and replacements complete and monthly repetitive

Question 22: What is the frequency of creep elongation inspection?

	Choice	Response Percent	Response Total
1	Monthly	17.5%	11
2	Anually	11.1%	7
3	When we encounter problems	27.0%	17
4	At end of coil life	6.3%	4
5	Never	28.6%	18
6	Comment	9.5%	6

Total Respondents:	**63**

Comments: 1 x Quarterly
5 x During Maintance/Furnace outage

Question 23: Please indicate which type of radiant coil outlet temperature measurement you apply and indicate the average life?

Total Respondents: 63

	less than 6 months	7 - 12 months	1 - 3 years	3 - 10 years	we never replaced or repaired them	Response Total
Thermowell insert	0.0% (0)	12.7% (7)	43.6% (24)	36.4% (20)	7.3% (4)	**55**
Skin thermocouple	0.0% (0)	11.1% (2)	16.7% (3)	50.0% (9)	22.2% (4)	**18**

Comment: Skin Thermocouple provide some longer life time than thermocouple inserted in a thermowell. Supposedly life time of thermocouple is mostly determined by aging/drift rather than mechanical damage.

Question 24: What radiant coil spares do you keep on hand?

	Choice	Response Percent	Response Total
1	None, we only order coils for planned coil replacments	0.0%	0
2	Loose tubes and fittings	60.0%	39
3	Pre-assembled radiant coils	38.5%	25
4	Comment	1.5%	1

Total Respondents:	**65**

Comment: We keep both tubes and fittings and a complete coil set

Question 25: Disposal - do you have a metal recycle program?

	Choice	Response Percent	Response Total
1	Yes	75.4%	49
2	No	24.6%	16

Total Respondents:	65

5 comments has been sorted into "Yes"

Question 26: What kind of decoking technique is applied?

	Choice	Response Percent	Response Total
1	Steam-air decoking w/o treatment (i.e. directly to atmosphere)	18.8%	12
2	Steam-air decoking through the firebox	26.6%	17
3	Steam-air decoking to a coke seperator/drum/cyclone	51.6%	33
4	Steam only (online decoking)	1.6%	1
5	Comment	1.6%	1

Total Respondents:	64

Comment: Steam-air decoking to a cyclone and then through the firebox

Question 27: Do you experience any of the following failures?

Total Respondents: 64

	Yes	No	Comment	Response Total
Erosion in decoke piping	81.3% (52)	18.8% (12)	0.0% (0)	**64**
Erosion of firebox refractory (after decoking to the firebox)	73.7% (14)	21.1% (4)	5.3% (1)	**19**

Comment: Minimal damage

Comment: Erosion of decoke piping as well as erosion of refractory are widely experienced.

Question 28: What type of primary quench exchangers do you have? A primary quench exchanger is the exchanger located closely to the furnace radiant coil outlet often referred to as (primary) Transfer Line Exchanger (TLE). In some of the newer plants several quench exchangers are used in series. In this survey we are not addressing issues with the secondary (or tertiary) exchangers.

	Choice	Response Percent	Response Total
1	Linear double pipe exchanger	28.6%	18
2	Multi-tube double pipe exchangers	31.7%	20
3	Conventional shell and tube exchangers	38.1%	24
4	None, we have a direct water quench	0.0%	0
5	Comment	1.6%	1

Total Respondents:	63

Comments: Both, Linear double pipe & Conventional shell and tube exchangers

Question 29: What is the orientation of your primary quench exchanger?

	Choice	Response Percent	Response Total
1	Horizontal	10.2%	6
2	Vertical	86.4%	51
3	Comment	3.4%	2

Total Respondents:	59

Comments: 1 x Both Horizontal and Vertical
1 x n/a

Question 30: What is the frequency of unscheduled furnace outage due to quench exchanger failure? Please answer this question on a per furnace basis, i.e. if you have 10 furnaces and on average year you have a failure shutting down two furnaces each year, the reply shall be once per 5 years.

	Choice	Response Percent	Response Total
1	More then once per year	7.8%	5
2	Once per 1 - 3 years	10.9%	7
3	Once per 3 - 5 years	20.3%	13
4	Once per 5 - 10 years	25.0%	16
5	Less than once per 10 years	15.6%	10
6	Never	20.3%	13

Total Respondents:	**64**

1 comment ("quench exchanger is controlled once per year to avoid unscheduled outages") has been sorted into "Never"

Question 31: How often do you mechanically clean your primary quench exchanger?

	Choice	Response Percent	Response Total
1	With every furnace decoke cycle	1.6%	1
2	With every 2 - 4 furnace decoke cycles	25.0%	16
3	With every 5 - 7 furnace decoke cycles	25.0%	16
4	With more than 7 cycles	12.5%	8
5	At a maintenance outage only	31.3%	20
6	Never	4.7%	3

Total Respondents:	**64**

1 comment ("once in 5-6 years") has been sorted into "At a maintenance outage only"

Question 32: Do you experience the following failures? Please select "yes" or "no":

					Response for Linear Double Pipe Exchanger		
	Yes	**No**	**Response Total**	**Comments**	**Yes**	**No**	**Response Total**
Inlet tube sheet erosion	39.7% (25)	60.3% (38)	**63**	2x n.a. has been sorted into "No"	17.6% (3)	82.4% (14)	**17**
Erosion on the quench exchanger outlet fitting	28.6% (18)	71.4% (45)	**63**	1x extensive corrosion on outlet cone has been sorted into "Yes"	35.3% (6)	64.7% (11)	**17**
Tube blockage	68.3% (43)	31.7% (20)	**63**	1x blockage on cone/tube/tubesheet sorted into "Yes"	29.4% (5)	70.5% (12)	**17**
Failure of tube to tube sheet weld	46.7% (28)	53.3% (32)	**60**	1x n.a. has been sorted into "No"	33.3% (5)	66.7% (10)	**15**
Other weld failures	16.7% (10)	83.3% (50)	**60**		25.0% (4)	75.0% (12)	**16**
Refractory failure	50.0% (31)	50.0% (31)	**62**	1x n.a. has been sorted into "No"	25.0% (4)	75.0% (12)	**16**
Water side fouling	17.0% (9)	83.0% (44)	**53**	1x fouling not expected not sorted in	0.0% (0)	100% (15)	**15**
Water side corrosion	35.0% (21)	65.0% (39)	**60**	1x frequent failures after 15 years has been sorted into "Yes"	13.3% (2)	86.7% (13)	**15**
Tube erosion	25.8% (16)	74.2% (46)	**62**		13.3% (2)	86.7% (13)	**15**
Bowing of tube sheet	6.8% (4)	93.2% (55)	**59**	1x n.a. has been sorted into "No"	6.7% (1)	93.3% (14)	**15**
Distortion of coke deflector	43.9% (25)	56.1% (32)	**57**	3x n.a. has been sorted into "No"	5.9% (1)	94.1% (16)	**17**
Gasket or seal failures	37.0% (20)	63.0% (34)	**54**		28.6% (4)	71.4% (10)	**14**
Outer tube in double pipe exchanger or outer shell in shell and tube exchanger	7.7% (4)	92.3% (48)	**52**	2x n.a. has been sorted into "No"	6.7% (1)	93.3% (14)	**15**
Gas inlet reducer (for linear exchangers Only)	37.5% (6)	62.5% (10)	**16**	only answers for linear exchangers listed	37.5% (6)	62.5% (10)	**16**

Question 33: Rank the three most severe quench exchanger failures?

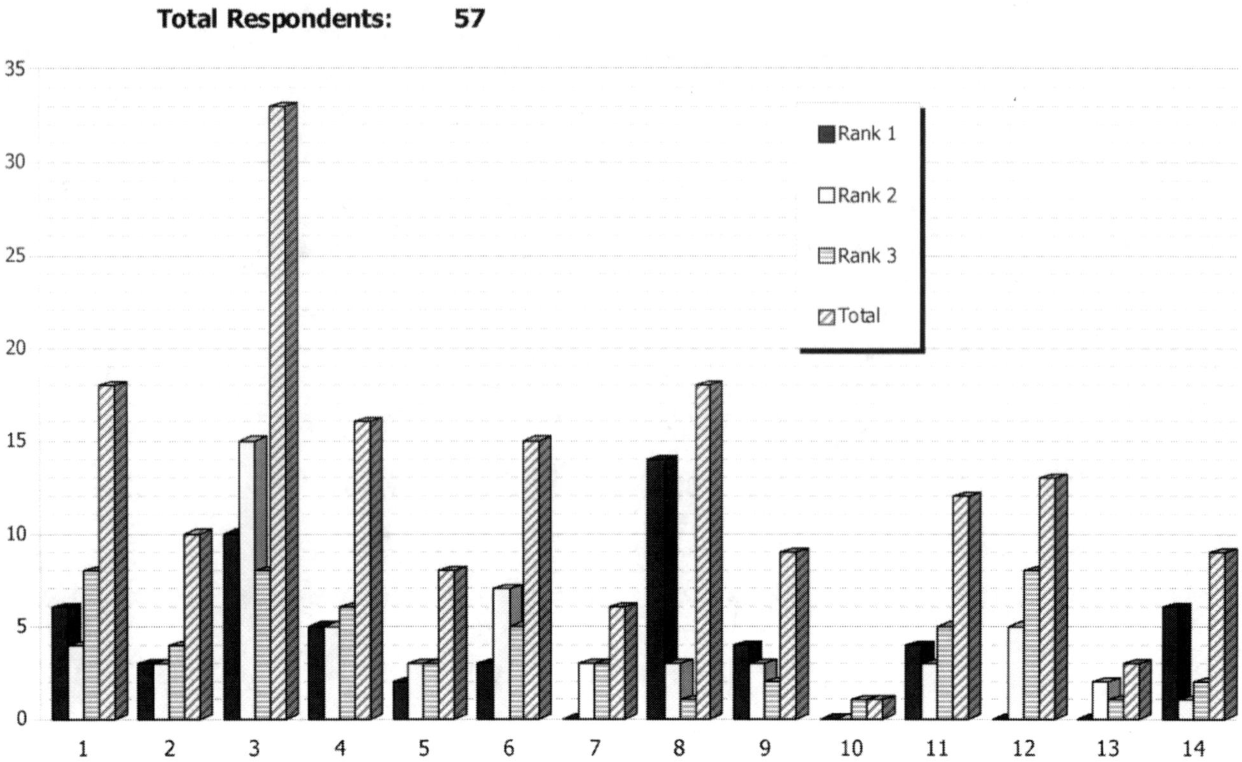

1 = Inlet tube sheet erosion
2 = Erosion of the quench exchanger outlet fitting
3 = Tube blockage
4 = Failure of the tube to tube sheet weld
5 = Other weld failures
6 = Refractory failures
7 = Water side fouling
8 = Water side corrosion
9 = Tube erosion
10 = Bowing of the tube sheet
11 = Distortion of the coke deflector
12 = Gasket or seal failures
13 = Failure of the outer tube in double pipe exchanger or outer sheel in shell and tube exchanger
14 = Gas inlet reducer (for linear exchangers only)

Comment: The most severe problem is tube blockage. The majority of the respondents (19 = 57.6%) are cracking only gaseous feedstocks. Surprisingly, even five respondents experience blockage that have linear quench exchangers (see Question 32). All of them list tube blockage under the three most severe failures (3 times rank "1"), Except for "bowing of tube sheet" all failure mechanisms are listed multiple times. Evident is that, if taking place, water side corrosion becomes a severe problem as it is mostly ranked "1".

Question 34: Please indicate which types of inspections you have performed on your quench exchangers. Mark all that apply.

	Choice	Response Percent	Response Total
1	Visual	84.4%	54
2	Boroscope	57.8%	37
3	Shell internal inspection	9.4%	6
4	UT on tube sheet	57.8%	37
5	Iris wall thickness check on tube sheet	14.1%	9
6	UT on gas outlet channels	15.6%	10
7	Hydro-test or pneumatic test	87.5%	56
8	RT radiography on tube sheet	9.4%	6
9	PT dye penetrant on tube to tube sheet welds	51.6%	33
10	MT magnetic particle on tube sheet	25.0%	16
11	Eddy current on tube sheet	6.3%	4
12	Hardness test	10.9%	7

Total Respondents:	64

Comments: 1 x Boroscope inspection after pickling of new (replaced) quench exchangers
1 x n/a

Question 35: Select the repairs that are performed. Mark all that apply.

Select the repairs that are performed. Mark all that apply.

	Choice	Response Percent	Response Total	Corrected *) Response Percent
1	Complete in-kind replacement of the exchanger	36.5%	19	28.4%
2	Weld overlay on the tube sheet	46.2%	24	35.8%
3	Weld overlay on the outlet fitting	15.4%	8	11.9%
4	Tube replacement	50.0%	26	38.8%
5	Partial tube replacement	42.3%	22	32.8%
6	Replacement of tube sheet segments	21.2%	11	16.4%
7	Comment	25.0%	13	19.4%

Total Respondents:	**52**
Total Non-Respondents:	**15**

Comments: 7 x no repairs, no repairs yet, no repairs in last five years or n.a.
6 x specific repairs not listed under choices 1-6

*) Note: It is assumed that those who did not respond, did not perform repairs, hence the corrected Response is based on 67 Total Respondents

Question 36: Please indicate what types of spares you are keeping for your primary quench exchanger

	Choice	Response Percent	Response Total
1	Complete quench exchangers	33.9%	21
2	Inlet chambers with refractory installed	50.0%	31
3	Tubes	69.4%	43
4	Spare tube sheet sections	16.1%	10
5	Coke impact devices	33.9%	21
6	Inlet reducer (for linear exchangers)	16.1%	10
7	Gaskets	67.7%	42
8	Comment	3.2%	2

Total Respondents:	**62**

Comments: 1 x complete double pipe segment also (for linear quench exchangers)
1 x no spares

Question 37: Have you ever replaced your complete quench exchangers due to "end-of-life" and if so please indicate the number of operating years prior to replacement? If you did replace your quench exchanger for reasons of a technology upgrade or because of operational upgrade please mark so and do not mark the age of the exchanger prior to replacement.

	Choice	Response Percent	Response Total
1	We replaced the quench exchangers due to technology upgrade	8.1%	5
2	Less than 5 years	1.6%	1
3	5-10 years	16.1%	10
4	10-20 years	17.7%	11
5	More than 20 years	9.7%	6
6	Never	46.8%	29

Total Respondents:	**62**

Question 38: What type of refractory do you use in your gas inlet chamber? Mark all that apply

	Choice	Response Percent	Response Total
1	Chemical bond	23.3%	10
2	Hydraulic bond high alumina (90% or higher)	11.6%	5
3	Hydraulic bond medium alumina (approx 50%)	20.9%	9
4	Needle reinforced refractory	30.2%	13
5	Anchored refractory	25.6%	11

Total Respondents: *)	**43**

Comments: 7 x n.a.
1 x 60% Al_2O_3 with wire mesh has been sorted into choices 3 + 4
1 x Thermbond has been sorted into Chemical bond

*) Note: Respondents that replied "n.a." are not counted

Question 39: How would you characterize your convection arrangement?

	Choice	Response Percent	Response Total
1	Natural draft with a stack for each furnace or set of furnaces	9.5%	6
2	Induced draft fan for each furnace or set of furnaces	74.6%	47
3	Flue gas ducted from various furnaces to a common stack	11.1%	7
4	Comment	4.8%	3
	Total Respondents:		**63**

Comments: 1 x 50% of furnaces to common stack, 50% to stack per set of furnaces
2 x balanced draft

Question 40: What is your average stack temperature?

	Choice	Response Percent	Response Total
1	Below 300 °F (149 °C)	21.7%	13
2	300 - 400 °F (149 - 204 °C)	45.0%	27
3	400 - 500 °F (204 - 260 °C)	20.0%	12
4	500 - 600 °F (260 - 315 °C)	8.3%	5
5	Above 600 °F (315 °C)	5.0%	3
	Total Respondents:		**60**

Question 41: Do you experience fouling in the convection section?

	Choice	Response Percent	Response Total
1	Internal	11.9%	7
2	External	42.4%	25
3	Internal and external	28.8%	17
4	No fouling	16.9%	10
	Total Respondents:		**59**

Question 42: How often do you clean the tube/fin externals of the convection section?

	Choice	Response Percent	Response Total
1	1 - 3 years	6.6%	4
2	4 - 5 years	16.4%	10
3	6 - 10 years	21.3%	13
4	Never	54.1%	33
5	Comment	1.6%	1

Total Respondents:	**61**

Comments: 2 x "n.a." are sorted into "Never"
1 x "once since start-up"

Question 43: What is the average effect on stack temperature after cleaning of the convection section?

	Choice	Response Percent	Response Total
1	0 - 50 °F (0 - 28 °C)	61.5%	16
2	50 - 100 °F (28 - 56 °C)	30.8%	8
3	100 - 150 °F (56 - 83 °C)	3.8%	1
4	More than 150 °F (83 °C)	3.8%	1

Total Respondents: *)	**26**

*) Note: Respondents that never performed convection cleaning, are not counted

Question 44: Do you use soot blowers?

	Choice	Response Percent	Response Total
1	Yes	5.2%	3
2	No	94.8%	55

Total Respondents:	**58**

Question 45: What is the average effect on stack temperature after sootblowing?

	Choice	Response Percent	Response Total
1	0 - 50 °F (0 - 28 °C)	66.7%	2
2	50 - 100 °F (28 - 56 °C)	0.0%	0
3	100 - 150 °F (56 - 83 °C)	0.0%	0
4	More than 150 °F (83 °C)	0.0%	0
5	Comment	33.3%	1

Total Respondents:	3

Comment: do not measure

Question 46: Where do you experience fouling on the inside of the convection tubes?

	Choice	Response Percent	Response Total
1	Nowhere	48.1%	26
2	In the boiler feed water preheat coil	0.0%	0
3	In the steam superheat coil	0.0%	0
4	In the HC preheat coil	20.4%	11
5	In the HC + Steam coil	27.8%	15
6	In the shock rows	3.7%	2

Total Respondents:	54

Comments: 1 x "in crossover only", 1 x "not significant" are sorted into "Nowhere",
3 x "in the HC + Steam coil, especially at the mixing point/zone of steam and feed"

Question 47: What metallurgy do you use in your convection section?

	Not applicable	Incoloy	Stainless steel	Low alloy steel	Carbon steel	Response Total
Feed preheat coil	3.8% (2)	5.7% (3)	9.4% (5)	11.3% (6)	69.8% (37)	**53**
Boiler feed water coil	17.3% (9)	1.9% (1)	1.9% (1)	13.5% (7)	65.4% (34)	**52**
Dilution steam superheat coil	12.0% (6)	12.0% (6)	22.0% (11)	32.0% (16)	22.0% (11)	**50**
High pressure steam superheat coil	16.7% (8)	4.2% (2)	27.1% (13)	52.1% (25)	0.0% (0)	**48**
Mix preheat coil	1.9% (1)	26.9% (14)	46.2% (24)	19.2% (10)	5.8% (3)	**52**
Shock tubes	19.1% (9)	51.1% (24)	27.7% (13)	2.1% (1)	0.0% (0)	**47**

Question 48: Please indicate whether you have experienced the following failures on your convection tubes:

	Yes	No	Comment		Response Total
Tube ruptures	59.6% (34)	40.4% (23)	0% (0)	1 comment: corrosion with carburization sorted into "Yes"	**57**
Erosion	38.2% (21)	50.9% (28)	10.9% (6)	6 comments: unknown 2 comments: "on bends" sorted into "Yes"	**55**
Tube support failure	66.1% (39)	33.9% (20)	0% (0)		**59**
Loss of extended suface (fins or studs)	38.9% (21)	55.6% (30)	5.6% (3)	3 comments: n.a.	**54**
Corrosion	46.0% (23)	54.0% (27)	0% (0)		**50**
Stress corrosion cracking	24.5% (12)	75.5% (37)	0% (0)		**49**
Premature stress failure	12.2% (6)	87.8% (43)	0% (0)	1 comment: manufacturer defect sorted into "Yes"	**49**

Question 49: How often do you clean the inside of the convection tubes?

	Choice	Response Percent	Response Total
1	1 - 3 years	12.5%	8
2	3 - 5 years	4.7%	3
3	5 - 10 years	7.8%	5
4	Never	68.8%	44
5	Comment	6.3%	4

Total Respondents:	**64**

Comments: 1 x "once in plant life time only", 1 x "occacionally", 1 x "when problems are detected", 1 x "decoking"

Question 50: What type of burner do you have?

	Choice	Response Percent	Response Total
1	Side wall burners	50.8%	32
2	Floor burners	14.3%	9
3	Both side wall and floor burners	34.9%	22

Total Respondents:	**63**

Question 51: What type of burners do you have in your furnace? Please mark all that apply if you have various kinds

	Choice	Response Percent	Response Total
1	Pre-mix radiant wall burner	67.7%	42
2	Pre-mix radiant wall burner with fuel staging	9.7%	6
3	Raw gas burner	9.7%	6
4	Staged fuel burner	21.0%	13
5	Staged air burner	1.6%	1
6	Burners with internal flue gas recirculation	9.7%	6
7	Burners with external flue gas recirculation	0.0%	0
8	Burners with steam injection	0.0%	0
9	Comment	3.2%	2

Total Respondents:	62

Comments: 1 x steam injection installed, 1 x vortex burners (radiant wall) also

Question 52: How many burners are installed per furnace?

	Choice	Response Percent	Response Total
1	Less than 20	1.6%	1
2	21 - 40	7.9%	5
3	41 - 60	7.9%	5
4	61 - 100	33.3%	21
5	101 - 150	25.4%	16
6	more than 150	23.8%	15

Total Respondents:	63

Question 53: What type of fuels are fired?

	Choice	Response Percent	Response Total
1	Gaseous fuel only	100.0%	63
2	Liquid fuels only	0.0%	0
3	Gaseous and liquid fuels	0.0%	0

Total Respondents:	63

Question 54: Please identify other burner options you may have

	Yes	No		Response Total
UV scanners	30.9% (17)	69.1% (38)		**55**
Ionisation rods	3.6% (2)	96.4% (53)		**55**
Other flame supervision	3.6% (2)	96.4% (54)	1 comment: flame supervision by pilots sorted into "Yes"	**56**
Automated remote ignition	21.4% (12)	78.6% (44)		**56**
Automated burner damper	20.0% (11)	80.0% (44)		**55**
Forced draft burners	20.0% (11)	80.0% (44)		**55**
Preheated air	12.7% (7)	87.3% (48)		**55**
Gas turbine integration	3.6% (2)	96.4% (53)		**55**

Question 55: Please identify the frequency of burner maintenance due to following causes

	More than once per week	Once per week	Once per month	Once per year	Less than once per year	Never	Response Total
Tip plugging	8.1% (5)	8.1% (5)	12.9% (8)	53.2% (33)	16.1% (10)	1.6% (1)	**62**
Tip damage	1.6% (1)	0.0% (0)	11.3% (7)	53.2% (33)	30.6% (19)	3.2% (2)	**62**
Flashback	5.2% (3)	0.0% (0)	10.3% (6)	19.0% (11)	19.0% (11)	46.6% (27)	**58**
Burner tile cracks	0.0% (0)	0.0% (0)	0.0% (0)	41.0% (25)	44.3% (27)	14.8% (9)	**61**
Damper problems	0.0% (0)	0.0% (0)	1.7% (1)	16.7% (10)	43.3% (26)	38.3% (23)	**60**
Pilot problems	0.0% (0)	4.8% (2)	2.4% (1)	19.0% (8)	28.6% (12)	45.2% (19)	**42**
Problems with UV scanners	2.7% (1)	0.0% (0)	16.2% (6)	13.5% (5)	13.5% (5)	54.1% (20)	**37**
Problems with ionisation rods	0.0% (0)	0.0% (0)	4.0% (1)	4.0% (1)	0.0% (0)	92.0% (23)	**25**
Failure due to mis-assembly of the burner	0.0% (0)	0.0% (0)	0.0% (0)	9.3% (4)	39.5% (17)	51.2% (22)	**43**

Note: Comments that this is not applicable / not installed are not counted

Question 56: How often do burner failures require a furnace shutdown? Again please answer on a per furnace basis. If you have 10 furnaces and you have to shut-down 5 furnaces on average each year your answer is once per 2 years

	Choice	Response Percent	Response Total
1	Once per 0 - 3 months	0.0%	0
2	Once per 4 - 12 months	0.0%	0
3	Once per 1 - 5 years	3.2%	2
4	Never	96.8%	60

Total Respondents:	**62**

Question 57: Please indicate which type of refractory you use in the radiant section of your furnace

	castable	brick	ceramic fiber	Response Total
floor	61.3% (38)	58.1% (36)	0.0% (0)	62
lower side walls (near floor burners)	20.0% (12)	76.7% (46)	8.3% (5)	60
upper side walls	6.8% (4)	62.7% (37)	32.2% (19)	59
arch	32.7% (19)	32.7% (19)	53.4% (31)	58

Question 58: If you use ceramic fiber blankets, please indicate which problems you encounter

	yes	no	Response Total
Shrinkage	80.5% (33)	19.5% (8)	41
Erosion	63.4% (26)	36.6% (15)	41
Failure of Tie-back anchors	59.6% (28)	40.4% (19)	47

Question 59: Please indicate which problems you encounter with brick refractory

	Yes	No		Response Total
Collapse	44.0% (22)	56.0% (28)	2 comments: "due to anchor failure" sorted into "Yes"	50
Breakage	79.7% (47)	20.3% (12)	6 comments: "partially distributed" sorted into "Yes"	59
Erosion due to burner impingement	49.0% (25)	51.0% (26)		51
Erosion due to decoke effluent impingement	34.9% (15)	65.1% (28)		43
Failure of tie-back anchors	57.1% (32)	42.9% (24)		56

Note: Comments that this is not applicable are not counted

Question 60: When performing repairs to brick refractory, what typical measures are used?

	Choice	Response Percent	Response Total
1	Localized areas are patched/repaired	65.4%	34
2	Wall sections are replaced as an assembly	26.9%	14
3	Wall sections are replaced and anchoring systems are renewed	42.3%	22
4	Don't know	0.0%	0
5	Comment	3.8%	2

Total Respondents: 52

Comments: 2 x n.a.
1 x only small repairs as needed sorted into choice 1

Question 61: What do you feel is the typical life expectancy of the following refractories?

	< 2 years	3-5 years	5-10 years	10-15 years	More than 15 years		Response Total
Ceramic blanket	5.3% (2)	23.7% (9)	39.5% (15)	21.1% (8)	10.5% (4)	1 comment: "as long as bricks stay intact" sorted into "5-10 years" as this is the expected life for bricks from the repondent	38
Ceramic Modules	0.0% (0)	10.0% (4)	37.5% (15)	40.0% (16)	12.5% (5)		40
Castable	0.0% (0)	4.2% (2)	33.3% (16)	20.8% (10)	41.7% (20)		48
Brick	0.0% (0)	2.0% (1)	28.0% (14)	34.0% (17)	36.0% (18)		50

Question 62: Do you perform thermography on the furnace casing to identify refractory issues?

	Choice	Response Percent	Response Total
1	Yes, on a regular scheduled basis	20.0%	11
2	Yes, prior to a maintenance turnaround	5.5%	3
3	Yes, on a as needed basis	40.0%	22
4	No	34.5%	19
5	Comment	0.0%	0

Total Respondents:	55

Question 63: Following a typical radiant coil replacement, how much time is required for refractory repairs?

	Choice	Response Percent	Response Total
1	2 additional days	10.9%	6
2	3-5 additional days	25.5%	14
3	6-10 additional days	32.7%	18
4	11-15 additional days	10.9%	6
5	More than 15 days	7.3%	4
6	Comment	12.7%	7

Total Respondents:	55

Comments: 3 x no additional days
2 x one additional day max.
1 x refractory is not repaired during coil maintenance
1 x no experience

Question 64: On average how many days per year are your furnaces out of operation due to maintenance activities?

	Choice	Response Percent	Response Total
1	0 - 6 days	17.9%	10
2	7 - 12 days	30.4%	17
3	13 - 18 days	17.9%	10
4	19 - 24 days	17.9%	10
5	24 - 30 days	8.9%	5
6	More than 30 days	7.1%	4
	Total Respondents:		**56**

Question 65: What are the 3 most important causes for furnace outage? Please use the primary reason for bringing the furnace off-line and disregard the other activities that may occur because of the convenience of doing them during the furnace outage.

Total Respondents: 56

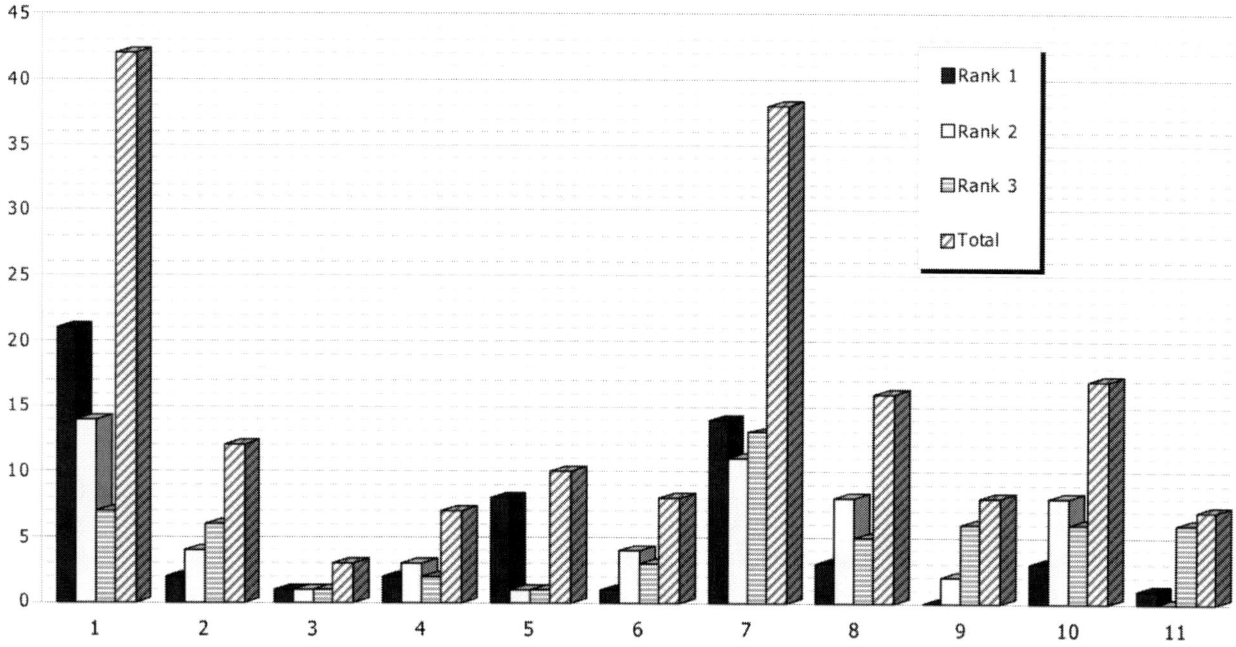

1 = Radiant coils (recoils, coil repair, coil plugging)
2 = Refractory
3 = Burner System (burner, damper, tile)
4 = Fans
5 = Decoke piping systems
6 = Other piping systems
7 = Quench exchanger cleaning
8 = Quench exchanger repair
9 = Steam system (drum, relief devices etc.)
10 = Instrumentation
11 = Other

Comment: There is clear evidence that two areas dominantly cause furnace outages: The radiant coils and cleaning of quench exchangers.

Thanks

Most important to the success of this survey was the support and participation of all the operating companies. The survey team would like to thank all the engineers involved in responding to the survey questionnaire.

We also thank all the members of the Maintenance & Reliability sub-committee for their review of the questionnaire as well as the survey report. A special thanks goes to the Maintenance Group of the European Ethylene Producers Committee. Their support in distribution of the survey was critical.

Steam Turbine Extraction APC Application

Eric S. Matthews

Control Systems Engineering Group

Equistar Chemicals, LP, a LyondellBasell company

Abstract

Significant challenges exist when trying to operate the steam system of a major chemical complex in the most efficient manner. One challenge is the constantly changing steam demand from different operating units. Each unit has its own level of operational autonomy which varies the steam demand. Another challenge is the distribution of responsibility for the operation of major steam-intensive equipment across multiple board technicians. For example, the boilers may be the responsibility of Board-A, while the two largest steam turbines are separately operated by Board-B and Board-C. It is difficult for technicians spread across multiple boards to manually control a complicated steam system in the most efficient manner. More than likely, the board technicians find an operating region within their comfort zone, which is a measurable distance from maximum efficiency. It is increasingly important to implement automated control strategies to overcome these challenges as utility costs continually increase. Equistar's Chocolate Bayou Chemical Complex has developed an Advanced Process Control (APC) application to operate the complex's steam system more efficiently by maximizing the extraction steam flows from the two largest turbines while monitoring key constraints in the turbines and steam letdown stations. Ultimately, the application reduces the cost of steam for all units inside the complex.

Introduction

Operating costs for U.S. chemical manufacturers have increased substantially over several years. The variable cost of steam produced from natural gas is of considerable importance. The spot price for natural gas has varied between US$5 and US$13 per million BTUs between 2004 and 2008 [(1)]. With costs varying so widely, manufacturers are focused on decreasing energy consumption. Advanced Process Control (APC) applications are well known for delivering benefits from energy savings and increased production. Equistar's Steam Turbine Extraction APC Application directly targets savings in the steam distribution system with the potential for increased plant throughput.

Operating Challenges

Variations in Steam Demand

Opportunities for energy savings within the steam system at Equistar's Chocolate Bayou Chemical Complex (the Complex) may have similarities with other major chemical complexes. The Complex consists of a liquids-cracking olefins unit, with gasoline separation, an aromatics unit and C4 processing units. Excluding the olefins unit, there are six different operating units with substantial operational autonomy. These six units can individually shutdown or start-up for a variety of reasons while the olefins unit continues to run. Aside from the shutdown of whole units, there are other activities that have a measurable impact on the Complex's steam system. Activities, such as pyrolysis furnace decokes, start-up or shutdown of pump turbines and steam to the flares, are also of consequence to the steam system. Operating adjustments are required for each change in steam demand in order to maintain maximum energy efficiency.

Distribution of Operating Responsibilities

The control of all operating units at the Complex is performed in a single, centralized control room. There are multiple board operators in the control room controlling the olefins unit as well as the other units. Communication is good among board technicians at the Complex. However, the important handles for controlling the important steam system variables are distributed throughout the boards in the control room. The Cracked Gas Compressor (CGC), the Propylene Refrigeration Compressor (PRC) and the boiler/utilities are controlled by different board technicians.

Description of Steam System

Nominal 1250 Psig Steam Header

The highest level of steam in the Complex is nominally 1250 psig. The pyrolysis furnaces and the boilers produce steam at 1250 psig. The 1250 psig steam has only two users: CGC and PRC turbines. Both compressors are operated on governor control and in suction pressure control. The suction pressures of both compressors are utilized in an APC and real-time optimization control scheme primarily involving the process side of the compressors.

Nominal 450 Psig Steam Header

The 450 psig steam header is fed by steam letdown from the 1250 psig header. Additionally, the CGC extracts steam into the 450 psig header. The flow of extraction steam from the CGC is determined by the pressure of the cavity inside the CGC. An increased extraction pressure in the cavity forces more steam out of the CGC turbine into the 450 psig steam header. More simply explained, increased extraction pressure means increased extraction flow from the turbine into the header.

Nominal 150 Psig Steam Header

The 150 psig steam header is fed by steam letdown from the 450 psig header. Additionally, the PRC extracts steam into the 150 psig header. The flow of extraction steam from the PRC is determined in a similar manner to the CGC: increased extraction pressure means increased extraction flow from the turbine into the header.

Nominal 50 Psig Steam Header

The 50 psig steam header is fed by steam letdown from the 150 psig header. There are two pressure letdown stations designated PIC050A and PIC050B. The letdown station primarily used is PIC050A. PIC050B is normally closed.

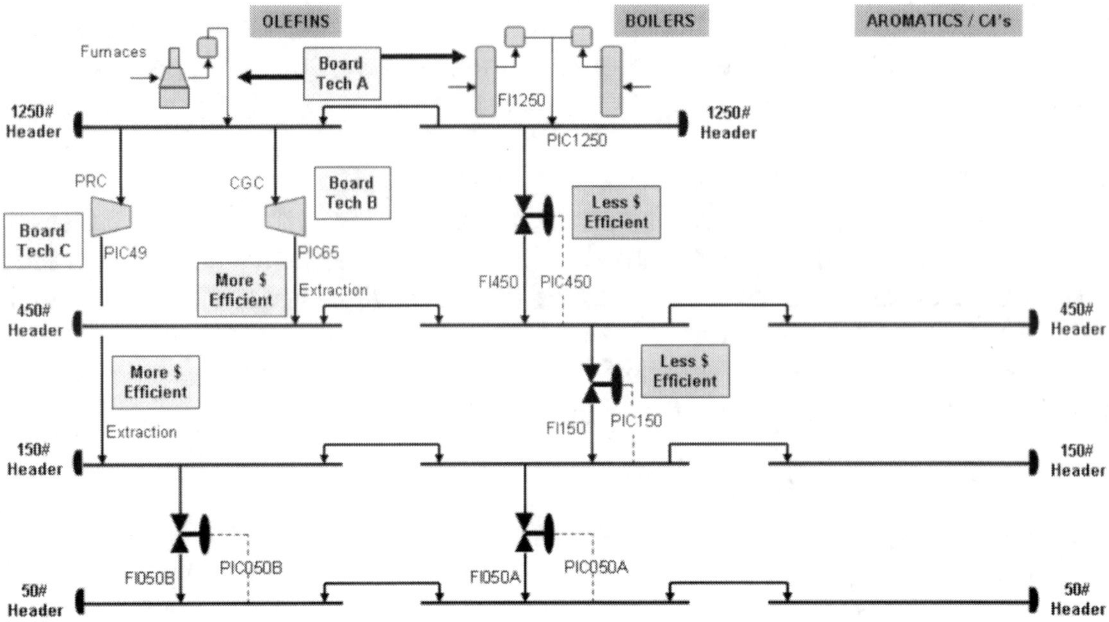

Complex Steam System Overview

Operating Objectives

Maximize Extraction Steam Flow

There are several important operating objectives to consider. When the compressor turbine is not against a constraint because of high plant throughput, it is more energy efficient to letdown steam to the 450 and 150 psig headers via turbine extractions than to letdown the steam through the throttle valves. Use the CGC as an example. When the 1250 psig steam is routed through the CGC turbine and extracted into the 450 psig header, work is accomplished inside the turbine. When 1250 psig steam is simply throttled down to 450 psig, no work is accomplished. Even more energy is wasted when 1250 psig steam is throttled to 450 psig and again throttled to 150 psig to satisfy the user demand at that level. So maximizing extraction steam from both the CGC and PRC turbines is an important operating objective.

Maintain Minimum Flow Through Throttle Valves

Another important operating objective is to maintain a minimum flow rate through the throttle valves at the 450 and 150 psig levels. The minimum flow can also be represented by a minimum valve opening. If the throttle valve is allowed to close for a significant period of time, there could be several adverse effects on steam quality. Header pressure would not be controlled. Desuperheater temperature controllers would not control well. Condensate could accumulate at the throttle valve. So it is important to maintain a minimum flow through the 450 and 150 psig throttle valves.

Maintain Minimum Boiler Load

The application should ultimately reduce the boiler load. A minimum boiler load is specified to prevent boiler instability and boiler trips.

Prevent Saturation of Turbine Extraction Valves

The turbine extraction valves, also known as the V2 racks, should be maintained with a minimum opening. Steam to the backend of the turbine helps cool the turbine.

APC Application Design

Constraints (CV's)

The constraints in the APC application are listed below.

- Boiler load, FI1250.PV
- Flow through 1250-to-450 psig throttle valve, FI450.PV
- Opening of 1250-to-450 psig throttle valve, PIC450.OUT
- Flow through 450-to-150 psig throttle valve, FI150.PV
- Opening of 450-to-150 psig throttle valve, PIC150.OUT
- CGC turbine extraction valve (V2 rack), PIC65.OUT
- PRC turbine extraction Valve (V2 rack), PIC49.OUT

Manipulated Variables (MV's)

The manipulated variables in the APC application are listed below.

- CGC turbine extraction pressure setpoint, PIC65.SP
- PRC turbine extraction pressure setpoint, PIC49.SP
- 150-to-50 psig throttle valve pressure setpoint, PIC050B.SP

Custom Programming for Extraction Pressure Controllers

In order to utilize the CGC and PRC extraction pressures as APC MV's, custom programming was required in the DCS. P-I-D control of the extraction pressures is not performed in the DCS. P-I-D control of the extraction pressures is performed in the compressor governors. Inside the DCS, the extraction pressure setpoint (SP), process variable (PV) and output (OP) can be viewed only. The SP, PV and OP are simply values read from the governor. A traditional P-I-D control block did not exist in the DCS to which the operator or an APC application could write a SP. Operator control of the extraction pressure SP is performed by clicking on DCS graphical buttons. There is a graphical button to raise the SP for each compressor and a graphical button to lower the SP. This graphical button sends a "1" or "0" to the respective governor, where the SP is changed. The longer the graphical button is depressed, the larger the change in SP. However, the exact amount of change of the SP can not be controlled precisely. It is often an iterative process to find the correct SP.

In order to implement the APC application, a traditional control block is required to which an operator or the APC application can write a SP. Custom programming was performed by the project team in the native DCS language. The custom program utilized a newly created P-I-D block for each extraction pressure. The P-I-D block accepted an operator/APC application entered value as the SP. The PV for the new P-I-D block was the SP value read from the governor. The custom program automatically "pressed" the graphical buttons so that the SP read from the governor matched the desired value entered by the operator or APC application. The programming is unique and the APC application could not be implemented without it.

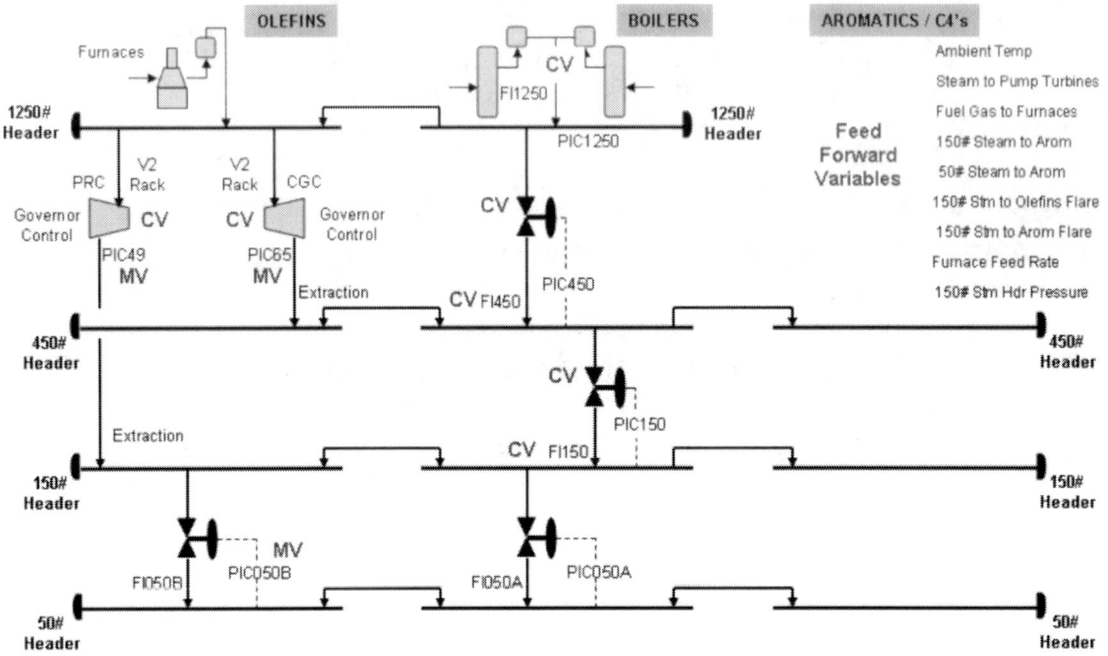

Design of Steam Turbine Extraction APC Application

Disturbance Variables

The disturbances that have an impact and can be quantified are mentioned below.

- Ambient temperature
- Steam flow to pump turbines (ones that have flow measurements)
- Fuel gas flow to furnaces
- Furnace feed rate
- 150 psig steam flow to Aromatics/C4 Units
- 50 psig steam flow to Aromatics/C4 Units
- 150 psig steam flow flare at Olefins Unit
- 50 psig steam flow to flare at Aromatics/C4 Units
- 150 psig steam header pressure setpoint, PIC050A.SP

APC Application Results

The APC Application designed to maximize extraction steam flow from the turbines of the CGC and PRC was commissioned on November 6, 2008. The Complex continued to operate normally until December 18, 2008 when the Olefins Unit was temporarily idled. From November 6 to December 18, the application was online more than 99% of the time. In order to illustrate the benefits delivered by the application, averaged hourly data was collected from January 1, 2008 to December 18, 2008. Unrepresentative data was removed, such as plant upsets and shutdowns (i.e., Hurricane Ike). All graphs were developed using a uniform approach described below.

- Process data prior to APC commissioning are plotted using a darker blue color
- Process data after APC commissioning are plotted using a lighter blue color
- A line representing the average is plotted through the pre- and post-APC data.
- The average line plotted through pre-APC data is designated 100% of normal.

Reduced Letdown Flow Through Throttle Valves

The normal, average letdown flow through the 1250-to-450 psig throttle valve prior to APC is designated 100% of normal. After APC implementation, the letdown flow has been reduced to 50% of normal.

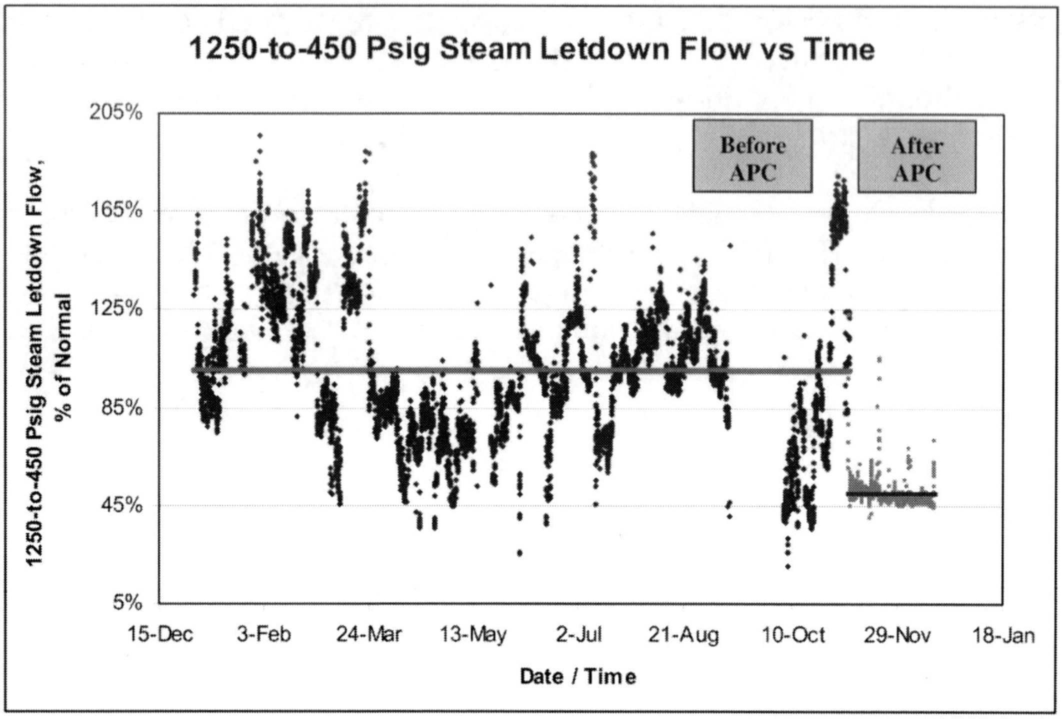

1250-to-450 Psig Steam Letdown Flow (FI450.PV) versus Time

The normal, average letdown flow through the 450-to-150 psig throttle valve prior to APC is designated 100% of normal. After APC implementation, the letdown flow has been reduced to 48% of normal (a reduction of 52%).

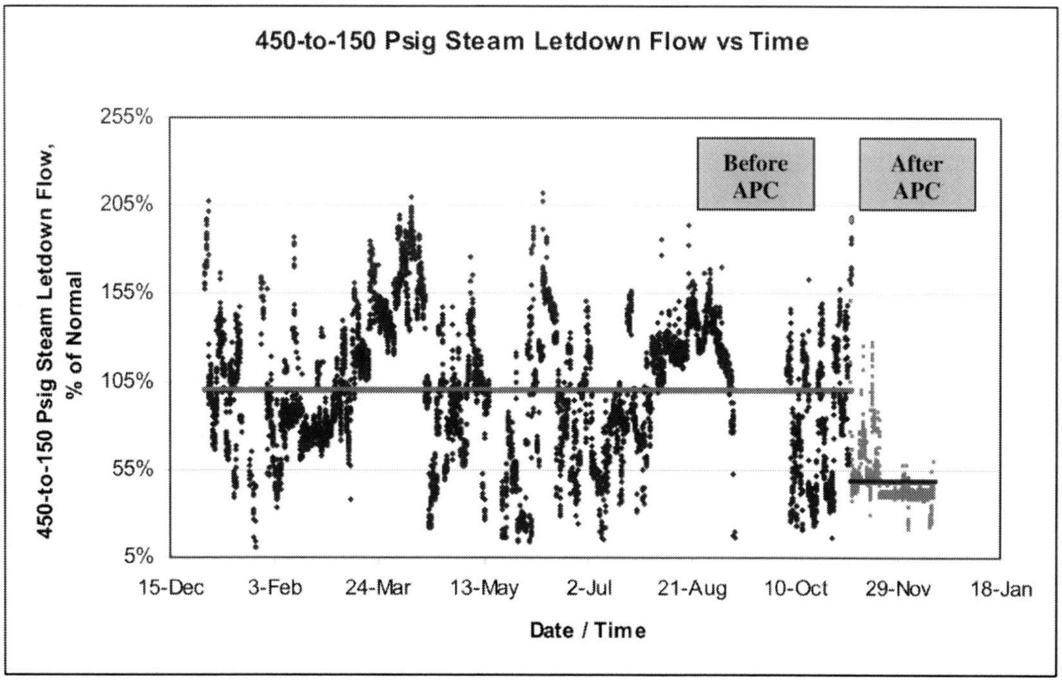

450-to-150 Psig Steam Letdown Flow (FI150.PV) versus Time

Reduced Boiler Load

The normal, average boiler load prior to APC implementation is designated 100% of normal. After APC implementation, the boiler load has averaged 93.5% of the normal load. This is a 6.5% reduction in boiler load.

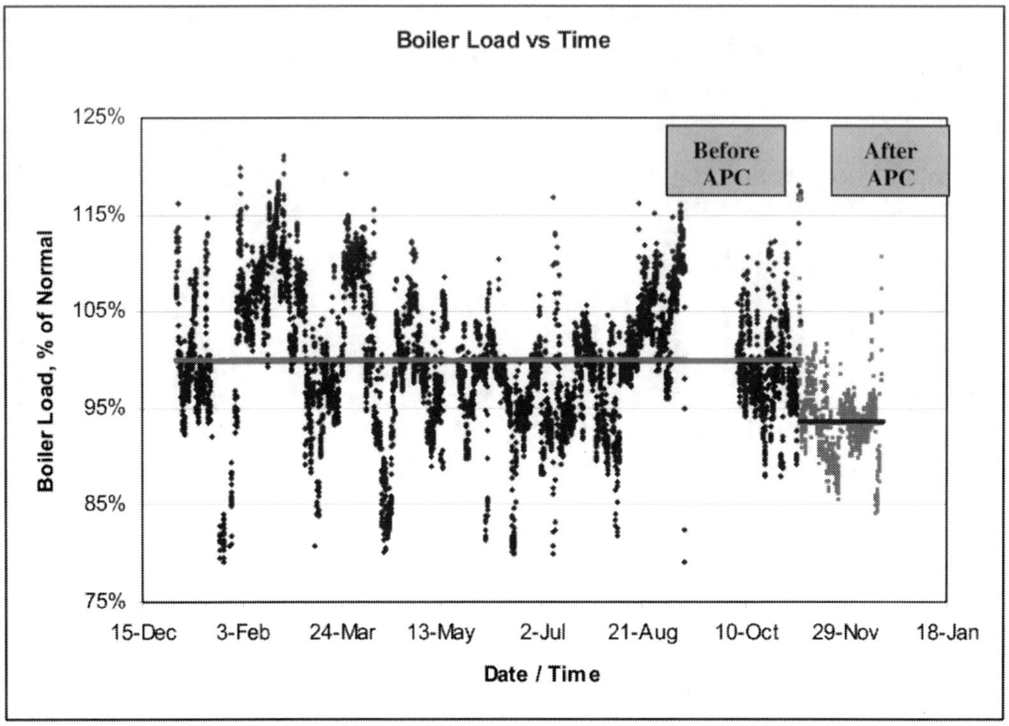

Boiler Load (FI1250.PV) versus Time

Availability of Benefits

The Steam Turbine Extraction APC Application is explicitly designed to reduce energy consumption. It also has the potential to increase plant throughput, which will be discussed later. Comparable benefits are available to the APC application regardless of whether the operating units are running at high or low rates. As is true in all manufacturing facilities, the board technician has a comfort region in which he prefers to operate. For the Complex steam system, the comfort factor is introduced to the letdown flows of the throttle valves. The technician prefers to have a higher letdown flow through the throttle valve so it does not have to be so closely monitored. The technician's comfort factor is utilized regardless of whether the plant throughput is high or low. Fortunately, there is very little non-linearity in this system so the models remain valid at essentially all plant rates or boiler rates. Therefore, the application's energy saving benefits are comparable under all process operating conditions.

Strategy for Implementing Costs in the Linear Program

In order to configure the linear program (LP), costs are required for the manipulated variables (MV's). Based on the step test data and knowledge of the operation, it was concluded that preferentially maximizing the PRC extraction pressure provides more benefit than maximizing the CGC extraction pressure. As revealed in the APC model relationships, PRC extraction to the 150 psig header reduces both throttle valve letdown flows to the 450 and 150 psig headers. CGC extraction to the 450 psig header only reduces the throttle valve letdown flow to the 450 psig header. A partial representation of the model matrix is provided below.

Table 1

Manipulated Variables (MV's)	***Controlled Variables (CV's)***	
	FI450.PV, 1250-to-450 Letdown Flow	FI150.PV, 450-to-150 Letdown Flow
PIC65.SP, CGC Extraction Pressure SP Positive move in MV	- Negative response in CV	(empty) No Response in CV
PIC49.SP, PRC Extraction Pressure SP Positive move in MV	- Negative response in CV	- Negative response in CV

Note: PRC extraction pressure is maximized first. The CGC extraction pressure is maximized second.

After the APC application was commissioned, the PRC extraction pressure has operated higher on average (from 160 to 166 psig or 6 psig higher). This is consistent with the LP strategy that was implemented. The CGC extraction pressure has operated lower on average after the APC strategy was commissioned (from 471 to 468 psig or 3 psig lower). Prior to the APC application, preferentially considering one extraction pressure over the other had not been implemented. This can be seen in the following two plots of the extraction pressures.

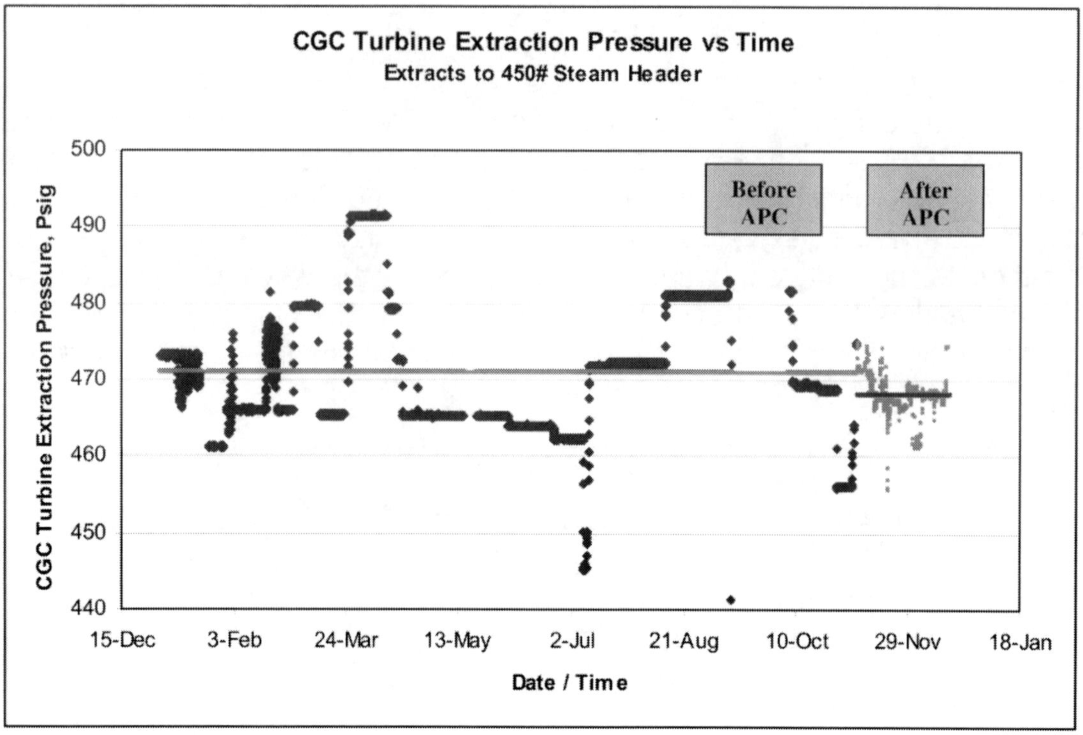

CGC Turbine Extraction Pressure (PIC65.SP) versus Time

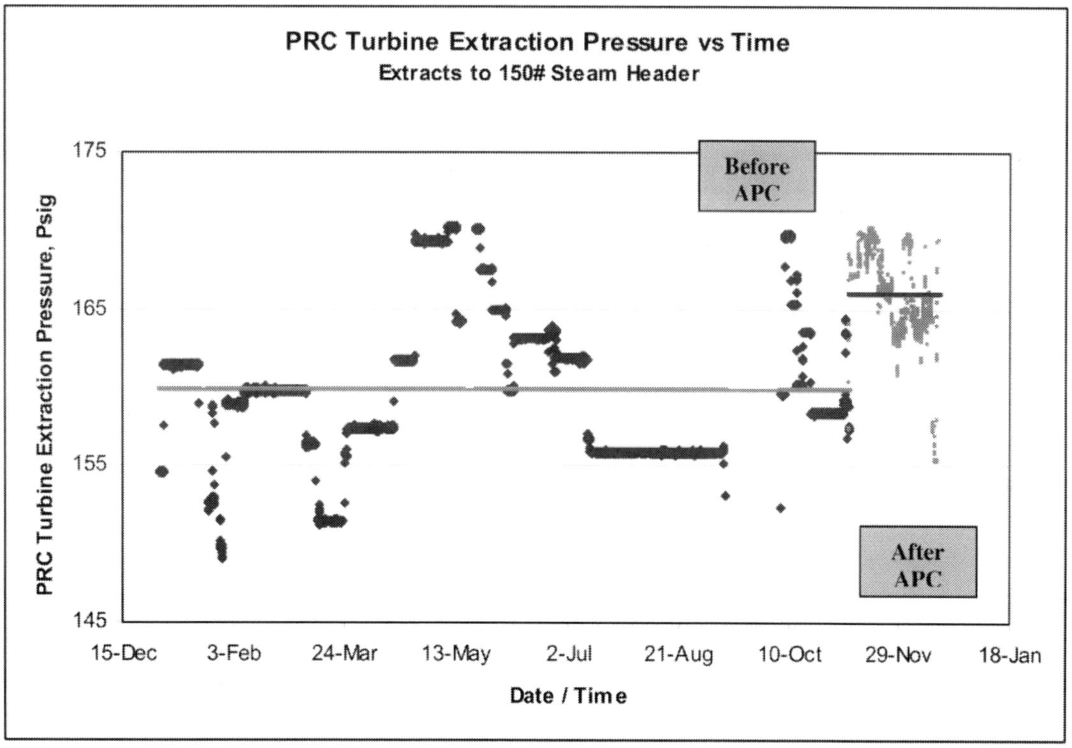

PRC Turbine Extraction Pressure (PIC49.SP) versus Time

Potential for Increasing Plant Throughput

The premise of this publication is that it is more efficient to distribute steam to lower level users via turbine extraction rather than through throttle valves as long as the turbine has spare capacity. However, maximum turbine power is generated by sending more steam to the condensing section. If the turbine is against its capacity limit and throughput becomes more important than efficiency, then it may not be possible to extract as much steam. This operating case must be considered in the context of this APC application.

The PRC on the Complex's olefins unit is considered to have more than sufficient capacity – it is oversized. The CGC, however, has limited plant throughput by operating at maximum speed. With this understanding of unit equipment and constraints, it can be recognized that the APC Application has the potential to directionally increase plant throughput. The PRC turbine extraction pressure is maximized first. Maximizing the extraction increases the efficiency of the steam system, but borrows from the PRC turbine capacity. This is acceptable in the case of the PRC because it has spare capacity. The CGC turbine extraction pressure is maximized second. And the data shows that the APC application has actually lowered the extraction pressure of the CGC. This implies that more steam is sent to the condensing section of the turbine creating marginally more capacity in the CGC. It can be estimated that the efficiency achieved by the APC application has not sacrificed throughput. Instead, there is the probability that marginally more plant capacity has been created in addition to efficiency. This hypothesis will be studied when the Complex is operating at full rates again.

Planned Enhancements

Utilization of Secondary 150-to-50 Psig Letdown (PIC050B)

Under most operating scenarios, the 150-to-50 psig steam letdown station, designated PIC050B, is closed. Pressure control for the 50 psig header is performed with PIC050A. PIC050A is frequently operated at a fixed pressure that is as low as possible, but still satisfies the weakest reboiler on the header. Periodically, the PRC turbine extraction pressure reaches its upper MV limit while the 450-to-150 psig letdown flow is not at its minimum. This means that more steam can be extracted from the PRC turbine into the 150 psig header if more driving force (differential pressure) can be created. Additional differential pressure can be created by (a) further increasing the PRC turbine extraction pressure or (b) lowering the 150 psig header pressure. Raising the PRC turbine extraction pressure is not an option because it is already at its upper MV limit. Historical operating data reveals that opening PIC050B lowers the pressure in that particular area of the 150 psig header. Physically in the field, this part of the header is also where the PRC turbine extraction steam ties into the header. Therefore, additional differential pressure is created so additional steam flow is extracted from the PRC turbine. This strategy utilizing the secondary header pressure controller will be employed when valve mechanical issues are resolved with PIC050B.

Introduce More Turbines as Disturbance Variables

The Complex has many pump turbines using 450 psig steam. These pump turbines exhaust into the 150 or 50 psig headers. The vast majority of these pump turbines do not have a flow measurement for steam to the turbines. While operating near the minimum letdown flows, taking one of these pump turbines offline creates a significant, but unmeasured disturbance. The disturbance causes the 450-to-150 psig steam letdown flow to swing low. A vector was manufactured to represent the time of the pump switch and the amount of steam flow used by its turbine. Currently, relationships exist for major pump turbines that introduce a significant disturbance into the system. If flow meters are installed on pump turbines, then these measurements can readily become Disturbance Variables. The existing relationships that have been built can be incorporated into the online model.

References

1. Federal Reserve Bank of St. Louis website, www.research.stlouisfed.org, January 2009.

PETRO 2 ETHYLENE PLANT APC APPLICATIONS

Brian O'Rourke
Principal APC Engineer
Westlake Chemical Corporation
Kenneth L. Praprost
APC Engineer
Matrikon International, Inc.

Abstract: The Petro 2 ethylene plant produces polymer-grade ethylene, chemical-grade propylene, and by-products from an ethane/propane feedstock. Many APC applications were commissioned several years ago, including MPC controllers for each of the five furnaces plus controllers for hot and cold side fractionation. Over the years, the performance and benefits of the APC applications had deteriorated and some controllers were no longer used. After the plant turnaround in 2007, a new project was launched to revive the APC applications. Matrikon software tools were purchased for monitoring and tuning PID loops, for monitoring APC controllers, and for automated step testing and modeling. These tools were utilized to upgrade APC models for the C2 Splitter, C3 Fractionator, Demethanizer, Deethanizer, and the five furnaces. The APC applications have been successfully re-commissioned and performance monitoring tools indicate high online service factors with improved performance. Severity controls have also been re-commissioned with new models and the addition of feedback updates from online analyzers. The APC applications allow the plant to maximize production when market conditions are favorable and to reduce energy costs.

1. Introduction

In the last few decades, APC technology has created tremendous value for the refining and petrochemical industries. An APC controller can push the process unit to a more profitable region without violating operation constraints. APC projects can often provide significant benefits resulting in quick payback periods. Refineries and petrochemical plants worldwide have implemented APC controllers for their major processing units, however, the large benefits cannot always be maintained as control performance may degrade over time due to changes in the process, instrumentation, feedstocks, operators, etc. Operators may turn off an APC application in situations when performance is not acceptable. If that trend continues, then the APC application may remain off for longer periods of time as operators lose confidence in the controller. Therefore, the maintenance of existing APC applications becomes a primary issue in order to sustain the benefits.

As the number of installed APC applications increases with limited process control engineering resources, the use of improved technology tools helps offset the required time investment to maintain existing applications and to implement new applications. Technology tools exist to monitor performance of existing APC applications and the performance of underlying regulatory control loops. Problems with regulatory control loops can be identified and performance can be improved through the use of control tuning technology and instrumentation maintenance and repair. When MPC models need to be updated, new technology tools also exist to automate and expedite plant step testing and to improve the quality of the identified models. These various technology tools assist process control engineers in re-commissioning APC applications.

APC re-commissioning efforts also commonly involve modifications or extensions to the existing designs. For example, new constraints may need to be added due to changes in plant operating conditions. New analyzers may also be available to include in the design and thereby improve the predictions and performance.

After a plant turnaround in 2007, Westlake Chemical initiated a new project with Matrikon to help revive and re-commission various APC applications in their Petro 2 ethylene plant in Sulphur, Louisiana. This project included the C2 Splitter, C3 Fractionator, Demethanizer, Deethanizer, and the five cracking furnaces. Matrikon software tools were purchased for monitoring and tuning PID loops, monitoring APC controllers, and for automated step testing and modeling. This paper summarizes the methodology used, issues encountered, and the results obtained.

1.1 Westlake Chemical Petro 2 Plant

The Petro 2 ethylene unit was built in the late 1990's with an original rated capacity of 1.0 billion lb/yr in Sulphur, Louisiana by ABB-Lummus for Westlake Chemical Corporation. It consists of process units and supporting facilities to produce polymer-grade ethylene, chemical-grade propylene, raw pyrolysis gasoline, and by-products.

The major sections of the plant include the following:

- Feed Separation, Pyrolysis Module, Feed Saturation, Quench System
- Charge Gas Compression, Acid Gas Removal, Charge Gas Drying, Dryer Regeneration
- Charge Gas Chilling, Demethanization, Hydrogen Purification
- Deethanization, Acetylene Hydrogenation, Ethylene Fractionation, Depropanization, MAPD Hydrogenation, Propylene Fractionation, Debutanization
- Ethylene Refrigeration
- Propylene Refrigeration
- Caustic Pre-treatment
- C4 Hydrogenation

The plant areas for which APC had been applied at Petro 2 include feed separation, the five cracking furnaces, charge gas compression, cold-side fractionation, refrigeration, and hot-side fractionation. The APC consists of several DMCplus multivariable controllers from Aspen Technologies.

For the cold side fractionation sections, there was a large DMCplus multivariable controller, which included six sub-controllers: CGC (Charge Gas Compressor), MRC (Methane Refrigeration Compressor), DM (Demethanizer), DE (Deethanizer), C2F (C2 Splitter), and REF (Ethylene Refrigeration Compressor & Propylene Refrigeration Compressor). The Cold Side controller was used, but operated with active sub-controllers for only the MRC and DE. The C2 Splitter sub-controller was not used and it was indicated that this controller's performance had never been satisfactory. The first goal of the project was to re-commission the C2 Splitter APC.

The hot side fractionation sections include three DMCplus multivariable controllers for the Feed Splitter, Depropanizer, and C3 Fractionator. The C3 Fractionator DMCplus controller was not being used. This controller had been used at times in the past, but was often turned off when the product propane composition increased too quickly toward the specification limit.

The hot side also included DMCplus multivariable controllers for each of the five cracking furnaces. The furnace DMCplus controllers were typically used, but in order to improve performance, especially with plans for more frequent changes to feedstocks, it was decided to retest one furnace of each type: SRT-5 and sweep coil. The furnace retesting and re-commissioning was executed as phase 2 of the project. Severity control had not been used for a long period of time due to various issues. To provide more reliable composition data for severity control, the sampling system for one of the effluent analyzers was upgraded. Updates were subsequently implemented for three of the other furnace effluent analyzers (and the 5th analyzer and sampling system are on order). Therefore, a major objective of the 2nd phase of the project was to re-commission the furnace severity controls.

The Deethanizer and Demethanizer sub-controllers of the Cold controller were also to be updated in the 2nd phase of the project.

2. APC Re-commissioning Methodology

The project methodology for APC re-commissioning needs to include several tasks:

- Examination of the process, instrumentation, and existing APC designs
- Detection and correction of regulatory control problems
- Step testing and model identification
- Re-commissioning with new models, tuning, calculations, logic
- Monitoring APC performance

These tasks must be executed efficiently in order to limit project costs; otherwise re-commissioning projects cannot always be justified. At the Westlake Chemical Petro 2 unit, Matrikon's Control Performance Monitor (CPM) technology was utilized to support many of the re-commissioning tasks as follows:

- CPM PID Monitoring module to detect regulatory control problems
- CPM Tai Ji PID module to improve regulatory control tuning
- CPM Tai Ji MPC module to provide automated step testing and model identification
- CPM MPC Monitoring module to monitor the APC applications

The Aspen Technologies' Aspen Watch software was also used for monitoring APC performance.

2.1 APC Design Changes

The first task in APC re-commissioning is to examine the existing APC designs with respect to the variables, operating objectives and constraints and how the design is affected by any changes to the process, instrumentation, feedstocks, or operating procedures. Such changes often require new models to be developed, new constraints to be added, etc. New measurements may be available to utilize as disturbance variables (DVs) or controlled variables (CVs). New manipulated variables (MVs) can be considered, particularly for designs that have not been working for long periods of time.

At Petro 2, the design that encountered the most changes was the C2 Splitter. During the turnaround in 2007, a new condenser was added to the C2 Splitter overhead to provide more cooling capacity. Work was also performed to improve the composition analyzer for the Deethanizer overhead / acetylene converter feed. The acetylene converters provide the feed to the C2 Splitter, therefore the ethane composition from this analyzer could be used as a DV in the C2 Splitter design to represent its feed composition.

The original controller design included reflux and product draw manipulated variables to control the reflux drum level and ethylene product composition. In normal operation, there is a level control PID loop for the reflux drum which adjusts the reflux flow to maintain the level. Also in normal operation, the ethylene product draw setpoint is adjusted by a ratio controller with operators defining the setpoint for the product draw to reflux ratio. During the re-commissioning pre-tests, it was determined that these normal operating procedures provided more stability and the CV responses were more consistent and predictable than with the original APC design. Therefore, it was decided to use the product draw ratio as an MV in the new APC design by leaving the ratio controller in AUTO and leaving the reflux drum level PID controller in AUTO to adjust the reflux.

The C2 Splitter is affected by the refrigeration systems as the PRC provides reflux condensing and reboiling and the ERC provides side reboiling. The bottoms reboiler propylene flow loop is in Manual because if the valve is opened too much, the reboiler condensate drum level cannot be maintained as flow blows through with loss of reboiler capacity. In the original APC design the reboiler valve was an MV, but operators were responsible for maintaining the bottoms reboiler level by adjusting the high limits for the reboiler valve. In the new design the reboiler condensate drum level was added as a constraint variable (CV).

For the C3 Fractionator it was recommended to retest and develop new models since the existing controller was not performing well and was not being used. This tower is also affected more than the C2 Splitter by furnace feedstock changes which were expected to occur more frequently. There were only a few changes made to the controller design. A condenser bypass valve was added during the shutdown. Operators adjust the bypass valve when the cooling water valve is outside its nominal operating range and cannot control the overhead pressure, so the bypass valve was added as a DV to account for its effect on the cooling water valve position. For this tower, the product draw and reflux ratio were kept as MVs to control the propylene product composition and reflux drum level. The product draw to reflux ratio was added as a CV to ensure that it remained within reasonable limits.

For the cracking furnaces it was decided to retest one furnace of each type. The main design change was to include temperature balancing of the coil outlet temperatures whereas the existing design was based on pass flow balancing. Temperature balancing helps provide more consistent conversion across the coils.

Improvements to the effluent analyzer sampling systems provided more consistent severity calculations which are based on the feed and effluent analyzers. Therefore, the severity models could include more reliable analyzer feedback in a model update strategy.

2.2 Regulatory Control Loops

Before the start of MPC step testing, it is important to check the performance of the regulatory control loops. Matrikon's CPM PID module provides several reporting methods to easily detect problems and prioritize maintenance for poorly performing regulatory control loops. The software reports numerous performance indices and highlights when values are outside of thresholds. Two key performance indices are the relative performance index (RPI) and the oscillation index. The RPI value is used to indicate the respond speed of the controller relative to the benchmark value (required settling time). A value close to 0 indicates poor control, whereas a value close to 1 indicates good controller performance. A value greater than 1 indicates an aggressive controller. The oscillation index indicates the degree of oscillation in the PV, with values greater than 0.4 indicating significant oscillation.

For the C3 Fractionator, the CPM summary reports were examined and indicated low

RPIs and significantly large oscillations for several control loops. The oscillations occurred during almost all conditions, but were worse when operating with four of the five furnaces. The reports also indicated a fairly common oscillation period of approximately 20 minutes for many of the loops.

The bottoms level loop and the overhead accumulator level loop had the largest oscillation indices and seemed to be the likely sources of the interactions and oscillations throughout the column. The bottoms level PID loop is part of a double cascade with a duty controller and the quench water flow controller. Figure 1 shows a detailed report for the quench water flow loop. These oscillations had propagated throughout the column.

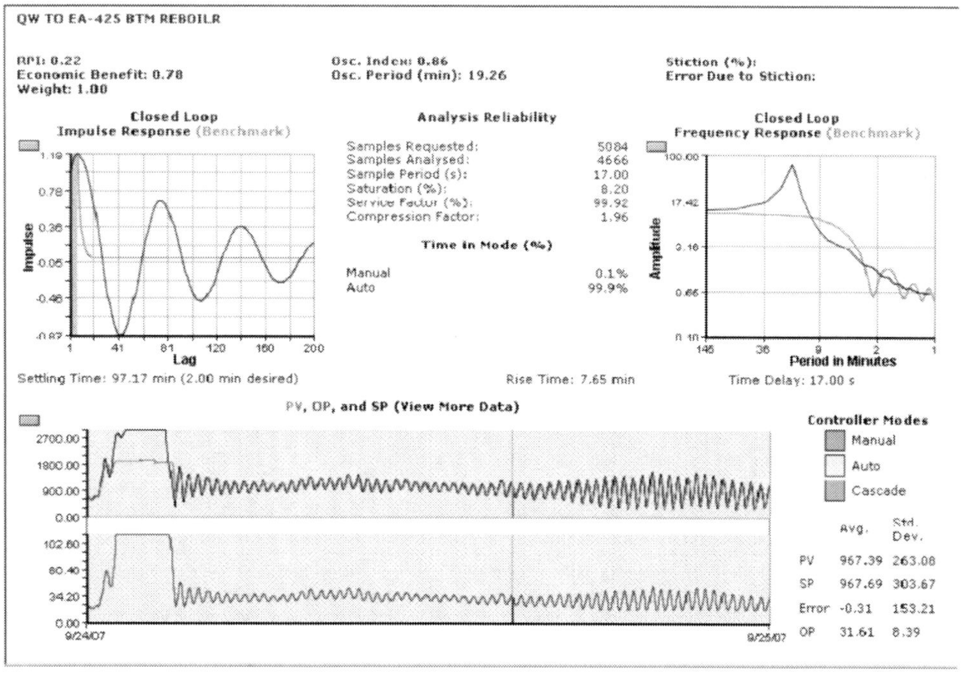

Figure 1: Reboiler Quench Water Flow Loop Detailed Report

From analysis of the reports, it seemed that this double cascade was increasing the magnitude of the oscillations, especially because the duty controller and quench water flow controller have similar speeds of response (an inner slave loop of a cascade should be several times faster than the outer master loop). The reports also indicated that the changes to the setpoint of the quench water flow loop were contributing heavily toward its PID error. The quench water setpoint is driven by the duty controller. Therefore, it was proposed to remove the duty controller and have the bottoms level output cascade directly to the quench water flow controller setpoint. The bottoms level PID loop was also retuned to provide slower integral reset action using Matrikon's Tai Ji PID tuning tool.

Figure 2 shows a new detailed report for the quench water flow loop indicating significantly reduced oscillations and lower PID error. Using the summary reports to compare results from one week before to one week after the PID changes were implemented, the RPI for the bottoms level loop increased from 0.25 to 0.54 and the oscillation index decreased from 0.81 to 0.33. The RPI for the quench water flow loop increased from 0.26 to 0.38 and the oscillation index decreased from 0.72 to 0.37. Also, the RPI for the overhead accumulator level loop increased from 0.27 to 0.56 and its oscillation index decreased from 0.80 to 0.44. In addition to reducing the oscillation magnitudes, the oscillation periods of the other control loops have been altered so there is no longer a common oscillation mode affecting the column.

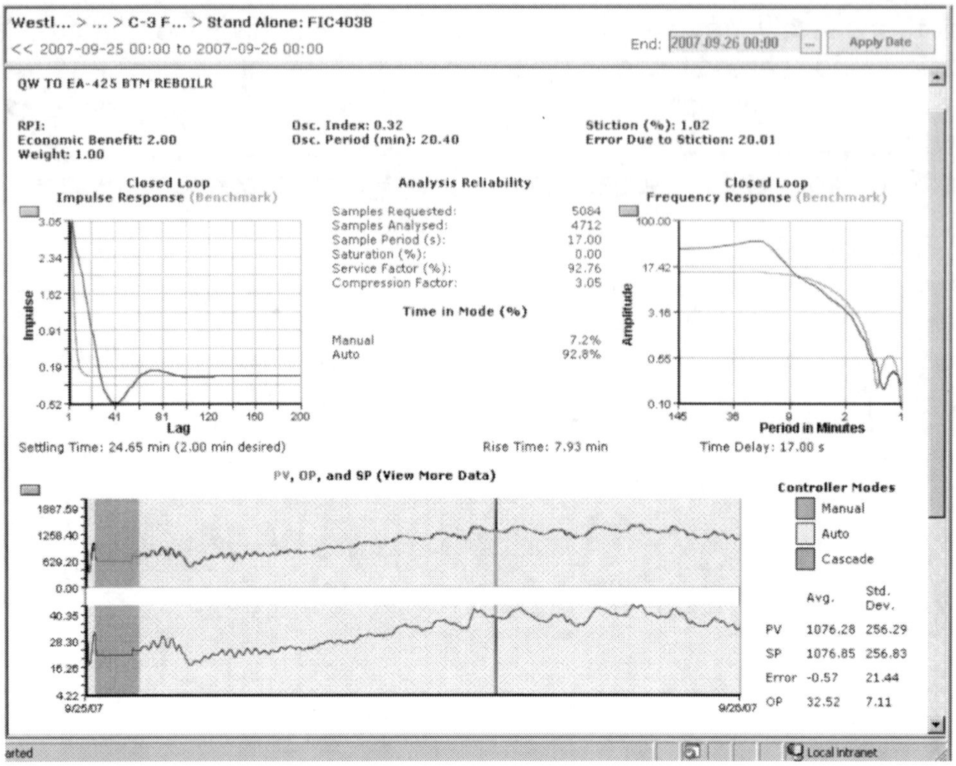

Figure 2: Reboiler Quench Water Flow Loop Detailed Report after PID Modifications

The advantage of the loop monitoring tool is that it automates the data collection, calculations, and reporting to expedite the performance assessment and diagnostic process. This helps reduce the pre-testing time and it is not necessary to perform manual setpoint changes to evaluate regulatory controller performance. Consistent monitoring of controller performance on a daily basis reduces maintenance costs and helps to optimize plant operations.

2.3 Step Testing and Modeling

Matrikon's Tai Ji MPC module offers significant advantages to reduce the time needed for step testing and modeling and provides a less intrusive testing environment. The Tai Ji technology calculates a step testing pattern that contains simultaneous uncorrelated moves in the independent variables as opposed to moving one variable at a time.

The software provides automated step testing and monitoring 24 hours a day. Step testing can be conducted in either open-loop or closed-loop. Closed-loop testing implies that some or all of the process CVs are controlled by an MPC controller or by PID loops. During closed-loop testing, the existing DMCplus controller is utilized to keep the CVs within constraint limits while a Tai Ji bias signal is added to each of the manipulated variables.

For the cracking furnaces and Deethanizer at Petro 2, the step testing was conducted in closed-loop control to provide the safest testing environment. For the C2 Splitter and C3 Fractionator, the step testing was conducted in open-loop since the operators had not been running those controllers. For these tests, step sizes were selected to try to prevent constraint violations. Operators could turn off the test if necessary. Step testing was performed throughout both the day and night shifts, with night shifts running completely in automatic mode, without engineering supervision.

With the Tai Ji multivariable modeling component, initial modeling passes could be performed during the step testing to evaluate the model qualities (graded A, B, C, or D). The model qualities help guide the step testing to a faster completion. For example, if models from some independent variables have poorer qualities, then the step sizes for those variables can be increased to enhance the test data for subsequent identifications. Figure 3 shows a Tai Ji step response plot with the model grade.

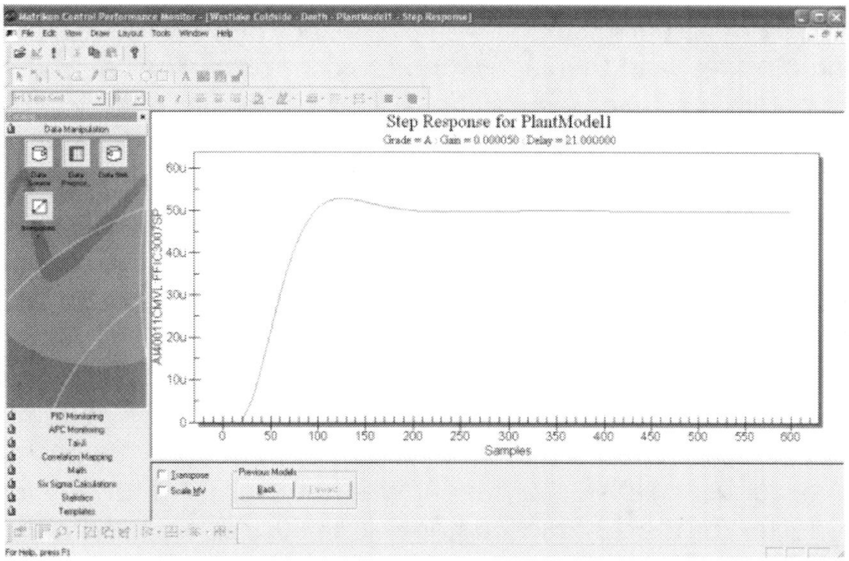

Figure 3: Step response plot with model grade

Some of the new models identified for the C3 Fractionator had slower dynamics than the original models, particularly for the composition analyzers. Gain changes were also identified for the analyzers and for some of the constraint variables.

The models for the C2 Splitter were very different than the original models because of design changes affecting the manipulated variables as described in the APC design sub-section.

For the furnaces, two sets of models are now used for the different furnace types. Several new models had similar gains and dynamics to the previous ones. However, there were some changes in the model gains for excess O_2 and for several valve constraints.

The advantages of the automated step testing and advanced modeling technology can be summarized as follows:

- Step-testing 24 hours a day and with multiple variables significantly reduces the total testing time and provides a richer set of data for model identification.
- Operators are not as stressed with the automated step-testing method and can focus on operating the plant.
- APC engineers can monitor the step test and make test adjustments without intrusion to the operators' daily work routine.
- The ability to quantify model quality with the identification software results in more accurate final models.

The inferential models for furnace severity control were also updated. New data was generated from Spyro runs for various sets of operating conditions. Different models were created for each furnace type. The new models, data validation, and analyzer feedback strategies were implemented in the Aspen IQ software.

2.4 APC Re-commissioning and Performance Monitoring

The APC Re-commissioning stage involves placing the new models online, possibly with some initial tuning changes, and then activating the controller. The controller performance is then examined to determine subsequent actions, such as further modifications to the models and tuning. Various methods are utilized to assess performance, including: trending tools, operator comments, and APC performance monitoring packages, such as Matrikon's CPM MPC Monitoring module and Aspen Technologies' Aspen Watch software. When there is a problem, the performance monitoring tools can help determine the source. For example, model quality indices may indicate that some models still need to be improved, or statistics on limits and effective service factors may indicate that operators have set some MV limits to be too restrictive, or performance indices may indicate that some tuning changes are needed, etc.

The service factor for the C2 Splitter MPC is 89.9% when all five Petro 2 furnaces are operating, which is when the greatest benefits can be achieved. In the past, this controller's performance had never been satisfactory and it was not ever used for any long periods of time. The service factor for the C3 Fractionator MPC is 69.1%. This controller had not been used at all for at least 6 to 9 months prior to the start of the re-commissioning project.

For the furnaces, the MPC service factor must be measured relative to the percent of time each furnace is up and running (i.e. not in decoke or shutdown). When averaging the results for four of the five furnaces, the MPC relative service factor is 97%. One of the furnaces is experiencing mechanical problems with a damper, so it is not running in APC until the damper is replaced. The service factor for the De-ethanizer sub-controller remains very high at 99%.

Figure 4 shows trend plots for the C2 Splitter APC. The ethane composition in the ethylene product is controlled close to its high limit to maximize ethylene production. The MPC high limit for ethane is selected by operators to provide a small buffer from the pipeline specification limits. The tight impurity specification and the slow composition dynamics make it very difficult for operators to control near the high limit for any period of time without APC.

The controller is also satisfying the low limit for the bottoms reboiler condensate level to prevent the reboiler vapor from blowing through.

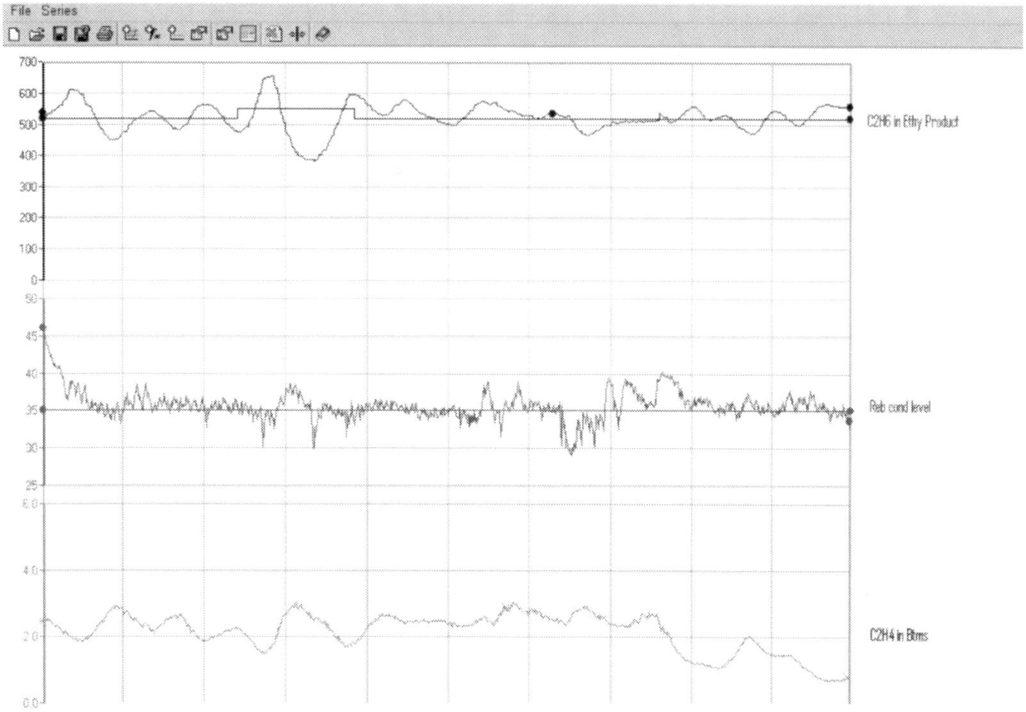

Figure 4: C2 Splitter APC Trends

Figure 5 shows trend plots for the C3 fractionator APC. The propylene production and reflux MVs are displayed together with the analyzer CVs. The propane impurity in the propylene product is controlled at its high limit and the propylene composition in the bottoms is held close to the low limit (set to prevent MAPD buildup in the column) while satisfying other constraints.

The MPC typically provides tight control over the C3 fractionator for long periods of time. However, there are periods when the controller is turned off due to large disturbances in the column operating conditions resulting from changes to furnace feedstocks and conversion targets. Such changes occur more frequently now because of the desire to reduce operating costs. Efforts have been made to increase the controller robustness to feed changes by modifying some of the models and tuning.

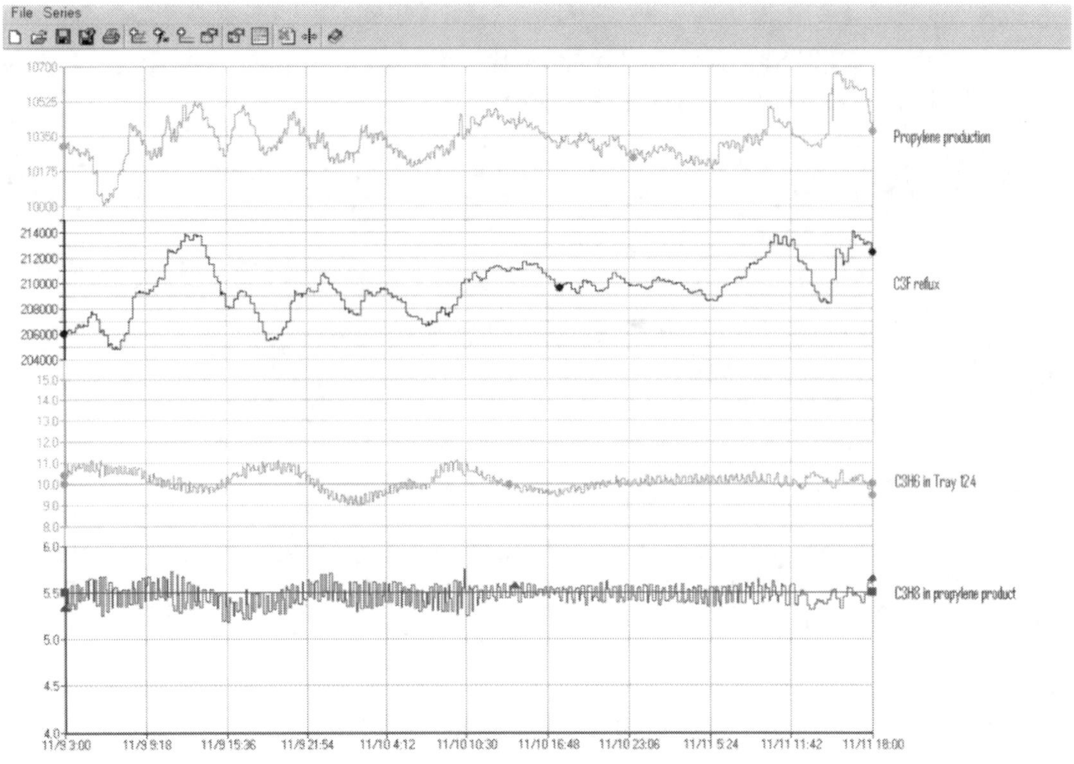

Figure 5: C3 Fractionator APC Trends

Figure 6 shows a detailed daily report from CPM's MPC Monitoring module for the furnace 3 severity CV. Several statistics are reported, including the service factor, average and peak violation, and steady state (SS) limit give-up (% of time SS target was outside of high/low limits). The report also includes trends displaying the high and low limits, SS target, PV, and unbiased prediction. Severity is controlled to a target, so the high and low limits are equal. The trend plots and statistics indicate good control of the furnace severity. The average violation is just 0.07% and the peak violation is 0.66%.

A model quality index and phase index are presented on the upper right side. The model quality index represents the SS prediction capability of the models for this CV (<1 implies under-predicting, >1 implies over-predicting). The phase index indicates the dynamic prediction capability of the models (1 for in-phase, 0 for 90 degrees out of phase, -1 for 180 degrees out of phase). The model quality index versus time is plotted on the bottom right, indicating good SS prediction capability. The report also includes a plot of the Autocorrelation of the prediction error and frequency response plots for the PV and unbiased prediction.

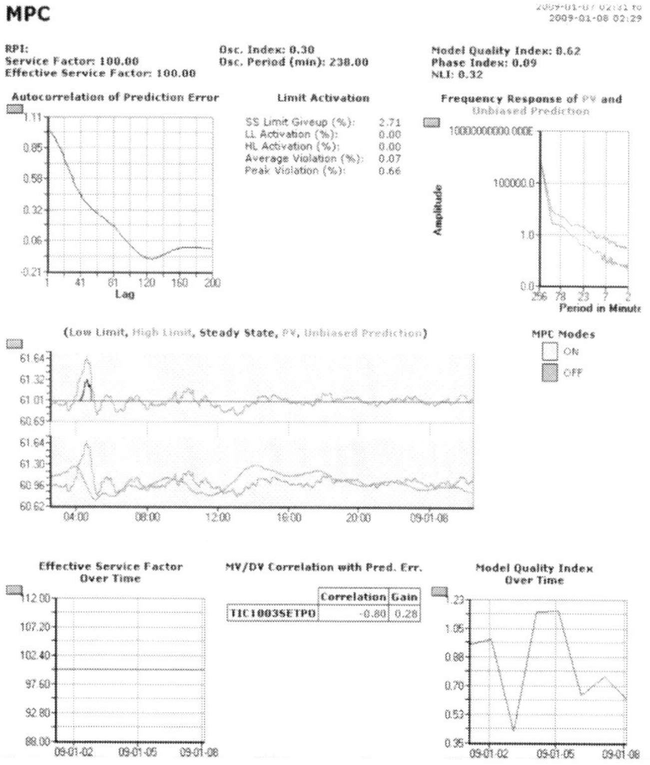

Figure 6: MPC Monitoring Report for Furnace 3 Severity CV

For comparing energy savings with APC, it must be based on data with very similar operating conditions and there should be reasonably long periods of time with and without APC. For the C2 Splitter, Table 1 shows a comparison of the data from Oct 27th to Nov 3rd, 2008. The ethylene fraction in the De-ethanizer overhead (C2 Splitter feed) is almost the same for the periods when APC was on versus off. The same holds true for the ethylene production relative to the ethylene in the feed, where ethylene in the feed is estimated as: De-euthanize overhead flow multiplied by the ethylene feed fraction. The reflux to product ratio is reduced by 4.04% with APC on.

Table 1: C2 Splitter APC Energy Savings

	Reflux to Product	Product to C2H4 in Feed	C2H4 Fraction in Feed
APC Off	5.323	0.915	65.88
APC On	5.108	0.915	65.92

To compare energy savings for the furnaces, a new metric was defined to measure BTU/LB of ethylene for each furnace. This metric uses the total firing divided by the furnace ethylene production, where ethylene production is calculated as the furnace hydrocarbon feed multiplied by the severity fraction. Table 2 shows a savings of approximately 2.0% with APC On versus APC Off for the BTU/LB of ethylene metric for furnace 3. This was calculated for the period from Oct 29th to Nov 1st, 2008 as the furnace 3 APC was off for a period of time and then turned on with very similar conditions for the feedstock and severity. The furnace APC applications save energy by reducing excess O_2 to the low limit for emissions, reducing the wall-to-hearth firing ratio, and reducing variations in severity and operating closer to target.

Table 2: Furnace APC Energy Savings

	BTU/LB of Ethylene Savings	C3H8 Fraction in Feed	Severity
APC Off	--	0.233	65.1
APC On	1.98%	0.231	64.9

3. Conclusion

Several APC applications have been successfully re-commissioned at the Westlake Chemical Petro 2 Ethylene plant. This paper has described the re-commissioning methodology, which included modifying APC designs, improving regulatory controls, step testing and modeling, tuning, and monitoring performance. The APC design changes were defined to account for changes to the process and feedstocks, to add variables for new or modified instrumentation, and to develop more robust models. The paper also described the use of Matrikon's Control Performance Monitor software to help expedite several of the re-commissioning tasks. The CPM technology includes modules for monitoring and tuning of regulatory control loops, automated step testing, multivariable model identification, and MPC performance monitoring.

The operators now consistently use APC applications that had been turned off, including the C2 Splitter, C3 Fractionator, and furnace severity controls. The APC applications allow the plant to maximize benefits by increasing production and reducing energy costs, while satisfying operating constraints.

A METHODOLOGY FOR REAL-TIME AUTOMATED PROCESS MONITORING IN ETHYLENE UNITS.

Unmesh Taskar

APC & Optimization Engineer

Chevron Phillips Chemical Company, LP – Sweeny complex

Abstract: Hi fidelity process models that are used for online optimization are employed to generate real time information related to plant operations and assets. Key performance and economic parameters were identified for each process equipment as well as for the overall unit. The capability of on-line optimization models was enhanced to compute these parameters during regular optimization cycle. The results are extracted into an Excel based reporting tool in an automated fashion. Some Parameters have the capability to be trended to show their variation over time. Economic impact of important process variables is also generated and presented in this report. The easy to use Excel graphical front end makes it very intuitive tool to be used by Process, Production, and Control Engineers and Operators. This tool can be used for identification of opportunities, problem identification and enhanced productivity.

1.0 Introduction

In an era of high energy and feedstock costs leading to continued tightening of profit margins in the ethylene business, pressures on chemical/production engineers are increasing. The importance of ensuring that all major assets in the plant remain working as efficiently as possible has grown. Also, high emphasis on improving safety and reliability requires these assets to keep operating at consistent performance level.

A typical ethylene production unit involves furnaces cracking various feeds into ethylene and several products at high temperatures. The cracked-gas mixture is later separated into ethylene and various components through cryogenic separation and fractionation. Thus, an ethylene unit includes several different types of process equipment and unit operations like steam cracking furnaces, compressors and turbines, distillation columns, etc. The performance of these process equipments can vary over time. The degradation in performance can significantly affect the "bottom-line" of the whole unit. Understandably, plant management is interested in closely monitoring the performance of these assets. Process monitoring is also driven by the need to understand the economic impact of some of the key variables to ensure that right decisions are being made to optimize the unit.

This paper outlines the development of a Microsoft® Excel-based performance monitoring tool at Chevron Phillips Chemical Company's Sweeny Olefins Complex. The tool is designed to help plant and process engineers in tracking the performance of the overall ethylene unit as well as some key pieces of equipment. Ultimately, the tool aims to present the relevant information in such a way as to facilitate the process of informed decision making and engineering analysis. The process of compiling this information into an Excel-based process monitoring report is almost completely automated.

2.0 Issues in Process Performance Monitoring

Challenges in Effective Process Monitoring

Because of the large number of process equipments involved in an ethylene unit, various key performance characteristics are required to be tracked. These parameters need to be regularly computed and historized in order to effectively monitor the performance. Also, the economic impact of under-performing assets needs to be frequently quantified to help plant management make more informed decisions about day-to-day operation.

In recent years, many olefins units have implemented advanced multivariable process controllers as well as real-time optimizers. These automated tools are intended to drive the unit towards a place of more profitable operation. However, regular effort is needed by engineers to make improvements, understand and explain that the plant is being steered in the correct direction.

Because of the constraints on time and engineering resources supporting the plant, it is often a challenging task to satisfy these requirements in a timely fashion. This can result in unexpected performance issues appearing without warning. It could also mean missing potential opportunities to improve the economic performance of the unit.

Effective process monitoring also suffers from a lack of a unified platform where all the information can be pulled together into one single interface with analysis. Typically useful information is distributed among different groups like process, controls, reliability, etc. Consequently, some of this information does not get communicated on a timely basis to take the required corrective action. Another issue faced in effective process monitoring is the number-intensive nature of the reports generated which provides little direct help in analysis.

Requirements of Process Monitoring Report.

The concerns mentioned above clearly illustrate the need to pursue creation of an efficient process monitoring tool. The tool should be able to address some of the issues listed above. For that reason, some basic criteria need to be met for any process monitoring effort. Obviously, it needs to capture the performance characteristic information of the key pieces of equipment in the plant. However, to be really effective for analysis, the information needs to be accompanied by historical trends and comparisons to design. This would help detect performance deterioration over time. In some cases, it would be quite advantageous for planners and decision makers if this information was supplemented by economic analysis. For example, what is the cost of deterioration of the heat transfer coefficient from design for a compressor inter-stage cooler in today's economics? Would potential savings from cleaning the exchanger and restoring the heat transfer coefficient to the design justify the cost of repair?

It is also quite important that the report be relatively quick and easy to generate on a regular basis, preferably almost completely automated. An often overlooked aspect in most engineering reports is the aesthetic aspect of the presentation of the data. In recent years, Microsoft® Excel has become a versatile and familiar interface used by most engineers. Information presented with Excel as a front-end interface would be easy to manage, display and automate. Excel therefore was the interface of choice to implement the process monitoring report.

3.0 Design

The design phase of the report included charting out, in detail, the information relevant to plant performance monitoring. At the conclusion of the design phase of the report, it was decided that it would be organized into following four categories.

1. Plant Overview

This should be designed to give a quick overview of important metrics tracked by plant management on a regular basis. This section should contain following information.

a. Energy Consumption – An index or a measure that captures the performance of the unit from an overall energy perspective would be an important parameter to track. This index should be accompanied by a historic trend to show variation in plant energy performance over time.

b. Economic Performance – Net costs of producing ethylene vs. the price of ethylene denote operating margins which should be displayed in the overview page. This information should be updated at least weekly with the latest economics. This should be split further between the feed cost component and the utilities component. A more important number is the cost of making additional production. Frequently, the incremental cost is higher than the overall cost. In the current environment of tight margins, this information can help management make vital decisions such as which type of feed is most profitable, etc.

c. Material Balance Table – Getting 100% material balance closure is a frequent issue faced by many ethylene plants. The effort consumes a lot of time for plant engineers. The material balance table should list all the feed and product meters and preferably provide some tool for hunting imbalance. For example, which meter is likely to be the "most wrong". Also, plant imbalance should be trended to track any improvement or deterioration over time.

d. Fuel Gas Balance Table – Just as with the material balance, the energy balance closure around the fuel gas system is closely watched by the plant management and should be included.

e. Steam Table – Steam consumption is another key component of the energy equation in an ethylene unit. The table should list major steam flows like steam produced in furnaces, steam consumed in major compressors, extraction, import and letdown flows.

f. Costs and Profit table – A table depicting $ values of all feed, product and utility import/export streams would add valuable information.

2. Furnaces

In every ethylene plant, furnaces are the most important pieces of equipment from the point of view of the energy consumed as well as impact on the product yields. Furnace performance monitoring forms an important aspect of overall unit performance monitoring. The Furnace section of the report should contain the following information.

a. Furnace Efficiency – Furnace efficiency indicates the performance of thermal cracking furnaces from an energy perspective. Furnace efficiency can be separated further into convection section efficiency and radiant section efficiency. Historical trends should accompany the furnace efficiency tabulation. Deterioration in convection efficiency could indicate convection section fouling, while declining radiant efficiency could indicate higher losses in the firebox. An economic impact computation accompanying the furnace efficiency would be quite helpful in decision making about future furnace maintenance work. Another statistic that could be presented is heat fired in every furnace per weight of ethylene and propylene produced in the effluent of that furnace. These parameters will further aid the analysis of a poor overall energy performance.

b. Furnace Overview Table – A furnace section should contain a table listing current key operating variables like conversion, feed rate and type, steam-to-hydrocarbon ratio, coil outlet temperature, days on line, etc. In addition, it should include certain other calculated numbers like coil pressure drop as well as calculated conversion of a co-cracking feed, wherever applicable.

c. Exhaust Oxygen analysis – Excessive oxygen content in the stack indicates a higher than necessary presence of air in the furnace stack and may lower furnace efficiency. Economic impact analysis of higher exhaust oxygen, which considers the effect of excess air on fuel gas consumption as well as steam generation, should be provided. This could provide guidance on the need to adjust damper position or ID fan speed, etc. Convection section leaks can lower steam production and reduce furnace efficiency. Calculation of percent air leak along with associated economic impact figures would also be beneficial.

d. Transfer line heat exchanger (TLE) heat transfer coefficients – TLEs recover heat from hot effluent gases and generate steam. Deteriorating transfer line heat exchange could reduce steam generation and indicate coking. A table listing TLE heat transfer coefficients, accompanied by historical trend, is necessary for effective monitoring of TLE performance. Also, the economic impact of deterioration of TLE heat transfer coefficients from the design case would be helpful in planning TLE maintenance.

e. Furnace feed versus conversion trade-off analysis – Ethylene furnaces are frequently limited by constraints in the back-end compression and separation section. In this scenario, how the furnaces should be operated for maximum economic benefits is a question confronting daily operations. Higher conversion means lower furnace feed and vice versa. An economic analysis of the feed versus conversion trade-off in the face of a typical back-end constraint will provide valuable clues to answer the dilemma.

3. Compressors and Turbines

Next to furnaces, compression of hot effluent gases and energy for providing refrigeration for separation columns consumes the most energy in an ethylene plant. Ethylene plants have three major multi-stage compressors: cracked gas, ethylene refrigeration, and propylene refrigeration. In addition, there may be turbo expanders and compressors as well as booster compressors. The compressor and turbine section should be designed to contain following information.

a. For each stage of the three major compressors there should be tables for flow, speed, polytropic efficiency (actual versus design) and polytropic head (actual versus design). This data must be supplemented by historical trends. An economic impact analysis for deviation from design should be done. There should also be a mechanism for drawing attention to any minimum recycle flows if they are open. A table listing the surge margin percentages for each stage would also add valuable information.

b. Steam turbine-driven compressors are the most common in ethylene plants. Data for extracting- and condensing-side efficiency should be tabulated along with their historical trends. A historical trend of design versus actual power, coupled with economic analysis, should be present.

c. For the cracked gas system cooling water inter-stage heat exchangers, heat transfer coefficients and pressure drops, wherever available should be plotted along with historical values. Similar information for the propylene system refrigeration cooling water condenser should be presented. Decay in these parameters from the design case is reflected in higher power consumption in compressors. The economic analysis could capture the cost in this scenario.

d. For the cooling water system, cooling water supply temperature and approach to dry and wet bulb temperature could be listed.

4. Fractionation Section

Cracked gases are fractionated into various components and products in the back-end separation section of an ethylene unit. Most of these distillation columns operate at high pressures and require low-temperature refrigerant for the condensers. Apart from the product purity specifications, optimum cut points in the distillation columns are determined by the trade-off between cost of the product lost as impurity from one end of the column and the cost of recovery. For example, the optimum ethylene impurity specification for the bottom end of the ethane-ethylene fractionator is determined by the relative magnitude of the loss of ethylene product versus cost of refrigeration required to recover that ethylene. A sensitivity analysis that lays out in detail the effect of the column cut points on important plant variables like, product value, utilities cost, etc. would be a valuable tool. With the sensitivity analysis one can isolate the effects of one decision variable on a number of dependent variables. This could lead to a better understanding of where to operate these columns for maximum profits.

The ultimate aim of the design described above was to make available to plant engineers an effective plant performance analysis tool. Obviously, further engineering analysis will be required to use this information to arrive at the final conclusions and recommendations.

4.0 Implementation

Based on the design requirements, it was clear that significant computational effort would be needed to accomplish the desired tasks. The only practical strategy would be to build a simulation model of the entire ethylene unit that could read online measurements and compute the parameters and information needed for engineering analysis.

Chevron Phillips Chemical Company's Sweeny ethylene complex has implemented online, real-time optimizers (RTO) based on Aspen Plus® optimization technology in all of their ethylene units. Over the last few years, real-time optimization has proved to be a highly applicable technology, particularly to the ethylene manufacturing process. Many ethylene plants around the world have employed rigorous online models for the purpose of real-time online optimization. It was realized that these online optimization models lend themselves quite suitable to meet our purpose of performance monitoring for the following reasons,

a. The rigorous, first-principles-based nature of the RTO models meant that a lot of engineering know-how was already built into the models. The model utilizes the SPYRO® kinetic model technology for predicting furnace yields. It contains performance maps for all stages of compressors and turbines. The separation columns use detailed tray-to-tray distillation models with suitable equation of state. The fully-integrated model covers the steam system, fuel gas system, refrigeration systems, etc. In a nutshell, a number of performance parameters of interest are either already being calculated inside the model or can be computed easily with minor modifications.
b. The models run continuously, with online data access, and are usually maintained up-to-date
c. The models contain current economic data, making the job of performing economic analysis much easier.

With these considerations, using RTO models as the basic data provider for the process monitoring report was an obvious choice. It was then a matter of following these steps to accomplish the report generation.

1. Since RTO models are primarily configured to perform online optimization, certain model modifications were carried out to calculate the desired numbers

required for the report. Some examples include calculation of air leakage in furnaces, incremental cost of ethylene, etc.

2. Aspen Plus® Open Equation Solver™ retains the Jacobian matrix information at the end of a converged solution for the entire model. Jacobian contains derivatives of all dependent variables such as profit, ethylene production, etc., with respect to all independent variables (furnace conversion, etc.). This is quite useful in presenting economic impact analysis for various independent variables. This Jacobian information was extracted by the Aspen Plus® sensitivity generation tool. The data contained in the sensitivity matrix was then maneuvered to present relevant economic figures.
3. In order to present historical trends for a number of parameters of interest, data calculated from the RTO models needed to be archived. For this purpose, several DCS tags were created to receive updated numbers at the end of every RTO solution. These tags were then archived into the plant data historian. The Excel process monitoring tool can automatically plot the data contained in the data historian.
4. Specific output files generated at the end of RTO solutions are stored with a date and time attribute. Excel macros can easily load this file using a user-specified time stamp and populate the cells with relevant information at the click of a button.

Currently, the report is being generated on a regular basis and findings are discussed among the engineers supporting the unit. The parameters of interest are being archived in the data historian. In fact, building a historical database of these important parameters can, in itself, be considered to be a useful byproduct of the process monitoring effort. Another notable benefit of the process monitoring has been realized on the real-time optimization side as well. Process monitoring reports display important information that is not as easily and frequently viewed otherwise. This has led to the improvements in the on-line model and higher confidence in the real-time optimization of the unit.

The layout of the current process monitoring report is presented in pictures below. Due to limitations of space and format only part of the actual report is displayed.

ETHYLENE UNIT : OVERVIEW

DATE	16-Nov-2008
TIME	12:03:41

ENERGY INDEX		
ETHYLENE PROD		
OPERATING MARGIN		

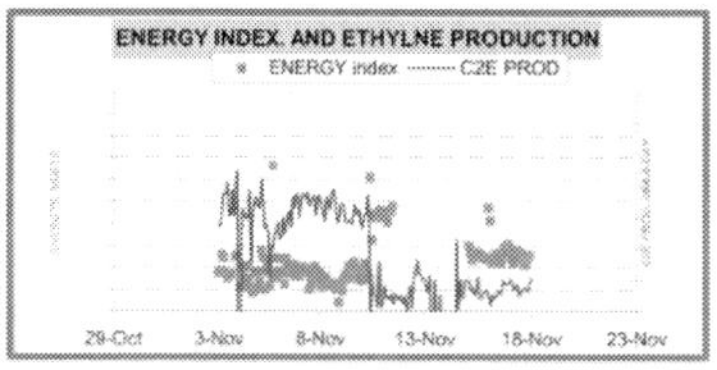

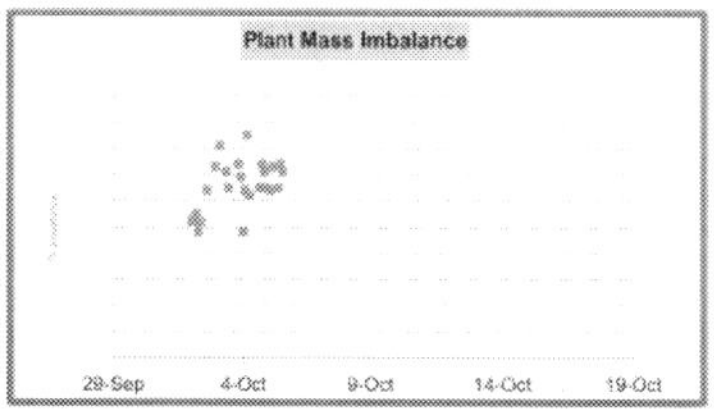

MATERIAL BALANCE

FEEDS (MLB/HR)	PLANT	MODEL
VAPOR FEED 1		
VAPOR FEED 2		
LIQUID FEED 1		
LIQUID FEED 2		
TOTAL		

Unit ZZ →

PRODUCTS (MLB/Hr)	PLANT	MODEL
ETHYLENE PRODUCT		
PROPYLENE PRODUCT		
C4 PRODUCT		
C5 PRODUCT		
GASOLINE		
TAIL GAS		
TOTAL		
IMBALANCE	100.00	100.00

FUEL GAS BALANCE

	FG VOL FLOW		HEATING VALUE			ENERGY	
	PLANT	MODEL	PLANT	MODEL		PLANT	MODEL
IMPORT	MMSCFD	MMSCFD	BTU/SCF	BTU/SCF		MMBTU/D	MMBTU/D
TAILGAS FROM COLDBOX					TAILGAS FROM COLDBOX		
OFFGAS					OFFGAS		
FUELGAS IMPORT					FUELGAS IMPORT		
TOTAL							
EXPORT							
TAILGAS EXPORT					TAILGAS EXPORT		
FUEL GAS TO FURNACES					FUEL GAS TO FURNACES		
TOTAL							
IMBALANCE						100.00	100.00

STEAM BALANCE

LEVEL 1 STM	PLANT	MODEL
IN	MLB/HR	MLB/HR
FURNACE 1		
FURNACE 2		
FURNACE 3		
FURNACE 4		
TOTAL		
OUT		
CGC TURBINE		
LEVEL 1 LETDOWN		
TOTAL		

LEVEL 2 STM	PLANT	MODEL
IN	MLB/HR	MLB/HR
LEVEL 2 IMPORT		
CGC EXTRACTION		
LEVEL 1 LD TO LEVEL 2		
TOTAL		
OUT		
ERC TURBINE		
PRC TURBINE		
LEVEL 2 LETDOWN		
OTHER LEVEL 2 USERS		
TOTAL		

PROFIT TABLE

FEEDS	$/HR
VAPOR FEED 1	
VAPOR FEED 2	
LIQUID FEED 1	
LIQUID FEED 2	
TOTAL	
PRODUCTS	**$/HR**
ETHYLENE PRODUCT	
PROPYLENE PRODUCT	
C4 PRODUCT	
C5 PRODUCT	
GASOLINE	
TAIL GAS	
TOTAL	
UTILITIES	**$/HR**
FUEL BURNT IN FURNACES	
LEVEL 2 STEAM IMPORT	
TOTAL	
OPERATING MARGIN	$/HR

Plant Overview

ETHYLENE UNIT FURNACE SUMMARY

DATE 18-Nov-2008
TIME 12:03:41

FURNACE OPERATION

	FEED TYPE	Online Days	TOT HC FEED	Conversion	COT	FG P	XOT	STM/HC RATIO	STACK T	COIL OUT P	COIL DP
		DAY	MLB/HR	CONV-PCT	F	PSIG	F		F	PSIG	PSI
FURNACE 1											
FURNACE 2											
FURNACE 3											
FURNACE 4											
		Total =	0.00								

FURNACE EFFICIENCY

	Radiant Efficiency	Convectn Efficiency	Overall Efficiency	Total Heat Fired	Stack Loss	BTU Fired / LB C2= & C3=
	PCT	PCT	PCT	MMBTU/HR	MMBTU/HR	BTU/LB
FURNACE 1						
FURNACE 2						
FURNACE 3						
FURNACE 4						

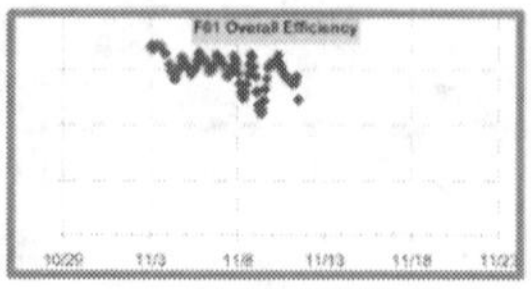

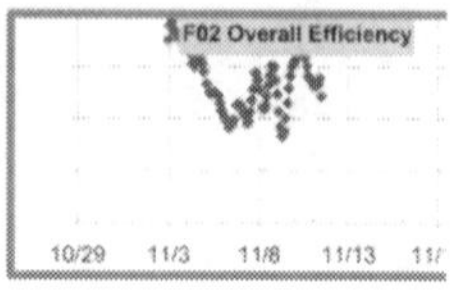

Furnace Effluent Mass Flows

	Ethylene	Propylene	H2+CH4	C4
	MLB/DAY	MLB/DAY	MLB/DAY	MLB/DAY
FURNACE 1				
FURNACE 2				
FURNACE 3				
FURNACE 4				
TOTAL				
BACKEND C2= LOSS				
BACKEND C3= LOSS				
BACKEND C4 LOSS				

Furnace Effluent Composition

	Ethylene	Propylene	H2+CH4	C4
	WT%	WT%	WT%	WT%
FURNACE 1				
FURNACE 2				
FURNACE 3				
FURNACE 4				

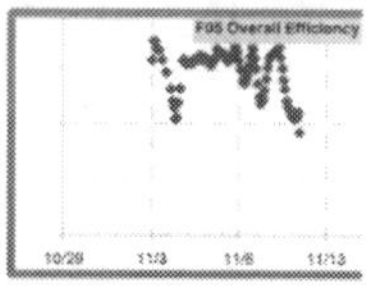

Convection Section Air Leak

	Stack O2 Dry basis	Arch O2 Wet basis	AirLeak %	Net Benefit if AirLeak Frac = 0
	MOLPCT	MOLPCT	PERCENT	$/hr
FURNACE 1				
FURNACE 2				
FURNACE 3				
FURNACE 4				

TLE HEAT TRANSFER

	TLE Ht Xfer Coeff East
	BTU/HR-SQFT-R
FURNACE 1	
FURNACE 2	
FURNACE 3	
FURNACE 4	

Recommended Furnace Conversion Moves in Various Scenarios

	*Scenario 1		**Scenario 2	
		Economics		Economics
Furnace 1 conversion				
Furnace 2 conversion				
Furnace 3 conversion				
Furnace 4 conversion				

Economics column indicates additional profit for 1% move in conversion in the direction indicated

Furnaces

ETHYLENE UNIT COMPRESSOR SUMMARY

DATE	18-Nov-2008
TIME	12:03:41

Cracked Gas compressor Efficiency and Head Data

		Stage 1	Stage 2	Stage 3	Stage 4
Efficiency Act	UNITLESS				
Efficiency Exp	UNITLESS				
Head Act	FT-LBF/LB				
Head Exp	FT-LBF/LB				

Cracked Gas Compression system

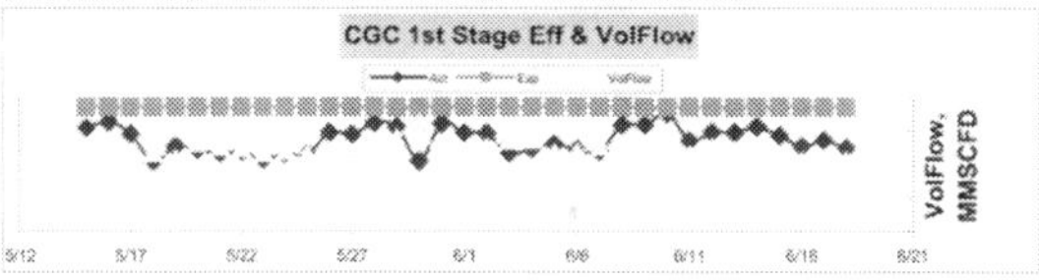

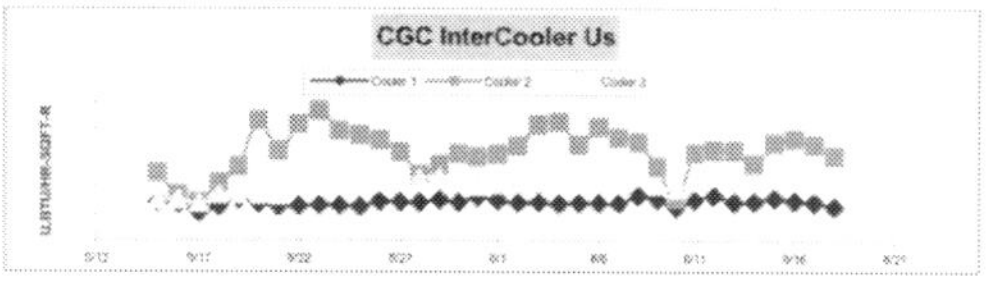

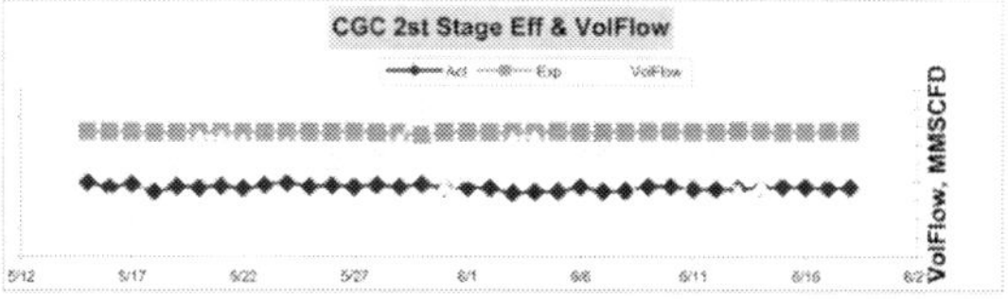

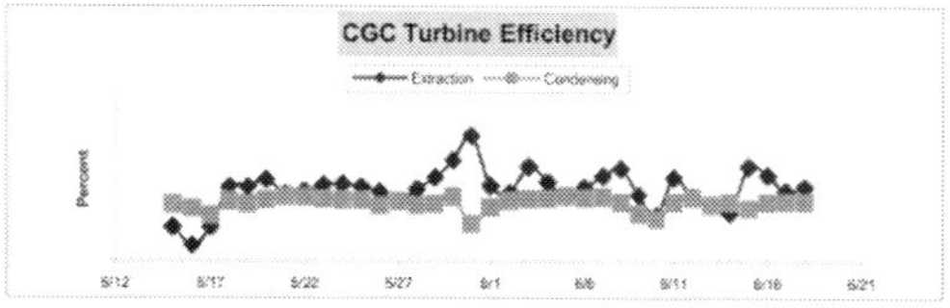

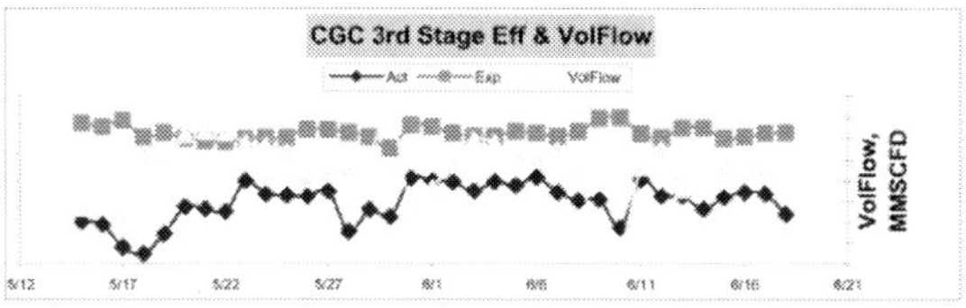

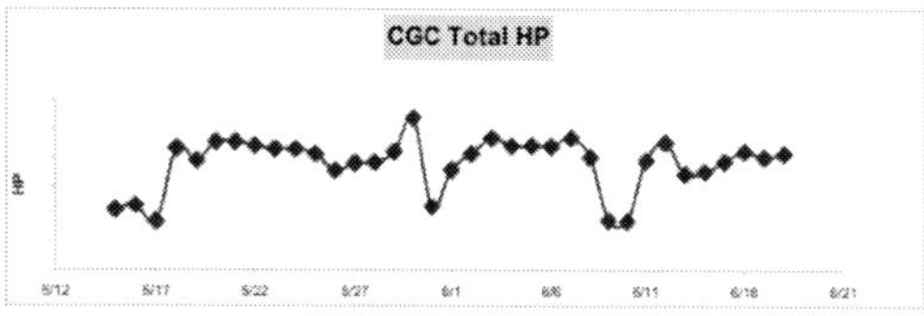

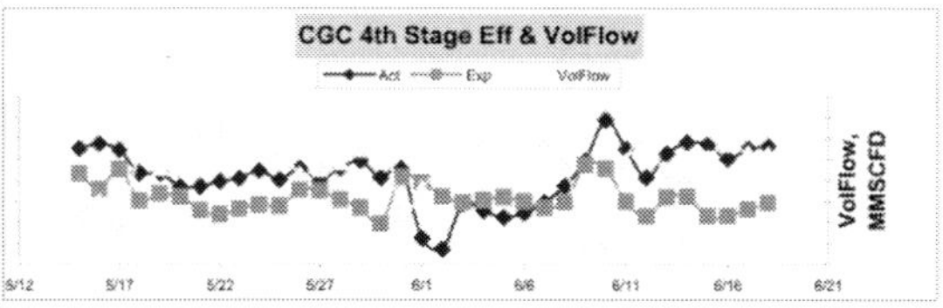

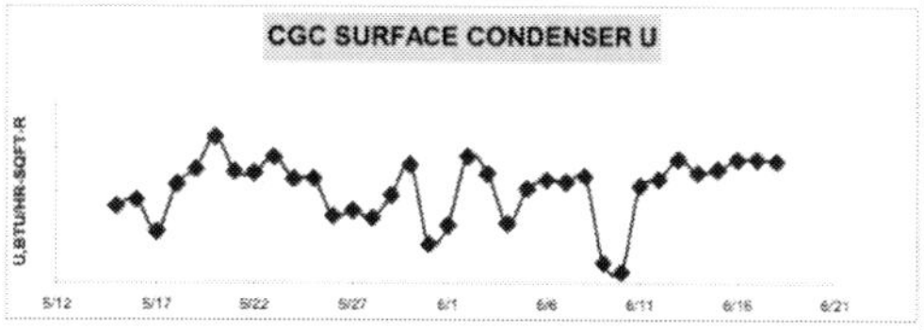

Heat Transfer Coeffcient Analysis

	Current	Design	Delta	450# STM Import Gain	Steam saving*	Vacuum improve
	BTU/HR-SQFT-R	BTU/HR-SQFT-R	BTU/HR-SQFT-R	MLB/HR	MLB/HR	
Cracked Gas Surface Condenser U						
Intercooler 1 Heat Trans Coeff						
Intercooler 2 Heat Trans Coeff						
Intercooler 3 Heat Trans Coeff						
Ethylene Comp Surface Cond U						
Propylene Comp Surface Cond U						

* Projected steam saving if heat exchanger operates at design U

Cooling Water Economics

	CW Change	HP Saving
	5	HP
Cracked Gas comp HP		
Ethylene comp HP		
Propylene comp HP		

Approach to wet bulb T	
Approach to dry bulb T	

Compressors and turbines

5.0 Conclusion

A tool for generating real-time process monitoring data was implemented at the Chevron Phillips Chemical Company's Sweeny olefins complex. Information contained in hi-fidelity process models that are used for online optimization was leveraged successfully. The capability of online optimization models was enhanced to compute these parameters during regular optimization cycle. An Excel graphical front-end makes report generation easy and more readable. Ultimately, the aim was to provide information to be used by Process, Production, and Control Engineers for identification of opportunities, problem identification and enhanced productivity. Process monitoring has also benefited real-time optimization efforts.

6.0 Acknowledgements

Author would like to acknowledge the contribution of the following people who helped during conception and implementation of the process monitoring methodology.

1. Richard Prickett, Technical manager at Chevron Phillips Chemical Company.
2. Sandeep Lal and Seung Sung, co-workers at Chevron Phillips Chemical Company.
3. Viral Maniar, Aspire Global solutions.

Cooperative, Distributed Model Predictive Control for Managing Resource Coupled Constraints

Brett T. Stewart
Dept. of Chemical Engineering, University of Wisconsin - Madison
James B. Rawlings
Dept. of Chemical Engineering, University of Wisconsin - Madison
Gabriele Pannocchia
Dept. of Chemical Engineering, University of Pisa

Abstract

In chemical plants, production often involves a series of unit operations interconnected through material, energy and information flows. Traditionally, in plants for which centralized control is judged impractical or unmanageable, the unit operations are controlled in a decentralized way so that closed-loop interactions between controllers are unmodeled. It is well known, however, that if these interactions are strong plantwide performance is poor.

Cooperative model predictive control (MPC) has been proposed as a method for coordinating multiple optimization-based controllers. In this control strategy, input trajectories are passed between controllers, each optimizing a plantwide objective. It has been shown that cooperative MPC is provably stable and is plantwide optimal at convergence. A necessary assumption for optimality is that each subsystem's inputs are constrained independently. Removing it gives suboptimal performance. This assumption is not valid, however, for plants in which the constraints between controllers are coupled. This situation arises if a common resource must be optimally shared between unit operations or in plants with strong inter-subsystem interactions and output constraints.

We show that including the coupled inputs in each controller eliminates the sub-optimality and gives good plantwide performances. We also propose an auxiliary optimization that enhances cooperative MPC to manage coupled constraints. This constraint manager decouples these constraints by computing the optimal inner box contained within the plantwide feasible space. The optimization proceeds asynchronously and can be terminated at a suboptimal iterate. We show the augmented optimization does not weaken the cooperative MPC stability properties and is plantwide optimal at convergence. We provide some examples showing performance properties.

1 Introduction

Model predictive control (MPC) has become one of the most popular forms of control in industry (Qin and Badgwell, 2003). In MPC, the inputs are determined via an online optimization, taking into account current operating conditions, such as disturbances or economic fluctuations. Its success is attributed to its ability to explicitly handle constraints and maintain stability. MPC has been implemented in many applications, particularly in the petro-chemical industry for controlling critical systems in chemical plants and refineries.

For plantwide control, it is often more favorable to use several MPC controllers instead of a single controller. Utilizing centralized control implies all control decisions are the result of a single optimization, and therefore the performance of the plant is reliant upon this optimization. Many practitioners view centralized control as impractical for these robustness reasons. Alternatively, decentralized control divides the plants into subsystems connected via material, energy, and information flows. Traditionally, the inter-subsystem interactions are treated as disturbances (Sandell Jr., Varaiya, Athans, and Safonov, 1978). Decentralized control techniques try to detune the controllers to minimize closed-loop interactions between subsystems, but it is well-known that decentralized control is unstable for highly interacting subsystems.

Between centralized and decentralized control techniques, there is distributed control. These methods maintain the topology of decentralized control, but, because they explicitly model the inter-subsystem interactions, give centralized-like stability properties. Cooperative control is a type of distributed control in which each subsystems solves a distributed optimization problem for the entire plant (Venkat, Rawlings, and Wright, 2006). This method has been shown to be plantwide stabilizing, even when the subsystems are highly interactive. Moreover, in the limit of infinite iterates, cooperative control converges to centralized control performance. For these reasons, we consider cooperative control in this paper.

This paper is primarily concerned with an extension of cooperative control to plants with coupled input constraints. In Section 2, we give a description of cooperative model predictive control. Section 3 motivates the problem of coupled constraints and offers two solution methods, including examples showing performance. Finally, we state conclusions.

2 Cooperative model predictive control

Cooperative model predictive control is a method of distributed control in which each subsystem's controller has the same plantwide objective function. Each controller participates in a plantwide optimization problem, taking into account not only how it can exploit the state of the plant, but also how it may potentially affect the other subsystems in the plant. The favorable stability properties of cooperative control are a result of this feature. In this section, we give a technical description of cooperative MPC.

2.1 Models

We assume linear time-invariant models such that each input affects each output. For each subsystem i to each subsystem j, there exists a discrete time model. For the interactions between subsystem i to subsystem j

$$x_{ij}(k+1) = A_{ij}x_{ij}(k) + B_{ij}u_j(k)$$

in which $x_{ij} \in \mathbb{R}^{n_{ij}}$, $u_j \in \mathbb{R}^{m_j}$, $A_{ij} \in \mathbb{R}^{n_{ij} \times n_{ij}}$, and $B_{ij} \in \mathbb{R}^{n_{ij} \times m_j}$. We assume that each (A_{ij}, B_{ij}) is stabilizable.

For each subsystem i the output is a linear combination of these states

$$y_i(k) = \sum_{j=1}^{M} C_{ij}x_{ij}$$

in which $y_i \in \mathbb{R}^{p_i}$ and $C_{ij} \in \mathbb{R}^{p_i \times n_{ij}}$. We assume that each (A_{ij}, C_{ij}) pair is detectable.

2.2 Objective function

For each subsystem i, there exists an objective function measuring the deviation of its states and inputs away from zero.

$$V_i(\mathbf{x}_i, \mathbf{u}_i; x_i(k)) = \frac{1}{2} \sum_{t=k}^{k+N-1} x_i(t)^T Q_i x_i(t) + u_i(t)^T R_i u_i(t) + x_i(k+N)^T P_{fi} x_i(k+N)$$

in which $x_i = [x_{i1}^T \cdots x_{iM}^T]^T \in \mathbb{R}^{n_i}$, $Q_i \in \mathbb{R}^{n_i \times n_i}$, $R_i \in \mathbb{R}^{m_i \times m_i}$, and the terminal penalty $P_{fi} \in \mathbb{R}^{n_i \times n_i}$. The arguments for the objective function, $\mathbf{x}_i \in \mathbb{R}^{Nn_i}$ and $\mathbf{u}_i \in \mathbb{R}^{Nm_i}$, are the concatenated vectors of the state and input trajectories. We assume each (A_{ij}, Q_{ij}) pair is detectable, in which Q_{ij} is the jth block diagonal of Q_i. We define the plantwide objective as a convex combination of each subsystem objective

$$V(\mathbf{x}, \mathbf{u}; x(k)) = \sum_{i=1}^{M} \rho_i V_i(\mathbf{x}_i, \mathbf{u}_i; x_i(k)) \qquad \text{s.t.} \quad \sum_{i=1}^{M} \rho_i = 1, \quad \rho_i > 0 \quad \forall i$$

in which $\mathbf{x} \in \mathbb{R}^{Nn}$ and $\mathbf{u} \in \mathbb{R}^{Nm}$ are the concatenated vectors of all state and input trajectories in the plant.

2.3 Constraints

For cooperative MPC, it is assumed that each subsystem's inputs have a separate, convex feasible space. For each subsystem i this feasible space is described by the polytope

$$D_i u_i \leq d_i$$

in which $D_i \in \mathbb{R}^{nc_i \times m_i}$ and $d_i \in \mathbb{R}^{nc_i}$. In the next section we explore the implications of expanding this assumption.

2.4 Cooperative target calculation

The target problem is solved in a similar way to the regulation problem. Each subsystem i solves for the its steady state in the following optimization. For iterate l

$$\begin{aligned} u_{s,i}^{0(l)} = \arg\min_{u_{s,i}} & \sum_{i=1}^{M} (y_{s,i} - y_{sp,i})^T Q_{s,i} (y_{s,i} - y_{sp,i}) + (u_{s,i} - u_{sp,i})^T R_{s,i} (u_{s,i} - u_{sp,i}) \\ \text{s.t.} \quad & (I - A_{ij}) x_{s,ij} = B_{ij} u_{s,j} + B_{d,ij} d_j(k) \quad \forall i, j \\ & y_{s,i} = \sum_{j=1}^{M} C_{ij} x_{s,ij} + C_{d,i} d_i(k) \quad \forall i \\ & D_i u_{s,i} \leq d_i \\ & u_{s,j} = u_{s,j}^l \quad \forall j \neq i \end{aligned}$$

in which $u_{sp,i} \in \mathbb{R}^{m_i}$ and $y_{sp,i} \in \mathbb{R}^{p_i}$ are externally provided setpoints and $u_{s,i} \in \mathbb{R}^{m_i}$, $x_{s,ij} \in \mathbb{R}^{n_{ij}}$ and $y_{s,i} \in \mathbb{R}^{p_i}$ are steady states for subsystem i. The vector $d_i(k) \in \mathbb{R}^{p_i}$ is the estimate of the disturbances affecting subsystem i at time k with $B_{d,ij} \in \mathbb{R}^{n_{ij} \times p_i}$ and $C_{d,i} \in \mathbb{R}^{p_i \times p_i}$. In practice we often set $B_{d,ij} = 0$ for $j \neq i$ so that disturbances are modeled to affect only local outputs. We employ a convex combination to update the iterate

$$u_{s,i}^{l+1} = w_i u_{s,i}^{0(l)} + (1 - w_i) u_{s,i}^l$$

in which w_i is the selected weighting of subsystem i in the plant such that $\sum_{i=1}^{M} w_i = 1$. The choice of w_i is a tuning parameter available to the practitioner. In general, a choice of a bigger w_i means that the inputs in subsystem i are more critical to the operation of the plant.

2.5 Cooperative dynamic optimization

For the dynamic problem, we represent the plant as deviation variables away from the calculated steady state. At each sample time k, every subsystem i solves the following optimization problem. For iterate l

$$\begin{aligned} \mathbf{u}_i^{0(l)} = \arg\min_{\mathbf{u}_i} & \; V(\mathbf{x}, \mathbf{u}; x(k)) \\ \text{s.t.} \quad & x_{ij}(t+1) = A_{ij} x_{ij}(t) + B_{ij} u_j(t) \quad \forall i, j \\ & D_i u_i(t) \leq d_i \quad \forall t = k, \ldots, k + N - 1 \\ & \mathbf{u}_j = \mathbf{u}_j^l \quad \forall j \neq i \end{aligned}$$

in which $\mathbf{u}_i = [u_i(k)^T, \ldots, u_i(k + N - 1)^T]^T$. The iterate update is a convex combination of each of these solutions

$$\mathbf{u}_i^{l+1} = w_i \mathbf{u}_i^{0(l)} + (1 - w_i) \mathbf{u}_i^l$$

Using this optimization, it can be shown that $V(\mathbf{x}^{l+1}, \mathbf{u}^{l+1}) \leq V(\mathbf{x}^l, \mathbf{u}^l)$ and $(\mathbf{x}^l, \mathbf{u}^l) \to (\mathbf{x}^*, \mathbf{u}^*)$ as $l \to \infty$.

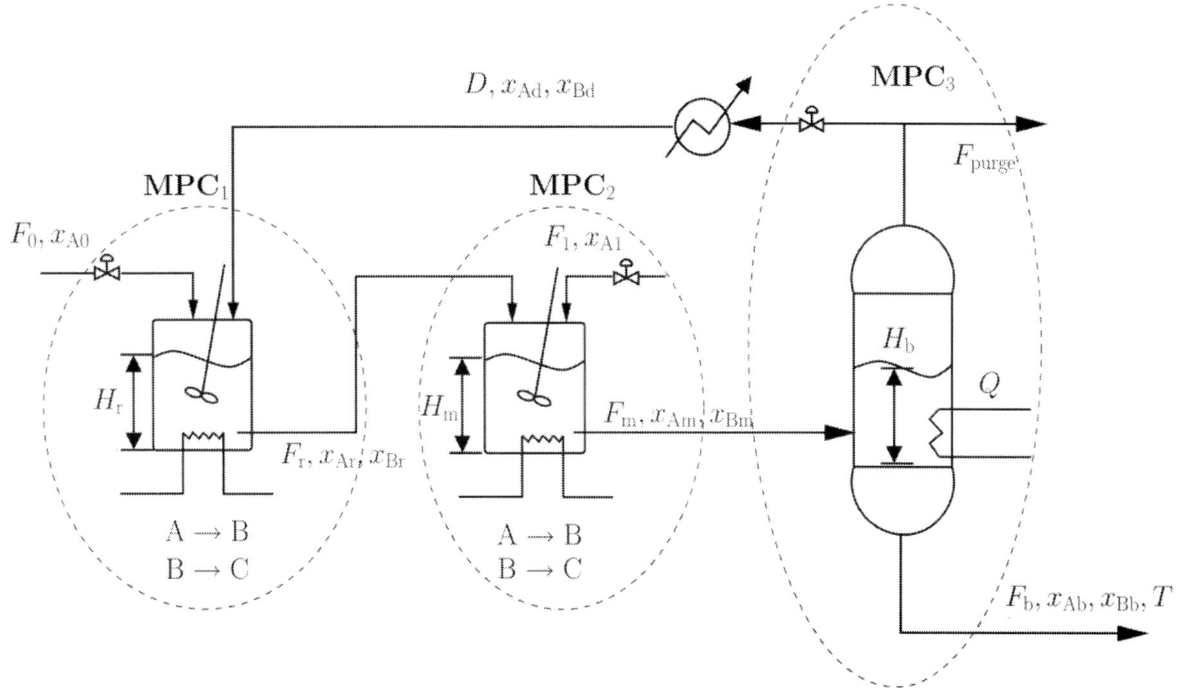

Figure 1: Two reactors in series with a separator and recycle loop. Each unit is controlled via a separate MPC.

3 Coupled constraints

In the previous section, we assume that each subsystem's inputs have separable constraints so that the feasible region does not depend upon the states or inputs of any other subsystem. This condition implies that each subsystem's inputs do not physically interact in anyway, e.g., the position of one valve does not mechanically limit the position of any another subsystem's valve. It also implies there are no physical or safety limitations spanning subsystems. In many cases this assumption is not valid. For instance, subsystems drawing the same resource are constrained by the resource's availability. It is impossible, however, for each subsystem to request all of a shared resource and receive it, so in practice there is a constraint that the total requests must not exceed the total availability. Such constraints are coupled constraints.

In general, the plantwide constraints are of the form

$$\begin{bmatrix} D_1 & & & \\ & D_2 & & \\ & & \ddots & \\ & & & D_M \\ D_{C1} & D_{C2} & \cdots & D_{CM} \end{bmatrix} \begin{bmatrix} u_1 \\ u_2 \\ \vdots \\ u_M \end{bmatrix} \leq \begin{bmatrix} d_1 \\ d_2 \\ \vdots \\ d_M \\ d_C \end{bmatrix} \tag{1}$$

in which the last row of constraints are the coupled constraints.

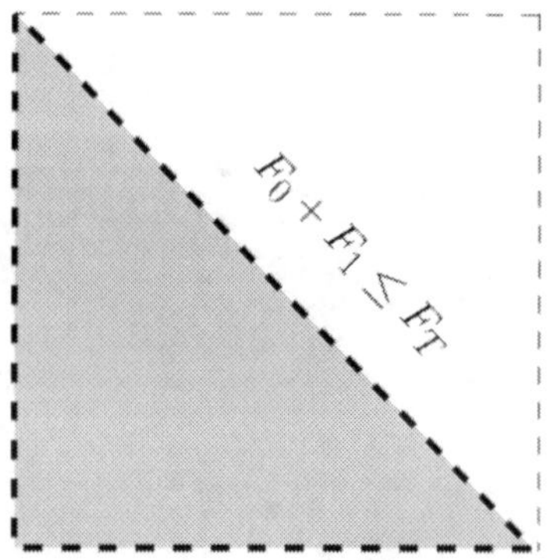

Figure 2: Geometry of coupled constraints.

3.0.1 Example

Consider the example in Figure 1. The reactant is distributed between the two reactors with feed rates F_0 and F_1. Usually the feeds draw from the same feed stock. In this case, there exists a limiting rate F_T defining the maximum feed available for distribution. This upper limit constrains the sum of the feed rates

$$F_0 + F_1 \leq F_T$$

This constraint is coupled because F_0 and F_1 are decision variables for different MPCs. This constraint defines a diagonal half-plane in the (F_0, F_1) space (see Figure 2).

3.1 A naive strategy

If we wish to allow coupled constraints in the optimization, then one method is to explicitly include these constraints into the cooperative MPC problem as hard constraints. In this strategy, the cooperative target calculation includes the following constraint in subsystem i

$$D_{Ci} u_{s,i} \leq d_C - \sum_{j \neq i}^{M} D_{Cj} u_{s,j}^{l}$$

in which the subscript $j \neq i$ implies $j \in \{1, \ldots, i-1, i+1, \ldots, M\}$. A similar constraint is included in the regulation problem.

These constraints, however, can cause the cooperative optimization to get stuck at a suboptimal point. This behavior is because the point may appear to be optimal to each subsystem when it reaches a point on the boundary of the coupled constraint. For instance, see Figure 3. This figure represents a target calculation in which the setpoint u_{sp} is unreachable due to constraints. The feasible region contains a diagonal, and therefore coupled, constraint. Suppose the cooperative optimization is initialized on the boundary of the coupled constraint at $u^0 = (u_1^0, u_2^0)$. Subsystem 1 optimizes u_1 (in the horizontal direction). In this direction the negative gradient points out of the feasible region, therefore the u_1 component stays the same upon optimization. For subsystem 2, the negative gradient also points out of the feasible region, and hence the u_2 component also does not move. Therefore if the cooperative optimization is initialized at point u^0 it cannot find a more optimal point. By

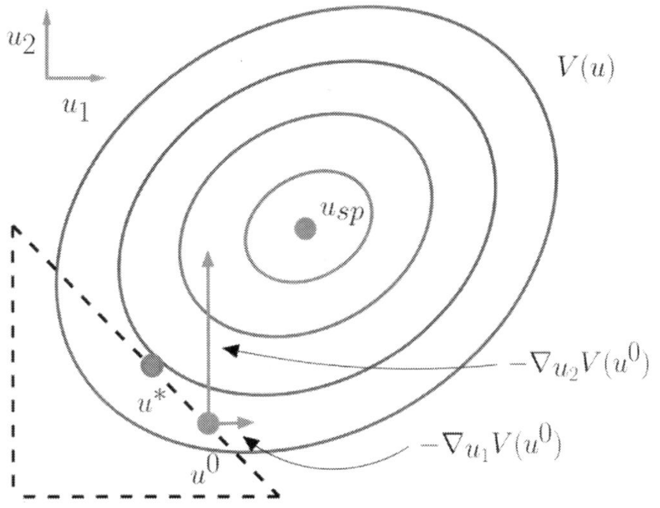

Figure 3: The coupled constraint causes the cooperative optimization to get stuck at a suboptimal point

inspection, however, u^0 is not the optimal. The point u^* is feasible and has the lowest cost. By relaxing the assumption of separable constraints, the optimality property is lost.

3.2 Infeasible solution methods

Current methods optimally solving distributed optimization problems with coupled constraints involve relaxing the coupled constraints and including them in the objective function. Then some type of penalty function method, e.g. augmented Lagrangian, is used to solve the optimization problem (Cheng, Forbes, and Yip, 2007; Kuwata and How, 2006; Necoara, Doan, and Suykens, 2008). These strategies, however, are not interior point methods, and therefore do not guarantee feasibility of the coupled constraints until convergence.

Feasibility of the coupled constraints, however, is an important property for robust and safe plantwide performance. Resource constraints specify hard physical limits for the plant. Other coupled constraints may be needed to ensure the plant stays within safety margins. Controllers that ask the plant to operate in areas violating these constraints can give unstable performance. Convergence of any cooperative optimization method takes time[1]. If convergence is required for feasibility and the optimization has not converged by the time an input is needed, then either an open-loop input or suboptimal input must be used. In either case the input may be infeasible. This motivates the use of methods that can be terminated early and remain feasible.

3.3 Input augmentation

One method that ensures the inputs are plantwide feasible is to augment each subsystem's decision variables with the inputs coupled via constraints. We repartition the plantwide

[1] In some cases, convergence of a cooperative optimization takes more time than the equivalent centralized optimization.

constraints from equation (1) to form the following

$$\begin{bmatrix} \tilde{D}_1 & & & & \\ & \tilde{D}_2 & & & \\ & & \ddots & & \\ & & & \tilde{D}_M & \\ & & & & \tilde{D}_C \end{bmatrix} \begin{bmatrix} u_{U,1} \\ u_{U,2} \\ \vdots \\ u_{U,M} \\ u_C \end{bmatrix} \leq \begin{bmatrix} \tilde{d}_1 \\ \tilde{d}_2 \\ \vdots \\ \tilde{d}_M \\ \tilde{d}_C \end{bmatrix}$$

in which $u_{U,i} \in \mathbb{R}^{\tilde{m}_i}$ are the inputs unaffected by coupled constraints in subsystem i and $u_C \in \mathbb{R}^{\tilde{m}_C}$ are all inputs affected by coupled constraints. Here, an input is "affected" by a coupled constraint not only by being directly involved. If an input is in the same row of D_i as an input included in a coupled constraint, then it is also included in u_C. In the case for all inputs being affected by coupled constraints, $u_C = [u_1^T, \ldots, u_M^T]^T$ and $u_{U,i} = \{\emptyset\}$ for all i. We define $u_{C,i} \in \mathbb{R}^{\tilde{m}_C}$ to be the copy of u_C in subsystem i such that

$$u_C = \sum_{i=1}^{M} \alpha_i u_{C,i}$$

in which α_i is a user specified convex weighting (i.e. $\sum_{i=1} \alpha_i = 1$ and $\alpha_i > 0, \forall i$). The constraints for subsystem i are then

$$\begin{bmatrix} \tilde{D}_i & \\ & \tilde{D}_C \end{bmatrix} \begin{bmatrix} u_{U,i} \\ u_{C,i} \end{bmatrix} \leq \begin{bmatrix} \tilde{d}_i \\ \tilde{d}_C \end{bmatrix}$$

Notice these constraints are not coupled between subsystems.

With these definitions, we implement cooperative MPC on the augmented system. In the optimizations described in sections 2.4 and 2.5, we replace the following parameters

$$u_i \leftarrow \begin{bmatrix} u_{U,i} \\ u_{C,i} \end{bmatrix} \quad D_i \leftarrow \begin{bmatrix} \tilde{D}_i & \\ & \tilde{D}_C \end{bmatrix} \quad d_i \leftarrow \begin{bmatrix} \tilde{d}_i \\ \tilde{d}_C \end{bmatrix}$$

Partitioning the model is done is similar to the constraint partitioning

$$A_{ij} \leftarrow A_{ij} \qquad B_{ij} \leftarrow \begin{bmatrix} \tilde{B}_{ij} & \alpha_i \tilde{B}_{i,C} \end{bmatrix} \qquad C_{ij} \leftarrow C_{ij}$$

in which $B_{i,C} \in \mathbb{R}^{\tilde{m}_i \times \tilde{m}_C}$. And finally the optimization variables are partitioned

$$Q_i \leftarrow Q_i \qquad R_i \leftarrow \begin{bmatrix} \tilde{R}_i & \\ & \tilde{R}_C \end{bmatrix}$$

in which $\tilde{R} \in \mathbb{R}^{\tilde{m}_C \times \tilde{m}_C}$ is the matrix of penalties for u_C.

3.3.1 Example

Consider the plant from the previous example, in which we have included the constraint $F_0 + F_1 \leq F_T$ as before. At time $t = 0$, there is a setpoint change in H_r that is unreachable. The control performance is shown in Figure 4. Explicitly using the coupled constraints in

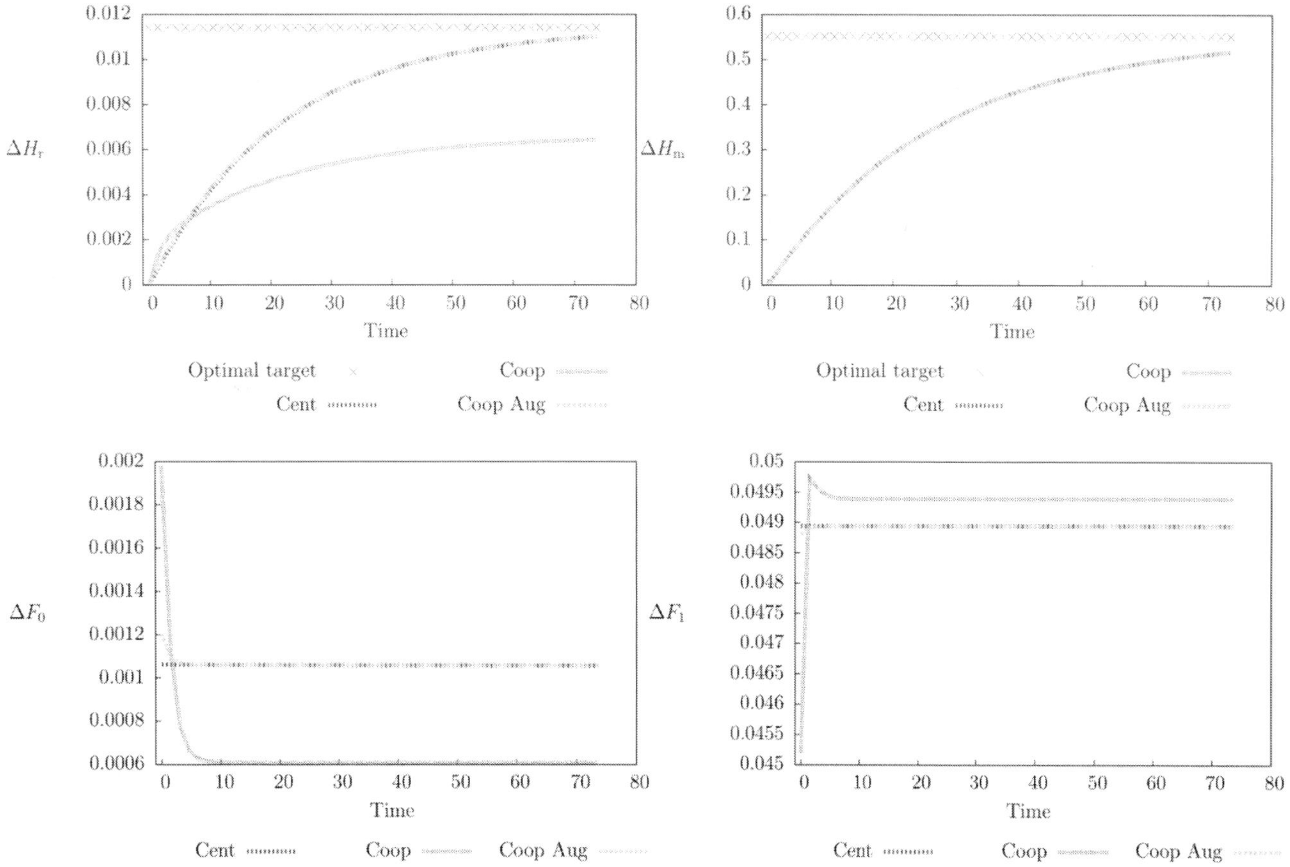

Figure 4: Performance of two reactor with separator and recycle. Coupled inputs are included between reactors. Cooperative control alone gives steady-sate offset. Augmenting the inputs (Coop Aug) gives performance similar to centralized.

	Cost	Performance loss
Centralized MPC	15.391	-
Cooperative MPC (1 iterate)	20.656	30.7%
with augmented inputs	15.402	0.071%

Table 1: Performance comparison of input augmentation example.

cooperative control gives steady-state offset. This suboptimal performance is caused by the coupled constraint. Augmenting the inputs, however, gives almost no performance loss (see Table 1).

Notice that the decision variables space for each subsystem increases in size from m_i to $\tilde{m}_i + \tilde{m}_C$. When the coupled constraints are sparse, such as when there are only a few shared resources in the plant, this increase in size does not adversely affect the computation time for each subsystem. If the coupling is non-sparse, however, then each subsystem's optimization approaches the size of the centralized problem. In the next section, we consider this case.

3.4 Constraint management

Coupled input constraints are also present in plants with output constraints. This occurs because each subsystem is a function of inputs from other subsystems. Placing limits upon outputs ultimately constraints the inputs. Therefore plants with inter-subsystem interactions and output constraints have coupled input constraints. Plants taking advantage of energy integration and material recycle have many inter-subsystem interactions. Utilizing the input augmentation method for these densely coupled plants results in each subsystem augmenting its decision variable space with almost every input. In this case, the number of decision variables may be too large for implementation. Alternatively, an auxiliary centralized calculation managing the constraint partitioning can be employed. [2] This auxiliary optimizer, here called the constraint manager, finds the optimal constraint partitioning such that the plantwide constraints are both feasible and uncoupled.

For a given set of coupled constraints $Du \leq d$, partitioned as in equation (1), consider the set

$$\hat{\Omega}(\hat{d}) = \{u : D_i u_i \leq \hat{d}_i, \sum \hat{d}_i = d\}$$

in which $\hat{d} \in \mathbb{R}^{Mn_C}$ is the vector of right-hand-sides for each subsystem's constraints. This set is constructed such that $\sum D_i u_i \leq d$ and is therefore plantwide feasible. Each $\hat{d}$ characterizes a feasible inner box for the original plantwide constraints. If we constrain the cooperative MPC problem to this set, then for each $\hat{d}$, there exists a resulting point of attraction $u(\hat{d})$. For instance, consider the example in Figure 3. Two feasible inner boxes and their resulting point of attraction are shown in Figure 5. Finding $\hat{d}^*$ such that $u^* \in \hat{\Omega}(\hat{d}^*)$ gives an implementable set of constraints that guarantees the cooperative MPC is optimal upon convergence.

The constraint manager runs in parallel with the cooperative MPC algorithm and solves the following centralized optimization

$$\begin{aligned} \min_{\hat{d}} \quad & V(x, u(\hat{d})) \\ \text{s.t.} \quad & \sum \hat{d}_i = d \\ & u(\hat{d}) = \arg\{\min_u V(x,u) \quad \text{s.t.} \quad D_i u_i \leq \hat{d}_i \quad \forall i, \quad \text{and model equations}\} \end{aligned}$$

in which x and u are the corresponding parameters for the cooperative target and dynamic problems. Finding iterates, however, is a difficult. The constraint manager solves a bilevel

[2] Instead of needing M almost centralized optimizations, we need only one centralized and M small optimizations.

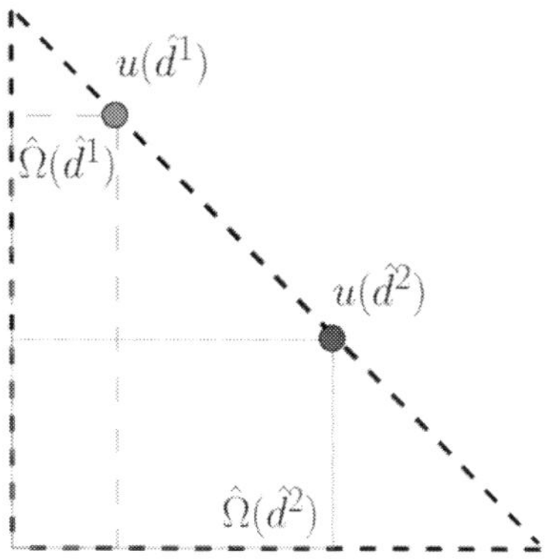

Figure 5: Constraint manager chooses sequence of inner box constraints $\{\hat{\Omega}(\hat{d}^k)\}$ forming point of attraction sequence $\{u(\hat{d}^k)\}$ so that $u(\hat{d}^k) \rightarrow u^*$.

optimization problem with quadratic objective functions in both the upper and lower problems. In general, merely checking first-order local optimality conditions for these problems is NP-hard because the lower problem is not differentiable (Vicente, Savard, and Júdice, 1994). Therefore, we employ a trick. Because both the upper and lower objective functions are the same, we can switch their order. We solve the lower problem via a feasible algorithm and generate a $\hat{d}$ vector that could have produced the lower problem solution

$$\begin{aligned} \min_{\mathbf{u}} \quad & V(x,u) \\ & \text{s.t.} \quad D_i u_i \leq d \quad \forall i \\ & \text{and model equations} \\ & \hat{d}(u) \in \mathcal{D}(u) = \{\hat{d} : D_i u_i \leq \hat{d}_i \quad \forall i\} \end{aligned}$$

in which $\mathcal{D}(u)$ is the set of $\hat{d}$ defining the inner box for which u is feasible. Because the set $\mathcal{D}(u)$ is not usually a singleton, we pick the $\hat{d}$ corresponding to the inner box with maximum feasible volume. If this optimization is solved via a feasible method as a suboptimal MPC problem (see Scokaert, Mayne, and Rawlings (1999)), then the suboptimal sequence of right-hand-sides can be injected into the plant without loss of stability, and the sequence converges to optimality. Again consider the example in Figure 3. Two iterates and the feasible inner boxes generated by the constraint manager are shown in Figure 6.

3.4.1 Example

The constraint manager is used on the same system as in the previous example. The performance of the constraint manager is shown in Figure 7. For this example, the cost lost is almost zero (see Table 2).

4 Conclusions and future work

In this paper we have shown a technical description of cooperative control and motivated the problem of coupled input constraints. We have shown two methods, input augmentation

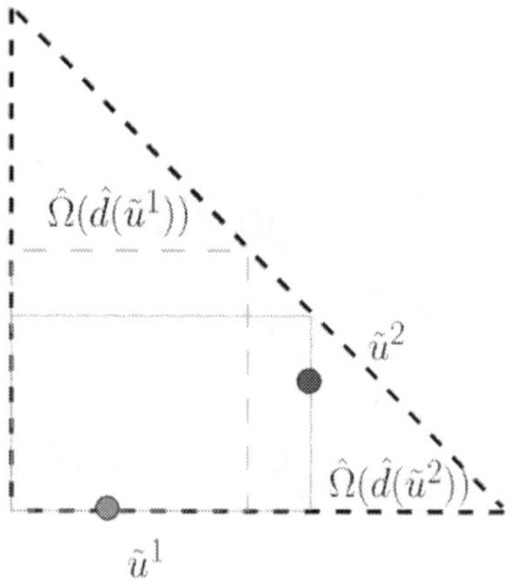

Figure 6: The constraint manager solves lower problem for candidate $\tilde{u}^k$ and chooses corresponding $\hat{\Omega}(\hat{d}(\tilde{u}^k))$ that maximizes the feasible volume.

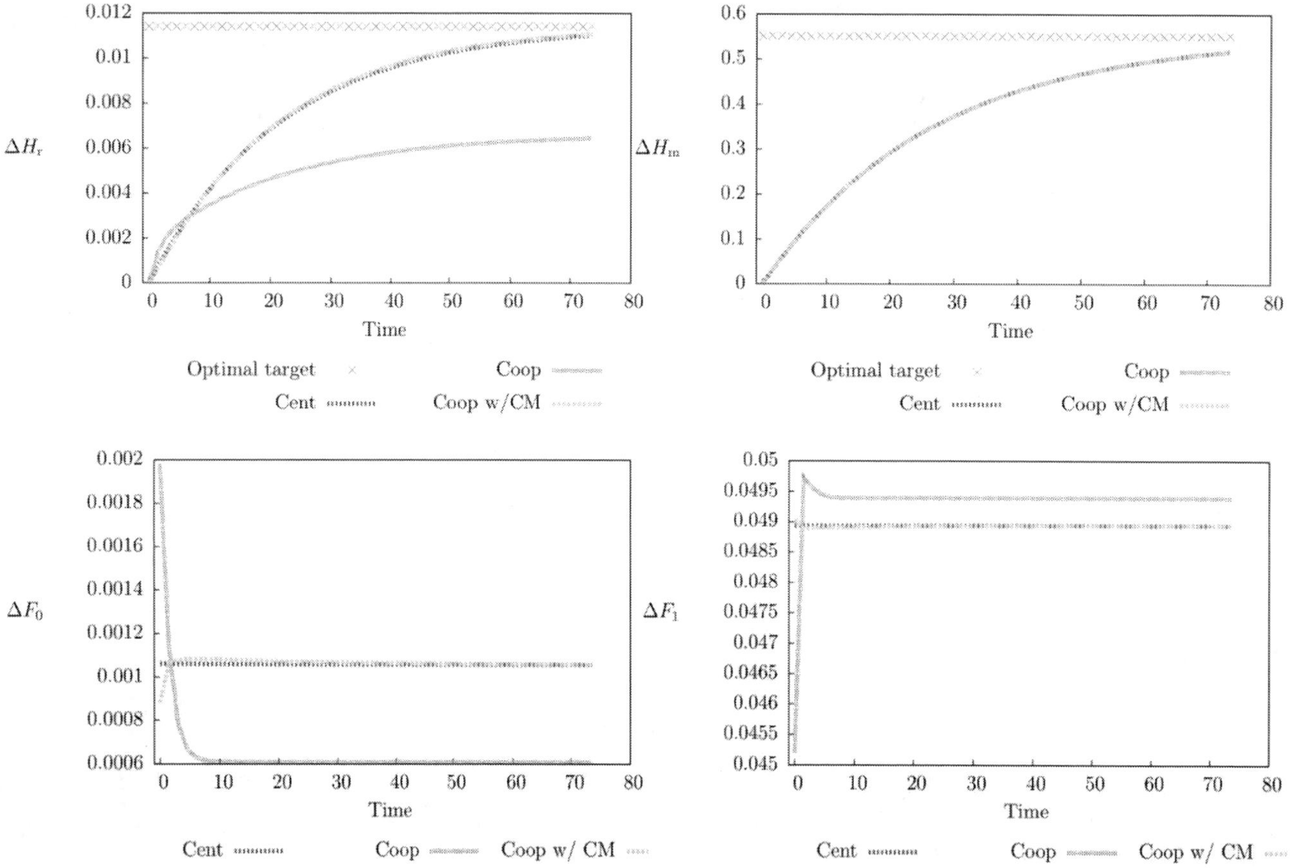

Figure 7: Performance of two reactor with separator and recycle. Coupled inputs are included between reactors. Using the auxiliary constraint manager (Coop w/CM) eliminates steady-state offset.

	Cost	Performance loss
Centralized MPC	15.391	-
Cooperative MPC (1 iterate)	20.656	30.7%
with constraint management	15.417	0.17%

Table 2: Performance comparison of input augmentation example.

and constraint management, each achieving centralized-like plantwide performance in the presence of these constraints without steady-state offset. An example is presented showing these features. As shown by this example, the performance of the two methods is similar. To extend this work, more examples are needed to show the advantages and disadvantages of each method. Moreover, there is a trade-off between density of constraint coupling and optimization time when selecting between the methods. More examples are needed to show this trade-off.

References

R. Cheng, JF Forbes, and WS Yip. Price-driven coordination method for solving plant-wide MPC problems. *J. Proc. Cont.*, 17(5):429–438, 2007.

Yoshiaki Kuwata and Jonathan How. Decentralized cooperative trajectory optimization for UAVs with coupling constraints. In *Proceedings of the 45th IEEE Conference on Decision and Control*, San Diego, California, December 13-15 2006.

Ion Necoara, Dang Doan, and J.A.K Suykens. Application of the proximal center decomposition method to distributed model predictive control. In *Proceedings of the IEEE Conference on Decision and Control*, Cancun, Mexico, December 9-11 2008.

S. Joe Qin and Thomas A. Badgwell. A survey of industrial model predictive control technology. *Control Eng. Prac.*, 11(7):733–764, 2003.

Nils R. Sandell Jr., Pravin Varaiya, Michael Athans, and Michael Safonov. Survey of decentralized control methods for larger scale systems. *IEEE Trans. Auto. Cont.*, 23(2): 108–128, 1978.

Pierre O. M. Scokaert, David Q. Mayne, and James B. Rawlings. Suboptimal model predictive control (feasibility implies stability). *IEEE Trans. Auto. Cont.*, 44(3):648–654, March 1999.

Aswin N. Venkat, James B. Rawlings, and Stephen J. Wright. Stability and optimality of distributed, linear MPC. Part 1: state feedback. Technical Report 2006–03, TWMCC, Department of Chemical and Biological Engineering, University of Wisconsin–Madison (Available at http://jbrwww.che.wisc.edu/tech-reports.html), October 2006.

L.N. Vicente, G. Savard, and J.J. Júdice. Descent approaches for quadratic bilevel programming. *J. Optim. Theory Appl.*, 81(2):379–399, 1994.

Best Practices to Support and Sustain Advanced Process Control and Optimization Projects in Remote Regions

Rogier Pouwer

Advanced Solutions Leader

Honeywell International

Steve Johansen

Manager, Services Marketing & Business Development

Honeywell International

Brendan Sheehan

Chemical Marketing Manager

Honeywell International

Abstract: This paper identifies potential challenges and pitfalls of supporting advanced control (APC) and optimization projects in remote regions and then describes best practices that help avoid them. These best practices consist of building a comprehensive support and maintenance framework that:

- Makes better use of remote monitoring technology
- Uses technology tools to more clearly identify problems and then prioritize resolution efforts
- Leverages the use and training of local resources
- Layers support services to enable fast response to simple problems with clear escalation channels
- Structures support contracts to goal both the supplier and end user to jointly support installed applications.

Introduction

This paper presents best practices and methodologies to support advanced process control (APC) and optimization projects in remote regions. With the majority of new ethylene capacity being built in the Middle East and China, more and more APC and optimization projects are being implemented in regions where is it is difficult to find and retain skilled resources to support the installed applications. If support issues are not explicitly addressed, application benefits can rapidly decline. This leaves the end user with a complex application that has been expensive to implement and now is not producing the returns that were expected.

Successful support of APC and optimization applications is not something that just happens after the project is complete. In order to be successful, application maintenance aspects need to be planned and addressed very early in the project cycle so that maintenance and support preparations can be worked in parallel with the project delivery efforts.

In Honeywell's experience, support and maintenance challenges arise in all project phases that are broadly categorized in this paper as:

- Pre-project
- Project implementation
- Post-project

Failing to address support factors in any of these stages will significantly lower the probability of sustaining long-term benefits from installed advanced control and optimization applications.

In each of these project phases, we will identify support challenges and pitfalls and then outline support best practices and methodologies that can be implemented to minimize their impact on the long-term supportability of the application. What you will find is that long-term APC and optimization success depends on a consistent and methodical layering of support components that must occur before, during, and after implementation.

Pre-Project Support Considerations

Challenges During Project Justification and Project Award

Decisions in the pre-project stage can directly impact the long-term success of supporting and sustaining advanced control and optimization projects. Prior to project award, the following factors commonly result in future support and maintenance challenges:

- Insufficient consideration of lifecycle support factors
- Overestimation of resource capabilities and/or expected results
- One-sided or non-definitive support agreements

Even though the project has not yet begun, any one of these factors can substantially increase the risk of not being able to support and sustain the APC and optimization benefits.

Pre-Project Challenge #1: Insufficient consideration of lifecycle support factors

Not giving appropriate focus to the long-term support implications at the time of project justification and award is probably the most common failure mode in advanced control and optimization pursuits. This oversight occurs for a variety of reasons:

- Support is commonly treated independently from the project effort
- The project is under cost pressure so non-capital expense items like support tools, training, and other maintenance related content is deferred
- The scope to support and maintain the applications is grossly underestimated

However, missing the opportunity to bring support considerations to the table at the time of project justification and contract award has broad repercussions with respect to laying a successful application support and maintenance framework.

Pre-Project Challenge #2: Overestimation of resource capabilities and/or expected results

Overestimation of internal resource capabilities and/or over promising of expected results will almost always lead to problems in future support of the application. Some common examples are:

- The bid is highly competitive and the end-user negotiates guarantees for the best possible performance
- The technology is very new which limits the supplier's ability to reliably predict/estimate end user benefits
- Technical skills required to maintain the applications and/or technology are underestimated.

Many customers in countries without sufficiently skilled internal resources use external suppliers for implementation of advanced control and optimizations solutions. Typical practice is to invite three or more companies to participate in a competitive bidding process in which each party provides a scope of work and an associated cost. To properly compare each company's bid, the customer will typically prepare a bid specification that describes what the scope of work will involve and what benefits are expected. To minimize risk and ensure that the project is justified, the customer will often ask the supplier to guarantee delivered benefits.

When selecting the best bid, a common outcome is to select the bid with the highest perceived value (guaranteed benefits less the projected cost.) However, this is where the above examples come into play.

In a competitive situation, a supplier may well be expected to guarantee more benefits than the end user can reliably expect to achieve. This is a calculated risk since it is not always easy to know the exact level of achievable benefits until the actual project is underway. If a supplier is comfortable with guaranteeing 3% throughput improvement, they may still accept a guarantee of 5% since the financial risk in offering the higher improvement level is typically defined as a fixed percentage of the contract value.

When the time comes to meet the guaranteed benefits, several scenarios can occur:

Scenario #1: The guaranteed benefit level is not achieved and the supplier ends up paying a contract penalty. However, because the project was justified on a higher expected ROI result, an increased support burden is usually placed on maintenance team to try to achieve the higher benefit level. In many cases, this is an unrealistic expectation because a well designed and implemented APC and optimization application will always push the plant up against multiple constraints. When active constraints are physical and/or hydraulic limitations, no amount of support effort is going to address the benefit shortfall.

Scenario #2: The guaranteed benefit level is achieved but is not sustainable on a consistent basis due to limits that may vary due to changing conditions such as ambient temperature, feed composition, catalyst poisons, and other transient factors. The long-term support and sustainability of the application will be continuously challenged to perform at the project sign-off level of performance. In the short term, applications can be tuned or modified to optimally run under controlled test conditions. However, frequent and active application changes may be unsustainable during the course of normal operations.

Scenario #3: The benefits are achieved and are sustainable, but only if higher skill levels are engaged in application support and maintenance. When new technologies are introduced into the market or when existing technologies are being applied in new ways, unforeseen problems tend to show up after exposure to the wide range of operating conditions. Only resources with sufficient understanding of both the application and the technology can efficiently troubleshoot and resolve application issues. When these resources reside within central engineering groups or external supplier organizations, significant delays can be introduced into the support process.

Pre-Project Challenge #3: One-sided or non-definitive application support agreements

In some cases, support responsibilities are defined and agreed to in the upfront project agreement that make the onus of responsibility for support and success of the project shift too far from one party to the other. Depending on how the contract is structured, the following problems may arise:

- The support contract may be too stringent and includes clauses that shift too much support responsibility to one party
- The support contract may be too specific and calls for performance indicators that may not reflect varying economic and operating objectives
- The support contract may be too vague and have no specific performance indicators or includes factors that do not directly relate to achieving benefits

A customer without sufficient in-house resources will often try to put all the responsibility of supporting the installed application on the supplier. The problem with this approach is that supplier resources can rarely spend as much time monitoring and interfacing with the application as an on-site resource. In addition, if no one on-site is both responsible for and can fix minor issues as they occur, the build-up of small unresolved problems over time can cause the application performance to quickly deteriorate.

The opposite problem may result if the contract is too vague. For example, some contracts may call for an application uptime metric to be maintained. Uptime is a useful performance indicator, as failure to stay online almost always means that there is some kind of problem. However, this metric is not independently sufficient to measure support performance since it is quite possible for an application to be running but adding no financial benefit to the customer. An example of this scenario occurs when manipulated variables are so tightly constrained that the application is not able to achieve any financially beneficial control objectives such as maximizing throughput or yield. In other cases, the application's control objective may not be aligned with the current economic conditions. Although the application is "ON," it is not pushing the plant in line with current business objectives.

Pre-Project Solutions: Maintenance Planning is Key

A major factor for improving the sustainability of advanced control and optimization benefits is ensuring that appropriate support and maintenance aspects of a new project are addressed early in the pre-project planning stage. Ideally, lifecycle application maintenance should be explicitly considered at the time that new projects are being evaluated and justification for funding is being considered. Lifecycle maintenance costs cannot be neglected in the project planning phase and honest evaluation of internal support capabilities need to be performed as part of the upfront project evaluation. A reality check should also be made on the proposed solution outcome. For example, if the outcome of a solution is expected to be a 5% ethylene yield increase, that result should be further validated against plant limits that might not be defined in the project bid specification (e.g. heater limits, hydraulic limits, available feedstocks, etc.)

Unfortunately, project goals are commonly misaligned with long-term support objectives due to the fact that projects are funded as capital expenditures whereas support and maintenance related costs are treated as operating expenses. When capital and expense funding is considered separately, project decisions regarding technology choice and implementation can easily be made that result in lower capital costs and faster project implementation. However, these same decisions can negatively impact the supportability and long-term profitability of the project. The better approach is to first define your support and maintenance plan and then factor those plans in when making project decisions.

In this early stage of project planning, the following best practices have been identified that proactively address support and maintenance:

- Define and communicate project support requirements
- Budget for support and maintenance resources

- Identify application support and maintenance ownership
- Define and set goals that drive desired internal and external behaviors

Best Practice #1: Define and communicate project support requirements

Communication of short-term and long-term support requirements is critical to ensure that project leaders have the right information at the right time to make the right business decisions. Regular communication ensures that all team members are aware of expected support and maintenance requirements. This knowledge can then be integrated into upfront project justification, budgeting, planning, and scheduling. In many cases, support requirements should directly determine the appropriate technology and project implementation choices that are made, not the other way around.

Support requirements typically fall into the following categories:

- Technical support and troubleshooting
- Technology updates and refresh
- Maintenance and support tools
- Performance monitoring and analysis
- Engineer and operator training
- Application maintenance

Plants that proactively define support plans around each of these categories have a much higher probability of sustaining their APC and optimization benefits once the project is implemented.

Best Practice #2: Budget for support and maintenance resources

After definition and communication of the support requirements, the next step is to ensure that a budget is put in place to address those needs. Some support and maintenance requirements such as "maintenance and support tools" and upfront "engineer and operator training" should be considered for inclusion within the original project scope. This allows for easier justification of the tools and training while also allowing the tools to be installed and used to baseline application performance.

Support components such as "technical support and troubleshooting" and "technology update and refresh" commonly involve service contracts and/or outsourced services with advanced control and optimization suppliers. Therefore, it is important that funds be planned and allocated to facilitate access to outside supplier expertise and knowledge if this is determined to be a necessary support requirement.

The last two budget elements, "performance monitoring and analysis" and "application maintenance" have become increasingly important to ensure successful support and maintenance outcomes. These items cover labor resources required for the execution of sustaining maintenance tasks such as application tuning, model maintenance, and performance monitoring activities. The decision in this case involves either dedicating internal resources on these tasks or outsourcing these activities to an outside supplier.

Best Practice #3: Identify application support and maintenance ownership

Organizational ownership for application support and maintenance should be identified early and must be aligned with the support budgets described previously. Early identification of support and maintenance ownership provides a key opportunity for the support owner to engage the project team and ensures a smooth transition between project team and the support owner. This process also allows the support owner sufficient time to prepare for the maintenance effort which commonly involves internal staffing or outsourcing decisions.

Best Practice #4: Define and set goals that drive desired internal and external behaviors

Although somewhat early at this phase of the project, some preliminary thought should be given to the identification of key performance indicators (KPI's) which represent the overall objectives of the project. These metrics can then be used to define and set business goals to drive sustained performance long after the project has been completed.

In cases where performance metrics are clearly defined, project performance guarantees are sometimes structured into the initial project scope. This is particularly true in advanced control and optimization of olefins plants where input costs are immediately available and the desired output, yield of ethylene, is easily discernable.

Identification of appropriate KPI's also helps the customer recognize which supplier solution is most likely to meet their needs because the application's objective should be aligned to meeting the identified KPI's. It also enables the customer to recognize when competing bids are similar even if the offered guaranteed performance metrics vary. If the objectives of different supplier applications are similar then they should push to the same set of constraints and the benefits achieved will be similar. In this case the customer should focus more on other aspects of the proposal such as how support is provided rather than the specific guaranteed benefits.

Project Implementation Support Considerations

Challenges During Project Design & Execution

Once the contract has been awarded, a project team typically comprised of both internal and external supplier resources is assigned with the objective of designing, implementing, and commissioning the application. Ideally the team would be comprised entirely of local resources, but more often than not it is a mixture of experienced resources from a central office, external supplier resources from an established region, and less experienced local resources.

During the project phase the following challenges are commonly encountered:

- Difficulty with obtaining and retaining experienced lead resources
- Inconsistent skill levels within locally available resource pools
- On-site work by skilled resources is constrained

Project Implementation Challenge #1: Difficulty with obtaining and retaining experienced lead resources

For experienced consultants and implementation staff, projects in remote regions are often not their preferred assignments. If a sufficient pool of available work exists in other geographies, getting suitable experienced resources to lead projects in remote regions can be very difficult.

Project Implementation Challenge #2: Inconsistent skill levels within locally available resources pools

In every region, there are varying degrees of technical skills available in the discipline of advanced control and optimization. However, the quantity and quality of those resources can vary dramatically region to region.

External suppliers with a global reach will often have established local offices in high growth regions such as the Middle East, China, and India in anticipation of expanded business opportunities. However, it takes time to hire suitable local engineers and then provide them with sufficient implementation experience to become fully independent with current technologies. Even then, as new technologies are rolled out, their ability to implement, maintain and support applications can diminish over time.

Suppliers are not the only ones that have trouble finding suitable technical resources in remote regions. For example, when APC and optimization projects have been awarded in the Middle East, the client engineer assigned to the project is commonly an external contractor. In fact, many of the skilled technical resources working for manufacturing companies in the Middle East are contractors, either on specific contract to the Middle East entity or on assignment from a large multinational corporation. As a result, the project may start the with one client engineer as the primary contact but that lead role may transition to others as contractors come on and off assignments.

These inconsistencies in skill levels and rotating project members can lead to sub-optimal controller designs that are much more difficult to understand and support after the project is completed.

Project Implementation Challenge #3: On-site work by skilled resources is constrained

While much of the design work can be done from local offices, APC and optimization projects and their follow-on support requirements typically require considerable on-site work effort. However, the amount of time allocated to on-site work effort is usually constrained by resource availability and cost factors.

Without sufficient on-site testing and validation, it is very likely that post-project support will have to manage and address operating scenarios that were very different from those evaluated during the project.

Project Implementation Solutions: Maintaining a Support Focus

Once a project implementation has begun, the support and maintenance requirements must also be coordinated, scheduled, and executed in parallel with the project execution. In fact, support and training tasks should be directly integrated as sub-tasks within the overall project delivery plan to ensure that visibility to these maintenance requirements is not lost.

Project implementation as well as post-project success relies heavily on the following:

- Local Support Ownership –to champion the application, understand and convey its objectives to others, and can internally articulate its business value.
- Local Operator Engagement – to ensure that operating personnel can describe and articulate the application's value as it applies to them.
- Local Support Capability – to enable the performance of day-to-day maintenance tasks and provide the ability to assist with higher level maintenance activities.
- Local Support Procedures – to enable a fast response to simple problems and identify escalation channels for resolving more complex issues.

As you can see from this list, the key word in all of these success factors is "local." It is also important that all of these factors are in place prior to project completion.

Best Practice #5: Local Support Ownership

The best approach to achieve local support ownership for an advanced control and optimization application is to engage the support organization as an integrated function of the project execution team. Knowledge transfer and understanding of the installed application is assured under this model and the likelihood of the support organization actively maintaining and championing the application success after project implementation is greatly increased.

The scope of this role can be very large or very small depending on how much maintenance activity is planned for this function. However, regardless of how much support and maintenance activities are performed and executed directly by the local support organization, identified local support resources must be accountable to:

- Represent the on-site coordination point of all support and maintenance activities
- Follow-up and drive the execution of problem resolution activities
- Internally communicate the value of the applications and the maintenance functions that are supporting them

Best Practice #6: Local Operator Engagement

Local operating personnel are always an important element of the success or failure of an advanced control and optimization project. Their operating role at the interface between the process and the advanced controls gives them the ultimate authority to use or not to use the application. Even when the application is in use, operating staff have full control of how much operating flexibility they transfer to the advanced control and optimization solutions.

If the operators fully understand the application and can see direct benefit, they are likely to actively use the application to its full benefit. They are also more prone to report identified issues and help the owning support team to resolve those issues.

If operators don't fully understand an application and cannot attribute any direct benefits, they will be much less likely to actively use the application or will only use the application in a limited fashion. Identified problems will not get reported and operating staff will not proactively help the support team to resolve issues.

Therefore, it is critical that operating staff are actively engaged in the implementation activities and get training in both the operating objectives of the application as well as how to operate the unit both with and without the advanced controls in place.

Best Practice #7: Local Support Capability

Local engineering and maintenance staff are the first line of defense when operators encounter problems or misunderstandings with advanced control and optimization applications. Therefore, these organizations must maintain some level of local expertise and understanding to address minor issues that arise in day-to-day operations.

The best opportunity to develop this on-site support expertise is during the project implementation. Local support engineers should interact and engage with the project team as much as possible to understand the design, the basic engineering configuration, and the objectives of the application. If the project team is leveraging external suppliers and on-line connections to support the project implementation, the support team should be looking to leverage similar resources and on-line connections to maintain the applications.

The final level of local expertise that is required will vary depending on the plant's support plans. At a minimum, local engineering and maintenance staff must be able to answer operating questions regarding the application. They should also be able to perform basic application troubleshooting and execute minor engineering configuration changes. At the other extreme, local engineering and maintenance staff that are fully trained in the design, implementation, operation, and maintenance of advanced control and optimization applications may effectively take on all maintenance and support activities.

Regardless of which expertise level applies to your facility, maintaining some level of local application expertise builds operator confidence in using the applications and enables a fast and efficient support response.

Best Practice #8: Local Support Procedures

Local support procedures represent the second line of defense when application support problems arise that are outside of the capability of local engineering and maintenance personnel. During project implementation, support escalation paths must be defined for the following support categories:

- Operating issues
- Software maintenance

- Engineering design / implementation
- System/application integration

Defining these support procedures prior to project completion will ensure a smooth transition to post-project support by enabling a fast response to simple application problems while explicitly defining clear channels for resolving more complex issues.

Local support procedures commonly involve internal and external support entities including central engineering groups, application software suppliers, control system suppliers, contract engineers, and system integrators. Therefore, it is important to establish these links prior to project hand-off.

Post-Project Support Considerations

Challenges – The Project is Over, Now What?

Many of the challenges that were discussed with respect to implementation are similarly related to the challenges that are introduced during the post-project phase:

- Limited access to the original project team
- Limited or no post-project support budget
- Decline in local application interest
- Communications, time zone, and language barriers

Post-Project Challenge #1: Limited access to the original project technical lead(s)

Now that the project is completed and handed over, the original project team is usually dissolved and its members are re-assigned elsewhere. Once the technical lead and key team members are actively working on other projects, it can now be very difficult to access these key resources to assist with any support and maintenance activities.

Post-Project Challenge #2: Limited or no post-project support budget

Once a project is successfully completed, project budgets on both the client and supplier sides of the transaction are typically closed-out and are no longer available to support post-project activities. Suppliers will struggle to authorize any unfunded work supporting a project that has been officially completed. End users will face similar constraints on their ability to fund additional time and material scopes unless these expenses were planned and budgeted well in advance of the project close-out.

Post-Project Challenge #3: Decline in local application interest

In many cases, the local engineer who championed the original project may also move on to other projects or switch focus to other priorities. Maintaining the original application quickly becomes a secondary priority as exciting new challenges are identified.

Provided that no one is complaining, the application may receive little or no further attention from either operations or engineering. Without active support and maintenance, advanced control and optimization applications can quickly reach a point where the applications are no longer delivering any significant benefits.

Post-Project Challenge #4: Communications, time zone, and language barriers

Even in today's world of global communication networks, connecting experienced resources in one region to a client site in another, still presents challenges to be overcome. Language, time-zones, and limited access to communication mediums like the internet can all be barriers that prevent effective interaction and communication. Even though the internet now has a very broad reach, there are still some regions in China and India where access is severely limited. In other regions like the Middle East, internet access is often highly restricted, particularly when it involves the Process Control Network (PCN.)

Because of these limitations, it can be very difficult for a support team to gain remote access to an application and/or facilitate and oversee work being performed by local resources.

Post-Project Solutions: Delivering on Lifecycle Support Promise

All the activities described thus far were in preparation of final project hand-off. The goal now is to sustain and incrementally improve the lifecycle benefits associated with the installed advanced control and optimization implementation. Once the applications are in the hands of the process operators and support responsibility has been transitioned away from the projects organization, long-term success usually depends on the following:

- Consistency of ownership
- Training refresh
- Application currency and relevance
- Maintenance and support enablers
- Organizational alignment

If appropriate attention is given to these five post-project support factors, there is a high probability of sustaining long-term benefits from advanced control and optimization in process facilities.

Best Practice #9: Consistency of Ownership

In an ideal world, technical members of the original project team will remain involved in the long-term support of the installed application. Under this model, there are proven cases of sustained optimization benefits. In one example, the technical lead on an ethylene plant optimization project retained long-term support responsibility for the application. Because he understood the technology and championed the application's success, he was able to continuously achieve application benefits for over 10 years. However, this situation of a technical lead taking on long-term support is not the norm and other solutions must be identified that address consistency of ownership.

The next best alternative is to have a clearly defined support owner. As support roles change over time, transition of this responsibility must be explicitly assigned.

In cases where external suppliers or engineering contractors were used to implement advanced control and optimization applications, continuing ties to those organizations and/or individuals also allows for improved consistency of ownership over the application lifecycle. Even when the supplier's lead engineer is no longer accessible, external suppliers are usually in the best position to provide consistent ownership due to their intimate understanding of the technology, access to project design documents, and their larger pools of expert resources.

Consistency of ownership ensures that pertinent information related to the original design, implementation, and application objectives is retained and shared over time.

Best Practice #10: Training Refresh

Periodic refresh training ensures that operating, engineering, and maintenance staff skills are sustained at appropriate levels to maintain local operations and support capabilities. The intended goals are to:

- Address the loss of application knowledge through retirement and attrition
- Reinforce proper application use under different operating conditions
- Increase local support capabilities to enable new functionality and obtain increased benefits

The scope of refresh training on an annual basis can vary substantially from a couple of hours to more than a week depending on individual plant requirements. Regardless of the scope, refresh operator and engineer training at some level should be considered as requirement to maintain on-site skill levels.

Best Practice #11: Application Currency and Relevance

Application currency and relevance are two related aspects of lifecycle application maintenance. Currency specifically refers to the application software release and its interrelationship to other hardware, software, and operating systems used to implement the solution. Relevance specifically refers to the applicability of the solution under current business and economic conditions.

A major pitfall in the long-term support of advanced control and optimization application is ironically the "if it isn't broke, don't fix it" mentality. Once an application is installed and running, most end users will only focus on the costs of maintaining application currency and relevance (e.g. license cost, engineering cost, application downtime, etc.) However, the risks and associated cost of *not* maintaining application currency and relevance is usually significantly higher (e.g. higher license cost, higher engineering costs, mismatched application objectives, more application downtime, and lost opportunities to leverage new application functionality.)

Maintaining application currency allows for increased support flexibility and ensures that installed applications are viable and supportable. Maintaining application relevance ensures that application objectives continue to match business and economic objectives.

As such, both application currency and relevance should be reflected in metrics used to measure the performance of the support teams.

Best Practice #12: Maintenance and Support Enablers

Over the last several years, there has been tremendous progress made in the area of maintenance and support enablers with the end goal of addressing global resource constraints and support challenges with implementing advanced control and optimization applications in remote regions. This best practice is further broken down into the following subcategories:

- Maintenance tools
- Remote services
- Outcome-based service contracts

Maintenance Tools are local software applications that are designed to increase maintenance staff productivity, lower the required experience level to perform certain maintenance functions, and/or provide application performance metrics and analysis to assist with maintenance and performance improvement efforts.

Use of these types of enablers is particularly useful when on-site resources do not have high levels of expertise. However, the same tools are also useful to customers with high levels of expertise as it greatly improves the productivity of on-site experts.

Specific to advanced control and optimization applications, maintenance tools are now available that can assist with identifying problem models and help optimize step testing and model tuning resolution efforts without necessarily requiring an advanced control expert to be directly involved. Similarly, tools that constantly monitor and present application performance data can be used to quickly identify application issues and prioritize resolution efforts.

Remote Services enable either one-way or two-way on-line access to advanced solutions data and/or user interfaces. Over the last several years, acceptance of these connections has become more and more widespread as end users have gained confidence with security aspects of these connections and have found that the benefits of these services can be significant.

At the lowest level of service, a one-way, on-line connection is established that allows only data to be transferred from the end user site to a remote location where data analysis, expert review, and reporting can occur. In this model, benefits are incrementally achieved through proactive notification of issues, more efficient and timely support responses, and expert input on resolution actions.

At the highest-level of service, a two-way on-line connection is established that can enable broad and bidirectional flexibility with respect to the types of objects and interactions that are allowed through the on-line connection. In this model, the same benefits noted for the unidirectional connection apply with additional benefit possibilities including real-time troubleshooting, on-line application maintenance, and more.

Remote capabilities offered by external suppliers and internal central engineering groups are expanding everyday with available performance management services for control loops, control system alarms, advanced process control, computer networks, as well as comprehensive process optimization solutions that encompass unit performance, energy management, and equipment health monitoring.

As the complexity of installed advanced control and optimization solutions rises, the need to interact with multiple types of experts quickly becomes apparent. Remote services provide the most cost effective approach to augment on-site capabilities with this expertise.

Outcome-based Service Contracts are a relatively new enabler that can be used to drive the long-term sustainability of advanced control and optimization benefits. Many projects contain performance clauses that are executed at the end of the project to ensure that the application performs to design expectations. In cases where outside suppliers are involved, these clauses may invoke risk penalties or performance rewards. However, this same approach can also be extended into lifecycle support strategies for applications.

If selected performance metrics are easily measured and cannot be manipulated by either the end user or the outside supplier, defined performance criteria can be used to drive sustained performance on a recurring basis through guaranteed service contracts with joint participation of the end user and outside service suppliers.

One word of caution though, these types of support and maintenance contracts must be structured such that both the supplier and the end user are equally goaled to maintain benefits. This cooperative service model keeps both the end user and external supplier interested and focused on keeping the applications performing optimally.

When contracts are not structured in this manner, problems can quickly arise. In a recent example, Honeywell was actively engaged in remotely supporting advanced control applications at a client site. Plant management's expectation was that the applications would be actively maintained under the support contract. However, Honeywell's support scope was effectively limited to monitoring the applications and making improvement recommendations back to the client organization. In this particular case, it was up to the site personnel to act on those recommendations but site personnel were not appropriately goaled for that activity. In this example, Honeywell was providing valid support recommendations, the client was not acting on those recommendations, and benefits were not being realized. As a result, the value of the support contract was being questioned. The solution in this particular example was to re-engage the plant's management team and align site goals and objectives to match the overall application support objectives.

Best Practice #13: Organizational Alignment

This last success factor is really a composite of all of the other best practices that have been described throughout this paper. As these lifecycle support solutions are implemented, conflicts with current business practices and organizational interactions are bound to arise and will need to be addressed accordingly within the end user organizations.

For example, maintenance of large scale advanced control and optimization applications can easily cross historical boundaries within organizational hierarchies that require different behaviors and goals to be defined for various business and operating departments. Support success can also require new relationships and personal networks to be established. For example, enabling remote services will require close coordination between the engineering department and possibly the corporate information technology group. In many cases, these organizational linkages do not currently exist and must be established as they are identified within the lifecycle support framework.

Lifecycle support solutions also tend to shift maintenance activities from a reactive support model to a proactive support model. However, this shift commonly conflicts with reactive operating methodologies where resources tend to be focused on immediate issues rather than proactive maintenance activities like maintaining application currency and relevance.

As was noted in the outcome-based service contract example, simple awareness and communication of these issues to appropriate levels of management will allow them to be addressed in a logical manner.

Conclusions

As we have noted throughout this paper, there are many challenges involved with sustaining advanced control and optimization benefits. These challenges are apparent in almost every advanced control and optimization project but can be particularly acute when projects are implemented in remote regions or at sites with limited advanced control and optimization expertise. The challenges can be introduced during all phases of the project from pre-award to final project sign-off with root causes ranging from improper justification criteria, poor technology choice, organization misalignment, technical resource limitations, setting wrong client/supplier expectations, and more.

This paper discussed best practices and lifecycle support methodologies that address many of the support and maintenance challenges required to sustain advanced process control and optimization benefits in remote regions. A key takeaway from this paper is that lifecycle application support must be planned for and then actively addressed in parallel with project delivery activities.

APC KPI Aggregation

Jim Hackney
Senior Advisor
Equistar Chemicals, LP, a LyondellBasell company
Rob Hawkins
Senior Advisor
Lyondell Chemical Company, a LyondellBasell company
Jeff Varner
Principal Control Systems Technologist
Equistar Chemicals, LP, a LyondellBasell company

Abstract: Key Performance Indicators (KPIs) can be an important facet of the maintenance of any system, as long as the KPIs are properly conceived, well defined, and understood. Similar to a process control feedback loop, KPI values are a measurement of the health of the system. The targets for the KPIs are similar to set points, and the system will drive towards the KPI targets, if not beyond. As the well-know quotation states, "What you measure is what you get." If you use the wrong KPI, you will tend to drive the system to that KPI anyway.

Typical performance KPIs measure the performance of a controller or subcontroller at a given time. To measure the performance of a controller over a period of time, all the controllers in an area, or all the controllers in an area over a period of time; special care must be given to make sure the aggregated KPI reflects the true performance of the system.

This paper outlines methods we have used to aggregate individual controller KPIs so as to monitor plant-wide APC performance, allocate resources and focus attention in olefins plants.

Introduction

Key Performance Indicators (KPIs) are used by people to indicate whether something is going right or wrong and generally provoke some action if something is wrong. It is similar to a process control feedback loop. KPIs can be critical to focusing the organization's actions. Some organizations may tie KPIs to personal, group, or operating unit performance and/or bonuses. In our organization, management at all levels looks at KPIs in various areas, such as safety, environmental, reliability and several others, including an Advanced Process Control (APC) on-stream time KPI for entire sites.

Last year we presented a paper on a multi-site APC and Real Time Optimization (RTO) effort[1]. At the end of that project several KPIs were set up within an olefins "communication portal." The portal displays site-by-site KPIs with drill down capability.

Once we started developing and coding the KPIs it became apparent that aggregating them into site-wide values was not as straightforward as one might imagine.

KPI Philosophy

Whether we realize it or not, we all use KPIs, since in a general sense a KPI is any measure we use for feedback about what an organization, group or individual does. Because KPIs are developed to trigger action, it is very important that they reflect the desired results. A KPI for management may be the same as that used by engineering, operations, or maintenance personnel; however, the action triggered will generally be different. Front-line personnel usually want specific, detailed information that help them pinpoint problems and ultimately solutions. Management needs to know how resources are to be allocated; if systems are not working well, perhaps more resources need to be applied. Further, management can tie KPIs to individual or group performance goals which can affect pay, bonuses, and promotions. What can be calculated quickly and easily may give way to stricter requirements for the KPI when it is used for such purposes. The types of KPIs discussed in this paper are not an individual's KPIs that give detailed information, but are those that get distributed up, down and sideways in the organization.

Who owns the KPI? One answer is: anyone who uses the KPI. One person may generate the KPI from the data available and send it out to a group of people. The group consists of two main subgroups: users and receivers. Receivers get the information but generally do nothing with it. The users however can be triggered to action. If the trigger is activated, the user can ignore it, call someone who can do something about the problem, or drill down to get more information before deciding what to do. The user can also pass the KPI along, vertically or horizontally across the organization. Although there should be a common, agreed upon trigger point, each individual user will be moved to action using his or her own trigger point. Eventually the KPI reaches one or more people in the organization who may act on the KPI, but generally do not pass it upward. One can view those people as the ultimate owners.

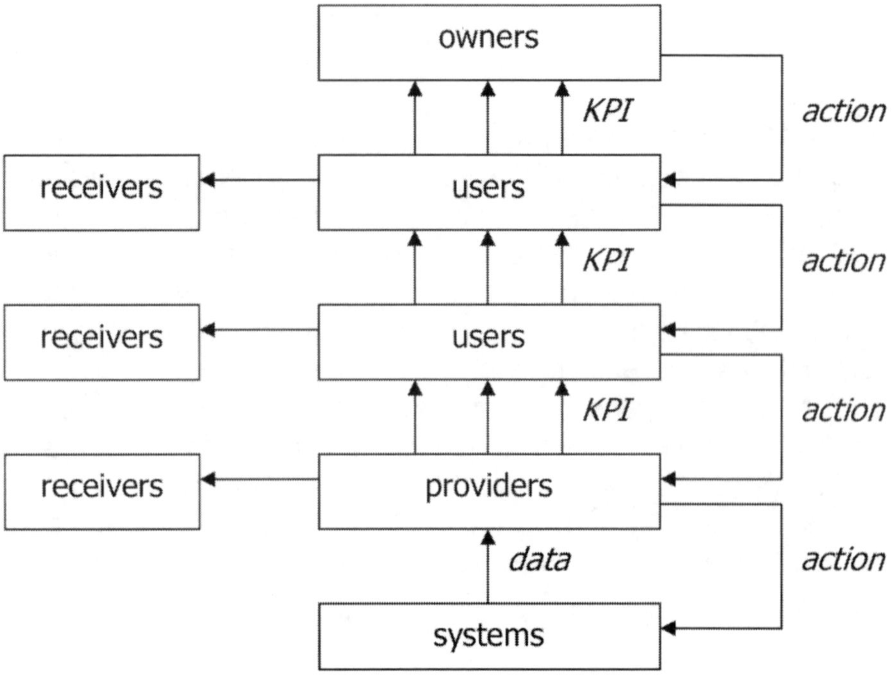

Figure 1. The flow of a KPI through an organization, leading to action by users and owners.

It is a good idea to normalize or scale the KPI so that different KPIs can be compared and even aggregated. Scaling 0 to 100% is a natural range. The trigger target must also be scaled to the same range, which can lead to different KPIs having different trigger points. To avoid this problem, scaling can be based on the percent of trigger; one can then have a secondary trigger indicating that the KPI is approaching the target value.

There are many other aspects of KPIs that can be mentioned:

- The KPI should reflect some goal or goals of the organization.
- The meaning of the KPI should be understandable and understood.
- The KPI trigger should be set so that action should be taken when the KPI exceeds the trigger. The trigger point should be standardized and agreed upon.
- For credibility, the KPI should be automated as much as is reasonable.
- Drill-down capability should point to the cause of the problem. Management is not necessarily interested in drilling down to the cause but is more interested in making sure the resources are applied to fix the problem. Therefore, drill-down capability is more useful for the practitioner than the manager. This is not to downplay the usefulness of drill-down capability, just to emphasize who would normally use that capability.
- Think carefully about what you count towards the KPI and what you exclude. Make sure it is understood by the KPI users.
- False positives should be minimized. Like an alarm, a KPI that frequently exceeds its target and yet requires no action to be taken will soon be ignored.
- The basic components to calculate the KPIs should be historized.

- No KPI is fool-proof in giving a completely accurate and easily interpreted picture of the state of the controller. Each one has its strengths and weaknesses.

There are some properties specific to APC KPIs, especially those to be aggregated:

- The purpose of APC KPIs is to give the user a sense of the state of health of the controller. For more details, one needs to either drill down to more information or talk to someone who can give the details and reasons for deviations, and attest to the state of the controller.
- Aggregated APC KPIs are generally for management's use or to compare site-to-site performance; in that sense they are owned by management. They are not intended to be triggers for action by the control engineer or specialist, as they should already be aware of the issues and problems with their controllers. Some of the ways management might use these KPIs include checking if more resources might be required, shifting or refocusing resources, and tracking long-term trends.
- Consider the scope of each KPI. For example, should the KPI reflect the performance of the whole controller, or should you report at a subcontroller level (for each distillation tower, for example).
- What should be reported if the controller is down? For example, if a utilization KPI is used, what should be reported when the controller is off? One answer is to report 0% utilization, since the controller degrees of freedom cannot be used when the controller is down. However, an equally valid philosophy is to report an undefined utilization value, especially if a separate on-stream time KPI is also presented. Thus, the utilization KPI does not measure the same aspect of performance as the on-stream KPI. If only one KPI is to be used, then the utilization can reflect both the utilization when the controller is on and when the controller is off. The definition should be clearly communicated to all users.

Aggregation

By aggregation, we mean combining KPIs into a single KPI. Figure 2 illustrates that aggregation can occur across different KPIs, across different processing units, across time periods or any combination of the three.

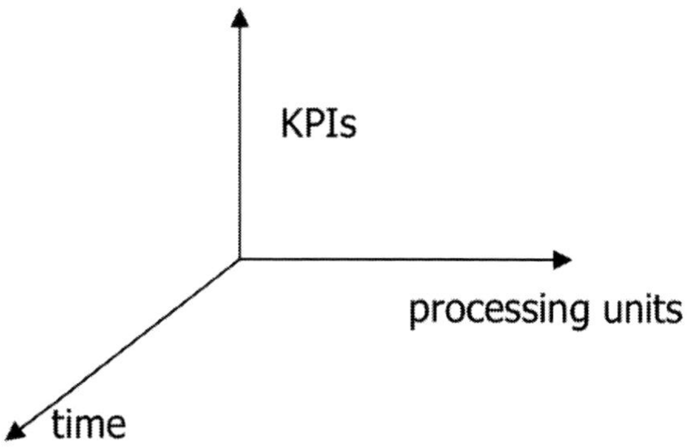

Figure 2. Three dimensional aspect of KPI aggregation.

Aggregation can be done a posteriori or continuously. A continuous calculation lends itself to more automation, but is more difficult.

Aggregation in Time

Time aggregation involves combining the KPIs calculated over one time frame into a KPI over a longer time period. Calculating the KPI using historical data once over the longer time period is the cleanest and simplest way, but we have seen situations where that is not feasible. For example, to improve efficiency our communication portal stores data daily in its own database. When the user requests data for a multi-day time period, the stored daily values have to be aggregated into a single value. The key to proper aggregation is to weight the time periods by the number of valid data points the individual time periods represent.

$$
\begin{aligned}
KPI &= \sum_{i=1}^{N} K_i w_i \\
w_i &= \frac{n_i}{\sum_{j=1}^{N} n_j}, \quad \sum_{k=1}^{N} w_i = 1
\end{aligned}
\qquad (1)
$$

Note that the weighting factors are normalized; that is, they sum to one.

As an example, we compute the APC percent on-stream time KPI only for when the process is available for the APC application to run. When the process is not available the percent on-stream KPI is then missing or undefined. In order to avoid manual corrections of the KPI, the flag for process availability should represent when the application should and should not be on. It can be tied to a minimum feed rate to a distillation column or a minimum fuel gas rate to a furnace. It might also be based on whether or not an APC model is available for the current conditions. Again, think carefully about what is excluded. For example, is the process considered available during commissioning of a controller? During step testing? If key instruments are not available due to maintenance? Whatever your criteria are, they need to be clearly stated and agreed upon.

The key for the undefined KPI data is that those periods should not contribute to the time-aggregated KPI; if the entire period has no valid values, the KPI should not be 0, but rather undefined.

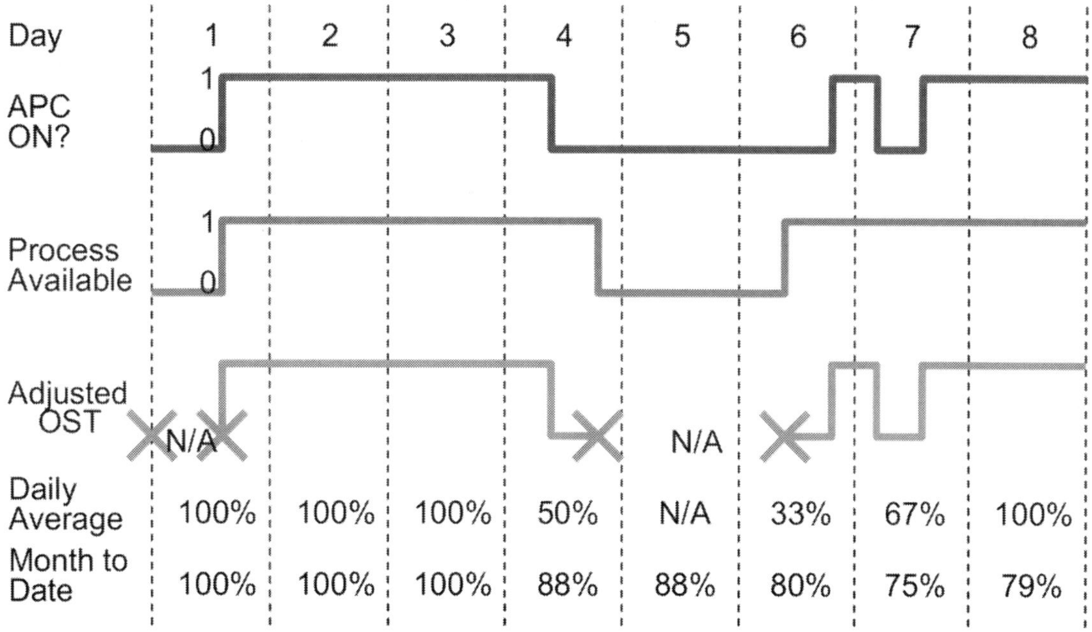

Figure 3. Illustration of APC On-Stream KPI. Eight days of data are shown. The APC on-stream, process availability and adjusted on-stream flags are historized, along with the number of valid data points per day. Note the gap in the adjusted OST when the process is not available for APC.

Aggregation Across Applications

For management use KPIs are usually aggregated across APC applications into a site-wide KPI. If your KPI is at the subcontroller level, the subcontroller KPIs can be aggregated into a controller KPI and then further aggregated into a site KPI. Equation 1 can be used for application aggregation as well. Again, the key is proper weighting of the KPIs. There are several possible weightings one can use.

- *Economically-based weighting* – Here the economic benefit generated by the individual (sub)application is judged against the other (sub)applications and a weighting factor is determined. Remember that the weighting factors must be normalized.
- *Size-based weighting* – A weighting factor based on the size of the application is used. We prefer to use the number of manipulated variables. When you are aggregating continuously, use the actual number of MVs that are available *by design*. Some MVs may be turned off *by design* under certain circumstances. Thus, the number of MVs used in the weighting should be historized as well. If you are aggregating a posteriori and the number of MVs available is not historized, you can use a fixed number of MVs. An operator turning an MV off is not sufficient to reduce the number of MVs available by design. Another sized-based weighting could be total number of variables (MVs+CVs+FFs),
- *Other weighting* – There can be any number of ways to weight the KPIs, including heuristic-based using perceived importance or complexity, or uniform weighting (simple averaging).

Aggregation Across KPIs

Different KPIs can be aggregated into a single composite KPI to indicate the health of a *system*[2]. This can take many forms, from a composite KPI number to an image like a carpet plot displaying all of the individual KPIs simultaneously. When computing a composite KPI number, one has to make sure that the scaling is such that the individual values can be combined and that the proper weighting is used.

Continuous or Batch?

Should the KPI be calculated continuously in time or as a batch job when the KPIs are to be published? The batch process is relatively easy and does not require much automation. But, it also requires more human intervention. If there are many ad hoc adjustments required every reporting period, the batch method may be necessary.

If most adjustments can be captured continuously, like in the process availability for the APC on-stream time, then calculating the KPI continuously makes sense. Setting up the calculations can be tedious, but you can automate that process as well. For example, we have developed internal code that automatically detects controllers and loops through the APC applications to calculate some of our KPIs and their aggregations. However, unless a "replay" mechanism is available, it may be difficult to make corrections or adjustments to the historical data when continuously calculating the KPIs. Automated KPIs give them credibility because there is a well-understood algorithm for the calculation and the results are not subjectively adjusted after calculation.

Example: APC On-Stream Time

As an example, we will examine the aggregation over time and subcontrollers of the on-stream time (OST). We will use the following nomenclature:

i = index to applications
k = index to time steps
OS = raw on-stream flag (0 = OFF, 1 = ON)
OA = adjusted (to process availability) on-stream flag (0 = OFF, 1 = ON)
PA = process availability flag (0 = not available, 1 = available)
M = number of degrees of freedom (# MVs available by design)
N_t = number of data points over the time period
N_a = number of applications to aggregate

At each cycle we aggregate across subcontrollers:

$$\overline{OA}_k = \frac{\sum_{i=1}^{N_a} OS_{i,k} PA_{i,k} M_{i,k}}{\sum_{i=1}^{N_a} PA_{i,k} M_{i,k}} \qquad (2)$$

This would be for aggregating subcontrollers to the controller level. Note that any subcontroller where $PA=0$ is excluded from the aggregation. We could also aggregate the controllers to a unit level, to a site level, etc. For rollups at these levels you can use a weighted average, weighted by the total degrees of freedom of each sublevel.

Next we aggregate data across a time period. Here we will do a simple average.

$$\overline{OA} = \frac{\sum_{i=1}^{N_t} \overline{OA}_k}{N_t} \qquad (3)$$

N_t is the number of valid points (not missing, unavailable, or bad values) during the period. If $N_t = 0$, then $\overline{OA}$ is undefined for the period. Equation 3 applies to the subcontroller and the other application aggregated values.

Although there are many ways to compute these values in real time, there are tradeoffs between complexity, memory and computational efficiency. Table 1 is a list of tags in the historian that might be used for continuous calculation of the aggregation of the APC on-stream KPI. The total number of tags is surprisingly high. For example, for a single subcontroller to compute the adjusted, daily, month-to-date, yearly, and year-to-date values would require at least 21 tags.

Table 1. Possible list of tags in historian for aggregating the on-stream KPI. N_2, N_3, etc. are the number of tags at each level of application aggregation (number of controllers, number of units, number of sites, etc.). More can be added; for example, the process availability flag can be aggregated as well.

Tag	**Definition**	**Number**
OS	Raw on-stream flag	N_a X # time aggregate levels
PA	Process availability flag	N_a
OA	Adjusted on-stream flag	($N_a + N_2 + N_3 +$...) X # time levels
Num	Numerator in Equations 2,3	($N_a + N_2 + N_3 +$...) X # time levels
Den	Denominator in Equations 2, 3	($N_a + N_2 + N_3 +$...) X # time levels

Conclusion

KPIs have become important tools for both practitioners and managers. For management, aggregating the KPIs over both time and applications is important. Something as simple as calculating an average KPI for a site over a time period turns out to be not so simple. There are many details associated with automating the aggregations. A very important part is how undefined values are handled. It is surprising how many historian tags are required.

With the greater emphasis on KPIs, potentially tying them to employee compensation, the accuracy of the aggregation must be given a fairly high priority. Hopefully our paper has shed some light on the issues involved and suggested some possible solutions.

References

1. Hackney, James (2008) "An Optimization Program in Lyondell," 20th Ethylene Producers Conference, April.
2. Gerold, Jane (2004) "Calculating a Composite KPI," *AutomationWorld*, June, p.13, http://www.Automationworld.com/research-771.

Decreasing Emissions and Improving Performance at Dow's LHC-7 Ethylene Plant in Freeport, TX

Bernardo M. Corripio
Cracking Furnace Technology Leader
The Dow Chemical Company

Susan T. Belveal
Texas Operations Hydrocarbons Improvement Leader
The Dow Chemical Company

Abstract: Starting in 2000, increasingly stringent air quality requirements from the Texas Commission on Environmental Quality (TCEQ) for ozone non-attainment areas forced Dow to consider various options for reducing NOx emissions at its Freeport, TX facilities, located about one hour south of Houston and included in the Houston-Galveston non-attainment region. This site was originally started up in 1941 and has facilities of varying ages, including ethylene plants started up in 1973 and 1995. The older ethylene plant, LHC-7, was a prime candidate for NOx reduction. This paper describes the site-wide study to reduce NOx emissions, the various options for equipment replacement, modification and shutdown considered and the improvements made to the LHC-7 ethylene plant to reduce NOx emissions. These improvements included replacement of a gas turbine driver for the propylene refrigeration compressor with a steam turbine and construction of an all-new furnace section to replace the existing furnaces. Economic, energy efficiency and NOx reduction results of the work will be described.

Regulation of NOx

During combustion, nitrogen in the air and/or fuel is thermally oxidized into nitric oxide (NO) or nitrogen dioxide (NO2). These compounds are referred to as "NOx". The Federal Clean Air Act of 1990 established ambient air quality standards for six pollutants, ozone being one. All states are required to be in attainment under those standards. States having non-attainment areas are required to submit a State Implementation Plan (SIP) to the Environmental Protection Agency (EPA) detailing how attainment will be reached. The Houston, Galveston, and Brazoria (HGB) areas have been designated as "non-attainment – severe" by the EPA for the Federal 1 Hour Ozone Standard.

The Dow Chemical Company has multiple sites in the HGB area, with each site consisting of multiple production facilities. In these facilities, various NOx sources were identified, including boilers, process heaters, stationary gas turbines, and pyrolysis furnaces, among others. Given the magnitude of impacted equipment and facilities, Dow's strategy was to address the regulatory compliance for the entire region, rather than on a site by site or facility by facility basis. A project team was assembled to complete abatement feasibility studies for all emissions sources. A third party was consulted to supplement Dow's limited experience with NOx abatement feasibility studies.

A total of twenty feasibility studies, grouped by Dow business area, were completed. Each study included a determination of the current emissions, technology alternatives to attain emission limits, cost estimates for each alternative and a recommended alternative. A variety of abatement options were evaluated such as low NOx burners, Selective Catalytic Reduction (SCR), steam injection, scrubbers, etc. These options were compiled across the facilities and sites and evaluated to select the most cost effective abatement projects to meet the required 90% NOx reduction per the SIP. Two projects were selected for the LHC-7 ethylene plant located in Freeport, TX.

LHC-7: A Case for Change

LHC-7 was a grandfathered facility and therefore did not have an air permit. However, due a capacity expansion project an air permit was developed. The expansion project was subsequently canceled but the obligation to meet the requirements of the air permit remained. In order to maintain the long term operation of the facility, it was recognized that significant investment would be needed to meet the air permit requirements as well as to address plant reliability concerns. Additionally, LHC-7, like all ethylene plants, has extensive internal energy integration and significant integration with the site power plant due to its large generation and use of steam and fuel. This made the plant a natural source of potential improvements to the NOx reduction scope.

A "refurbish vs. shutdown" 20 year economic assessment was completed with commercial and technical input. The refurbishment scope included options to upgrade existing equipment or replace equipment to meet air permit requirements in addition to scopes necessary to extend the lifetime of the facility. The refurbishment option was the most economically viable alternative. The refurbishment scope included the following NOx abatement projects:

1. Replace gas turbine driver for propylene refrigeration compressor with a steam turbine
2. Replace original ten furnaces with new furnaces having ultra low NOx burners and provisions for future Selective Catalytic Reduction (SCR) NOx abatement

Gas Turbine Replacement and Site Optimization

This project replaced the gas turbine and the helper steam turbine and gearbox used to drive the propylene refrigeration compressor with a new 50,000 hp steam turbine driver. The new turbine is driven by 1250 psig steam supplied by the site power plant; it has a controlled extraction of 235 psig steam and exhausts 30 psig steam. This project also eliminated the waste heat boiler associated with the gas turbine. New steam lines from the power plant and steam distribution piping for the lower steam levels were included in the scope.

Important alternatives to the replacement of the gas turbine with the new 50,000 hp steam turbine were:

- Installation of Selective Catalytic Reduction (SCR) with steam injection system on the existing gas turbine.
- Replacement with a smaller steam turbine (25,000 hp).
- Replacement with a 25,000 hp or 45,000 hp electric motor.

The economics of these options were inferior to that of the selected option. The economic drivers for the evaluation included:

- Improvements in ethylene plant and site power generation efficiency
- Reliability improvements. The project allows LHC-7 to run longer between major turnarounds; the turnaround interval was formerly set by the 4-year limitation of the gas turbine.
- Capital reduction.

The turbine replacement was delivered reduced NOx emissions of over 340 tons/year. The project design was coordinated with the other NOx abatement projects, including the furnace replacement and work at the site power plant to optimize capital spending and benefits. The benefits delivered by the combined projects included

- Increased ethylene production for TX-7 (>70 MMlb/yr);
- Reduced natural gas consumption of the Freeport site by (>230 MMBtu/hour);
- Improved ethylene plant and site power generation efficiencies (net 7% improvement);

- Elimination of one-time maintenance spending for repairs to the gas turbine (>$2MM) and annualized overhaul costs for the gas turbine (>0.5MM/yr).

One example of the design optimization was the choice of steam generation level for the new furnaces. The existing furnaces at LHC-7 generated 600 psig steam. The option of 1250 psig steam generation was considered carefully in light of the added high pressure steam demand to drive the new turbine. It turned out to be more cost-effective for the site as a whole to implement the new furnaces with 600 psig steam generation despite the lower local efficiency for the new LHC-7 furnaces.

Furnace Replacement

This project replaced the existing ten LHC-7 furnaces with five furnaces. Nameplate plant capacity did not change and steam generation remained at 600 psig. The new furnaces have induced draft fans with low NOx burners.

To avoid an extended outage, the new furnaces were constructed several hundred feet away from the existing furnace area. The scope of the project included the piperack and piping to connect the furnaces into the existing plant and the addition of an electrical power supply and Motor Control Center.

The new furnaces reduced NOx emissions by more than 300 tons/yr and significantly upgraded the furnace technology. Benefits include conversion energy reduction, increased yield to desired products, increased ethylene production, and maintenance savings greater than $2 MM/yr. The project will significantly extend the life of the LHC-7 cracker.

An important alternative to the complete furnace replacement was the minimum investment case. This case involved replacing burners to meet NOx regulations, replacing convection sections to resolve furnace health issues, and adding significant instrumentation to comply with Dow safety standards. Despite a lower initial investment cost, this option provides none of the above mentioned furnace technology gains, energy savings, or maintenance savings. The new furnaces option provided far more value creation for Dow.

Also, as previously mentioned, high pressure steam generation, additional steam generation from tertiary quench exchangers, and various plot locations were considered. The chosen alternative had the highest NPV creation relative to the minimum investment case. An earlier alternative included significant expansion of the plant capacity as well as NOx reduction, but was determined to be too costly.

An additional investment was made to the convection section to allow future addition of SCR technology in case environmental regulations call for even greater NOx reductions in the future. This investment also served as a hedge in case the low NOx burners did not reduce

NOx as planned; however, the burners performed as advertised and delivered the promised NOx reductions.

Summary

Dow's NOx reduction effort at its Freeport, Texas facilities is an important example of meeting stringent new environmental regulations while maximizing value to the company. By examining the synergies across operating units and being willing to spend more capital than the minimum required to attain emission limits, Dow implemented a solution which meets the requirements while using capital effectively and allowing room both for further growth of the site and more stringent future requirements. The ethylene plant proved to be fruitful ground for project synergies due to extensive internal integration and with the site utilities driven by its large generation and use of steam and fuel.

THE EFFECTS OF CAUSTIC TOWER OPERATIONS AND SPENT CAUSTIC HANDLING ON THE ZIMPRO® WET AIR OXIDATION (WAO) OF ETHYLENE SPENT CAUSTIC

Clayton Maugans, Ph.D.
Siemens Water Technologies

Mike Howdeshell, Ph.D.
Siemens Water Technologies

Steve De Haan
Lummus Technology

Abstract: Ethylene plant spent caustic is frequently treated using a Spent Caustic Oxidation Unit (SCOU). Commonly, this is a medium pressure (~400 psig) Wet Air Oxidation (WAO) system. There is spent caustic storage for feeding the system. Events that cause excessive downtime in the WAO unit are not acceptable because the spent caustic storage capacity is limited. The hazardous and malodorous nature of the spent caustic makes large inventory unappealing; and ad-hoc storage difficult.

WAO system operations are affected by the composition of the spent caustic. Off-spec feed can cause disruptions or even damage to the WAO system. The design and operations of the caustic tower, gasoline wash, and spent caustic storage affect the spent caustic composition, which ultimately affects WAO operations. While such disruptions are rare, when they do occur the most common cause is excessive organic contamination (red oil) in the spent caustic.

This paper describes the WAO system and the spent caustic parameters that affect WAO operations. Red oil in particular affects WAO operations and the upstream handling, and techniques to best manage this material are discussed.

Introduction

This paper discusses the effects of caustic tower operations and spent caustic handling on the reliability and performance of a wet air oxidation (WAO) system. A description is given of the WAO system and the feed parameters that affect its operation; the caustic scrubbing tower design and operations; and spent caustic storage and handling. The paper closes with recommendations for operating techniques for best reliability of the WAO system.

Wet Air Oxidation: purpose and designs

Spent caustic is intensely malodorous, toxic, and highly alkaline. It contains high chemical oxygen demand (COD), some total organic carbon (TOC), and high total dissolved solids (TDS).

These properties make it difficult to treat by conventional biological treatment. Ethylene plants often have a spent caustic oxidation unit, which is a wet air oxidation (WAO) system to treat the wastewater by oxidizing sulfides, mercaptans, and a majority of the COD. This allows the oxidized spent caustic to be further treated by conventional biological treatment prior to discharge.

There are three design temperature ranges for spent caustic oxidation.

Low temperature WAO: (100 - 150 °C)

This WAO design is more common in older ethylene plants. This is a relatively low cost system and is fabricated from inexpensive alloys such as carbon steel. The design achieves some bulk COD destruction and destroys much of the sulfide content, but sulfide and mercaptan destruction is incomplete. Overall performance and reliability are relatively poor. Effluent is typically high in COD and reduced sulfur (mostly as thiosulfate) and may still be malodorous. Fouling, caused by organic molecule polymerization, creates an organic scale and fouling rates in these older designs are high.

Medium temperature WAO: (200 - 230 °C)

This WAO design is more typical for new ethylene plants. These units achieve high gross COD reduction with complete sulfide and mercaptan oxidation. TOC concentration is not affected much, but the nature of the TOC is changed, which makes for low fouling rates and better bio-treatability. Effluent is low in COD, contains low levels of thiosulfate, and is no longer malodorous.

High temperature WAO: (>250 °C)

This is the most effective (and expensive) form of WAO. Performance is similar to medium temperature WAO, with the addition that TOC is also greatly reduced. High temperature WAO is unnecessary and not used for ethylene plants because the TOC of ethylene spent caustic is rarely greater than a few thousand ppm. This design is commonly used for refinery spent caustics that have a high TOC content.

WAO plant description

Medium temperature (200 – 230 °C) WAO systems are the most commonly built today and this paper will focus on that application. Figure 1 is a block flow diagram for a typical medium temperature WAO system. Typical spent caustic composition is shown in Table 1.

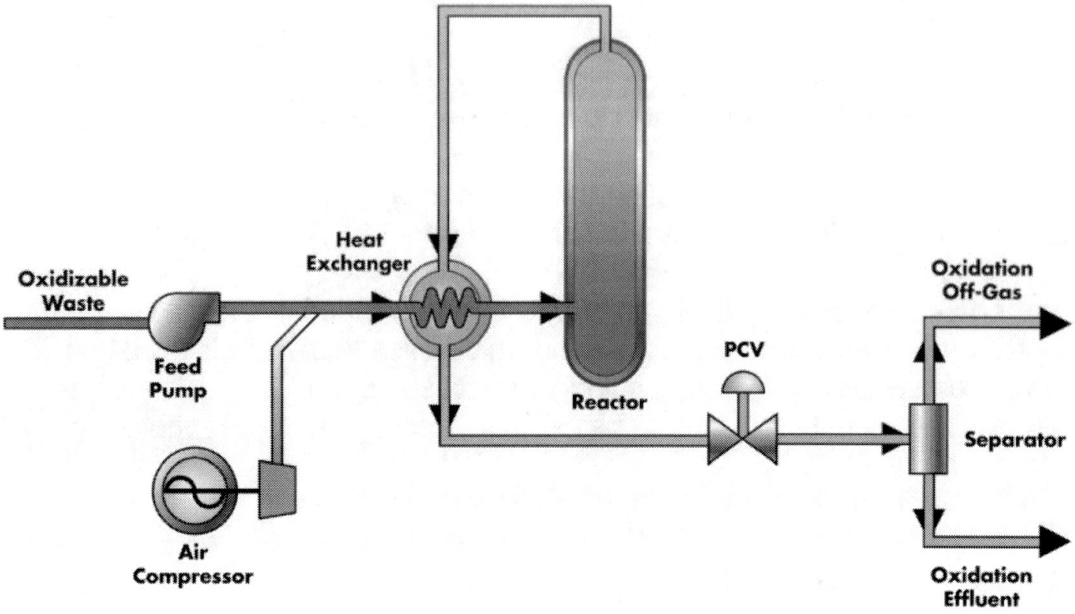

Figure 1. Typical Zimpro® WAO Block Flow Diagram.

Table 1
Typical Ethylene Plant Spent Caustic Composition (1)

Compound	Concentration Range (wt%)
NaHS	0.5%-6%
Na_2CO_3	1%-5%
NaOH	1%-4%
NaSR	0%-0.2%
Soluble oil	50-150 ppm
TOC	50-1500 ppm
Benzene	20-100 ppm

In this design, a high pressure pump is used to pressurize the spent caustic. Compressed air is combined with the pressurized spent caustic and the fluid passes through the feed/effluent heat exchanger (F/E HX). The F/E HX heats the fluid and initiates the oxidation reaction. The feed passes into the WAO reactor, where the oxidation takes place. The hot liquor and off-gas pass out the top of the reactor and back to the F/E HX, where the fluid is cooled. The effluent is depressured across a control valve and the off-gas is separated from the oxidized liquor in a separator. The liquid effluent is discharged or passes to the plant biological facility for final polishing. The off-gas is typically sent to an afterburner.

The reactor is a bubble column. Oxygen transfers from the gas to the liquid phase as bubbles rise through the liquid full reactor to satisfy the oxygen demand of the reaction. WAO reactions at 200 °C are shown below (2):

1. $NaHS + NaOH + 2\ O_2 \rightarrow Na_2SO_4 + H_2O$
2. $NaSR + NaOH + O_2 \rightarrow Na_2SO_4 + RCOONa$ (unbalanced)
3. $R + NaOH + O_2 \rightarrow CH_3COONa$ (unbalanced)
4. $R + NaOH + O_2 \rightarrow NaOOCCOONa$ (unbalanced)
5. $R + NaOH + O_2 \rightarrow Na_2CO_3 + H_2O$ (unbalanced)

The dissolved oxygen reacts with the sulfide compounds to produce sodium thiosulfate, which further oxidizes to form sodium sulfate (reaction 1). As shown in Table 1, over a thousand ppm of TOC may be present. At 200 °C organic compounds undergo only partial oxidation (reactions 3 and 4). The COD load is lowered and approximately 0-10% of the TOC content is completely mineralized (reaction 5).

Normally the equipment can operate for several years between cleanings or other maintenance. The process is operated with sufficient air to maintain about 5% residual oxygen in the off-gas, assuring an oxidative environment throughout. Metallurgy is selected to withstand this environment.

Parameters that affect WAO operations

Feed COD and sulfides

The COD concentration in the spent caustic is affected by the caustic tower operations and the furnace sulfide and carbonyl production rates. Sulfides make up the bulk of the spent caustic COD content. The WAO system is designed to utilize the exothermic heat release from the oxidation reaction. Low COD spent caustic (<20,000 ppm) will require more utility (steam) consumption in the WAO system, but otherwise will not have any negative impact. High COD spent caustic must be diluted or processed at a reduced flow rate in order to avoid exceeding the oxygen capacity of the air compressor. Otherwise, high COD will cause the off-gas oxygen concentration to become low. The off-gas is continuously monitored and safety interlocks shut down the unit when the oxygen is too low.

Without the interlocks, the WAO system could operate with insufficient oxygen for sulfide oxidation, creating poor quality effluent. With insufficient oxygen, the TOC will not be oxidized and polymerization/coking can occur, which will cause fouling. In addition, the metallurgy is selected to withstand an oxidizing environment, so rapid corrosion can occur under anoxic conditions.

TOC/red oil

In ethylene plants, WAO systems are designed to process a certain maximum content of TOC – usually just a few thousand ppm. The composition of the ethylene gas stream, the way the caustic tower is operated, and the way the spent caustic is stored can result in oil contamination of the spent caustic. This can cause pockets of pure oil (oil slugs) to enter the WAO system – greatly exceeding the design maximum. Oil slugs have high COD and a slug will quickly exceed the capacity of the air compressor, causing the system to become oxygen deficient. When this happens, the off-gas oxygen monitoring safety interlock should trip the WAO unit, to prevent damage to the system. Since there is some delay before the trip, frequent oil excursions will result in coking and plugging of the F/E HX.

Caustic

The metallurgy of the WAO system is selected to withstand oxidative conditions at high pH. Caustic is consumed in the above reactions. Typically there is sufficient surplus caustic in the feed to maintain alkaline conditions in the effluent.

Sodium carbonate

Sodium carbonate is a pH buffer and results from CO_2 gas adsorbed in the caustic tower. Some carbonate is also formed from the TOC oxidation reaction (reaction 5), though only a small portion of the TOC is oxidized fully to CO_2. Carbonate helps regulate the pH in the WAO system and is beneficial to the operation of a WAO unit. Essentially all of the CO_2 and H_2S in the ethylene gas stream are captured in the caustic tower – so there is little that can be done at the caustic tower to affect carbonate mass flow in the spent caustic.

Chlorides

High chloride concentration often comes from using poor quality caustic in the tower. High chloride concentration is corrosive to the standard WAO metallurgy. WAO systems are typically designed for a maximum chloride concentration of 1,500 ppm.

Iron

Spent caustic may have iron present as iron sulfides. This is oxidized in the WAO system to iron oxide and sulfate. Usually the concentration is only a few ppm, - a low enough level that the iron oxides are soluble enough to pass through the system or form only a trivial amount of precipitation. When the iron content in the feed is high, the iron will form an insoluble scale and mechanical or chemical cleaning will be needed. High iron contamination can come from using poor quality caustic, acid, or water or from systems that are experiencing high corrosion in the upstream carbon steel components (e.g. in the caustic tower).

Hardness

Because the pH is high, spent caustic typically has little hardness. Most hardness scaling issues come from the use of poor quality service water for dilution and start-up/shut-down. Any hardness in dilution or start-up water that is fed to the unit will come out on the heat transfer surfaces as scale. Ideally, the hardness of feed service water should be well under 25 ppm as $CaCO_3$, though even that will eventually cause scale.

Caustic tower additives

Special chemicals are sometimes added to improve operations of the caustic tower (e.g. antifouling agents to prevent organic fouling). These chemicals and reaction products will pass with the spent caustic. In general this appears to maintain, if not improve, the operability of the WAO system as well. In rare cases, a new additive could be tested in an older caustic tower to solubilize or otherwise remove existing fouling. This can cause a large increase in the TOC and COD content of the spent caustic, which could affect WAO operations.

Upstream operations and their affects on the WAO plant

Ethylene handling

Ethylene gas contains CO_2, H_2S, mercaptans, and other organic molecules. The caustic scrubbing tower is used to adsorb and remove these contaminants. Before entering the tower, the ethylene gas is compressed and cooled to a temperature approach to cooling water. The liquid water and condensed hydrocarbons are separated from the vapor in a drum. The vapor feed to the caustic tower is then reheated by about 5 to 10 °C, to control organic compound condensation in the tower.

Insufficient reheating results in excessive condensing of organic compounds in the tower, which could increase the volume of red oil in the spent caustic that must be handled. Excessive reheating is inefficient.

Caustic tower scrubbing – overview

The caustic tower is a vertical gas/liquid contactor that is pressurized and operated between 30 and 50 °C. The typical caustic tower (Figure 2) has 3 to 4 stages, with the top stage the water-wash stage, the second stage the strong-caustic stage, and the bottom stages the intermediate- and weak-caustic stages. The caustic from each stage cascades down to the next stage. In the bottom stage, most of the free caustic has been consumed and the weak caustic is loaded with sulfides, carbonates, and hydrocarbons. A portion of the weak liquor is recirculated in the bottom stage and the remainder is discharged as spent caustic.

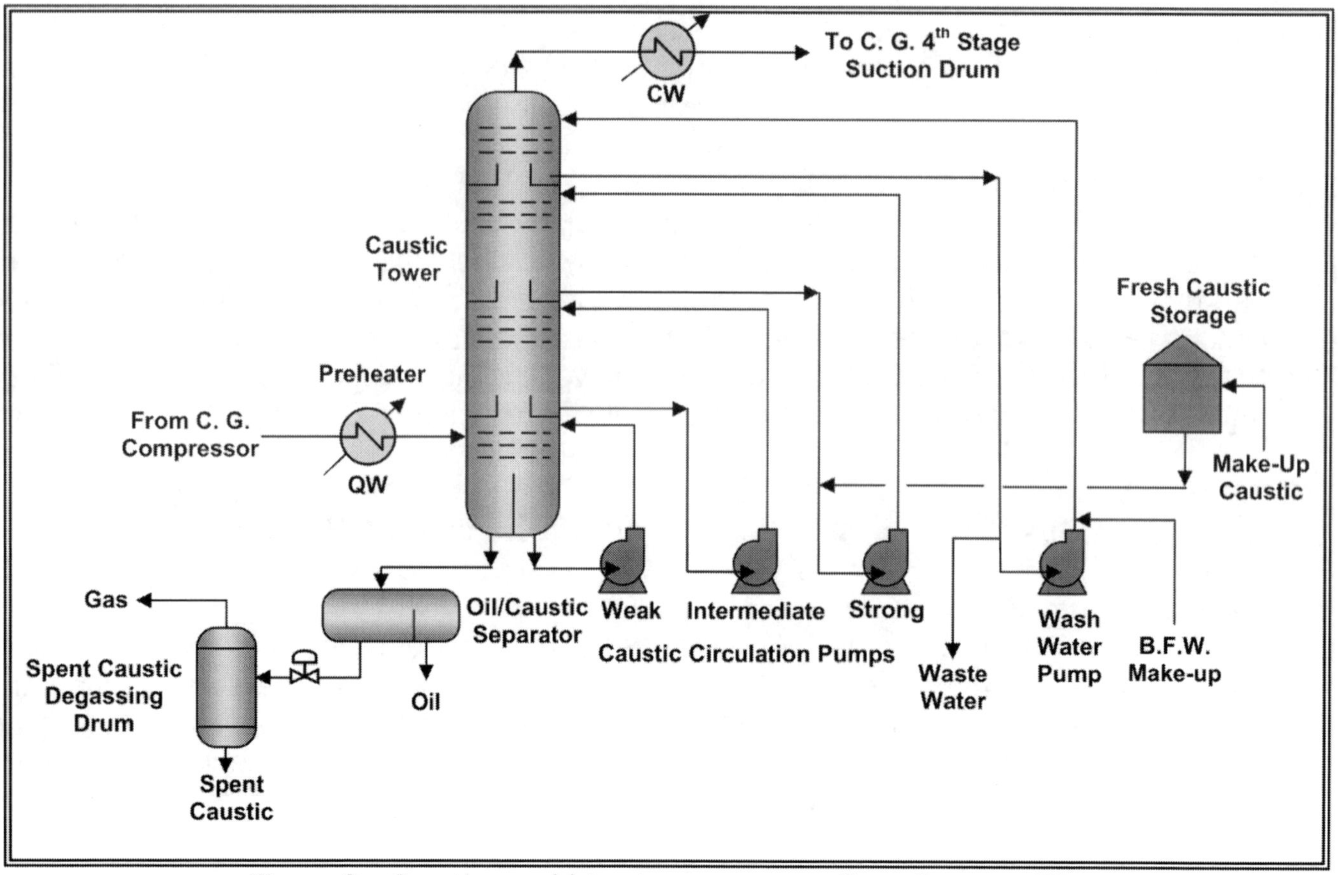

Figure 2. Caustic scrubbing tower process flow diagram (3)

Caustic tower design and operation

Most caustic towers are segregated into stages, with each stage having a liquid reservoir at the bottom. Gas/liquid contacting is enhanced by recirculating the caustic from the reservoir to the top of that stage. A portion of the reservoir is cascaded down to the next stage.

Floating on top of each caustic reservoir may be a layer of hydrocarbon oil, sometimes called "red oil". The tower should be designed to avoid retaining oil since excessive residence time worsens polymerization. The intermediate section sumps should therefore be designed with a standpipe for down flow to the next section. The standpipe allows the oil floating on the caustic to exit to the next stage, thus minimizing residence time. The bottom section sump is designed to allow skimming of the oil by the operators. Oil removed on a routine and frequent basis is typically light, easy to separate, and less prone to fouling than aged oil. The polymerization reaction is temperature dependant, so the warmer the tower, the faster the red oil formation rate. This is another reason that excessive reheating of the feed should be avoided.

Some towers do not have reservoirs with circulation pumps between the stages. Without the reservoirs, the red oil cannot be removed from within the tower. This can lead to a heavier hydrocarbon load on the tower and in the spent caustic.

After the lowest stage, the spent caustic collects in the bottom of the tower. In some towers, a separate tap in the bottoms is used to remove the floating red oil layer, so that it does not continue to polymerize. Others may drain the bottoms completely on a periodic basis. Most towers also send the mixed spent caustic to a gravity-separation drum to skim the oil.

Red oil: polymerization and fouling

Polymer formation is the most common cause of fouling in caustic towers. The entrained oils formed from the polymerization can upset the WAO operations as well.

The polymerization reaction results from the aldol condensation of acetaldehyde. Acetaldehyde and vinyl acetate (an acetaldehyde forming compound) are formed in the ethylene process and adsorbed into the caustic. Contaminated cracker feedstocks that result in the formation of these compounds have the undesirable side effect of more polymer formation in the tower.

The polymerization reaction can generally be written as:

1) $CH_3COH + NaOH \leftrightarrow Na^+ \, C^-H_2COH$
2) $Na^+ \, C^-H_2COH + CH_3COH \rightarrow \rightarrow$ polymer

The rate of reaction increases with temperature and acetaldehyde concentration. The initial, smaller polymer remains soluble in the caustic. As the polymer grows, it forms a light insoluble oil that floats on top of the aqueous phase. In this form, it can be separated by a simple skimming step. Some residual will remain and that portion will continue to react and form heavier polymers, eventually forming solids that may become entrained or suspended in the caustic. The larger polymers are more difficult to remove because they do not float and may adhere to metal surfaces.

Depending on the extent of polymerization, dehydration, and other contaminants, the polymer can appear to be red, yellow, green, or other colors. Typically it is red, and so the oil layer is usually referred to as "red oil". The red oil will absorb other organics from the cracked gas as well as corrosion products from the tower, and thus increase in volume. Some heavy organic compounds in the cracked gas will condense and remain in the tower. These compounds also make up the total composition of red oil.

Red oil that is not well managed in the tower will exit with the spent caustic. If not removed, this oil can ultimately pass to the WAO system as high COD slugs.

Spent caustic handling and gasoline washing

As shown in Figure 2, the discharged spent caustic passes to a separation drum where caustic, entrained gasses and entrained oils are separated by gravity. The spent caustic then passes through a depressurization valve and into a degassing drum where the evolved gases from depressurization are removed. The spent caustic is then routed to a gasoline washing step or to a storage tank.

The "gasoline" wash is usually pygas (from the quench water tower), steam cracked naphtha (SCN, from the primary fractionator), or some other stream or combination. Figure 3 shows a gasoline wash configuration, though the location and configuration of gasoline washes can vary.

The gasoline wash removes entrained oils as well as some dissolved organic compounds. The loaded gasoline from the wash may also be contaminated with some caustic, which is removed by a separate water-wash step. The used gasoline can be disposed by burning for fuel value or blending into one of the liquid streams from the primary fractionator or quench tower.

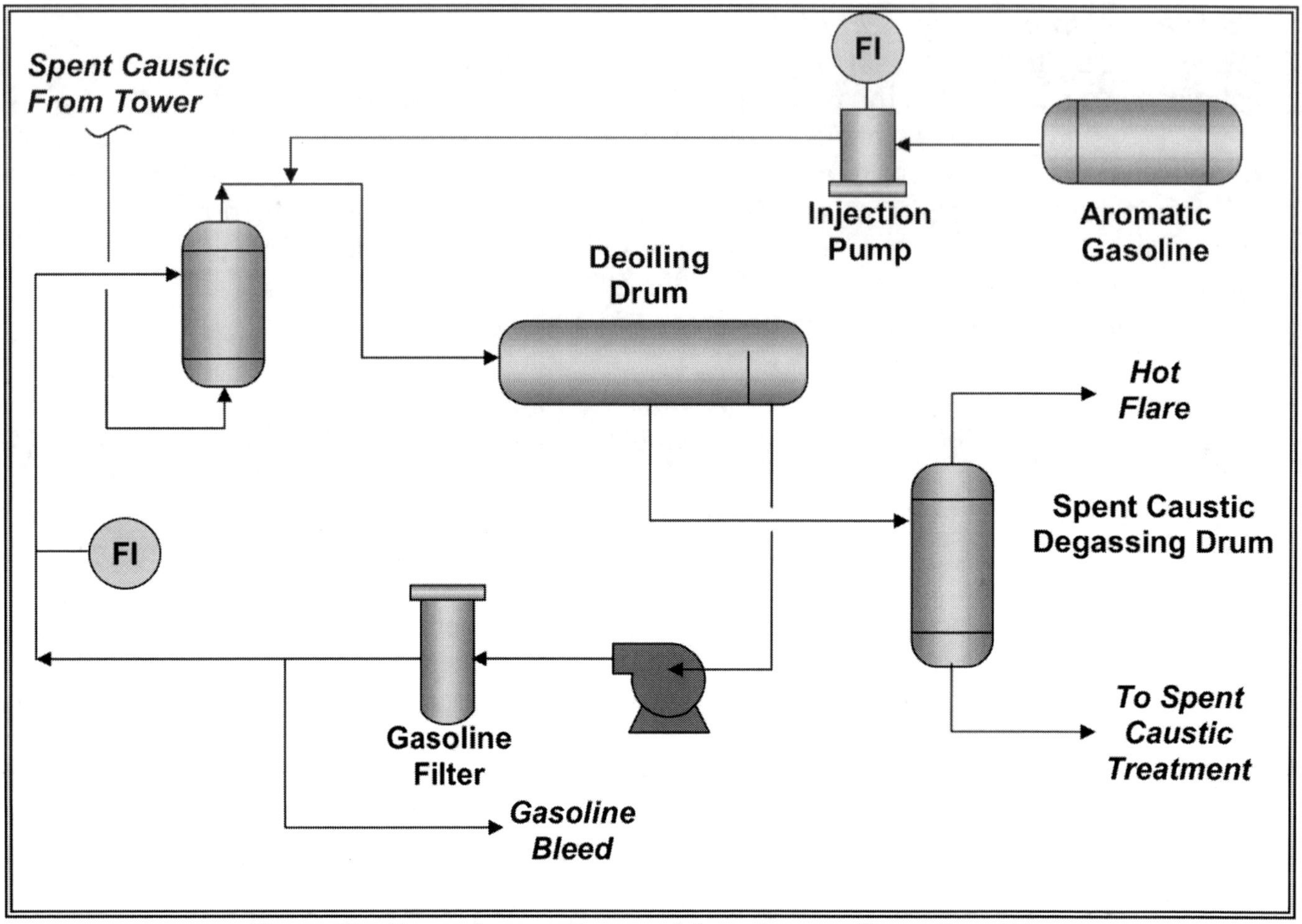

Figure 3. Example gasoline washing step configuration (4)

Spent caustic storage

The spent caustic is transferred to a storage tank. Residence time in the storage tank is typically 24 hours or longer. This tank serves as an equalization tank and is the feed tank for the WAO system. The configuration and operation of this tank can have an impact on the WAO system operations.

Even after a gasoline wash, the spent caustic will contain some red oil precursors. These compounds will continue to polymerize in the storage tank and form more red oil. Oxygen can promote these reactions, so the storage tank should be kept under an inert gas atmosphere. If the oil layer is not removed, it will continue to polymerize and form denser layers that may become entrained in the spent caustic. This will cause oil slugs to enter the

WAO system and result in WAO plant upsets. The hazard is concentrated slugs of red oil entering the feed line.

Spent caustic tanks are equipped with provision for removing the floating oil layer. Most common is a floating skim line, which floats on top of the caustic phase to skim the oil, but floating skimmers are not always reliable. The flexible piping becomes less flexible after years of use. Eventually the floating line may be abandoned and fixed side taps on the tank can be used instead. This can be effective; however, it is not certain if channeling (i.e. "rat-holing") is occurring, where the local red oil is removed, but the floating layer across the entire diameter of the tank may not be. The most effective form of skimming is thought to be "floating cover" style tanks, with a skim line attached to the bottom of the floating cover.

Over several years of operation, a heavy organic sludge will build in the bottom of a non-agitated tank - formed from the aged red oil - and can have a depth of 3 feet or more. As this depth increases, it is possible for slugs of the oil sludge to be drawn into the WAO feed tap. The location of tank inlet lines can also have an affect, as the risk of drawing an oil slug is increased if an inlet stream disturbs the bottom oil sludge. Typically the spent caustic draw line is located several feet from the bottom of the tank. Periodically cleaning the tank to remove the sludge build-up will reduce the risk of heavy oil slugs entering the WAO system.

Conclusions

WAO systems are used for spent caustic treatment and are both reliable and effective. However, WAO reliability can be hampered by off-spec feed, which is affected by the upstream processing and handling of the spent caustic. For maximum up-time, spent caustic management should be done in accordance to the following recommendations:

1. The ethylene gas is cooled and then reheated prior to entering the caustic tower.
2. Low chloride (<500 ppm Cl^-) caustic is used in the caustic tower.
3. Skim loops and/or antifoulant additives are used in the caustic tower to manage the red oil as it forms.
4. A gasoline wash is used with the fresh spent caustic to remove entrained oils and red oil precursors.
5. The spent caustic is stored in a nitrogen blanketed storage tank; the liquor is skimmed daily to remove freshly formed oils; the tank is periodically drained and the accumulated sludge is removed from the bottom; and the feed tap to the WAO system is at least 3 feet above the floor of the tank.
6. Boiler feed water or other low-mineral water is used for all dilution and flush streams.

References

1. Ellis, C.; Lawson, R.; Brandenburg, B. (1994) "Wet Air Oxidation of Ethylene Plant Spent Caustic", *6th Annual Ethylene Producers Conference*

2. Grover, R.; Gomaa, H. (1993) "Proven Technologies Manage Olefin Plant's Spent Caustic", *Hydrocarbon Processing,* September, pp 61-69.

3. Reid, J. (1998) "Introduction to Sulfur Compounds and Sulfur Removal as Applied to Olefin Unit Design and Operation", *10th Ethylene Producers' Conference*, pp 713-755.

4. Matthews, R. (1997) "Performance update : Low Pressure Wet Air Oxidation unit at Grangemouth, Scotland", *Environmental progress*, vol. 16, no1, pp. 9-12.

Optical Remote Sensing of Fugitive VOC Losses

Steven H. Ramsey

Principal Consultant

ENVIRON

Ram A. Hashmonay, Ph.D.

Principal Scientist

ARCADIS

Abstract: Recent years have shown great advances in the detection and measurement of fugitive volatile organic compound ("VOC") and other emissions using optical remote sensing ("ORS") devices. Use of these new techniques in the U.S., Canada, and Europe shows surprising quantities of VOC being lost by the processing, distribution, and consumption of petroleum and petroleum byproducts. New field studies indicate that unexpected levels of fugitive losses are routinely emitted from storage tanks, pumps, pipes, cooling towers, wastewater separators and the like. These emissions are extremely important for two reasons: (1) they are end-products of the refining process, which means they are both volatile and toxic; and (2) they occur throughout the national energy network ranging from well-head to refining to distribution to storage and retailing. These uncounted emissions could have a strong impact on environmental control strategies and emission inventories. Many of these losses are currently uncounted in the emission inventories and thus are lost for offsets and trading. There are also significant implications with respect to compliance with federal and state rules and New Source Review permitting. Significant ORS advances have occurred in the characterization of fugitive emissions from a variety of non-refinery sources such as landfills, animal feeding operations, and other industrial facilities. The following will be discussed: (1) review of Solar Occultation Flux (SOF), Differential Absorption LIDAR (DIAL), and Radial Plume Mapping (RPM) measurement methods and findings over the past year and how those findings relate to estimation methods currently in use; (2) review ongoing and imminent ORS studies; and (3) present options available to government and industry on how to use these studies to improve estimation – and control – of VOC fugitive emissions.

Review of Optical Remote Sensing Technologies

Introduction to ORS

Environmental optical remote sensing, or ORS, involves the use of scientific instruments to detect the presence of pollutants of interest and, in certain cases, measure the concentration of those pollutants in the ambient air. Some of the ORS techniques allow for direct measurement of emission rates from sources that are not amenable to measurement using traditional stack testing methods. Examples of these sources include equipment leaks, storage tanks, delayed cokers at petroleum refineries, wastewater treatment plants, landfills, and industrial flares.

ORS takes advantage of the fact that different pollutants will absorb radiation – and radiate when heated – at different wavelengths across the electromagnetic spectrum. Many of the most common pollutants of interest absorb radiation in the infrared ("IR") portions of the spectrum. If the spectrum for the pollutant(s) of interest is known, one can use ORS to "look" for those compounds.

There are three basic categories of ORS technologies:

- Those that can detect the presence of a compound but don't give any quantitative information and don't provide any information on speciation. Examples include passive infrared ("IR") cameras that are widely used in finding leaks at industrial facilities.
- Those that can provide compound speciation and concentration data, but not how much is present. Examples include open-path ("OP") sensors used to detect the presence and concentration of harmful or explosive gases.
- Those that can measure mass flux, or emission rate, of specific compounds.

Following is a brief discussion of each of the three technologies in this last group: Solar Occultation Flux ("SOF"), Differential Absorption LIDAR ("DIAL"), and Vertical Radial Plume Mapping ("VRPM").

Solar Occultation Flux

SOF is a passive infrared method based on the measurement of IR intensity spectra of the sun from a mobile vehicle in combination with point measurements with a parallel IR system coupled to a White-cell. In order to obtain the flux from a particular source, the vehicle is driven in such as way that the detected solar light traverses across the plume of air pollutants. The flux is then obtained as the integrated sum of the retrieved path averaged concentrations, multiplied by the wind speed. Wind speed and wind direction are measured using high masts and GPS balloon sondes.

The SOF technique has been used for over five years in Sweden to quantify emissions from petroleum refineries. It was also used to measure reactive VOCs in the Houston Ship Channel area during the second Texas Air Quality Study ("TexAQS II"). SOF measurements are conducted in the mid-IR (2.5-8 μm) range. Therefore, this technique can be effectively used to detect and quantify many VOCs including alkanes and olefins (but not aromatics), as well as certain inorganic pollutants such as nitrogen dioxide and sulfur dioxide.

SOF is limited in that it only works in daytime under sunny conditions as the plume needs to be between the SOF receiver and the sun. SOF also gives better results if there is some distance between the source of emissions and the location where the measurement is taken. The uncertainty of the results is typically 20 to 50 percent. Measurements conducted near the source are associated with larger uncertainties as the plume is closer to the ground and the wind field is more disturbed. To avoid the uncertainties due to turbulence and mixing, measurements are usually made at some downwind distance, usually between 0.5 to 2 km, to avoid uncertainty due to turbulence and mixing. Wind speed and direction are the largest known sources of error for the SOF technique. To reduce this fluctuation, the time spent in the plume is minimized by driving the SOF vehicle at speeds ranging from 30 to 35 mph. A diagram of the SOF technique is shown in Figure 1.

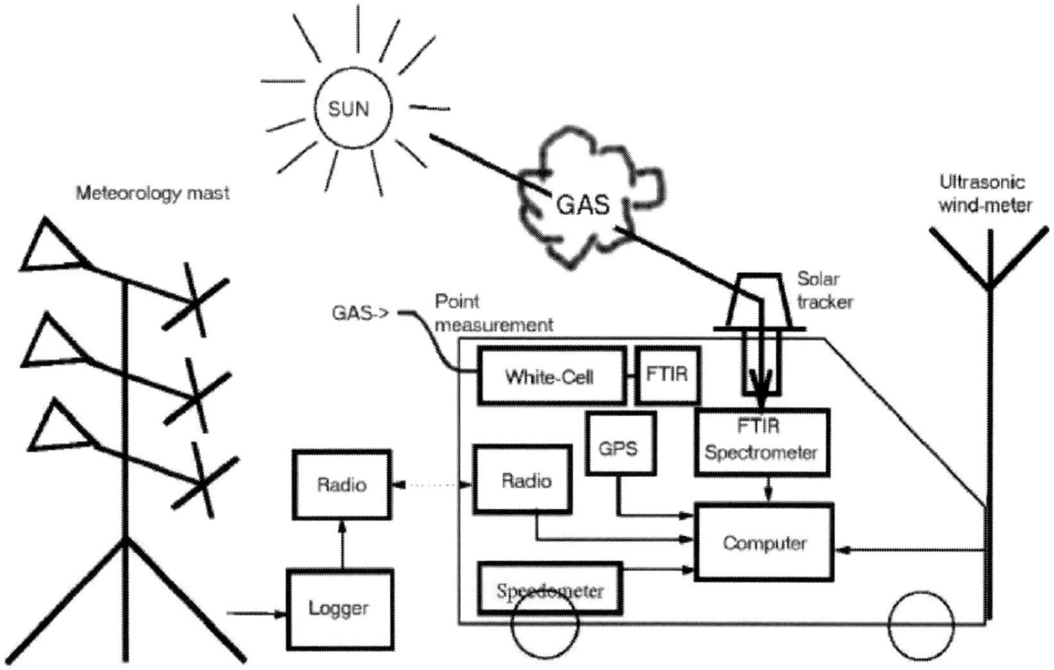

Figure 1. SOF System
(Source: www.FluxSense.se)

There is a single commercial provider of SOF: FluxSense of Göteborg, Sweden.

Differential Absorption LIDAR

DIAL is based on the Differential Absorption LIDAR principle. LIDAR is an acronym for Light Detection and Ranging and is a range finding device which uses light as the ranging medium instead of the radio or microwaves used in conventional radar devices. DIAL is a development of the simple LIDAR which uses light (normally tunable laser light) spectroscopically. That is, with the knowledge of the electromagnetic absorption characteristics of a specific gas, the source laser can be tuned to the wavelength of a defined absorption feature and can be used to make a measurement of the concentration of the gas in the laser path. In a 'differential' mode this concentration measurement can be made more accurately as the absorbed signal is compared with a second signal generated from laser light at a wavelength just off the defined absorption feature. If the two wavelengths are kept close spectrally and are sent into the atmosphere close temporally, they will both be scattered by the atmosphere in the same way. The difference in the powers of the two signals recorded by a detector gives a direct measurement of the amount of target gas in the laser path. If sufficient laser output energy is available, it is possible to use the small fraction of light backscattered to a detector (housed with the source) from atmospheric aerosols etc. to make the concentration measurements. A pulsed laser source will allow the return signals to be time-gated according to their position of backscatter thus giving the system a range resolved concentration measurement capability. The DIAL technique is shown graphically in Figure 2.

Figure 2. DIAL Technique
(Source: www.spectrasyne.ltd.uk)

There are currently two commercial providers of land-based, mobile DIAL technologies: Spectrasyne and the UK National Physical Laboratory ("NPL"). Both are based in the UK but have conducted surveys in the U.S. Spectrasyne's DIAL system employs two lasers: an ultraviolet (UV) laser to measure certain criteria pollutants (NO_X, SO_2 and O_3) and/or light aromatic compounds, such as benzene, toluene and xylenes; and an IR laser to measure methane and/or total hydrocarbons. To speciate emissions, Spectrasyne manually places sampling pumps and absorption tubes at various locations around the area being surveyed.

The tubes are sent to third-party laboratories for analysis. NPL's DIAL system employs a single laser that can be tuned to operate in the UV, visible or IR spectral regions. The DIAL analysis gives a good estimate of total hydrocarbon emissions. NPL improves the analysis by combining the DIAL measurements with the results of a gas chromatography ("GC)" analysis of the emitted gases.

With respect to validation of the DIAL technique, in-field validation measurements carried out by NPL include measurements of a known source of methane which were within 10% of the actual emitted value. For aliphatic hydrocarbons, the measurements were within 12% of the actual emissions. For toluene, the measurements were within 15% of the actual value and for VOC emissions, the measurements were within 8% of the actual value calculated by standard point method. The NPL DIAL has also been validated using the OP calibration facility developed at NPL. This is a 10 m long windowless calibration cell which can generate controllable test gas regions for testing the sensitivity, accuracy and spatial resolution of DIAL and other OP techniques. The NPL DIAL was also used in parallel with a USEPA DOAS system and canister samples (analyzed in a laboratory at a later time) for side-by-side comparisons of benzene measurements during recent studies for the Texas Commission on Environmental Quality ("TCEQ"). The three measurement techniques showed good correlation.

Spectrasyne carried out its first mass emission validation study in 1991 at Spadeadam, Cumbria in the UK, a defense site. In this study, DIAL measurements were carried on a controlled release of methane, 200 m-400 m downwind. Agreement between release measurements and DIAL was within 10%. The NPL DIAL (then owned and operated by Shell Oil) was also involved in this study. Another mass emission study was carried out by Spectrasyne in collaboration with the European Oil Industries Association for Environment, Health and Safety ("CONCAWE"), using a river barge tanker vent as a single emission source. In these studies, the maximum deviation in DIAL measurements was 15%.

Vertical Radial Plume Mapping

VRPM uses an open-path, Path-Integrated Optical Remote Sensing ("PI-ORS") system in multiple beam configurations, coupled with wind measurement, to measure emission fluxes. The VRPM methodology was first proposed in late 1990s and attempted to mimic and provide a lower-cost alternative to DIAL. Any scanning PI-ORS system that can provide path-integrated concentration ("PIC") data can be used in a VRPM application. These include Open-Path Fourier Transform Infrared ("OP-FTIR") spectroscopy and Ultra-Violet Differential Optical Absorption Spectroscopy ("UV-DOAS"). The OP-FTIR and UV-DOAS technologies are widely used due to their capability of simultaneous chemical detection for a large number of gas species of environmental interest. The VRPM methodology has been validated by measuring emission fluxes from controlled gas releases simulating area sources and is described in U.S. EPA Other Test Method 10 ("OTM-10") which was made available for use in July 2006. This method has been applied and demonstrated at various sites for measuring emission fluxes.

The VRPM methodology uses a vertical scanning plane downwind of the source to directly measure the gaseous flux. Two different beam configurations of the VRPM methodology are recommended: the five-beam (or more) and the three-beam VRPM configuration. Figure 3 illustrates the setup for these two VRPM beam configurations. In the five-beam (or more) configuration, the ORS instrument sequentially scans over five path determining components ("PDCs"). PDCs are typically retroreflecting mirrors for most ORS systems but can be also light sources or detectors. Three PDCs are along the ground-level crosswind direction (beams a, b, and c in Figure 3), and the other two are elevated on a vertical structure (beams e and f in Figure 3). The additional beam (d) in Figure 3 is for 6-beam configuration, which provides better spatial definition of the plume in the crosswind direction. In the three-beam configuration, the ORS instrument sequentially scans over three PDCs. Only one beam is along the ground level (beam c or d in Figure 3) and the other two are elevated on a vertical structure (beams e and f in Figure 3).

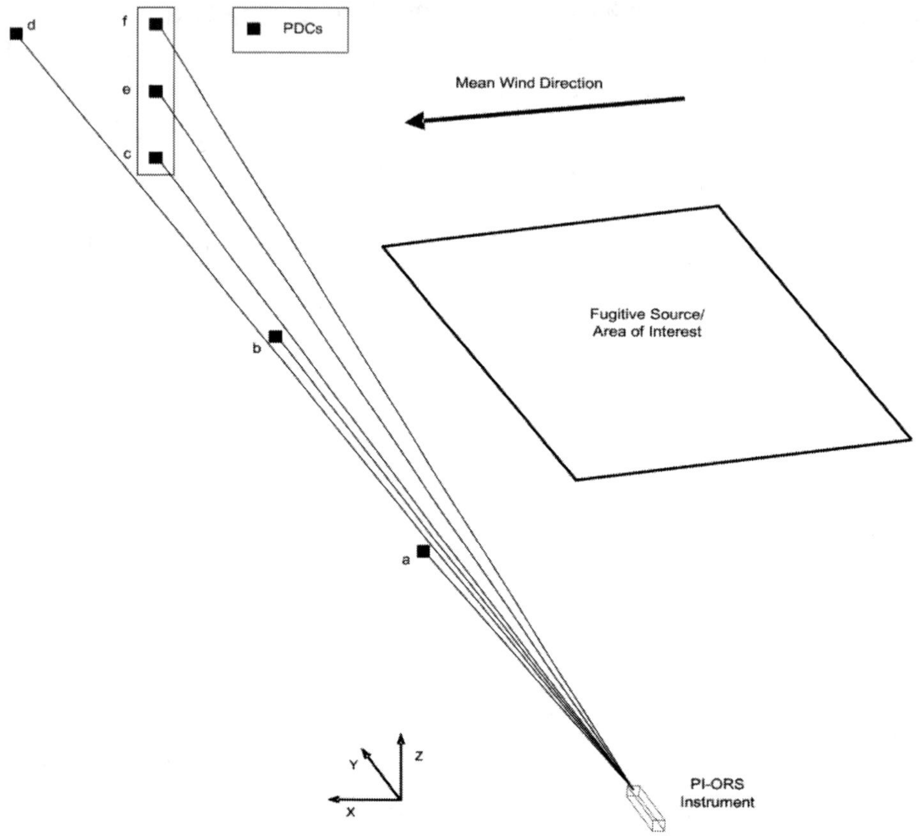

Figure 3. VRPM Configuration

PIC data are collected over time, completing many cycles through the defined beams of each configuration. Once the PIC for all beam paths are averaged with the predetermined grouping of cycles for the gas species of interest, the VRPM calculations make use of the PIC information to reconstruct a plume map in the vertical downwind plane using the smooth basis function minimization ("SBFM) approach. Once the parameters of the function are found for a specific run, the VRPM procedure calculates the concentration values for every square elementary unit in a vertical plane. Then, the VRPM procedure integrates the values, incorporating wind speed data at each height level to compute the flux.

ARCADIS is the licensed provider of the VRPM technology.

Summary of Recent Studies

SOF

Over the past eight years, FluxSense has conducted approximately 50 SOF surveys. These include:

- Since 2001, FluxSense has conducted annual SOF surveys of emissions from three Swedish refineries as part of the KORUS program. Measurements have been made for total alkanes.
- Since 2001, FluxSense has conducted annual SOF surveys of two Swedish chemical plants. Measurements have been made for total alkanes.
- Participation in TexAQS II during the summer of 2006. Measurements were made for alkanes, olefins, NO_2 and SO_2. The TexAQS survey clearly indicated the value of SOF as a screening technique for rapidly identifying areas of elevated emissions of target compounds.
- During 2005 and 2006, measurement of emissions from bitumen refineries in Sweden.

In addition to the work already completed, the Houston Advanced Research Center ("HARC") is looking to fund a SOF study that would be conducted at the same time as a DIAL study, allowing a side-by-side comparison of the technologies.

DIAL

Much of Spectrasyne's work has been conducting fenceline surveys of industrial facilities in Northern Europe, looking for emission plumes, and providing information to the facilities on measured emission rates and potential sources of the emissions. Since 1992, Spectrasyne has conducted approximately 150 DIAL surveys at a variety of industrial sites including petroleum refineries, chemical plants, storage terminals, natural gas plants, flares, barges and ships and airport operations. More recently, Spectrasyne has conducted several surveys in North America, including:

- A Canadian petroleum refinery (2005),
- A Canadian oil sands production facility (2005), and
- A Texas petroleum refinery (2007).

Discussions of some of their work can be found on their website: www.spectrasyne.ltd.uk.

Since acquiring their DIAL system from Shell, NPL has conducted a handful of DIAL surveys, including several in the U.S.. These include:

- A Texas petroleum refinery and an independent storage terminal during 2007,
- A municipal landfill during 2008, and
- A coke oven battery in the Midwest during 2008.

In addition to this work, NPL been commissioned to by the City of Houston to conduct benzene surveys at a Houston Ship Channel petroleum refinery during 2009. They may also be working with the TCEQ to evaluate flare performance sometime during 2009.

VRPM

VPRM has been used in a number of recent air quality management research and commercial applications including:

- Measurement of mercury emissions from a chlor-alkali facility using a UV-DOAS instrument.
- Measurement of VOC and methane emissions from landfills, combined animal feeding operations (CAFOs), Superfund and Brownfield sites.
- Measurement of hydrogen fluoride emissions from phosphate processing operations.
- Measurement of multiple hydrocarbon emissions from refueling stations.
- Measurement of multiple hydrocarbon emissions from petroleum refinery and petrochemical manufacturing operations.

Many of these studies have been conducted in cooperation with the U.S. EPA.

Findings of Potential Interest & Implications

Between 1988 and 1999, Spectrasyne conducted six DIAL surveys of total hydrocarbon emissions from a refinery in Göteborg, Sweden. Their findings are shown in Figure 4.

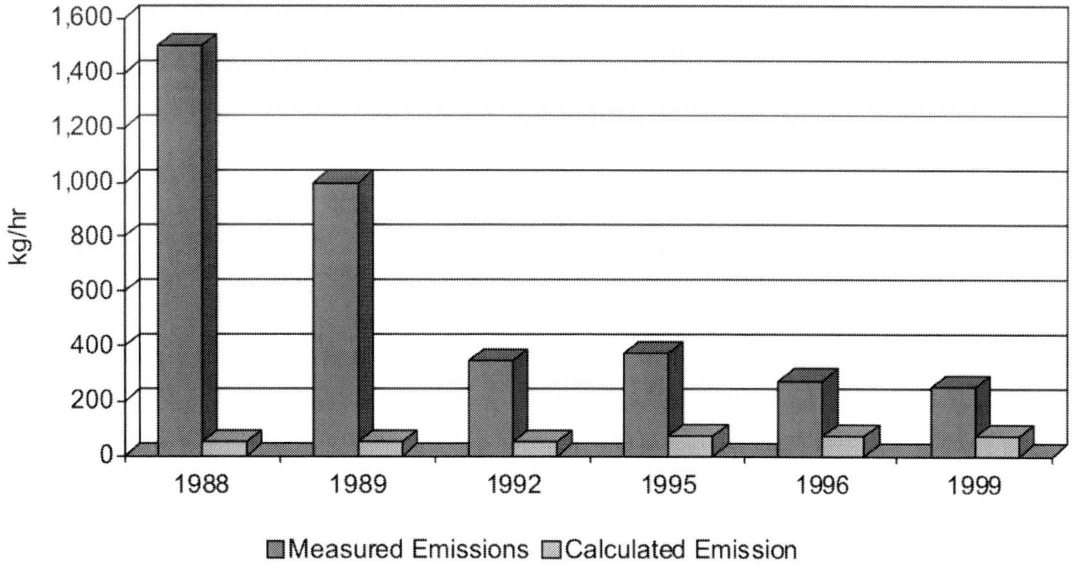

Figure 4. DIAL Survey Results

As reported by Spectrasyne, most of the emissions at this refinery originated from product storage, crude oil storage, process fugitives and, to a lesser extent, wastewater management. This refinery does not have a delayed coker, another process unit that Spectrasyne has determined to be a potentially large source of unaccounted or under-accounted emissions.

The purpose of the surveys was to provide information to the refinery that could be used to improve refinery operations and reduce emissions. As shown, the program was very successful, with measured emissions reduced by over 80% between 1988 and 1999. Two interesting items of note:

1. Even as actual emissions, as measured by DIAL, were reduced dramatically, calculated emissions using accepted emission factors were essentially unchanged.
2. While actual emissions, as measured by DIAL, had been reduced by over 80% during the course of this work, they were still approximately 3-times the calculated values.

These findings may be indicative of one or more of the following:

- The DIAL survey results do not accurately reflect actual emissions.
- Short-term emissions, as measured by DIAL, are not indicative of long-term emission rates. [While a rigorous statistical analysis has not been performed by the authors, one would think that measurements consistently showing measured short-term emission rates much greater than calculated long-term emissions would constitute sufficient evidence that long-term emissions are being underestimated.]
- The refinery was not using the emission factors properly.
- The emission factors represent an oversimplification of refinery operations and, as a result, may not able to consistently estimate emissions from real-world refinery operations accurately.

The DIAL survey findings from the Göteborg Refinery are generally consistent with ORS surveys conducted at other refineries. For example:

- Environment Canada has worked with private industry on several occasions to conduct DIAL surveys of petroleum refining and oil sands production operations. Publication by Environment Canada of the findings of a 2005 petroleum refinery survey confirmed what was already known from years of study in Europe: actual hydrocarbon emissions from petroleum refineries are typically 10 to 30 times greater than estimates obtained using approved emission factors. The main sources of underreported emissions tend to be crude and product storage tanks, wastewater treatment operations, delayed cokers, and equipment leaks – all of the things that cannot be directly measured using traditional stack testing methods but can be directly measured using ORS technologies.
- In August 2007, the TCEQ commissioned NPL to conduct a series of studies at a Texas City refinery and independent storage terminal. The findings included: a) emissions from the delayed coker, while larger than those calculated using approved emission factors, were not as large as those reported for the Canadian refinery; b) emissions from some crude oil storage tanks were much larger than were expected or calculated using approved emission factors; and c) emergency flares were shown to achieve much lower than expected control of routine emissions during normal operations.

In addition to indicating that over-reliance on emission factors may result in substantial underreporting of emissions, the surveys seem to indicate that individual facilities have unique emission fingerprints that may not be conducive to using a one-size-fits-all approach to estimating emissions.

The findings from the facility-specific surveys are mirrored by airshed-wide assessments. As presented in the findings report for TexAQS II, while there are indications that ethylene emissions from industrial sources in the Houston area decreased by approximately 40% between 2000 and 2006, measurements indicate that reported emissions for 2004 underestimate actual 2006 emissions by one to two orders of magnitude.

There are some fairly significant consequences if, in fact, use of accepted emission factors results in chronic underestimation of emissions.

Air Quality Management and Planning: Air quality management organizations utilize emission inventories prepared and submitted by regulated entities as key inputs to the planning process. If this information is incorrect, planning efforts may result in flawed and ineffective strategies.

Regulatory Programs: Many regulatory programs (e.g. National Emission Standards for Hazardous Air Pollutants for Source Categories, and Title V Federal Operating Permits) have emission threshold triggers. Flawed emission estimation techniques could result in improper determinations of applicability – and associated risk of enforcement and penalties.

New Source Review: The entire New Source Review ("NSR") permitting program assumes an underlying accuracy in the emission estimation process. If emission estimates are not accurate, the integrity of the NSR applicability determination, the Best Available Control Technology ("BACT") review, and ambient air quality impacts analysis will be affected.

Liability & License to Operate: Neighboring communities will not be pleased if they learn that industrial facilities in the vicinity have been underestimating, potentially significantly underestimating, emissions and, thus, community exposure to air pollutants.

Conclusions and Recommendations

ORS technologies have been conclusively demonstrated as valuable tools to industry in efforts to improve characterization and quantification of fugitive VOC emissions. However, application of ORS has not been widespread – at least not in the U.S. – and findings from ORS surveys have not been used to improve emission factors and the emission estimation process followed by the vast majority of industrial facilities. To improve this state of affairs, the authors recommend the following:

- ORS technologies, especially DIAL, need to be generally more available and at a lower cost. Having only two DIAL suppliers – both based out of the UK – and costs as high as $500,000 for a one month facility survey are significant barriers to widespread use. The solution would seem to be establishing a commercial DIAL business in the U.S.
- More ORS surveys need to be conducted and the information from those surveys needs to be made available for use in improving emission estimation techniques. We believe

that, for the most part, industry is willing to participate in this effort were it not for concerns about use of ORS survey findings as credible evidence in enforcement proceedings. Regulatory agencies need to think this process through, finding ways to provide incentives and remove barriers to industry doing the right thing.

- EPA needs to continue with their efforts to develop standard methods and guidance for use of ORS technologies. Until this is done, there will continue to be questions about the validity of ORS survey findings.
- EPA should develop a master plan for formulating better emission factors using ORS survey findings. This could include conducting their own ORS surveys specifically for the purpose of improving emission estimation techniques and taking advantage of ORS surveys conducted by others. Included in the latter category would be findings from ORS surveys conducted as a result of consent decree requirements.

References

1. Chambers, Allan K., M. Strosher, T. Wootton, J. Moncrieff, and P. McCready, "Direct Measurement of Fugitive Emissions of Hydrocarbons from a Refinery," *J. Air & Waste Manage. Assoc.* 58:1047–1056, 2008.

2. Cowling Ellis B., C. Furiness, B. Dimitriades, D. Parrish, Final Rapid Science Synthesis Report: Findings from the Second Texas Air Quality Study (TexAQS II)," TCEQ Contract Number 582-4-65614, August 2007.

3. EPA, "Optical Remote Sensing for Emission Characterization from Non-point Sources," USEPA Other Test Method Protocol, OTM-10, 2006.

4. EPA, "Evaluation of Fugitive Emissions Using Ground-Based Optical Remote Sensing Technology," USEPA Report EPA/600/R-07/032, February 2007.

5. Gronlund, R.; M. Sjoholm, P. Weibring, H. Edner, S. Svanberg, "Elemental Mercury Emissions from Chlor-Alkali Plants Measured by LIDAR Techniques;" *Atmos. Environ.* 2005, *39,* 7474-7480.

6. Hashmonay, R.A., "International Applications of OTM-10 in Chemical and Petroleum Industries," 2nd International Workshop on Remote Sensing of Emissions: New Technologies and Recent Work, Research Triangle Park, NC, April 2008.

7. Hashmonay R.A., D.F. Natschke, K. Wagoner, D.B. Harris, E.L. Thompson, and M.G. Yost " Field Evaluation of a Method for Estimating Gaseous Fluxes from Area Sources Using Open-Path Fourier Transform Infrared," *Environ. Sci. Technol.*, 35, 2309–2313, 2001.

8. Hashmonay, R.A., and M.G. Yost, "Innovative Approach for Estimating Fugitive Gaseous Fluxes Using Computed Tomography and Remote Optical Sensing Techniques," *J. Air Waste Manage. Assoc.*, 49, 966-972, 1999.

9. Hashmonay, R.A., R. Varma, M.T. Modrak, R.H. Kagann, R.R. Segall and P.D. Sullivan, "Radial Plume Mapping: A US EPA Test Method for Area and Fugitive Source Emission Monitoring Using Optical Remote Sensing," Y.J. Kim and U. Platt (eds.), *Advanced Environmental Monitoring,* Spring 2007.

10. Mellqvist, J., "Industrial Emission Measurements Using the Solar Occultation Flux Method," 2nd International Workshop on Remote Sensing of Emissions: New Technologies and Recent Work, Research Triangle Park, NC, April 2008.

11. Modrak, M. T.; R.A. Hashmonay, R. Kagann, "Measurement of Fugitive Emissions at a Region I Landfill;" EPA-600/R-04-001; U.S. Environmental Protection Agency, Research and Development, Work Assignment No. 4-003, January 2004.

12. Moncrieff, J., T. Wootton, "Reflections on 20 years of DIAL VOC Measurements in the Oil and Gas Industries," 2nd International Workshop on Remote Sensing of Emissions: New Technologies and Recent Work, Research Triangle Park, NC, April 2008.

13. Nettles, R. "Texas Commission on Environmental Quality Differential Absorption LIDAR Project, Summer 2007, Texas City, Texas," 2nd International Workshop on Remote Sensing of Emissions: New Technologies and Recent Work, Research Triangle Park, NC, April 2008.

14. Thoma, E. D., R.C. Shores, E.L. Thompson, D.B. Harris, S.A. Thorneloe, R.M. Varma, R.A. Hashmonay M. T. Modrak, D.F. Natschke, H.A. Gamble, "Open Path Tunable Diode Laser Absorption Spectroscopy for Acquisition of Fugitive Emission Flux Data," *J. Air and Waste Manage. Assoc.* 55, 658-668, 2005.

15. Walmsley, H.L.; S. J. O'Connor, "The Use of Differential Absorption LIDAR to Measure Atmospheric Emission Rates at Industrial Facilities, " P*roceedings of the A&WMA's International Conference on Optical Sensing for Environmental and Process Monitoring,* Pittsburgh, PA, 1996, p 127.

16. Yost, M.G.; R.A. Hashmonay, "Mapping Air Contaminants Using Path-Integrated Optical Remote Sensing with a Non-overlapping Variable Path Length Beam Geometry," U.S. Patent 6,542,242, 2003.

Underestimated Emissions Inventories

Alex Cuclis
Research Scientist
Houston Advanced Research Center

Abstract: A number of studies in the U.S. and Europe indicate that there are substantial errors in emission inventories of volatile organic compounds (VOCs) for refineries and chemical plants. Although significant reductions of VOCs have occurred over the past three decades, some reports suggest that emissions are underestimated by an order of magnitude or more. There are technologies used mainly in Europe that can provide substantially improved emissions inventories measuring VOCs with remote sensing tools instead of the current practices that are based on engineering estimation techniques.

Introduction

Significant reductions have been made in VOC emissions throughout the United States over the last three decades. There is, however, a growing body of evidence that emissions factors developed 20 years ago, which were based on engineering estimates for plant operations, underestimate VOC emissions by an order of magnitude or more.[1,2,3,4,5] These discrepancies need to be resolved to develop the most cost effective control strategies for meeting federal ozone standards as well as identifying the best strategies for addressing emissions of air toxics and greenhouse gases. Currently neither ethane nor methane are part of the VOC inventories developed in the U.S. and the American Petroleum Institute estimates that more than 40% of the CO_2 that is released from the energy intensive oil refining industry can come from noncombustible sources, like the venting that occurs from cat crackers and hydrogen plants.[6] How will chemical engineers identify, quantify and audit GHGs and other emissions in the next 20 years? This paper discusses several case studies that illustrate the differences between VOC emission inventories and values that have been obtained using several different direct measurement techniques and the uncertainties associated with the different approaches. Also included is a discussion of variable and routine emissions. To what extent do measured concentration spikes at ground-level monitors represent emission events (intermittent releases due to start-ups, shutdowns or upsets) at facilities as opposed to changes in wind direction and meteorology that brings a plume to ground-level for a brief period of time? When should focus be placed on "emission events" from process vents, delayed cokers, relief valves and flares or on routine emissions from process areas, storage tanks, wastewater facilities, etc.?

Many other factors add to the complexity of the problem. Since ozone has been a major concern in Houston, the focus has been placed on highly reactive VOCs (HRVOCs) like ethene and propene. Houston is where much of the U.S. production of these chemicals occur. However, as concerns grow about air toxics like 1,3 butadiene (which is also an HRVOC), benzene, etc., other issues become more important. Recently, another VOC and air toxic, formaldehyde, has garnered substantial attention, because it acts as a supercharged catalyst for ozone production by providing chemically active radicals that promote ozone formation.

This paper will provide a brief description of how VOC emission inventories are generated for refineries and chemical plants, identify some of the reasons why the inventories may not include all emissions, how emissions have been measured and compared to inventories and identify the concerns raised regarding the comparison of measurements to emission inventories. This will be followed by a discussion of three case studies: VOC emissions in Europe from 1988 to the present with Differential Absorption LIDAR (DIAL) and Solar

Occultation Flux (SOF), the Texas Air Quality Study 2000 (TexAQS 2000) and the follow-up study in 2006 (TexAQS II). DIAL and SOF have been used in many different types of facilities however the case study in this paper focuses on the surveys performed in refineries. The TexAQS 2000 and TexAQS II field campaigns found abundant information on the meteorology and chemistry of the atmosphere but the case studies here will focus on the findings related to ethene and propene, since they are very reactive and can rapidly create ozone.

Estimating VOC Emissions at Refineries and Chemical Plants

In the U.S., a VOC inventory includes all organic compounds except for methane, ethane and a few other compounds that are considered non-reactive in terms of their ability to form ozone. To build a VOC emissions inventory for a petrochemical facility, the following sources are typically considered:

Blowdowns
Flares
Catalytic Crackers
Cooling Towers
Fugitive Emissions
Marine Vessel Loading
Process Heaters
Tanks
Vacuum Distillation
Wastewater Treatment
Other

The procedures for developing these inventories are provided on the EPA Technology Transfer Network (TTN) Clearinghouse for Inventories and Emission Factors (CHIEF) website at: http://www.epa.gov/ttn/chief/. Typically an emissions factor is developed that represents the amount of released chemical in relation to the throughput and the type of equipment. According to the EPA website, "In most cases, these factors are simply averages of all available data of acceptable quality, and are generally assumed to be representative of long-term averages for all facilities in the source category (i.e., a population average)." [7] This is an important point to note; because some claim that emissions measurements are merely a snapshot at a particular time and are not representative of annual emissions. However, the emissions factor used is in effect a single number of an average of many measurements that is seldom based on site specific measurements. The general equation for emissions estimation is:

$$E = A \times EF \times (1 - ER/100)$$

where:

- E = emissions;
- A = activity rate;
- EF = emission factor, and
- ER =overall emission reduction efficiency, % [7]

Sources of Error in VOC Emission Estimates

Some of the common reasons why the emissions estimates may not reflect the actual emissions are as follows:

1. Emissions estimates typically assume the equipment is "well-maintained." The equipment may be poorly maintained or not maintained at all, and emit much higher emissions that when it is in good condition. Over 50% of U.S. refineries have shutdown since 1980, and those that remain have had to operate on tightly constrained budgets. Chemical plants have seen similar cost pressures.

2. Default parameters may be used instead of actual numbers. For example, a facility may assume a low vapor pressure for crude oil tanks when crude oil often contains a substantial portion of very light components. This practice has the potential to substantially underestimate emissions from these tanks.

3. The emission factor that is applied is not accurate for a specific application due to differences in design or operation.

4. There may be sources inside facilities that have emissions, but there are no emissions estimating techniques to capture them.

5. There may be emissions in unexpected places due to operator error (e.g., a valve was inadvertently left open or a blind flange was not installed, etc.).

The case studies provided below indicate that direct measurements of VOCs indicate VOC fluxes that are often more than an order of magnitude higher than reported in the emissions inventories. When the Texas Natural Resource Conservation Commission (TNRCC), the previous name for the Texas Commission on Environmental Quality (TCEQ), first found out about these problems after Texas Air Quality Study 2000, a field campaign in the Houston area in 2000, they hired a firm to investigate the problem. The resulting report concluded that emissions from industrial sources of VOC submitted to the TNRCC have been calculated using the methods approved by TNRCC and EPA. The report identified that the majority of errors were in methods and quality assurance of the data.[8]

The U.S. EPA has performed their own analyses and acknowledges that there are substantial problems with the way emissions inventories are built. Another concern that EPA has identified is that "Emissions factors, intended for use in developing emissions inventories, have been inappropriately used for key environmental decisions beyond their intended use."[9]

Comparison of VOC Emission Inventory Estimates to Measured Values

There are two commonly used approaches to compare emission inventory estimates to measured values:

1. Direct measurement using advanced analytical techniques such as Differential Absorption LIDAR (DIAL) or Solar Occultation Flux (SOF).

2. Comparing ratios of the hydrocarbons of interest to nitrogen oxides with ground-level monitors or samples taken by aircraft.

Figure 1. Techniques for measuring VOC emissions.

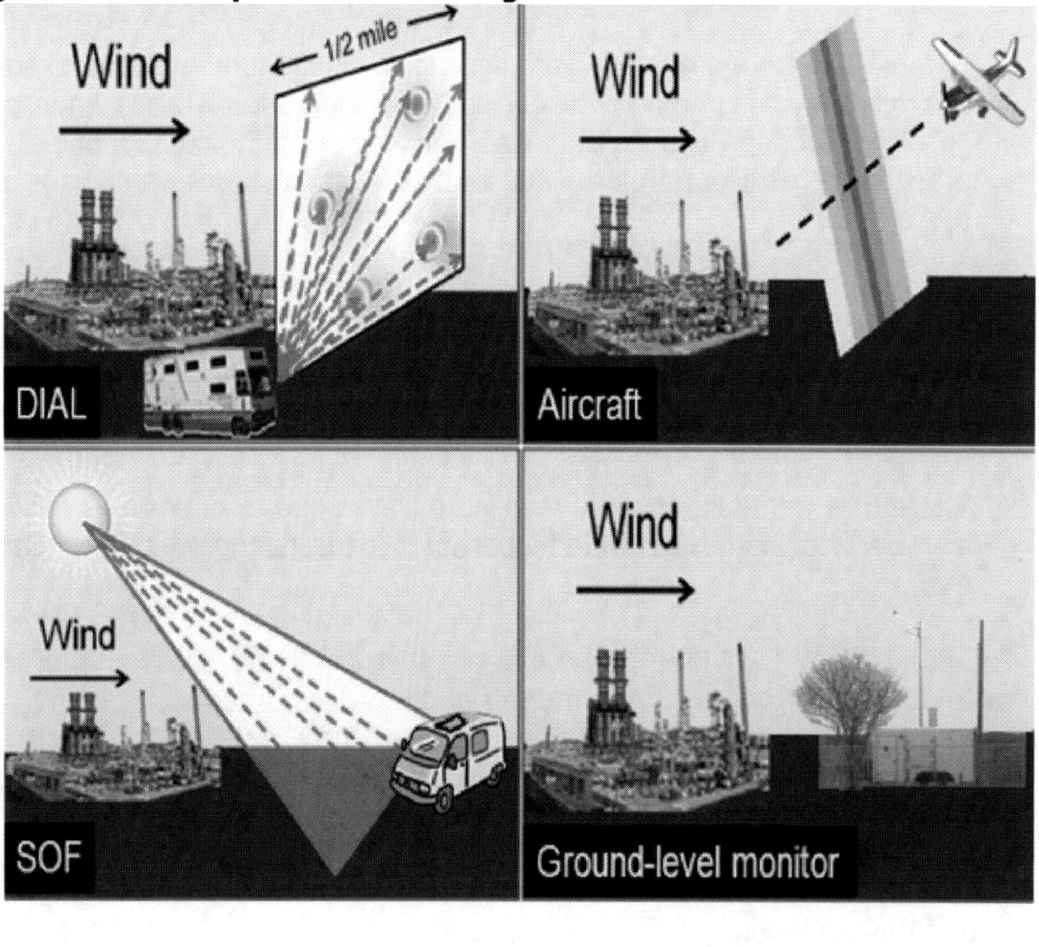

Details of the principles and history of the DIAL and SOF analytical techniques can be found elsewhere.[4,10,11] The key to these direct measurement techniques (shown in the two left panes in Figure 1) is that in both cases they obtain a 2-dimensional concentration map of the atmosphere. DIAL typically measures an upwind and downwind vertical plane and uses the wind speed to estimate a VOC flux across an area. Its range is limited to about half a mile and as a result it is typically used inside facilities to measure emissions of specific areas. SOF can perform some tasks similar to DIAL, however it cannot get range resolved concentrations – or identify specific concentrations in space. Instead SOF generates an average concentration along the measurement path. When the path is changed as the vehicle moves, the changing measurements are integrated into a different type of concentration map that is converted to a vertical plane perpendicular to the wind direction. These measurements are then converted into VOC flux numbers or VOCs in lbs/hour, which are compared to emissions inventories. SOF has also been used to measure emissions from outside of facilities to generate VOC flux numbers. [10] Since SOF uses the sun as its light source, measurements can only occur during daylight.

In order to achieve the best results, DIAL and SOF techniques rely on returning to the same location on several different days when the wind is blowing in the same direction, in order to get an average value of emissions. Emissions will vary and are dependent on wind speed, however, even the lowest measurements are typically several times higher than the reported inventory. Since 1988, Spectrasyne (a privately owned company that was previously part of BP research), has performed DIAL studies. They have amassed a significant amount of experience and data by performing 30 surveys of refineries over the past 20 years. Although many DIAL and SOF studies have been performed at other facilities, refineries have been selected to provide examples. It should be noted that every refinery is different in feedstock, size and complexity among other things. Despite these differences it is worth providing the results that were found in this representation of a "typical" large, complex refinery. [12]

Figure 2. Typical distribution of VOC emissions at a complex refinery.

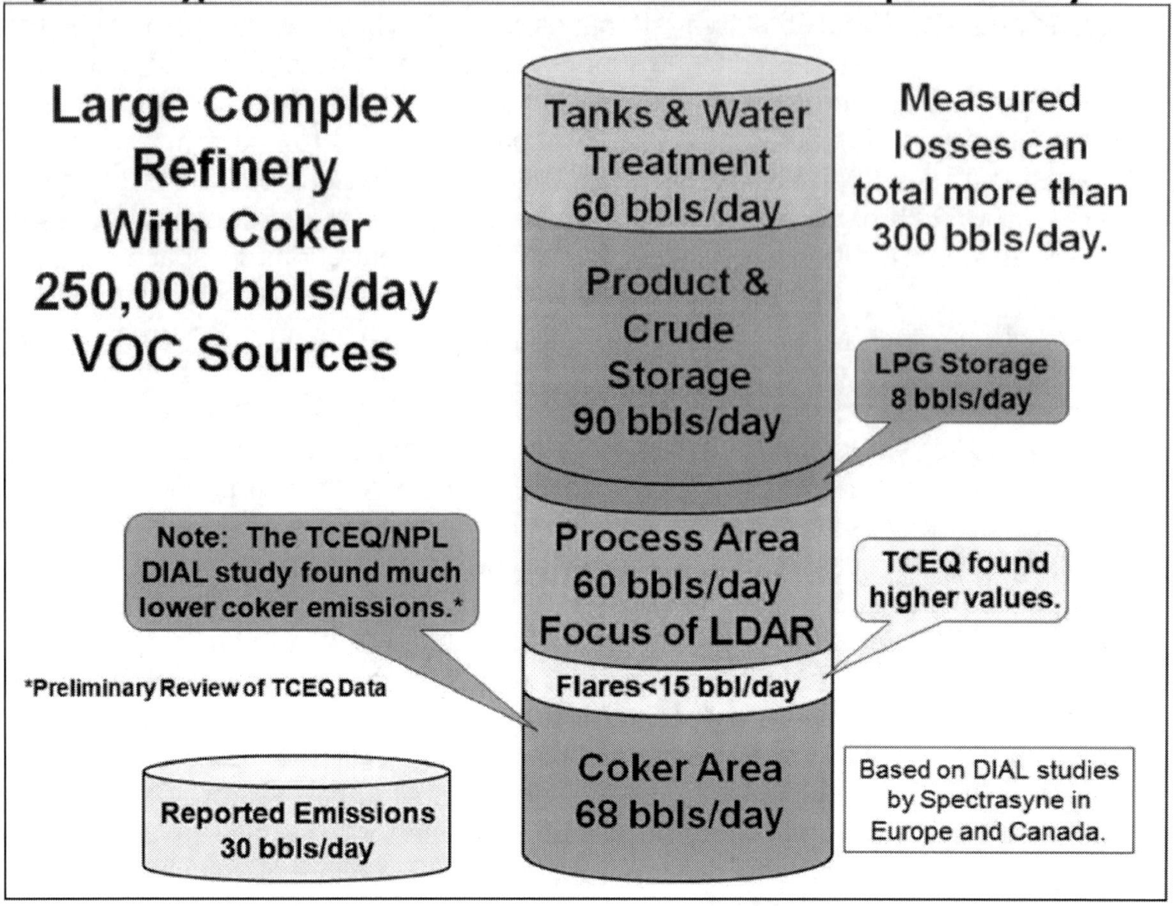

Although the reported emissions from a 250,000 barrels/day refinery may be the weight equivalent of 30 barrels/day (a barrel is approximately 300 pounds), the measured emissions are typically 10 times that amount, or a loss of 300 barrels worth of crude per day in a refinery that processes 250,000 barrels/day. Based on Spectrasyne's experience in Europe and Canada, storage tanks tend to be the biggest problem. When product, crude and intermediate tank emissions are combined they can represent about 50% of a refinery's emissions. Other areas like delayed cokers can emit very high amounts of VOCs, including HRVOCs. In Spectrasyne's experience, flares have not been too much of a problem and generally contribute less than 5% of the total VOC emissions.

It is difficult to determine how well these results reflect what is happening at U.S. refineries, although there have been over 35 comprehensive DIAL and SOF studies performed at refineries in Europe. To date, no comprehensive studies have been performed in the U.S. The TCEQ performed the first U.S. regulatory agency sponsored DIAL study in 2007 and performed some tests inside a refinery and at other facilities. [13]

Another method used to evaluate VOC emissions is to compare measured hydrocarbon/ NO_x concentration ratios to the reported inventory VOC/NO_x ratios. One ratio commonly used is non-methane organic compounds or NMOC to NO_x (NMOC would include ethane, whereas reported VOCs in the U.S. do not). For example, suppose that for a given area the reported NMOCs is "X" tons/year and the reported NO_x is "Y" tons/year. First the mass numbers are converted into moles. For the NMOCs this requires knowing the mass amounts of the individual species, selecting the appropriate molecular weights. Then the total number of moles can be calculated. The same is true of NOx. For simplicity sake assume that X tons of NMOC is equal to Z moles of NMOC, and Y tons of NOx is equal to 10 Z. The reported NMOC/NOx molar ratio would be Z divided by 10 Z or 0.1. Now assume the measured concentration of NMOC downwind of the area is 100 ppb and the measured NO_x at the same location is also 100 ppb. The measured NMOC/ NO_x ratio would be 1.0. If the 1:1 concentration ratio is consistent over a long period of time and representative of the area in question, and if the molecular weight distribution of NMOC emissions is consistent, then if 100 moles of NO_x are reported, then one would expect 100 moles of NMOCs to be emitted as well. However if the reported NMOC/NOx molar ratio works out to be 0.1, then one would expect for every 100 moles of NOx emitted only 10 moles of NMOCs. The expected amount of moles based on measurements is a factor of 10 times higher than the expected amount of moles based on the inventory. Converting the numbers back from moles to mass may generate a different factor.

Care must be taken when calculating these ratios to use the same units throughout. When measurements are made the results are typically reported in percent volume or percent molar concentration. However the inventories are typically reported in units of mass such as 100 tons of NO_x. Converting the mass inventories of VOCs in to mole fractions requires knowledge of molecular weights of the specific species. This conversion is not a trivial task and should be reviewed carefully.

The reason that NO_x is chosen for ratio calculation is that point source NO_x emissions inventories are considered to be reasonably accurate. This is backed up by measurements in aircraft which suggest the emissions inventories are accurate to +/- 50%.[14] That may sound high, but having an error of 50% that could go either way is much better than having a low bias which says reported NMOCs are off by 1000 – 10,000% or a factor between 10 and 100.

Ground-level monitors have also been used to calculate NMOC/ NO_x ratios. These may be useful in a general way, for example, when comparing ground level monitor ratios near petrochemical areas to other monitors; however the ratio is prone to errors near petrochemical facilities, because it is only one point in space as opposed to a two dimensional reading. Ground monitors are valuable in that they identify the concentration of compounds where people are breathing, however it is not clear how well they can represent the air space that passes by them.

One improvement is to take several measurements along a line. This is the technique used when an aircraft takes samples at several points downwind of a petrochemical facility, as has been done in the TexAQS 2000 and TexAQS II field studies and in other areas. One assumption that is made in these studies is that the concentration along the vertical path of the planetary boundary layer (the layer of the earth's atmosphere where weather occurs) is relatively constant. The measurements taken horizontally help to confine the size of the pollution plume. The TexAQS studies in Houston have focused on ethene/NO_x ratios and to lesser extent propene/NO_x ratios. Some of the concerns about the assumptions and methods used in this analysis are discussed in the following section.

Criticisms of Comparison Methods between Estimations and Measurements

There are two basic criticisms of comparing measurements to inventory estimates:

1. Measurements are typically a snapshot in time and/or space and inventories are developed for annual reporting.

2. Emissions are variable in time and space.

Although measurements are typically a snapshot in time and space, the growing number of VOC measurements at petrochemical facilities and studies performed have garnered sufficient evidence for a number of different groups of scientists to conclude that the VOC emission inventories are underestimated by as much as an order of magnitude and maybe more.[1,2,3] It would be an error to take the highest reading and extrapolate it to the annual average, just as it would be an error to get the lowest reading and extrapolate it. The results of direct measurement techniques show that emission inventories for NO_x are generally reasonable, but the consistent low bias of the reported VOC emissions is indicative that there are discrepancies that need to be resolved. Specific case studies of DIAL and SOF measurements in Europe (and one in Canada) from 1988-2008, TexAQS 2000 and TexAQS II are provided in the sections that follow. When direct measurements are made with 2-dimensional techniques substantial variability is observed, but the observed variability occurs above a baseline of emissions that is several times higher than the reported emissions (see Figure 3).[15]

The validity of DIAL and SOF measurement techniques is widely accepted as indicated by a recent report by CONCAWE, the European equivalent to the American Petroleum Institute. However, there are a couple of notable exceptions. One concern is the error due to changes in wind speed and other parameters. The other concern is about extrapolating measurements from a relatively short measurement basis and extrapolating it to annual results.[11] However, one Swedish refinery that used DIAL during the 1990's and SOF since 2005 reports their annualized emissions on the basis of a SOF survey. [16]

Specific critiques that have been voiced about the aircraft data have been that the sampling has preferentially occurred during peak emission time periods, or during emission events, or that the flight path of the aircraft is more often at an altitude that crosses a relatively high concentration of VOCs in the plume that is extrapolated throughout the vertical dimension of the planetary boundary layer. Researchers respond that they take many samples and the average provides a good statistical representation of what the emissions are. They also note that when compared with the SOF measurements any individual measurement may be off by as much as a factor of 2, however, there is not a bias, either SOF or the aircraft measurements can be higher. In either case, both the SOF and the aircraft measurements are substantially higher than would be predicted by the inventory. [1, 15] The error bars in figure 3 indicate the range of measured values found by repeated sampling at the same location downwind of some facilities. The inventory numbers are substantially lower than even the low end of the range of measurements.

Figure 3. Comparison of ethene emissions from Mt Belvieu, Houston, measured during TexAQS II by SOF and airborne measurements by NOAA (de Gouw) [15]

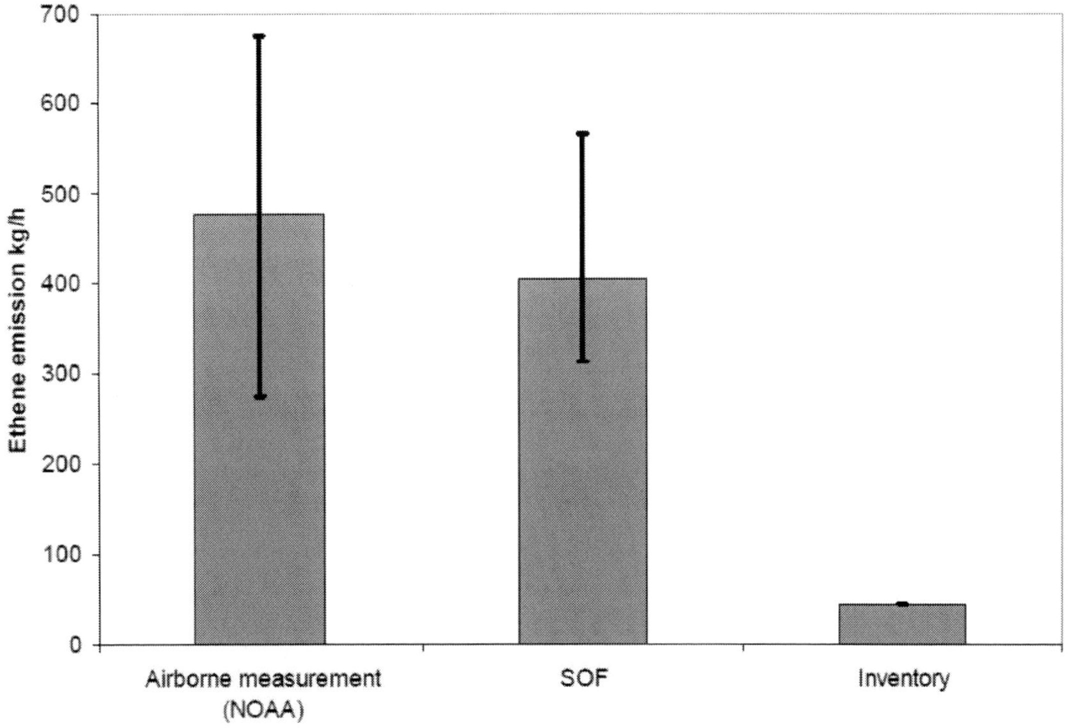

Case Study: Europe 1988 – 2008

DIAL is a technique that has been used to measure VOC mass flux at refineries and chemical plants since 1988. DIAL has been used as a technique since the 1960's, but in 1979 BP started a research team to develop an application specifically to locate and quantify VOC emissions. BP worked with National Physical Laboratories (NPL) in the U.K. and had an operational system by 1988. The system was operated at many BP refineries and chemical plants in Europe through 1992. BP then sold the technology to the employees who developed and operated it. They formed a new company called Spectrasyne that continues to operate today. NPL built their own DIAL system as did Shell Global Solutions in the U.K (with NPL's assistance). Shell shutdown their system after operating for about 6 years. NPL purchased and refurbished the Shell system and continues to operate it today.

In the late 1990's, a professor at Chalmers University in Sweden, Johan Mellqvist, developed SOF, a technique which uses a mobile FTIR and the sun as a light source. By driving with the FTIR detector across areas near VOC sources, SOF builds a 2-dimensional map of emissions. Corrections must be made due to the angle of the measurement path from the sun to the instrument, but when translated to a plane perpendicular to the wind, a mass flux of VOCs can be measured.

Numerous DIAL and SOF surveys have been made at refineries, chemical plants, oil production sites, landfills and other VOC and SO2 sources. Below is a graph that summarizes the results from 35 DIAL and SOF surveys at refineries over the past 20 years. The plot shows VOC emissions in barrels for every 100,000 barrels of throughput.

Figure 4. Emissions measured at 35 refineries. [17]

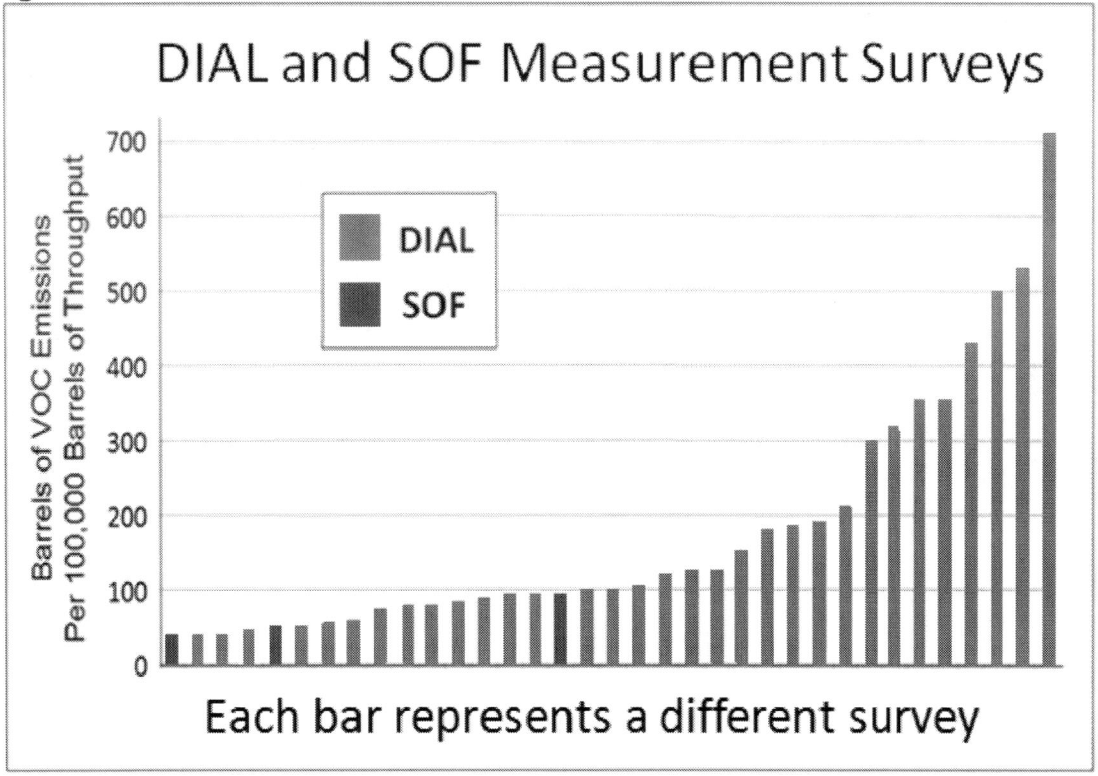

These DIAL and SOF surveys show a median emission rate for refineries of 120 barrels for every 100,000 barrels of throughput. Naturally refineries differ significantly depending on size and complexity, but this is meant to show a rough depiction of what the measured emissions are. These measured amounts are about 5-20 times higher than would be expected on the basis of the reported VOC emissions inventory. By way of comparison the nine refineries in the Houston area report about 12 barrels worth of emissions for every 100,000 barrels of throughput,[18] or about 1/10th of what has been measured in Europe. The lowest emitting refineries in Europe are typically the one that have been through several DIAL and/or SOF surveys and have taken steps to fix emission problems. Although criticized for not being representative statistically one would expect that at some point in time the measured emissions would be less than reported, but in 35 studies that has not ever happened.

Comprehensive refinery measurements like these have not yet been performed in the U.S. The reasons are in part practical and in part regulatory. No one in the U.S. owns these technologies, at least not in a way that can be applied to measuring emissions at refineries and chemical plants. As a result, the equipment must be shipped from Europe to the U.S. in order to perform the survey increasing the costs. The other consideration is that in the U.S. regulations require companies to fix any item they find is out of compliance. If they bring in a new technology that identifies emissions are a factor of 10 or more higher than they have been reporting – or are permitted for – they may be obligated to spend many millions of dollars to fix the problems that are found.

Case Study: TexAQS 2000

During the summer of 2000, over $20 million dollars was spent by more 250 scientists and engineers, using five airplanes and numerous ground instruments to perform an extensive field campaign to characterize the pollution in Houston area. The results of the "TexAQS 2000" field campaign were summarized in Accelerated Science Evaluation of Ozone Formation in the Houston-Galveston Area: Emissions Inventories[19] (referred to here as the TexAQS 2000 report). This report provides the following examples to argue that emissions are substantially higher than reported:

Excerpt 1: Prior to 1992 NMOC emissions were 5.9 times higher than reported.

> *"On average predicted NMOC concentrations in Houston are 5.9 times lower than observed values."*[20]

This was a statement made by the EPA in a review of data from the mid 1980's but prior to 1992 looking at ratios of NMOCs to NO_x in the Houston area.

Excerpt 2: A 1995/6 study also says NMOC emissions were substantially higher than reported.

> *"The measured ratios of NMOC to NO_x ranged from 4-12, and the measured ratios were consistently larger than those from the emissions inventory."*[21,22]

Although not specifically stated in the TexAQS 2000 report, the plots which follow the above quote imply that the observed NMOC values are several times higher than what would be expected from the emissions inventory (See Figures 6 and 7 in the TexAQS 2000 report[19]).

Similar comparisons have been made to compare estimated and measured ratios of ethene and propene to NO_x. However there is a discrepancy among researchers in the TexAQS 2000 report.

Excerpt 3: Aircraft measurements indicate alkene emissions were higher than reported.

> *"Airborne measurements of the concentrations of hydrocarbons and NO_x have been compared to inventory data. Ratios of measured concentrations of propene to measurements of NO_x and measured concentrations of*

ethene to measurements of NO_x (all measured by aircraft in apparently well mixed plumes) have values of about 1. For sources at Chocolate Bayou, Freeport and Sweeny ratios of alkenes to NO_x using data from the TNRCC point source inventory are 0.3, 0.1 and 0.2 respectively (T. Ryerson et al, 2001)."[19]

The data are attributed to a presentation by T. Ryerson in August, 2001[23], however the presentation which lists the "0.3, 0.1 and 0.2" numbers are for total NMOC/NO_x ratios, not alkene/NO_x ratios. Ryerson later published the following results in the Journal of Geophysical Research in 2003[6].

Table 1 (adapted from Ryerson, et. al.) [6]

Complex	Tabulated Ethene/NO_x	Measured Ethene/NO_x	Tabulated Propene/NO_x	Measured Propene/NO_x
Sweeny	0.05	3.6	0.03	2.0
Freeport	0.05	1.5	0.01	0.5
Chocolate Bayou	0.08	2.0	0.10	4.0

The measured ratios of ethene to NO_x and propene to NO_x range from 0.5 – 4.0, with a median value of 2.0. The tabulated or inventory numbers reflect an ethene to NO_x and propene to NO_x ratios in the range of 0.01 to 0.10. These are the values used by Ryerson, et. al.[24] to estimate that ethene and propene emissions are off by a factor between 10 and 100. Wert, et. al. [25] found similar results. The order of magnitude difference between the alkene ratios reported in the TexAQS 2000 report and the Ryerson and Wert reports may explain the differences expressed in the reporting of the deviations in measured and reported amounts of either a factor between 3 and 10 (according to the TexAQS 2000 report) and a factor between 10 and 100 (according to the Ryerson and Wert reports).

Excerpt 4: Does photochemical modeling agree with measurements?

This discrepancy brings up another issue identified in the TexAQS 2000 report based on this statement:

The two approaches commonly used to evaluate the accuracy of emission estimates are to (1) employ photochemical models to relate emission estimates to observed concentrations and (2) to examine ratios of NMOC concentrations to NO_x or NO_y concentrations. Both of these approaches have been employed and yield similar results. [19]

If the focus is kept not on propene/ NO_x and ethene/NO_x ratios, the measured data versus the reported data indicate errors of 10-100. However according to the TexAQS 2000 report, regional photochemical modeling indicated that imputing the ethene and propene inventories by a factor of approximately 6 would bring the ozone levels predicted by the model in agreement with measured ozone values. The apparent discrepancy that the measured values imply reported emissions are off by a factor between 10-100 but the model can function by making an adjustment of only 6 has not been resolved.

Excerpt 5: Variability can be related to either plant conditions or meteorology.

> *Ground observations, taken over a period of several years, and other data, suggest that emissions exhibit significant temporal variability. The implication of this finding is that NMOC/NO_x ratios will be variable, and individual plumes may have very different ozone formation potentials at different times. The extent the observed variability is caused changes in emissions or changes in meteorological conditions such as wind direction, vertical mixing and atmospheric stability, has not been determined.* [19]

Variability is a very important issue to resolve. If the errors in the inventory can all (or mostly) be ascribed to emission events, it effectively says that the errors occur because "It was a bad day at the refinery (or chemical plant)." If, however, there are continuous emissions causing substantial discrepancies in the inventory then "Every day is a bad day at the refinery (or chemical plant)."

Prior to TexAQS 2000 most people believed that the emission inventories were reasonably accurate. After TexAQS 2000 a much higher awareness was placed on VOC emissions, especially emission events. Newly proposed rules that focused on highly reactive VOCs (HRVOCs) from flares and cooling water towers encouraged more activity to reduce emissions prior to the implementation of the required monitors. Another technology development, the IR Camera, began to be used in the industry and helped to identify and reduce emissions as well. By the time TexAQS II began in 2005 and 2006 substantial reductions had been achieved as discussed in the next section.

Case Study: TexAQS II

A follow-up study to the TexAQS 2000 campaign occurred during 2005 and 2006 called TexAQS II. The intensive period of the campaign occurred during the months of August and September 2006. The initial findings of TexAQS II were published by TCEQ on their website in a document called "The Final Rapid Science Synthesis Report: Findings from the Second Texas Air Quality Study (TexAQS II)". [1] Scientists who reviewed the data indicated that emissions of ethene had dropped by about 40%, but were still higher than expected from the inventory by a factor of 10 or more. Some of the key excerpts from the report related to VOC emissions inventories are provided below.

Excerpt 1: Alkene/NO_x Emission Ratios

Table 2 Alkene/NO_x Emission Ratios[1]

Table C1. Ethene/NO_x emission ratios for petrochemical complexes determined from measurements in 2000, 2002, and 2006, compared with emission inventories.

	Ethene/NO_x Emission Ratios				
	Inventories		Measured		
	1999 [a]	2004 [b]	2000 [c]	2002 [c]	2006
Sweeny	0.05	0.019	3.6	1.7	0.5
Freeport	0.05	0.030	1.5	0.62	0.32
Choc. Bayou	0.08	0.048	2.0	1.2	0.62

[a] TNRCC emission inventory.
[b] TCEQ point source emission inventory with 1999 VOC speciation.
[c] Ryerson et al. (2003.)

The 2006 inventory was not available at the time of this report, so comparisons were made to the most recently available inventory in 2004. Concentrations of both ethene and NO_x decreased substantially. The ratios of ethene to NO_x dropped significantly during this time period as well. NO_x concentrations dropped, but not as much as ethene concentrations.

Excerpt 2: Four different analyses indicate ethene emissions dropped by about 40%.

Thus, analyses based on four different measured parameters (ethene/NO_x emission ratios in plumes of petrochemical facilities, the ambient distribution of ethene concentrations, the ambient distribution of formaldehyde, and long-term auto-GC ethene measurements) have all found evidence for a significant decrease in ethene emissions. The unanimity of these four analyses increases our confidence that a significant decrease in HRVOC emissions from Houston area increases our confidence that a significant decrease in HRVOC emissions from Houston area petrochemical facilities has actually occurred in the period between 2000 and 2006. Considering all of these analyses, the best estimate is that ethene emissions have decreased by about 40% (i.e., a factor of 1.7) between 2000 and 2006. The much more limited evidence available for propene suggests that emissions of this HRVOC may have decreased similarly. [1]

Weather conditions were significantly different in 2000 than in 2006, but even when those considerations are taken in to account substantial reductions were observed indicating that the plans for reducing emissions were successful despite problems with the emissions inventories. Some of the reasons for the declines include the increased attention to and reduction emission events at facilities and the use of the IR Camera to identify and reduce emissions. Also, HRVOC rules, which were not yet in effect by 2006, but were in the process of being implemented, likely encouraged facilities to make in preparation for the new rules that would impact the management of emissions from flares and cooling water towers.

<u>Excerpt 3:</u> Emissions do not drop to zero when the planes fly away.

> *Measurements of ethene emission fluxes from petrochemical facilities during TexAQS 2006 indicate that the 2004 TCEQ point source database underestimates these 2006 emissions by one to two orders of magnitude. Repeated sampling of the same petrochemical facility showed that the ethene emission flux remained constant to within a factor of two.* [1]

One concern raised about the aircraft data in 2000 was that the aircraft took too few measurements of the same location and essentially used the highest values to generate the emissions numbers. For this reason an aircraft was instructed to go back to the same location several times when the wind was blowing in a specified direction. They did find variation (see Figure 3), but even at the lowest level of the measurements the emissions were many times higher than would be anticipated from the inventory.

<u>Excerpt 4:</u> Solar Occultation Flux (SOF) and Aircraft (WP-3D) got similar results.

> *Flux estimates from the SOF and WP-3D measurements were compared for different sources (Sweeny, Freeport, Texas City) and generally agreed within a factor of 2, with no systematic difference as to which technique was higher or lower.* [1]

The aircraft measurements take samples every 5 seconds across a plume and assume a constant vertical concentration. In some cases, the aircraft transect will catch samples that are higher than the average vertical concentration and at other times lower than the average. SOF measurements count all of the ethene molecules between the sun and the SOF solar collector located on top of the van and integrate along the path that the van drives, making adjustments to correct for the angle of measurement. Some researchers speculate that the SOF numbers are more accurate than those determined from the NOAA aircraft samples, since SOF does an integrated analysis between the ground level and the boundary layer. The measurements agreed within a factor of 2 and in all cases with no bias of either the aircraft or the SOF measurements being high or low. All of the measurements, including the lowest measurements, were close to an order of magnitude higher than the reported emissions.

Excerpt 5: Continuous hourly emissions of ethene are greater than most emission events

Murphy and Allen (2005) investigated the role of large, accidental releases of HRVOC in ozone formation in the HGB area. They identified 763 HRVOC emission events in a one-year period (31 January 2003 to 30 January 2004). More than half of these events released less than 1000 lbs (454 kg) total HRVOC. Thus, the Mont Belvieu complex, as an example of petrochemical facilities in the HGB area, routinely emits more ethene each hour than the total released in most of the individual accidental release events considered by Murphy and Allen. These facilities represent a significantly larger source of HRVOC and more substantial contribution to ozone formation than indicated by current emission inventories.[1]

This addresses the focus on emission events and suggests that the continuous emissions from facilities are a significant part of the problem in forming ozone.

Excerpt 6: Despite reductions ethene emissions are still underestimated by an order of magnitude.

In summary, the discrepancy between the HRVOC emission measured in the field and those included in inventories has improved to some degree between the 2000 and 2006 TexAQS studies. The measured emissions have decreased by about 40%, while the inventoried emissions (at least as indicated by the limited Special 2006 Hourly Emission Inventory) have increased by about a factor of 2. However, in spite of this progress, the latest available emission inventories still underestimate ethene emissions by approximately an order of magnitude.[1]

There has been success with reductions of emissions, but emission inventories have been around for over 20 years and it still appears that as of 2006 the measured ethene emissions are about 10 times higher than expected from the inventories. A follow-up analysis performed after the special inventory for 2006 was complete confirmed that measured levels ethene and propene were approximately 10 times higher than expected from the inventory.

Impact of the IR Camera

The IR Camera, developed in the early 2000's, has been an effective way for facilities to identify VOC emissions that were otherwise undetected. The camera is relatively inexpensive (about $100,000) and can be used by plant personnel. It can identify leaks of many different kinds of VOCs very quickly simply by looking through the camera and viewing the shady clouds of emissions that emanate from various leak sources yet cannot be seen with the naked eye. The drawbacks are that the camera cannot see all peaks and its response levels to different compounds vary by two orders of magnitude depending on the compound. The detection limits are around 250 ppm-meters for the compounds that are easiest to identify. In contrast SOF and DIAL have detection limits of about 1 ppm-meter, suggesting that these techniques are 2-5 orders of magnitude more sensitive than the IR Camera. The IR Camera cannot quantify emissions.

One of the outcomes of the camera use has been the identification of large, continuous emission releases. Although the camera does not quantify emissions, several facility owners who have fixed leaks have calculated that the savings from recovered product would be valued at over $1,000,000 per year from a single leak. That equates (assuming 40 cents/pound of VOCs) a release of over 1200 tons of VOCs, more than is reported by most U.S. refineries during any given year. This and other stories from those familiar with the findings of the IR Camera, as well as the measurement taken using DIAL and SOF indicate that many of the inventory problems are related to continuous emissions and not just variable emission events.

Resolving these problems are not as easy as one might expect. In one example a "million dollar leaker" (a leak from a single source that is costing the facility $1 million per year in product losses to the atmosphere) may be as simple as closing a valve that an operator left open. However that million dollar leaker may be the result of a tank that is more than 50 years old, has a foundation that is not cracked or leaking but damaged so badly that the tank is no longer round and the floating roof cannot seal against the wall. The only solution may be to tear down and rebuild the tank at a cost of $30 million. It is not difficult to see why some facilities may not want to purchase a $100,000 IR Camera that can "save" them $1,000,000 from a single leak.

Impacts on Developing Strategies for Reducing VOC Emissions

To develop effective control strategies there must be substantial evidence about what are the Best Available Control Strategies (BACT) or the Maximum Achievable Control Strategies (MACT), however identifying those strategies is difficult if emissions estimates are too low by an order of magnitude. Improving emissions inventories and locating specific sources of VOC emissions by encouraging the use of more direct measurement techniques including DIAL, SOF and the IR Camera and other available techniques such as FTIR and GCs essential. Continuous Emissions Monitors should be considered where the technology can be applied.

Conclusions and Recommendations

A greater emphasis must be placed on counting the molecules in the air. Measurements provide an audit of performance and can identify previously unknown emission sources as well as operator and maintenance errors. Addressing the issues associated with variable emissions from things such as blowdowns, flares, relief valves and delayed cokers is important, but addressing the continuous emissions from leaking sources in the process or storage tank areas is important too. The errors in the emission inventory cannot be dismissed as merely a function of variable emissions, as if it were a matter of having good and bad days at the refinery (or chemical plant) versus every day being a bad day – at least as far as routine measurements go.

If a new emissions accounting system provides more accurate, *but much higher emissions,* then consideration must be given to the impact to implementation of the new accounting system and its impact on existing and new permits, any existing VOC taxes or penalties and the competitive edge to facilities that are not placed under similar constraints. Care must be taken to identify the appropriate ways to implement policies that improve the accuracy of VOC emissions inventories without causing harm to those who take extra measures to reduce emissions. For example, if facilities volunteer to perform more direct measurements and find that emissions are costly to fix, then the greenest facilities may incur the greatest cost, and if the greenest facilities go bankrupt, then everyone loses.

The recommendation of this paper is for chemical engineers to form a workgroup that identifies all of the incentives and disincentives for using 2-dimensional direct detection and measurement techniques, such as the IR Camera, DIAL and SOF to address VOC, air toxics and greenhouse gas issues. This workgroup should also identify how to address all of the resulting regulatory issues and the impacts working in an international market. The workgroup should produce recommendations for action on this problem and submit to the Administrator of the EPA, the EPA Clean Air Scientific Advisory Committee, the Environment Czar, the White House Council on Environmental Quality and the appropriate personnel at the state agencies.

References

1. Cowling, Ellis B., Furiness, Cari, Dimitriades, Basil, Parrish, David, and Estes, Mark. "Final Rapid Science Synthesis Report: Findings from the Second Texas Air Quality Study (TexAQS II)", prepared for the Texas Commission on Environmental Quality, August 31, 2007. Accessed on October 16, 2008 from http://www.tceq.state.tx.us/assets/public/implementation/air/texaqs/doc/rsst_final_report.pdf.

2. Beatty, George, Olaguer, Eduardo, Sattler, Melanie, Klaus, Chris, Alvarez, Ramon, Carmichael, Greg, Carter, William, Dickerson, Russ, Hildebrand, Susana, Merrill, John, Mobley, David, Parry, Rice, Pinto, Joseph and Scire, Joseph. "Texas Environmental Research Consortium Strategic Research Plan 2007-2009," November 2007. Accessed on February 10, 2009 at http://www.tercairquality.org/AQR/About/SRP

3. De Gouw, Joost, Hekkert, S. te. Lintel, Mellqvist, J., Warneke, C., Atlas, E.L., Fehsenfeld, F.C., Fried, A., Harren, F.J.M., Holloway, J.S., Lefer, B., Lueb, R., Meagher, J.F., Parrish, D.D., Patel, M., Pope, L., Richter, D., Rivera, C., Ryerson, T.B., Samuellson, J., Walega, J., Washenfelder, R.A., Weibring, P., and Zhu, X. "Airborne Measurements of Ethene from Industrial Sources Using Laser Photo-Acoustic Spectroscopy," to be published by Environment, Science and Technology in March 2009.

4. Chambers, Allen and Strosher, Mel. "Refinery Demonstration of Optical Technologies for Fugitive Emissions and for Leak Detection," Prepared for Environment Canada, Ontario Ministry of the Environment and Alberta Environment, November 1, 2006.

5. Mellqvist, Johan, Johansson, John, Samuelson, Jerker, Rivera, Claudia, Lefer, Barry and Alvarez, Sergio. "Comparison of Solar Occultation Flux Measurements to the 2006 TCEQ Emission Inventory and Airborne Measurements for the TexAQS II," November 7, 2008, accessed from www.tceq.state.tx.us/assets/public/implementation/air/am/contracts/reports/da/20081108-comparison_solar_occultation_flux_measurements.pdf.

6. Shires, Theresa M. and Loughram, Christopher J., "Compendium of Greenhouse Gas Emissions Inventories for the Oil and Gas Industry," prepared for the American Petroleum Institute, February 2004. Accessed on February 10, 2009 from www.**api**.org/ehs/climate/new/upload/2004_**COMPENDIUM**.pdf.

7. EPA website downloaded from http://www.epa.gov/ttn/chief/ap42/index.html on February 4, 2009.

8. L. Chinkin, D. Coe, "Ground Truth Verification of Emissions in the Houston Ship
Channel Area: Revised Final Report", Report prepared for TNRCC, Aug 2002.

9. Milligan, Patrick, Martinsky, Frank, Good, Kevin and Nelson, Bill. "EPA Can Improve Emissions Factor Development and Management," Report No. 2006-P-00117, prepared for the EPA Office of Inspector General, March 22, 2006.

10. Mellqvist, Johan, Samuelsson, Jerker, Rivera, Claudia, Lefer, Barry and Patel, Monica. "Measurements of Industrial Emissions of VOCs, NH3, NO2 and SO2 in Texas using the Solar Occultation Flux Method and Mobile DOAS," prepared for the Texas Environmental Research Consortium, August 2007. Accessed on February 10, 2009 from http://files.harc.edu/Projects/AirQuality/Projects/H053.2005/H053FinalReport.pdf.

11. CONCAWE. "Optical Methods for Remote Measurement of Diffuse VOCs: Their Role in the Quantification of Annual Refinery Emissions," Report no. 6/08, June 2008.

12. Moncrieff, Jan. "Reflections on 20 Years of DIAL Measurements in the Oil and Gas Industries, presentation at the Houston Area Monitoring Group meeting, 3055 Timmons, Houston, Texas on September 26, 2007.

13. Robinson, Rod and Gardner, Tom. "Measurements of VOC emissions from petrochemical industry sites in the Houston area using Differential Absorption Lidar (DIAL) during Summer 2007," draft for comment submitted to the Texas Commission on Environmental Quality by National Physical Laboratories, February 8, 2008. Accessed on February 25, 2009 from www.tceq.state.tx.us/assets/public/implementation/air/am/contracts/reports/oth/20080206-dial_test_report.pdf

14. Ryerson, T.B., Trainer, M., Angevine, W.M., Brock, C.A., Dissly, R.W., Fehsenfeld, F.C., Frost, G.J., Goldan, P.D., Holloway, J.S., Hübler, G., Jakoubek, R.O., Kuster, W.C., Neuman, J.A., Nicks Jr., D.K., Parrish, D.D., Roberts, J.M., Sueper, E.T., Atlas, E.L., Donnelly, S.G., Flocke, F., Fried, A., Potter, W.T., Schauffler, S., Stroud, A.J. Weinheimer, B.P. Wert, C. Wiedinmyer, R.J. Alvarez, R.M. Banta, L.S. Darby, V. and Senff, C.J. "Effect of petrochemical industrial emissions of reactive alkenes and NO_x on tropospheric ozone formation in Houston, Texas," 2003: *J. Geophys. Res.*, **108**(d8), 4249, doi:10.1029/2002JD003070.

15. Mellqvist, Johan. "Surveillance of Industrial VOC Leaks by SOF-TCT and FLIR," presented to the Texas Environmental Research Consortium's Science Advisory Committee on November 21, 2007.

16. Shell Raffinaderi AB, "MILJÖREDOVISNING enligt EMAS 2007" http://www.shell.com/static/se-sv/downloads/about_shell/miljoredovisning_raff_2007.pdf accessed on October 17, 2008.

17. Cuclis, Alex. "Policy Implications of Measuring Point Source VOC Emissions," submitted to Environment, Science and Technology, December 2008.

18. Texas Commission on Environmental Quality (TCEQ). 2004 State Summary of Criteria Pollutants Sorted by Company Name, September 20, 2006. http://www.tceq.state.tx.us/assets/public/implementation/air/ie/pseisums/03ss_co.pdf accessed on November 10, 2008.

19. Allen, David, Durrenberger, Cyril and Texas Natural Resource Conservation Commission, Technical Analysis Division, "Accelerated Science Evaluation of Ozone Formation in the Houston-Galveston Area: Emissions Inventories," Version 3.0, February 5, 2003.

20. Baugues, K. "Further Comparisons of Ambient and Emission Inventory HMOC/NO_x Ratios". Presentation at 85th Annual Meeting & Exhibition, Air & Waste Management Association, June 1992.

21. Korc, M., Jones, C., Chinkin, L., Main, H. and Roberts, P. "Use of PAMS data to evaluate the Texas COAST Emission Inventory", Final report prepared for U. S. EPA, December 1995.

22. Main, H., Roberts, P., and Korc, M. PAMS Data Analysis Workshop; Illustrating the Use of PAMS Data to Support Ozone Control Programs. Dec. 1996.

23. Ryerson, T. et al, 2001. "Contrasting effects of different anthropogenic source types of ozone formation in Houston". Presentation to the Texas Natural Resource Conservation Commission at the TexAQS Workshop, in Austin, Texas, August, 2001.

24. Ryerson, T.B., M. Trainer, W.M. Angevine, C.A. Brock, R.W. Dissly, F.C. Fehsenfeld, G.J. Frost, P.D. Goldan, J.S. Holloway, G. Hübler, R.O. Jakoubek, W.C. Kuster, J.A. Neuman, D.K. Nicks Jr., D.D. Parrish, J.M. Roberts, and D.T. Sueper, E.L. Atlas, S.G. Donnelly, F. Flocke, A. Fried, W.T. Potter, S. Schauffler, V. Stroud, A.J. Weinheimer, B.P. Wert, C. Wiedinmyer, R.J. Alvarez, R.M. Banta, L.S. Darby, and C.J. Senff, 2003: Effect of petrochemical

industrial emissions of reactive alkenes and NO_x on tropospheric ozone formation in Houston, Texas, 2003: *J. Geophys. Res.*, **108**(d8), 4249, doi:10.1029/2002JD003070.

25. Wert, B.P., M. Trainer, A. Fried, T.B. Ryerson, B. Henry, W. Potter, W.M. Angevine, E. Atlas, S.G. Donnelly, F.C. Fehsenfeld, G.J. Frost, P.D. Goldan, A. Hansel, J.S. Holloway, G. Hübler, W.C. Kuster, D.K. Nicks Jr., J.A. Newman, D.D. Parrish, S. Schauffler, J. Stutz, D.T. Sueper, C. Wiedinmyer, and A. Wisthaler, 2003: Signature of terminal alkene oxidation in airborne formaldehyde measurements during TexAQS 2000. *J. Geophys. Res.*, **108**(d3), 4104, doi:10.1029/2002JD002502.

74e Carbon Legislation Update

Chris Berendt
Pace Global Energy Services
Fairfax, VA 22033
Chris.Berendt@Paceglobal.com

Abstract:

Mr. Berendt will present to the conference an update on the current status of the U.S. Carbon Markets. Examining the development and interplay of state, regional, and federal actions (Legislative and Administrative), he will focus on the timing and design of the pending carbon regulations. From carbon compliance requirements impacting operating margins to mandatory GHG inventory reporting and opportunities to influence carbon market design, Mr. Berendt will detail the shifting carbon regulatory landscape through the lens of operational impact. To conclude, he will give an overview of emerging carbon management best practices in the industrial sector.

Additives to Prevent Coking in Ethylene Furnaces

Girish Srinivas
Program Manager
TDA Research Inc.
Jeffrey R. Engel
Engineer
TDA Research Inc.
David T. Wickham
Vice President
Reaction Systems LLC
Michael E. Karpuk
President
TDA Research Inc.
James Nabity
Principal Engineer
TDA Research Inc.
Steven C. Gebhard
Senior Scientist
TDA Research Inc.

Abstract: TDA Research Inc. has identified several additives that block the formation of the metastable carbide intermediates on high alloy metal surfaces during the thermal cracking of n-heptane. The additives work by forming strong, thermally stable surface compounds upon which filamentous coke cannot form. These compounds are stable at temperatures in excess of 1000°C (1832°F), and in the case of one of our more recent formulations, also resist attack by sulfur. We have conducted laboratory experiments using hot metal tubes to determine how well our additives suppressed coke formation and found that under conditions that would otherwise result in rapid coking, our additives completely blocked filamentous coke formation when used at concentrations as low as 10 ppm. In tests done using stainless steel tube sections, 761 mg of coke was formed in 8 hours when our additive was not used. In contrast, only 58 mg coke was formed in 16 hours when 10 ppm of our additive was added to the n-heptane feedstock. In addition, microscopic examination showed that the coke that did form was not filamentous. We also found that pretreating the stainless steel tube sections with 10 ppm of our additive for four hours prior to use, filamentous coke formation was suppressed for up to 40 hours after the flow of additive was stopped. We also tested Inconel 625 and Incoloy 800HT tube sections at 900°C, with and without one of our coke suppression additives and for both alloys, the presence of our additive effectively eliminated the formation of filamentous coke. Finally, we tested several alloys and additives using JP-7 jet fuel as the feedstock.

Introduction

Coke Formation in Ethylene Production

Ethylene is the largest volume building block chemical used in the petrochemical industry. It is also the largest volume organic commodity chemical, with a total annual U.S. production capacity of about 40 billion pounds (2, 3). It is commonly produced by thermal cracking of ethane, LPG, naphtha, and in some cases heavy gas oils at temperatures near 900°C (1652°F) at pressures ranging from 1.5 to 3 atm. The reaction occurs in the gas phase through a free radical mechanism (1-3).

Unfortunately, the conditions used in ethylene production also promote coking and the accumulation of carbonaceous deposits on the walls of the pyrolysis tubes. Coke buildup creates an insulating layer that decreases heat transfer across the reactor walls and also increases pressure drop through the reactor. Eventually, the furnace must be taken off line to be decoked using steam and air, and these interruptions in production are costly.

Coke can form by several different mechanisms. Filamentous coke forms through a catalytic mechanism in which the iron and nickel in the metal reactor walls react with the hot hydrocarbon feed and catalyze the formation of carbon filaments (4, 5). Coke can also form through gas phase reactions, involving free radical intermediates. Understanding the mechanism responsible for coke deposition is critical to identifying approaches to mitigate the problem.

Fortunately, it is easy to determine what mechanism is responsible for the coke because each type of deposit looks quite different when viewed under a scanning electron microscope (SEM). Catalytic or filamentous coke, as its name implies, forms in the shape of long filaments approximately 0.25 to 1 μm in diameter (left side of Figure 1). On the other hand, condensation coke deposits that form through gas phase reactions appear more often as small spherically shaped particles (right side of Figure 1).

Under conditions encountered in ethylene production furnaces, the coke formed is almost entirely filamentous. Accordingly, coke mitigation technologies currently employed involve either the use of tubes that have been pre-coated with an inert material or the addition of compounds to the feed that reduce the catalytic activity of the iron and nickel metals. Unfortunately, neither approach is 100% satisfactory. The inert layers in pre-coated tubes eventually become depleted and new expensive tubes must be installed. In addition, welds formed during the installation are uncoated and can cause coke formation. Current additives, which consist primarily of sulfur or phosphorus compounds, only reduce the rate of coke formation by 10–20% and are expensive to use (6, 7). Even with sulfur added to the feed, ethylene furnaces must shut down every 40 – 100 days or so to remove coke (2).

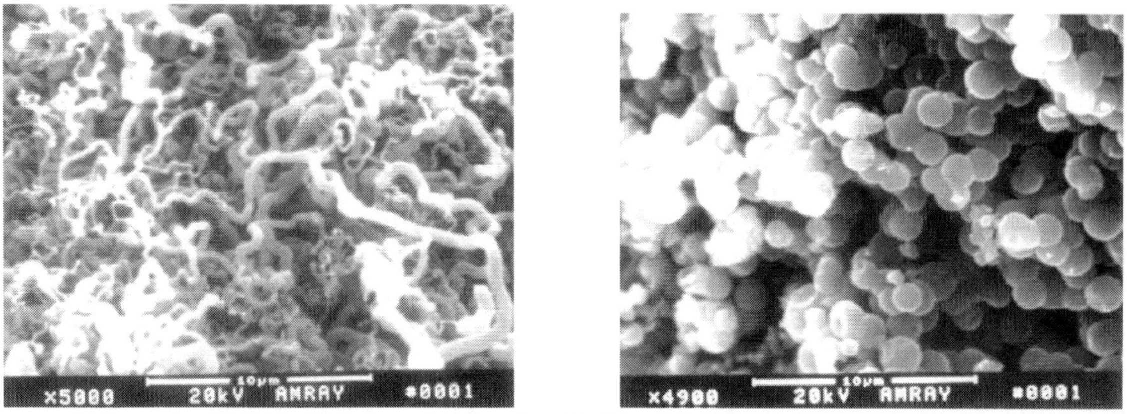

Figure 1.. SEM of filamentous coke (left) and condensation coke (right).

The Mechanism of Filamentous Coke Formation

The mechanism for filamentous coke formation has been studied previously (4, 8-13). A simplified schematic of the process is shown in Figure 2. The initial step in both processes is the catalytic decomposition of the hydrocarbon feedstock on the hot metal surfaces, leaving a layer of reactive carbon on the reactor walls or heating coils (Step 1). After the reactive carbon has been formed, it can migrate into the metal lattice and form iron carbide (Fe_3C) and possibly nickel carbide (Ni_3C) (4, 12-13) (Step 2). Since both of these carbides are metastable species, they tend to decompose under these reaction conditions into graphitic carbon and the respective metal (Step 3). The formation of graphitic carbon represents a very strong thermodynamic driving force for the reaction (12, 13). Once graphitic carbon has been deposited within the bulk metal, it creates pressure within crystal structure, causing some small metal particles to become dislodged. In a pyrolytic environment, the dislodged crystallites continue to catalyze carbon formation, sometimes forming a carbon filament on the back side of the metal particle (13).

Step 1 Step 2 Step 3

$$C_nH_{2n+2} \rightarrow C + Fe + Ni \rightarrow Fe_3C + Ni_3C \rightarrow Fe + Ni + C_{(graphite)}$$

Figure 2. Steps involved for filamentous coke formation.

The mechanism illustrated in Figure 2 shows that iron and nickel carbides are intermediates in the coke formation process. Thus, our approach to blocking filamentous coke formation was to identify additives that will preferentially react with the iron and nickel, preventing the metals from forming iron and nickel carbides.

Formation of Iron and Nickel Carbide Under Pyrolytic Conditions

Iron carbide and nickel carbides are thought to be important intermediates in the coke formation process. This suggestion is supported by thermodynamic calculations, which are illustrated in Figure 3. The results show that under pyrolytic conditions almost all of the iron and about half of the nickel have been converted to the carbide form. This indicates that the initial formation of both iron and nickel carbide is strongly favored under these reaction conditions and supports the conclusion that these carbides likely are important reaction intermediates in the coke formation process. This result is consistent with previous work in the area and also supports the mechanism proposed in Figure 2.

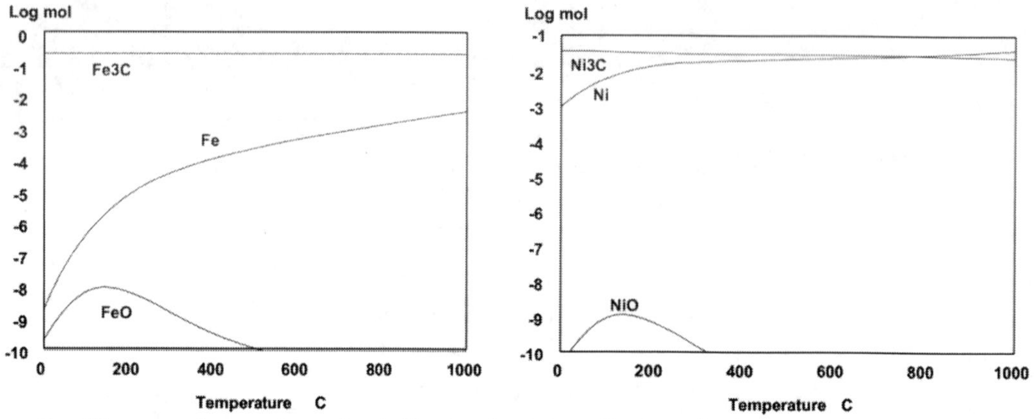

Figure 3. Thermodynamic calculations showing favored phases of iron and nickel in a pyrolysis environment.

Equilibrium Calculation Method

Because hydrocarbon pyrolysis is done at high temperatures (ca. 800°C), thermodynamic equilibrium calculations are useful in predicting what phases will be stable under specific conditions. For example, when Fe, Ni and n-heptane are used as the starting materials, and the composition of the final equilibrium product mixture is calculated, substantial amounts of iron and nickel carbides are predicted to form (Figure 3). The *surface* chemistry of the metal tubes can be simulated by starting with a large excess of hydrocarbon compared to the metal; for example, 100 moles of n-heptane, one mole of Fe or one mole of Ni. This simulates the surface conditions because relative to the surface of the tubes, there is a very large excess of hydrocarbon available with which to react. While thermodynamically, most of the carbon in the heptane ends up as elemental carbon, the Fe and Ni react to form Fe_3C and Ni_3C. Tiny amounts of FeO and NiO are predicted to be stable at low temperatures and these compounds arise because we included 70 ppm of oxygen that naturally dissolves in the hydrocarbon feed when it is exposed to ambient air (at high temperatures the dissolved oxygen reacts to form CO_2). Also, note that the y-axes in the graphs in Figure 3 are logarithmic so that the concentrations of unreacted metals is extremely low, indicating that essentially all of the 'surface" Fe and Ni react to form the carbides.

Figure 3 also "predicts" that the carbides should form at low temperatures. While the carbides would form if possible, the kinetic rates of the carbide forming reactions are, for all practical purposes, zero at lower temperatures and the carbides don't form. At the high temperatures used in hydrocarbon pyrolysis however, the formation of metal carbides is kinetically possible, so the thermodynamic predictions become more useful.

The equilibrium calculations were done using the computer program HSC Chemistry for Windows (Outotec). HSC is a reasonably priced program that runs very quickly on an ordinary PC. Because the calculations are thermodynamic, the results are independent of any chemical reactions, and therefore, they are good for predicting the most stable compounds that are energetically allowed for a particular starting composition, temperature and pressure.

The input to the HSC program includes the species of interest (the program has a very large database or organic and inorganic compounds, radicals, ions, etc.), their starting concentrations, the pressure and the temperature. The program then iteratively calculates the concentrations of each species that minimizes the total free energy of the system. In the mode used here, the operator chooses the temperature range of interest (at constant pressure) and how many data points to calculate within that range (the program can do up to 200). The program then calculates the minimum the Gibbs free energy composition at each temperature. The output is given in both graphical and tabular form.

HSC Chemistry derives its name from H (enthalpy), S (entropy) and C (specific heat). HSC uses the following relationships to calculate the heat (ΔH), entropy (ΔS) and the change in free energy (ΔG) of the reaction, from which the equilibrium constant is calculated. These types of calculations would be extremely tedious if done by hand.

The enthalpy is calculated from: $H_T = H^0_{298K} + \int_{298}^{T} Cp\,dT + \Sigma H_{tr}$ where H_T is the enthalpy at the temperature of interest, H_{298} is the standard enthalpy of formation at 298.15K, C_p is the heat capacity and H_{tr} is the heat of transformation which is added when there is a phase change such as melting or boiling. The entropy is calculated from: $S_T = S^0_{298K} + \int_{298}^{T} \frac{Cp}{T} dT + \frac{\Sigma H_{tr}}{T}$ where S_T is the entropy at temperature T and the other variables have meanings similar to the ones used in the enthalpy calculation. The Gibbs free energy at constant temperature and pressure is: $G_T = H_T - TS_T$. The heat, entropy and free energy of a reaction are then calculated from the following relationships: $\Delta H_{reaction} = \Sigma \Delta H_{reactants} - \Sigma \Delta H_{products}$, $\Delta S_{reaction} = \Sigma \Delta S_{reactants} - \Sigma \Delta S_{products}$ and $\Delta G_{reaction} = \Sigma \nu_i G_{reactants} - \Sigma \nu_i G_{products}$ where ν_i are the stoichiometric coefficients of the reaction. The equilibrium constant is then calculated from $\Delta G_{reaction}$ and is given by: $\Delta G_{reaction} = RTlnK_{eq}$, where R is the gas constant, T is the temperature and K_{eq} is the equilibrium constant. Figure 3 shows the output when starting with Fe, Ni and n-heptane.

TDA Approach to Preventing Filamentous Coke

Our approach to eliminating filamentous coke is to add a compound that reacts with iron and nickel preventing the metals from forming iron and nickel carbides.

In order to evaluate the feasibility of this approach and identify candidate additives for testing, we carried out thermodynamic calculations (similar to those described above) to evaluate the stability of various nickel and iron compounds that could be formed by the addition of a suitable additive. The results of these calculations indicated that this approach was valid. We found that selenium should form very stable compounds with iron and nickel, effectively preventing either metal from forming a carbide (14).

Figure 4 shows the phases that are stable in the presence of 10 ppm selenium for n-heptane cracking at the same temperature, pressure and concentrations that were used in the calculations that generated Figure 3. Even at very low concentrations, selenium converts all of the nickel and iron on the metal surface into nickel and iron selenides (NiSe and FeSe), eliminating carbide formation (even at 1000°C the concentration of Fe_3C is almost 1000 times lower than that of FeSe). This occurs simply because the selenides are much more stable than the carbides. In addition, Figure 4 shows that the selenide phases are stable at pyrolysis temperatures. As a result, the addition of a selenium-containing compound should block the formation of iron and nickel carbides, which in-turn should prevent filamentous coke formation. Because the selenides are stable at temperatures in excess of 1000°C (1832°F), it might be possible to use very low concentrations of a selenium compound in the feed to prevent filamentous coke formation on a continuous basis. Alternatively it might be possible to add the selenium compound periodically to passivate the reactor tubes rather than having to add Se continuously.

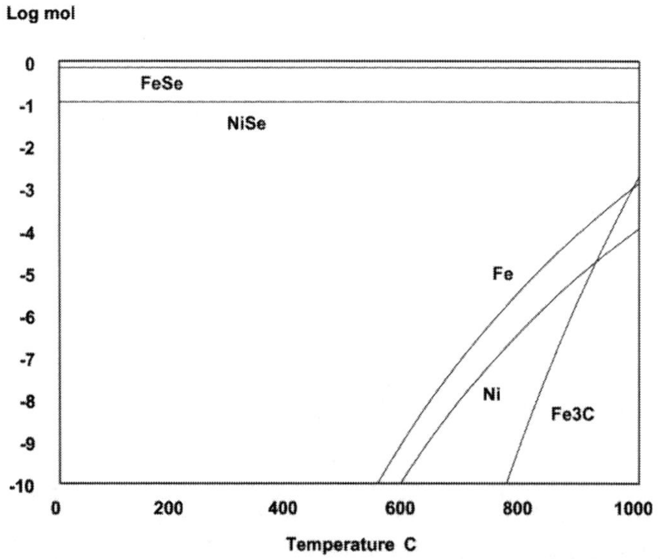

Figure 4. Preferential formation of iron and nickel selenide under pyrolysis conditions.

Experimental Methods

A diagram of the test apparatus is shown in Figure 5. The three primary feed flows: the primary feedstock, the mixture of additive in the feedstock, and water are all controlled separately. The three liquid flows are introduced at the desired pressure and flow rate with separate high pressure liquid pumps. All of the individual flow streams pass through separate preheaters before they are combined in the heated test section. The test section will consist of a 12-in length of ¼-in OD tubing. For these tests we used 304 stainless steel, Inconel 625 and alloy 800HT tubing, Because temperature control down the length is a critical parameter, we monitor the temperature profile by spot-welding thermocouples approximately every three inches down the length of the reactor. A Lindberg tube furnace is used to heat the test section to temperatures up to 950°C. We use the last thermocouple (TC-5) for temperature feedback to control the oven power. The flow stream exiting the reactor is combined with a flow of ambient temperature nitrogen, which rapidly quenches the products. The entire system is controlled with a PC running a process control program. The rate of coke deposition was determined by the weight gain in the tubing test section as a function of time (the tubes were periodically removed, rinsed with acetone to remove tar, dried, weighed and then placed back in the apparatus. At the end of each test, the tubes were sectioned longitudinally and examined under a scanning electron microscope.

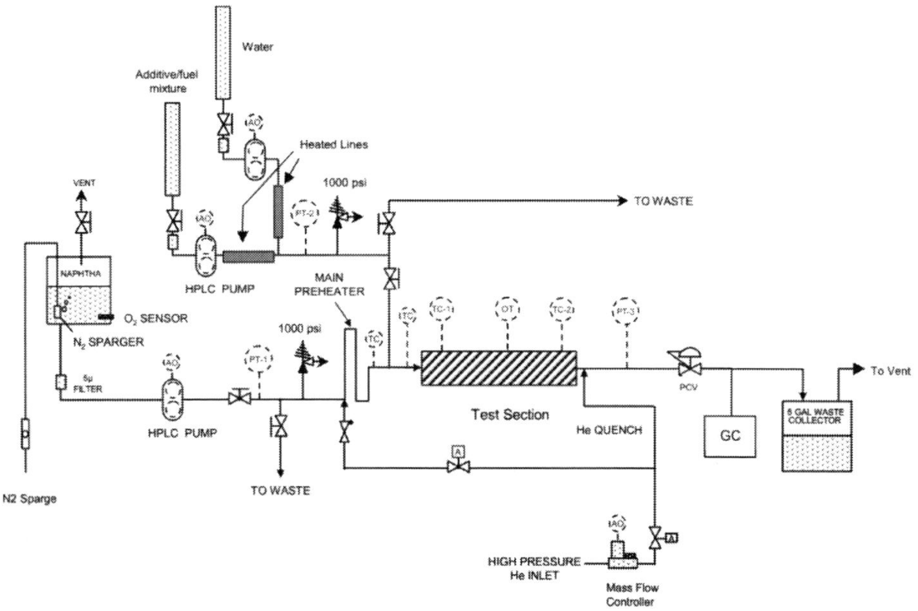

Figure 5. Process and instrumentation diagram of the test apparatus

Test Results

To measure the effect of the Se additive on reducing filamentous coke formation in metal tubes, we conducted tests in which we flowed the feedstock through a ¼-in section of tubing (initially 304 stainless steel) maintained at high temperatures with and without the selenium- additive. In each test, we started with a new pre-weighed tube, mounted it in an automated test apparatus, flowed a model compound naphtha feedstock (n-heptane) while maintaining the test section at the desired temperature (measured by a thermocouple mounted on the outer wall of the tube). At the completion of the test, we removed the tube, rinsed it with acetone (to remove any pyrolysis oil or tar), dried it at 110°C, and then reweighed it to determine how much coke was deposited inside the tube. We also removed the coke and examined it under a scanning electron microscope (SEM) to determine if the coke was catalytic (filamentous coke) or thermal (gas phase condensation coke) had formed. When used, selenium was added in the form of diphenyl selenide (Figure 6) at concentrations of 10 and 30 ppm as Se.

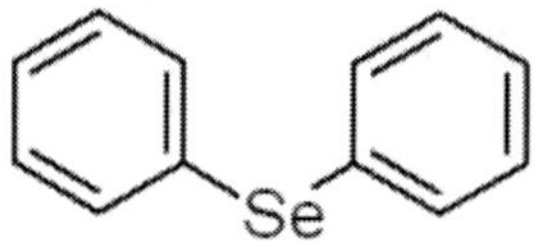

Figure 6. Diphenyl selenide

In the first series of tests, we used stainless steel tubing at high pressure in order to accelerate the coke deposition rate. Under these conditions, we found that maintaining the wall temperature at 685°C (1265°F) resulted in measurable coke deposition rates in a period of eight hours without additive. We also conducted tests under conditions more representative of those used in ethylene production. We carried out one series of tests using a mixture of feed and steam along with an Inconel 625 test section at a temperature of 750°C (1382°F). We also carried out tests using alloy 800 HT at a temperature of 900°C (1652°F). Both alloys were readily available for laboratory testing. In future work, we will examine the effectiveness on the 25Cr-35Ni cast alloy compositions that are now used in most ethylene plants.

Results with Stainless Steel Test Sections

Initially, our search for a coke suppression additive was based on projects for the U.S. Air Force and NASA for their hypersonic aircraft programs. In those applications, we were examining the use of thermal fuel cracking as an endothermic heat sink for cooling aircraft components when the speed of the aircraft exceeds about Mach 5. At these speeds, the heat sink requirements are too high to be accommodated by fuel vaporization alone. The cracked products are used as fuel for the engine. For this reason, we initially did tests with 304L stainless steel tubing sections. While, stainless steel is obviously not used in ethylene furnaces, the results obtained from our tests with stainless steel suggested ways to prevent coke in high Ni-Cr alloys that are used in ethylene production.

Figure 7 is a summary of the results we obtained in the tests conducted with stainless steel test sections at high pressure. High pressure (20 bar) was used to simulate the conditions during hypersonic aircraft cooling. In the tests without additive, there was very little weight change in the tube up to a period of six hours. The figure shows an accumulation of 41 and 24 mg of coke for periods of five and six hours respectively. However, by seven hours coke began forming at a substantial rate; we obtained 145 mg coke after seven hours, and by eight hours the amount of coke rapidly accumulated (243, 371, and 761 mg). These results show that there is an induction period at the beginning of the tube life where very little coke forms, but that after this period, the rate of coke formation increases rapidly, providing significant quantities of coke after seven hours of operation. Moreover, when we looked at the coke under the SEM, it had the characteristic filamentous appearance (Figure 8).

When we conducted tests where diphenyl selenide was added. Figure 7shows that coke formation was almost entirely prevented. In five separate tests with the Se additive, very little coke formed even when we extended the run time to 16 hours. In three eight hours tests with either 10 or 30 ppm of added Se, we obtained 21, 37 and 40 mg coke, values that are far lower than the amounts of coke formed in the absence of selenium. In addition, in a 16-hour test with only 10 ppm Se, we obtained 58 mg coke, again much lower than that obtained in eight hours without additive. Finally, SEM analyses performed on the small amounts of coke that were collected when Se was present showed no evidence of filaments (Figure 9); the coke that was deposited had the characteristic morphology of coke produced by gas phase condensation reactions. These results show that the selenium additive was very effective at reducing filamentous coke formation, which accounts for the majority of the coke formed under ethylene production conditions.

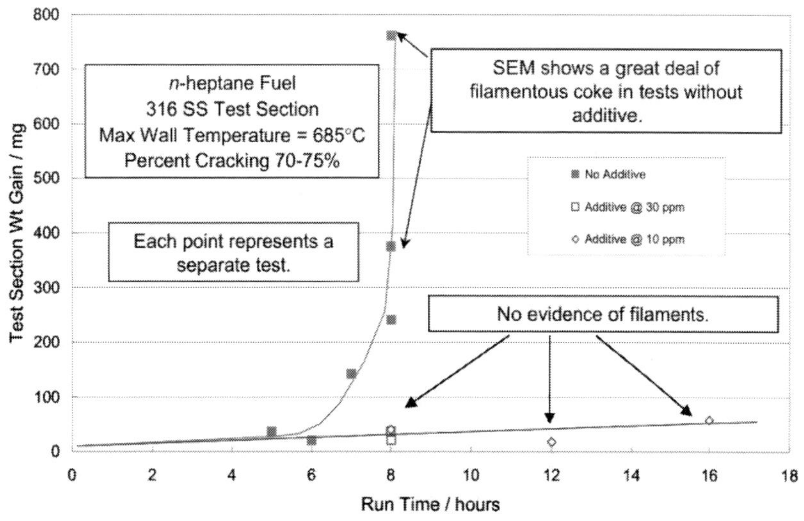

Figure 7. Quantity of coke accumulated in tests with and without additive.

Because it would be less costly to use an additive as a pretreatment or on an intermittent basis rather than continuously (and much less expensive than using tubes that have been modified using specialized metallurgical coatings), we conducted a test in which we pretreated the test section by injecting 10 ppm of diphenyl selenide for the first four hours of the test only. We then stopped the diphenyl selenide flow and monitored the amount of coke that accumulated as the tube was run without any additional additive. At the end of each test interval, we removed the reactor (heated tubing section) to determine how much coke had accumulated by measuring its weight gain. We then reinstalled the same tube section and continued the test for the next time interval.

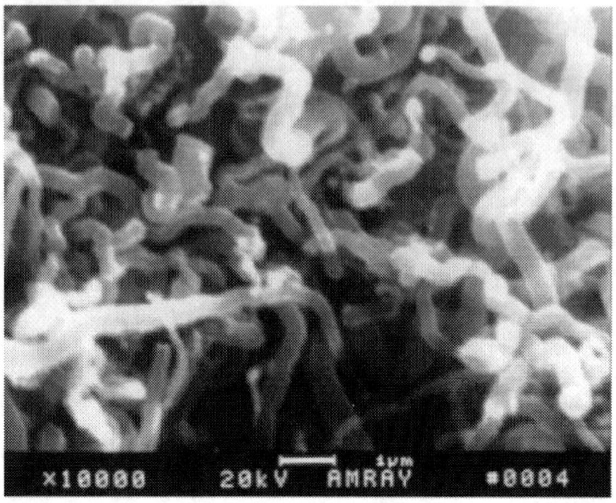

Figure 8. SEM photograph of coke that formed without additive.

The results of this series of tests are shown in Figure 10. We have included the results obtained with no additive for comparison. The figure shows that at the first inspection period of 16 hours, the total quantity of coke that had accumulated was only 15 mg. This is significantly less than the values measured after eight hours of operation when no additive was used. After 28 hours of time the total carbon accumulated is only 43 mg. We conducted SEM analyses on this coke and found no evidence of filaments. Finally, at 40 hours, the weight of coke is 88 mg.

Overall, these results show that very low concentrations of the diphenyl selenide greatly reduces the rate of filamentous coke formation in hot stainless steel tubes. We obtained very little coke when we added either 10 ppm Se on a continuous basis or when we added it for only the first four hours of a 40 hour test. Moreover, SEM images of the small quantities of coke that did accumulate when the additive was present, showed no trace of filamentous coke (only condensation coke).

Figure 9. SEM image of coke formed when Se was present.

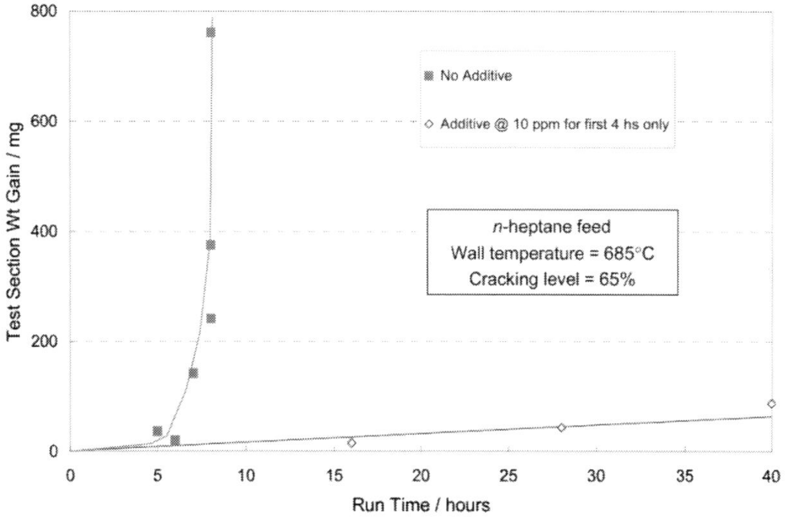

Figure 10. Coke formed after 4 hour pretreatment.

Results obtained with Inconel Tubes

We also carried out several tests at under conditions more representative of those used to produce ethylene. We reduced the pressure to three atmospheres, added water at a molar ratio of 1:3 water to feed, and used a test section fabricated from Inconel 625. This alloy contains 58% Ni, 20-23% Cr, 8-10% Mo along with some Fe, Nb, Ta, Al and other elements. Again our goal was to identify conditions where we formed measurable quantities of filamentous coke within the tube and then compare the quantities formed with and without additive. Unfortunately, when we maintained test section at the maximum allowable wall temperature of 750°C (1382°F) for 24 hours without additive, we were only able to collect about 10 mg carbon, a value that is very close to our detection limit. However, when we removed the material from the reactor and completed SEM analyses of carbon, we did find that it had the characteristic filamentous structure.

Figure 11 shows that the carbon filaments collected in the Inconel tube have a diameter of approximately 0.1 to 0.2 microns. In addition when we examined the inside of the reactor tube at the conclusion of the test, we found that it had a dull appearance, characteristic of carbon coated on the surface.

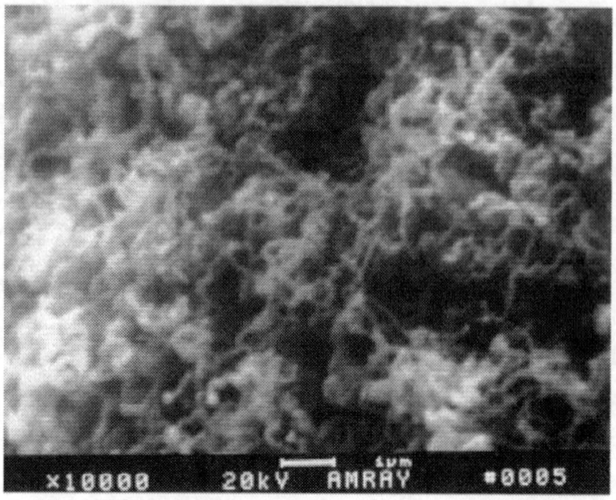

Figure 11. Filamentous coke from an Inconel tube.

We then conducted a test under identical conditions with 10 ppm of diphenyl selenide contained in the feed. At the conclusion of this test, we were not able to remove any carbon from the reactor, indicating that the additive completely suppressed the formation of filamentous coke, at least for the duration of the experiment. In addition, the inside of this tube had a very shiny appearance, supporting the conclusion that no carbon filaments were present on the surface.

After the tests were concluded, we cut sections from the two tubes, one that was used with additive and one used without additive. Photographs of the inside walls of each tube are shown in Figure 12. The section on the left was run without the additive and the photograph shows that the inside surface has a dull appearance indicating the presence of carbon. On the other hand, the surface of the tube on the right side, which was used with Se additive, was smooth and shiny, indicating that no carbon had accumulated during the test.

These tests are still very encouraging. They have shown that the selenium additive, when present at 10 ppm, prevents filamentous coke in a high temperature alloy reactor, under conditions more representative of those used in ethylene production.

Results Obtained with 800 HT Test Sections

We then obtained an alloy that is more representative of the alloys used in the pyrolysis tubes in ethylene furnaces, 800 HT. The tests were also done at higher temperatures, 900°C (1652°F). At this temperature we found that measurable quantities of coke could be obtained in relatively short times, less than one hour after reaching temperature (we did not add steam in these tests in order to accelerate the coke formation rate).

The results we have obtained to date are shown in Figure 13. Without additive, we obtained weights of 1650 and 1450 mg carbon inside the test section during the testing sequence, or an average of 1550 mg in the two runs. We then conducted two tests with the selenium additive. In these tests, we pretreated the test section with additive and then conducted the remainder of the test without further addition. We pretreated the tube with 30 ppm Se at 550 psi and 500°C for one hour. We then lowered the pressure, discontinued the additive, and increased the wall temperature to 900°C (1652°F).

Figure 12. Inconel tube run without additive (left) and with additive (right).

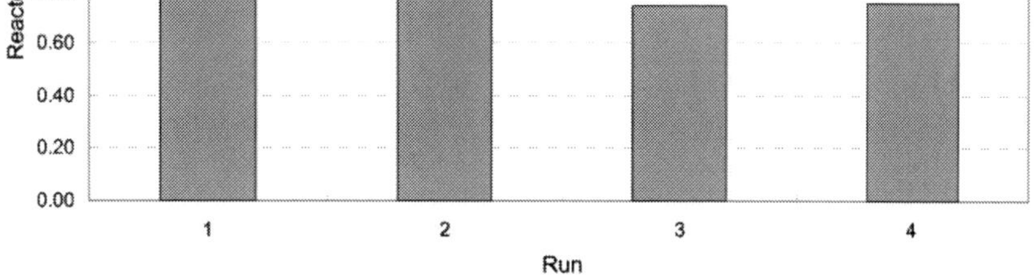

Figure 13. Coke accumulation with and without additive at 900°C (1832°F).

Figure 13 shows that the pretreatment caused a substantial reduction in total coke deposition. Following the pretreatment, we obtained 740 and 750 mg coke in the test section or about one half of the coke that accumulated when no additive was used. Although this is a substantial reduction, the effect that the additive has on filamentous coke, as determined by SEM, may be much greater.

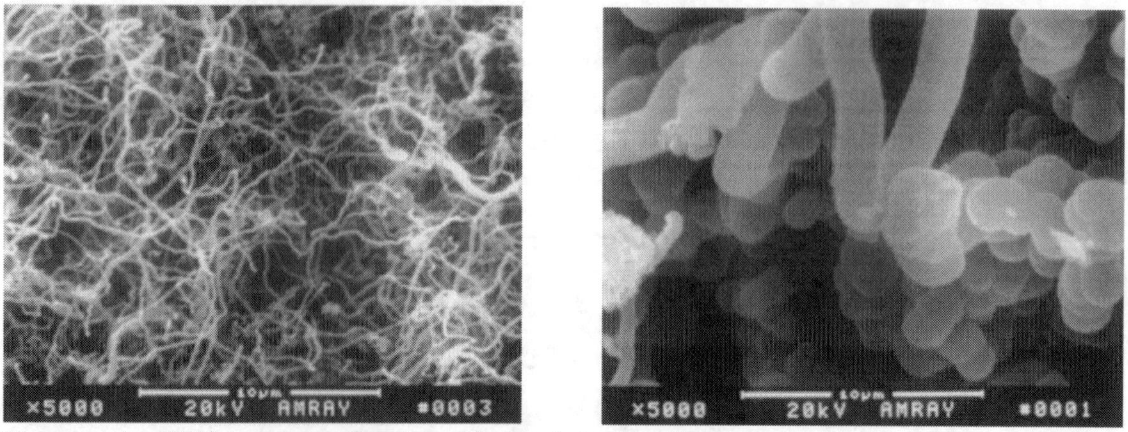

Figure 14. SEM images of coke obtained during the tests with n-heptane cracking in an 800 HT tube without additive (left side) and with additive (right side).

We obtained samples of coke formed during each test and examined them under the SEM. A photograph of the coke collected without additive is shown on the left side of Figure 14, while an image of the coke formed with additive is presented on the right side of the figure. In both cases, we obtained multiple samples and the images shown are representative of all the samples collected. Figure 14 shows that the coke formed without additive is comprised almost entirely of very well defined filaments that are appear to be less than one micron in diameter. Thus, as in the other series of tests, we conclude that without the selenium additive, essentially all of the coke deposited in the test section is catalytic or filamentous coke.

On the other hand, the photograph on the right side of Figure 14 shows that the coke deposit when the Se additive is present has a much different morphology than the coke obtained without additive. There are very few well-defined filaments like those in the left photograph. In addition, the figure shows the appearance of many spherically shaped deposits, which are characteristic of condensation coke. Because we are operating at relatively high residence times and at very high temperature, we are not surprised to see more condensation coke. However, the presence of condensation coke, which is included in the overall weight gain, may be masking the effectiveness of the additive is suppressing the formation of filamentous coke. In fact, it appears from the SEM images that the additive has almost completely eliminated filamentous coke formation. It is likely that if we were able to conduct tests at higher flow rates (lower residence times), we would be able to reduce the interference from condensation coke. Based on the SEM images, this would result in a more significant weight reduction when the additive is present.

Summary and Conclusions

In summary the results obtained under a number of test conditions have shown that the addition of low concentrations of diphenyl selenide stops filamentous coke formation in stainless steel, Inconel 625 and Alloy 800HT during n-hexane cracking. In tests with stainless steel reactors under accelerated aging conditions, we found that 10 ppm Se reduced the quantity of coke from over 700 mg in eight hours to less than 60 mg in 16 hours. In addition SEM images show that the coke collected with additive contained no traces of filamentous coke. We also found that the addition of 10 ppm Se blocked coke formation in a test section made from Inconel 625 and with a feed containing 33% steam. Finally, in tests conducted at 900°C (1652°F) with alloy 800 HT, the additive produced over 50% reduction in total coke accumulation. More importantly, SEM images indicate that very few filaments were present, again showing that the additive blocks filamentous coke.

We believe that the selenium additive functions by reacting with the iron and nickel in the metal surface, preventing the formation of iron and nickel carbides, which are important intermediates in the filamentous coke formation process. In addition, thermodynamic calculations suggest that the additive should prevent carbide formation at temperatures in excess of 1000°C (1832°F).

The work described in this paper was done to meet the needs of NASA for endothermic fuel cracking for hypersonic aircraft cooling, nevertheless, we feel that our additives could be very useful to the ethylene industry for inhibiting coke formation and therefore, increasing the time between turnarounds and decoking operations, which would potentially lower the operating costs of the ethylene plant. Our additives can be used as a pretreatment, or supplied continuously to the feed is very low concentrations (10 – 30 ppm) to provide coke inhibition. The low concentrations keep additive costs low.

References

1. Kniel, L.; Winter, O. and Stork, K. (1980) Ethylene, Keystone to the Petrochemical Industry, Chemical Series/2, Marcel Dekker, New York

2. Sundaram, K.M.; Shreehan, M.M. and Olszewski, E.F. (2001) "Ethylene." In Kirk Othmer Encyclopedia of Chemical Technology, Vol 10. 593-631.

3. Zimmermann, H. and Walzi, R. (2007) "Ethylene," in Ullmann's Encyclopedia of Industrial Chemistry, 1-58.

4. Vaish, S. and D. Kunzru (1989). "Triphenyl Phosphite as a Coke Inhibitor During Naphtha Pyrolysis," *Ind. Eng. Chem. Res.* 28, 1293-1299.

5. Reyniers, G.C., Froment, G.F., Kopinke, F.D., and Zimmerman, G. (1994). "Coke Formation in the Thermal Cracking of Hydrocarbons. 4. Modeling of Coke Formation in Naphtha Cracking" *Ind. Eng Chem Res.*, 33, 2584-2590.

6. Tong, Y. and M.K. Poindexter (1999). "Method of Inhibiting Coke Deposition in Pyrolysis Furnaces", U.S. Patent Number 5,954,943

7. Woerde, H.M., S. Barendregt, F. Humbolt and C. Brun (2002). "Mitigate Coke Formation – A new anticoke technology sustains a noncatalytic layer on cracking tubes that reduces coking", *Hydrocarbon Processing*, March 2002, 45-50.

8. Bartholomew, C.H., Agrawal, L.K., and Katzer, J.R., "Sulfur Poisoning of Metals", *Adv. Catal.*, 31, 135-242 (1982).

9. Butt, J.B. and Petersen, E.E., (1988) Activation, Deactivation, and Poisoning of Catalysts, Academic Press.

10. Baker, R. T. K., Yates, D. J. C., and Dumesic, J. A. (1982). "Filamentous Carbon Formation Over Iron Surfaces," in Coke Formation on Metal Surfaces, *ACS Symposium Series 202*, 1.

11. Reyniers, M.F.S.G, and G.F. Froment (1995). "Influence of Metal Surface and Sulfur Addition on Coke Deposition in the Thermal Cracking of Hydrocarbons", *Ind. Eng Chem Res.*, 34, 773-785.

12. Grabke, J.J., (1995) "Metal Dusting of Low- and High-Alloy Steels," *Corrosion*, 51, No. 9, 711-720.

13. De Jong, K.P. and J.W. Geus (2000) "Carbon Nanofibers: Catalytic Synthesis and Applications," *Catal. Rev.-Sci. Eng.* 42(4) 481-510.

14. Wickham, D.T., Engel, J.R., Karpuk, M.E. (2002). "Methods for Suppression of Filamentous Coke Formation", U.S. Patent No. 6,482,311 B1.

HYDROGEN TRANSPORT MEMBRANE TECHNOLOGY FOR ETHYLENE PRODUCTION

Uthamalingam (Balu) Balachandran
Materials Scientist
Argonne National Laboratory

Tae H. Lee
Materials Scientist
Argonne National Laboratory

Steve E. Dorris
Materials Scientist
Argonne National Laboratory

Carl A. Udovich
Chemist/Heterogeneous Catalysis
Argonne National Laboratory

Christopher L. Marshall
Group Leader/Heterogeneous Catalysis
Argonne National Laboratory

Abstract: We have demonstrated in the laboratory a new way to make ethylene via ethane dehydrogenation using a dense hydrogen transport membrane (HTM) to drive the unfavorable equilibrium conversion. The HTM drives the conversion of ethane by selectively removing hydrogen that is produced by the dehydrogenation reaction. Preliminary laboratory experiments using small disk-type HTMs show that the new approach can afford ethylene yields above the thermodynamic equilibrium limit, at high ethane conversion, while eliminating production of greenhouse gases in the reactor section. Moreover, the new approach results in a much simpler product slate than that obtained from today's pyrolysis technology. This simpler product slate affords an opportunity to lower the cost of the "back end" purification train. We present here preliminary results from ethane dehydrogenation experiments and describe the development of HTMs suitable for ethylene production.

Introduction

Ethylene, the largest volume commodity chemical manufactured today, is currently produced by pyrolysis of ethane in the presence of steam. Thermodynamically limited in terms of its equilibrium conversion, this process requires substantial energy input. Pyrolysis of ethane also produces significant amounts of greenhouse gases (CO and CO_2) in the reactor section, because carbon directly contacts steam. In a new approach for producing ethylene [1, 2], a dense hydrogen transport membrane (HTM) selectively removes hydrogen during the dehydrogenation of ethane, thereby shifting the equilibrium toward ethylene production. This process is shown schematically in Fig. 1.

Preliminary laboratory experiments showed that this approach gives ethylene yields above the thermodynamic equilibrium limit, even at high ethane conversion. Because the HTM prevents carbon-containing species from contacting air or steam, the production of greenhouse gases is avoided in the reactor section, and a much simpler product slate is produced. The simpler product stream should significantly lower the cost of "back end" purification, which accounts for ≈70% of the capital cost and ≈50% of the energy consumption for a conventional ethylene production unit. Besides significantly reducing energy costs, the new approach should result in savings in feedstock and capital costs.

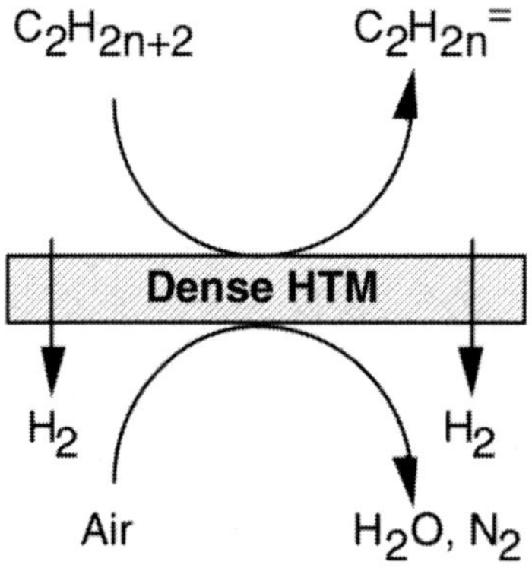

Fig. 1 Schematic of process for producing ethylene via dehydrogenation of ethane.

Experimental

A dense disk-type HTM in the form of a ceramic/metal composite was fabricated to produce ethylene by dehydrogenation of ethane at 850°C [3, 4]. The membrane for the dehydrogenation experiment contained 50-60 vol.% Pd with the balance being Y_2O_3-stabilized ZrO_2 (TZ-3Y). It was prepared from a mechanical mixture of Pd (Technic) and TZ-3Y (TOSOH) powders that were uniaxially pressed at ≈70 MPa in a stainless steel die. After sintering at 1400-1450°C for 5 h in air, the membrane had a thickness of ≈0.25 mm and a diameter of ≈20 mm. In the dehydrogenation experiment, ethane at ambient pressure was passed over one side of the HTM, while air flowed over the other side. The membrane was heated to ≈790-890°C. The hydrogen produced by dehydrogenation was removed and transported through the HTM to the air side of the membrane. The air provided the necessary driving force for transport of hydrogen through the HTM. The reaction between transported hydrogen and oxygen in air can provide some of the heat needed for the dehydrogenation reaction.

Because the ethylene production rate depends directly on the rate at which hydrogen is removed from the dehydrogenation zone, the HTM must deliver a high hydrogen flux for the new process to be effective. The hydrogen flux of HTMs was measured in tests using feed gases with known hydrogen concentrations. The HTMs were made as either a self-supporting sintered disk, like that in the dehydrogenation experiment, or a thin-film membrane supported by a porous ceramic substrate.

Thin-film membranes were prepared by mixing palladium and TZ-3Y in a solution of α-terpineol and isopropyl alcohol with polyvinyl butyral as a binder and dibutyl phthalate as a plasticizer. A paste was prepared to give 50 vol.% Pd in the final thin film. The concentration of isopropyl alcohol was varied to control the viscosity of the paste. To prepare porous substrates for HTM thin films, alumina hydrate powder was mixed with carbon powder (0-30 wt.%) as a pore former, and then the powder mixture was cold isostatically pressed ($\approx$70 MPa) in the shape of either a disk or a tube. The pressed specimen was then pre-sintered at 800-900°C in air for 5 h. Paste containing the membrane components was painted onto the surface of the pre-sintered porous substrate and then was sintered for 5 h in air at 1400-1450°C. Film thickness was less than 15 μm.

The hydrogen flux through an HTM sample was measured by affixing the sample (either a disk or a tube) to the Al_2O_3 tube in the assembly shown in Fig. 2. A seal formed during heating at 950°C, when spring-loaded rods squeezed a gold ring between the membrane and the Al_2O_3 tube. One side of the sample was purged with feed gas composed of 4-100% H_2/balance He, while the other side was purged by a sweep gas of 100 ppm H_2/balance N_2. All gases were at ambient pressure. The hydrogen flux was measured in the temperature range 400-900°C.

Gas flow rates (150 ml/min) were controlled with MKS mass flow controllers. A Hewlett-Packard 6890 gas chromatograph was used to measure the concentrations of He and H_2 in the sweep gas. The He concentration in the sweep gas was used to correct the hydrogen concentration in the sweep gas for leakage through the seal. The calculated hydrogen leakage rate was typically <10% of the total H_2 flux through the membrane.

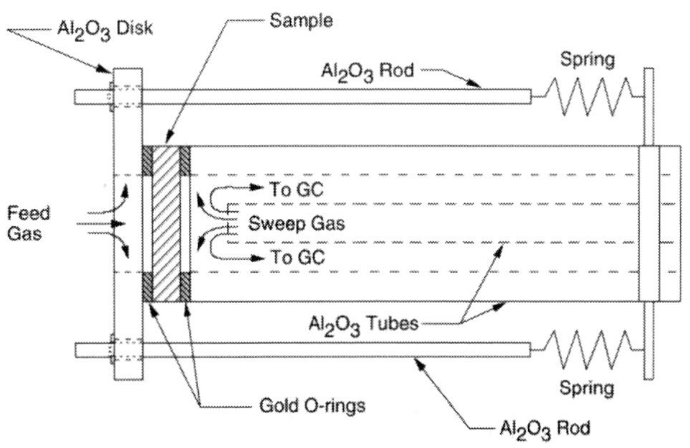

Fig. 2 Fixture for measuring hydrogen flux through HTM.

Results

Figure 3 shows the results from an ethane dehydrogenation experiment done at 790-890°C using a self-supporting HTM disk (thickness ≈0.25 mm). This experiment was conducted at BP, Naperville, IL. In this experiment, the hydrogen produced by dehydrogenation of ethane was removed and transported to the air side of the HTM, where it reacted with oxygen to form steam. The reaction between hydrogen and oxygen on the air side provides some of the energy needed for the dehydrogenation reaction and establishes a low partial pressure of hydrogen, which provides the driving force for hydrogen transport through the HTM. At 850°C and 1 atm pressure, equilibrium conversion of ethane limits the ethylene yield to 64%, whereas our experiment gave an ethylene yield of 69% with a selectivity of 88% under the same conditions. Coking was not a problem in runs extending over several weeks. These results are encouraging even though the HTM probably did not give a high hydrogen flux due to its relatively large thickness (0.25 mm).

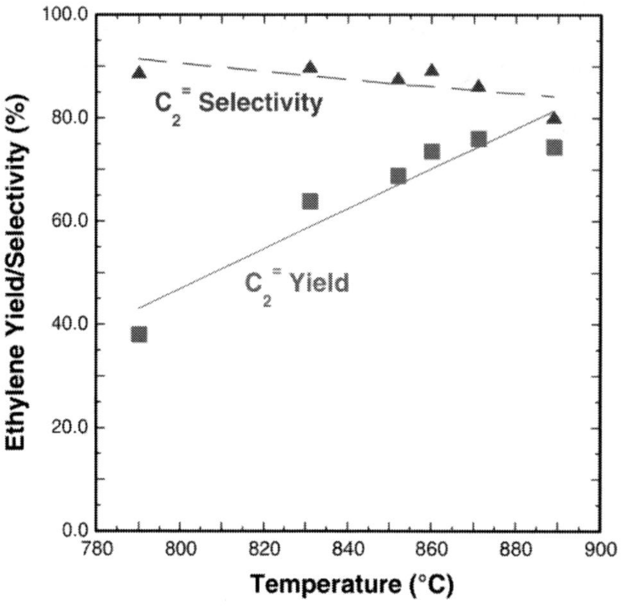

Fig. 3 Results from dehydrogenation of ethane at 850°C with 0.25-mm-thick HTM.

Increasing the hydrogen flux through the HTM would enhance the ethylene yield at lower temperatures with a reasonable residence time. Furthermore, reducing the operating temperature would suppress unwanted side reactions, eliminate NO_x formation on the air side of the HTM, and prolong the membrane's life. Figure 4 shows the hydrogen flux at 900°C versus the inverse of membrane thickness for HTMs like that used in the dehydrogenation experiment. The feed gas during these measurements was 80% H_2/balance He, and the sweep gas was N_2. The hydrogen flux varies linearly with the inverse of membrane thickness, indicating that the membrane's bulk properties control its flux in this thickness range. The HTM for the dehydrogenation experiment and the thickest membrane tested for Fig. 4 had a similar thickness (0.21-0.25 mm); therefore, their hydrogen transport properties are expected to be similar. Since the dehydrogenation experiment was conducted, the hydrogen flux through HTMs has been increased significantly by reducing their thickness.

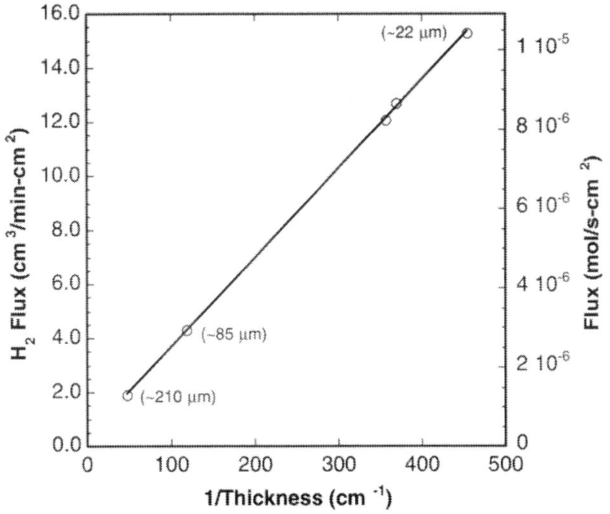

Fig. 4 Hydrogen flux at 900°C vs. inverse thickness of HTM. Feed was 80% H_2/balance He; sweep was N_2.

The data in Fig. 4 were obtained with self-supporting membranes that were prepared by polishing sintered disks. This fabrication technique is, of course, impractical for producing membranes with large surface areas, and samples made this way are very fragile. We thus sought to make thinner membranes with better mechanical integrity by a scalable method. To that end, we developed the paste painting technique for fabricating dense thin film membranes on porous supports. A cross-sectional view of a film made by paste-painting is shown in Fig. 5, where a dense ≈15-μm-thick HTM film is clearly visible on top of a porous alumina substrate.

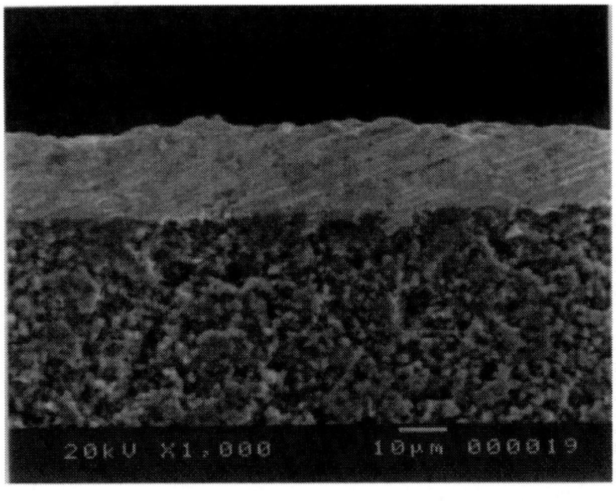

Fig. 5 Cross-sectional view of ≈15-μm-thick HTM film on porous alumina substrate.

Figure 6 shows the hydrogen flux for a ≈15-μm-thick HTM made by paste painting.

The flux (≈52 cm^3/min-cm^2) at 900°C is much higher than could be obtained with the 0.25-mm-thick HTM for the dehydrogenation experiment, based on thickness. The flux at 500°C (≈32 cm^3/min-cm^2) was even much higher than expected for a 0.25-mm-thick HTM at 900°C. These results suggest that dehydrogenation should be favorable at temperatures <800°C.

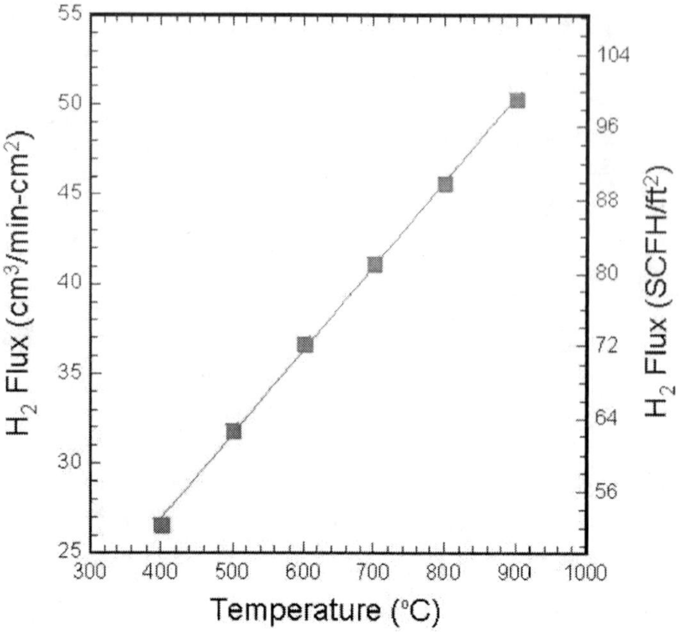

Fig. 6 Hydrogen flux vs. temperature for ≈15-μm-thick HTM film on top of porous alumina. Feed gas was 90% H_2/ balance He, and sweep gas was N_2.

In order for the new approach for producing ethylene to be practical, HTMs must be available in a shape that has a large active area, such as tubes; therefore, we adapted the paste-painting technique to fabricate dense HTMs on porous tubes. To this point, we have fabricated small (≈8-10 cm in length with outside diameter of ≈1 cm) tubes for demonstration purposes. Tubes made by this method typically contain a dense HTM film with a thickness of ≈25-60 μm. A tubular thin-film membrane with film thickness of ≈60 μm passed hydrogen at a rate of ≈190 cm^3/min at 900°C and ≈165 cm^3/min at 700°C.

Conclusions

We have demonstrated a new process to make ethylene via ethane dehydrogenation using a dense disk-type HTM to drive the unfavorable equilibrium conversion. This new process should reduce energy costs and overall costs, as well as greenhouse gases. Preliminary results obtained at 850°C with a relatively thick HTM show that this approach provides ethylene yields above the thermodynamic equilibrium limit. Because the ethylene production rate depends directly on the rate at which hydrogen is removed from the dehydrogenation zone, we developed a technique for fabricating thinner HTMs with significantly higher hydrogen fluxes. HTMs with higher hydrogen flux would enhance the process by allowing dehydrogenation to be performed at a lower temperature, which would suppress unwanted side reactions, eliminate NO_x production on the air side, and prolong membrane life. We realize what is accomplished so far is very preliminary, and commercialization of this new process will require several additional steps. At this time we are looking for industrial partner(s) to confirm the initial data and move this technology forward toward eventual commercialization.

Acknowledgments

Work supported by the U.S. Department of Energy, Office of Industrial Technologies Program and Office of Fossil Energy, National Energy Technology Laboratory's Hydrogen and Syngas Technology Program, under Contract DE-AC02-06CH11357.

References

1. U. Balachandran, U. S. Patent 7,329,791, issued Feb. 12, 2008.
2. U. Balachandran, T. H. Lee, S. E. Dorris, C. A. Udovich, C. G. Scouten, and C. L. Marshall, Proc. of Clean Technology 2008 Conference, June 1-5, 2008, Boston, MA.
3. S. E. Dorris, T. H. Lee, and U. Balachandran, U. S. Patent 6,569,226, issued May 27, 2003.
4. U. Balachandran, T. H. Lee, L. Chen, S. J. Song, J. J. Picciolo, and S. E. Dorris, Fuel, 85, 150, (2006).

AUTHOR INDEX

AUTHOR INDEX

American Institute of Chemical Engineers
3 Park Avenue
New York, New York 10016-5991

CURRAN ASSOCIATES INC
proceedings.com

21st Ethylene Producers Conference 2009

Topical Conference at the 2009 AIChE Spring National Meeting

Tampa, Florida
26 - 30 April 2009

Volume 2 of 2

ISBN: 978-1-61567-256-1

AIChE

21st Ethylene Producers Conference 2009

Topical Conference at the 2009 AIChE Spring National Meeting

Tampa, Florida
26 – 30 April 2009

Volume 2 of 2

ISBN: 978-1-61567-256-1

Printed from e-media with permission by:

Curran Associates, Inc.
57 Morehouse Lane
Red Hook, NY 12571

Some format issues inherent in the e-media version may also appear in this print version.

Printed by Curran Associates, Inc. (2009)

For permission requests, please contact AIChE
at the address below.

AIChE
3 Park Avenue
New York, NY 10016-5991

Phone: (203) 702-7660
Fax: (203) 775-5177

www.aiche.org

Additional copies of this publication are available from:

Curran Associates, Inc.
57 Morehouse Lane
Red Hook, NY 12571 USA
Phone: 845-758-0400
Fax: 845-758-2634
Email: curran@proceedings.com
Web: www.proceedings.com

TABLE OF CONTENTS

VOLUME 1

VOLUME 2

TABLE OF CONTENTS

VOLUME 1

VOLUME 2

TAKING THE PETRO OUT OF PETROCHEMICALS

OLD FEEDSTOCK OPTIONS REVISITED

Mark A. Wegenka

Senior Consultant

CMAI

Abstract:

One hundred years ago, most non-fuel industrial products, including chemicals were made from biomass (plants, trees and vegetables) or coal / coal tars. Since then, industrial chemical production has been dominated by petroleum based feedstocks. More recently, high oil and gas prices, dwindling domestic reserves and growing demand for transport fuels and chemicals in rapidly growing or large established economies such as the U.S., China and India have refocused attention back on these "old" feedstocks. Moreover, environmental concerns have resulted in mandated legislation for "renewable content" in fuels that are focusing efforts on improving the technologies and cost position for chemicals used as fuels (e.g. ethanol and biodiesel). Coal's abundance, relatively low pricing / price volatility and developments in technologies that improve its cost position and environmental performance are supporting trends. While coal's reputation as a "dirty" fuel in the West is slowing its penetration, realities suggest that it will become a larger share of the future global energy mix with corresponding impacts on chemical feedstocks. The implication for chemicals is becoming clearer. The political and technological advancements for coal and biomass will offer chemical producers alternatives other than traditional petroleum feedstocks. This paper will address the current issues facing chemical producers interested in diversifying and expanding by utilizing the two primary alternative feedstock alternatives to traditional petroleum feedstocks, biomass and coal. It will examine the economics associated with their use and explore the realities of the technologies available today and highlight the associated issues driving or inhibiting their use into chemicals. Finally, it will provide a brief economic snapshot and investment comparison of a coal-to-polyolefins facility in China vs. Middle East E/P cracking and liquids cracking in China.

In the case of biomass, interest has been soaring; probably too quickly. Economics are difficult with existing technologies, even at recent high energy and ethylene values. Developing cellulosic technologies offer the hope of using lower-value feeds, but yields are still relatively low. Chinese investment in coal based production of methanol (via gasification) and vinyls (via carbide / acetylene) is growing substantially. New, world-scale coal to methanol facilities coupled with derivative olefins and polyolefins are now also appearing. Our analysis concludes that MTO is cash-cost competitive with Middle East E/P economics delivered to China. Even with its relatively high CAPEX requirements, MTO has the best returns of all cases at a mine-mouth coal cost structure, while the liquids-based cracker has the highest cash cost structure and lowest returns in a high crude environment. However, MTO returns on market coal value are lower at current crude oil prices, demonstrating the risks to investment.

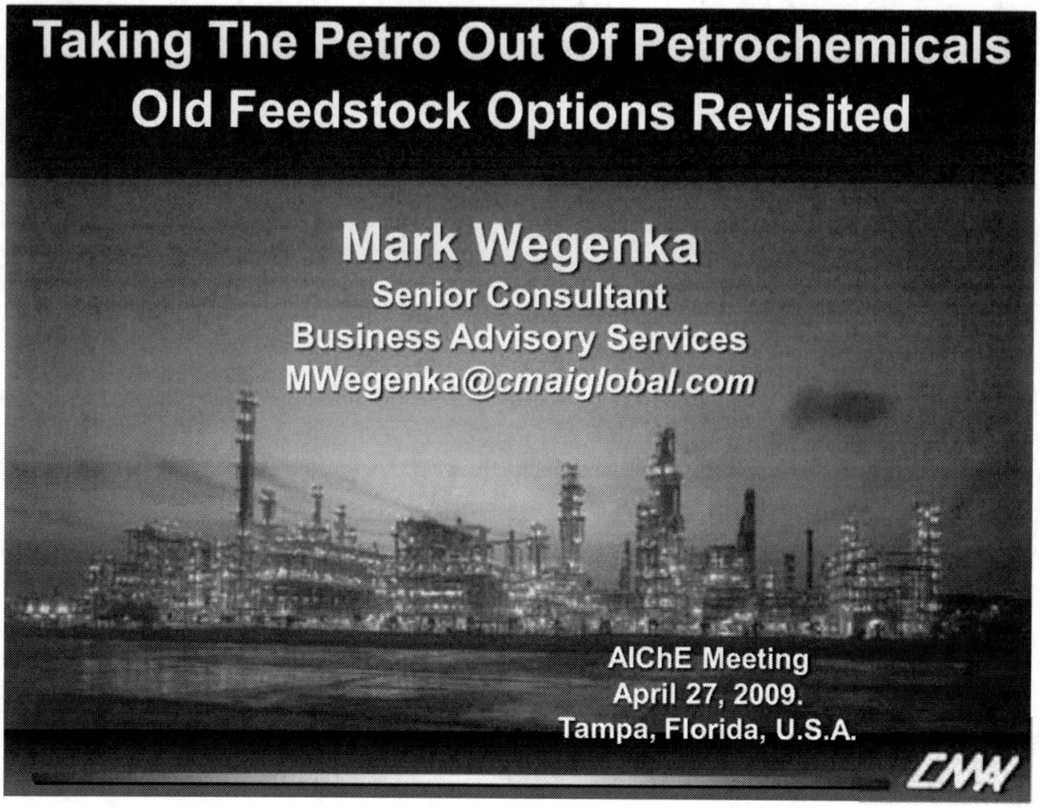
Taking The Petro Out Of Petrochemicals
Old Feedstock Options Revisited
Mark Wegenka
Senior Consultant
Business Advisory Services
MWegenka@cmaiglobal.com
AIChE Meeting
April 27, 2009.
Tampa, Florida, U.S.A.
CMAI

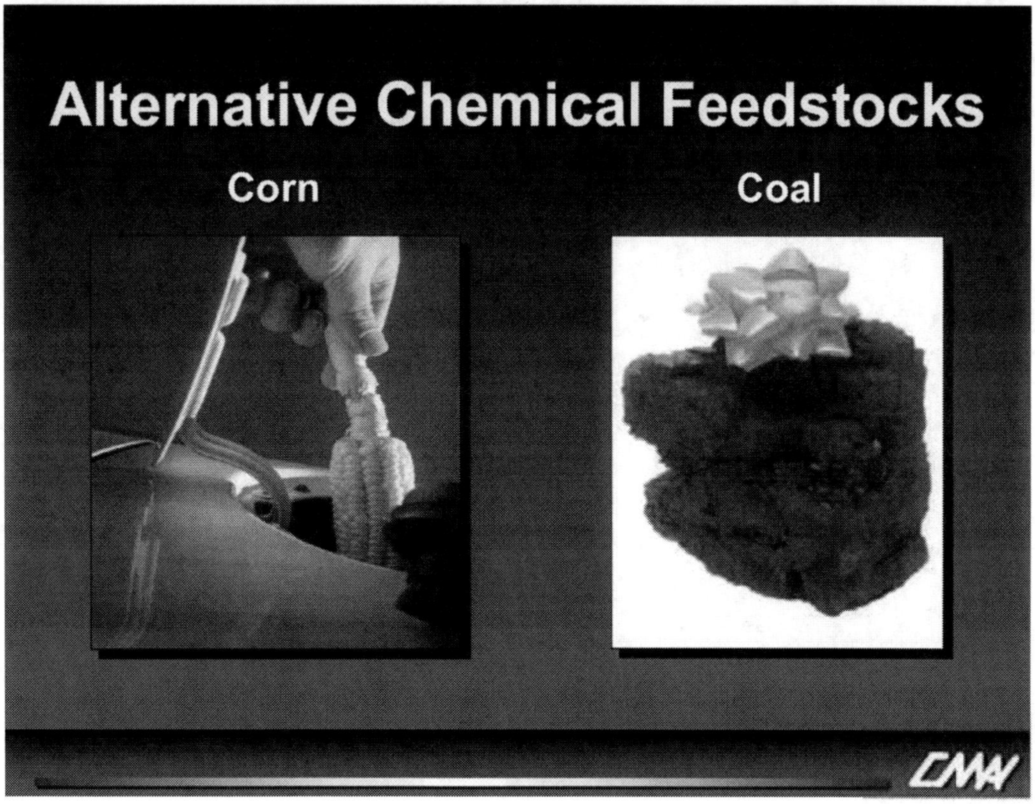

One hundred years ago, most non-fuel industrial products, including chemicals were made from biomass (plants, trees and vegetables) or coal / coal tars. By the mid 1900s, large scale oil discoveries and the increasing demand for transportation fuels led to the development of more economically efficient chemical production from petroleum (oil and gas). Since that point, aside from a brief respite during the oil shocks in the 1970s, industrial chemical feedstocks have been dominated by petroleum feedstocks. More recently, high oil and gas prices, dwindling domestic reserves and growing demand for transport fuels and chemicals in rapidly growing or large established economies such as the U.S., China and India have refocused attention back on these "old" feedstocks. Moreover, environmental concerns have resulted in mandated legislation for "renewable content" in fuels that are focusing efforts on improving the technologies and cost position for chemicals used as fuels (e.g. ethanol / biodiesel). Coal's abundance, relatively low pricing / pricing volatility and developments in technologies that improve its cost position and environmental performance are supporting trends. While coal's reputation as a "dirty" fuel in the West is slowing its penetration, realities suggest that it will become a larger share of the future global energy mix with corresponding impacts on chemical feedstocks. The implication for chemicals is becoming clearer. The political and technological advancements for coal and biomass will offer chemical producers alternatives other than traditional petroleum feedstocks. This paper will address the current issues facing chemical producers interested in diversifying / expanding by utilizing alternatives to traditional petroleum feedstocks.

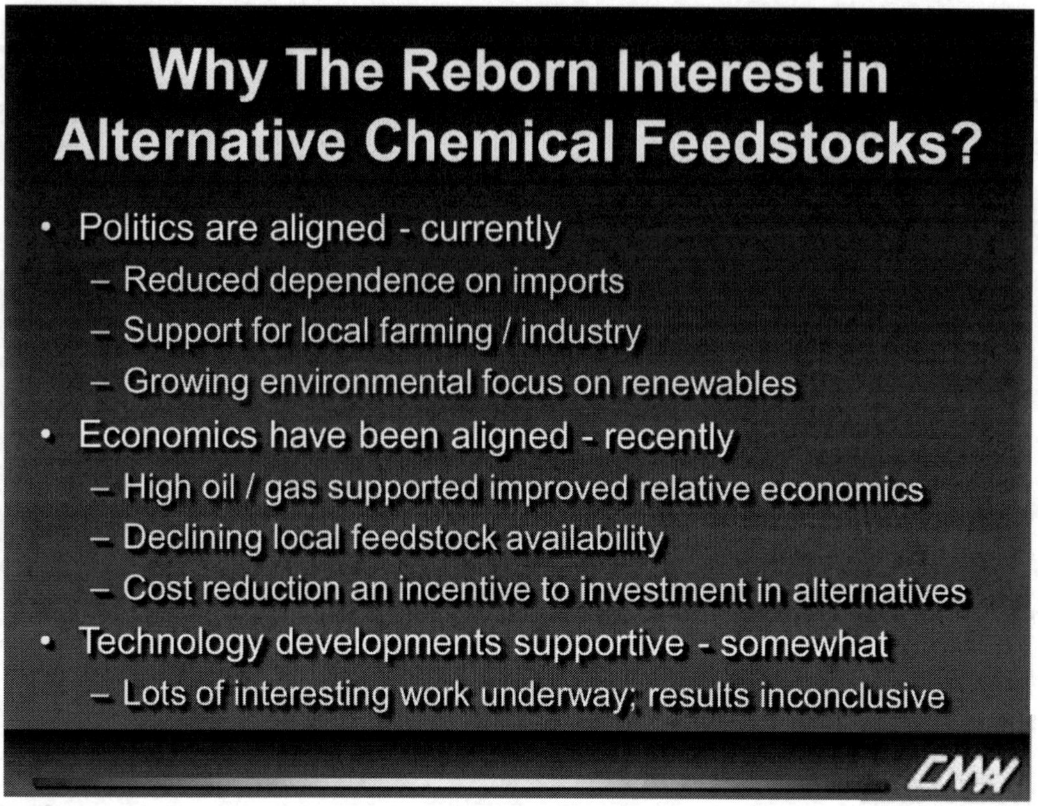

Surging interest in renewable, indigenous sources of fuels has created interest in the associated opportunities for the chemical industry. Politicians see a win-win with support from farmers, the environmentally conscious, and in hydrocarbon short countries, those who want to decrease reliance on imports. As a result, new legislation mandating renewable content and providing incentives in the form of subsidies, tax incentives and direct funding for research and application development has grown.

The current economic realities are also supportive. Not withstanding the recent declines, relatively high values of oil and gas and declining natural gas reserves in many established countries are combining at a time when chemical companies are looking for lower cost alternative feedstocks. Security of supply and local sourcing are also desired. As opportunities for growth are becoming more limited; alternative feedstocks present new options.

Technology developments promise future potential improvements and are focusing on reducing the most significant economic and technical barriers for implementation.

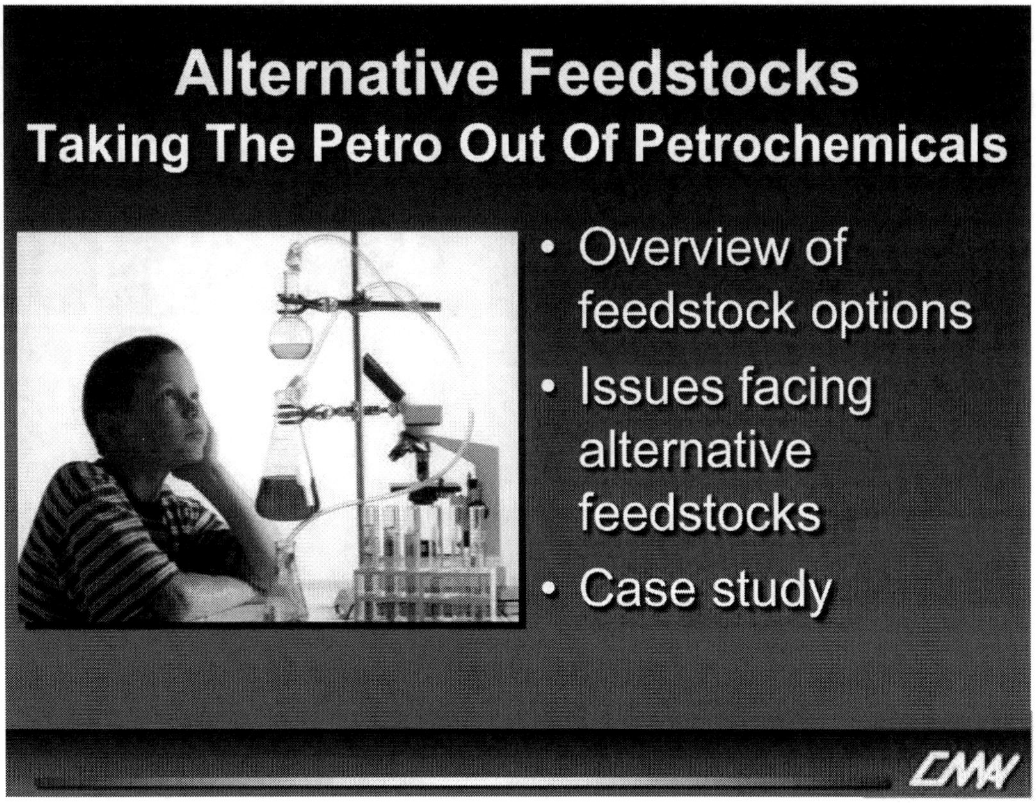

This paper will address the two primary alternative feedstock options; broadly falling into the categories of biomass and coal.

- It will examine the economics associated with their use and explore the realities of the technologies available today and highlight the associated issues driving or inhibiting use into chemicals.
- Finally, it will provide a brief economic snapshot and investment comparison of a coal-to-polyolefins facility in China vs. Middle East E/P cracking and liquids cracking in China.

Figure 1

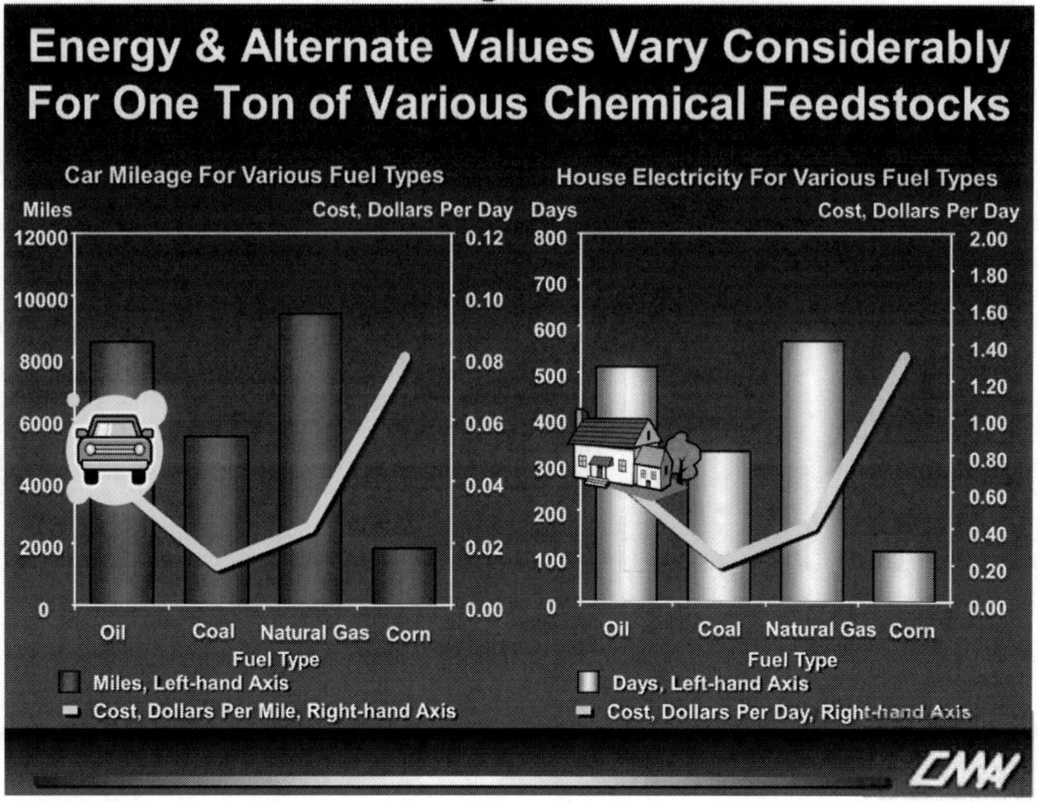

All materials have a basic intrinsic energy content. The bar graphs here show the amount of energy available in a ton of various energy containing materials into two of the primary energy demand sectors; namely transportation and electricity. The analysis uses average U.S. fleet mileage efficiency and average U.S. home power usage.

As shown, both the distance an automobile can drive and the time a house can be powered varies considerably with each of the different materials selected. Moreover, the cost associated with each of those materials is also quite variable.

Coal is by far the cheapest of the fuels, followed by natural gas. The energy content of a ton of corn (in its fermented form, ethanol) is higher in cost than crude oil. Note, this analysis does not include the effect of government subsidies, and it doesn't address the costs required to process each of these materials into the quality and form required for consumption. That cost of extracting the energy in a usable form is one of the reasons that the materials differ in the value of their raw form.

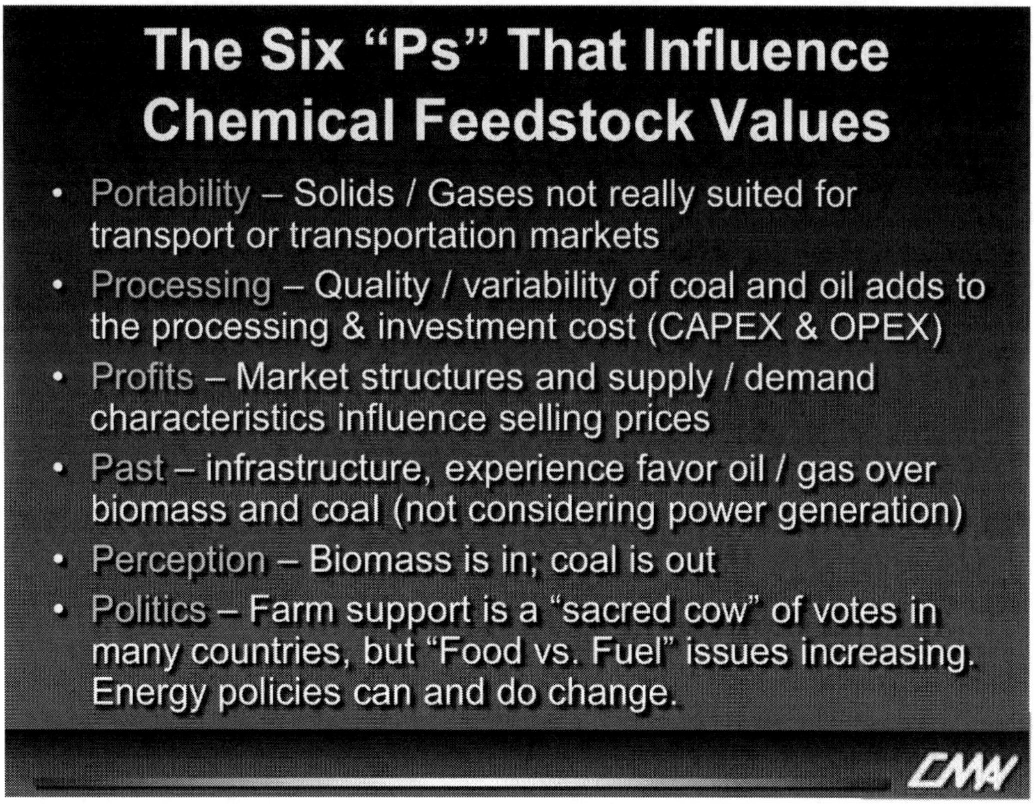

The actual use and cost of energy is predicated both on its fundamental intrinsic attributes, and the economic value of the material(s). The political and popular views of the impact of the energy on other important drivers such as the environment, risk in use, or impact on a country's economy also factor in.

An example that illustrates one aspect of this is a material's physical form. Since gases have a lower relative energy density than solid or liquids the transportation costs are higher per unit of energy contained. Solids do not easily flow, so likewise are more limited in the way in which they can be employed and cost more to transport.

Perception is also important. In the West, coal has historically been associated with a poor environmental footprint; while renewables, such as corn ethanol, have a more positive image (note this is not addressing whether the disparity in reputation should exist, just recognizing the reality that it does).

Figure 2

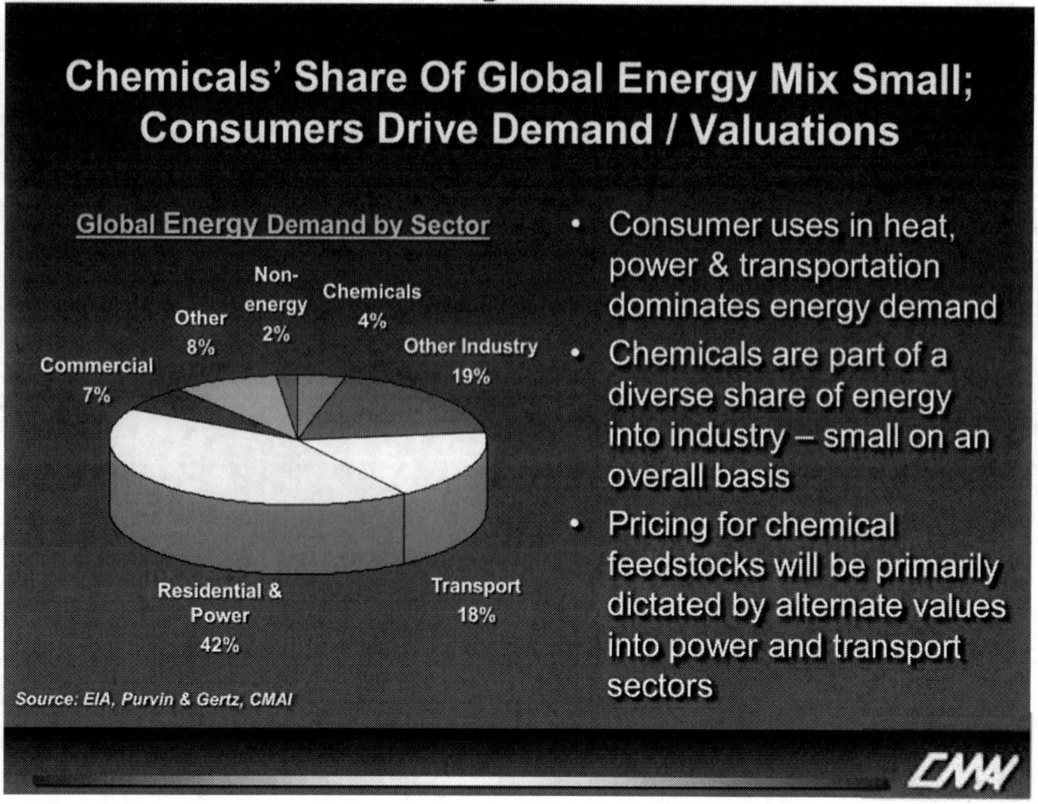

Consumer use in heat, power & transportation dominates the demand for energy. Chemicals represent only about 4% of overall energy demand. Pricing for chemical feedstocks will be primarily dictated by their alternate values into power and transportation sectors.

Figure 3

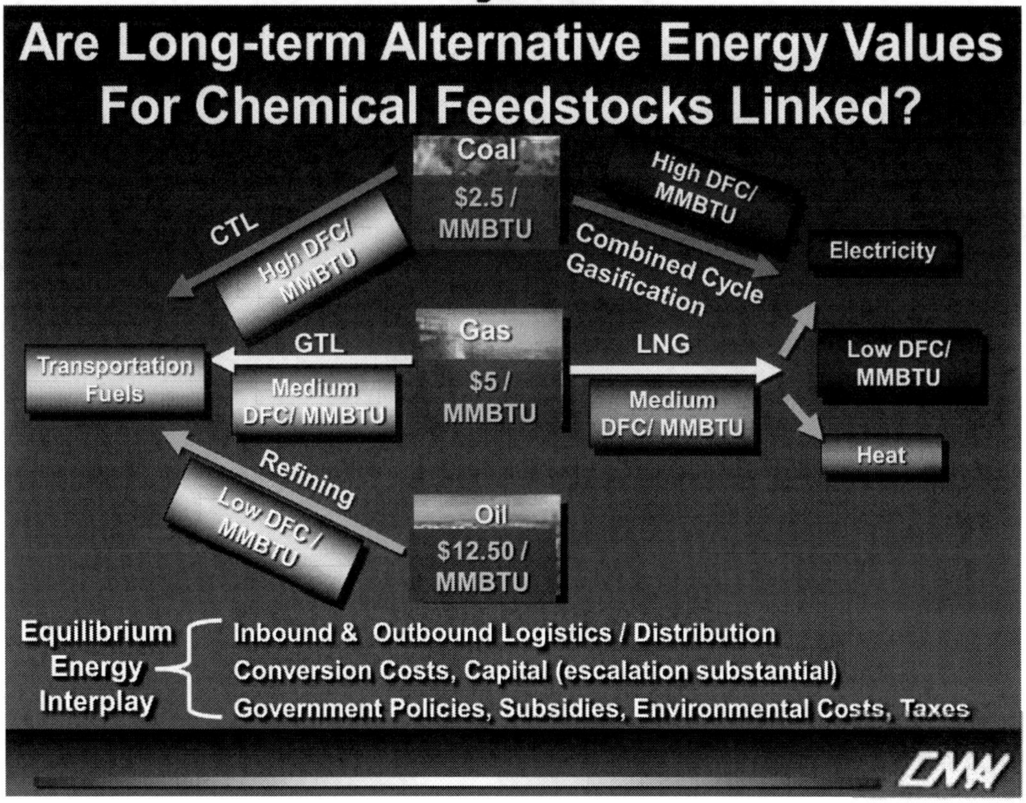

In many cases the value of chemical feedstocks is bounded by substitution between alternative energy uses. Technologies to convert between various energy forms have been in existence for some time. Lower CAPEX, logistics and operating costs and historically lower values for oil have prevented some of these technologies from becoming more widely practiced.

More recently, increasing scale and rising oil and gas values in regional markets have combined to increase the economic potential down the LNG / GTL / CTL / gasification chain, with more long-term potential for investment and energy inter-substitution. Recent Direct Fixed Capital escalation (DFC for coal and gas to liquid projects is generally high) and the bursting of the recent oil price bubble, along with uncertainty over how relative pricing will play out in the future, has slowed investment. OPEC's potential reaction to more substantive investment in non-oil routes to fuels is one over-arching risk to investment in these alternative pathways.

Figure 4

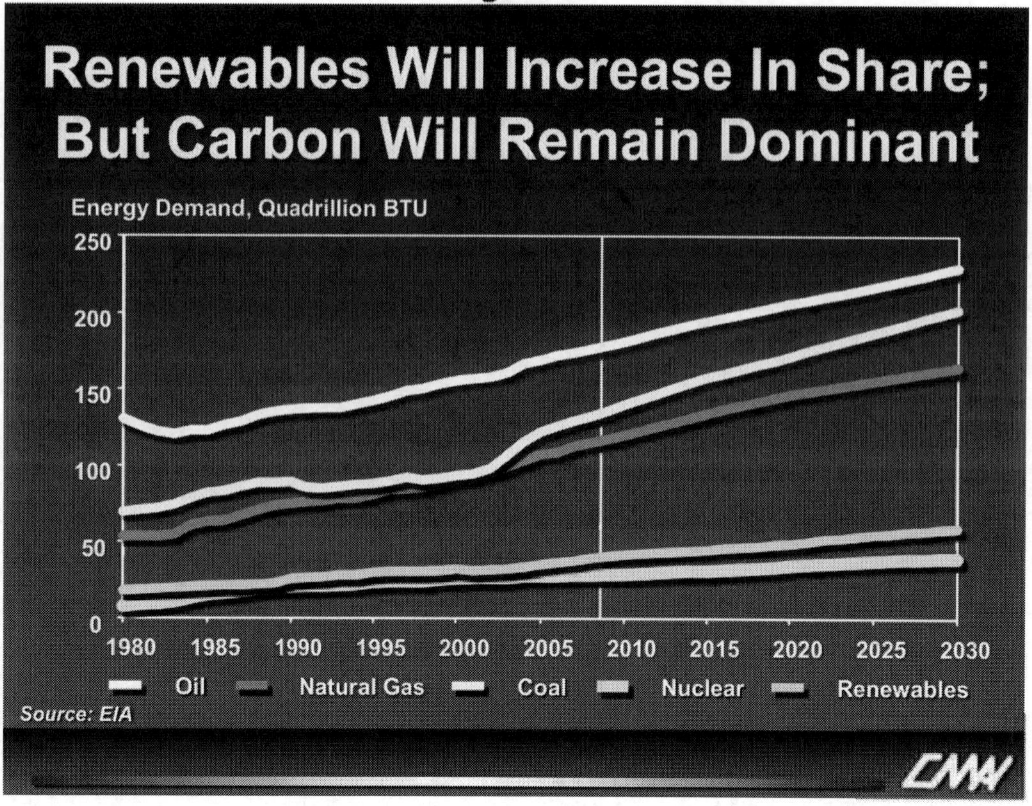

Hydrocarbons currently accounts for more than 85% of the global energy mix. Renewables (including biomass solar, wind and hydroelectric) constitute less than 10%. With expectations of reduced energy intensity (i.e. economies becoming more productive at using incremental energy); global energy demand is forecast to rise only about 2% per annum. Participation by the highly populous and emerging economies will drive this growth.

The future sources and exact breakdown of global energy demand is not only dependent on the actual rate of energy demand growth, but among other factors, relative costs, technology development and government policy. In reality, only establishing some general scenarios and trends is plausible.

These general trends are:
- carbon bearing fuels will continue to be the majority of energy utilized
- coal and natural gas are likely to increase their share
- renewables are likely to increase in share, but will remain a small fraction of the total .

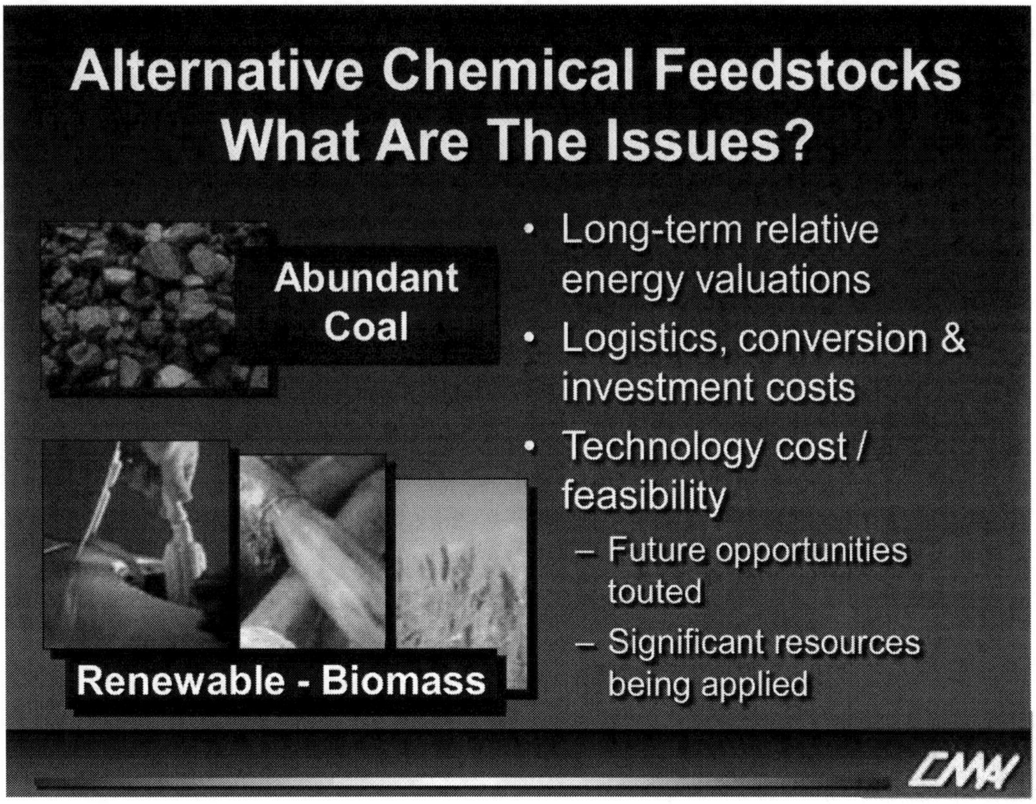

The issues for alternative chemical feedstocks are primarily these three:

1) Long-term relative energy valuations
2) Logistics, conversion & investment costs
3) Technology costs and feasibility

On this last point, significant resources are being applied, and progress is being made.

Figure 5

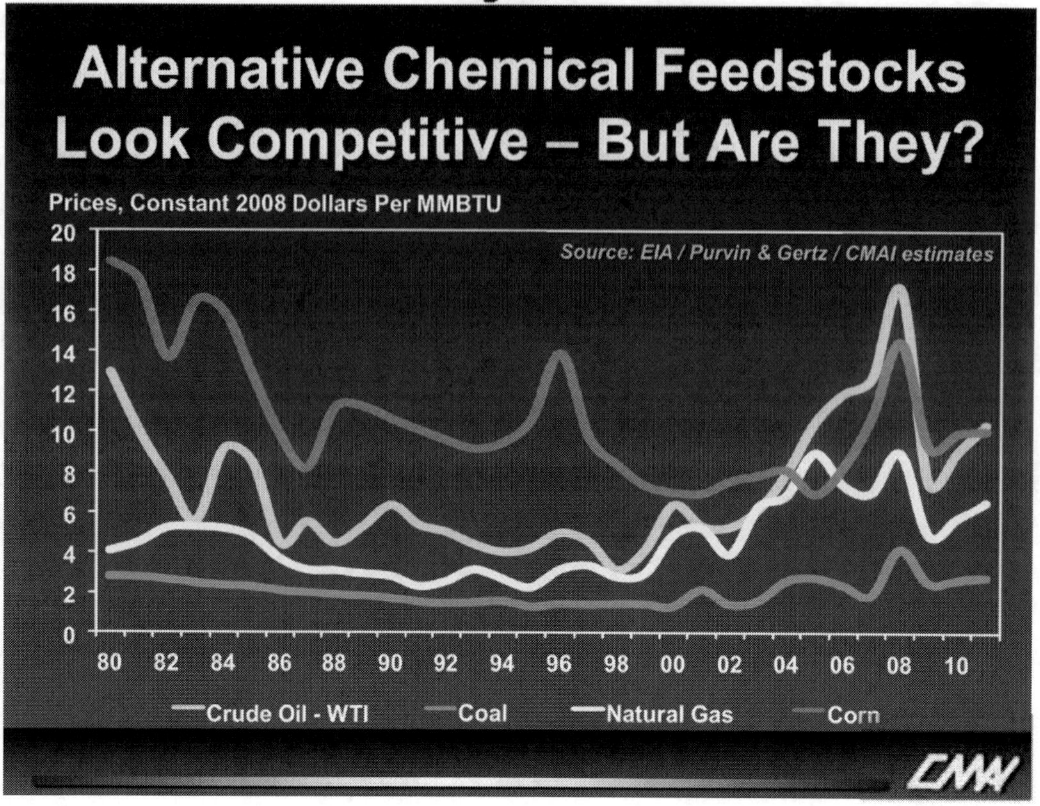

Costs for chemicals vary regionally. The primary driver is the cost of natural gas (for feedstock or electricity) and /or the cost of petroleum (naphtha, gasoline or gas-oil). The price setting regions for chemicals are typically based on petroleum feedstocks, though gas can and does influence them, especially regionally. As the graphic shows, in 1980 oil experienced a spike in prices much like the recent market increase, which generated initial support for the agriculture fuels programs of today – though today's interest is also driven by environmental issues.

Several additional points can be made by analyzing the relative price history:

(1) Oil / gas relative differentials closed, but have recently reopened
(2) Coal pricing has been relatively impervious to movements in oil and gas. Price stability is one of the appealing attributes for coal as a chemical feedstock.
(3) Coal pricing, in constant $2008 has been steadily declining until the recently, and is now much lower than oil and gas..
(4) Corn had become competitive with fossil fuels, but followed oil in its recent price rise, which generated a significant "Food versus Fuel" debate.

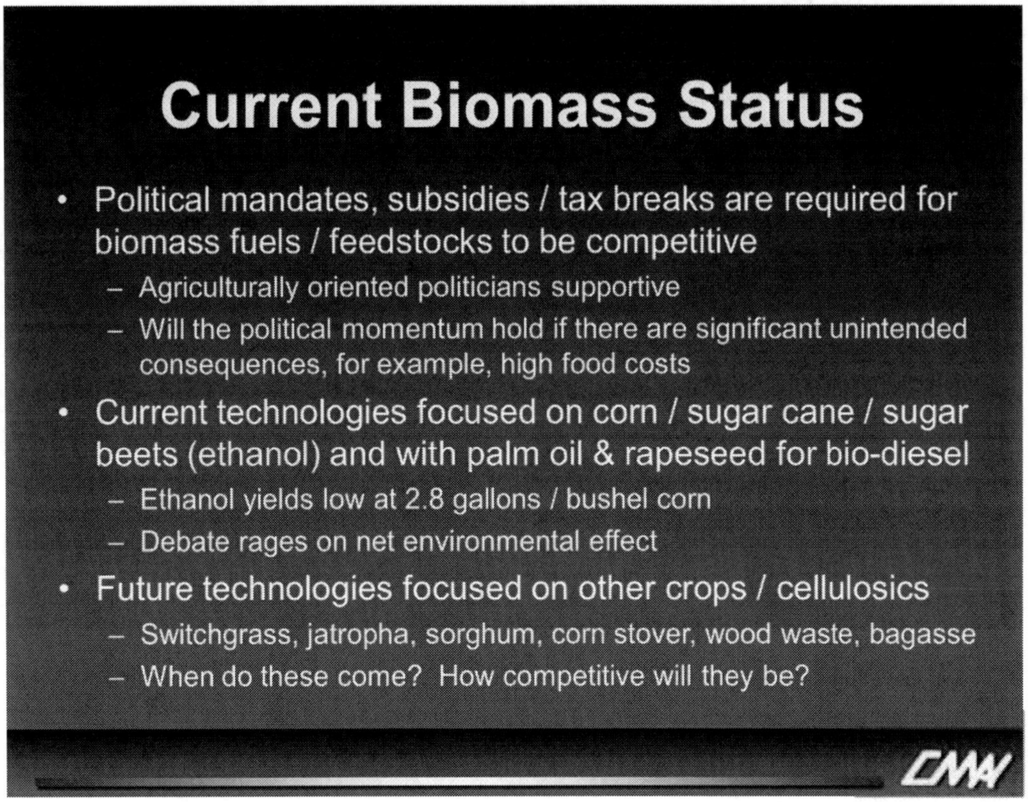

Political mandates, subsidies and tax breaks are required for biomass fuels and feedstocks to be competitive. Agriculturally oriented politicians are supportive, but will the political momentum hold if there are significant unintended consequences (for example, high food costs)?

Current technologies are focused on corn, sugar cane and sugar beets for ethanol, and on palm oil & rapeseed for bio-diesel. However, ethanol yields are low at 2.8 to 3 gallons / bushel of corn, and debate rages on the net environmental effect and net energy balance of ethanol.

Future technologies are focused on other crops and cellulosics, like switchgrass, jatropha, sorghum, corn stover, wood waste, bagasse, but when do these come and how competitive will they be?

Figure 6

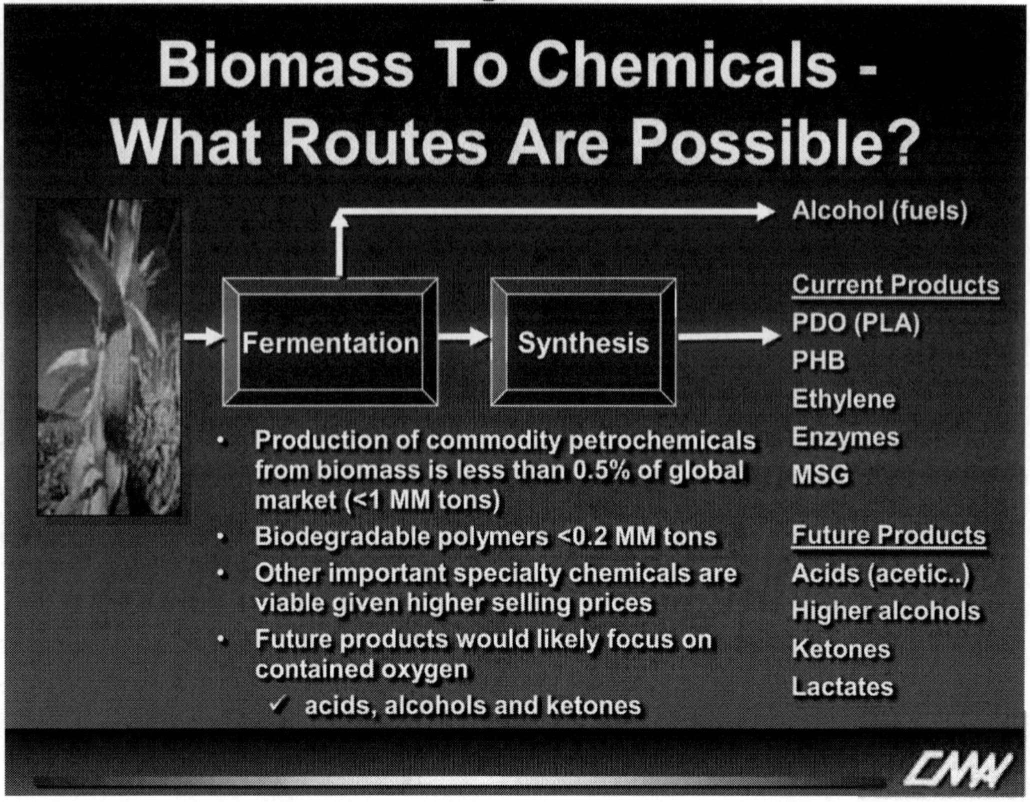

Production of commodity petrochemicals from biomass is currently less than 0.5% of the global market (<1 MM tons). Of that total, biodegradable polymers are <0.2 MM tons.

Some other important specialty chemicals are viable given higher selling prices, but future products would likely focus on contained oxygen, in acids, alcohols and ketones.

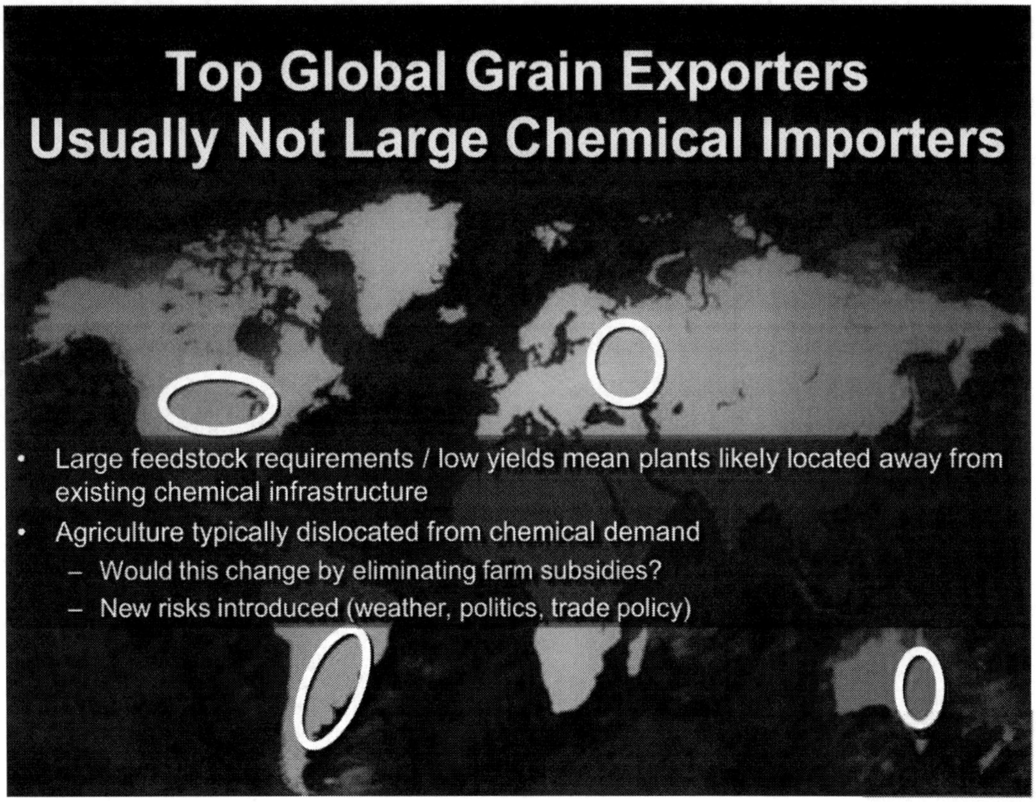

Because (currently) the yields are relatively low (with large amounts of biomass required per unit of output), to keep transportation costs low it is most likely that new facilities will be located close to the feedstock (crop) supply. In most cases, these are rural areas, and not near end consumers nor other chemical producers. This "de-clustering" will likely mean logistics cost penalties.

Except for the U.S. the top five grain exporters are either chemical exporters or only small net chemical importers and typically have small domestic markets With the current cost structure, exporting commodity biomass based material may be problematic.

There are also risks associated with connecting the energy and agricultural industries. Agriculture is one of the most controversial areas with respect to international trade and government policy. Policy change could dramatically impact the value proposition. Agricultural markets are highly unpredictable (weather, pests, crop substitution), with a corresponding risk to the cost structure.

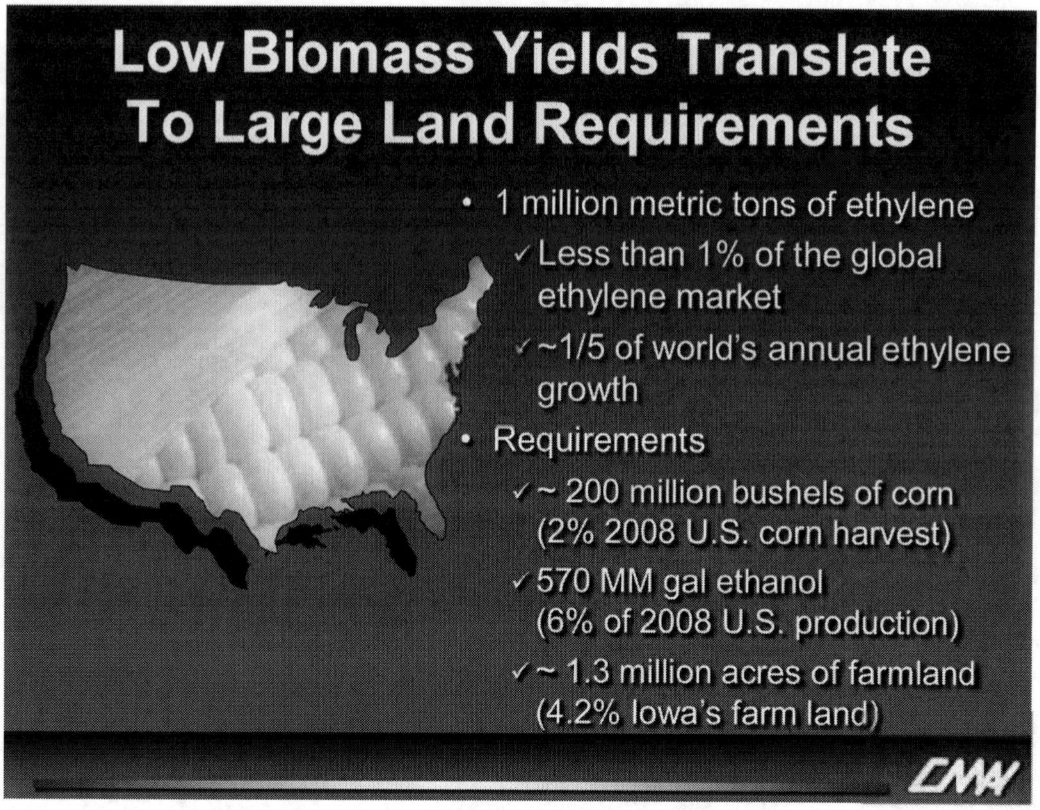

Low biomass yields translate into large land requirements for the production of chemicals.

One conventional ethylene cracker produces about a million metric tons of ethylene, which is less than 1% of the global ethylene market, and is only about 1/5 of the world's annual ethylene demand growth.

But this production volume requires:

- ~ 200 million bushels of corn (2% 2008 U.S. corn harvest)
- 570 MM gal ethanol (6% of 2008 U.S. production)
- ~ 1.3 million acres of farmland (4.2% Iowa's farm land)

Figure 7

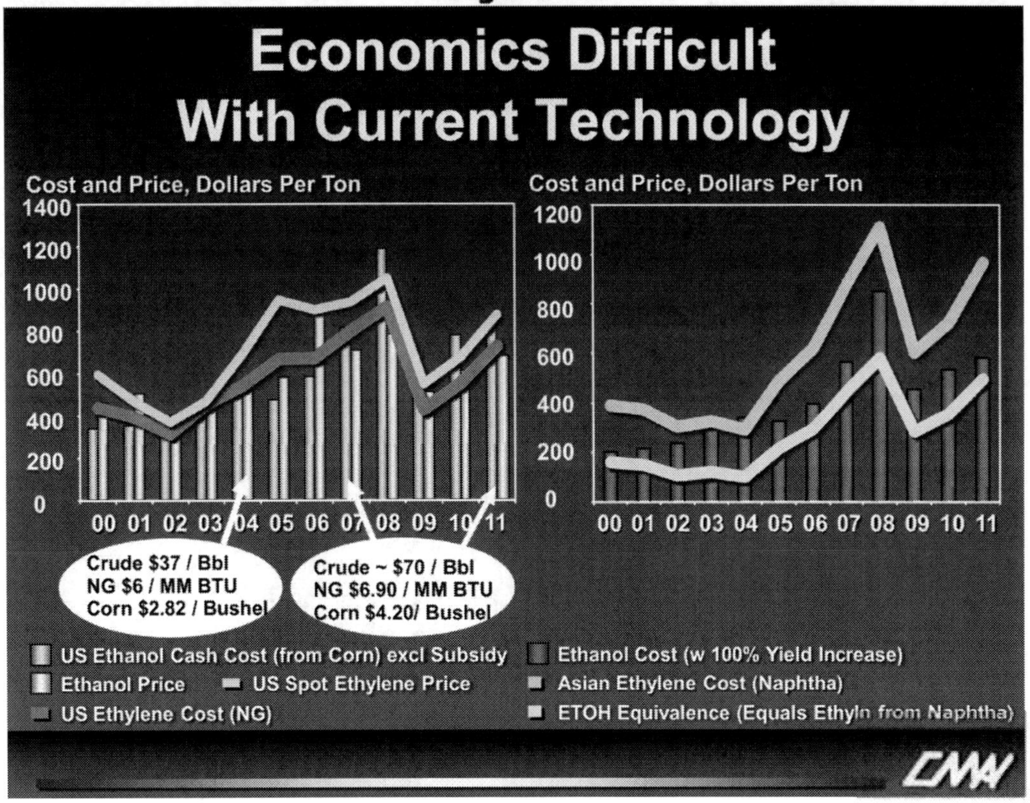

Ethylene is one of the chemicals often quoted as a possible biomass derivative. Dehydration of ethanol is an energy intensive process creating significant water by-product.

While ethanol dehydration is commercially practiced, hydration of ethylene to ethanol is also commercial. In fact, more ethanol is made from ethylene than vice versa. As the graph on the left shows, when natural gas and ethylene markets were weak (in 2001) the U.S. spot ethylene price was below the ethanol price. More recently, when natural gas and ethylene markets tightened, ethylene hydration economics suffered. Ethanol's rapid escalation in 2006 through 2008 with higher corn and fuel use closed the gap, but insufficiently.

The graph on the right compares the required ethanol price (blue line) needed to be competitive with naphtha based economics, assuming a theoretical reduction in corn costs by half (due, for example, to lower market prices for corn or improved yields). As shown, to be competitive at naphtha based economics at $40 crude, ethanol prices would have to move to near ten year lows, although with the higher oil forecasts of $70-$80/BBL after 2011 positive margins might be possible, if ethanol yields improve and government subsidies continue. Sugar based ethanol processes might be viable as well.

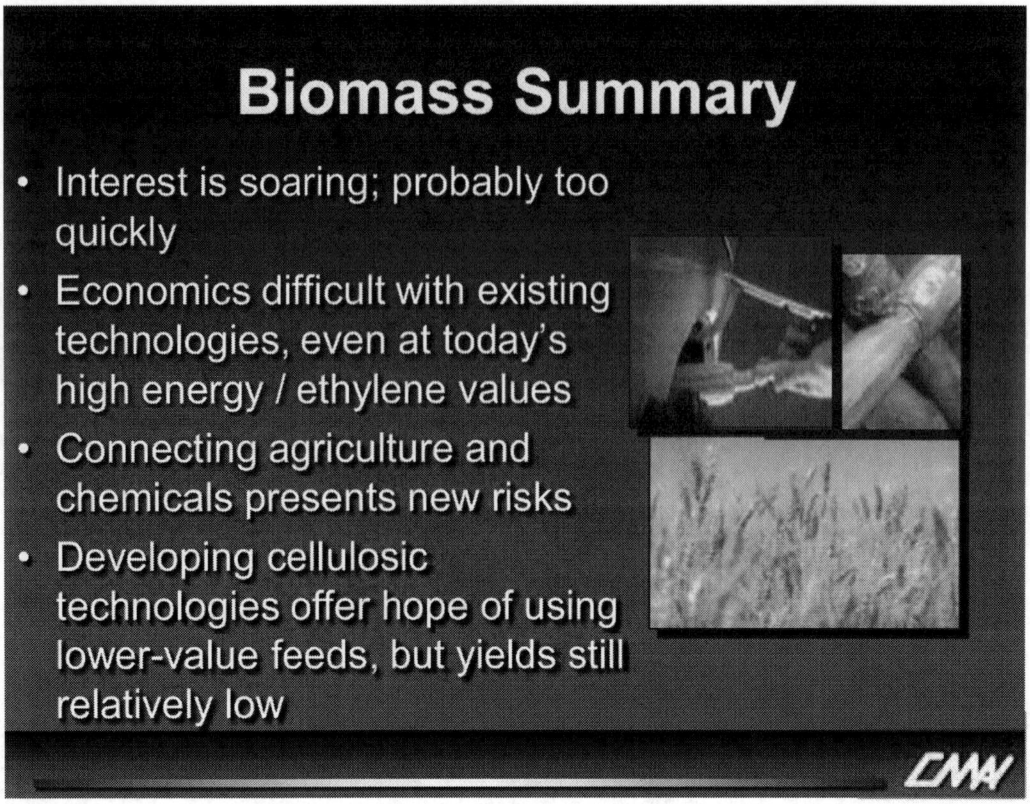

In summary, even with its problems, interest has been soaring in biomass, probably too quickly. Economics are difficult with existing technologies, even at the recent high energy and ethylene values. Developing cellulosic technologies offer the hope of using lower-value feeds, but yields are still relatively low. An enzyme breakthrough is needed, and they currently require political mandates and subsidies for positive economics.

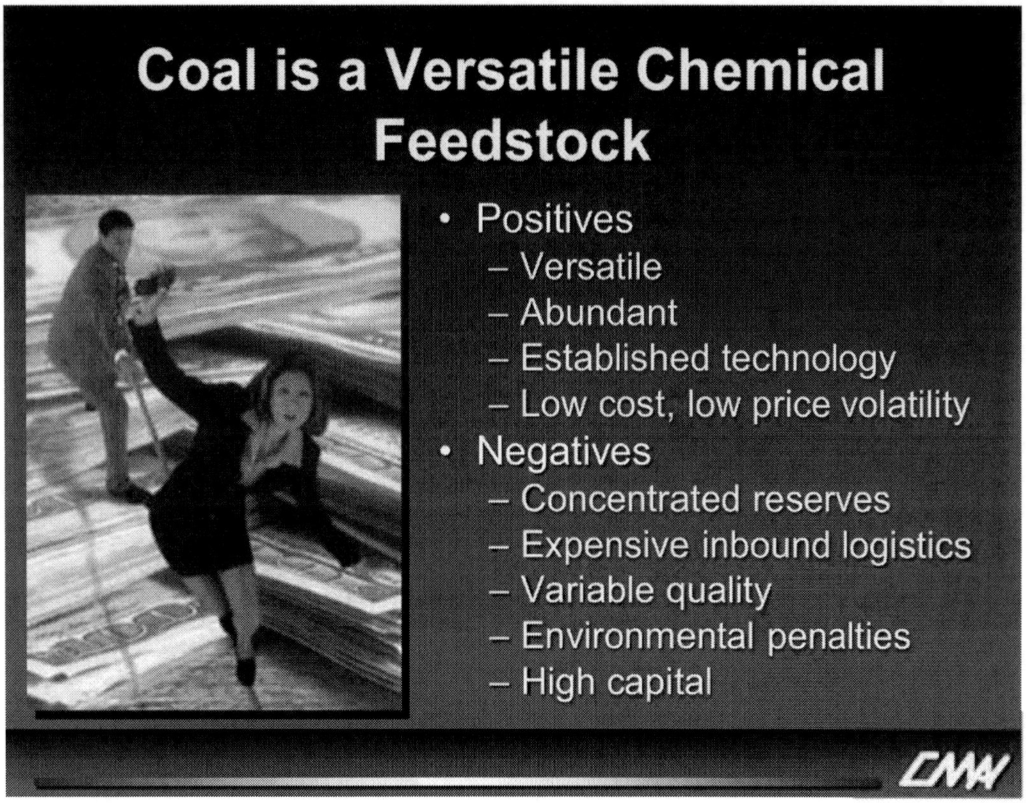

Coal is one of the oldest and most established energy sources. It is a versatile fuel, with proven, improving technologies allowing its intrinsic energy to be used in a number of areas. It is best suited to the generation of electricity. Coal is abundant with an estimated 200 years of globally proven reserves. Pricing has been tame, both on an absolute and relative basis.

Reserves are concentrated in a few countries. Transporting coal is expensive; rail operators have historically captured a significant portion of the delivered value. Coal quality and energy content are highly variable. Higher grades are used in coke production; lower grades may even be unsuitable for power and offer additional transportation challenges.

Because coal is an established fuel, most facilities were built with lower environmental standards. Coal mining was historically a visible environmental blight with a poor safety record. Coal generates high levels of solid waste, has lower energy production efficiencies, and produces more CO2. New more efficient abatement technologies are now available; but coal's reputation suffers from its legacy.

Figure 8

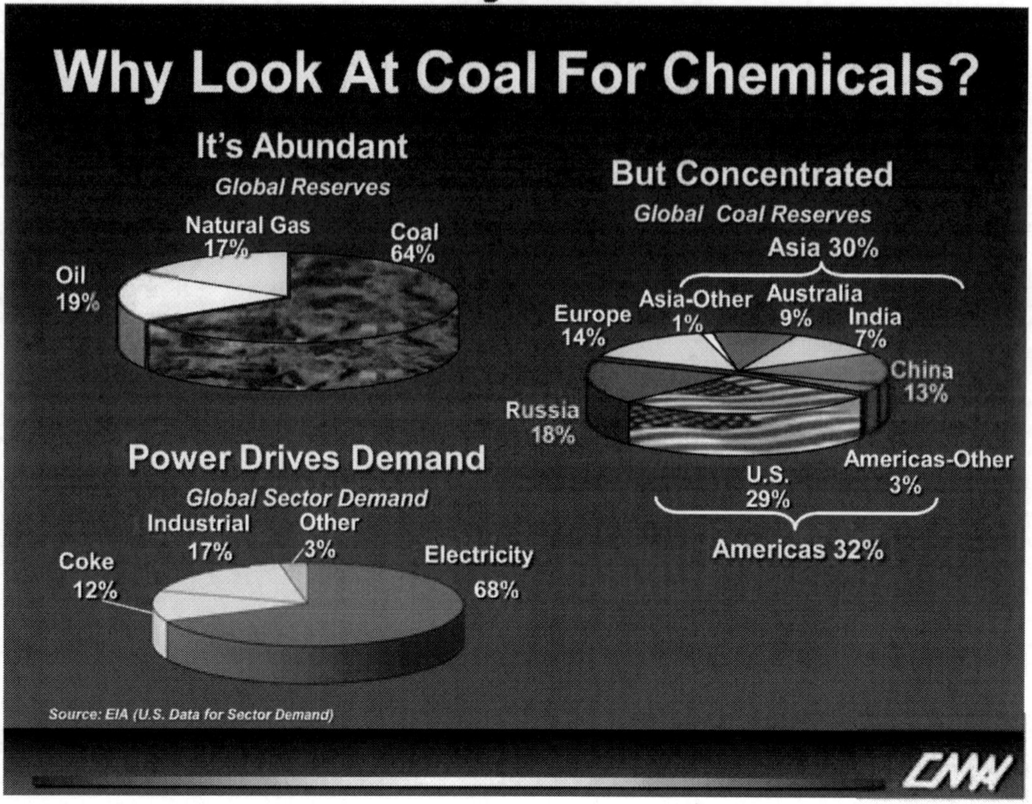

On an equivalent BTU basis, coal reserves are approximately four times higher than either natural gas or oil. Because of this, it is increasingly likely that coal will become a larger piece of the overall energy matrix.

However, similar to oil, the largest coal reserves are concentrated in only a handful of countries. The largest of these is the U.S., where much of the supply is controlled by the government as ~60% of the reserves exist on public land. Russia, China, India and Australia also have large reserves. China and India both have plans to exploit their reserves, especially given they are gas and liquid energy source deficient.

In most cases, coal's primarily application will be in the electricity segment as coal is most suited to conversion to power. However, a number of new projects to convert coal to chemicals and even synthetic crude or petroleum fractions are currently planned or under construction.

Figure 9

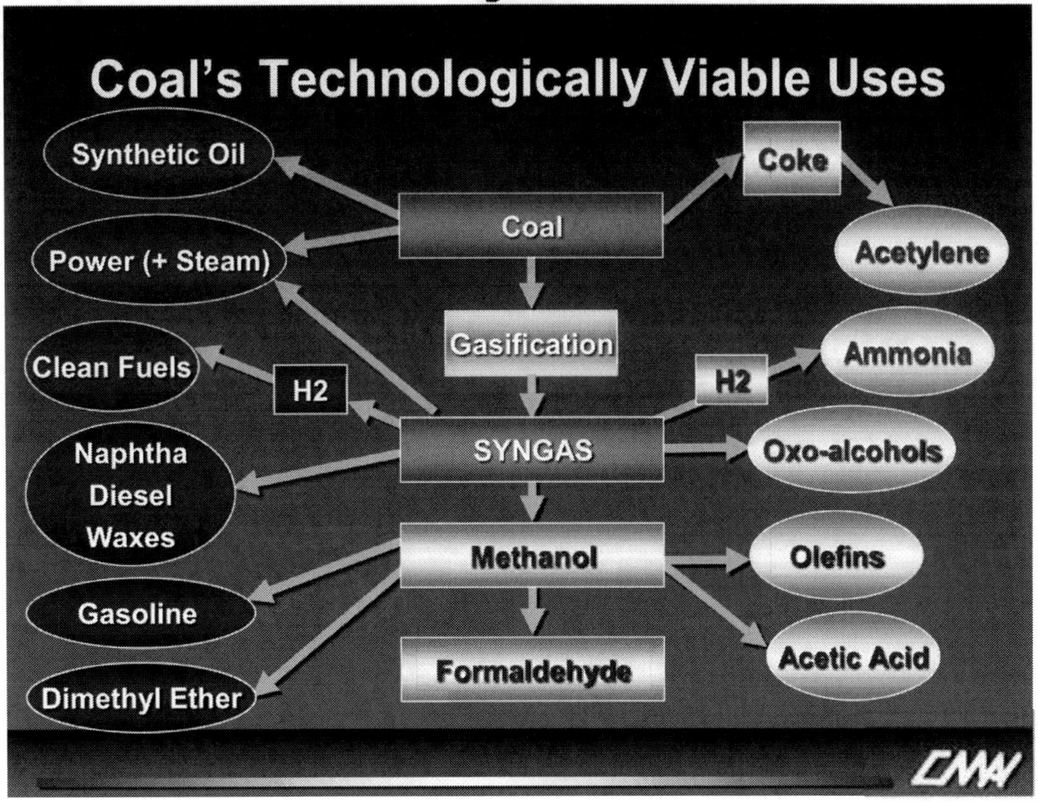

Coal is an extremely versatile material. It can be directly combusted to produce power and - this is currently its largest end use. It can be liquefied to synthetic petroleum or it can be anaerobically transformed to coke for smelting and calcium carbide. Calcium carbide can then be subsequently processed for production of acetylene, which competes against ethylene in the production of Vinyls.

Gasification / partial oxidation of coal is a technology growing in stature. Gasification can be used to generate synthesis gas (SYNGAS) – a mixture of hydrogen and carbon monoxide. SYNGAS has a high energy value and is a significant intermediate for production of clean synthetic fuel blend-stocks such as naphtha and diesel. Other chemicals manufactured from SYNGAS include oxo-alcohols, ammonia and methanol. Methanol can be used for production of gasoline, DME (a diesel substitute) and as co-feed for bio-diesel. Methanol is also a large established chemical intermediate, used in the production of formaldehyde and acetic acid.

Figure 10

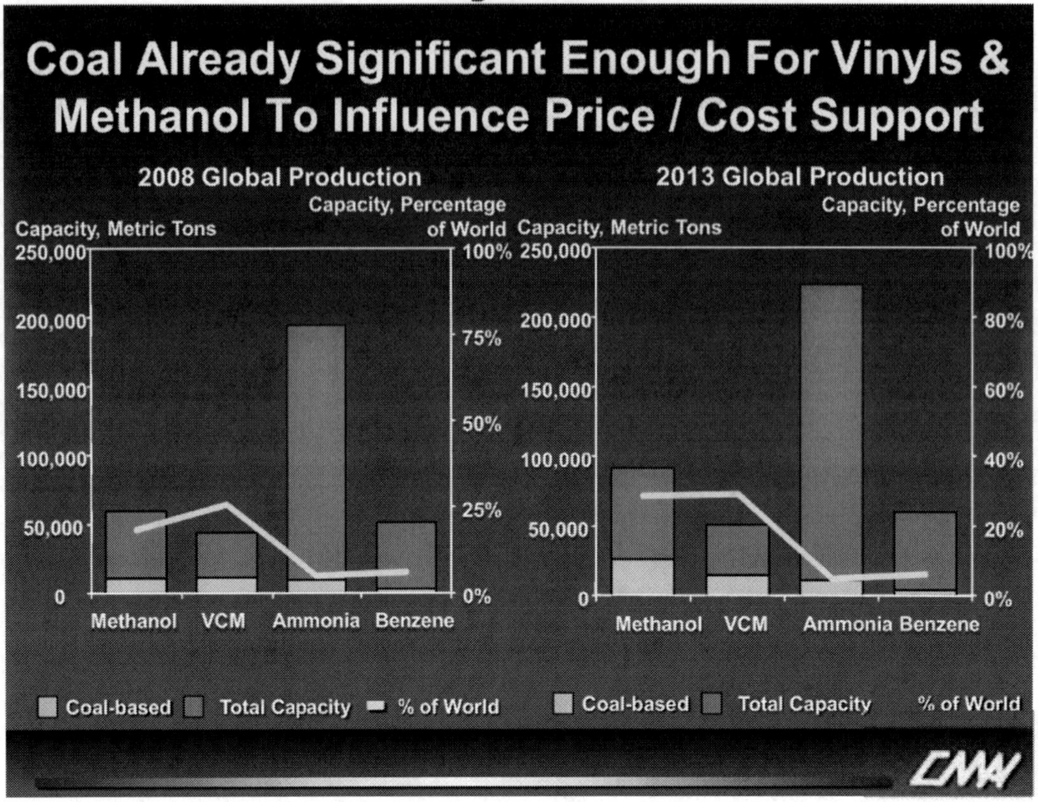

Chinese investment in coal based production of methanol (via gasification) and vinyls (via carbide / acetylene) is growing substantially. Growth in these products in China has greatly influenced Chinese PVC balances. Direct blending of methanol into fuel and strong growth in solvent and other uses has attracted significant investment. These smaller facilities (some integrated into coal, coke or other derivatives) were the result of unregulated growth in the rural provinces supported by measures to help rural growth initiatives. New, world-scale facilities coupled with derivative olefin / polyolefin are now also under construction.

Other chemicals based on coal are small and not very significant to the market. In the case of aromatics, the coal tars provide a feedstock from which higher valued co-products can come, but these are small. Ammonia from coal via SYNGAS is also practiced, but these facilities are generally focusing on the Chinese domestic market and represent a small share of overall world capacity.

Figure 11

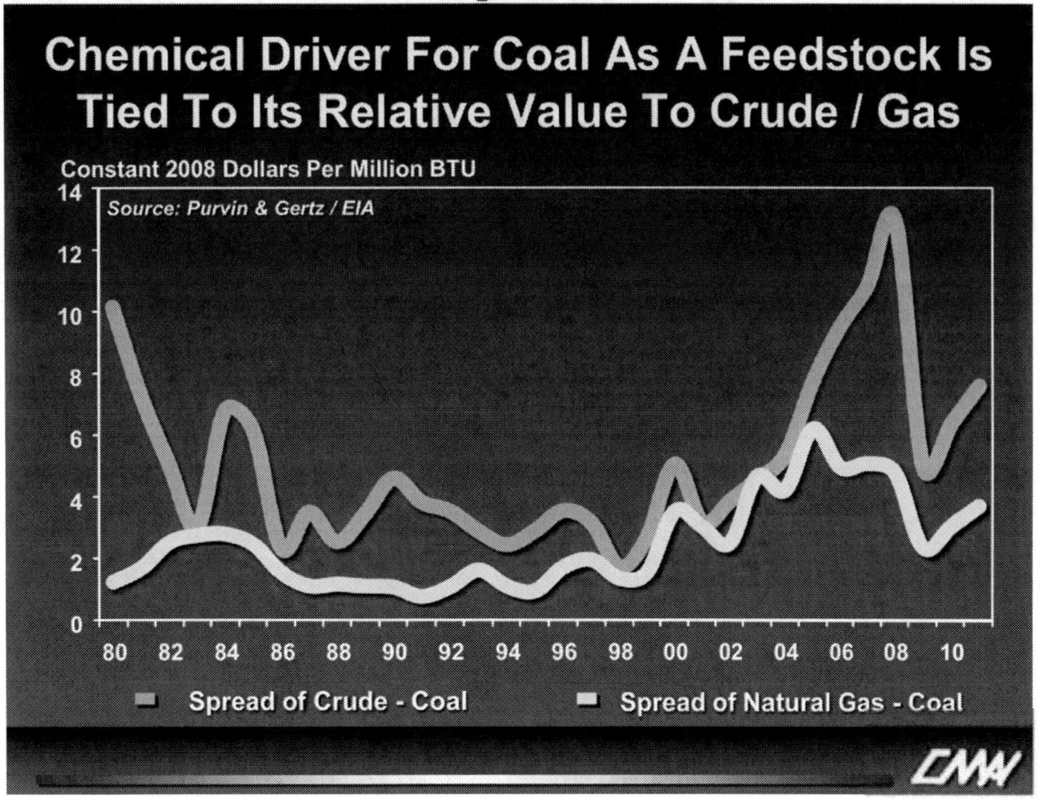

This graphic shows the constant 2008$ spread between gas and coal, and between oil and coal on a $ per MM BTU basis. As shown, the historical spread between natural gas and coal has been quite low. This was true even during the run up of oil in the early 1980s. While these trends have eased from recent lofty highs, they are expected to return going forward for those countries (such as China and the U.S.) that require imported LNG to meet demand.

While oil's relative value to coal was at these levels previously, spreads have not been this high since the early 1980s. Spreads are forecast to be at elevated levels in the future as well.

These economic drivers are important to those considering coal as a source of energy for chemicals, fuel or power, and to making decisions on long-term sources of feed.

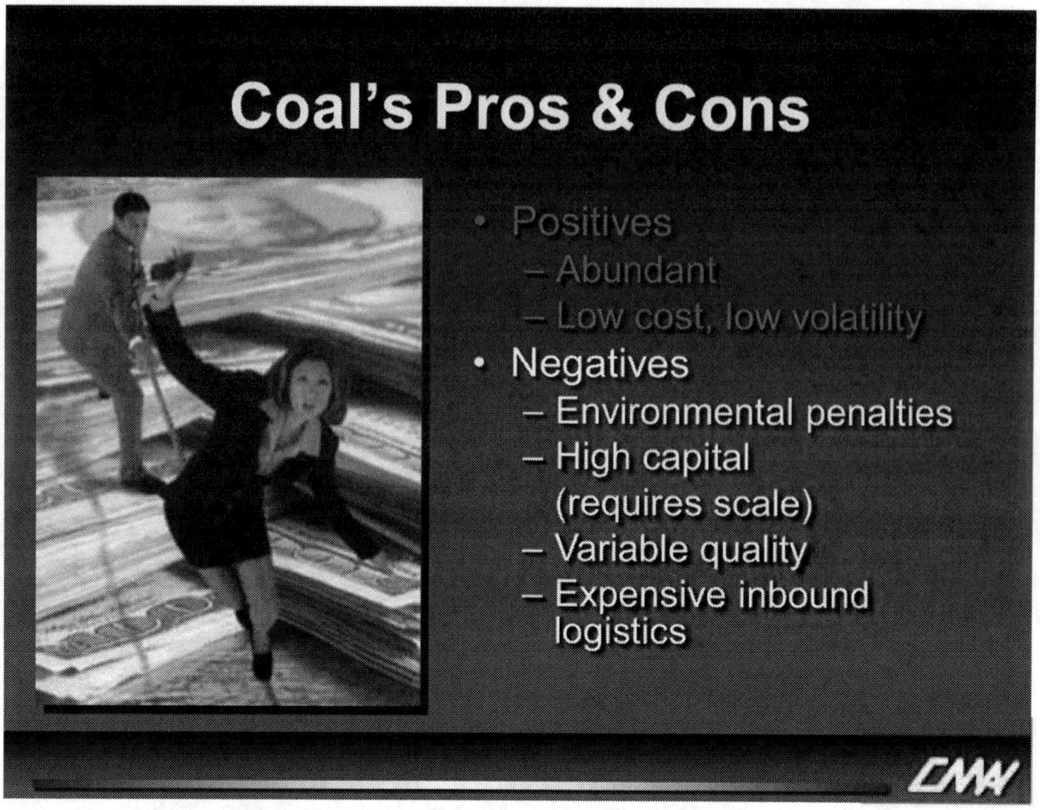

In addition to coal's positive characteristics, which include availability, low cost, and low price volatility, it has some negative ones as well, including its environmental profile, high investment requirements, variable quality, and relatively high logistics costs.

Figure 12

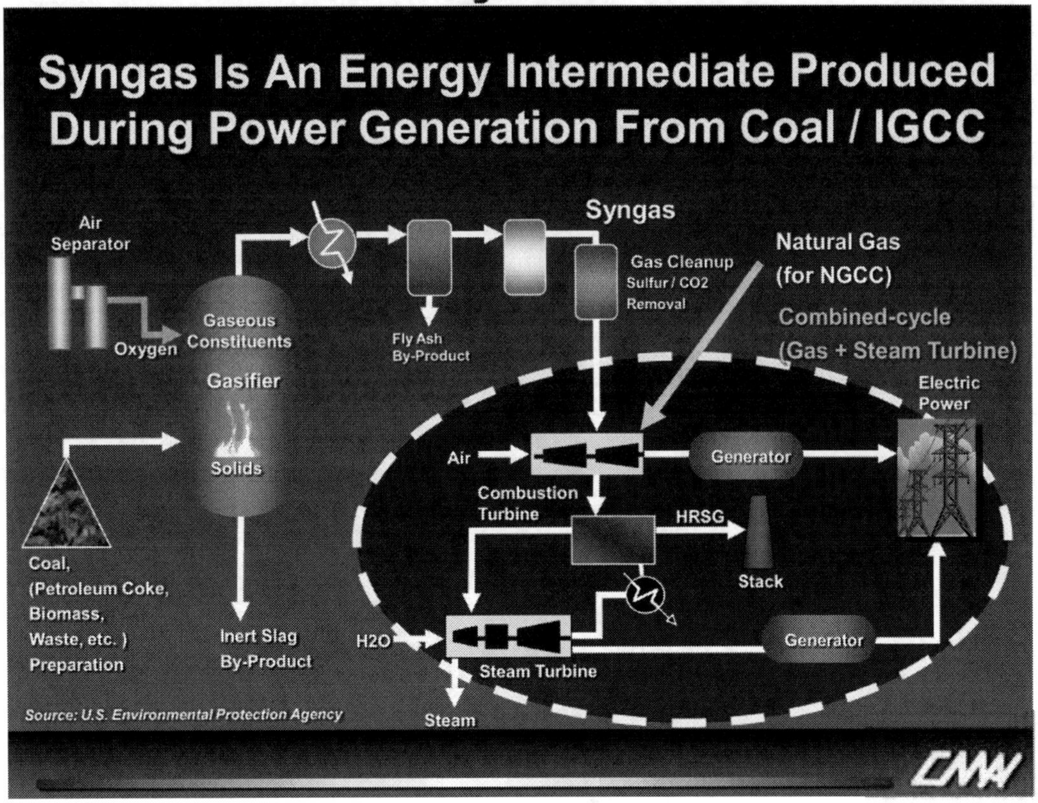

- Coal's use for power is either through direct combustion of pulverized coal or gasification. More recently gasification's (partial oxidation of coal) inherent advantages with respect to efficiency, particulates production and reduced effluent gas treatment have made it preferred. Gasification uses about 1/10 the oxygen vs combustion. As a result, the effluent gases are more concentrated in CO_2 and there is much less volume to treat. A typical gasifier flow is shown in the graphic. Oxygen is used to gasify coal in an oxygen-starved environment. The resultant SYNGAS mixture is cleaned and a water-gas shift reactor is used (if desired) to maximize H_2 before being sent to power generation (or alternatively, chemicals).

- Downstream of the gasifier, the operation is very similar to that of natural gas combined-cycle electricity generation. The use of a gasifier almost doubles the capital cost. However, natural gas combustion turbines are higher in efficiency, though fuel costs are higher and clean-up of CO_2 is more difficult.

Table 1

Coal's Competition With Natural Gas In Power Generation Via IGCC

Technology / Metric	PC (Pulverized Coal)	NGCC (Natural Gas Combined Cycle)	IGCC (Integrated Gasification Combined Cycle)
CAPEX	3X	1X	3X
Efficiency	35%	50%-60%	40%
Solid Waste	Base (Fly-ash)	NIL	+50% (slag)
CO_2	Base – Low conc	0.4x – Low conc	0.9x – High Conc
On-stream, %	70-80	85	70-80 Depends on Redundancy
Future CO_2 Recovery Cost	1X	1X	0.25X

The table on the left summarizes a comparison of coal in direct combustion vs gasification and natural gas. As shown, the largest differentials exist in capital cost and efficiency. With coal combustion, generally natural gas is favored – at fuel cost parity. This is the situation that existed in the U.S. through most of the 1990s and led to a large build-up of natural gas combined-cycle turbines (many have since gone bankrupt).

Outside of the U.S., those countries that were choosing between coal and natural gas, unless they had a very cheap domestic supply of gas, generally chose coal – though not necessarily via gasification. The decision to go with gasification over conventional combustion has to do primarily with the environmental issues around SOx / NOx and future expectations of required CO2 sequestration. Recovery of CO2 from gasification is much less expensive (both CAPEX and OPEX) so that impacts decisions most prominently in countries with higher environmental standards.

Figure 13

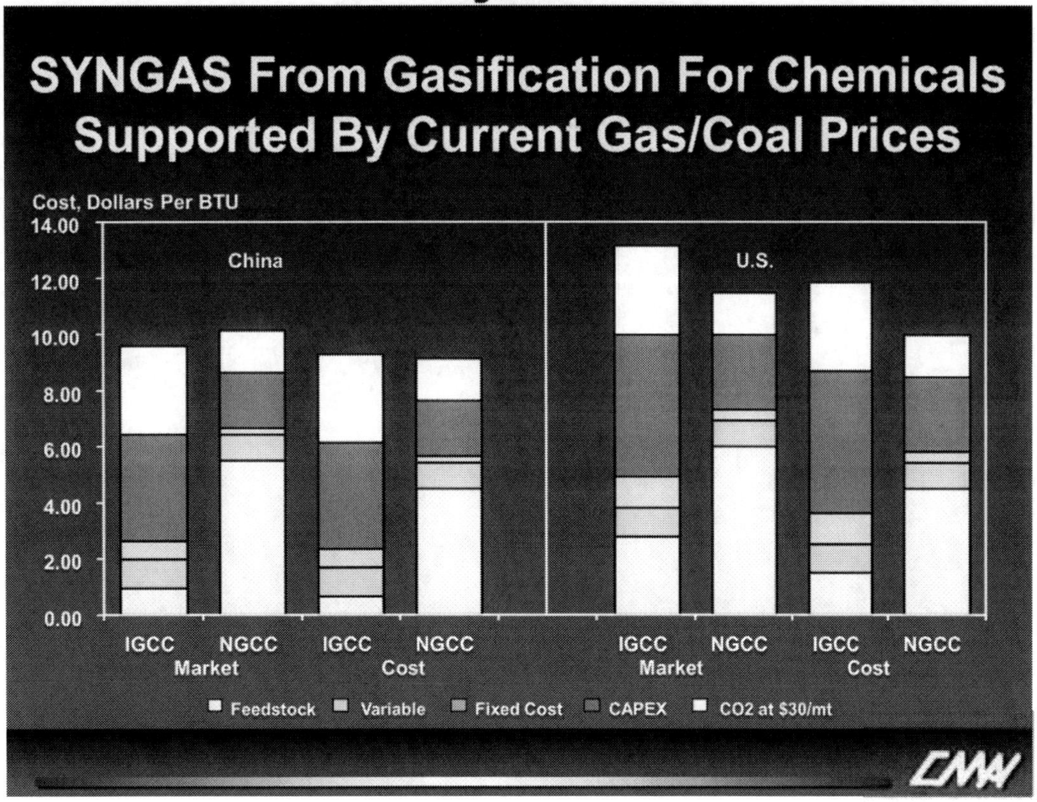

This analysis compares the economics of electricity generation via coal gasification vs natural gas when both are integrated into a combined cycle system. Assuming the same values for by-product steam in both cases, the economics show that the variable cost of power generation is much lower for coal. In China, where coal mine-mouth costs can approach $10.00 per ton ($0.40 per MMBTU) for power-grade coal, and CAPEX escalation has been more contained, gasification is a clear winner. In the U.S. the total return differences are not as apparent – though with the forward forecast of gas at $6 per MMBTU, coal is extremely close.

Offsets against the cash cost differences are primarily associated with the higher CAPEX and the "assumed" cap and trade carbon tax ($30 per metric ton CO2). These taxes are in place in the EU and have not yet "crossed the pond" on a federal basis, but they are expected to be implemented in the US and possibly even China. This analysis penalizes IGCC to some extent because while coal does have higher CO2 emissions, the incremental capital and OPEX required to recover the CO2 (and thus prevent paying the CO2 costs shown here) is much lower for IGCC vs NGCC. (Note: the capital here is shown without CO2 recovery).

Figure 14

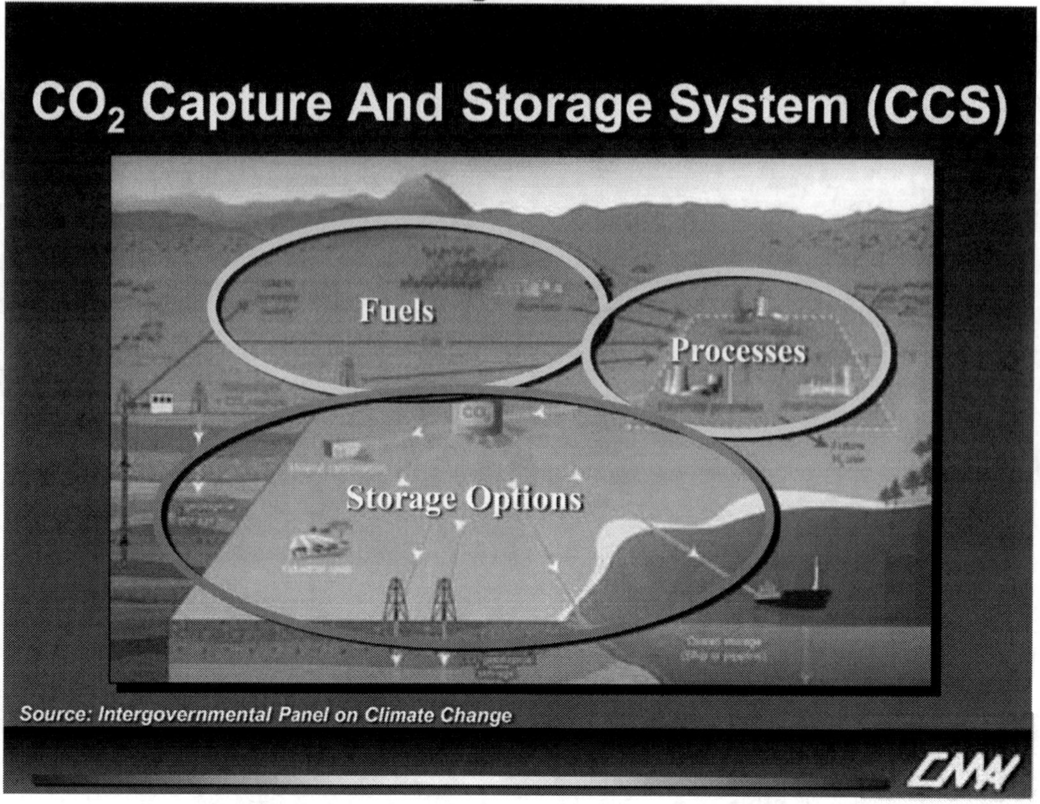

CO2 minimization via alternative fuel use, improved process efficiency, and storage options that sequester the CO2 are the focus of many government / research programs. Fuels can be coal, gas and oil, but also biomass which may lead to reduced or negative emissions.

Processes where CO2 is produced from these fuels is mainly electricity generation, but CO2 can also be captured from petrochemical plants, cement or steel plants.

Geological storage can be done in oil or gas fields, saline formations or in coal beds. Ocean storage can be done by direct injection in the water column, or by depositing the CO2 on the sea floor where it will be isolated from the atmosphere longer. Mineral carbonation is, in principle, also possible. However it is still expensive both in terms of costs and of energy use. The CO2 can also be used in industrial facilities, but the potential for CO2 reduction for this is small.

Figure 15

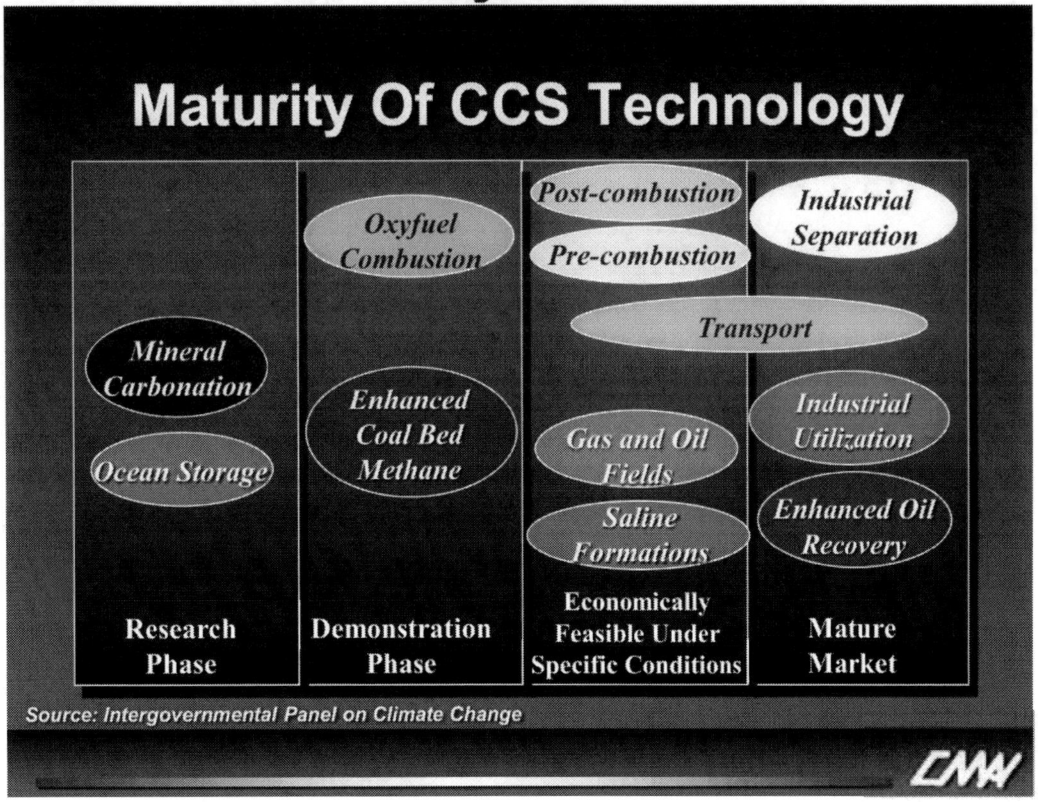

In this chart, each oval is placed in the sector representing the highest level of maturity currently seen for that component. Components of Carbon Capture and Storage are in various stages of development. Complete CCS systems can be assembled from existing technologies that are mature or economically feasible under specific conditions, although the state of development of the overall system may be less than some of its separate components. There is relatively little experience in combining CO2 capture, transport and storage into a fully integrated commercial size CCS system. The utilization of CCS for large-scale power plants (the potential application of major interest) still remains to be implemented.

Note that Enhanced Oil Recovery (EOR) is pictured as "mature market", but this is only true for EOR as a technology for enhanced recovery of oil. If EOR is considered as a CO2 abatement option, it is only "economically feasible under specific conditions".

Table 2

CCS Component Costs
Dollars Per Metric Ton CO_2

Capture		Transportation	Storage		Total *
Power Plant	\$15 - \$75	\$1 - \$8 per ton (250 Km)	Geological	0.5 - 8.0	6.5 - 225.0
Gas/NH_3	\$5 - \$55		Ocean	5.0 - 30.0	
Other	\$25 - \$115		Mineral	50.0 - 100.0	

Source: Intergovernmental Panel on Climate Change
** Costs may not be additive, depending on situation*

Costs for the various components of a CCS system vary widely, depending on the reference plant and the wide range in CO2 source, transport and storage situations. In most CCS systems, the cost of capture (including compression) is the largest cost component. Transportation of CO2 over 250 km costs between 1 and 8 \$/mtCO2.

Geological storage costs vary between \$0.50 and US\$8 per metric ton stored and exclude monitoring costs (< \$0.30 /mtCO2). The costs for ocean storage are very uncertain at this point, but are estimated to vary between \$5 and \$30 /mtCO2. Mineral carbonation costs are still very high.

Over the next decade, the cost of capture could be reduced by 20–30%, and more should be achievable by new technologies that are still in the research or demonstration phase. The costs of transport and storage of CO2 is expected to decrease more slowly as the technology matures further and the scale increases.

Figure 16

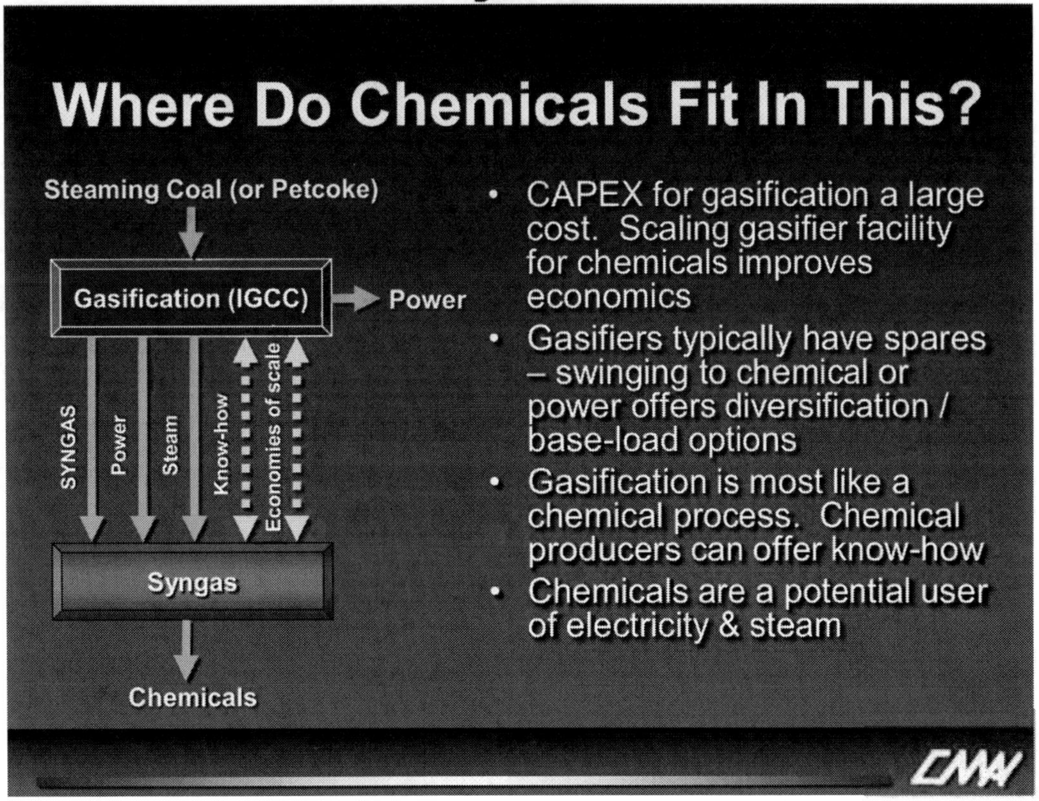

The CAPEX for gasification is a large cost. Scaling gasifier facilities to include chemicals improves their economics. Gasifiers typically have spares, so swinging to chemical or power production offers diversification and optimization of production.

In addition, gasification is more like a chemical process, and chemical producers can offer valuable know-how. Finally, chemical processes are a potential user of both electricity & steam.

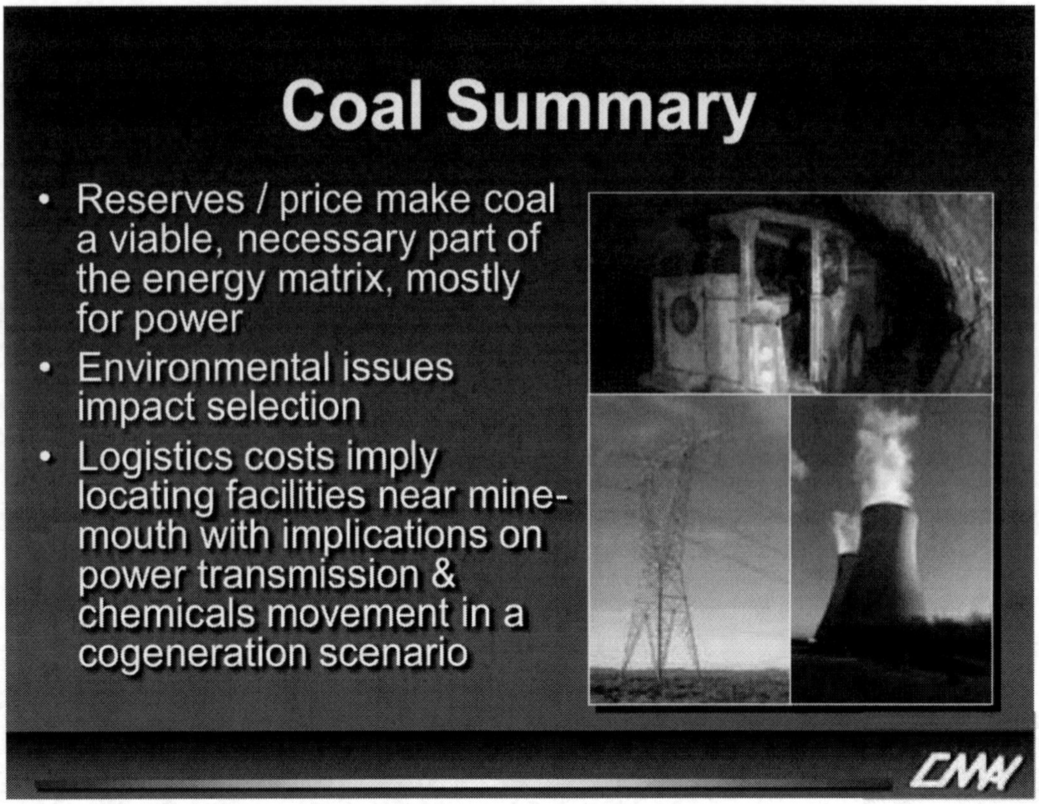

Coal's large availability and price characteristics make it a viable necessary part of the energy matrix, mostly for electricity production, but environmental issues affect its selection. Coal gasification especially shines when C02 recovery is required.

Logistics costs imply locating facilities near the mine-mouth, with implications on power transmission & chemicals movement in a cogeneration scenario.

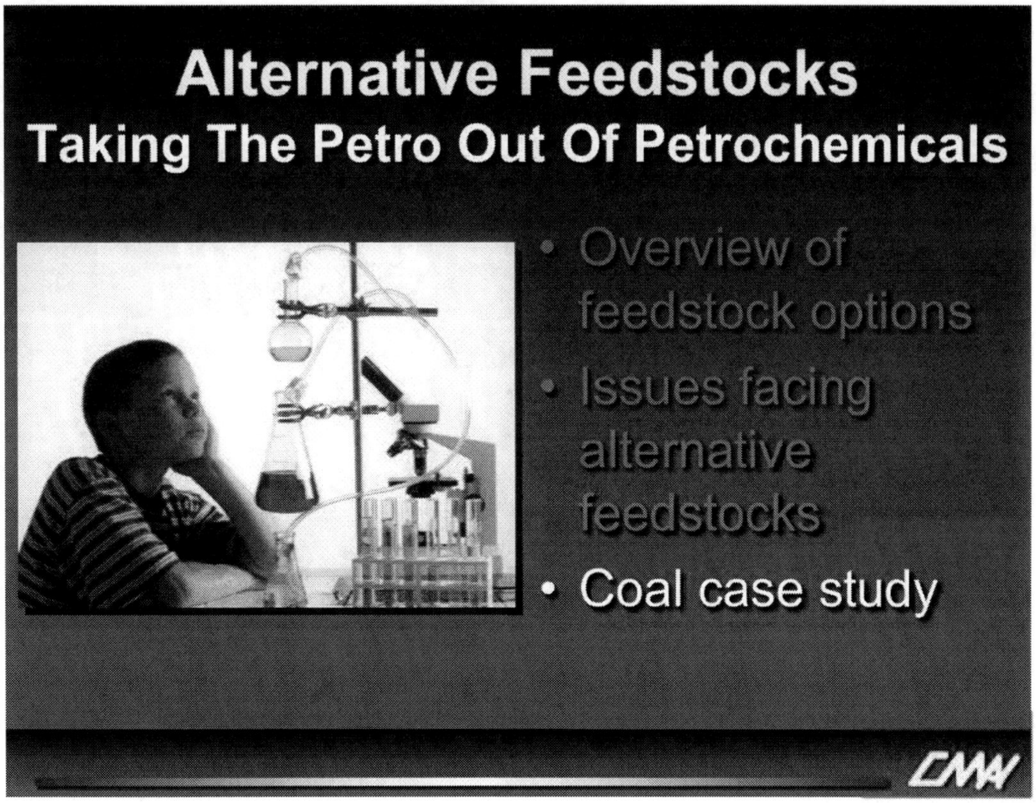

The following slides analyze and demonstrate the competitiveness and some of the issues associated with converting coal to chemicals. To bring out these points the analysis focuses on three different routes and locations to produce olefins and polyolefins for delivery into China.

These selections were made because olefins represent one of the largest and most important of the petrochemical building blocks and we believe represents a real-world scenario. In addition, technologies to convert coal into chemicals are in place with new processes being commercialized, and there is substantial interest in further investment once the technology risks have been minimized. It should be noted the numbers presented in here are generic and based on publicly available information. The results should be considered on a relative basis, rather than an absolute basis. Even so, important insights can be gleaned.

Figure 17

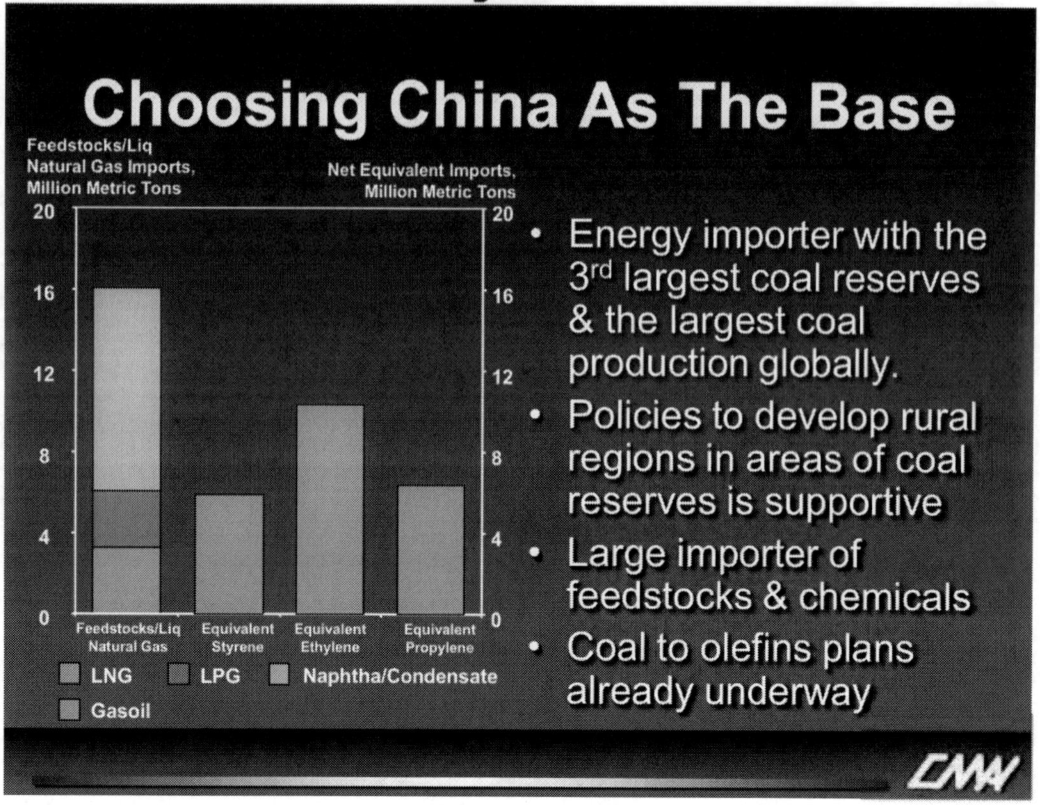

The reasons for choosing China as the basis for this study are simple:

China is the world's largest coal producer and has the third largest coal reserves globally. It is also a large importer of energy, feedstocks & chemicals. Government policies to develop rural regions in area of coal reserves are supportive. Finally, their coal to olefins plans are already underway.

Table 3

1.5 Million Tons of MTO Projects Approved by NDRC & Proceeding

MTO/MTP Projects	Locations	Process	Olefin Capacity (Thousand Metric Tons)	Estimated Startup Timing
Datang International Powe	Duolun, Inner Mongolia	MTP	500	2010
Shenghua Ningmei Group	Ningdong, Ningxia	MTP	500	2010
Baotou Shenghua	Baotou, Inner Mongolia	MTO	520	2010
Shenghua/Dow JV	Yuling, Shaanxi	MTO	1000	-

Three MTO projects have been approved for investment by the National Development and Reform Commission (China's central government organization regulating investment). They all involve the Shenghua group; a very large state-owned coal company with large reserves in Northwest China and Inner Mongolia.

Datang International's MTP project will be based on local coal, with technology from Lurgi for coal to olefin, and Unipol for olefin to polyolefin.

Shenghua Ningmei Group's MTP project will also be based on local coal resources. The MTP technology is Lurgi.

Shenghua Group's MTO project will source coal locally, too. Technology is from the Chinese Academy of Sciences (CAS).

The Shenghua/ Dow project has not yet been approved though a feasibility study has been completed and submitted to the central government.

Figure 18

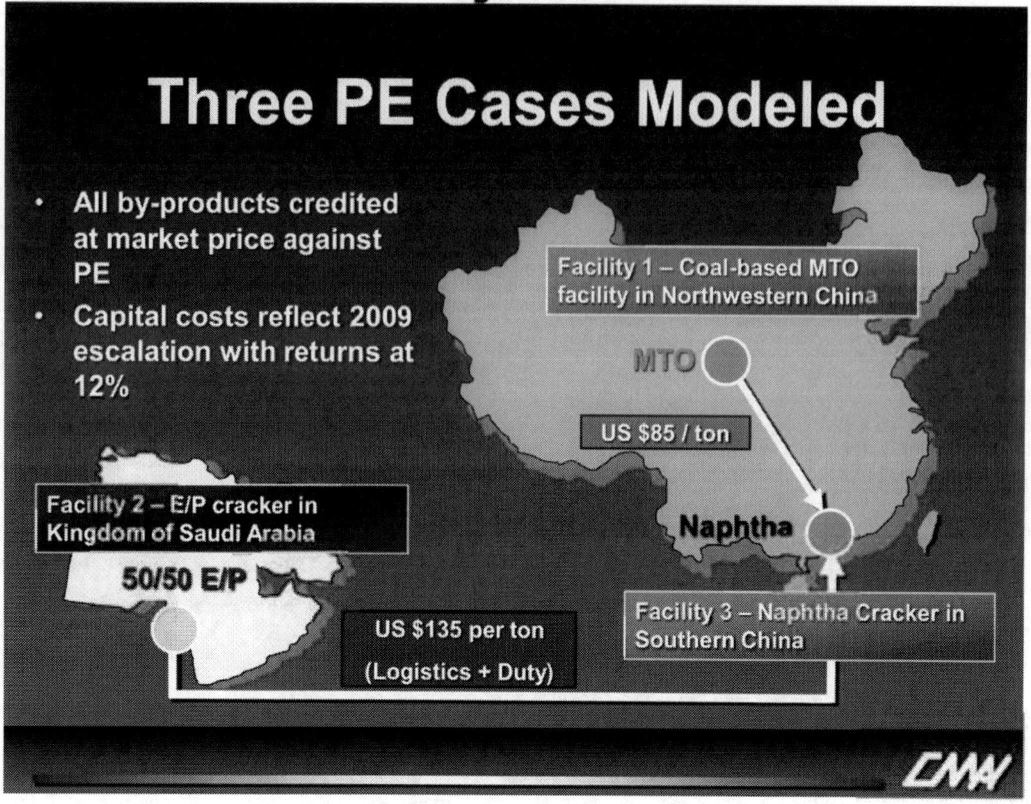

Three Polyethylene facilities have been modeled for this study.

Facility 1 is an MTO plant assumed to be located in Shaanxi province in NW China. The plant is a coal-based gasification facility using thermal grade coal (modeled at both cost and market) producing 1 million metric tons of polyolefin via a methanol intermediate. All by-products are sold on a local basis and all polyolefin product is assumed to be shipped by rail to Guangdong province and sold at market prices.

Facility 2 is an 800 KTA 50:50 ethane / propane cracker in Saudi Arabia with ethane valued at $0.75 per million BTU and propane valued at local pricing (based on a discount to exports). Propylene is valued at local market price (based on Asia netbacks) and all logistics costs for product that is ocean shipped from KSA to South China are included.

Facility 3 is a 900 KTA liquids cracker (full-range naphtha) in Guangdong province with all feedstock and co-product prices at local market. Polyolefins are produced and sold into the local market with local logistics included.

Figure 19

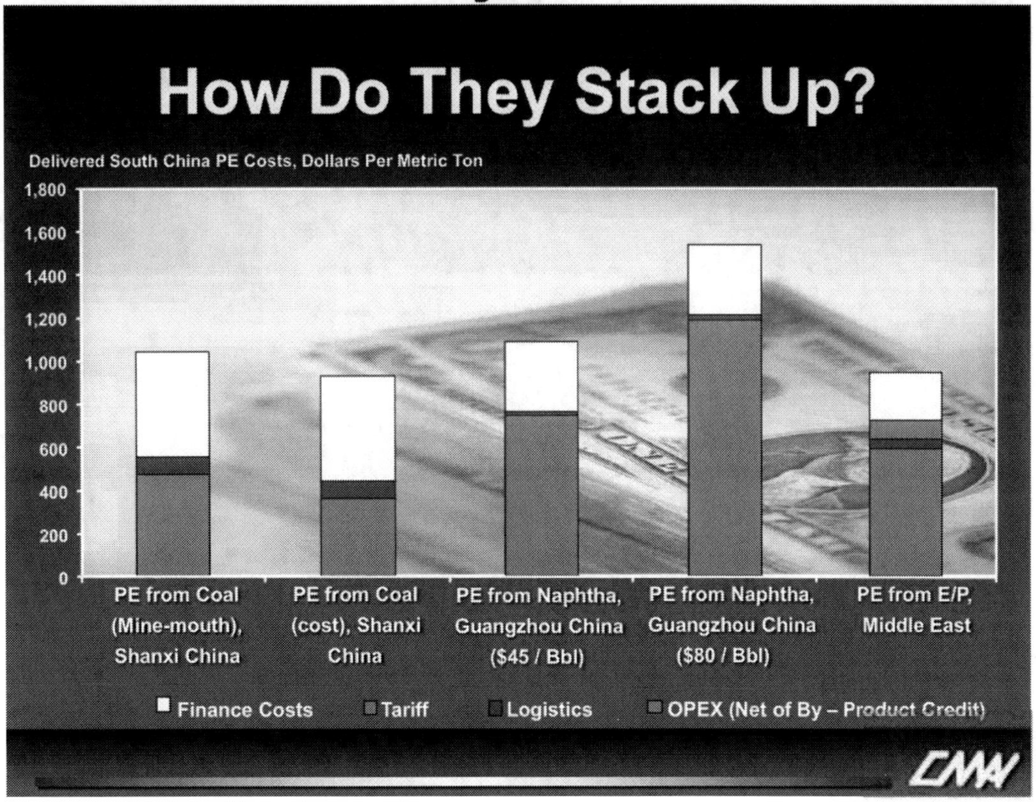

Production and capital cost analysis of these facilities show, perhaps surprisingly, that the best delivered cost is from the coal-based MTO facility running on coal at mine-mouth production costs (near $0.40 / MM BTU). The Middle East facility is a slightly higher cost; disadvantaged by PE tariffs and logistics costs (PE tariff costs are 6.5% under WTO). However, when the value of coal into the MTO facility is at coastal market netback, (about a 3-4x change), Middle East production is more competitive. The liquids cracker, at current forecast crude prices, has the highest cash cost, despite having the lowest logistics burdens.

Returns change the picture. On an overall return basis, at current crude prices, returns for PE from coal at cost and Middle East E/P cracker economics are at near parity. PE via coal is lower than naphtha cracking, but this position does not hold at lower crude prices. Gasification to methanol to olefins is capital intensive; even more so than steam cracking. Recent capital costs escalation is a more significant penalty for the coal process.

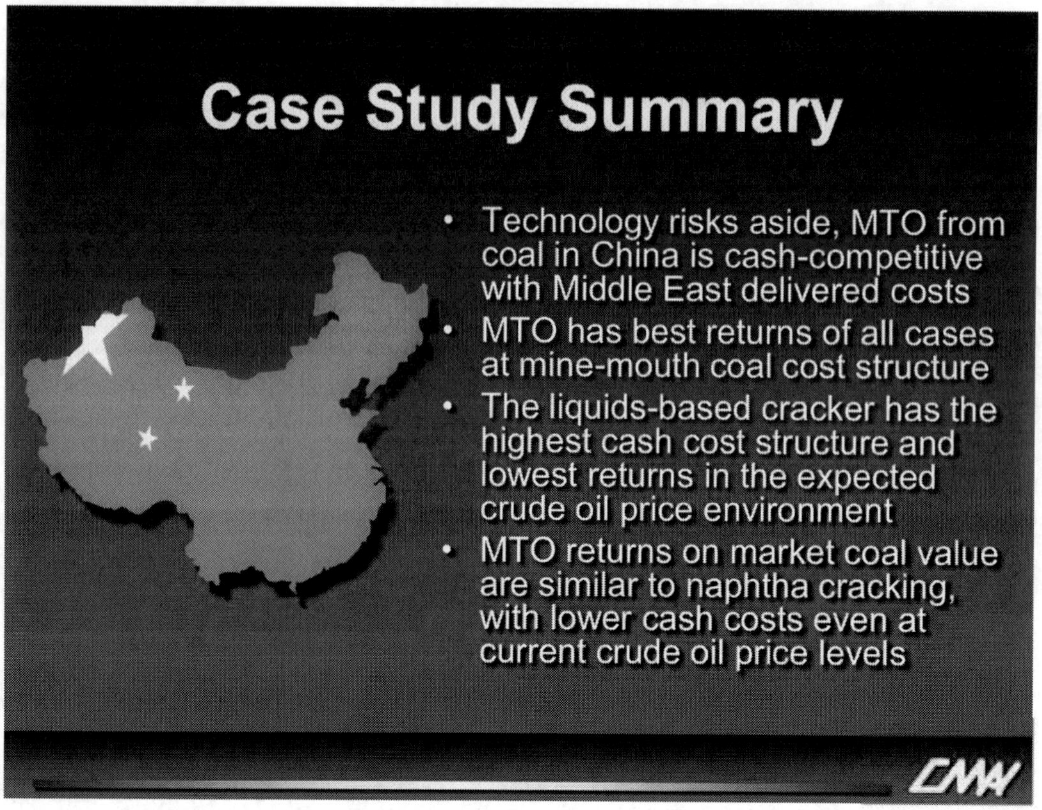

Technology risks aside, northwest China MTO from coal is cash-cost competitive with Middle East economics delivered into China.

MTO has the best returns of all cases at a mine-mouth coal cost structure.

The local liquids-based cracker has the highest cash cost structure and lowest returns in a high crude environment, but returns are similar to an MTO plant that is based on market priced coal at current crude oil price levels. However, the market based coal MTO plant's cash costs are below those of the naphtha cracker even at current energy prices.

Figure 20

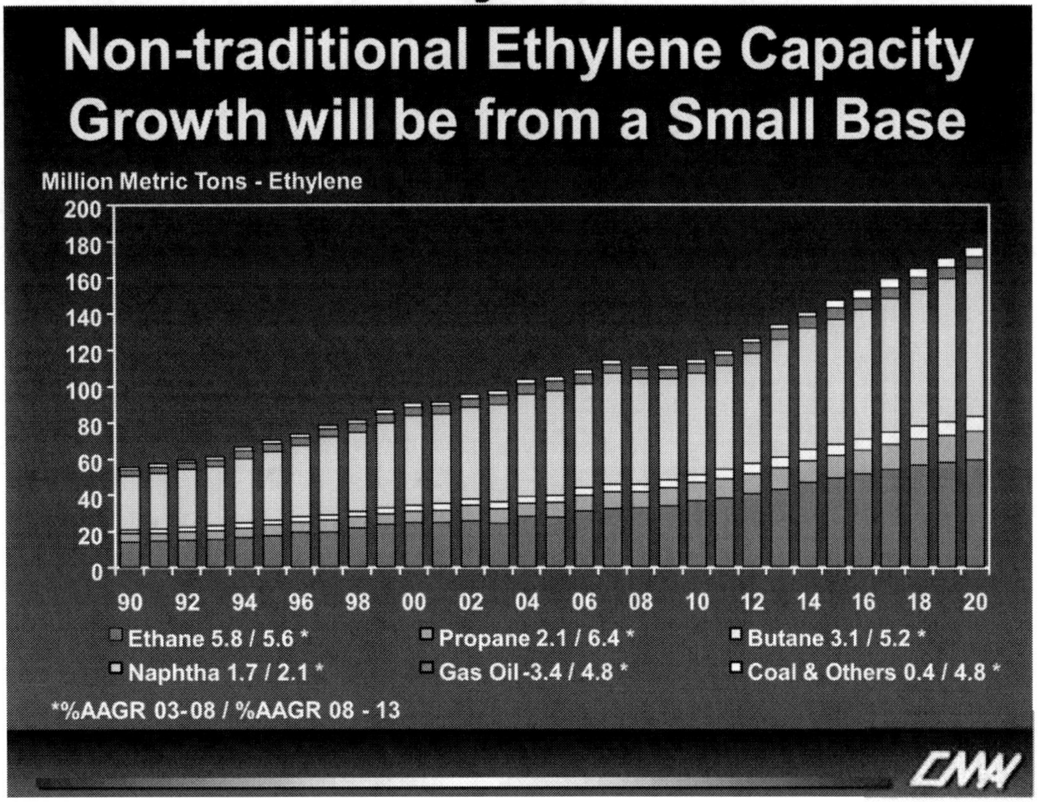

In spite of the favorable economics versus Asian naphtha cracking, and a fairly significant amount of development effort underway, the forecast growth of non-traditional ethylene feedstocks (which are represented by the small segment at the top of each bar on this chart) is not expected to appreciably change the feedslate mix over the next decade. Even though the growth rate is expected to be high, it will be from a very small base in 2010, and the existing capacity is very large.

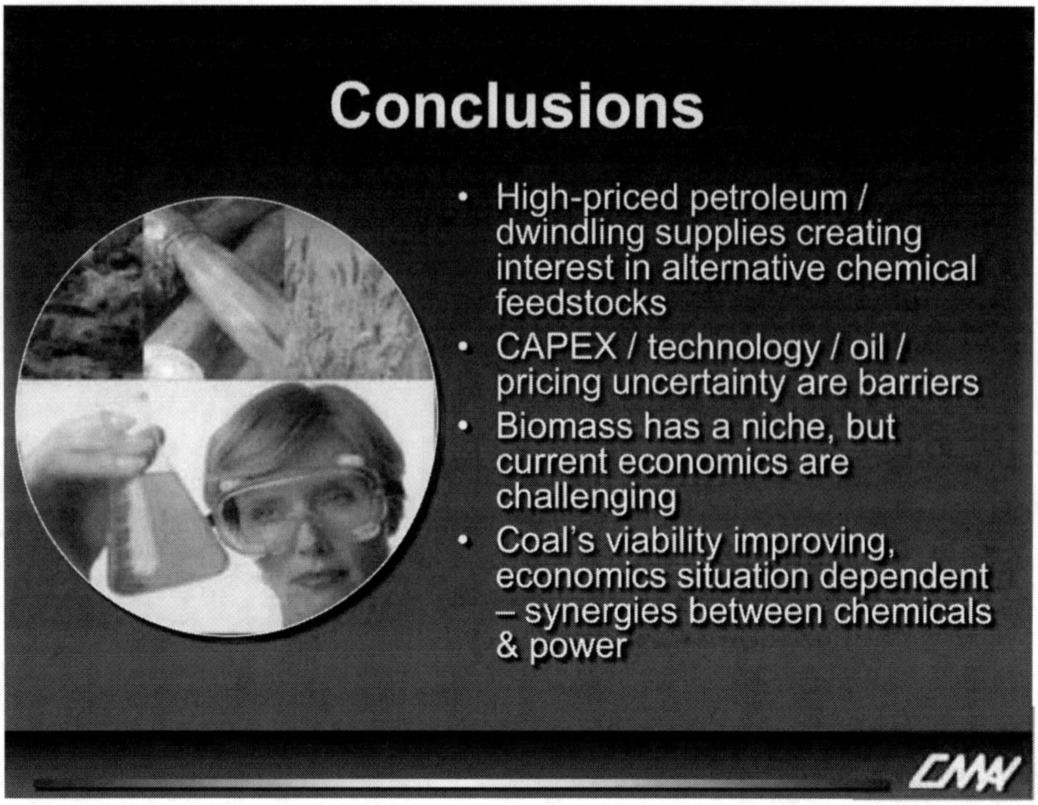

In conclusion, the recent experience of high oil prices and dwindling supplies have created an interest in alternative chemical feedstocks. However, high capital investment requirements, new technology risks and volatile oil prices are all barriers to entry. Biomass may have a niche, especially if high oil prices return, but current economics are not encouraging. Coal's viability is improving, primarily in China, where a large coal chemical production base already exists, and it appears there may be some synergies between the chemical and power industries with respect to coal gasification, as well as in CO2 capture and sequestration. As these clean coal technologies develop, they may provide more incentives to use coal as a chemical feedstock.

Application of Heat from High Temperature Nuclear Reactors for Process Furnaces

Daniel Barnett
Chief Engineer
Shaw Energy & Chemicals Group
Reiner Kuhr
Executive Consultant
Stone & Webster Management Consultants
Dan Mears
****
PBMR

Abstract: Ongoing work to develop, demonstrate and commercialize the next generation of high temperature nuclear reactors points to possible petrochemical plant applications to displace fuel gases used to directly fire furnaces in various hydrocarbon processes. Smaller scale nuclear reactors that produce heat at 750 to 900C via a high pressure helium loop can be integrated with process convective heat exchangers to reduce premium fuel consumption and associated emissions. The enhanced safety characteristics of these nuclear reactors are expected to enable their physical location near to process units, which will be important to minimize the cost of connective high pressure helium piping for heat delivery. A number of traditional direct-fired furnace applications are reviewed which can instead be satisfied by nuclear process heat. Some parametric economics and a review of likely implementation and technology development challenges are also discussed.

Low Emissions Burner for Ethylene Furnaces

Roger Poe
Research Associate
John Zink Company, LLC
Michael Claxton
Sr. Technical Consultant
John Zink Company, LLC
Charles Baukal
Director, John Zink Institute
John Zink Company, LLC
David Brown
Sr. Engineering Technologist
Shaw Energy & Chemicals Group

Abstract: A new burner (patent pending), referred to as the Halo™ burner, has been developed that incorporates the Coanda principle for controlling fluid flow, mixing, and stability. This aerodynamic effect is sometimes referred to as boundary layer attachment. The Halo™ burner has some unique properties that give it enhanced performance compared to other burner technologies. It can entrain more furnace gases than previous designs, where furnace gas entrainment has been limited by the lower flammability limit for the mixture of the fuel, air, and inerts. Increased entrainment helps minimize NO_x emissions. It has a significantly wider turndown range and more compact size compared to other ultra low NO_x burners. The flame quality is improved because of the increased gas velocities that help minimize interactions between adjacent flames. The flames are less easily disturbed because of the increased momentum. The flames also tend to be significantly tighter in diameter than comparable low NO_x burners due to the adhesion and flame forming properties of the Coanda affect. The tighter flame diameter, along with the increased momentum of unobstructed combustion air, minimizes flame issues typically associated with the longer flames needed in ethylene furnaces. The Coanda-formed combustion envelope greatly reduces the chance for flame roll over and the subsequent flame impingement on process tubes, known to shorten process tube life. This flame control also minimizes the chances for flame-to-flame interaction between adjacent burners. These interactions can significantly increase flame lengths, possibly even causing flames to coalesce into larger flames, which can lead to flame impingement and increased NO_x emissions. The heat flux profile is also more uniform, which has many operational benefits. The combustion efficiency is high and the CO emissions are low from startup to shutdown. These are benefits for burners used in ethylene service. Noise levels are also lower than comparable low NO_x burners. This paper discusses testing of a wall-fired Halo™ burner using a variety of fuels in a pilot-scale ethylene cracking furnace.

Introduction

Cracking furnaces often present significant burner challenges compared to process heaters, because of higher temperatures and more specific heat flux profile requirements. Minimizing pollution emissions is a general requirement for all industrial combustion applications.[1] New generations of burner technology continue to reduce emissions, but often impose operational limitations. For example, with some low NO_x burner technologies, low NO_x emissions may only be achievable over a relatively narrow range of operating conditions. A new burner technology has been developed using some advanced fluid dynamic mixing and control techniques, which minimizes both NO_x and CO emissions over a much wider range of operating conditions compared to other types of advanced burner designs. This new burner has a much wider turndown ratio and a more compact flame, which are both important advantages in cracking furnaces. A detailed discussion of the challenges, design and development of this burner is given elsewhere.[2]

Coanda Principle

A new burner (patent pending) has been developed that incorporates the Coanda effect[3] for controlling fluid flow, mixing, and stability. This aerodynamic effect is sometimes referred to as boundary layer attachment. Figure 1 shows a diagram of the Coanda principle. A fluid is supplied to a chamber with a small opening. A curved surface with a radius *R* within a specified range causes the fluid to be drawn against the surface, instead of being injected straight out of the nozzle. The fluid outlet velocity must be within a certain range and the surface must have the appropriate characteristics for this principle to apply. If the fluid velocity is too low or too high, the fluid will not attach to the surface.

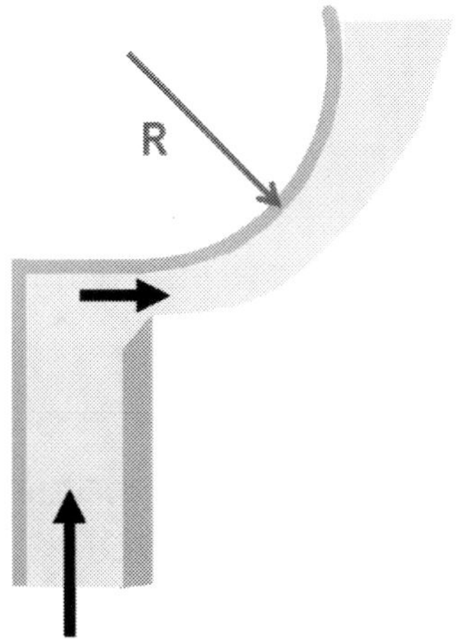

Figure 1. Coanda principle.

COOLstar Halo™ Burner

The Coanda principle has been implemented in a new burner development. A drawing of the new burner is shown in Figure 2a. Shown is a natural draft version of the burner. One of the key design features is the shape of the tile which has an outer Coanda surfaces integrated into the overall design to augment both mixing and stability within the burner. Figure 2b shows a process schematic for the operation of the Coanda burner. Ambient air is drawn into the burner throat by the draft in the furnace. A damper is used to control the air flow into the burner.

This burner has been extensively tested and its operation continues to be explored and documented. This paper discusses pilot scale testing performed on one version of this burner at the John Zink Research and Development Test Center in Tulsa, Oklahoma.

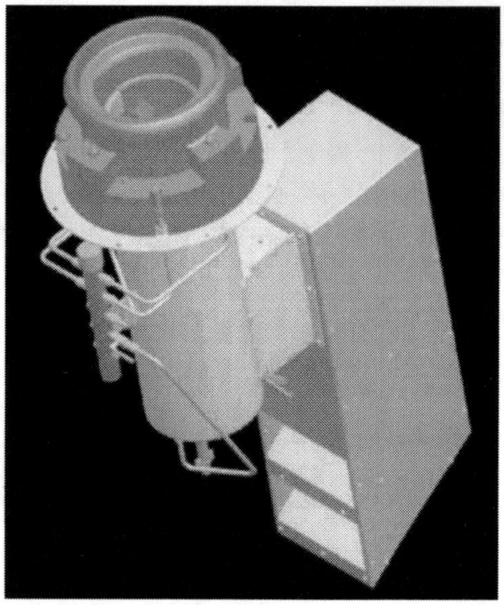

(a)

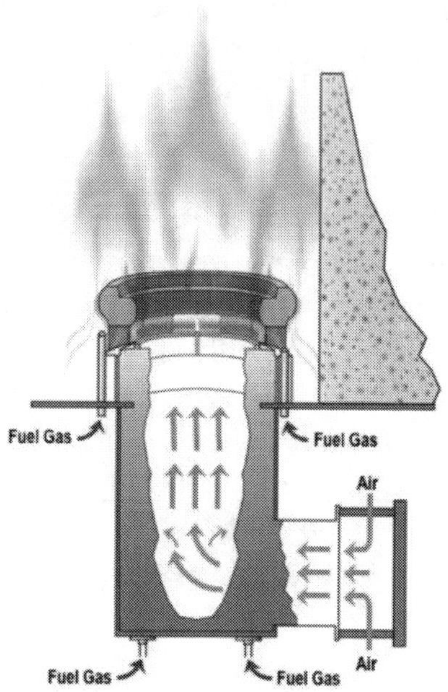

(b)

Figure 2. COOLstar Halo burner: (a) drawing and (b) process schematic.

Test Setup

The HALO™ burner (Figure 2) was vertically installed adjacent to the north wall, firing upward into a test furnace (see Figure 3) which is a rectangular heater with internal dimensions of 7 ft (2.1 m) wide, 5 ft (1.5 m) deep, and 30 ft (9.1 m) tall. The furnace is cooled by four single pass water tubes located along wall opposite the burner. The furnace temperature is regulated by insulating these tubes. For the tests discussed here, one of the middle tubes was un-insulated while the other three tubes were fully insulated. This resulted in furnace temperatures of approximately 2200°F (1200°C).

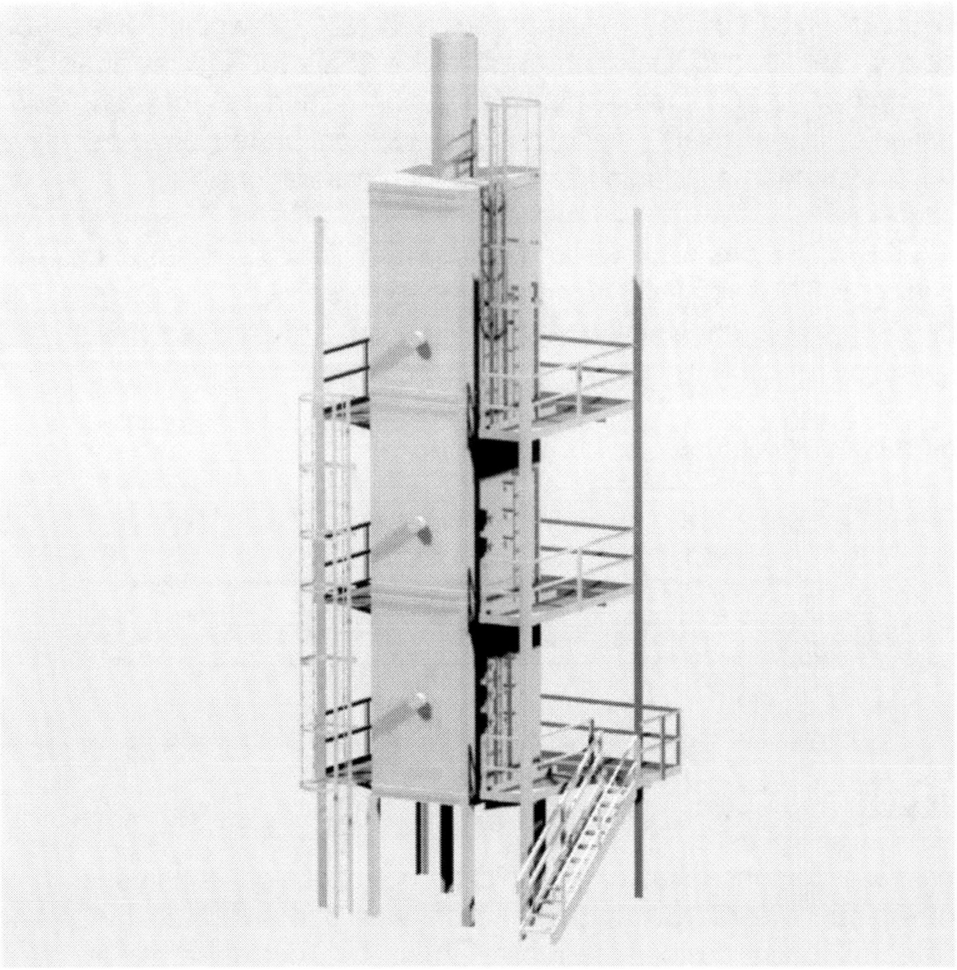

Figure 3. Test Furnace #4, John Zink Co. R&D Test Center (Tulsa, OK).

A velocity thermocouple (also known as a suction thermocouple or suction pyrometer) was used to measure furnace gas temperatures.[4] It consisted of a thermocouple recessed inside a radiation shield. An eductor rapidly aspirated furnace gas across the thermocouple. This configuration maximizes the convective heat transfer to the thermocouple while minimizing radiation exchange between thermocouple and its surroundings, thus assuring that the equilibrium is nearly that of the true gas temperature. The velocity thermocouple was mounted through the furnace wall approximately 16 ft (4.9 m) above the floor.

The furnace draft was measured with an automatic, temperature-compensated, pressure transducer as well as an inclined manometer connected to a pressure tap at the burner level. Fuel flow rates were measured using calibrated orifice meters, fully corrected for temperature and pressure. Given the measured fuel flow rates, the data acquisition system calculated the real-time total fuel flow rate and the total rate of heat release, based on the lower heating value (LHV) of the fuel.

State-of-the-art continuous emissions monitors (CEMs) were used to measure emissions species concentrations of NO_x, CO, and O_2.[5] Stack samples were continuously extracted at high flow rates to minimize response time. A continuous slipstream went to a condenser and sample conditioner which chilled the sample to 34°F (1°C) dewpoint. Once the moisture was removed, the dry, conditioned sample flowed to the CEMs for species analysis. NO_x, CO, and O_2 were determined using chemiluminescent, non-dispersive infrared, and paramagnetic detection analyzers, respectively. Each analyzer was calibrated once per day at a minimum with primary NIST standards including zero and span calibrations.

Multiple fuel compositions were tested as shown in Table 1. Natural gas from a pipeline in Tulsa (referred to as TNG) nominally consisted of 93.4% CH_4, 2.7% C_2H_6, 0.60% C_3H_8, 0.20% C_4H_{10}, 0.70% CO_2, and 2.4% N_2 by volume. High purity hydrogen (>99% H_2) and propane (>93% C_3H_8) were also used as components in some of the test fuels.

Table 1. Test fuel compositions.

Fuel	TNG (vol. %)	H_2 (vol. %)	C_3H_8 (vol. %)
A	34	66	0
B	30	70	0
C	57	40	3
D	50	0	50

Test Description

The burner was demonstrated under cold firebox, light-off conditions. The light-off and furnace warm-up fuel was TNG. Initial light-off was demonstrated by setting the burner's damper at the full open position and pre-setting the design draft at the furnace floor. Following light-off, the furnace draft was re-adjusted to the design of 0.45 – 0.50" WC (112 – 125 Pa) and the firing rate was ramped up to the normal heat release of 5.29 MMBtu/hr (1.55 MW). This ramp-up was accomplished over less than 30 minutes. At no time during the cold light-off or during temperature ramp-up to a furnace temperature of over 1900°F (1038°C) did CO exceed 150 ppmvd.

The burner was then switched to fuel A at the normal firing rate where the furnace was allowed to heat soak. The flame quality was observed and stack emissions were recorded. The firing rate was then increased until the stack monitor showed CO increasing (referred to as "CO break"), indicating the burner was short of combustion air. There was no observed instability in the burner, even at ~0% O_2 and 2500 ppmvd CO in the stack.

The furnace draft was then increased until the burner was again operating at approximately 10% excess air (~2% O_2, dry). The burner maintained stability at all times through the CO break and following when the air flow was increased.

The furnace draft and burner firing rate were then increased to the burner maximum rated capacity (6.35 MMBtu/hr or 1.86 MW), while maintaining the operational excess air. The flame quality was observed and the stack emissions recorded.

The firing rate and excess air were then reduced to the normal firing rate with the design excess O_2 of 2% and the furnace was put into automatic mode. While the control system maintained operational stability, a heat flux profile was collected.

After the flux measurements were collected, a water-cooled CO probe was used to check for measurable CO near the furnace cooling tubes. Measurements were taken at multiple elevations and up to 20" (~50 cm) past the tubes, into the firebox. The test furnace is 61" (155 cm) deep from the fired wall to the centerline of the furnace cooling tubes. Periodic small measurements of less than 50 ppmvd of CO were detected. However, most readings were less than 10 ppmvd CO in the vicinity of the cooling tubes.

The burner was then switched to fuel B. Normal, CO break from normal, and maxinum firing rates were demonstrated while observing flame quality and stability, and recording stack emissions.

This ended that day's demonstration points and the furnace was shut down. The furnace was left to cool over night, allowing for a cold start the next day. The next morning the demonstration began with cold light-off of the burner on fuel D which closely simulates the LHV and molecular weight of ethane, although the combustion characteristics are more difficult. Methane is a more stable hydrocarbon with a higher ignition temperature and propane has a lower vapor pressure.

After cold light-off on this simulated C2 fuel was demonstrated to have no problems, the fuel was switched to fuel A where cold light-off was demonstrated. The burner was again shut down, and cold light-off was demonstrated on fuel B. The burner was switched back to fuel A and over the next hour the firing rate was ramped up to the normal firing rate while CO was monitored. The maximum CO excursion was less than 450 ppmvd (uncorrected). The total CO shift was from approximately 100 ppmvd in a firebox environment of less than 800°F (430°C) up to 450 ppmvd and back down to less than 10 ppmvd in an 1800°F (982°C) firebox, with the spike lasting less than two minutes.

Maintaining the normal firing rate at 2% excess O_2 on fuel A, a longer water-cooled CO probe was used to map the flame boundary. By varying the sampling elevation and the probe insertion depth, the estimated flame boundary using 2000 ppmvd as the iso-surface was determined.

The burner was switched back to fuel B and the normal firing rate where CO break was demonstrated. Burner flame quality and stack emissions were recorded.

The final demonstration point was to establish the most stringent turndown condition. The burner was returned to the fuel A, the burner draft was set at 0.49 in. H_2O (122 Pa), and the damper was set to the normal operation position. The fuel flow was then reduced. Watching for any visual indications of instability or any monitored indications such as unburned hydrocarbons (UHCs) or draft pulsations, the burner's fuel input was reduced to 0.21 MMBtu/hr (0.062 MW) at a measured fuel pressure of 0.2 psig (~0.014 kg/cm^2) without any problems.

Test Results

The maximum corrected NO_x of 32 ppmvd at 3% O_2 (64.7 mg/Nm^3) was recorded at maximum rate at a furnace temperature of 2280°F (1250°C). CO emissions were a maximum of <2 ppmvd at 3% O_2 (2 mg/Nm^3) under normal operating conditions. The flame length was estimated at less than 15 ft (4.6 m) based on 1250 ppmvd CO iso-surface determined through probing. The flame diameter and length were estimated at 3 ft (1 m) and 15 ft (4.6 m), respectively based on visual observation of the heated wall. The maximum noise level was 77.4 dBA at 3 ft (1 m) from a single burner.

The heat flux profile in a cracking furnace is very important to maximize the process fluid throughput.[6] If there are hot spots in the profile, then either the firing rate may need to be reduced or the furnace shut down sooner because of increased coking. Figure 4 shows the measured heat flux profile.

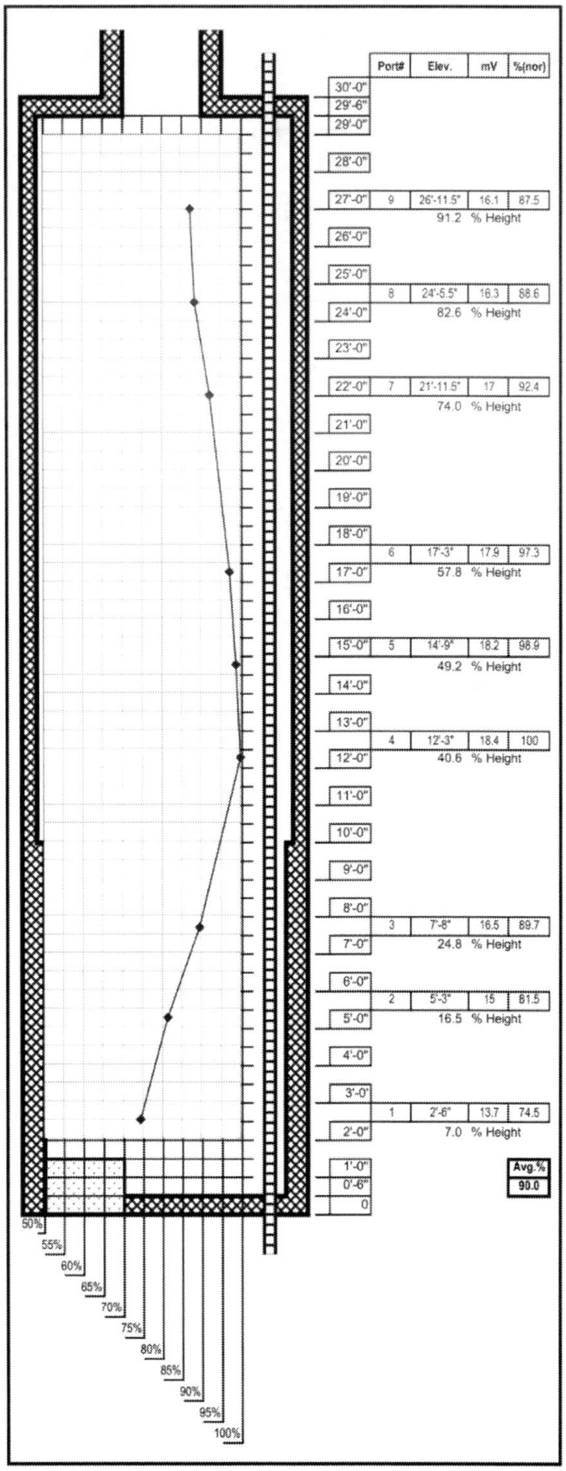

Figure 4. Measured heat flux profile.

Conclusions

A new burner has been developed which has some significant improvements over previous burners used in cracking furnaces. The heat flux profile is significantly uniform, the turndown is wide, the flame is stable over a wide range of operating conditions, the combustion efficiency is high from startup to shutdown, the burner is easy to maintain, and both the CO and NO_x emissions are low.

References

1. C. Baukal (ed.), *Industrial Burners Handbook*, CRC Press, Boca Raton, FL, 2004.
2. R. Poe, A. Patel, C. Baukal, and D. Wright, Advanced Combustion System for Cracking Furnaces, *Petroleum Technology Quarterly*, Vol. 12, No. 3, pp. 113-117, 2007.
3. H. Coanda, "Propelling Device," U.S. patent 2,108,652, February 15, 1938.
4. J. Gifford and J. Heinlein, "Combustion Controls," Chapter 12 in *The John Zink Combustion Handbook* (C. Baukal, ed.), CRC Press, Boca Raton, FL, 2001.
5. C. Baukal, *Industrial Combustion Pollution and Control*, Taylor and Francis, New York, 2004.
6. C. Baukal, *Heat Transfer in Industrial Combustion*, CRC Press, Boca Raton, FL, 2000.

MERT Performance and Technology Update

Michael Györffy
Research and Development Metallurgist
Kubota Metal Corporation
Makoto Hineno
Group Manager Material Research and Development Department
Kubota Corporation
Kunihude Hashimoto
Assistant Manager Informal Steel Casting Department
Kubota Corporation
Seong-Hoan Park
Deputy General Manager
YEOCHUN NCC Co. Ltd.
Min-Soo You
Manager
YEOCHUN NCC Co. Ltd.

Abstract: MERT (Mixing Element Radiant Tube) is a centrifugally cast radiant tube with a helical mixing element on the tube inside diameter for use in steam pyrolysis (cracking) furnaces. There are over 300 furnace installations of MERT technology worldwide in a wide range of furnace designs and tube diameters which achieve improved efficiency and productivity. MERT technology provides high heat transfer coefficient (up to 1.5 times that of bare tubes) by increasing process fluid mixing and break down of the thermal boundary layer on the tube internal surface. The improved heat transfer and mixing is achieved by a helical element integrally bonded to the tube inside diameter. Progressively, the design of this element has been optimized to reduce pressure drop, first through the use of our SLIT-MERT product and now with the latest design: X-MERT.

The enhanced thermal and flow properties imparted by X-MERT elements result in lower tube metal temperatures, lower coking rates and longer run times. Additional savings are achieved through a reduction in the number of decoke cycles. The high turbulence and mixing also improves cracking performance by reducing over-cracking conditions adjacent to the tube wall and provides a more uniform hydrocarbon temperature cross section. In order to achieve these benefits, there is some resistance to flow quantified as an increased friction factor. Compared to a standard tube of the same inside diameter, this friction raises the overall initial pressure drop in the cracking coil at the start of run. But due to the low coking rate of MERT technology, this pressure drop difference may not be significant by the end of run when compared to a bare coil. Nevertheless, any increase in coil pressure is considered detrimental to cracking yield performance, thus the development of lower friction X-MERT technology.

The heat transfer and friction characteristics of X-MERT tubes have been compared to that of standard bare tubes. This comparison has been performed using CFD (Computational Fluid Dynamics), experimental laboratory experiments, computer simulations and full-scale commercial installations. The commercial installations have confirmed the beneficial effects of X-MERT, including reduced tube metal temperature and reduced coking, leading to significantly increased run times. It has been further demonstrated that feed rate and/or severity increases can be consistently achieved without reducing the original run time, due to the high heat transfer capability of MERT. These have confirmed that the thermal efficiency of X-MERT is superior to bare tubes and while offering the lower pressure drop compared to previous generations of MERT technology.

This paper is an update of Kubota's MERT technology and describes the latest X-MERT geometry and performance.

MERT (Mixing Element Radiant Tube) is a centrifugal cast radiant tube with a helical mixing element on the tube inside diameter for use in cracking furnaces for ethylene production. The helical element induces turbulent swirl flow, thereby reducing the stagnant boundary layer of fluid. This increases the inside heat transfer to the fluid cross-section and achieves a more uniform temperature profile. This effect is shown in Figure 1. MERT is distinct from traditional internally finned tubes, in that it does not rely on increased surface area of the helical element in order to achieve improved heat transfer. [1] Furthermore, the mixing element does not reduce the effective inside diameter of the tube nor increase its weight significantly. These characteristics, in combination with the performance benefits, have resulted in the great popularity of MERT. Since 1996, MERT technology has been installed in over 300 commercial cracking furnaces, providing the benefits of improved heat transfer, reduced coking, reduced tube metal temperature, increased production rates and increased run times.[2, 3]

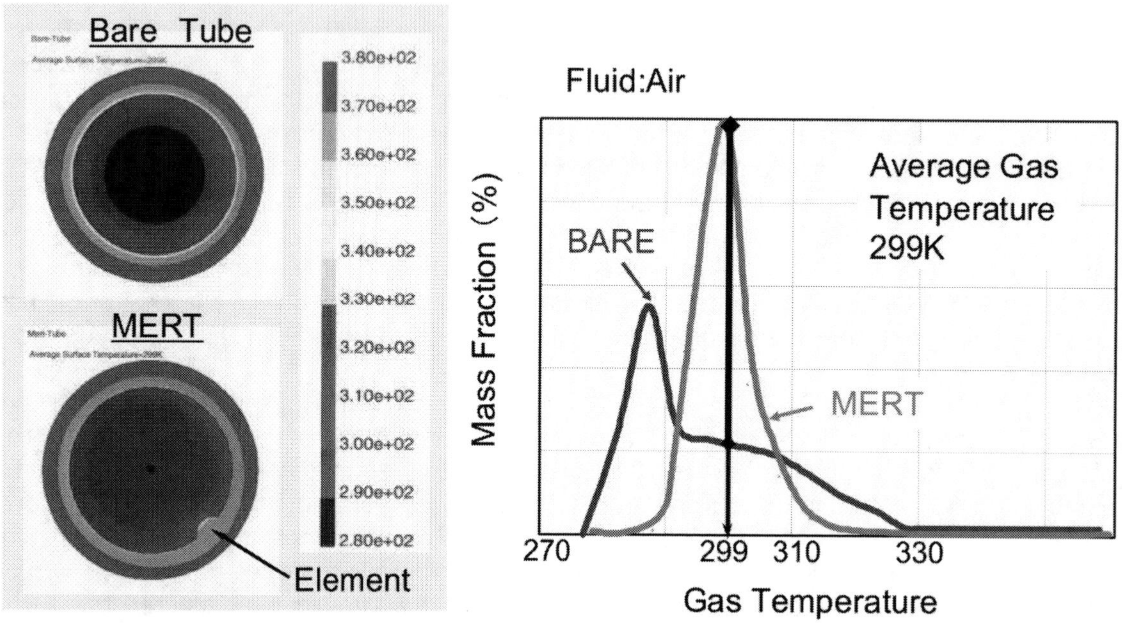

Figure 1: CFD analysis of boundary layer mixing and benefits to make uniform fluid temperature.

However, in some cases MERT could not be applied to a particular furnace design due to pressure drop limitations at the plant. As a result the number of MERT passes was reduced in order to accommodate the pressure drop limitation. To maximize the MERT effect on a furnace, the friction factor needed to be minimized relative to a constant high heat transfer coefficient. [4] In order to address these limitations, Kubota has developed the next generation of MERT technology, which provides an improved ratio of heat transfer to friction factor. This provides more opportunities to apply MERT technology in furnaces where there are pressure drop limitations or to provide increase process benefits. The new development has been validated in the laboratory and subsequently it was installed in a commercial furnace where MERT had previously been installed. This trial highlighted the benefits of the improved technology.

Improvements in the element design led to Slit-MERT which has an interrupted helical element. For a detailed description of Slit-MERT refer to the article by Matsueda et al. [5]. The next generation of MERT has improved upon this success and is called X-MERT. It is similar to the helical finned MERT, except that the element is applied intermittently and at a smaller pitch. The geometry differences between MERT, Slit-MERT and X-MERT are shown in Figure 2. As compared to the standard MERT tube and Slit MERT, the essential geometry is very similar, except for the increased number of small intermittent elements.

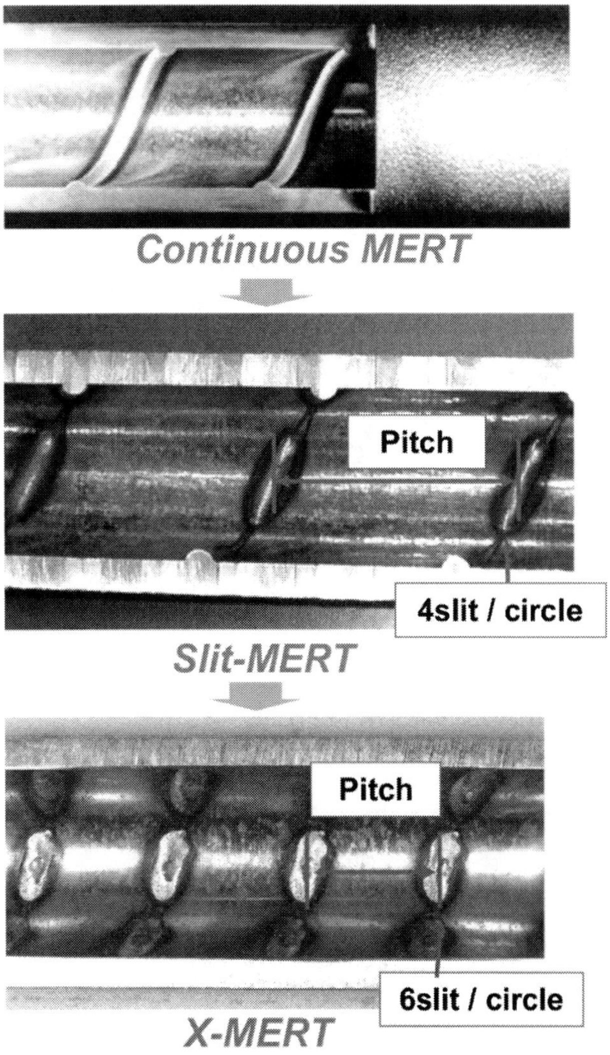

Figure 2: Photographs showing the difference between MERT, Slit-MERT and X-MERT geometries.

X-MERT was designed to decrease the resistance to flow while maintaining a high heat transfer coefficient, and improve upon the limits of Slit MERT. This is accomplished by decreasing the element height, but increasing the frequency of elements. This provides a more focused convective effect to the boundary layer of the process fluid, without producing excessive turbulence and resistance across the fluid cross-section. The helical alignment of the segments is maintained, maintaining the swirl mixing effect throughout the length of the tube. The concept of X-MERT is to reduce the pressure drop with a high frequency element geometry, while still maintaining heat transfer and tube metal uniformity and this design was optimized using computational fluid dynamics (CFD).

The fundamental properties of X-MERT were determined using the same laboratory apparatus as was used to characterize MERT, and Slit MERT and this has been described by Magnan et al.[4] Lab scale data allowed for Kubota to generate a data base relating tube geometry and fluid flow through the tube to the increased heat transfer and friction factor of the X-MERT tube. A summary is shown in table 1 and this data is then used in a cracking furnace simulation program, to apply X-MERT to address specific furnace geometries, pressure limitations and hot spots.

Table 1: Relationship of geometry to the pressure drop at constant heat transfer.

	Element Geometry	**Heat Transfer relative to Bare Tube**	**Pressure Drop relative to Bare Tube**
X-MERT		1.4	2.1
Slit MERT	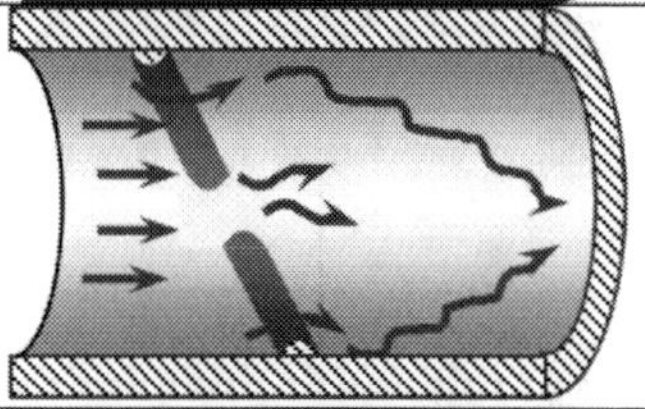	1.4	2.4
MERT	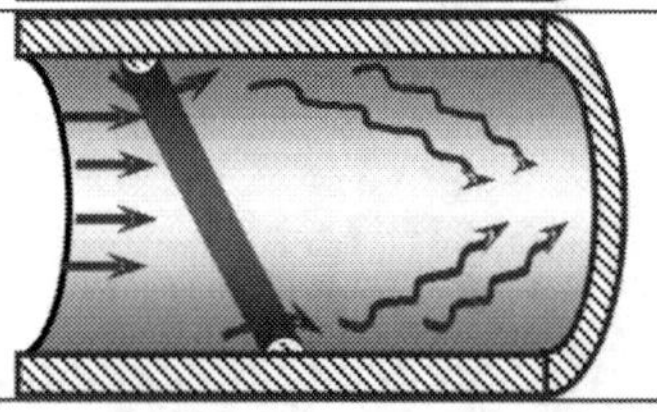	1.4	3.0

X-MERT is installed in a modified SRT-V furnace limited by tube metal temperature, and the X-MERT tubes replace bare tubes on both the coil inlet and the coil outlet. This geometry is shown in figure 3 and the feedstock is light naphtha. The first run is reported and the tube metal temperature (TMT) and pressure drop data are shown figures 4 and 5. For this first run under the same feed stock and P/E ratio, X-MERT allowed for increased feed rate and extension of run length due to the increased heat transfer and mixing effect of MERT technology. Laboratory data collected and applied to the furnace simulation agreed well with the actual observed results. This allows the design of the tubes to balance the end of run conditions so that TMT and pressure drop are both reached at approximately 100 days.

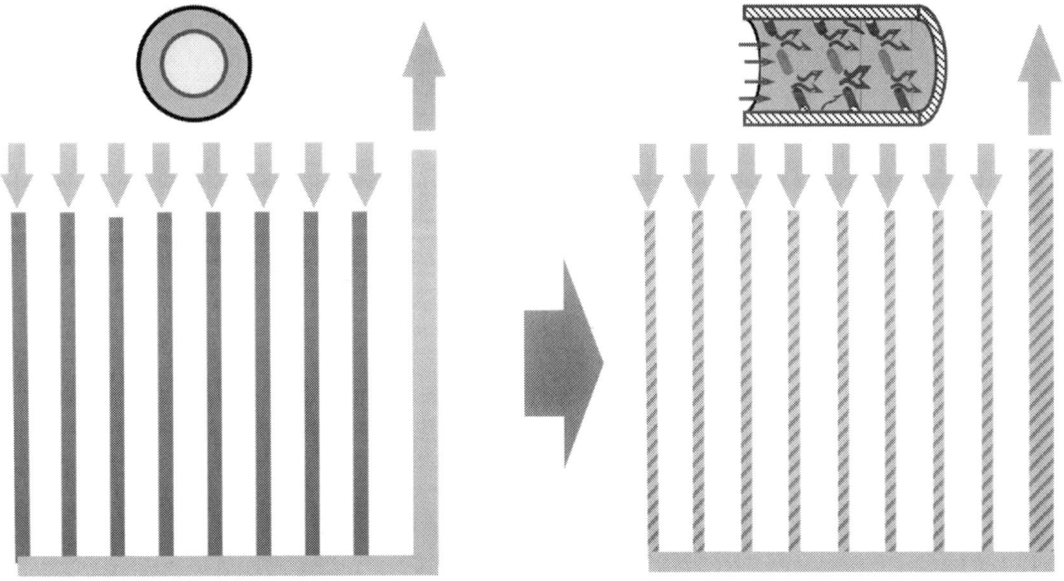

Original Coil Configuration, inlets and outlets are bare tubes.

X-MERT Coil Configuration, inlets and outlets are X-MERT tubes.

Figure 3: Commercial furnace application, modified SRT V coil - bare tube compared to X-MERT tubes.

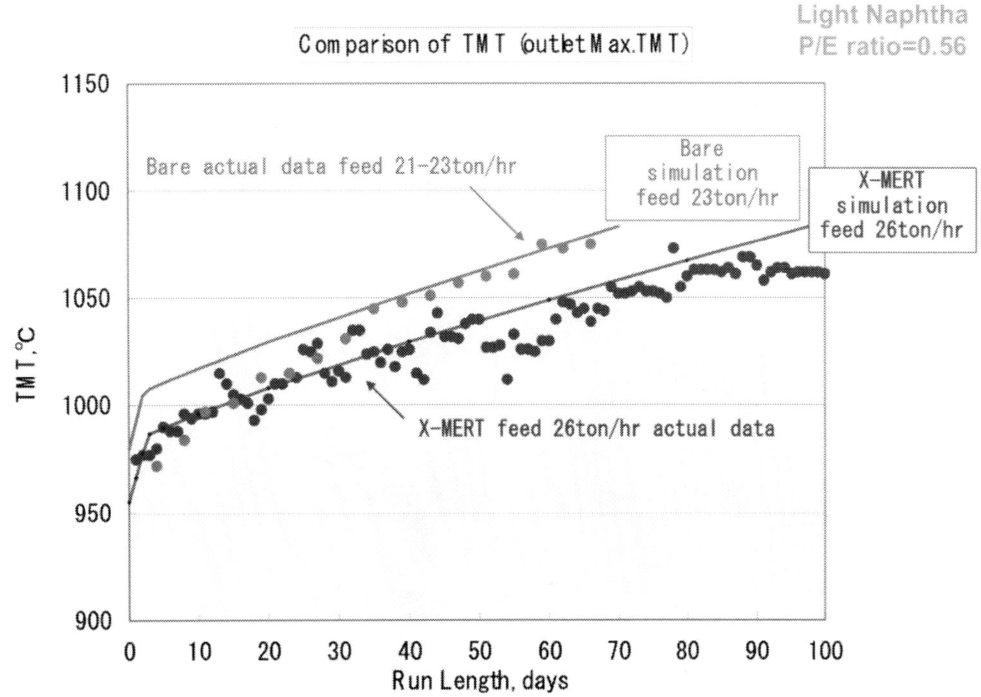

Figure 4: First run TMT data comparing bare tube performance at 23tons/hr with X-MERT performance at 26 tons/hr.

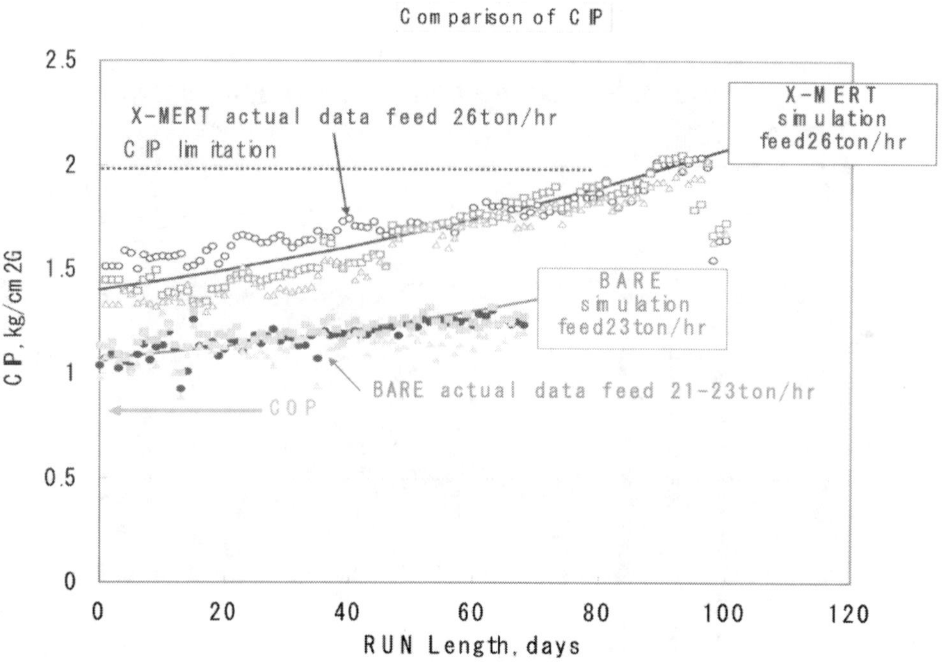

Figure 5: First run pressure data comparing bare tube performance at 23 tons/hr to X-MERT tube performance at 26 tons/hr.

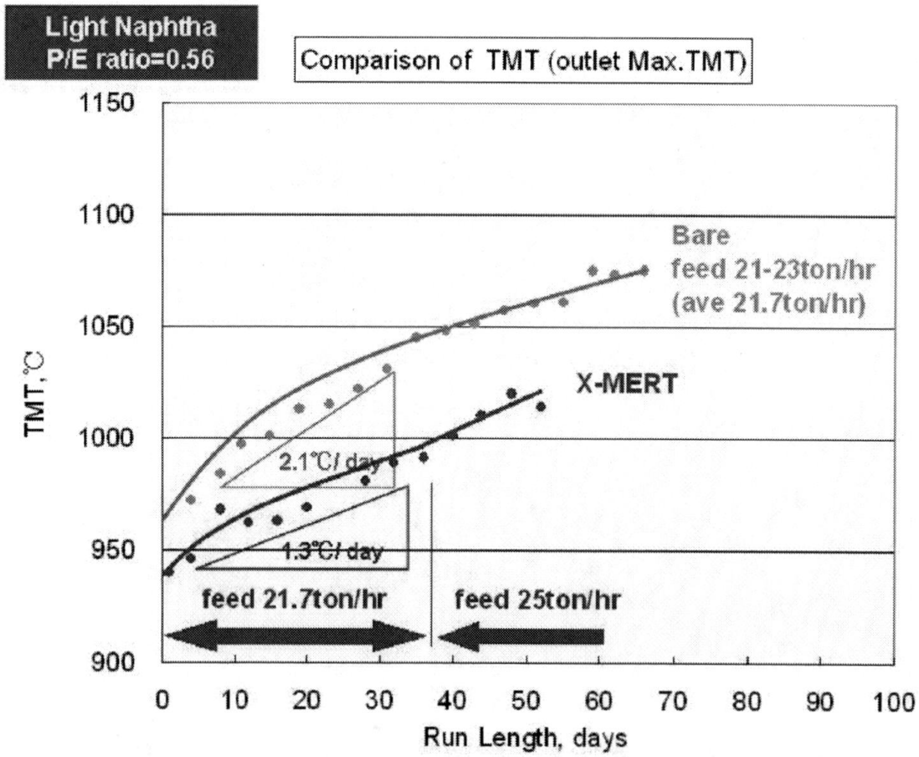

Figure 6: Ongoing second run data comparing bare tube to X-MERT performance at the same feed rate for the first 35 days.

Briefly summarizing the major observations from the commercial performance data, with X-MERT:

- The run time increased from 65 to 100 days.
- The feed rate increased from 23 tons/hour to 26 tons/hour.
- The tube metal temperature reduced by approximately 20 degrees C.
- The improvements were realized under the furnace coil inlet pressure (CIP) limitation.

For this same installation's second run, the feed rate was kept constant for the first 35 days, along with the feed stock and P/E ratio. The collected tube metal temperature data is shown in figure 6. The data clearly shows a start of run TMT reduction of 20°C. There is also a decrease in the rate of TMT increase with the X-MERT tubes and this is attributed to the increased fluid uniformity resulting is less coke formation.

The development of X-MERT is an improvement on the already successful MERT technologies by increasing the heat transfer to friction factor ratio of the cracking tube. This provides added benefits to the ethylene producer where TMT reductions can be achieved while minimizing the increased pressure drop penalties. Laboratory test data was consistent with the theoretical development of the new geometry and was successfully implemented in a commercial furnace installation. Kubota's next generation X-MERT is now available for manufacture and is being installed in several full furnace coil replacements. This technology provides still better radiant coil performance than MERT or Slit MERT and will build on the success of MERT technology.

References:

1. T. Torigoe, K Hamada, M. Furuta, M. Sakashita, K. Otsubo, M. Tomita, AICHE 1999 Spring National Meeting - Ethylene Furnace Technology and Operations at the 1999 EPC, "Mixing Element Radiant Tube (MERT) Improves Cracking Furnace Performance", March 16, 1999.
2. J. Magnan, K. Hamada, K. Otsubo, K. Mae, E. Sohn, ERTC Petrochemical Conference, "Improved Furnace Efficiency Using High Heat Transfer MERT tubes", Amsterdam, Netherlands, February 20-22, 2002.
3. J. Magnan, K. Hamada, K. Otsubo, E. Sohn, Petroleum Technology Quarterly, "Improved Steam Cracker furnace Efficiency", Summer 2002.
4. J. Magnan, K. Otsubo, S. Matsueda, M. Fujiwara, ERTC Petrochemical Conference, "Next Generation MERT Achieves Further Advances in Cracking Tube Performance." Vienna, 2004.
5. S. Matsueda, K. Otsubo, M. Fugiwara, M. Tomita, 8th ARTC Petrochemical Conference, "Next Generation MERT Achieves Further Advances in Cracking Tube Performance.", Kuala Lumpur, April 2005.

Effect of Operating Conditions and Feedstock Composition on Run lengths of Steam Cracking Coils

Dr. ir. Kevin M. Van Geem

Laboratory for Chemical Technology, Universiteit Gent

Ir. Steven P. Pyl

Laboratory for Chemical Technology

Universiteit Gent

Prof. dr. M.F. Reyniers

Laboratory for Chemical Technology

Universiteit Gent

Prof. dr. ir. G.B. Marin

Laboratory for Chemical Technology

Universiteit Gent

Abstract: Many variables affect the run length of industrial steam cracking furnaces such as feedstock type, process conditions, coil geometry and addition of additives. Incorporation of experimentally validated coking models in COILSIM1D extend its capabilities from predicting product yields to simulating run lengths for gaseous and liquid feedstocks. The coking model of Plehiers predicts the coking rates for steam cracking of light gasses, while the model of Reyniers et al. allows to simulate the coking rate of heavier feedstocks ranging from light naphtha fractions, over gas oils and heavy condensates. Both coking models account for the heterogeneous non-catalytic, or so-called asymptotic, coking only. Good agreement is obtained between simulated and observed run lengths of several types of industrial-cracking furnaces (Lummus SRT I, GK1) using either ethane or naphtha as feedstock. Moreover, the proposed methodology for predicting run lengths also allows to estimate the beneficial effect of additives on industrial run lengths. The application of Si- and S-containing compounds as additives for the suppression of coke formation can potentially double the run length of industrial steam crackers.

Introduction

A problem inherently associated with the construction materials used in ethene plants is their tendency to promote the formation of a carbonaceous deposit, so-called coke, on the wall of the reactor coil and the TLE. The accumulation of coke reduces the tube cross section, increasing heat transfer resistance and pressure drop. Depending on the coke deposition rate, the furnace must be periodically shut down for decoking. Cleaning operations are carried out either mechanically or by passing steam and/or air through the coils and TLEs to burn off the coke. The loss of the furnaces' availability due to decoking, the decrease of the olefin selectivity and the energy losses associated with the accumulation of coke on the reactor wall have important negative consequences for the economics of the cracking process (Ranzi et al. 1985). Moreover, carburization can lead to deterioration and/or damage of tube materials. Hence, coke formation and equipment fouling are a major challenge in steam cracking furnaces. Therefore, ethene producers have been and continue to be interested in technologies to reduce coke formation in order to achieve longer furnace run lengths.

Many efforts have been made by academic and industrial researchers to understand the coking problem. Based on various laboratory and industrial studies, several coking mechanisms have been proposed. In general, three coke formation mechanisms have been delineated: the catalytic mechanism, the free-radical mechanism, and the droplets condensation/tar deposition mechanism (Lahaye et al., 1977; Kumar and Kunzru, 1985; Kopinke 1988, Albright, 1988; Vleneyi, 1991; Kopinke, 1993; Zou et al., 1993; Cai et al., 2001). The catalytic mechanism, producing filamentous coke with nickel and iron acting as a catalyst, is mainly important when clean metal surfaces are exposed to the gas phase at high temperatures. Although the details of the mechanism of filament formation are still heavily debated in literature (Helveg et al., 2004; Anderson and Abild-Pedersen, 2007; Zhu et al., 2007; Abild-Pedersen et al., 2006; Lamouroux et al., 2007; De Jong and Gues, 2000; Hofmann et al., 2005), it is generally accepted that it involves a series of surface reactions, leading to the decomposition of the hydrocarbon and resulting in the formation of surface carbon atoms, followed by a sequence of solution, diffusion and precipitation of carbon.

The free-radical mechanism involves reactions between the solid coke and entities in the vapor phase, including free radicals, olefins, aromatics, etc. (Wauters and Marin, 2001; Van Geem et al., 2004). As radical coke formation decreases rapidly with decreasing temperature, it can be expected that this mechanism mainly contributes to coke formation in the radiant coil and at the entrance of the TLE where the temperature is still high.

The droplet condensation mechanism applies when heavy polynuclear aromatics are either present in the feed or, are formed as a result of secondary chemical condensation reactions, such as Diels-Alder reactions. These aromatics either condense directly on the wall or, they condense in the bulk gas phase, and then subsequently collect on the wall. This mechanism is mainly important when cracking heavier feedstocks, such as gas oils, vacuum residue, and bitumen, and in places where gases are cooled such as the TLE.

Excessive coke deposits inside the furnace coils can negatively impact plant safety and productivity. Coke is known to have a tendency to deposit heavily where the diameter of the coil changes and in the coil's bents (Froment, 1990; Laheye et al., 1988; Plehiers et al., 1990). Along the reactor coil the wall temperature profile varies strongly because the firing nozzles are placed at the bottom and side walls of the furnace, creating a nonuniform flux profile. The nonuniform distribution of temperature and pressure makes coke deposit heavily on certain sections of the coil. Those sections are subjected to particularly high temperatures by the insulation effect of the coke and so-called "hot spots" can arise. Because hot spots decrease the durability of the coils and increase the coke growth rate, they can jeopardize safe and continuous operation of the furnace.

A major problem in a practical furnace system is that one cannot measure the coke thickness directly without cutting the coils of the furnace. In addition, the uneven distribution of coke makes measurement at a single point not representative. However, the coke thickness can be estimated via simulation of the product yields, the tube skin temperature profile, and the pressure profile if an adequate simulation model is available. This modeling strategy can be used in combination with modern decoking scheduling strategies such as the Proactive Scheduling Strategy of Lim et al. (2009) to determine the optimal operation conditions. In the next paragraphs an overview is given about the capabilities of the LCT software for simulating run lengths for both gas fed and liquid fed industrial furnaces. Also the influence of additives can be accounted for and will be discussed.

COILIM1D

Steam cracking is a non-isothermal, non-adiabatic and non-isobaric process. Hence, the 1 dimensional reactor model equations consist of the transport equations for mass, momentum and energy. The steady state continuity equation for a SEMK component j in the process gas mixture over an infinitesimal volume element with cross sectional surface area Ω, circumference ω and length dz is:

$$\frac{dF_j}{dz} = \left(\sum_{k=1}^{n_r} \upsilon_{kj}\, r_k \right) \Omega \qquad (1)$$

with F_j the molar flow rate of SEMK component j, $r_{V,k}$ the reaction rate of reaction k, and υ_{kj} the stoechiometric coefficient of SEMK component j. The energy equation is given by:

$$\sum_j F_j\, c_{pj} \frac{dT}{dz} = \omega\, q + \Omega \sum_k r_k (-\Delta H_k) \qquad (2)$$

with q the heat flux to the process gas, c_{pj} the heat capacity of component j at temperature T, $\Delta_f H_k$ the standard enthalpy of formation of species k, $R_{v,k}$ the net production rate for species k. The momentum equation accounting for friction and changes in momentum is given by:

$$-\frac{dp_t}{dz} = \alpha \left(\frac{2f}{d_t} + \frac{\zeta}{\pi\, r_b} \right) \rho_g\, u^2 + \alpha\, \rho_g\, u \frac{du}{dz} \qquad (3)$$

with p_t the total pressure, α a conversion factor, f the Fanning friction factor, ρ the density of the gas mixture, r_b the radius of the bend, d_t the diameter and v the velocity. The boundary conditions of the model equations are:

$T=T_0 \qquad C_j=C_{j0} \qquad p=p_0 \qquad (z = 0)$

Integration of the last two model equations is only required when the temperature and/or pressure profile are not imposed. Based on the reactions, and reaction rate coefficients, the production rate $r_{v,k}$ of each component j by reaction k, can be expressed as a function of the concentrations of the involved species. The resulting set of continuity equations forms a stiff set of non-linear first order differential equations. The stiffness of the set of equations originates from the large difference (several orders of magnitude) of the eigenvalues related to the molecular species on the one hand and the radical species on the other hand. The stiff solver DASSL (Li and Petzold, 1999) is used to integrate the stiff set of differential equations.

If coke formation needs to be accounted for at any time and position the following continuity equation needs to be added to the set of differential equations:

$$\frac{\partial C_{coke}}{\partial t} = r_c \tag{4}$$

With r_c the rate of coke formation. The position dependence is obtained through the temperature and partial pressure dependence of the rate of coke formation.

Coking models

At the operating conditions prevailing in industrial cracking units, the largest amount of coke formed during the run length results from the heterogeneous, non-catalytic coke formation (Reyniers et al., 1994). When modeling the coke formation phenomena in the cracking coil the focus has therefore been put on this mechanism. Two types of models have been developed at the LCT over the years; i.e. semi-empirical models (Plehiers et al., 1989; Reyniers et al., 1994) and a model based on elementary reactions (Wauters and Marin, 2001). Although the model based on elementary reactions should have a wider applicability it is currently too time consuming to be used in an industrial environment. The coking model of Plehiers (Plehiers et al., 1989; Plehiers and Froment, 1990) is developed for predicting coking rates for steam cracking of light hydrocarbon feedstocks. The coking rate is then calculated via the following equation:

$$r_C = A_{C2H4} \exp\left(-\frac{E_{a,C2H4}}{RT}\right) + A_{C3H6} \exp\left(-\frac{E_{a,C3H6}}{RT}\right) \tag{5}$$

The model of Reyniers et al. (1994) allows to simulate the coking rate of heavier feedstocks ranging from light naphtha fractions up to heavy condensates. This model considers 12 reactions in parallel for describing the formation of coke. The rate of coke formation is then expressed as:

$$r_C = C_{H2} C_{CH4} \sum_{i=1}^{12} k_i C_i \tag{6}$$

Both coking models have been developed based on a huge number of reference experiments performed either on the pilot plant setup or one of the bench scale setups. The pilot plant setup has been extensively discussed in previous publications (Dhuyvetter et al. 2001, Wang et al. 2007) The analysis section has been recently upgraded by adding a unique GC×GC including both a Time of Flight Mass Spectrometer (Tof-MS) and a Flame Ionization Detector (FID) that allows to obtain an even more detailed quantitative and qualitative analysis of the cracking effluent and feedstock composition. Two new bench scale installations allow to study coke formation under conditions similar to those encountered in the cracking coil and in the transfer line exchanger (Van Geem et al., submitted).

Ethane cracking furnace

To perform a complete furnace simulation several options exist with COILSIM1D. COILSIM1D can be combined with Fluent (Stefanidis et al., 2008), where Fluent provides the heat flux profile from simulating the fire box. Another possibility is to obtain the heat flux profile from one of the in house developed software codes FURNACE (Heynderickx and Nozawa, 2005) or FLOWSIM (Stefanidis et al., 2006). Van Geem et al. (2008) showed that the desired reactor outlet conditions (e.g. COT or P/E-ratio) can also be used as input. In order to solve the resulting two point boundary condition problem, the shooting method (Meade et al., 1996) is applied in an iterative procedure. This allows the program to generate the inlet pressure and the heat flux profile corresponding to the requested cracking severity indices at the reactor outlet. This last option is used in the current paper for an ethane cracking furnace. The main characteristics of this furnace are given in Table 1.

Predicting the run length of a furnace requires solving the preceding set of differential equations. To do so, the run time is increased in a stepwise manner. Incremental pseudo steady state is assumed for the coking since the main cracking reactions are much faster than the coke formation. This means that the rate of coke formation may be considered constant in each time interval. For the present simulation runtime increments of 100 hours were found to be appropriate. For each time increment a COILSIM1D simulation is performed shooting at the specified outlet condition, i.e. an outlet ethane conversion of 64%.

Table 1: Furnace, reactor geometry and process conditions for ethane cracking furnace

Furnace	
Furnace length	9.60 m
Furnace height	12.85 m
Furnace width	1.80 m
Thickness refractory material	0.25 m
Thickness insulation material	0.05 m
Reactor coil	
Type	swaged coil
Number of reactors	4
Number of passes	8
Reactor length	100.96 m
Internal reactor diameter (passes 1-6)	0.124 m
External reactor diameter (passes 1-6)	0.140 m
Internal reactor diameter (passes 7-8)	0.136 m
External reactor diameter (passes 7-8)	0.152 m
Wall thickness	0.008 m
Ethane flow rate per reactor coil	2800 kg h^{-1}
Steam dilution	0.365 kg/kg
CIT (a)	656 °C
COP (b)	1.95 10^5 Pa
Composition of the feed	
Ethane	100.0 wt%
Outlet specification	
Ethane conversion	63.4 %

Performing this kind of simulation is straightforward with the new GUI developed for COILSIM1D. One can now easily specify the reactor geometry, feedstock composition and necessary outlet conditions. Figure 1 shows the resulting outlet composition simulated under initial conditions.

Achieving a constant conversion level requires a constant total heat input to the process gas. As more coke is formed, see Figure 2, the external tube wall temperature rises and with it the average flue gas temperature because in order to maintain the heat supply to the process gas, the firing duty has to be increased with 5%.

For the simulations the coil outlet pressure was kept fixed in time as was measured in the industrial plant. Due to the formation of the coke layer the total pressure drop over the reactor increased with 3.7 10^4 Pa, see Figure 3, which corresponds well with the industrially observed value of 4.0 10^4 Pa.

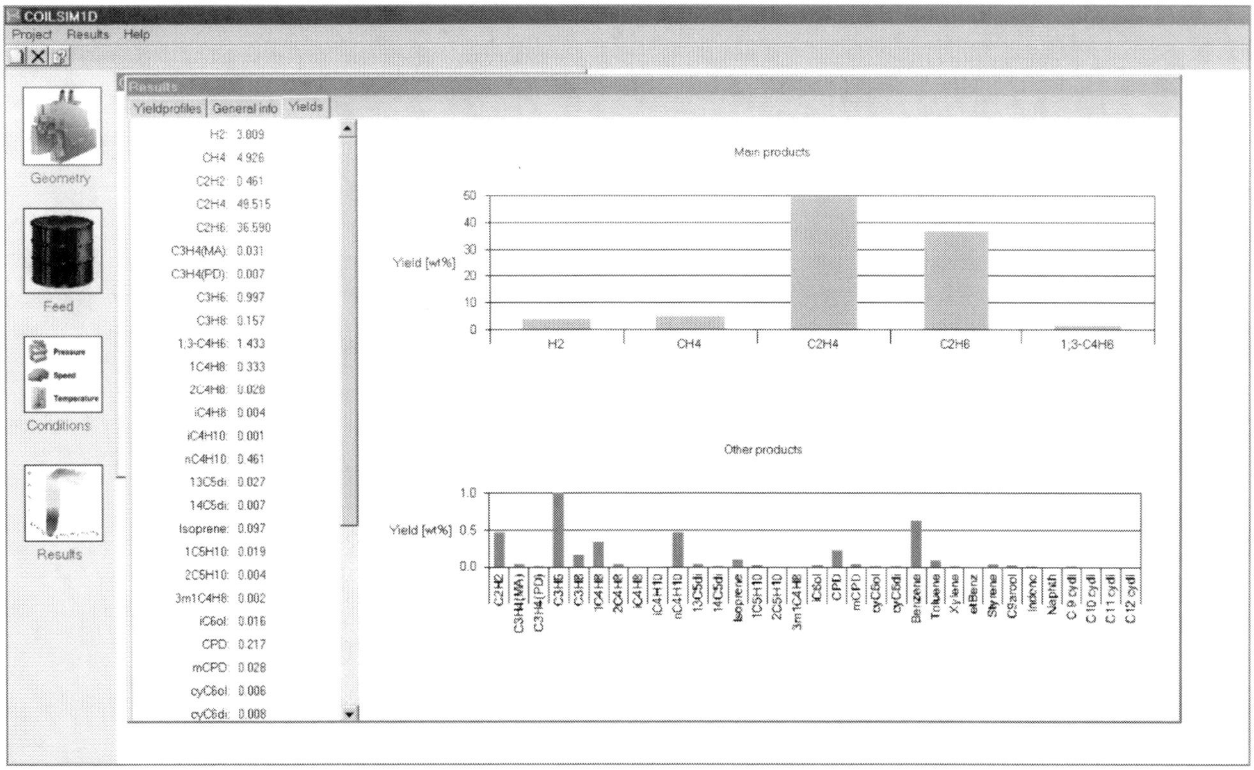

Figure 1. Simulated outlet composition under the initial conditions for an ethane furnace, i.e. no coke deposited on the reactor wall [CIT = 656°C, X_{ethane} = 63.4 wt%, COP = 1.95 10^5 Pa]

The maximum in the coke thickness is reached in the last U bend of the coil and amounts to 13.0 10^{-3}m. The peaks correspond with the bottom of the furnace, where the flue gas temperature reaches its highest values. The valleys correspond with the top of the firebox, where the temperature is lower and hence the coking rate is relatively lower than in the bottom. The average rise of the maximum temperature is 1.75 °C a day. The total run length is determined based on the maximum allowable tube skin temperature, i.e. 1054.4°C. This limit is reached after 1350 hours, which is in perfect agreement with the actual run length of the industrial coil of 1344 hours.

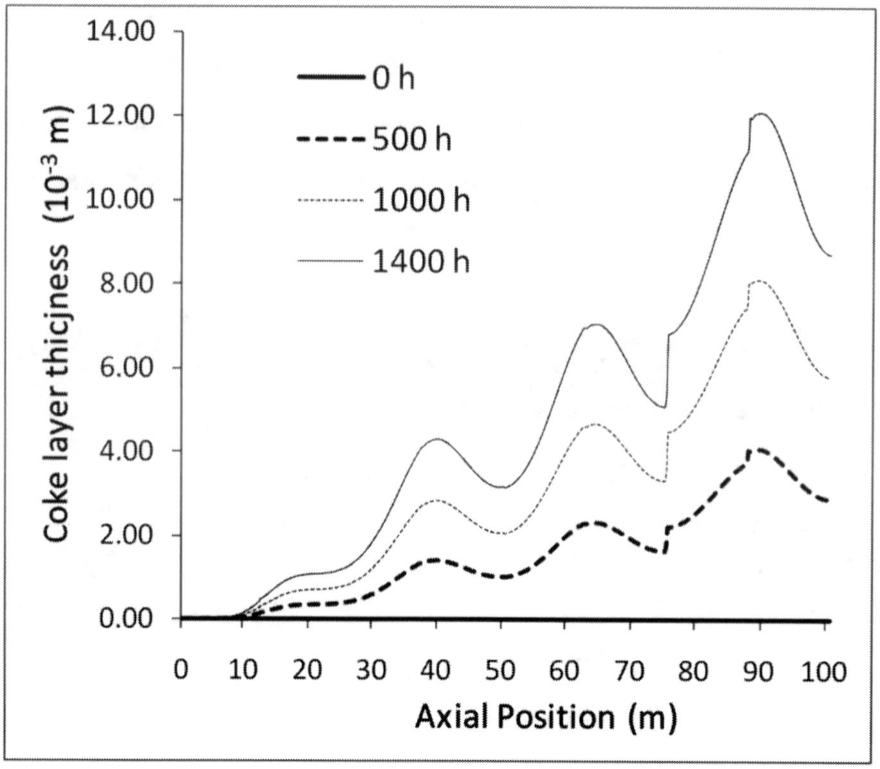

Figure 2. Simulated coke thickness as function of time on stream for an Ethane furnace

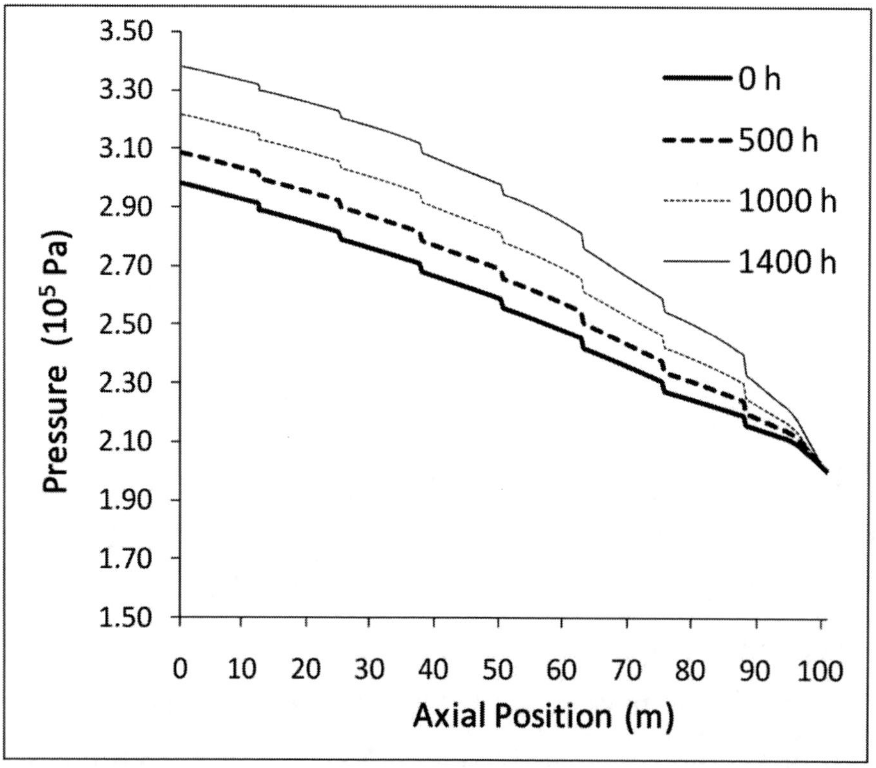

Figure 3. Simulated pressure profile as function of time on stream for an Ethane furnace

Naphtha cracking furnace

The same methodology as the one described in the previous chapter can also be applied for a naphtha cracking furnace. Table 2 summarizes the characteristics of the naphtha furnace. For the feedstock only a limited number of commercial indices are available, and hence, the detailed molecular composition needs to be reconstructed first. The method of Van Geem et al. (2007, 2008) is used. The resulting detailed PIONA weight fraction distribution is given in Table 3.

Table 2: Furnace, reactor geometry and process conditions for ethane cracking furnace

Reactor coil	
Type	Splitcoil
Number of reactors	8
Number of passes	6
Reactor length	53.89 m
Internal reactor diameter (passes 1-4)	0.080 m
External reactor diameter (passes 1-4)	0.096 m
Internal reactor diameter (passes 5-6)	0.114 m
External reactor diameter (passes 5-6)	0.130 m
Wall thickness	0.008 m
Ethane flow rate per reactor coil	2785 kg h^{-1}
Steam dilution	0.70 kg/kg
CIT (a)	620 °C
COP (b)	1.45 10^5 Pa
Feedstock characteristics	
PINA analysis (wt%)	
n-praffins	37.0 wt%
isoparaffins	33.0 wt%
naphthenes	18.0 wt%
aromatics	12.0 wt%
ASTM D86 BP (wt%)	
IBP	36.0
50%	94.5
FBP	161.0
Specific density	0.706
Outlet specification	
Ethylene yield	28.6 wt%

Table 3. Reconstructed feedstock composition

(Wt%)	C_3	C_4	C_5	C_6	C_7
Paraffins	0.00	3.36	10.72	7.86	5.12
Isoparaffins	0.00	0.00	6.48	7.29	5.81
NACC[1]	0.00	0.00	0.89	4.59	4.05
Cyclic C_5	*0.00*	*0.00*	*0.89*	*1.74*	*2.70*
Cyclic C_6	*0.00*	*0.00*	*0.00*	*2.84*	*1.35*
Aromatic	0.00	0.00	0.00	0.56	1.42
TOTAL	**0.00**	**3.36**	**18.09**	**20.30**	**16.39**
(Wt%)	**C_8**	**C_9**	**C_{10}**	**C_{11}**	**C_{12}**
Paraffins	4.09	3.51	1.97	0.38	0.00
Isoparaffins	4.73	4.09	3.90	0.70	0.00
NACC[1]	3.92	4.56	0.00	0.00	0.00
Cyclic C_5	*2.35*	*2.28*	*0.00*	*0.00*	*0.00*
Cyclic C_6	*1.57*	*2.28*	*0.00*	*0.00*	*0.00*
Aromatic	3.95	6.07	0.00	0.00	0.00
TOTAL	**16.69**	**18.23**	**5.87**	**1.08**	**0.00**

Figure 5 shows the evolution with time of the external wall temperature profile for the considered reactor. The maximum observed in this profile should be watched carefully during operation, since there is a limitation imposed by the metallurgy. In the present case the limiting value is reached after 2000 hours or 83 days. The simulated value is in excellent agreement with the observed plant value of 85 days.

[1] NACC: Non-aromatic cyclic compounds with a 5-membered ring, a 6-membered ring, and bicyclic compounds

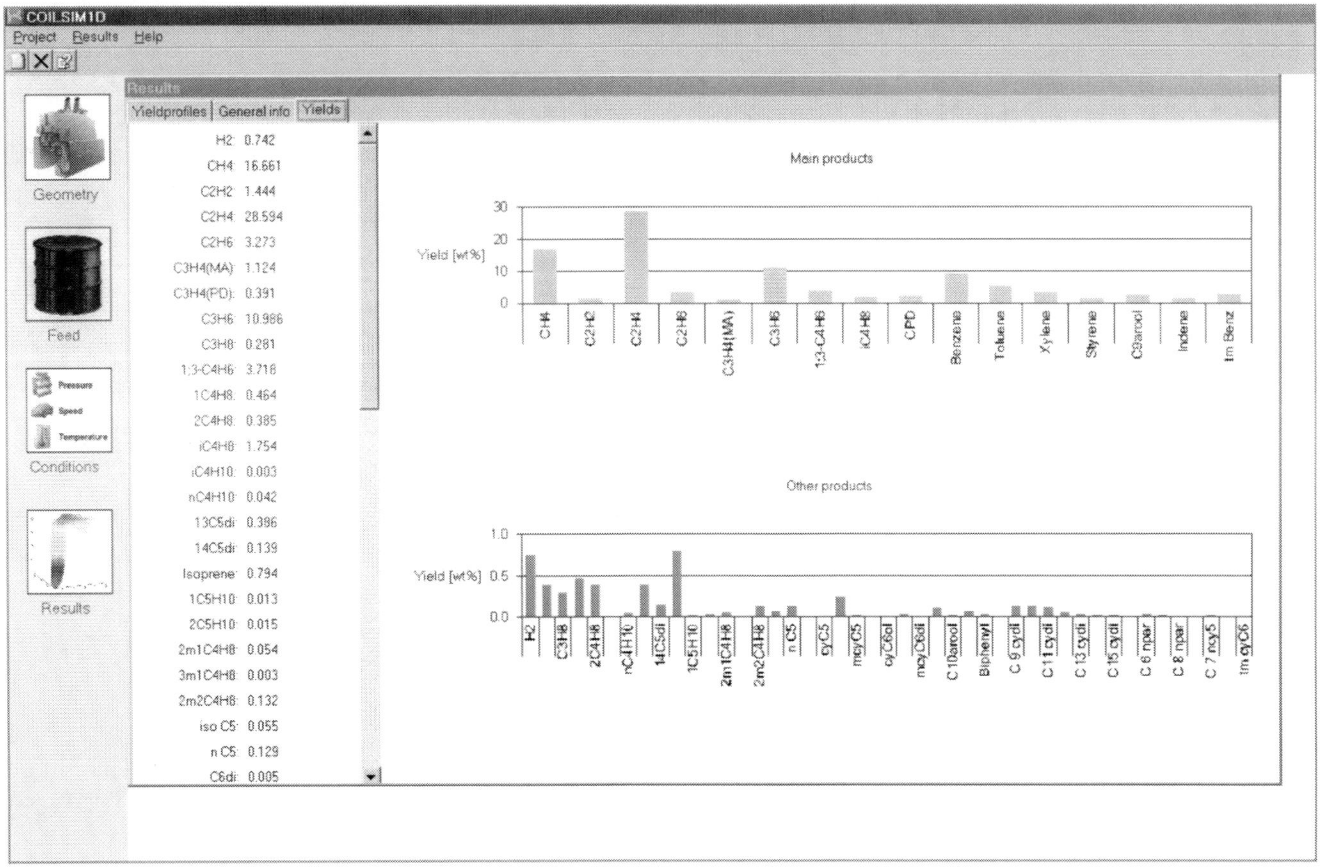

Figure 4. Simulated outlet composition under the initial conditions for an ethane furnace, i.e. no coke deposited on the reactor wall [CIT = 620°C, Ethylene yield = 28.6 wt%, COP = 1.45 10^5 Pa]

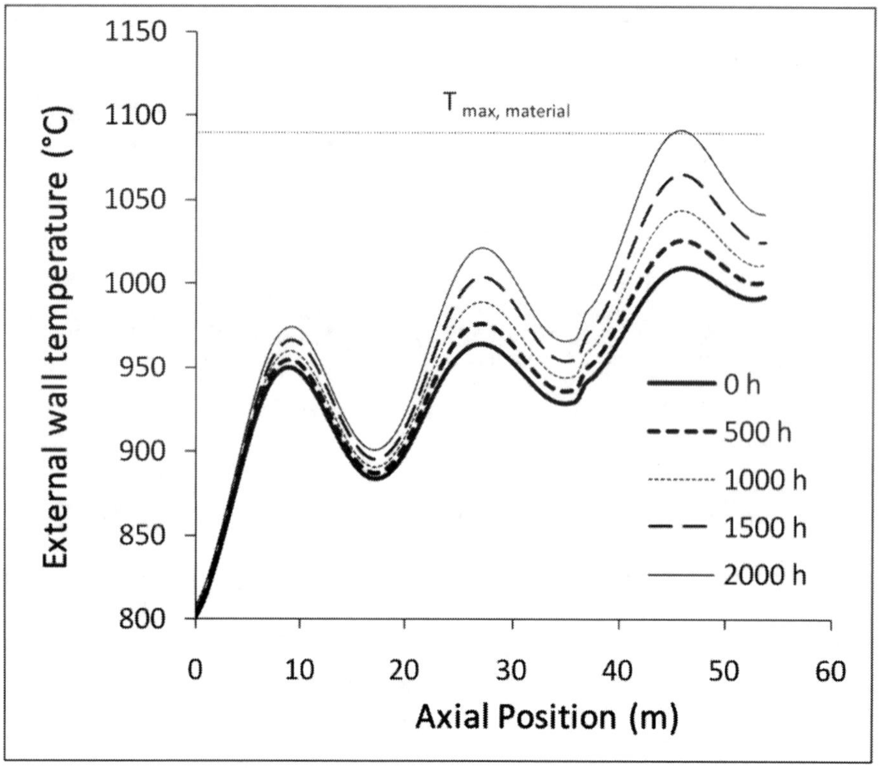

Figure 5. External wall temperature profile as function of time on stream (naphtha furnace)

Additives

The influence of the combination of two Si-containing additives, BTMS and TEOS, with DMDS on the steam cracking of hexane is first experimentally evaluated in a CSTR set-up (Wang et al., 2008). Application of the Si additives does not cause a significant change in the conversion of hexane and the product selectivities. For the combination of Si-pretreatment with presulfidation, the coke formation during the steam cracking of hexane is found to be affected by the amount of DMDS, the temperature and the duration of the presulfidation (Wang et al., 2008). Under these optimal presulfidation conditions, coke deposition is minimal and the combination of Si-pretreatment and presulfidation results in a decrease in the rate of coke formation by 32%. The results observed in the CSTR set-up are confirmed in coking runs performed in the pilot plant set-up for a set of ethane cracking experiments. Si-pretreatment + presulfidation + continuous addition of 2 ppm-wt DMDS reduces the coke formation by approximately 70% and the CO production by more than 90%. The suppressing effect of the

Si-pretreatment in the pilot plant is assigned to the deposition of a thin layer of SiO2 on the inner surface of the reactor acting as a physical barrier and preventing direct contact between the process gas stream and the alloy.

COILSIM1D allows to take in to account the effect of dosing additives by assuming that that the effect of the additives is temperature independent. Hence, the influence of the use of the additives on the amount of coke deposited can be simulated by multiplying the pre-exponential factors in equation (5) with a temperature independent reduction coefficient. Comparison between the measured amount of coke in the pilot plant and the one simulated with COILSIM1D indicates that the coking rate is reduced with approximately 65 % in the experiments upon Si/S-pretreatment followed by a continuous addition of 2 ppm of DMDS with the ethane feed.

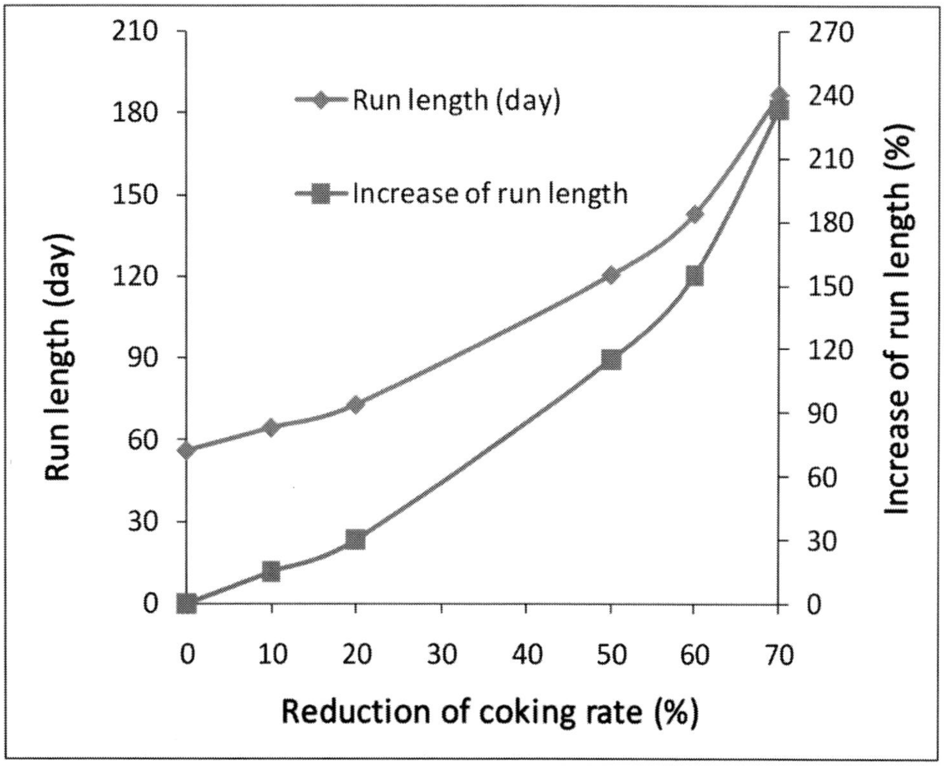

Figure 6: Simulated run length of an industrial ethane furnace as a function of the uniform reduction of coke formation [CIT = 656°C, X_{ethane} = 63.4 wt%, COP = 1.95 10^5 Pa]

These results can be extrapolated to estimate the maximal achievable run length in an industrial ethane furnace pretreated and operated under the same conditions. The run length as a function of the uniform reduction of the coking rate is given in Figure 6. A given reduction in the coking rate results in a more than proportional increase of the run length. For instance, a reduction in the coking rate of 50% increases the run length from 56 to 120 days; the increase in run length amounts to 115%. Application of the Si/S-containing additives at the optimal conditions in the pilot plant set-up decreases the coke formation by approximately 65%. Assuming that the same reduction of the coking rate can be obtained in an industrial ethane cracker, its run length can reach 186 days, i.e. an increase by 233%. Simulation of the run length for an industrial ethane cracker indicates that the run length can at least be doubled if the reduction of the coking rate reaches 65%, as is observed in the pilot plant tests. The use of a combination of Si- and S-containing compounds offers a promising method to effectively reduce coke formation in steam cracking of hydrocarbons.

Conclusions

Mechanistic considerations and experimental information were used to develop coking models that allow predicting the coking rate for both gaseous and liquid feedstocks. These coking models can be combined with a reactor model and a furnace model to simulate the run length of industrial steam cracking furnaces. The growth of the coke layer is accurately simulated, and so is the evolution of the external tube skin temperature profile. The simulated and experimental run lengths agree within 5%. COILSIM1D allows to take in to account the effect of dosing additives by assuming that the effect of the additives is independent of the temperature.

Simulations of this kind can be used to optimize furnace operation, including decoking scheduling, for various feedstocks and operating conditions. They can be used as a guide for the adaptation of the operating variables, aiming at a prolongation of the run length of the furnace.

References

1. Albright, L. F.; Marek, J. C. (1988) Mechanistic Model for Formation of Coke in Pyrolysis Units Producing Ethylene. *Ind. Eng. Chem. Res.*, 27, 755-759.
2. Velenyi, L. J.; Song, Y.; Fagley, J. C. (1991) Carbon Deposition in Ethane Pyrolysis Reactors. *Ind. Eng. Chem. Res.*, 30, 1708-1712.
3. Andersson, M.P.; Abild-Pedersen, F. (2007) Carbide induced reconstruction of monatomic steps on Ni(111) – A density functional study. *Surface Science* , 601, 649–655.
4. Abild-Pedersen, F.;Nørskov, J.K.; Rostrup-Nielsen, J.R.; Sehested, J.; Helveg, S. (2006) Mechanisms for catalytic carbon nanofiber growth studied by ab initio density functional theory calculations.*Physical Review B*, 73.
5. Cai, H.; Krzywicki, A.; Oballa, M. C. Coke Formation in Steam Crackers for Ethylene Production. *Chem. Eng. Process.(* 2002), 41, 199-214.
6. De Jong, K.P; Geus, J.W. (2000)Carbon Nanofibers: Catalytic Synthesis and Applications *Catal. Rev.-Sci. Eng.*, 42(4), 481–510.
7. Dhuyvetter, I., Reyniers, M.F., Froment, G.F. and Marin, G.B. (2001) The Influence of Dimethyl Disulfide on Naphtha Steam Cracking. *Ind. Eng. Chem. Res.,* 40, 4353-4362.
8. Froment, G. F. (1990) Coke formation in the thermal cracking of hydrocarbons. *Rev. Chem. Eng.*, 6 (4), 293–928.
9. Harutyunyan, A. R.; Tokune, T. (2005) Liquid as a required catalyst phase for carbon single-walled nanotube growth. *Applied physics letters*, 87.
10. Helveg, S.; Lopez-Cartes, C.; Sehested, J.; Hansen, P.L.; Clausen, B.S.; Rostrup-Nielsen, J.R.; Abild-Pedersen, F.; Nørskov, J.K. (2004) Atomic-scale imaging of carbon nanofibre growth. *Nature*, 427, 426-429.
11. Heynderickx, G.J., Nozawa, M. (2005) Banded gas and nongray surface radiation models for high-emissivity coatings. *AIChE J.*, 51 (10), 2721-2736.
12. Hofmann S, Csanyi G, Ferrari AC, Payne MC, Robertson J. (2005) Surface diffusion: The low activation energy path for nanotube growth. *Physical review letters*, 95, 3.

13. Kopinke, F.-D.; Zimmermann, G.; Reyniers, G. C.; Froment G. F. (1993) Relative Rates of Coke Formation from Hydrocarbons in Steam Cracking of Naphtha. 2. Paraffins, Naphthenes, Mono-, Di-,and Cycloolefins, and Acetylenes. *Ind. Eng. Chem. Res.*, 32, 56-61.

14. Kopinke, F.D. G. Zimmermann, S. Nowak (1988) On the Mechanism of Coke Formation in Steam Cracking – Conclusions from results obtained by Tracer Experiments, *Carbon*, 26, 117-124.

15. Kumar, P.; Kunzru, D. (1985) Kinetics of Coke Deposition in Naphtha Pyrolysis. *Can. J. Chem. Eng.*, 63, 598-604.

16. Kukovitsky, F.; L'vov S.G.; Sainov, N.A. (2000) VLS-growth of carbon nanotubes from the vapour. *Chemical Physics Letters* , 317, 65–70.

17. Lahaye, J.; Badie, P.; Ducret, J. (1977) Mechanism of carbon formation during steam cracking of hydrocarbons. *Carbon*, 15 (2), 87-99.

18. Lamouroux, E.; Serp, P.; Kalck, P. (2007) Catalytic Routes Towards Single Wall Carbon Nanotubes. *Catalysis Reviews*, 49,341-405.

19. Heejin Lim, H.; Choi, J.; Realff, M.; Lee, J.H.; Park S. (2009) Proactive Scheduling Strategy Applied to Decoking Operations of an Industrial Naphtha Cracking Furnace System, *Ind. Eng. Chem. Res.* In press

20. Mora, E.; Harutyunyan, A.R. (2008) Study of Single-Walled Carbon Nanotubes Growth via the Catalyst Lifetime. *J. Phys. Chem. C*, 112, 4805-4812

21. Plehiers P.M., Reyniers G.C. and Froment G.F. (1990) Simulation of the Run Length of an Ethane Cracking Furnace, *Ind. Eng. Chem. Res.*, 29, 636-644.

22. Plehiers P.M. and Froment G.F. (1989) Coupled Simulation of Heat Transfer and Reaction in a Steam Reforming Furnace, *Chemical Engineering Communications*, 80, 81-99.

23. Ranzi, E.; Dente, M.; Pierucci, S.; Barendregt, S.; Cronin, P. (1985) Coking simulation aids on-stream time, *Oil Gas J.*, 49- 52.

24. Reyniers G.C., Froment G.F. Kopinke F.D. and Zimmerman G. (1994) Coke formation in thermal cracking of hydrocarbons. 4. Modeling coke formation in, naphtha cracking, *Ind. Eng. Chem. Res.*, 33, 2854-2850.

25. Stefanidis, G.D., Van Geem, K.M., Heynderickx, G.J., Marin, G.B. (2008) Evaluation of high-emissivity coatings in steam cracking furnaces using a non-grey gas radiation model, *Chemical Engineering Journal.*

26. Stefanidis, G.D., Merci, B., Heynderickx, G.J., Marin, G.B. (2006) CFD simulations of steam cracking furnaces using detailed combustion mechanisms. *Comp. & Chem. Eng.,* 30 (4): 635-649,.

27. Van Geem, K. M.; Heynderickx, G. J.; Marin, G.B. (2004) A Comparison of One and Two-dimensional Reactor Models for Steam Cracking: Effect on Yields and Coking rate. *AIChE J.*, 50, 173 – 183.

28. Van Geem, K.M., Reyniers, M.F. and Marin G.B. (2005) Two Severity Indices for Scale-Up of Steam Cracking Coils, *Ind. Eng. Chem. Res.*, 44, 3402-3411.

29. Van Geem, K.M., Hudebine, B., Reyniers, M.F., Whal, F., Verstraete J.J. and Marin G.B. (2007) Molecular reconstruction of naphtha steam cracking feedstocks based on commercial indices. *Comp. & Chem. Eng.,* 31, 1020-1034.

30. Van Geem, K.M., Reyniers, M.F., and Marin G.B. (2008) Challenges of Modeling Steam Cracking of Heavy Feedstocks. *Oil & Gas Journal. Rev. IFP.*

31. Wang, J.; Reyniers, M.F.; Marin, G.B. (2007) Influence of dimethyl disulfide on coke formation during steam cracking of hydrocarbons. *Ind. Eng. Chem. Res.*, 46, 4134-4148.

32. Wang, J.; Reyniers, M.F.; Van Geem, K.M.; Marin, G.B. (2008) The Influence of Silicon and Silicon/Sulfur-Containing Additives on Coke Formation during Steam Cracking of Hydrocarbons. *Ind. Eng. Chem. Res.*, 47, 5, 1468-1482.

33. Wauters, S.; Marin, G. B. (2002) Kinetic modeling of coke formation during steam cracking. *Ind. Eng. Chem. Res.*, 41(10), 2379-2391.

34. Zhu, Y-A; Dai, Y-D.; Chen, D.; Yuan, W.K. (2007) First-principles study of C chemisorption and diffusion on the surface and in the subsurfaces of Ni(111) during the growth of carbon nanofibers. *Surface Science*, 601, 1319–1325.

35. Zou, R.; Lou, Q.; Mo, S.; Feng, S. (1993) Study on a Kinetic Model of Atmospheric Gas Oil Pyrolysis and Coke Deposition. *Ind. Eng. Chem. Res.*, 32, 843-847.

LUBE OIL IN ETHYLENE REFRIGERATION MACHINE

Jared T. Simpson (Author)
Chemical Engineer
Eastman Chemical Company, Longview, TX
G Michael McLeod (Co-Author)
Assistant Superintendent, HCC-3 Operations
Eastman Chemical Company, Longview, TX

Abstract: Following a plant turnaround in mid-2007, the ethylene refrigeration compressor was restarted. During restart, the compressor began to surge. The compressor was shut down due to high amps with an apparent seal failure. The older design of the wet seals allowed approximately 400 gallons of lube oil to leak into the refrigeration system. The seal was replaced and the majority of this oil was removed using a toluene wash followed by several washes with a mix of n-propanol and methanol. The compressor was restarted without issue and ran nine months until a power failure tripped the plant.

During the power failure the refrigeration system warmed allowing the stagnant oil to move into undesired locations. Upon startup, the plant experienced poor heat transfer and high pressure drop in the cold box. This resulted in heavy ethylene losses from the demethanizer, and below target production due to pressure drop concerns. When normal operating conditions could not be met the refrigeration system was warmed and cleaned using toluene to disperse the oil and chased with methanol several times. This effort had little effect on improving heat transfer in the cold box. The plant was shut down to inspect the demethanizer and clean the cold box. During this shutdown the first, second, third stage suction drums and demethanizer condenser were cleaned by circulating toluene and chased with methanol. The plant was brought online and heavy losses to fuel were still occurring. To help increase heat exchange an additional exchanger was put online. This exchanger has cracked gas on the tube side and ethylene refrigeration on the shell side. Using this exchanger cooled the demethanizer feeds resulting in better tower performance and lower pressure drop. For weeks following this restart of the compressor, the shell side of the exchanger was blown down several times a shift to remove any oil. This greatly improved heat transfer in the cold box, but did not restore optimal performance.

The plant continues to operate with lower heat transfer in the cold box and higher than normal losses to fuel. Several actions were taken to reduce the potential for this failure during startup. Startup procedures were revised to ensure the seal pots are unblocked when putting compressor oil system into service and starting the refrigeration compressor. A shutdown strategy was developed to prevent damage due to prolonged compressor surging. The attempts to remove the oil resulted in increased downtime and loss of production. This paper will further describe the mechanisms of failure and provide more detail on the activities taken to remedy the problem and concepts to reduce the severity.

Oil Contamination and Removal

The cracking plant described in this paper makes use of wet seals for the propylene and ethylene refrigeration compressors. The refrigeration system is an open-loop system and spans across the fractionation section of the plant. The low point in the system is a core exchanger that transfers heat from the refrigeration system to cracked gas on its way to fractionation. Figure 1 shows how oil contamination ends up the cold box and how loss of heat transfer increases losses in the demethanizer.

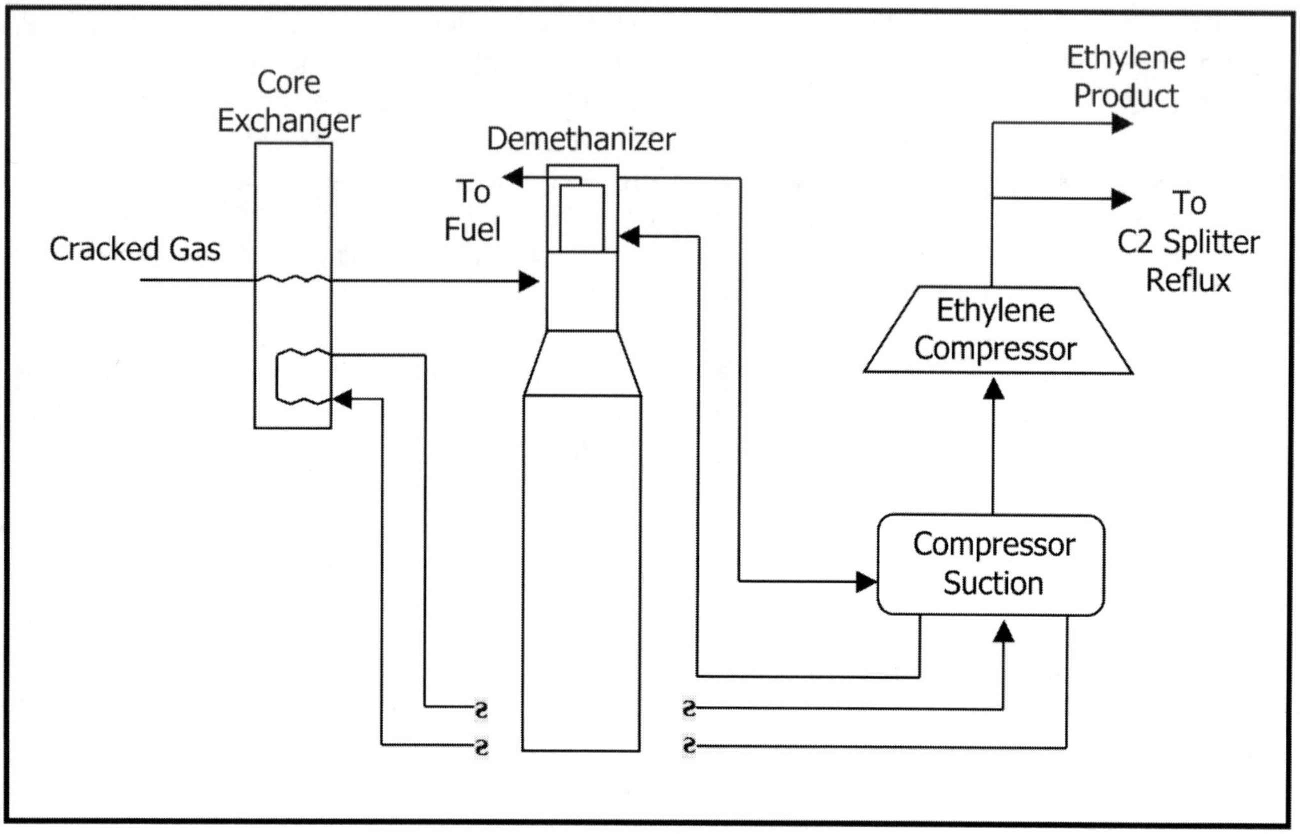

Figure 1 – Diagram of Open Loop Refrigeration

Following the cracking plant turnaround, the ethylene refrigeration compressor was put online and ran smoothly. Approximately two hours into run the differential pressure on the C2 splitter increased by 2 psig. The intermediate reflux pump for the tower was put online and the differential pressure on the tower dropped. The amps of the ethylene compressor increased but stabilized. Thirty minutes passed until the levels in the first and second stage suction drums began to fluctuate and the compressor began to surge. This caused the compressor amps and discharge pressure to swing. The kickback controllers were left in manual to keep them from chasing the surge. Three hours into the run the auxiliary oil pump kicked on and bearing vibrations increased above scale. This was believed to be the result of the compressor surging and the seal pots being blocked in. Five hours into run the compressor shutdown due to high amps. During the startup/shutdown there was no damage to other equipment except the seal to the ethylene compressor.

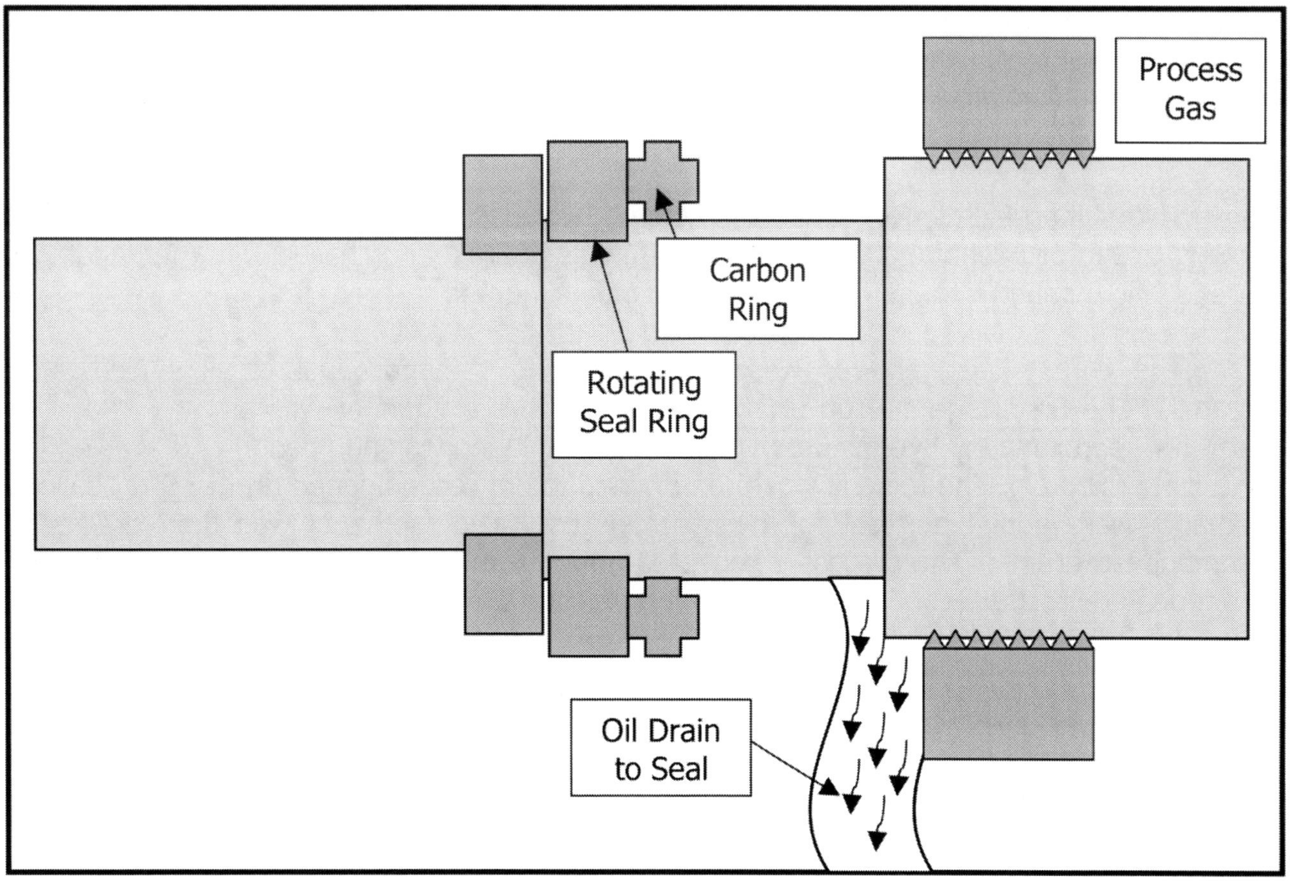

Figure 2 – Diagram of Wet Seal

When the seal was damaged approximately 400 gallons of oil contaminated the ethylene system. The contamination spread from the cold box to the C2 splitter reboilers. While the seal was being replaced, the free oil was removed from the system. The sight glasses for the C2 splitter, the C2 splitter reflux drum, the first, second, and third stage suction drums were all isolated. Eight drums of toluene were pumped via portable air pump into the C2 splitter reflux tank. The tank soaked for thirty minutes while nitrogen was bubbled at the base. The toluene was let down into the third stage suction drum and allowed to soak for thirty minutes. The reflux tank was blown dry and 400 gallons of methanol were pumped into the tank. An additional twelve drums of toluene were added to the third stage suction and allowed to soak. After thirty minutes it was let down into the second stage suction. The methanol from the reflux tank was let down into the third stage suction and 600 gallons of methanol were added. After a settling period the toluene in the second stage was let down into the first stage. The methanol from the third stage was then let down into the second stage. After the toluene in the first stage settled for 30 minutes in was pumped to the C2 splitter bottoms pump suction.

The bottom's pump of the C2 splitter was off and blocked in. The toluene from the first stage suction drum was pumped into the C2 splitter bottom's pump suction line and pumped up the column and allowed to set for thirty minutes. After the settling period the bottom's pump, pumped the toluene into the column until pump suction was loss. Nitrogen was blown through the column during the cleaning process. The toluene was allowed to fall through the column and settled in the column's reboiler. The toluene was than pumped into a storage container and vacuumed trucked out. Methanol was added to the column in the same manner.

The sight glasses for each tank/column were isolated and filled with toluene and allowed to set for twenty minutes. The toluene was drained and chased with methanol. Then toluene was pumped into a storage container and vacuumed trucked out.

The core exchanger was washed by isolating the individual ethylene refrigeration streams. The first stage suction drum flow to the core exchanger was blocked in and blow cased for approximately five minutes. One drum of toluene was pumped in and allowed to set for twenty minutes. The toluene was then drained into a reconditioned drum. Sixty gallons of methanol were added and set for twenty minutes. The methanol was than drained into reconditioned drums. The process is repeated until the methanol out is oil free. This method was repeated for the second stage suction stream through the core exchanger.

All bleeds to the ethylene refrigeration system were blown down until dry. The process of replacing the seal of the compressor and cleaning the refrigeration system added an additional seven days to the shut down. The plant was restarted without any issues and normal operating conditions were achieved. The plant ran for nine months until a power interruption shut down the refrigeration compressors and warmed the refrigeration system.

Loss of Heat Transfer in Cold Box

After the power interruption the plant restarted but poor heat transfer in the cold box was observed. Feed temperatures to the demethanizer were fifteen to twenty degrees warmer than normal. During the power outage the oil that had not been removed during the original cleaning thawed and flowed into the core exchanger. The oil plugged the refrigeration flow causing lower heat transfer in the exchanger. This increase in feed temperatures to the demethanizer caused the column to have poor separation. The overhead ethylene losses to fuel tripled.

The ethylene refrigeration streams through the core exchanger needed to be isolated and cleaned. Each pass was blocked in, one at a time, and filled with toluene. The toluene was allowed to settle for 20 minutes and then drained into a reconditioned drum. After draining, the toluene was chased with methanol. This process was repeated until the methanol contained no oil residue. After several cleanings of the core exchanger the tower still did not see an increase in separation. A gamma scan of the demethanizer found the top section of the tower to have several trays of flooding, a three tray vapor space, and flooding for three trays above the vapor space. Tray damage was suspected, and the plant was shut down to inspect the tower.

While the plant was down the first, second, third stage suction drums, and demethanizer condenser were isolated and cleaned. A 2000 gallon bulk tank of toluene was delivered to the cracking plant. Temporary piping was connected to the suction drums and the demethanizer condenser. The toluene was circulated from the demethanizer condenser to the suction drums for a period of 24 hrs. The toluene was than pumped back into the toluene bulk tank. The system was than filled with methanol and blown down into drums. The process of filling the system with methanol and blowing the system down was repeated until there was no oil residue in the methanol. Several gallons of oil were removed from the process. Before startup the ethylene refrigeration system was blown until dry. Figure 3 shows a picture of a double block and bleed assembly that was used to relieve line pressure and ensure the line was dry before disconnect.

Figure 3 – Double Block and Bleed Assembly

The plant was restarted without issue, but poor heat transfer in the cold box and high ethylene losses to fuel continued. Several attempts were made to clean the ethylene refrigeration system over a period of two months with the plant online. Each cleaning yielded marginal improvements in heat transfer and tower performance. To help increase the heat transfer an additional exchanger was placed on line. Figure 4 shows the placement of this exchanger in the refrigeration system. This exchanger has cracked gas from the second feed drum to the demethanizer on the tube side and ethylene refrigeration on the shell side. Oil that accumulated was removed via blow down several times a shift. This addition vastly improved heat transfer and tower performance, but did not fully restore performance to normal operations.

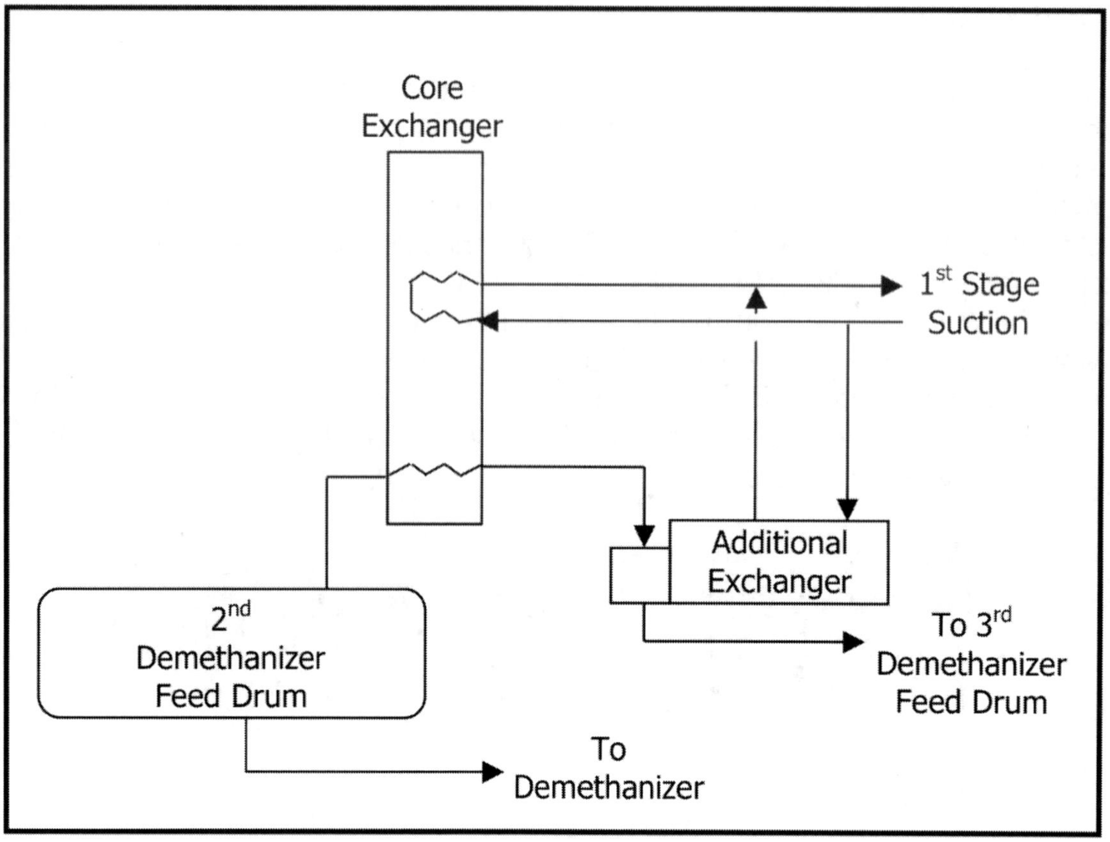

Figure 4 – Addition of Exchanger to Refrigeration System

Changes and Current Operations

Several changes were made to lower the chance of this type of failure. Startup procedures were revised to ensure seal pots are unblocked prior to putting compressor oil systems into service. A start-up checklist was created that includes seal pots and correct buffer gas alignment. The parts list was updated to use only original equipment manufacture's seals and parts. A shutdown strategy was developed to prevent prolonged surging. The strategy focused around a time limit it for startup where the startup would be aborted and the cause of the surging identified prior to attempting another startup. A high seal differential alarm was investigated, but found to not be necessary. The low seal differential alarms are critical and already on the system.

Procedures are in place to reduce the probability and severity of this problem. It encompasses a differential pressure being maintained between the seal and the sour gas side of the compressor. If the differential pressure is lost or becomes too low suggesting large amounts of oil could leak into the system, it is shut down and isolated as quickly as possible. The contaminated equipment will receive several washes with a solvent to disperse the oil and chased with chemical that can remove the solvent, yet not freeze if residual remains.

Since system performance degrades with each shutdown it has been concluded that very small amounts of oil still remain in the system and accumulates in low spots (i.e. core exchanger) where they cannot be removed by available means with the plant operational. As a result, during each shutdown the refrigeration system is warmed and the core exchanger is cleaned with toluene and chased with methanol.

Operational Issues and Solutions with Brazed Aluminum Heat Exchangers

Melissa Wiese, Production Engineer
Equistar Chemicals, A LyondellBasell Company

Brazed aluminum heat exchangers are used widely across the petrochemical industry. Their high heat transfer area and small size make them especially ideal for use in cold fractionation and refrigeration sections of olefins plants. A brazed aluminum exchanger can provide 6-10 times greater heat transfer surface area density per volume at 25-50 percent lower initial cost than shell and tube heat exchangers. (Chart, 2006, Lower Capital Cost section) However, the compact design and delicate construction of these exchangers can cause unique operational issues and constraints over their conventional counterparts. Nonetheless, with careful engineering and operation, these issues can be overcome or be significantly mitigated.

Core exchangers, or brazed aluminum exchangers, are unique in that they allow heat exchange from multiple streams within the same exchanger. Therefore, design of these exchangers is a complex process and must be done in close conjunction with the chosen manufacturer. Most importantly, the design is created with the following limits in mind: pass-to-pass temperature differential and chill-down rate. The purpose of these limits is to preserve the mechanical integrity of the exchangers by avoiding failures caused by thermal stresses. This complex analysis is typically completed by the core exchanger manufacturer.

A core exchanger is made up of multiple small layers of fins and plates. Each layer must be hand assembled and coated with flux before being brazed. Contamination during assembly can cause premature failure of the brazed joints, so extra care is made to keep the assembly area free of dirt and grime. After brazing is complete, the core exchanger emerges as one solid piece. At this point, any internal issues are permanent and cannot be repaired by disassembly. Using special welding techniques, the cores are then outfitted with headers and flanges and then tested before being shipped to a customer.

While very efficient, these exchangers are susceptible to unique failure modes over conventional shell and tube heat exchangers. The overriding number of failures comes primarily from thermal fatigue. Equistar's plant in LaPorte, Texas suffered a failure just prior to their turnaround in 2007. An external hole developed in one of their core exchangers at a parting sheet, the metal that separates one layer from another. The leak was detected with a LEL meter and was clearly visible after the insulation was taken off. The failure was attributed to two primary root causes: high pass-to-pass temperature differentials in excess of the design parameters (50°F for liquid to gas and 100°F for gas to gas passes) and high thermal ramp rate upon chill-down in excess of design (1.8°F / min).

In September 2008, work done by operations and engineering at Equistar's two olefins plants in Channelview, Texas (OP-1 and OP-2) confirmed that operation within the design conditions is possible under startup conditions. Each of the two nearly identical

Lummus plants has a set of two core exchangers in the demethanizer feed chilling section. As the active passes in the core exchangers on startup are gas phase, this group was able to use the more generous guideline for pass to pass temperature differential, but was still held to the 1.8°F / minute chilling rate limit. During startup, refrigeration was introduced to each shell and tube exchanger in the demethanizer chilling train at normal levels per the startup procedure. The engineering group identified temperatures on the outlet of each pass of the core exchanger to monitor during startup. Their analysis revealed that the temperatures on the outlet gave them the most accurate view of the bulk metal chilling rate in the core exchangers over the inlet temperature. The rate of temperature drop was determined to be overwhelmingly a function of gas flow rate through the exchangers and not the temperature or level of the refrigerant. Once the cores were chilled down as much as possible with only refrigerant, introduction of demethanizer reflux drum liquid, known at the Channelview plant as medium pressure methane, to the cores was made to chill the cores down even further. From this point on, the startup sequence proceeded as normal.

This procedure was tried in both units in succession during the startup from Hurricane Ike. The first plant to try this new procedure, OP-2, was successful in keeping within design constraints but for a few minor incidents. The trial in OP-1, a few days later, was deemed a complete success. With the adoption of this new procedure, the Channelview plant can now control much of their future thermal fatigue failures.

Thermal fatigue, however, is just one of a handful of failure modes associated with brazed aluminum exchangers. The next most prevalent failure Equistar has seen has been caused by mercury embrittlement. In the 1980's the Chocolate Bayou plant in Alvin, Texas suffered a core exchanger failure due to mercury embrittlement. Mercury entered the plant through heavy liquid feedstocks. To combat this issue, the Chocolate Bayou plant installed a carbon bed downstream of the furnaces to catch mercury. The OP-2 plant in Channelview installed a layer of silver impregnated mole sieve in their process gas dryers for the same purpose. Testing has revealed that both systems have been successful in removing mercury from the process gas stream and have helped to protect the core exchangers from further damage.

The largest issue that has plagued the Equistar's olefins plants has been a lack of understanding of brazed aluminum exchangers; their manufacture, design, constraints, issues and solutions. Starting in 2008, a number of presentations on this subject have been made to members of the engineering community throughout the US operations. These presentations, coupled with further discussions in best practice teams in the past year have ignited a sense of urgency and ingenuity to successfully solve these unique operational problems. As this information is relayed to the operations staff, ownership over this issue continues to grow and Equistar's staff continues to minimize the odds of a premature exchanger failure.

Reference

Chart, Inc. (2006) *Brazed Aluminum Heat Exchangers,* Retrieved February 18, 2009 from http://www.chart-ind.com/app_ec.cfm

Energy Management System for Ethylene Complex

Edward C. Patterson

Process Engineering Specialist – Utilities & Energy

Shell Chemical LP – Norco Chemical Plant

Abstract: Energy costs typically dwarf other operating expenses for an ethylene plant. In addition, fuel consumed in furnaces and boilers contributes to greenhouse gas (GHG) emissions. Shell, like many companies, has goals to reduce both operating costs and GHG emissions.

Tracking and reporting some aspects of plant performance (e.g., simple yield reporting) are relatively straightforward and calculations typically involve only a few meters or other process variables. Due to heat integration between process units and the complexity of energy flows into and out of units within an ethylene complex, it is difficult to get a clear picture of overall energy performance. Energy benchmarks and the approach to performance monitoring are often ad-hoc and customized. This can lead to problems maintaining the calculations and reporting tools. It can also make it difficult to understand relative performance between even similar units.

Shell's Norco Chemical Plant developed an Energy Management System (EMS) to provide real-time reporting of energy consumption and serve as a standard platform for managing energy performance. This paper will provide an overview of the EMS, Shell's approach to developing the system, the benefits of the EMS, and potential barriers to developing and maintaining an effective EMS.

Introduction

Not unlike other ethylene plants, energy ranks as a significant component of total operating costs for the Norco Chemical Plant. Improvements of a few percent can result in substantial improvement to contribution margin. Reducing energy consumption is also well aligned with our goals to reduce greenhouse gas emissions.

Given a large portfolio of energy improvement opportunities, it became obvious that we needed a tool to track progress and sustain performance improvements. This became one of the key drivers for development of our Energy Management System (EMS). This paper will provide an overview of the EMS, details on development of the tool, and the benefits we have observed.

Starting Point for Energy Management

While we had completed energy studies in the past and implemented improvements, we lacked effective tools for assessing whether the improvements were being sustained. It was recognized that energy is an area where potential profits from past improvements can easily leak away without careful monitoring. In addition, due to the large number of energy, utility, and waste heat streams flowing into, out of, and between units, it is difficult to gain a complete picture of overall Site energy performance.

Prior to development of the EMS, our approach to energy management was generally reactive in nature. While we had energy goals, calculation and reporting of energy performance was by quarterly reporting to the corporate center and much less frequent external benchmarking. By the time these results were available, months of poor performance may have accrued and catch-up on energy goals was difficult or impossible. In addition, these metrics provided little in the way of diagnostic tools to allow engineering and operations staff to diagnose problem areas. It was obvious that we needed a real-time tool to monitor changes in energy performance and allow our staff to drill-down into the causes of performance shortfalls. We also expected to learn from changes in energy performance and identify new opportunities for improvement.

While our Energy Team saw a definite need for an EMS, we knew that gaining approval from Site leadership to develop the tool would take a compelling business case. The first step toward developing the EMS was an alignment session with Site leadership to:

- review our current energy gap, past efforts, and the deficiencies in past energy performance tracking and reporting,
- present a business case for why an EMS was needed,
- summarize resources required to build the system,
- outline what is needed to support the system: maintenance of energy critical instruments, demonstrated management support for energy goals, and integration of the tool into operations and business decision-making processes; and finally,

- gain approval to build the EMS.

Although our Site leadership was supportive of developing an EMS, they were not interested in taking on a large technical effort to build an elegant, rigorous, and sophisticated tool. We had several examples within the company of EMS-like systems that failed because they were either too complex for the users or too difficult to understand and maintain. The systems eventually fell into disuse. Our Site leadership approved development of the EMS, but also provided clear success criteria for the project. They wanted an EMS that was simple, "fit for purpose", and easy to use and maintain. We were very successful in delivering such a system within three months, using a very small project team.

Overview of the EMS

The EMS was implemented within our existing Raddical Insight™ based process information system. The EMS consists of a hierarchy of displays that progress from an overview of Site energy performance to summaries of individual unit performance to specific equipment based energy gaps and, finally, down to operating targets that influence energy performance. This is illustrated in Figure 1.

The Site Overview, Unit Overview, and Sub-Unit KPI displays show energy performance in terms of defined Key Performance Indicators (KPI's). For the Site Overview and Unit Overview displays, we chose to use "specific energy consumption" (e.g., BTU/LB or BTU/B of product) as the KPI. The KPI's are displayed against the agreed target or goal for that KPI. The cost gap or cost of not meeting the target is also displayed. The cost gap is calculated in $USD using the current fuel price. This was done to assure that EMS users understand the current cost impact, but also have a measure of actual energy performance that is unaffected by fuel price fluctuations.

Site Overview Display

The Site Overview display is shown in Figure 2. This display is intended for use by Site managers, the Site Energy Focal Point, and others having a role in improving and auditing total Site energy performance. The graphical trend at the top of the display shows the overall Site KPI, BTU's per LB of high value chemicals (HVC), versus the Site goal for this KPI. The numerical scaling on the y-axis for the graphs has been omitted intentionally for this publication. The graphs in the actual EMS displays would show scales in engineering units. They are also auto-scaled, so that they remain appropriate for the numerical range of the data. As mentioned earlier, the cost of the performance gap in $USD is trended and displayed below the KPI graph.

At the bottom of the display are graphs displaying the individual Unit KPI's versus the agreed targets for those units. Clicking on the buttons labeled with the unit names will bring the user to the overview display for that unit. An example Unit Overview display is shown in Figure 3. The Unit KPI trends at the bottom of the Site Overview display are intended to help the user identify which unit or units are driving changes in the overall Site performance. For example, Figure 2 shows that the step change in Site energy performance correlates with a similar change in the GO-1 & OP-1 area. The GO-1 & OP-1 Unit Overview display can be used to further drill-down into the cause of the step change.

Unit Overview Display

The Unit Overview display is intended for use by the unit manager and engineering and operations staff assigned to the unit. It focuses on the energy performance for a specific unit(s). It is structured similar to the Site Overview display. The top of the display shows unit energy performance versus the target and also the cost of the gap in energy performance. At the bottom of the display, there are trends that break the unit into sections and display the Sub-Unit energy KPI's defined for those sections. Similar to the Site Overview display, the buttons above the Sub-Unit KPI trends will bring the user to the Sub-Unit KPI display. An example of one of these displays is shown in Figure 4.

Sub-Unit KPI and Driver Displays

The Sub-Unit KPI displays show equipment or system based energy KPI's and trends. Figure 4 shows KPI's developed for a set of furnaces. These include: specific energy consumption per pound of feed and stack heat losses. A gap cost is also displayed for the deviation from target. Clicking on the Driver button will display the Energy Driver display (Figure 5) which shows specific process variables and targets that can be used to drive improvement in the energy KPI. In this case, stack O2 and temperature are displayed, along with other variables that affect energy performance.

Energy Flow Displays

Energy Flow displays are also provided for each unit. Figure 6 is an example of one of these displays, which shows the energy flows into and out of a unit. These flows make up the overall energy balance for the unit, as defined in the Unit Energy KPI. The Energy Flow displays are available in both native flow units (e.g., MLB/HR) and in energy units (MMBTU/HR).

Development of the EMS

Defining Energy Balances & Energy KPI's

The first step in developing an EMS is to define and gain agreement on the Site and Unit Energy KPI's. In our case, we decided to use a KPI that is based on specific energy consumption: BTU/LB of high value chemicals (HVC). After agreeing on the KPI, we then needed to define what streams would be included in the Site and unit energy envelops. We also had to define the products that would be included in the product slate. Fortunately, we had done previous work to define Site and Unit Energy KPI's. The output of that work was a four page document that clearly defined what streams would be included in the Site and Unit KPI's. The document was reviewed and approved by technical staff and management stakeholders. Having an approved KPI definition document helped streamline development of the EMS.

Careful consideration should be given toward selecting Energy KPI's. Ideally, the KPI should be something that is aligned with Site goals or business strategy. In our case, we chose BTU/LB HVC. For other Sites, BTU/LB of ethylene, BTU/LB of ethylene plus propylene, BTU/LB of feed, or other metrics might be more appropriate.

Prototype

When we started development of the EMS, we ran into a roadblock in the initial design of the system. The intended users had never seen such a system and really did not have a good picture of what an EMS should look like or what it could do for them. To stimulate the creative process, we developed a prototype EMS for one of the operating units. The prototype included just a few displays: a prototype Unit KPI display, a Sub-Unit KPI display, and a Driver display. We reviewed the prototype with a group of users. It was enough to demonstrate the variability in energy performance, show the size of the cost gap, and demonstrate how the displays might be structured. After the prototype review, we had a good list of suggestions for improvement and features to add, proposed KPI's, a proposal for dividing the unit into sections, etc. The prototype took about 40 staff-hours to develop. This small investment proved very valuable in jump-starting EMS development. Once we settled on a concept for the EMS, the remainder of the development proceeded quickly.

Capitalize on Synergy with Other Sites / Plants

After development of the EMS pilot, it was shared with staff at Shell's Deer Park Chemical Plant and the technical advisor in our corporate center who supports ethylene operations at both Sites. Both the Deer Park Plant and our corporate center expressed interest in the system. As a result, we undertook a parallel path toward development. Project Teams were established at both Sites. The goal was to develop a template system that could be easily replicated at other cracker sites globally.

This parallel effort provided many benefits. Ideas and learnings from one Site could be shared with the other Site. Weekly teleconferences were established between the project leads at each Site to share progress and learnings. More frequent ad hoc discussions took place between team members at the Sites to standardize the EMS structure and approach. The synergy between the Sites resulted in a product that was better than what either Site would have developed in isolation. It is estimated that this approach reduced development time by at least 20 percent, by sharing standard KPI's, displays, etc. It is recommended that other developers look for opportunities to cross fertilize ideas between similar Sites and between similar plants at a single Site. It can streamline development and result in a better product.

System Development

Most of the calculations within our EMS are simple ratios, steam enthalpy calculations, and conversions of utility flows from native flow units (e.g., MLB/HR) into energy flow units (MMBTU/HR). Other calculations such as heat loss from furnace stacks were based on correlations, as opposed to rigorous calculations involving fuel and flue gas compositions and enthalpies. This 80/20 approach helped improve the speed of calculations and displays and minimized the amount of time required to build the system. We believe that we can achieve substantial improvement in our energy performance using these simple calculations versus rigorous process models or more detailed analysis.

Giving some prior thought to how the EMS should be structured within your existing process information system can streamline development, eliminate rework, and improve maintenance of the final product. Some things to consider are:

- Overall display structure or hierarchy
- How to subdivide process units into sections or Sub-Units
- How the user will drill-down from overview displays to details on actual energy drivers. Develop a drill-down that works for your plant.
- What common links or features do you want on various types of displays?
- Calculation point naming convention, naming of displays, folder structure for storing and organizing displays, etc. This makes it easier to have several people working in parallel, while delivering a system that looks like it was built by one person.

Our system was developed over a three month period with a project team staffed by about 1.5 people. The team included: our Site Energy Focal Point, a summer engineering undergraduate intern, and a consulting engineer from Shell Global Solutions (US) Inc.

Benefits from the EMS

We have observed numerous benefits from the EMS, both during development and in using the final product. This section outlines some of the benefits we have seen.

Validation of Existing Instrumentation and Process Information System

During development of the EMS, we uncovered and corrected numerous problems with instrumentation and our existing process information system displays. Examples include: broken meters, incorrect meters, problems with meter descriptions, analyzer problems, errors in our existing displays and calculations, etc. Correcting these errors helped improve the accuracy of energy flows billed to the operating units.

Improving the accuracy of energy flows between operating units might be considered balloon squeezing by some. I.e., simply shifting costs from one unit to another, while not achieving real savings. However, improved accuracy can yield substantial benefits for Sites that exchange energy with 3rd parties or JV units. It can also improve the accuracy of data supplied for external industry energy benchmarking studies and internal business planning models.

Faster Detection of Instrument Errors

The EMS has resulted in quicker detection and identification of errors in flow measurements and other instruments. In the past many of these errors would have gone undetected or would have been discovered only after a shift in performance was finally detected via trends from less frequent reports (e.g., graphs in monthly cost reports or other performance reporting). Faster detection allows errors to be corrected more quickly and reduces rework in cost accounting systems, environmental reporting, etc.

Energy Savings

Use of the tool has allowed us to detect changes in energy performance and reduce energy costs. Examples include:

- Increased awareness of the cost of parking furnaces in hot steam standby. We now avoid doing this.
- Increased awareness of the impact of operating changes and control system deficiencies that result in increased kickback / recycle on the large compressors. Work has been initiated to reduce these kickbacks.
- Awareness of changes in energy performance due to business plan changes. These effects are being included in our business planning models as we learn from the EMS.
- Detected large amount of steam being consumed by a unit idled for maintenance. Leakage was thought to be minimal until reported by EMS. The steam was blocked-in and substantial savings were achieved.
- Increased focus on furnace stack heat losses due to gap cost reporting. Work is underway to reduce stack $O2$'s, clean furnace convection sections, etc.
- Unit steam header pressure management – One unit has been able to save substantial steam costs by managing steam header pressures to avoid the need to letdown from higher pressure levels.
- EMS provides demonstrated "before and after" evidence of the benefits of energy activities already in the pipeline.
- EMS standardizes the way engineering and operations staff monitor energy performance. Additional measures have been added as we learn about the way energy is used in our units.
- Standardizes energy performance calculations and avoids loss of individualized tools when people change assignments.

Using this tool, we expect to improve our energy performance by 3-5% over the next several years.

Conclusion

This paper described our approach to developing and implementing an EMS. It is our belief that substantial energy savings can be achieved with a relatively simple EMS that can be quickly implemented at a Site. Development can be streamlined by developing the EMS in parallel at similar Sites or for similar units at the same Site. Displaying the cost of the gap in energy performance in monetary terms is an effective tool for changing a Site's culture toward managing energy costs. We have already achieved substantial savings using the EMS tool and expect to improve our energy performance by 3-5% over the next several years by continuing to use and enhancing the tool.

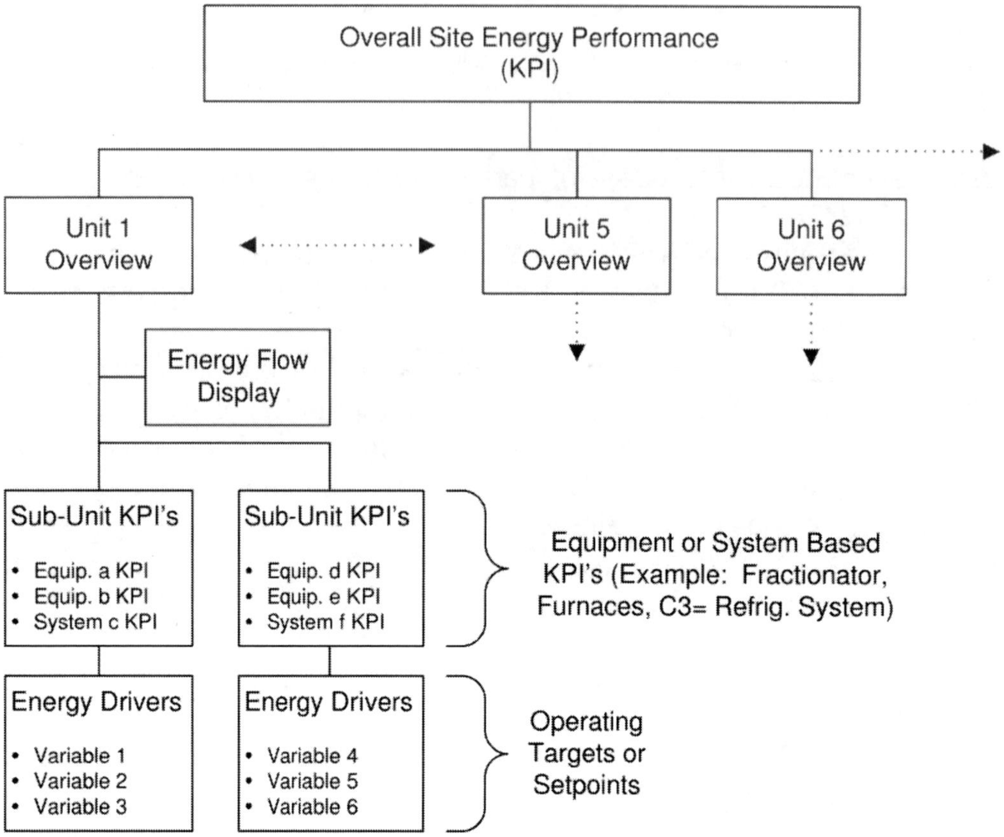

Figure 1 – EMS Display Hierarchy

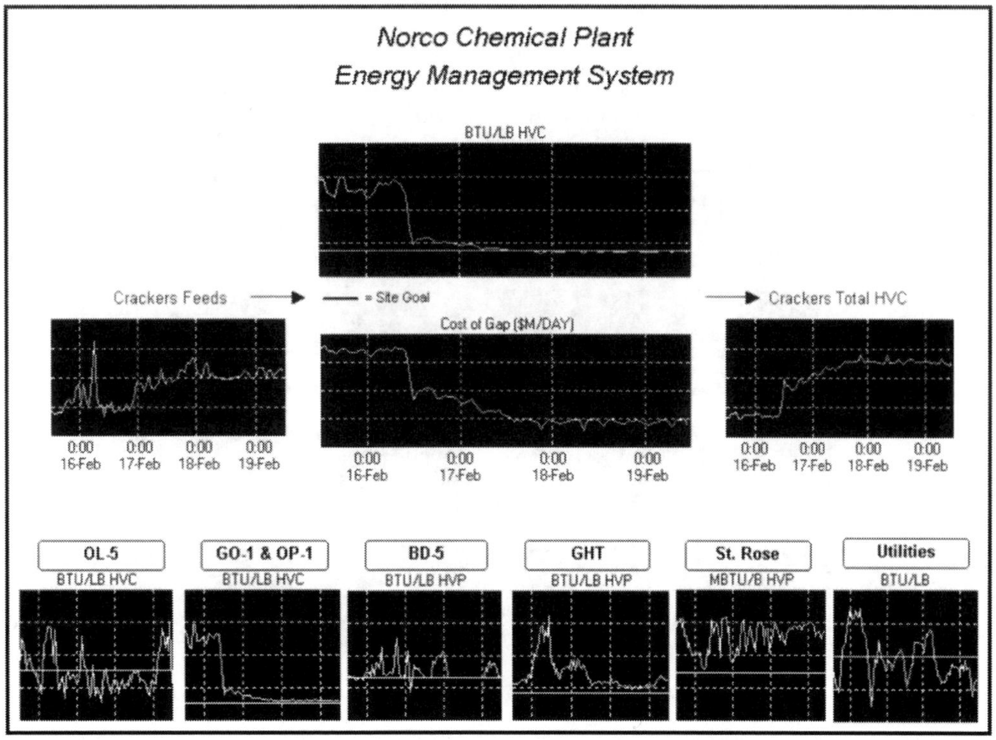

Figure 2 – Site Overview Display

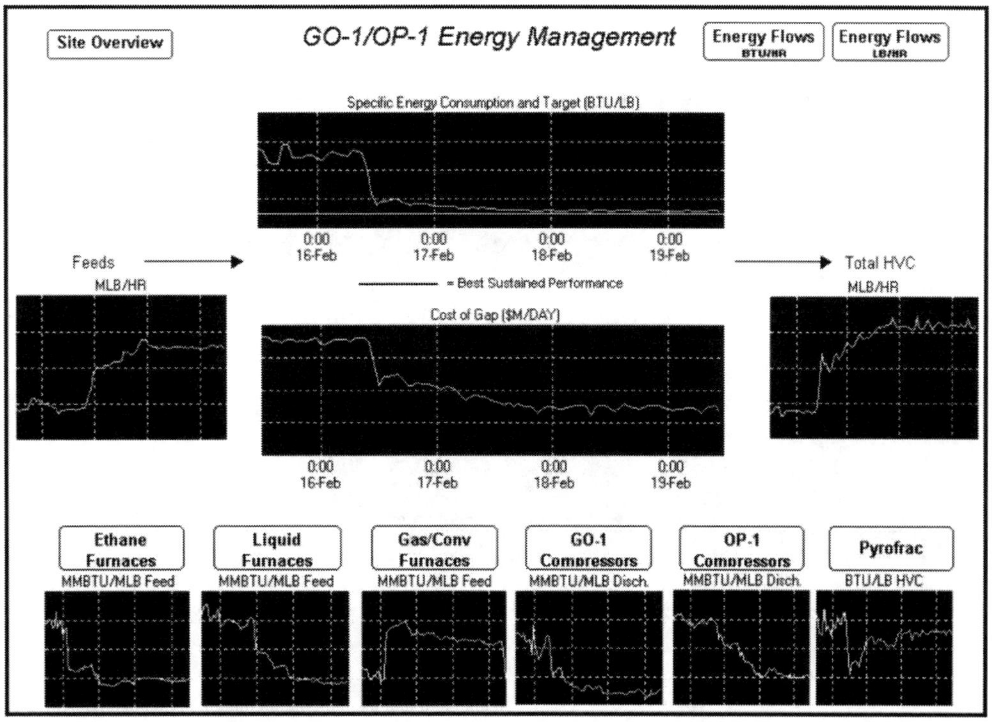

Figure 3 – Unit Overview Display

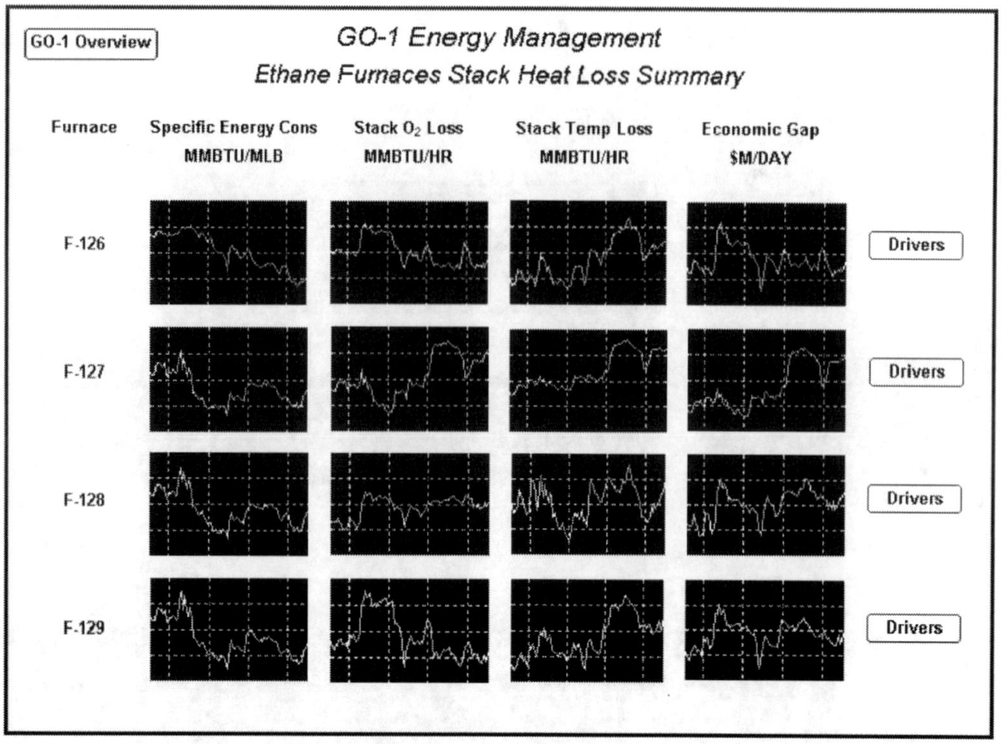

Figure 4 – Sub-Unit KPI Display

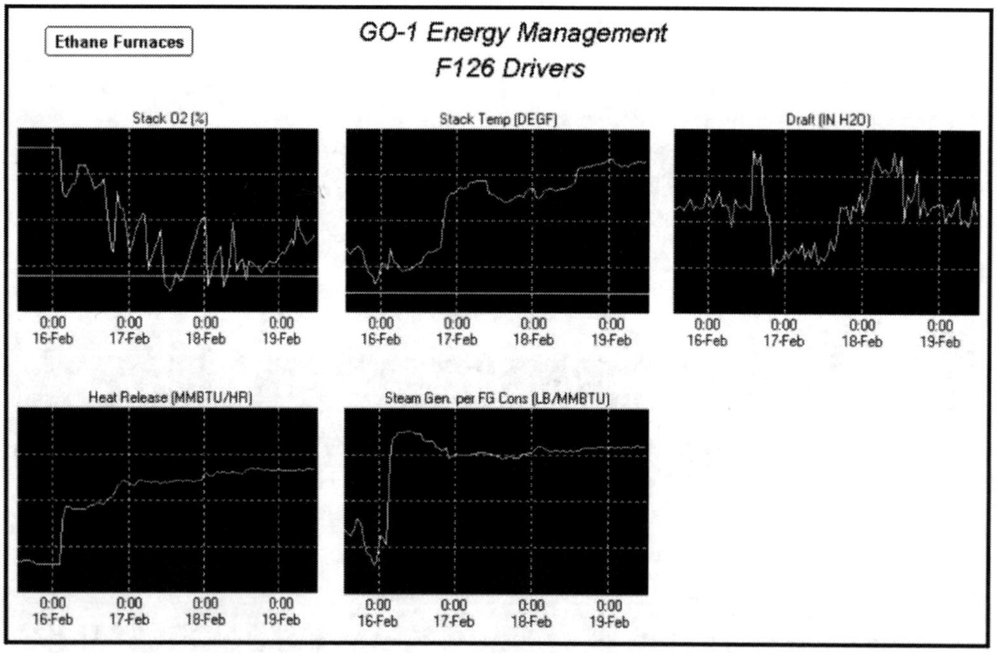

Figure 5 – Energy Driver Display

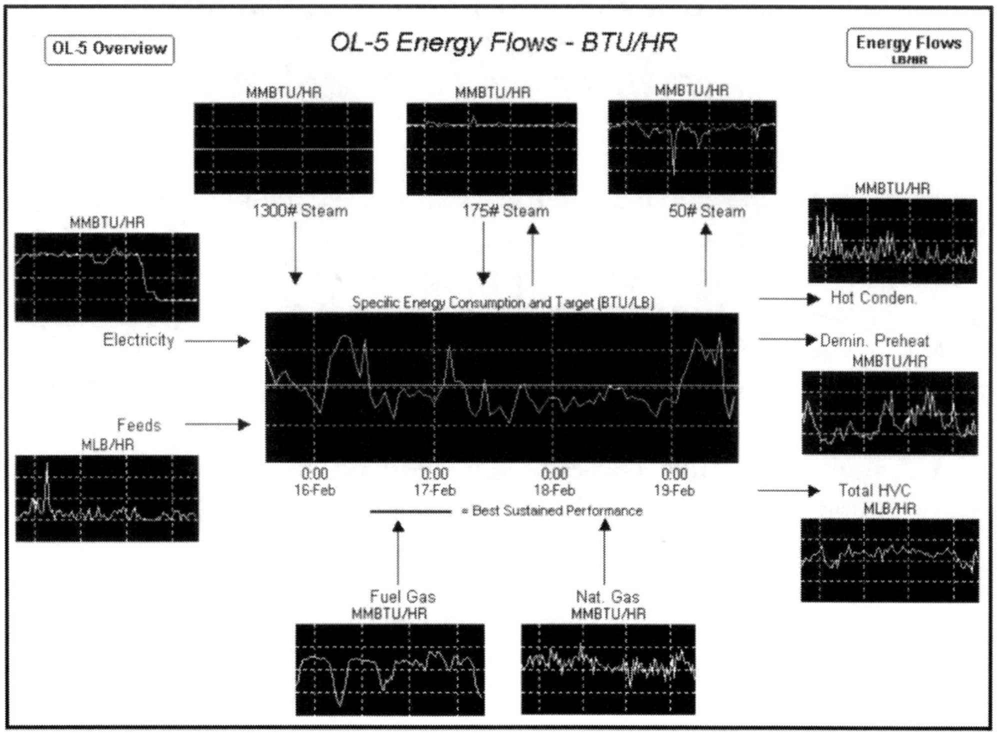

Figure 6 – Energy Flow Display

Operator Use of Mobile Technology

Michael J. Kennedy
Principal Engineer/Human Factors, Control Systems & Electrical Engineering
Equistar Chemicals, LP, a LyondellBasell company

This paper will provide an overview of the Operator Handheld Computer Project that is currently being executed within our company. The scope of this corporate project involves replacing the use of outside operator paper reading sheets with handheld computers. The project has two primary objectives. The first objective is to increase the effectiveness and efficiency of the outside field operators by integrating the use of handheld computers into their job activities. These activities include process and equipment data capture, data trending, and process equipment troubleshooting and maintenance. The second objective is to establish a system where this collected data can be readily reviewed and analyzed by other process support personnel. Anticipated benefits include strengthening the existing Operating Discipline process, improving process equipment reliability, facilitating required environmental reporting, electronically capturing corrective action best practices, and facilitating unit optimization efforts. This paper will include a summary of the project's background, objective, scope, justification, critical success factors, and implementation plan. Examples of actual benefits observed to date will also be included.

Operator Handheld Computer Project Plan

Objectives

This project has two primary objectives. The initial objective is to increase the impact of our outside field operators upon improving process equipment reliability and reducing maintenance costs. The second objective is to establish a more consistent and disciplined process to review non-DCS plant operating data. This is accomplished via making the captured process data readily available for timelier trending and analysis by plant field operators and other plant support personnel.

Scope

The primary scope of this project is to replace the use of paper log sheets by the outside field operators with the use of handheld computers. These paper log sheets are called Operating Discipline (OD) Evaluation Sheets. They were designed to help the outside field operator gather important process information and evaluate the status of critical process equipment and systems. These sheets were designed to ensure that all plant equipment and processes are being adequately checked, thereby improving plant reliability and performance.

The OD Evaluation Sheets focus on the evaluation of the equipment and process. An OD Evaluation Sheet was developed for each process area in all sites. Normal operating ranges and limits are shown on the sheet for each process reading. A supporting Technical Basis document was also generated for each OD Evaluation Sheet, which explained why the normal ranges and limits were set as they were. If a reading was found to be outside the normal range, the outside field operator is required to take corrective action in order to bring the condition back within the normal range. The operator is also required to record any particular corrective action in a comment on the paper OD Evaluation Sheet. Example of OD Evaluation Sheet comments include:

- → The specific action taken to correct at time of evaluation
- → Communication made to board operator
- → Request troubleshooting
- → Submitted maintenance work notification

New OD Evaluation Sheets were printed daily and the formal rounds made by the outside field operators were made with OD Evaluation Sheets in hand. The operator was required to sign and date the Sheet upon completion. Any modifications to the OD Evaluation Sheets required approval of the Unit Process Specialist. Unit support personnel were assigned to review these OD Evaluation Sheets at some routine frequency to ensure that the above process was being followed.

Therefore, the goal of the OD process utilizing these paper Evaluation Sheets was "doing something the best way consistently every time". It involved:

- → How we monitor variables, trends, and performance metrics
- → How we collect, evaluate, and communicate operating information
- → How we identify and solve problems
- → How we prevent reoccurrence
- → How we optimize performance

→ How we capture knowledge

The bottom line benefits of OD include improvements in unit safety, environmental compliance, process equipment reliability, product quality, and cost efficiency.

By replacing these paper OD Evaluation Sheets with the use of a handheld computer system, we are able to enhance our Operating Discipline process. Using the handheld computer, the operator is able to view previous process data points and trends for each reading. When an "out-of-range" reading is entered, corrective action information is instantly prescribed. The corrective actions include the unit "best practices". The operator is then able to record actual corrective actions using the handheld computer. Another benefit of using this electronic system instead of paper is that various data reports can be automatically generated and delivered to plant support personnel. For example, instead of collecting and reviewing a pile of paper OD Evaluation Sheets at some frequency, the unit production specialist or engineer can now receive a report via the company email system listing the out-of-range readings only for a particular process area. These are referred to as "exception reports".

The project scope also included integrations with other internal information systems. Integration was established with SAP to allow the initiation of maintenance work order (MWO) notifications by the outside field operator using the handheld computer. Therefore, if the field operator observed a pump seal failure while taking process readings for that same pump, the operator can easily initiate a MWO using the handheld computer. A link was also made between this handheld computer system and the existing Aspen IP21 data historian system used at the plants in order for the plant support engineers to easily access any information collected with the handheld computers.

Justification

Therefore, the anticipated project benefits include the following:

→ Strengthen the Operating Discipline Process
→ Improve reliability of unit operations
→ Reduce process equipment maintenance costs
→ Facilitate required environmental reporting
→ Electronically communicate corrective best practices for out-of-range conditions
→ Improve the internal Maintenance and Operations Work Process

The anticipated financial criteria for this IT capital project included a 43% IRR with a 3.3 year payback. This payback projection was determined following internal and external benchmarking efforts. Internally, one particular site had begun using handheld computers two years ago. We looked to this site as our "proof of concept" site. The primary benefit that this site achieved with the handheld computers was the ability of operations personnel to more easily collect, collate, and trend outside reading data with DCS process data when analyzing unit performance and implementing continuous improvement opportunities. At this site, where an ethylene unit is located, approximately 2,000 cracking furnace field readings are now captured at a routine frequency using the handheld computer. This data is used to monitor the coking rates of the cracking furnaces. With this data available electronically, coking rates can be quantified quickly allowing prompt process adjustments on the cracking furnaces to maximize cracking furnace run lengths, resulting in fewer decokes and increased cracking furnace availability.

This projected payback was also supported via some external benchmarking, which was facilitated by the handheld software supplier used at the above site. The project team visited an external local refinery, which had been using handheld computers for the last two years. The actual benefits that this refinery achieved with the implementation of their handheld computers involved a significant reduction in rotating equipment maintenance costs. Similar maintenance cost reductions were also communicated to project team members during an annual handheld computer users' group conference, which is organized by one of the handheld software companies.

Project Assumptions

The following assumptions were defined during project planning:

1.) Internal company resources would be made available to staff the Project Team.
2.) The appropriate resources would also be made available to staff a Site Core Team at each site when needed.
3.) All implementation sites have an effective Operating Discipline process already established including the use of paper OD Evaluation Sheets noted above.
4.) Application software with all needed functionality is available.
5.) The internal IT organization implements all necessary application infrastructure on a timely bases.
6.) Internal IT personnel and other resources are also available to develop integration with SAP and Aspen IP21.

Technology

Following a survey and review of available technology, the Project Team selected the IntelaTrac mobile technology software that was available from SAT Corporation of Houston, Texas. The source for this same IntelaTrac software is now from Wonderware Mobile Solutions. Similarly, following a review of available hardware, the Project Team selected the Motorola Symbol MC9090 handheld computer shown below.

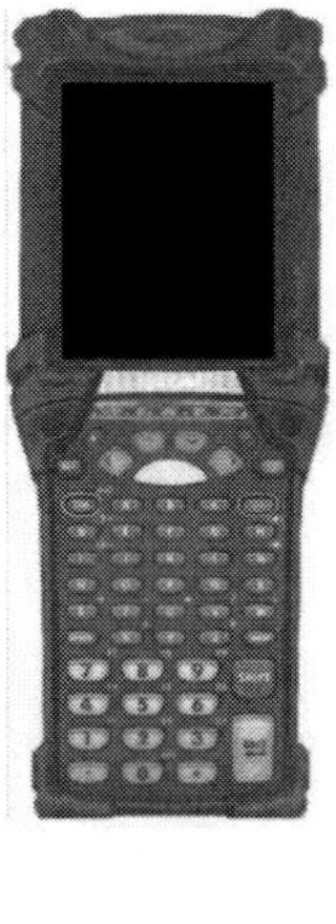

System Architecture

Following some discussion and review, the Project Team decided to implement a handheld computer system, which utilizes Network Handheld Cradles or "docking stations", instead of a wireless system. The system would operate on the company's internal business IT network. Sufficiently reliable and responsive communications would be required from the handheld computers to the Local Application "Sync" Server and to the IntelaTrac Database Server. The servers would be centralized to the greatest extent possible to accompany the responsiveness requirements of the application. The system architecture is illustrated below:

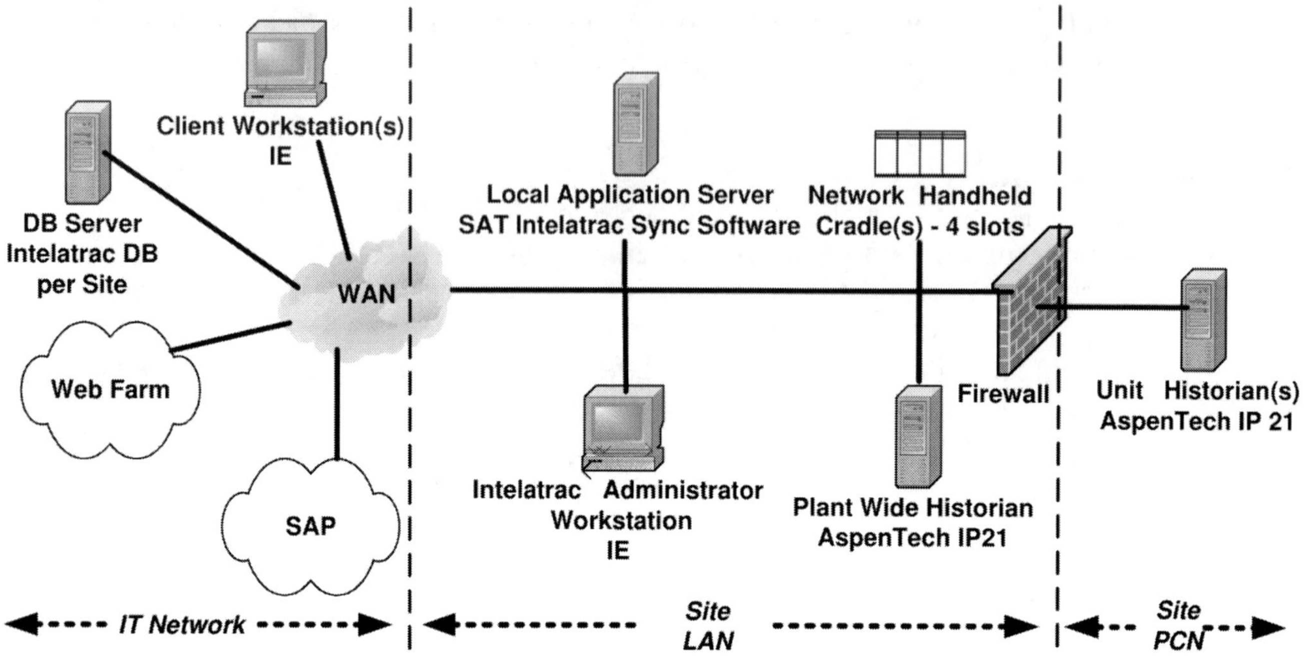

Deployment

This project would be implemented on a site-by-site basis at all 28 company sites over a three year period. Extra time was included in the project schedule for the initial project site to ensure that we had a reliable hardware and software system tested and established before moving forward to other locations.

A project communications schedule was defined, which was to be executed for each site. The Project Team would initially communicate the project plan to the site manager and staff approximately 6 months prior to site implementation. The site staff would then be requested to quickly identify the members of their Handheld Computer Site Core Team whose members would include:

→ **Site Management Sponsor** – provides link to site management, supports Site Core Team; usually Operations Supt.

→ **Site Champion** – leads the effort at the site, serves as the site handheld computer "cheerleader", responsible for tracking of initial implementation scope on site, responsible for continued development, approves changes/upgrades to the handheld computer electronic rounds, future member of the company's internal Handheld Computer Best Practice Team; usually an Operations Specialist

→ **Site Super User** – "maintains" the system, subject matter expert, initial point of contact for site system issues, back-up Area Super User for modifying electronic rounds; usually a Control Systems Specialist

→ **Area Super Users** – "builds" the initial electronic rounds based upon the original paper OD Evaluation Sheets, subject matter expert, implements approved changes and upgrades to the electronic rounds, train other operators in their process area; always outside field operators

→ **IT Support** – perform any required hard reboots of handheld computers, perform application reboots, maintain onsite spares of handheld computers, docking station cradles, and batteries; member of Site IT Client Services

Once the Site Champion at the site was identified, the Project Team would begin working with the individual regarding a project communications plan to the site operator workforce. The Site Champion would usually begin issuing Handheld Computer Project related newsletters or mass emails to site personnel, which explain the project scope and benefits. An actual handheld computer was often left with the Site Champion at this time to allow demonstrations with operations personnel.

Approximately 1-2 months prior to site implementation, a communications meeting would be conducted with all site engineering and support personnel. This would include those with operations, process engineering, production engineering, equipment reliability, maintenance, and/or environmental responsibilities. In addition to communicating the project scope to this audience, they would receive some preliminary training regarding access to this new electronic data. They would also be strongly encouraged to begin brainstorming how easy access to this outside operator round data could assist them in their day-to-day responsibilities.

Also, approximately 1-2 months prior to site implementation, project hardware and software installation would begin. For example, the location of the handheld docking cradles would be determined and any necessary Ethernet cabling would be installed. The exact number of necessary handheld computers would also be defined at this time. One handheld computer was provided for each outside operator job. The IntelaTrac software would also be loaded onto these handheld computers and onto a number of desktop PC's to be used for training and electronic round building.

The next major deployment step would be the Training Week for the above Site Core Team members, which included training on both the IntelaTrac software and handheld hardware. This training focused on how to build the particular electronic rounds for use on the handheld computer. Initially, the plan was to have this training provided by the software vendor at each of the sites. Following the implementation at the initial site, this plan was changed to where the required training at future sites would be provided by IntelaTrac trained members of the Project Team.

The trained Area Super Users from the site would then work together on a straight-day schedule in the same location to begin building the electronic rounds for their assigned locations of the process unit. By working together in the same room to build rounds, they could then easily communicate with each other for assistance and to ensure that the rounds were being generated in a fairly common format. After a draft set of electronic rounds were created, the Area Super Users would load these rounds onto the handheld computers and begin "testing" them in the field. In addition to testing the rounds, this time would also be used by the Area Super Users to begin introducing the handheld hardware device to their fellow field operators. Depending upon the number of available Area Super Users and number of electronic OD Evaluation Sheet rounds to be created, approximately 2-5 weeks were required for the Area Super Users to work full-time building, testing, and upgrading their new handheld electronic rounds.

When the new electronic rounds were completed and ready for use, the Go-Live phase would be planned and executed. The primary component of this phase was One-on-One training between the one of the Area Super Users and each of the outside field operators from that plant area. The Area Super User would then escort their trainee when he/she made their initial outside round using the handheld computer. The Area Super User could then immediately assist the trainee as needed. Each outside field operator was also given their own copy of a Handheld Computer Quick Reference Guide, which summarized the operation of the hardware.

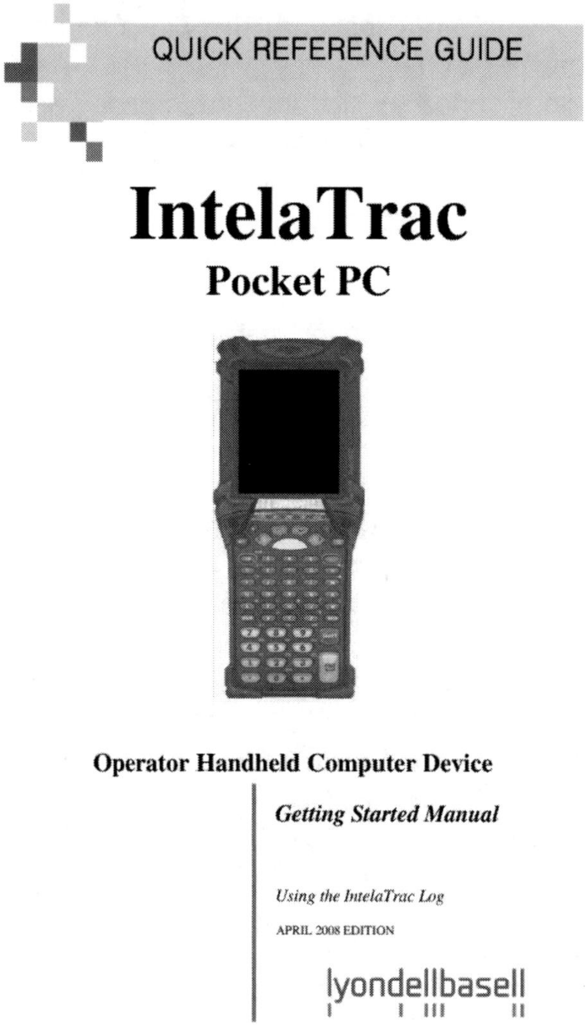

Project Team technical resources were readily available via telephone and email to each Site Core Team from their initial training week through the Go-Live phase. Project updates and any technical issues were also verbally communicated from each of the sites to the Project Team for follow-up during its weekly project conference call meetings, which were conducted during the entire project.

Expectations

Project expectations included the following:

1.) This had to be an "Operator Owned Project" for it to be successful. The outside field operators needed to begin viewing the handheld computer as another critical tool for their use.
2.) Similarly, operator buy-in and involvement is essential! The operators needed to become comfortable using the handheld computer. A process needed to be in place at each site to collect and respond to their input/feedback/recommendations regarding their handheld computers.
3.) Utilize in-house operator expertise as much as possible. We needed to determine and utilize the "best practice" corrective action on the handheld computers for all out of range conditions. This information was often available from the more experienced unit operators and/or unit support personnel.
4.) As described above, we expected that the use of these handheld computers would enhance our existing Operating Discipline process, thereby improving process equipment reliability and reducing maintenance costs.
5.) Since this process information would be available electronically, an expectation was that more of the operations support personnel would begin utilizing it as part of their unit decision making processes.
6.) A continuing goal would be to fully utilize this mobile technology. For example, peripheral monitoring equipment such as vibration probes and pyrometers can be used to communicate directly with these handheld computers.
7.) After the Initial Implementation scope of handheld computer applications were rolled out at a site, we expected the site to define other applications that may be specific to their location that are not completed each shift. Examples include safety shower inspection checklist, data around special heat exchangers to define extent of fouling, and lock-out/tag-out verification.
8.) Any handheld computer Best Practice, including the above special site applications for the handheld computers, would be shared between sites. This would be done via routine Handheld Computer Best Practice Team Meetings and also through the use of an internal computer site where all best practices relating to the use of handheld computers are collected.

Operator Handheld Computer Project Evaluation

Overall Performance

Handheld computers are now being used by all 1,500+ field operators at all major Lyondell sites worldwide. This project was completed during the first quarter of 2009 nine months ahead of schedule and 33% below the original approved project cost. Most of these project savings were due to the Project Team's early decision to form and utilize an internal Project Training Team for the initial handheld computer software and hardware training at each site. The initial project plan included utilizing the software vendor trainers for each site. After resolving some hardware and software issues with the vendors during the implementation at the initial sites, the system is currently operating very well.

Operator Buy-in

The Project Team accomplished its most critical initial success factor, which was to achieve **strong operator buy-in**. This was accomplished through a number of planned and executed project steps:

1.) This handheld computer hardware and software system is reliable and easy-to-use for the operators. Handheld battery life has not been an issue at any site. The battery is recharged when the computer is left in its docking cradle after the round data is collected and transferred to the IntelaTrac database. Sometimes, the software needs to be "rebooted", which can easily be accomplished by the field operator.

2.) When a site did encounter some technical issue with the handheld computer system, the Project Team members from our IT organization responded very quickly. They stayed on the problem and worked with the vendors as needed until the problem was resolved.

3.) The scope and potential benefits of this project were communicated to the operators at each site well before the handheld computers were brought on site. This was accomplished at some sites with a number of project communication newsletters given to the operators and at other sites via direct communications between the operators and the Handheld Computer Project Site Core Team members.

4.) A very good job was done in selecting the members of each Site Core Team. Naturally, if the project team leaders from the site were excited about this project, much of their enthusiasm was carried over to the field operators from that site.

5.) The Project Training Team was comprised of Control Systems Specialists and field operators from a number of internal sites. Because each Training Team trainer had a manufacturing plant background, they were able to relate and communicate well with the members of each Site Core Team. Therefore, trained operators, i.e., not engineers, were training other operators.

6.) After each Site Core Team member was trained on how to configure round sheets using the IntelaTrac software, they began building their actual rounds for the handhelds. When testing their newly configured rounds in the field using the handheld computers, they often interfaced with their fellow operators to begin showing them the handheld computer and its display of the same paper round questions. They were also able to receive any feedback and listen to any concerns.

7.) An effective training plan was generated by the Project Training Team and conducted consistently at each site with each Site Core Team. This plan included utilizing effective training materials and available technical support for the Site Core Team during round building.

8.) When the handheld computers were rolled-out at a site to all field operators, they would receive some initial general handheld computer training in a group setting. But, when they were to use the handheld computer for the very first time, they were escorted in the field for the entire round **one-on-one** with a member of that Site Core Team. Therefore, if any problem with the handheld round and/or computer operation were encountered, the operator would be able to get the issue quickly resolved via the Site Core Team trainer.

9.) When the rounds were configured for the handheld computers, we utilized only the existing paper round sheet questions. New round sheet questions were not added during this time. The goal was for the **only** change for the field operator was simply going from recording readings on paper to entering them on the computers.

10.) When the field operators from a particular site began using the handheld computers, they often recognized and suggested upgrades for the configured electronic rounds. Upgrades often involved the recommended Corrective Action for a particular out-of-range condition. Field operators also began suggesting that the handheld computer be used for other specific non-OD paper rounds that were being completed by the field operators at some frequency.

11.) When field operators began using the handheld computer, they began to recognize more rapid response by other plant support personnel to the process and equipment issues that they recognized and recorded in the field.

Key Performance Indicators

The following Key Performance Indicators (KPI's) were defined for monthly tracking and communication following the initial implementation at a site:

1.) **Bad Actors** – Total number of process equipment readings found out of range requiring operator intervention.

2.) **Operator Response** - % of out of range readings where the field operator recorded corrective action

3.) **Full Implementation/Utilization** – Total number of additional procedures created following the initial implementation at the site, i.e., those in addition to the OD Evaluation Sheets

4.) **SAP Maintenance Notification Interface** – Total number of maintenance work orders initiated in the field using the handheld computer

5.) **Total Reported Saves & Benefits** - Total one-time or continuous $ savings achieved or facilitated via use of the handheld computer

To date, the **Bad Actors** KPI is trending downward at many sites, which is desired. The chart below shows this monthly KPI for a number of polymer sites during 2008. The

Operator Response KPI also varies between sites and is usually in the 80-99% range. The average for all sites to date was 86% during 2008. The goal for this KPI is naturally 100%. Regarding the **Full Implementation/ Utilization** KPI, a total of 148 new rounds or procedures from all sites were built for use with the handheld computer during 2008. The goal is for each site to utilize this handheld computer as much as possible, as measured by this KPI. The **SAP Maintenance Notification Interface** KPI totaled 145 Maintenance Work Orders through 2008. The goal is to see this KPI increasing over time.

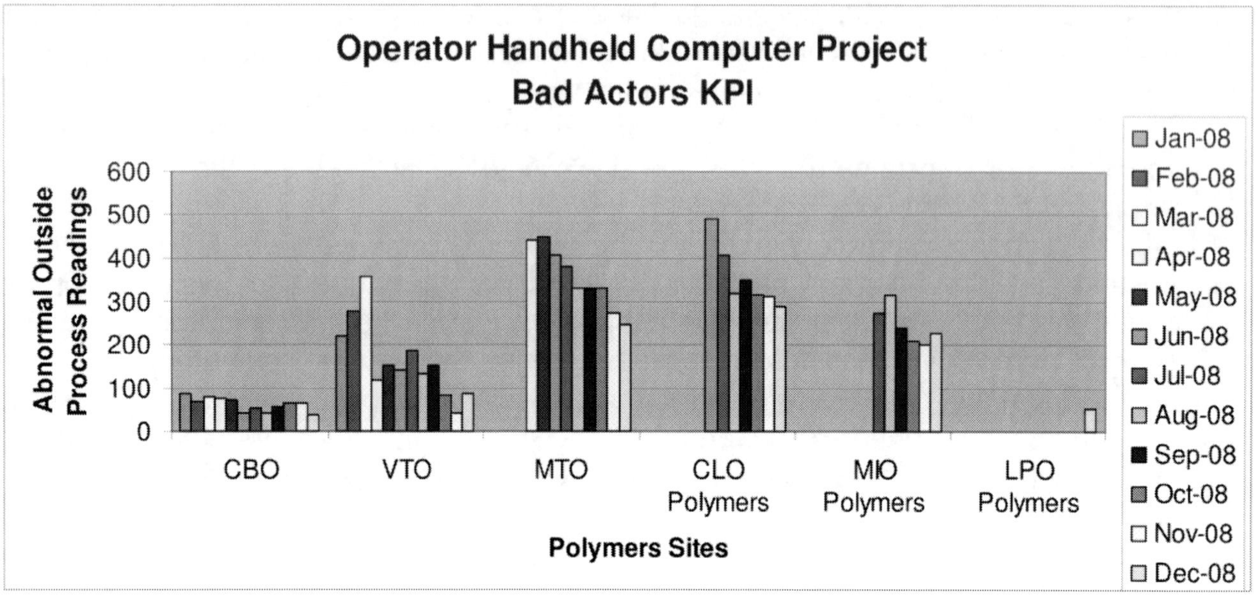

Regarding the **Total Reported Saves & Benefits** KPI, the total $ saves and benefits has resulting in this project paying for itself in approximately 1.3 years versus the originally projected 3.3 years. These benefits have been obtained from the enhancement of our existing Operating Discipline process resulting in many "early catches" of process equipment operating issues and therefore, reduced maintenance costs. Some $ benefits have also been obtained from the utilization of this newly readily available electronic process data by plant support personnel. A Handheld Computer "Success Story" log is being maintained to collect and share all significant handheld computer successes reported by the sites. Specific handheld computer success stories from various sites include the following:

1.) "The handheld computers had been in service only a few days when the utilities operator noticed that the reading on air compressor oil temperature was higher than the previous reading. The operator used the trending capability in the handheld to trend the oil temperature and saw that the oil temperature had increased ~ 5°F in the last few days causing the common systems alarm to come on. Using the IntelaTrac handheld, the operator identified the problem. The portable air compressor was put on line, the main air compressor shutdown and the oil cooler was pulled and cleaned. Without the IntelaTrac handheld, the main air compressor would more than likely have shutdown on its own causing low instrument air pressure and kicking in the nitrogen backup system, which is more expensive and would have caused operations to rush to make the changes. Instead the turn over to the temporary went smooth and easy. The total cost savings was about $10,000.00 (N2 cost and unscheduled maintenance). The big savings from our standpoint was that it prevented an incident that could have had negative affects on production. Loosing instrument air could have shut the plant down under certain circumstances and this cost could have been high. The fact we were able to prevent the incident by use of the handheld has greater value to us."

2.) "On the Handheld Computer Roll-Out day at the Clinton plant, I was checking with each of the High Density operators to make sure that the rounds were working well. I wanted to make sure the operating ranges were working and that we weren't getting unnecessary flags/generating unnecessary exception reports. On two of our recycle pumps, J 0612 & J 0611 (identical systems), we had some oil temps that were questionable. J 0612's oil temp was at 165F. J 0611's oil temp was running at 110F. The normal oil temp range on both pumps is 110 – 135F. The operator mentioned to me that the temp of J 0611 had been running on the cool side while J 0612's temp had been in the 160+ range for awhile. I spoke with the Unit Specialist in the hopes of determining a range that would work for both pumps. Once I brought the temperature situation to the Specialist's attention, the Specialist mentioned that we had recently changed the cooler on J 0611. Also, the temperature on J 0612 was noted as excessive. The "success story" that the use of the handhelds brought about was that on the very next day, the cooler on J 0612 was changed out. The oil temperature came down to a more acceptable range. Also, due to the oil temp range in question, engineering got involved. Engineering contacted the recycle pump vendor to make sure that the oil temp ranges were exactly what we needed to maintain pump reliability."

3.) "The reliability of the Process Gas Compressor wash oil system was very poor after start-up from the last turnaround. We discovered that several of the instruments were ranged incorrectly, there were missing pressure indications along the header that were necessary for monitoring, and the quills were completely plugged on the 2nd and 3rd stages. Operations used the hand written reports to record their readings for this system and they collected these in a pile that was boxed up and stored, but probably never reviewed by engineering. The IntelaTrac handheld computer system has improved our monitoring and reliability of the wash oil system for the PGC machine significantly. After operations included critical pressure indications within the system, I was able to correct the range issues on all of the flow meters for the wash oil system. The important readings, such as wash oil stage injection pressure and flow rate indicate whether the system is having problems. These readings are now compiled in the electronic historian for engineering to review and trend on a daily basis. It makes troubleshooting plugged quills, meter problems, and other issues very easy which allows us to fix problems before they become a major issue. For example, in June 2004, the PGC efficiency started to decline at an alarming rate over the next 3 months through September 2004. We figured out that the wash oil system was clogged with rust/scale that was limiting the flow to the machine allowing polymer to build up within the stages. This polymer degrades the efficiency over time causing us to use more energy to run the compressor. We lost approximately 5-7% efficiency per stage over those 3 months costing us around $300M - $500M annually in 900 PSIG Utility steam. We did some heavy solvent flushes and recovered some of this efficiency, but the loss of wash oil had caused some permanent damages that only the next T/A would repair. Losing wash oil also prevents us from putting our chemistry to the compressor that helps inhibit polymer growth and allow for longer life cycles between turnarounds. The IntelaTrac handheld computers allow these problems to be identified and tracked very efficiently today. I can keep an eye on my unit with minimal effort and make sure that we don't repeat the 2004 incident in the future. Currently, the IntelaTrac system has helped to identify several plugged quills that led to project money that installed a filtration unit in the wash oil system. With the IntelaTrac and filtration unit, the wash oil system has been without an incident for the past 11 months. This holds some major value for the next compressor run, which should avoid the significant efficiency losses from the previous incident."

4.) "We had a new technician working in my process area. While making rounds, the technician entered a lube oil temperature of 80°; the range is 90° to 100°. We hadn't had a chance to customize the default notes but when I saw the temperature reading on the report, we found that the linkage on the controller had come loose causing the steam valve to remain closed. It is possible with the cold front moving in; we may have lost the level in the Hyper compressor lube oil tank causing a trip."

5.) "Operations personnel for OP2 B shift utilized their IntelaTrac's to document a small tube leak in the radiant section of a cracking furnace in OP2. They also wanted to ensure that the appropriate support personnel are aware that the leak exists. It was assumed that everyone was aware of the leak and an incident report had been created to account for the emissions. That was not the case. Through the use of IntelaTrac Web Reports, the documented leak reached the appropriate personnel and an investigation took place. Historically, leaks in this section of the furnace are deminimis. However, if the leak was to continue and there wasn't engineering calculations documented within 14 days, this would have resulted in a Title V environmental deviation. An incident report has been created and a Title V deviation avoided. Thanks to everyone involved who helped resolve this matter."

Operator Handheld Computer Project Path Forward

New Success Factor

As reported above, the **initial** success factor that the Project Team focused upon was achieving strong **operator buy-in**, which was achieved. With the initial roll-out of the handheld computers completed to all sites included in the project scope, the success factor for this project has now changed. The new success factor is to continue to utilize this new handheld computer tool to **maximize the field operator's impact upon equipment reliability**. The primary step in achieving this is to audit and upgrade, if necessary, the existing round sheet questions and corrective actions on the handheld computers. Are the field operators looking at what needs to be looked at when conducting their daily rounds with respect to critical equipment reliability? Assistance from the site reliability engineers will be needed to answer this. Potential round question upgrades will initially focus on the questions involving process critical and high maintenance cost process equipment.

Monthly tracking and communication of the above KPI's will continue, which should encourage continued utilization of the handheld computer tool. Members from each Site Core Team comprise a Handheld Computer Best Practice Team, which meets on a regular schedule. All identified handheld computer best practices and any new applications can then be shared across all sites.

Resolution of Coke Problems in the Quench Oil Loop of Ethylene Plant

Rolf Ebus

General Manager

Facet Deutschland GmbH

Jan Doskocil

Ethylene Plant Manager

Unipetrol RPA

Jiri Pospisil

Process Engineer

Intecha Ltd.

Laura M. Lynne

Market Manager Refining

PECOFacet

Abstract: Recent advancements in the resolution of coke problems in the Quench Oil Loop of ethylene plants. The discussion of Operational Problems with Coke Deposition and their Solving. Operational benefits and cost savings based on Facet & Intecha's Coke Removal Unit implementation.

Foreword

The UNIPETROL's ethylene plant started up in 1980. The plant was very flexible from the point of view of processed feedstocks allowing the plant to crack LPG, primary naphtha, kerosene and atmospheric gas oil. This standard Lummus steam cracker with a nameplate capacity 450 kTA of ethylene was almost immediately converted to crack hydrocracker residue instead of kerosene. Since 1982 the primary plant feedstock has been a wide range naphtha and hydrocraker residue, both at high severity, causing the production of coke to be really tremendous and unfavourably impact the plant operation.

Until the political changes in 1989, the plant operated on average at sixty per cent of nameplate capacity with run lengths less than two years. Even at these conditions the quench oil loop was full of deposited coke. The free market demands soon pressured management to extend the time between turnarounds. Two serious accidents (total plugging of steam developer and destruction of primary fractionator's trays), both due to the heavy coke depositions, advanced the decision to solve the problem seriously and systematically.

The problem of coke deposition was solved by implementing the coke removal unit (CRU), the common technology of Facet Deutschland and Intecha companies during the plant shutdown in 1997. The coke removal unit has now operated for 12 years without any operational failure or problems, has removed approximately 250 tons of coke total, and completely solved the coke deposition problem. Its' unique feature – the isolation of dry, odorless coke as well as its' overwhelming reliability are the reason for presenting this paper.

Coke Deposition in the Quench Oil Loop

The deposition of coke is naturally characteristic for the quench oil loop equipment. The coke is produced in radiant coils of pyrolysis heaters either by catalytic mechanism or by the condensation of soots and tars (polycyclic aromatic structures). In either case the result is the same – the smaller portion of coke remains deposited on the walls of radiant coils whilst the majority of coke continues with pyrolysis heaters effluent further inside the hot section and presents a serious threat to equipment, plant performance and safety.

In order to describe the deposition of coke and its impact on quench oil loop performance in full detail, the picture 1 is provided on following page. As mentioned above, the coke produced during the pyrolysis inside the radiant coils is carried away with the heaters' effluent. The first pieces of equipment threatened by coke are quench oil fittings and gate valves. Before the CRU was started up, it was common this kind of equipment suffered mainly from erosion and the plugging of quench oil fittings. On the other hand, the problems with gate valves were not solved or significantly influenced by the CRU implementation.

Unfortunately, the primary fractionator in addition to its' designed purpose, acts as a big coke accumulator. The coke can be found practically everywhere inside in the middle and bottom part of the vessel. There are two most dangerous subjects – the deposition of coke on the trays and plugging of downcomers. In case of intensive coke deposition the downcomers are almost totally plugged and the destruction of trays takes place. In 1994 the amount of deposited coke in combination of plugged downcomers led to the destruction of three trays in the middle part of primary fractionator. The negative impact on production was significant – due to the fact the incident occured during the severe winter. Production had to be reduced to only 60 per cent of capacity for five months until the fall shutdown could be rescheduled for spring. Regardless of this incident, the bottom part of primary fractionator was always heavily plugged with deposited coke. The particles of coke formed here are very heavy conglomerates and sometimes several tens of centimeters in diameter. Taking into account the original size of coke particles the formation of such conglomerates caused not only accumulation but also a long and significant movement and mixing of slurry inside the bottom of vessel. These conditions caused huge damage to the primary fractionator internals and due to erosion.

The basket strainer filters as well as the impellers and internals of the quench oil loop circulation pumps are the most threatened equipment. The typical basket strainer damage is the rupture of the strainer due to the erosion and intrusion of coke conglomerates into the impeller of the circulation pump downstream the basket strainer. Before the implementation of CRU the average repair cycle of basket strainer was 1200 operating hours. Based on the size and hardness of coke conglomerate the damage to the quench oil circulation pumps varied from neglible scratching to the impeller to its total destruction together with sealing and pump chamber. The average repair cycle was almost about 8000 operating hours.

The next critical item were process (and low pressure) steam developers. The process steam developers are designed having the quench oil in the shell side so the coke deposited among the tubes along the whole tube bundle whereas the higher concentration of coke was typically close to the inlet of quench oil and at the steady pockets behind the bulkheads. Typically, the deposition of coke was so extensive it was practically imposible to remove the tube bundle from the heat exchanger. In 1993 one process steam developer plugged so quickly and intensively that the quench oil supply line internally damaged by erosion cracked and quench oil flowed out. Although the incident was solved without harm to operating personal, the plant had to be shut down and there was significant environmental pollution.

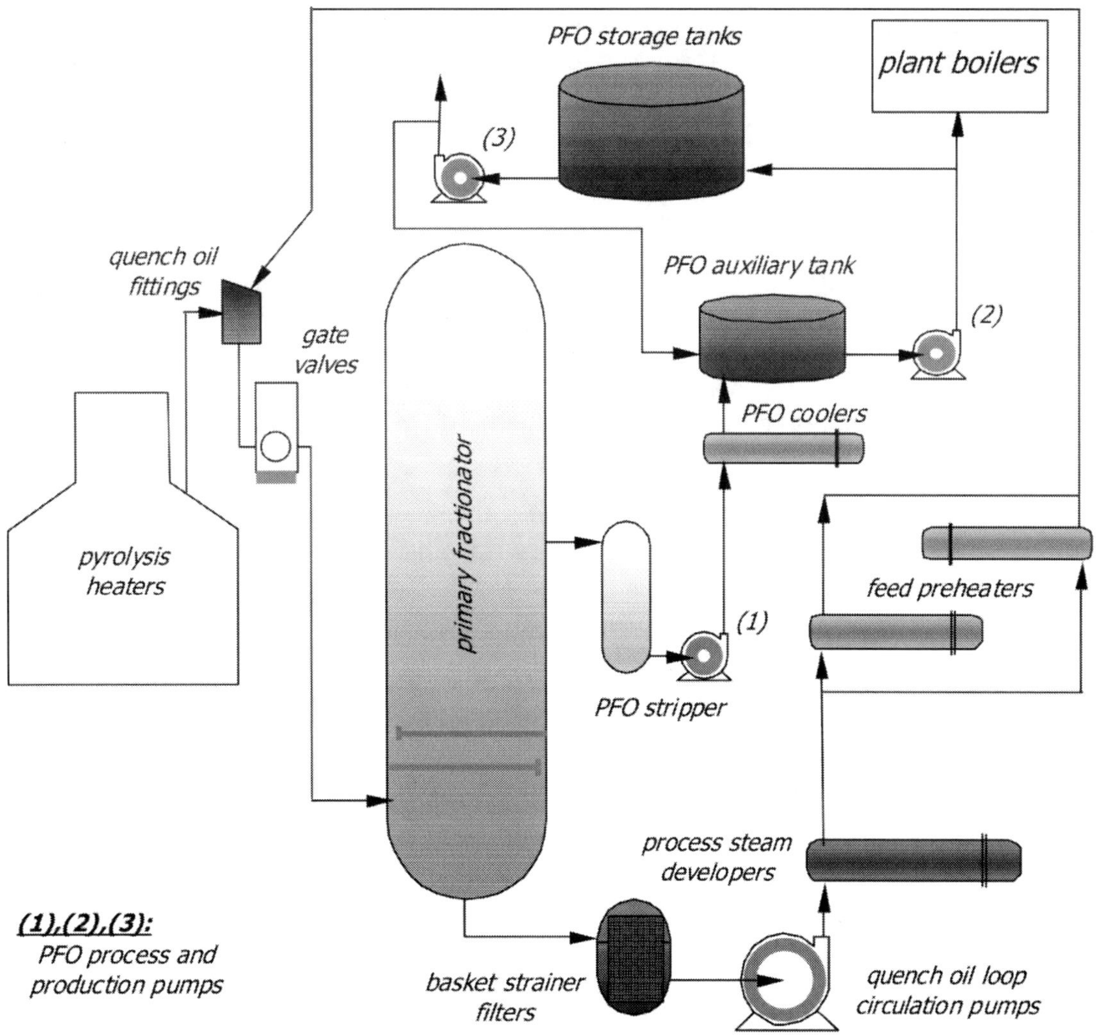

Picture 1 *The sketch of quench oil loop with the illustration of coke deposition and problematic areas*

Although the problems caused by coke deposition were not as severe as the problems in the primary fractionator or basket strainer filters, the quench oil loop itself, as well as heat exchangers (feed preheaters, PFO coolers) suffered. The deposition was typical for large diameter piping, sometimes non-operational (suction of spare circulation pumps) or in the pipings' dead ends.

The most impressive coke deposition took place in the main and auxiliary pyrolysis fuel oil storage tanks. At the end of a plant operational cycle the thickness of the deposited coke layer was more than one meter. As the contamination of tanks increased during the operation cycle, the problems started to occur also at the boiler house. The package of deposited coke from the top of the layer was sometimes suddenly carried with the pumped PFO and plugged the boiler house's burners. The loss of steam production resulted in a plant load reduction or even an emergency trip.

Generally, all problems described in detail above have one common source – the presence of a huge amount of small coke particles inside the quench oil loop. These particles interacted together, accumulated, gained in mass and settled down along the whole quench oil loop and finally deposited in the quench oil piping and other equipment. The common basket strainer filters were not capable of removing the coke from the loop so the coke deposits were found practically in every piece of equipment in the quench oil loop. But the plugging of equipment was only one feature of the problem. The most negative factor was erosion – not only from the point of view of reliability and economy but also from the point of view operational safety.

Solving the Problem

The proper solution of the problem had to comprise several objectives listed further below. The principle limitation for the design was the impossibility of acquiring a representative sample of a particle size distribution from the quench oil loop. Although not knowing the particle size distribution appeared as a complication, in fact the appropriate choice of system equipment made this factor negligible. The desired features of the proposed CRU system were

- **effectivness** – system had to be able to remove a majority of coke from the quench oil loop. The real removable portion had to defined by the characteristic of used equipment.
- **continual operation** – no breaks for cleaning or coke removing were allowed.
- **robustness and flexibility** – system had to be capable of operation at different throughputs of quench oil and also to operate with one spare vessel in each stage. This objective was due to the concern of vessel and internals damage caused by erosion. In reality, operating the CRU is practically without influence of erosion on equipment and all vessels are normally in operation.
- **simplicity of operation** – no special or complicated procedures were allowed. The system had to be designed operational friendly due to the unpleasant nature of quench oil. This meant the primary objective was to exlude any operation requiring direct handling of quench oil.
- **environment friendliness** – no waste generation was allowed. The only produced waste is dry, odorless coke, free of quench oil components, easy to handle and store.

The listed objectives could be fulfilled only by the implementation of a unique two stage system which was based on the combined operation of hydrocyclones and filters. This proposed approach was used due to the fact that widely used settlers were not capable of ensuring the removing of all coke present in the quench oil. They were big, did not operate continuously and last but not least the isolated coke was produced in the form of slurry – very complicated to handle and to treat.

The first CRU designed based on the above mentioned objectives consisted of four hydrocyclones and three candle filters (see picture 2). The system works in two stages. The first stage is realized by parallel operation of hydrocyclones. The whole stream of quench oil from the discharge of quench oil circulation pumps enters the upper part of hydrocyclones via the distributor. The tangential inlet nozzle ensures the required spiral movement of coke particles necessary for separation. Whilst the coke particles are consequently separated in the separation chamber, the coke-free quench oil leaves the hydrocyclone through the product outlet situated atop the vessel. The concentration of coke increase towards the bottom part of hydrocyclone and a coke-rich stream leaves the hydrocyclone through the outlet situated at the bottom of the vessel (see picture 3).

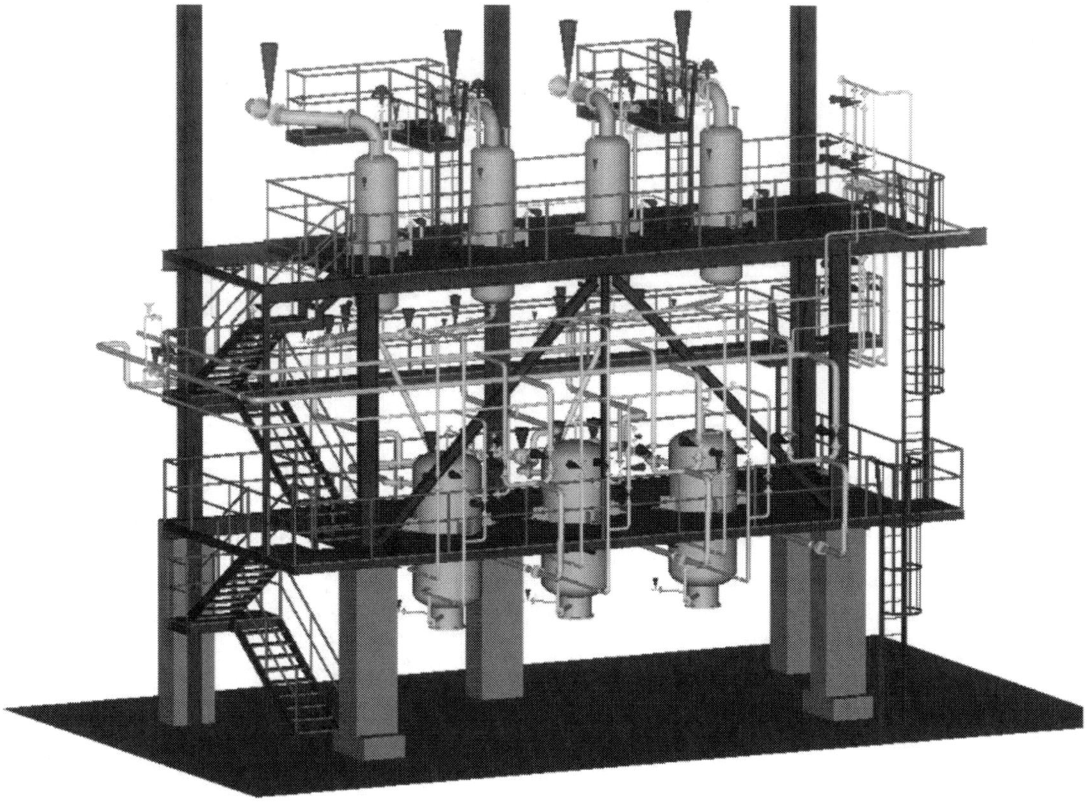

Picture 2 *The arrangement of CRU pilot plant at UNIPETROL's steam cracker, Litvínov, Czech Republic*

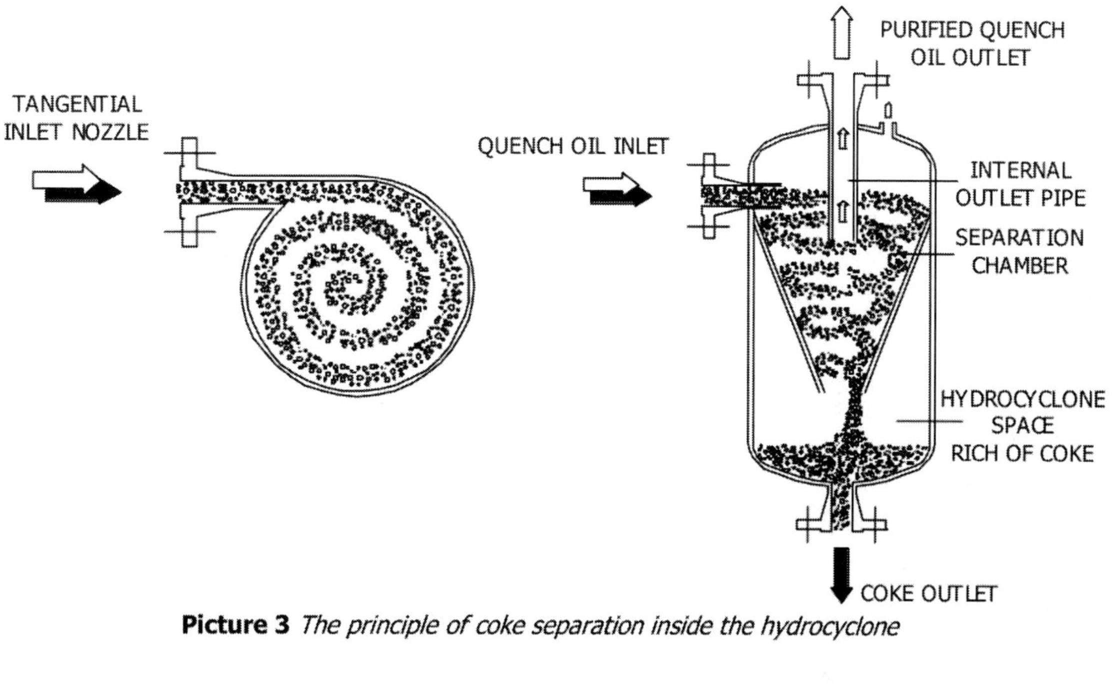

Picture 3 *The principle of coke separation inside the hydrocyclone*

The implementation of hydrocyclones made using filters feasible and more effective. Whilst the typical flow of quench oil through all hydrocyclones varies from 3800 to 4800 tons per hour, the total amount of the stream, rich with coke, to the filters is less than 60 tons per hour. This means approximately only one hundreth of total quench oil amount is filtrated although this stream carries practically all coke. The separation of coke this way is very efficient.

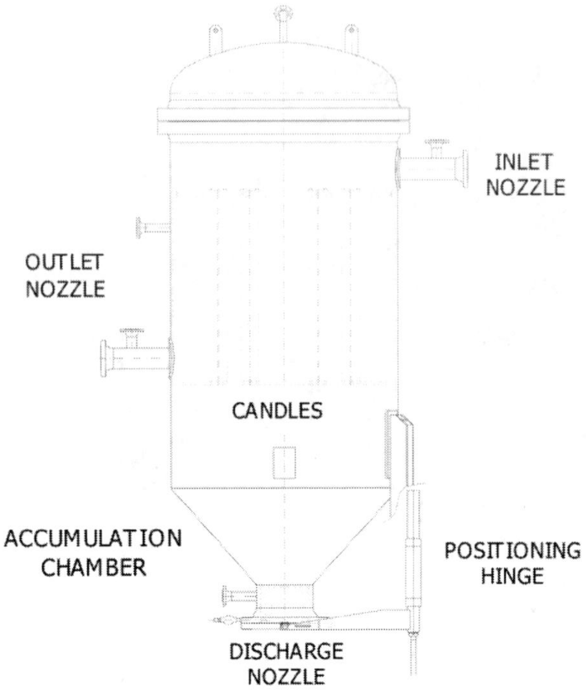

Picture 4 *The scheme of used candle filter*

The second, filtration stage is realized using the normal candle filters (see picture 4). Contrary to the hydrocyclones operation, only two of three filters are used. The remaining one is in stand by or it is prepared for cleaning/cleaned. Only the utilization of candle filters in combination with unique washing and cleaning procedure allowed Unipetrol to fully accomplish the desired objectives.

The candle filters contain twenty candles which remove coke from the quench oil. The mesh of the candles has a 500 μm diameter but different candles can be used based on desired separation efficiency.

Whatever the open architecture of the system is, the production of dry coke, is the consistent feature of the design. The capability of every realized coke removal unit could be very simply tailored by implementation of desired number of hydrocyclones and filters as well as by choice of appropriate type of both used vessels.

Operation of CRU

The operation of the coke removal unit can be described as a fully continuous process. The quench oil is led through the hydrocyclones and the coke is removed from the main stream of oil. The side streams of all hydrocyclones, rich with coke, are collected together and the stream through the common header enters the battery of filters. The number of operating filters depends on the coke production as well as the length of their operational period. The criterion of termination of the coke filter operating period is pressure drop which is measured along the battery of filters. The proper load for each filter is monitored by flow meters mounted for each filter.

As mentioned above, the unique features of the described coke removal technology is the production of dry and odorless coke as well as production of no waste (except the coke, of course). The production of this kind of coke is caused by the special washing procedure. The procedure is based on the sequence of washing and purging operations as well as on utilizitation of streams which can be found nearby at hot section of the plant (pyrolysis gasoline, fuel gas, process stream and nitrogen).

The filter has to be isolated from the process when the maximum pressure drop is reached. The quench oil is ëxpelled from the filter by the fuel gas directly to the primary fractionator. At this moment the accumulation chamber of the filter and candles are full of slurry. The removal of the remaining quench oil is accomplished by washing the filtration cake by pyrolysis gasoline. At the end of this step the coke is free of heavy hydrocarbons and small conglomerates are totally washed by pyrolysis gasoline. Also, the pyrolysis gasoline wash is sent to the primary fractionator. The next step of the procedure is purging the filtration cake using middle pressure process steam. Naturally, before the purging of steam, the pyrolysis gasoline is expelled to primary fractionator. The purging of coke using middle pressure steam is provided to the PFO stripper. When the purging of the filtration cake is over, the coke is totally free of hydrocarbons. The final drying process is realized using nitrogen vented to atmosphere. Finally, the nitrogen is closed, the discharging nozzle opened and the dry, odorless coke falls from filter by gravity.

The length of individual steps of described procedure was determined experimentally. Due to the fact the one filter is always non-operational there is no need to be in a hurry. Even in a situation when the quench oil compounds are not fully released from the filtration cake (e.g. if the washing procedure is not done properly) the filter is closed and individual steps of washing procedure can be repeated to obtain desired quality of coke.

Benefits and Savings

Although the main asset of CRU operation is a quench oil loop free of coke, the gains can be easily divided to two areas – operational benefits and cost savings.

- **operational benefits**

The performance of the CRU during one year of plant operation is illustrated on the picture 5 below. The chosen year, 2002, was not so pleasant from the stable operation point of view. The plant suffered three emergency trips during the year – the worst one on January, 15th caused by the loss of electric power. The next two, (August 16th and December 4th) were triggered by CGC turbine failure. The remaining trips lasted only four hours and all were caused by problems with the propylene refrigerant compressor. The operation of the plant is marked by the blue bar.

The graphical expression of amount and frequency of coke discharge in relation to plant operation shows several interesting features:

- after the emergency trips the frequency of filter discharge always increased. The typical cases are trips from the middle of January and August. Especially The period after the first trip is especially interesting. Due to the loss of electric power there was no possibility to decoke the furnaces. The coke from the walls of radiant coils was released due to the gradual cooling of coils and during the restoring of the normal plant operation (furnaces were brought back to hot stand-by and after it decoked) was accumulated in the quench oil loop. The picture allows estimating the amount of released coke to be almost 2 tons and the cleaning time to 3 weeks.

A similar situation occured after the trip of CGC in the summer. This proved the CRU provides the instant cleaning of quench oil loop in addition to the normal periodical one. The described function is very useful mainly from the accumulation point of view and is a primary contributor to operational excellence.

- the total amount of discharged coke strongly depends on type and amount of processed feedstock
- the amount of coke discharged on regular basis was 15.65 tons, trips have added extra 2.5 ton

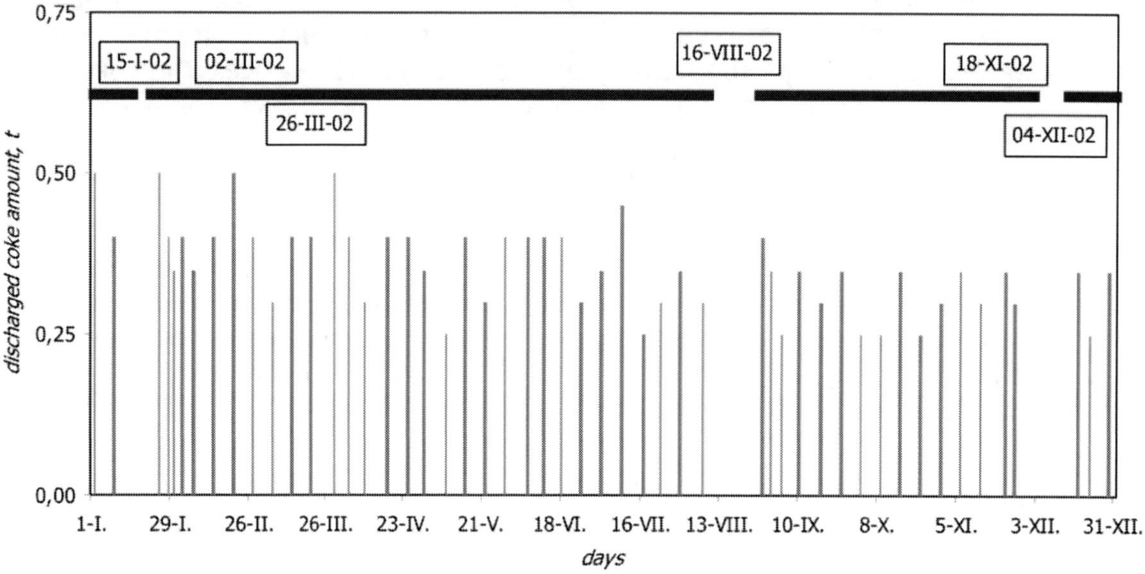

Picture 5 *The illustration of discharge periodicity and amount of discharged coke during the year 2002*

The coke free quench oil loop is much easier to operate. The operational excellence increased notably at the area of basket strainer filters and quench oil circulation pumps but other major parts of quench oil loop have been significantly impacted. The reason is very simple – the suppression of erosion. The comparison of required maintenance performed at different pieces of equipment (QOCP – quench oil circulation pumps) is given in following picture

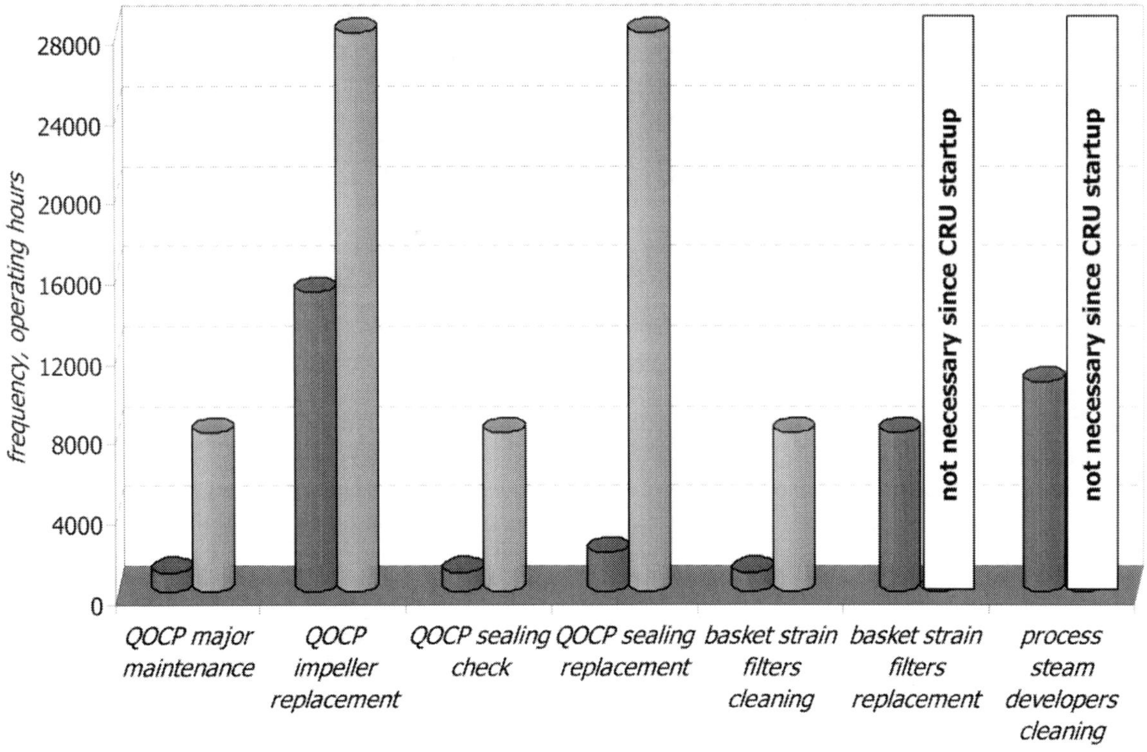

Picture 6 *The comparison of required maintenance performed*

The comparison provided above shows remarkable improvement of operational excellence. Several further benefits are also very important.

- the extension of the time between plant turnarounds from two to five years would have been impossible without the CRU
- the removal of coke from the quench oil loop has practically eliminated its deposition in storage tanks
- there is no leakage from the equipment due to the decrease of maintenance demands the tigtness tests and blinding is not so frequent
- no coke deposits in the primary fractionator, no problems with the fouling of downcomers and damage of the trays.

- **cost savings**

The presented brief cost savings estimation is based on following prepositions:

- the positive impacts on plant performance mentioned in the text above
- Extension of the time between plant turnarounds from two to four years (i.e. at least one cleaning of the primary fractionator had to be done without the operating CRU).
- the negative impact of a poorly operating primary fractionator (higher delta p) on plant capacity at the end of a run

- local maintenance and spare parts costs valid in the year 2008
- for simplicity the boiler area is exluded from the cost savings estimation, the influence of lower steam production is taken into account only in the case of process steam
- no costs for handling and storing the drycoke are take into account (they are negligible in comparison with costs for handling with slurry)

The cost saving estimation is given in the table 1 below. The costs savings are summarized for a four year run.

Table 1 *The cost savings estimation during one four years long turnaround cycle*

cost saving item	cost savings, Euros (€)	note
cleaning of primary fractionator	€ 1.500.000	preparation of the tower for cleaning and cleaning works
production losses (15 days)	€ 11.800.000	expressed as the loss of gross margin on production
plant load reduction	€ 42.000.000	cut of 10 % of normal load for 6 months - twice per 3 months due to the dP of primary tower
maintenance works	€ 4.700.000	all works except at primary fractonator area
process steam developers cleaning	€ 3.200.000	estimation based on average process steam developers operational cycle 10.500 hours
spare parts	€ 3.400.000	internals for basket strainers and pumps
process steam production	€ 4.800.000	based on price of imported steam
CRU maintenance, spare parts	€ 0	No failure, extensive repair or major maintenance works have been realized
total	**€ 71.400.000**	**estimated per a 4 years long turnaround cycle**

Conclusions

The CRU pilot plant at Unipetrol's steam cracker in Litvnov, Czech Republic has been in operation for 12 years. Since the start up in 1997 there have been no CRU failures, extensive repair or unusual maintenance. The reliability of technology is excellent.

The CRU is capable of removing practically all coke contained in the quench oil loop. There are no coke deposits, internals damage or product losses anymore. The implementation of the CRU allowed prolonging the turnaround cycle from two to more than 4 years. The estimated cost savings are at least $ 71.4 million USD for one four year uninterrupted production run.

The unique feature of the CRU is its' capability of producing dry, odorlesscoke, free of quench oil compounds. This possibility is due to the special washing and purging procedure developed for the Coke Process Filters. During the procedure only local utilities are used and no waste is produced. The dry powder coke presents no harm to the environment and it is easy to handle and store.

Finally, this two stage system consisting of hydrocyclones and filters could be tailored exactly to the user's needs. The final layout is done based on detailed local survey and incorporates the proper equipment number and size determination as well as utility availability and connections. The experience gained in the CRU pilot plant are used for design of new plants or revamp existing ones.

CAUSTIC TOWER OPERATION AT WESTLAKE CHEMICAL

David Payne
Principal Production Engineer
Westlake Chemical Corporation

Abstract: The Caustic Tower is one area most operating personnel and engineers prefer smooth operation with very few upsets. No one wants to deal with ethylene going out of spec on carbon dioxide, but it is mandatory to monitor the Caustic Tower operation by taking samples to verify proper operation. Unlike the first Westlake ethylene plant (Petro 1) that has a Caustic Tower with three circulating loops and uses trays, the second Westlake ethylene plant (Petro 2) utilizes a Caustic Tower design with two packed beds and eight bubble-cap trays at the top which can be more challenging to operate. Generally, industry prefers trays, which includes ripple trays, over packed beds for a Caustic Tower application. This report discusses the basics of red oil generation and how the red oil is removed from Westlake's Caustic Tower. It also contains other operating issues with the packed bed Caustic Tower found after further analysis that could lead to operational problems beyond red oil fouling.

1.0 Introduction

This paper discusses the general design features and monitoring program for the Caustic Tower at Westlake's Petro 2 ethylene plant located in Sulphur, Louisiana. A caustic tower monitoring program has been utilized to ensure that red oil polymer is kept under control. Also, a method to determine the proper red oil dosage is provided, and sampling program is discussed to monitor red oil polymer performance, which was established prior to the 2006 turnaround. Lastly, several upsets have occurred since 2006 startup, and there are additional operational challenges that have emerged. These challenges will be discussed.

The Petro 2 unit was shut down for a September, 2006 turnaround where the original packing in the Caustic Tower was removed after nine years of service. Modifications were made that included installation of new packing and new feed distributor trays designed for fouling service. Based on two years of data collection around red oil only, the lower packed bed was changed to one with more open area so it could handle higher solids.

After the startup of a clean Caustic Tower, the required red oil dosage was higher than anticipated. The Caustic Tower ran smoothly at the lower dosage until there was an upset late December, 2006, but no plausible explanation could be found at the time. The red oil inhibitor dosage was increased. Then the Caustic Tower ran fairly smoothly for the next two years after the turnaround. Again, unstable operations occurred after the Hurricane Ike startup in late September, 2008 and continued until the end of the year. After further review of the collected data, it was determined that red oil formation was not the primary mechanism for the excessive fouling at the Caustic Tower. Another proposed chemical mechanism is proposed that is consistent with the symptoms found in this upset and also the one that occurred in late December, 2006.

2.0 Caustic Tower Operating History

The Westlake Petro 2 ethylene plant was started up in November, 1997. The plant was designed to crack a combination of ethane, propane, and butane. The Caustic Tower is located between the third stage knockout drum and fourth stage suction drum of the crack gas compressor train. The Caustic Tower has liquid flow going down the tower with the crack gas flowing up the tower. The three different sections of the Caustic Tower are shown in Figure 1. The top section of the Caustic Tower has a water wash section consisting of three bubble-cap trays to absorb any caustic that is carried over from below in the caustic section. The strong circulating caustic section has five bubble-cap trays before the liquid drops to an upper packed bed feed distributor. The crack gas is moving counter-flow with the liquid flow through a number of vapor risers. The liquid caustic is distributed across the distributor on top of the 25 feet of Koch Glitsch (KG) IMTP # 40 random packing. On the strong caustic discharge side of the pump, the 20 wt % caustic, Boiler water for dilution, and red oil treatment chemical are added before entering the tower. The upper circulating flow is set at a design flow of 130,000 lbs/hr or 230-240 gpm. After dropping through the packing, the caustic accumulates at the bottom of the vessel and then drains through an 8-inch pipe to below the liquid level on the lower packed bed feed distributor tray. The vapor from the lower bed has four pipe risers to separate the vapor flow from the liquid.

The lower caustic circulating loop has a second injection point for the red oil inhibitor before it reaches the lower packed bed feed distributor tray. The lower bed circulating flow is set at its design flow of 130,000 lbs/hr. The lower packed bed has 25 feet of the KG # 40 packing. The maximum design for acid gases consisted of 45 lbs/hr of hydrogen sulfide and 274 lbs/hr of carbon dioxide. Normal loadings are less than 50% of the maximum design ones given. Virtually all of the hydrogen sulfide is reacted in the lower bed since its reaction is so much faster. Ideally, the upper packed bed removes the final 20 – 25 % of the remaining carbon dioxide.

The impurities in crack gas that lead to the formation of red oil polymers consist of compounds similar to acetaldehyde (CH_3CHO with MW 44.05), which is the main reference compound, and other similar oxygenated compounds. The acetaldehyde is generated at the furnaces at the cracking temperatures, and it is not increased in the crack gas compressor stages. If the acetaldehyde, or AA, is absorbed into the circulating caustic and contacts an inhibitor, its reaction to a longer polymer chain is retarded. Some of the acetaldehyde is not absorbed into the caustic, and it exits with the overhead gas leaving the Caustic Tower. If the AA absorbed into the caustic is not treated to stop its reaction with caustic, then the polymerization of the AA chain can continue, which requires time for the reaction to lead to a longer chain polymer. The amount of AA going overhead will increase if there is poor gas-liquid contact or the amount of inhibitor is too low. Initially, the short chain polymer will show up as a yellow solid when filtered out of the caustic, and as the chain length increases its color will transform toward an orange and then a dark orange color as the polymer chain increases. The double chain length polymer of acetaldehyde is crotonaldehyde ($CH_3CH=CHCH_3O$ with MW 70.09) and has limited solubility in the spent caustic. The longer chain polymers will migrate to the Pyrolysis Gasoline phase that floats on top of the caustic. There are five companies that supply inhibitors to retard red oil generation at the Caustic Tower; they are Baker, Chimec (German company), GE Betz, Dorf Ketal, and Nalco. The suppliers have treatment programs that will retard red oil generation and work well with adequate dosage for the application. The performance will be dependent on monitoring the tower for swings in acetaldehyde concentrations in the gas phase and polymer levels in the circulating loops.

The Caustic Tower design provides a vertical baffle that splits the tower bottom with roughly 65% of the volume to handle the lower circulating flow going to its pump. The floating Pyrolysis Gasoline phase along with the red oil overflows the vertical baffle into the Spent Caustic compartment. The drawing given in Figure 2 shows how the two phases are collected and removed from the bottom of the tower. It was found that the lower circulating loop had to be reduced to 60% of the design flow to reduce turbulence so that the organic phase could separate. To operate at the design lower circulating flows, the bottom of the Caustic Tower should have a larger diameter by at least 20% to provide adequate phase separation for the design circulating rates. On the Spent Caustic side of the baffle, the level control is typically set at 50 – 60% for normal blow down. The organic phase floats on top on the non-turbulent side of the baffle and continues to build up as it overflows from the other side of the baffle. To remove this organic layer, the DCS Board Operator closes the Spent Caustic outlet level control valve so that the level of Spent Caustic increases so that the organic phase collects in the red oil compartment. The outside operator drains this compartment to remove the organic phase and red oil polymer. The local sight glass shows the vertical quantity of the organic phase. Based on previous experience, the operator may have to remove two to seven batches of this small compartment. Several factors make this manual operation inconsistent, which include human factors, lighting in the area at night, plant operating issues in the operator's area, and cleanliness of the sight glass. Under the optimum conditions, the time for this operation should be 1 -1.5 hrs per shift if it done on a consistent basis each shift.

Another operating variable involves the fluctuating amount of the organic phase. With clean interstage coolers on the compressor third stage and an adequately sized preheater to add superheat to the crack gas, the organic phase is minimized and fairly consistent. The basic design does provide a small heat exchanger to add some superheat to the gas prior to entering the Caustic Tower. If this exchanger preheats the gas stream at best by 10° F, then less of the organic phase is condensed by the circulating caustic. If the overhead gas temperature leaving the Caustic Tower is greater than the temperature at the preheater inlet, then little condensation of organics is done while passing through the tower. After a few years of operation, the coolers and preheater became fouled, then there will be more organics condensed in the tower and required more manual red oil decants to remove the organic phase.

The condensed organics that float to organic phase should leave with the red oil decant. In the Westlake Chemical plant, this portion is sent to a tank for a lower grade of Pyrolysis Gasoline. Some of the condensed organics, which includes short chain red oil polymer are soluble in the Spent Caustic and exit with the continuous blow down Spent Caustic stream. This Spent Caustic is sent to a Pyrolysis Gasoline-Spent Caustic Coalescer, which is shown in Figure 3. Some Pyrolysis Gasoline from the Debutanizer bottoms, about 5-10 % of the Spent Caustic flow, is mixed with the Spent Caustic to separate the Pyrolysis Gasoline phase from the Spent Caustic as it moves through the packing section and baffles in the coalescer. The Spent Caustic stream should be free of a lighter organic phase. The polymer test, which will be described later in the paper, should show polymer level under control in the stream. It is sent to the Zimpro unit feed tank so that the sodium sulfide in the Spent Caustic can be oxidized to sodium sulfate. This conversion reduces odor issues and makes it a waste stream adequate for the biological treatment.

3.0 Caustic Tower Inspection and Modifications in 2006

After nine years of service, Westlake personnel found the crack gas interstage coolers and preheater exchangers severely fouled on the process gas side. The new tower design had to process higher crack gas rates with a lower tower pressure drop.

The inspection of the tower internals found significant solids and packing rings on the feed distributor trays. Both packed beds had similar issues, but the lower one and its distributor had more solids. The sides of the tower had significant solids, and solids were attached to most of the rings and showed many compacted rings. Also the pump suction strainer on the circulating pumps was collecting compacted Koch Glitsch IMTP #40 packing that apparently passed through the support plate. The upper section trays had deposited solids that were removed, and the trays returned to service. Solids were collected from the interstage cooler, preheater exchanger, and from the internals of the Caustic Tower. The solid material showed significant cross-linked polymers that are not soluble in any solvent. The material on the interstage coolers was gray to brown material and in the Caustic Tower had some gray and other darker reddish-brown solids.

The different sections of the Caustic Tower are shown in Figure 1. Both liquid distributor trays were replaced with one designed to handle fouling service. The circulating lower loop caustic enters the tower via an 8-inch pipe with eight smaller 1-1/2 inch pipes to spread the liquid across the liquid feed distributor tray. In close proximity of the circulating pipe is the 8-inch drain pipe that drops the liquid caustic solution from the upper section to the feed distributor tray for the lower packed bed. The liquid caustic solution drops under the liquid on the distributor tray. Ideally, the circulated weak caustic mixes with the stronger caustic on the distributor tray before the liquid flows through the feed tubes and drops on the lower packed bed.

The new feed tubes on the liquid feed distributor trays are 14-inch high tubes with a 45 degree cut at the top of the 5/8-inch BFW 16 tubes. The tube has an orifice located 1-1/2 inches above the distributor tray. The orifice size is designed to provide design flows when the liquid height is 6-8 inches above the tray. If the orifice is restricted, then the liquid level increases to maintain identical circulating flow. As it becomes more restricted, liquid height matches the top of the slanted cut at the top of the feed tube. This is an alternate route for the liquid to drip on the packing; the liquid flow this route is about double the normal route. It is important to prevent solids from being transferred by the circulating pump to the feed distributor tray to prevent this poor distribution. There were approximately 500 feed dip tubes to distribute the caustic across the packed bed. The feed tube is designed for fouling service but not with solids being on the distributor tray decking.

The original design had Koch Glitsch IMTP # 40 random packing in the upper and lower bed. Since the lower bed will be challenged with more condensed organics and heavier solids loading, it was decided to use the IMTP # 50 random packing that has more open volume. The larger # 50 packing is less efficient in removing carbon dioxide, which was a concern. The original plan was modified to use the # 50 only for the lower 50% of the lower bed, and the # 40 size for the upper half of the lower bed. The upper packed bed continued to use the original # 40 random packing. These modifications were installed in the fall of 2006.

The newly packed Caustic Tower started up late 2006, and it was noted that the inhibitor dosage was below the removal efficiency guidelines for acetaldehyde (AA) specified above. For two previous years prior to the tower packing change out, data was collected on AA in the inlet and outlet crack gas stream. The removal efficiency was consistently in the 60 – 75% but not up to the 80% level. With a clean tower and new packing, the required inhibitor dosage had to be increased by 50% to duplicate similar results for AA removal. The AA removal efficiencies at the higher inhibitor loading had some data above 80%. The inhibitor dosage was on the low side prior to a Caustic Tower upset late December, 2006. After a thorough review of plant operations by the author, operating personnel and by the inhibitor supplier, no logical explanation could be found. The lower circulating loop flow meter showed low values on the DCS, but the field meter showed normal flow. The upper caustic loop concentration stayed stable at high concentrations and was removing only 25 % of the carbon dioxide during this time period. The lower loop flow meter was corrected, but caustic concentrations did not change. A few days after the original upset, a brownish material appeared in the lower circulating caustic loop. The inhibitor dosage was increased after this incident.

The Caustic Tower performed well for 2007. There was over six months of continuous operation in 2007 until late December. The inhibitor dosage varied by 10% and the upper and lower circulating loops typically had polymer contents that stayed within the control targets. The polymer content would swing up occasionally but with a small increase of inhibitor it would drop to expected targets. Since the polymer level was in control, it was decided to improve the consistency of the red oil decant.

4.0 New Red Oil Decant Automated Operation

The red oil decant manual operation was not consistently done between shifts, and it required significant time for the outside operator. Since the polymer levels at the Caustic Tower were kept under control, a new red oil decant procedure was started in early 2008. The Spent Caustic level control valve maintained 55-60% level for a normal blow down to the Pyrolysis Gasoline-Spent Caustic Coalescer. The processing of Spent Caustic at the Coalescers is shown in Figure 3. To remove red oil, the DCS Board Operator opened the level control valve to empty the compartment to the Coalescer. The operation takes about an hour to drain it and then for the level go back to 55% level, and it is done once per shift. The polymer content of the Spent Caustic is still checked twice a week to ensure that there is no excess polymer going to the Zimpro unit feed tank. The operation is done consistently and eliminates the various factors involved with the manual red oil decant. Since the phase separation is not as critical, higher circulating rates, if needed, could be used in the lower bed.

The first coalescer has a section of packing rings roughly one foot in length in the horizontal vessel. The packing rings assist the Debutanizer Bottoms stream, or Pyrolysis Gasoline solvent, in contacting the Spent Caustic. The heavier Spent Caustic underflows the baffle and exits from its two compartments. The Pyrolysis Gasoline removes some of the organics from the Spent Caustic and floats to the top phase and overflows the baffle to its compartment. This stream is sent to the Water-Pyrolysis Gasoline Coalescer to rinse any caustic from the Pyrolysis Gasoline portion. Generally, the modification runs very smoothly and has caused no operating issues.

5.0 Caustic Tower Routine Sampling

The next section discusses the normal sampling and lab tests run to monitor the red oil polymer at the Caustic Tower. The tower inlet and outlet crack gas samples are collected in their sample gas bombs. The gas sample bomb is kept in the vertical position in case there is some moisture or condensable organics in the bomb because acetaldehyde, or AA, is water soluble. Then a fixed volume of gas from the sample bomb is collected and bled to a gas sample tube. Lab sample tubes made by Sensidyne read AA concentrations up to 140 ppm and are used to determine the inlet and outlet AA values. The expected inlet AA should be 32 – 40 ppm for maximum furnace operations, and the AA should be in the range of 22- 28 ppm for reduced rates. The removal efficiency for AA should be in the range of 60 – 80%. If sample results are less than 60%, then higher inhibitor dosage is required. Some data collected at the Westlake ethane cracker with a Caustic Tower with only trays gives AA on the inlet in the 32 – 36 ppm range, and its removal efficiency is between 60 – 70%.

Then samples should be gathered for the upper circulating loop, lower circulating loop, Spent Caustic outlet sample, and then the outlet flow from the Pyrolysis Gasoline-Spent Caustic coalescer. The samples should be visually examined for amount of organics and any high salt content in the caustic, which will make the sample cloudy in appearance. The weak caustic sample can have a greenish color if excessive sulfiding agent is being used. The appearance of foam should be noted, but it has not been common. Any differences in the normal appearance should be noted. Now the caustic samples are tested for polymer content. The amount of solid residue is not always all red oil polymer; it can contain some heavier Pyrolysis Gasoline portions at times. As the data history is collected, then swings in polymer content can show changes in red oil content. Sample size for the test can be as low as 20-30 ml range with extra care or as high as 1000 ml. Ideally, weigh each sample to obtain the sample size. A solvent similar to methylene chloride is used to rigorously mix with the caustic samples in a separatory funnel so that the solvent can extract soluble polymers and other organics from the caustic. The organic phase with the solvent is drained separately from the caustic. The solvent with polymers is evaporated so that only the dried residue is left. The weight of this residue is normally reported as ppm polymer, and a reasonable target is less than 500 ppm. The polymer content results should be tracked and compared to previous results. If there are two consecutive high results or an unusual event has occurred, then the inhibitor dosage is increased. This set of lab tests should show how well the red oil polymer is being controlled.

6.0 Other Caustic Tower Operating Issues

Earlier, it was mentioned that the Caustic Tower ran very smoothly for the last portion of 2007, and it continued most of 2008. After the plant startup after Hurricane Ike, the tower showed similar issues as seen in a minor upset in late April, 2008. The caustic was stacking in the upper bed and caustic was being carried over to the water wash section. Normally, the caustic concentration in this section is less than 1.0 wt %. When the water was tested, it ran over 3.0 wt %. The inhibitor treatment kept the polymer content within the normal range. The tower operations seemed to return to stable operation for a couple of weeks, and then the erratic operation would return.

The unit had a minor trip on 10/07/08, and the problems became much more frequent. On November 10, the lower caustic circulating loop had some dark brown solids in it. The solids were collected so that laboratory analytical tests could be performed. To locate trouble points in the tower, it was decided to perform several tower scans. One tower scan was done at the lower rates and one at the higher rates. The tower scan showed very poor liquid distribution in the lower packed bed, especially the # 40 packing section. High solids on the feed distributor tray could contribute to the poor distribution. Based on this finding, it was recommended that dual basket filters on the discharge of the circulating pumps be installed to keep the distributor tray free of solids. Significant solids on the lower distributor tray causes the drain tubes to not operate as intended, and excessive solids caused inconsistent drainage of caustic from the upper bed.

The red oil treatment program is not working was the first conclusion; the dosage should be increased. The question in my mind later became, "How could low crack gas rates cause the red oil polymer to get out of control? Is there a plausible explanation for this occurrence?" First, the red oil inhibitor dosage was maintained at a constant dosage even though the crack gas rates were reduced. The acetaldehyde, or AA, concentrations were lower than normal at 22 – 28 ppm, or even lower with high rates in October. The AA results for 2008 are given in Table 1. The effective dosage was increased due to the lower AA concentrations and the lower crack gas flow. The polymer testing did not show consistently higher polymer content. Lastly, the appearance of the brown material on 11/12/08 was not consistent with that of red oil polymer, and the lab analyses of the material indicated it was material from crack gas polymerization during the compression step. Similar type material in color was found only in the lower caustic loop during late December, 2006 when the crack gas rates were low. At that time, no plausible explanation was found. There was still no plausible explanation for red oil polymer to go out of control at low crack gas rates. The new question for the writer became, "What chemical and/or process mechanism could explain an increase of polymer solids while operating at low crack gas rates?"

7.0 Verification of Crack Gas Fouling

When the production rates are high, the crack gas is exposed to the higher compression temperatures, which the Westlake plant has kept below 205 ° F. Under these conditions, the crack gas compressor has little or no kickback flow, and the anti-foulant program is generally based on maximum rates. The components that can polymerize are under control since they are basically have only one pass through the compression steps and exit to be processed downstream. The only negative factor is the higher compressor temperatures, but the residence time for polymerization is minimized.

When the production rates are reduced so that the third stage compressor flow approaches its surge line, the kickback valves must open to allow more flow for the third stage. In 2006, the reduced rates occur for 30- 45 days with the average kickback flow being at 17,000 lbs/hr, which was 7.4 % of the forward flow. In late 2008, the reduced rates occur for 60 days with the average kickback flow being at 24,600 lbs/hr, which was 11.4 % of the forward flow. The third stage discharge flow is cooled and condensed liquid is collected at the knockout drum, before the kickback flow separates from the main flow. This recycled gas has its original carbon dioxide, hydrogen sulfide, and acetaldehyde loadings because the recycle stream has not reached the Caustic Tower. Some of the acetaldehyde, or AA, and possibly carbon dioxide may be removed with the condensed water phase at the first two stages. This removal would explain the lower levels of AA at lower crack gas rates. These components went through the cracking furnaces at temperatures above 1500 ° F so they are very stable. For the more reactive compounds in the crack gas, the only separation step is the cooling and condensing step at each compressor stage. At the lower crack gas rates, the compression step only reaches 200 ° F as compared to the higher temperature mentioned above. The coolers at the lower gas rate should condense more of the heavier compounds in the crack gas. The coolers and knockout drum are the only separation step to **remove all** of the reactive components before they are recycled back for another compression step. The writer is proposing that a small amount of the more reactive components is exposed to many more compression cycles than the average for the crack gas flow going forward. This small fraction is exposed to additional residence time at the 200° F to generate more free radical polymer chains than normally seen under maximum production rate conditions.

To study the fouling at the Caustic Tower and the crack gas compressor system, it was decided to track the pressure profile across the entire system, which is shown in Figure 4. The pressure profile started at the third stage discharge pressure to the fourth stage suction drum overhead pressure as shown in Figure 4. The compressor third stage flow then splits at the outlet of the knockout drum where the kickback valve can divert some flow to the first stage as needed. Then the crack gas goes to the Caustic Tower to remove acid gases, and its overhead goes to the fourth stage knockout drum, which has a DCS pressure reading. The operating period for analysis was from 11/10/06 to 4/08/07 with the system in service a maximum of 160 days. There were only two plant upsets during this time period. The new PDI model for the third stage system used the top two variables, which were third stage gas flow and kickback flow, had an excellent correlation coefficient of 0.932 and a standard error of 0.530 psid. The details for the correlation study are given in Table 2. The PDI model used the third stage discharge flow and its kickback, or recycle flow leaving the third stage knockout pot that returns gas to the first stage suction. The operating days was the third most significant variable. Also during the December, 2006 period there was low ethylene production rates, which led to higher than normal recycle flows. In late December, 2006 there was a major caustic tower upset mentioned above.

In early June, 2007, there was a 200-day operating period until 12/24/07 where there was no plant operating upset. When the correlations were run to check the PDI model, the third stage flow leaving the crack gas compressor and the kick back flow gave a very high correlation. The correlation study for this time period is shown in Table 3. The number of operating days dropped significantly in its importance for the correlation. The author thought the Caustic Tower performed better when there were no plant upsets. Using only the third stage flow, the correlation coefficient was 0.945 with a standard error of 0.452 psi.

Now the author went back to the model with two variables, or the first model, and used the model error or the difference of PDI measured minus PDI predicted to track the increase or decrease for the system over time. The PDI indicator would show the system change over time, which is shown in Figure 5. During the initial operations, the PDI indicator was below zero to the end of 2006. After the second plant operating upset on 1/15/2007, the PDI indicator took a step change to a positive 0.4 psid and continued on the positive side. As the PDI increases, there is additional pressure drop somewhere in the system. After the late May-early June, 2007 upset, the PDI indicator took another step up. The next time period shows the 200 consecutive operating day period. This data showed a gradual increase and stabilized with less than normal data scatter. The run ended with the plant upset in late December, 2007. There was no reduction of production for this December period. For 2008, the plant rates were at high levels until the upset the week of 4/22/08, and the interruptions from the two hurricanes in September, 2008. The plant started production after Ike on 9/22/2008, but there was no tower PDI reading to detect any abnormal Caustic Tower response. The model PDI indicator showed no major response during the initial plant startup until after a plant upset on 10/07/2008. A few weeks later, there was a major Caustic Tower upset on 11/12/2008 with the appearance of the dark brown solids in the lower circulating caustic loop. These solids were later analyzed to be cyclopentadiene molecule chains (C_5H_6 with MW of 66.10) plus some other components. Cyclopentadiene has a normal boiling point of 105.8 ^oF and polymerizes with the Diels-Alder mechanism at temperatures at the compressor discharge temperatures. The ethylene production rates were reduced from 10/27/2008 until 12/27/2008. The reduction was similar to what was done in November-December, 2006.

Since there was no PDI reading across the Caustic Tower during the startup after Hurricane Ike, the writer could not determine if there was an upset involving the unit shutdown and startup. The Caustic Tower PDI reading before the upset period typically read 1.0 – 1.6 psid for low and high gas flow conditions. It reads the PDI from below the bottom bed to above the top of the water wash section. It has valves located to measure PDI over different sections of the tower. The gas flow to the Caustic Tower was the most significant variable with a correlation coefficient of 0.494 and a standard error of 0.234 psi. The equation for the Tower PDI is 0.0000056 * Gas Flow to Tower – 0.155, psid. Then a similar PDI indicator was used for the Caustic Tower with it being the indicator being the Measured PDI – Calculated PDI. The time sequence of the Caustic Tower PDI Indicator is given in Figure 6. The Tower PDI trend did show a small increase during 2007, and it slowly decreased prior to the hurricanes interruptions in 2008. The main increase in the tower PDI occurred after the hurricanes. The time sequence for the Caustic Tower PDI shows a different pattern over time than the one for the third stage system shown in Figure 6.

The crack gas conditions from a plant upset has much more recycle flow than the low production conditions. The effect of the exposure can be much higher than the low production case, but the time duration is much shorter. Since the plant startup in late October, 2006, there have been eight to ten plant upset conditions. If the system pressure drop has increased by 3.0 psi, then each shutdown has added 0.3 – 0.375 psi to the system pressure drop. The sequence of operations between plant upsets gives the history of the PDI indicator in Table 5. The other factor involves the low production rate periods. The high kickback flow condition occurred for the initial startup on 10/31/06 through 12/2006 and the recent period that occurred from 10/27/08 – 12/27/08. These two periods accounted for 64 days with average kickback flows over 20,000 lbs/hr, or 52.4% of the all high kickback days of the 111 days. All other time periods had a total of 47 days out of 687 days, or 8.3% of the days. These findings are given for each time period in Table 5. The writer concluded the recycling of the crack gas due to low rates and from plant upsets was exposing a small population of reactive components to excessive residence time that formed polymer solids. These solids are later transported to the lower circulating loop of the Caustic Tower. If these solids reach the liquid distributor tray, then there will be serious operating issues at the Caustic Tower.

8.0 Conclusions from Study

The monitoring of the Caustic Tower is important to measure acetaldehyde loadings and removal to track the performance of the red oil inhibitor program. The polymer tests verify that the program is performing as expected. If the tower operation has an upset, it is important to visually look at the caustic samples to determine if the color of the solid particles are orange to red in color or tend to be brown in color so that the correct chemical mechanism is selected for the cause for the fouling. If the solid particles are brown in color, the source may be from the crack gas system and can be verified from lab tests looking for cyclopentadiene.

The chains of cyclopentadiene are not generated at the 95-105 $^{\circ}$ F operating temperatures seen at the Caustic Tower. It had to be generated at the crack gas compressor discharge where temperatures are high enough to make these polymer chains. It is suspected that the normal anti-foulant treatment for the crack gas compressor is inadequate for the unusual operating conditions when there is significant recycle. It is not fully understood how the polymeric material are transported to the Caustic Tower. When this material was found for both cases, the plant had upsets about two to three weeks prior to date.

The list of recommendations to improve the Caustic Tower operation is now listed below:

1) The plant operations should be reviewed to either minimize the recycle flow of the crack gas compressor and/or add additional anti-foulant dosage to control the polymerization for the recycle crack gas operations.
2) At some point, the Caustic Tower will require an outage to remove the solids from the distributor and probably replace the lower bed packing. Both caustic circulating loops should have dual filters installed to prevent the solids from reaching the liquid distributor tray.
3) The pressure drop for the third stage of the crack gas system will be monitored for future trends. The other stages will be added to the model so they can be monitored for future changes to monitor fouling in the system.
4) The red oil inhibitor dosage should be adjusted until the acetaldehyde removal efficiency reaches at least 60%.
5) If red oil polymer is kept under control, then the automated procedure to empty the Spent Caustic compartment located at the bottom of the Caustic Tower should be considered.

The subject matter of the original paper was expanded due to recent operating issues of the Caustic Tower. Several unexpected results were achieved by analyzing basic questions from the evidence found during the upset periods. There were many contributions made by several persons that assisted in completing this study on the Caustic Tower operations at Westlake Chemical plant. Their assistance was greatly appreciated.

Figure 1: Basic Packed Bed Caustic Tower

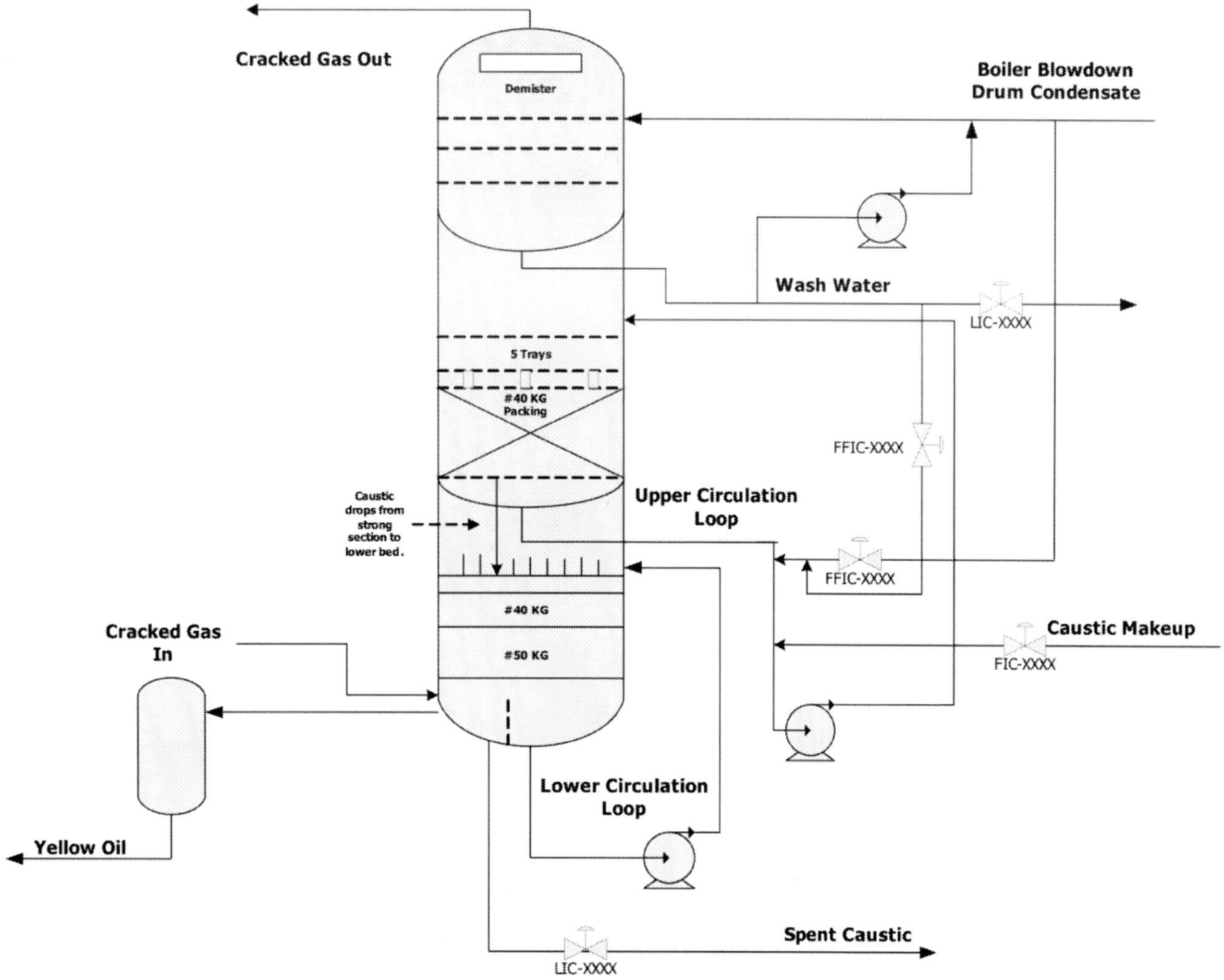

Figure 2: Caustic Tower Bottoms for Spent Caustic

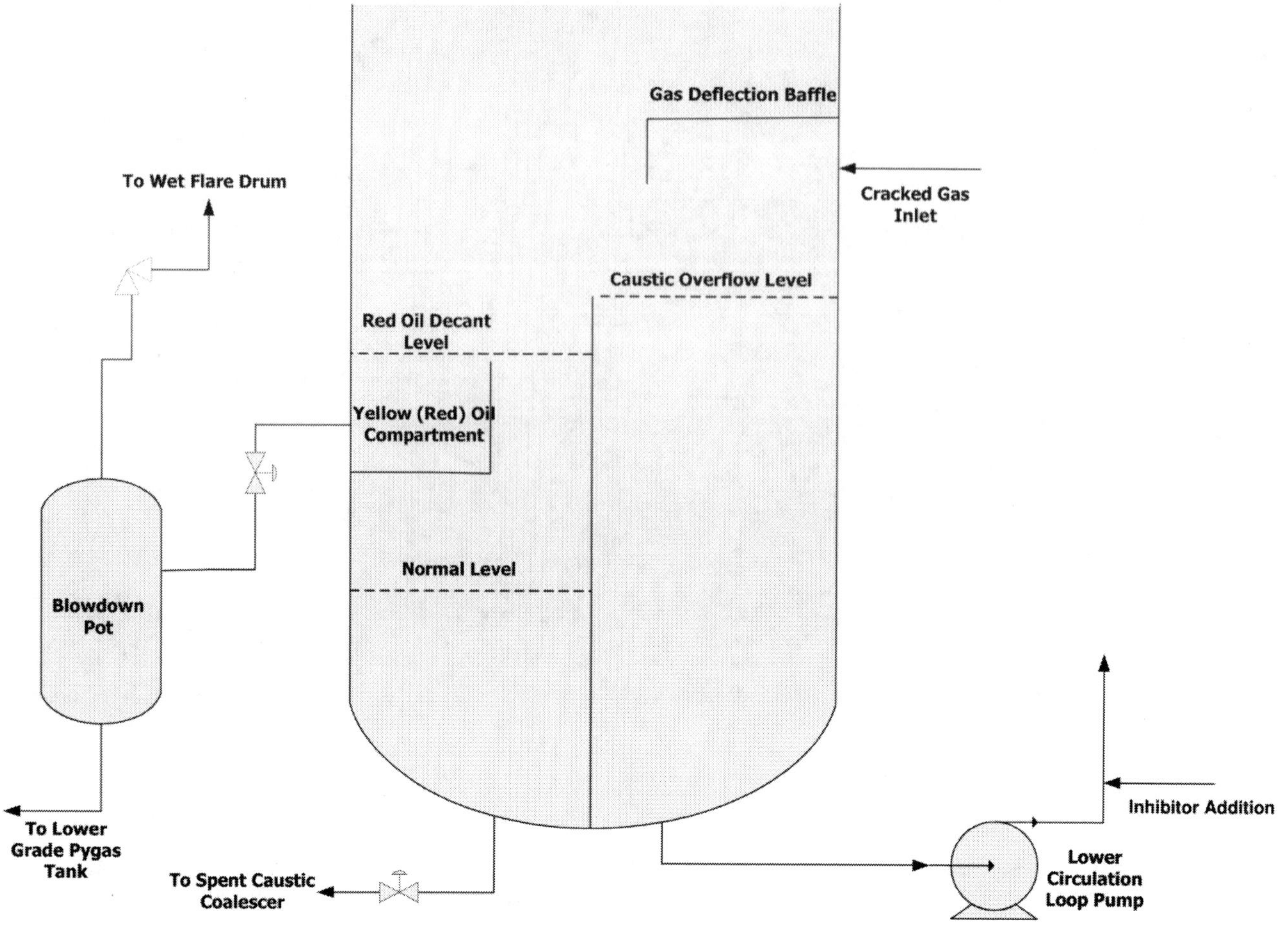

Figure 3: Spent Caustic Processing at Coalescers

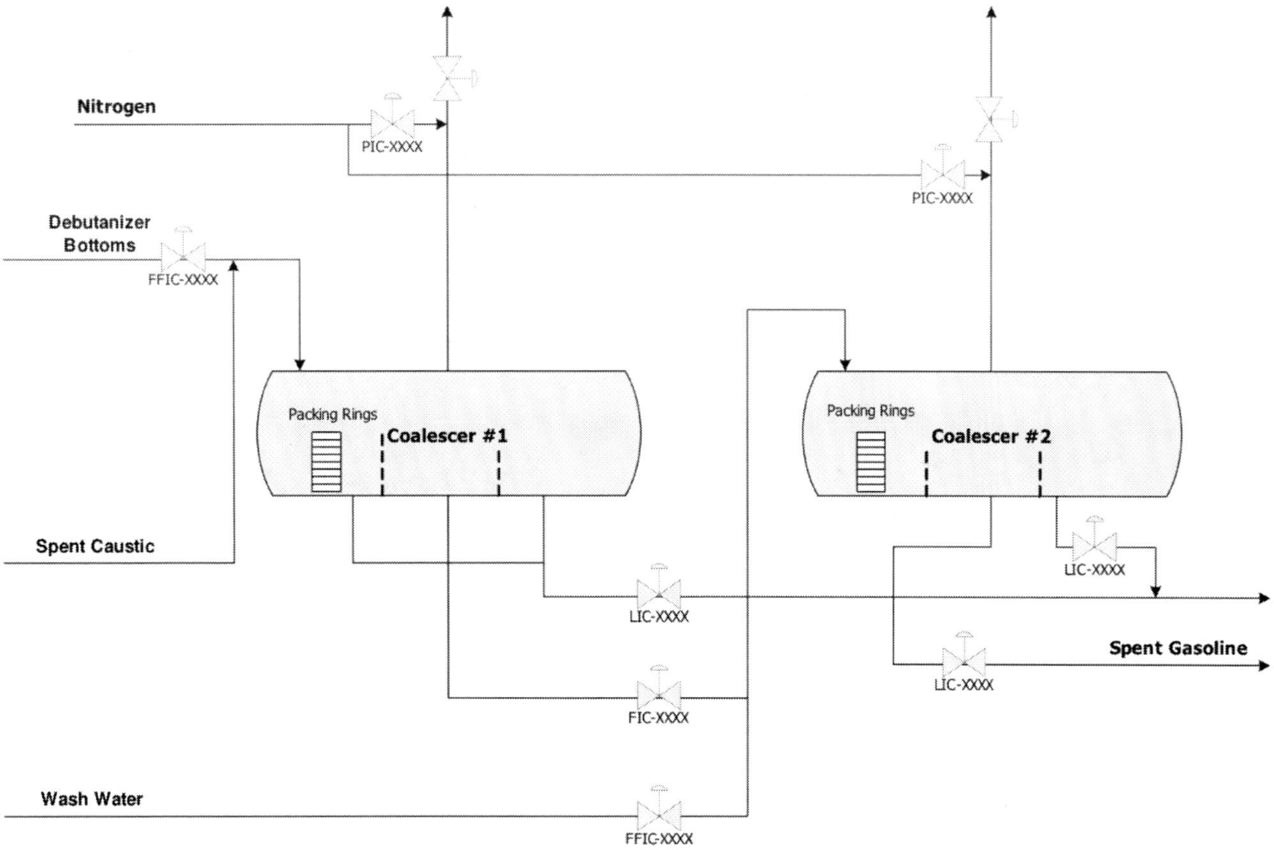

Figure 4: Crack Gas 3rd Stage Discharge to 4th Stage Suction Drum

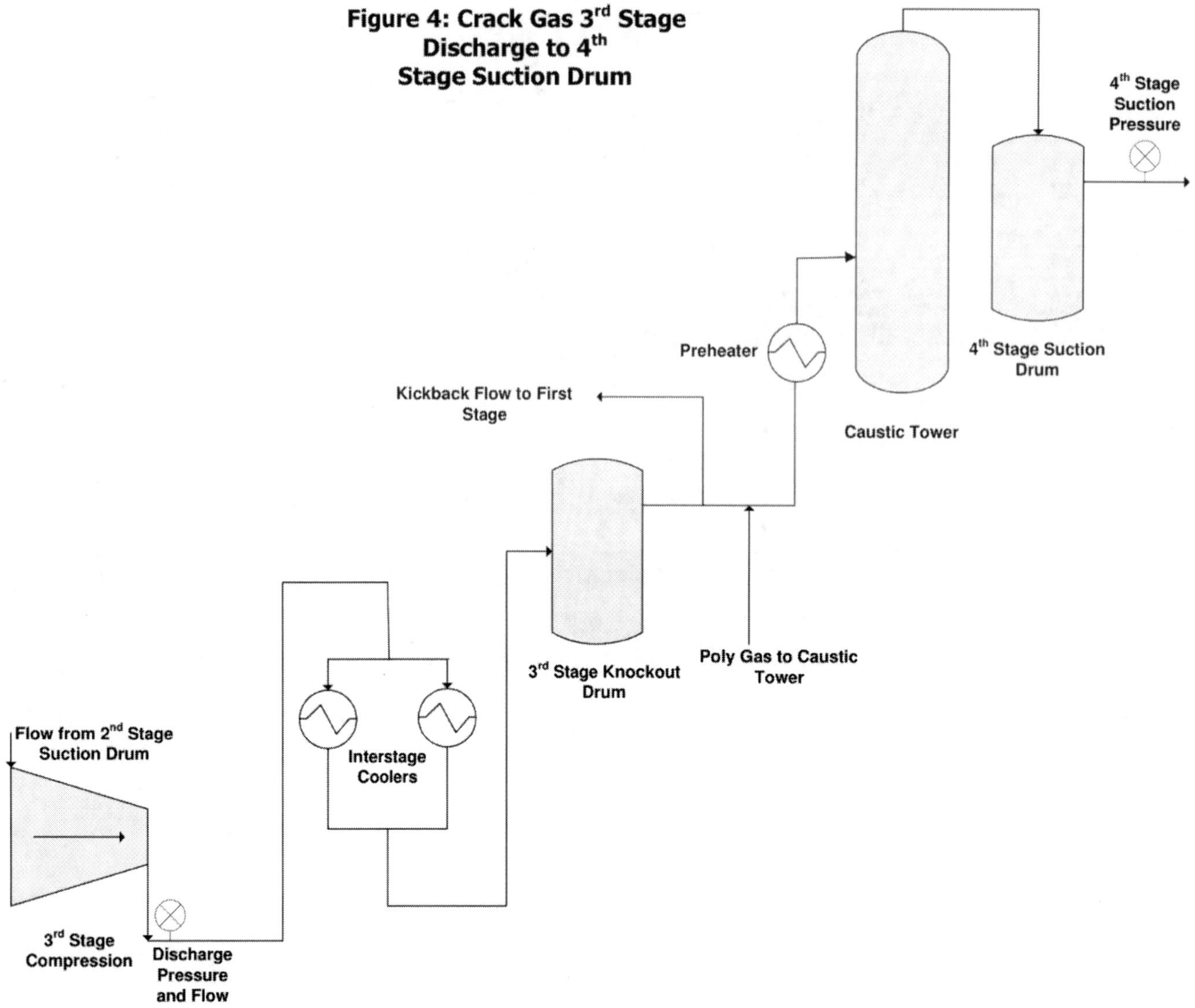

Fig 5 Model PDI Difference vs Run Days

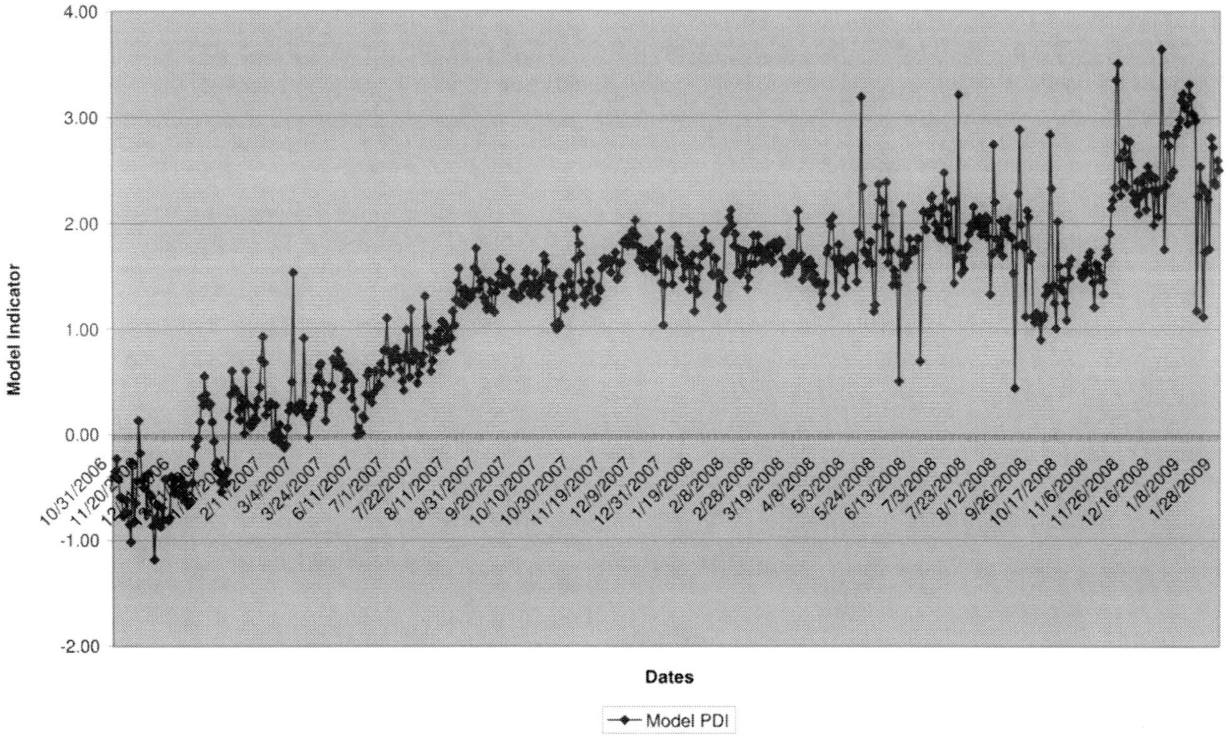

Fig 6 Tower PDI Difference vs Run Days

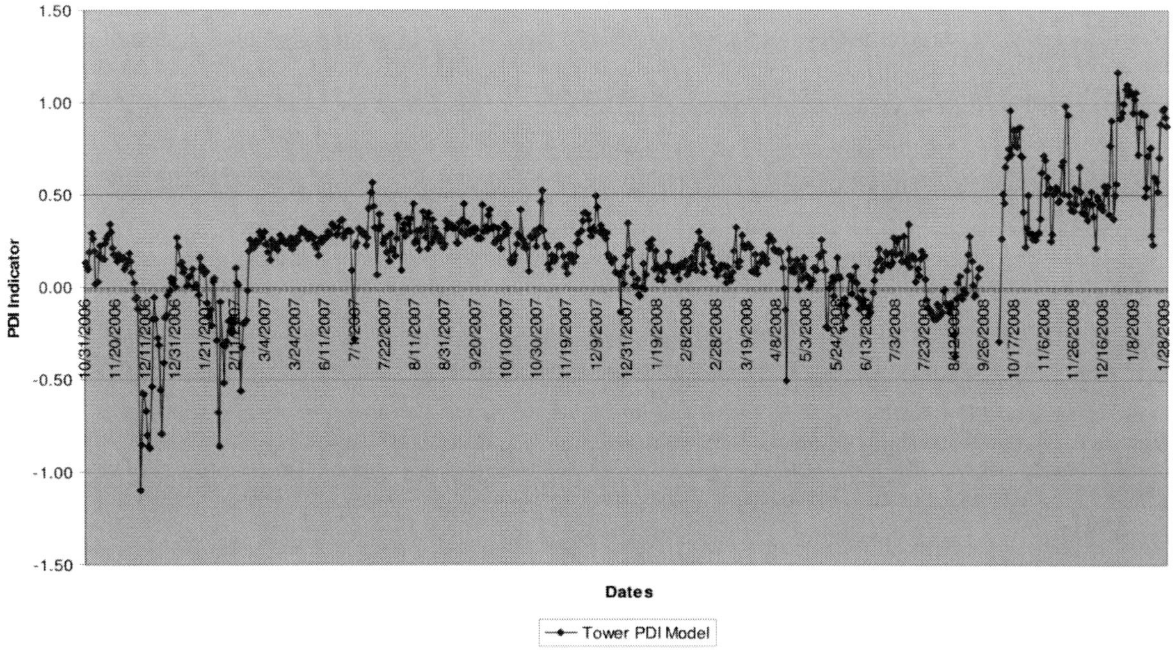

TABLE 1 Red Oil Potential Based on Acetaldehyde Loadings

Values shown are average values

Month Year	Acetaldehyde Inlet, ppm	Acetaldehyde Outlet, ppm	Removal Efficiency %
1Q, 2008	43.4	13.6	68.64%
Maximum Production			
April. 2008	39.9	11.5	71.16%
May, 2008	31.8	9.6	69.93%
June, 2008	26.4	8.2	68.91%
July, 2008	31.6	10.7	66.20%
August, 2008	29.6	9.8	67.09%
Sept- Oct, 2008	14.7	6.2	57.71%
Nov- Dec, 2008	26.8	7.9	70.51%

TABLE 2 CGC Model Shift of System Pressure Drop
3rd Stage Discharge to 4th Stage Suction

		Low Rates	High Rates	
Pressure Drop for	**EA-203 A/B**	4.7 psid	6.1 psid	Expected drop
Assumes equal gas flow to each cooler				
		Low Rates	**High Rates**	
Pressure Drop for	**DA-201**	1.0 psid	1.6 psid	Expected drop

Pressure Drop Model = **3rd Stage Discharge** to 4th Stage Suction Drum

Initially use data with clean system 11/10/2006 to 4/8/2007 **146 Operating Days**

Model evaluates five variables for correlation- listed in order of importance

PDI vs Summary of Results	**3rd Stage Flow lbs/hr**	**Kickback Flow Recycle Flow to 1st Stage lbs/hr**	**Run Days**	**3rd Stage Dischg Pressure psig**
	Regression Statistics	*Regression Statistics*	*Regression Statistics*	*Regression Statistics*
Correlation Coefficient R	0.8817	0.7602	0.5396	0.4791
R Square	0.7774	0.5779	0.2911	0.2295
Adjusted R Square	0.7759	0.5750	0.2862	0.2242
Standard Error	0.6890	0.9488	1.2295	1.2819
Count	146	146	146	146

	3rd Stage Flow plus Kickback Flow
Correlation Coefficient R	0.9321
R Square	0.8688
Adjusted R Square	0.8669
Standard Error	0.5309

PDI Calculated **= 0.0000799 * Flow 3rd - 0.000054 * Kickback - 7.043** Used model without days in it

TABLE 3 **3rd Stage Discharge to 4th Stage Suction**

Consecutive Run of 200 Days without a Single Plant Trip

Dates 6/8/2007 to 12/24/2007

PDI vs Summary of Results	**3rd Stage Flow** **lbs/hr**	**Kickback Flow** **Recycle Flow to 1st Stage** **lbs/hr**	**3rd Stage** **Dischg Pressure** **psig**	**Run Days**
	Regression Statistics	*Regression Statistics*	*Regression Statistics*	*Regression Statistics*
Correlation Coefficient R	0.9449	0.8454	0.5758	0.2536
R Square	0.8928	0.7146	0.3315	0.0643
Adjusted R Square	0.8923	0.7132	0.3281	0.0596
Standard Error	0.4521	0.7378	1.1292	1.3360
Data Points	200	200	200	200

PDI = 0.000104 * Flow 3rd Stage - 12.19 , psi

	3rd Stage Flow lbs/hr	**3rd Stage Pressure Drop** psid
Maximum Production	255000	14.3
Minimum Production	236900	12.4

TABLE 4 **CGC Model Shift of System Pressure Drop**
3rd Stage Discharge to 4th Stage Suction

PDI Calculated = 0.0000799 * Flow 3rd - 0.000054 * Kickback - 7.043

The new indicator would show trends over time showing when shifts up and down occur. It is compensated by the two flow variables.

Compare Indicator = Actual Measured PDI - Calculated PDI Value

Initially with clean system, indicator is showing a negative reading as psid.

Condition	**Flow 3rd Stage lbs/hr**	**Kickback Flow lbs/hr**	**System PDI, psid**	**Coolers PDI, psig**	**% Total PDI**	**Tower PDI, psid**	**Tower % Total**
Maximum Production In 2008 Actual	255000	5000	15.7	6.1	38.9%	2.4	15.3%
Maximum Production In 2006 with Calc PDI	255000	5000	13.1	5.24	40.0%	1.5	11.5%
Minimum Production In 2006	236900	24600	10.6	4.7	44.3%	1.0	9.4%

TABLE 5 CGC Model Shift of System Pressure Drop 3rd Stage Discharge to 4th Stage Suction

Time Periods Start	Time Periods End	Compare Indicator psid	Operating Days for Run	Run Days Days	Actual P_Drop psi	Kickback Flow 3rd Stage lbs/hr	Flow 3rd Stage lbs/hr	Comments	Occurrence of Kickback Flows > 20,000 lbs/hr Number	% of Days
10/31/2006	11/26/2006		26	26	12.04	13174	255060		6	23.1%
10/31/2006	12/3/2006	-0.59	34	34	12.00	12056	253933		6	17.6%
12/4/2006	**Plant Trip**									
12/24/2006	**Caustic Tower lower bed upset**		**Found brown material in lower bed**						16	39.0%
12/5/2006	1/14/2007	-0.32	41	75	11.19	16817	243556	Low Production		
1/15/2007	**Plant trip**									
1/16/2007	3/21/2007	0.30	63		13.07	7793	253302	**High Production** 3.045 MMlb/D avg	1	1.6%
1/16/2007	4/8/2007	0.35	81	156	13.18	6820	253278		1	1.2%
4/8/2007	**Plant turnaround**			5/09/2007		**Plant Startup from T/A**				
5/20/2007	**Furnace & Boiler trips**			5/31/07		**Ethylene Fractionator fire outage**				
6/7/2007				356	14.05	5950	252710	**Max Production** 3.364 MMlbs/day avg		
12/27/2007	4/22/2008	1.64	119	475	14.52	5815	253242		7	3.5%
4/28/2008	8/29/2008	1.85	123	598	14.12	8290	247999		11	9.2%
	Hurricane Gustav			**Hurricane Ike**					18	14.6%
9/24/2008	10/6/2008	1.58	13	611	14.48	13169	247999		1	7.7%
10/7/2008	**CGC slowdown**	11/12/2008		**Found brown material in lower bed**						
10/8/2008	10/26/2008	1.9	19	630	14.56	4643	255066		2	10.5%
10/27/2008	12/26/2008	2.14	57	688	12.67	24582	236908	Slow Production Recycle valve open	42	73.7%
12/27/2008	1/29/2009	2.60	34	722	16.24	3884	261512	**High Production** 3.185 MMlbs/day avg	1	2.9%

CAUSTIC TOWER OPERATION CONSIDERATIONS FOR EFFECTIVE PERFORMANCE

Charles Hammond, Ph. D.
Research Associate
Sasol North America Inc.
Vance Ham
Separations Technology Specialist
Process Equipment Technology
Shaw Energy & Chemicals Group

Caustic Tower operation is a key factor to the achievement of on-specification ethylene and propylene production. The tight limit on the allowable contaminant content levels driven by polymer producers necessitates close scrutiny and evaluation of the operation and performance of the Caustic Tower System.

Acid gases are created during the thermal cracking of any given steam-cracker feedstock. The preferred choice for removing these acid gases from the cracked gas is by absorption using dilute caustic soda. The Absorber, or Caustic Tower, is designed and operated to remove these acid gases; namely carbon dioxide (CO_2) and hydrogen sulfide (H_2S) and to a lesser extent some of the weaker acid gases such as mercaptans (R-SH).

This paper provides an overview of what the Caustic Tower process is, the acid gas and caustic concentrations and reactions of the process, and the key operating parameters. The impact of purchased caustic quality on performance is discussed, along with the handling of the fresh caustic solutions. Caustic Tower fouling and factors involved with its presence are described. Fundamentals of Caustic Tower controls and monitoring are provided to aid in the improvement of performance.

Introduction

The Cracking Furnace Section of an ethylene plant makes many by-products as it produces ethylene and other useful olefins. During the thermal cracking of an ethylene plant's given hydrocarbon feedstocks, acid gases such as hydrogen sulfide and carbon dioxide are pyrolysis by-products. These acid gas constituents of the total cracking effluent flowing to the ethylene plant Recovery Section must be removed sufficiently, so as to achieve an ethylene product and a propylene product (in the majority of ethylene plants) which are on-specification, and eliminate the potential negative effect of the acid gases upon downstream catalyst beds in the recovery process.

As these acid gases are very light, and are carried with the cracked gas stream within the cracked gas compression train, the preferred choice for removing the acid gases from the cracked gas stream is by absorption using a solution of dilute caustic soda. The Absorber, or Caustic Tower, washes the cracked gas with circulating caustic solutions, reacting the caustic soda with the acid gas compounds.

The Acid Gas Removal System, consisting of the Caustic Tower and its ancillary supporting equipment, is positioned in the basic process flow scheme, as shown in Figure 1, upstream of the final (fourth or fifth) stage of cracked gas compression.

Given the very important function that the Caustic Tower operation plays in the overall success of an ethylene plant, steps to ensure reliable operations of the entire Acid Gas Removal System must be taken. In a modern ethylene plant the caustic system has grown well beyond the Caustic Tower itself. Besides the Caustic Tower and its ancillaries to scrub the cracked gas (CG) clean of its acid gases, the spent caustic solution by-product of the acid gas scrubbing process must be considered as well.

Spent Caustic cleanup processing has expanded greatly over the last twenty years, consisting of pretreatment processes, stripper towers, coalescers, oxidation units, or incinerator units, etc. While the operation of the Caustic Tower may be good enough to get by, if the operation is not mindful of potential negative impacts the operating method and Caustic Tower performance may have on the downstream Spent Caustic handling equipment the results could be crippling to the entire ethylene plant manufacturing campaign. Thus, understanding and diagnosing Caustic Tower process performance and possible problems can be a key to maintaining effective ethylene plant operations.

But, before discussing the Acid Gas Removal System operations, the Operator needs to understand the generation of acid gases, in order to appreciate the prime factor in Caustic Tower operations – the concentration of acid gas constituents in the CG feed stream. Therefore, to get started, the process factors that influence and determine the CG acid gas content will be described.

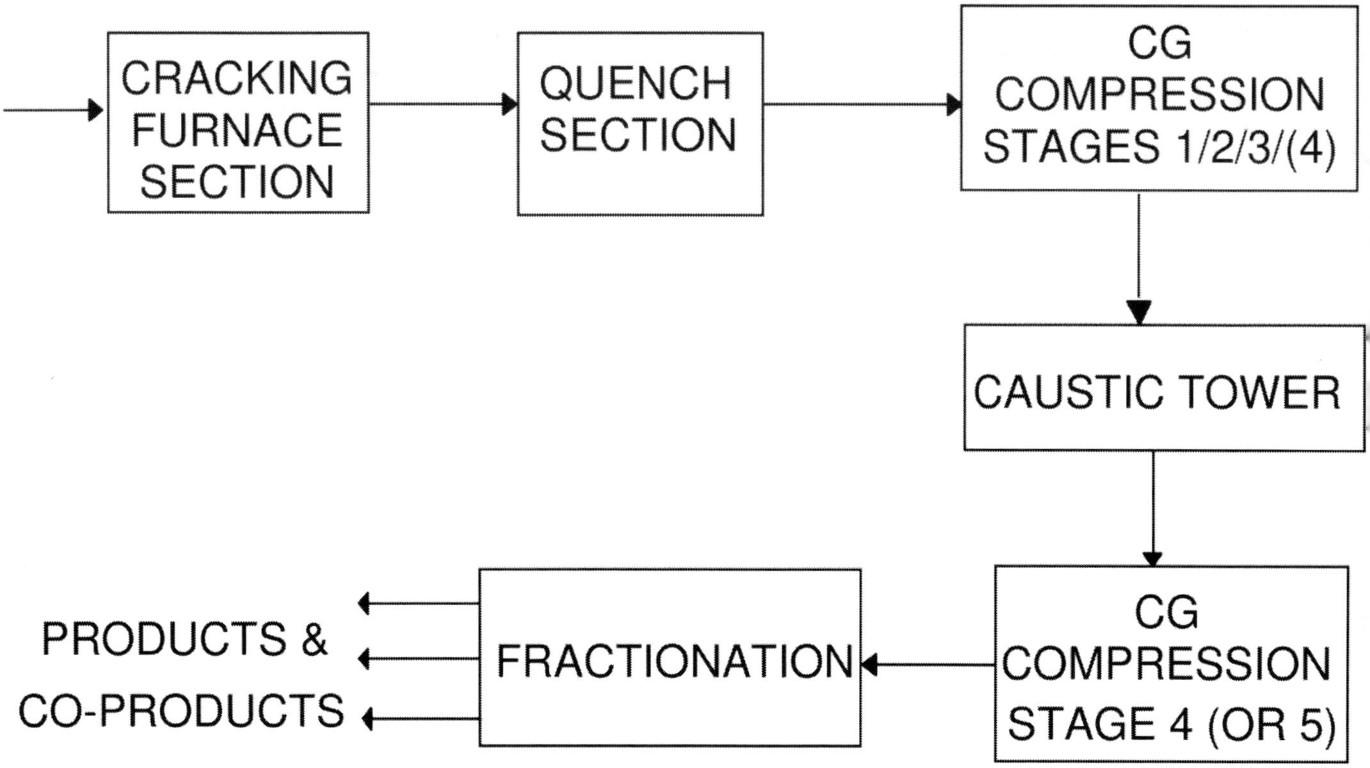

Figure 1 – Basic Ethylene Plant Flow Schematic

Source of Hydrogen Sulfide (H_2S)

The Caustic Tower cracked gas feed's H_2S content can be the result of two sulfur sources associated with the pyrolysis step in the ethylene plant production process. First, there is the sulfur which can be naturally occurring in an ethylene plant's fresh hydrocarbon cracking feed(s). Or, second, there is the sulfur containing compound that is deliberately added to a cracking furnace's hydrocarbon plus dilution steam feed mix, prior to the mix flowing into the furnace radiant coils. This second potential source for sulfur is also associated directly with the CO_2 production in the cracking furnaces as well, and this intertwined connection of CO_2 make in the pyrolysis step with H_2S yield will be explained later. But, before describing the H_2S yield from the cracking furnace area, the first sulfur source – the naturally occurring sulfur in the cracking feed – will be discussed so the reader can understand how this parametric factor affects the amount of H_2S that may be expected in the cracked gas.

Feedstocks and their Sulfur Content

Various types of fresh cracking feedstocks and their corresponding sulfur content include:

- LPG feedstocks used for gas cracking which typically contain very low sulfur levels. However, commercial propane and butanes may contain as much as 100 and 200 ppm sulfur by weight, respectively.

- Caustic treated C_3 LPG feedstock which can contain carbonyl sulfide (COS), in the range of 2 to 25 ppm by weight as sulfur (S), and mercaptan (CH_3SH), in the range of 5 to 10 ppm(S) by weight.

- Caustic treated C_4's feedstock which could contain light mercaptans in the range of 10 to 50 ppm(S) by weight.

- Light naphthas or natural gas liquids (NGL) which may contain heavy mercaptans and other organically bound sulfur species, and typically have total sulfur contents of about 100 to 150 ppm(S) by weight.

- Low sulfur full range naphthas with straight chain organic compounds and typically total sulfur contents less than 600 ppm(S) by weight.

- Atmospheric gas oil (AGO), heavy atmospheric gas oil (HAGO), and vacuum gas oil (VGO) feedstocks contain a variety of sulfur compounds at total sulfur contents that can range from 1000 ppm(S) to several weight percent, depending upon the crude source and upstream refinery distillation, hydro-processing and treating, and their boiling point ranges. These heavy feedstocks contain straight chain hydrocarbon compounds bound with sulfur, as well as thiophenes and higher order ring compounds, which are much more resistant to cracking than simple straight chain sulfur compounds.

- Field condensates (a prime cracking feedstock for many feed flexible liquids crackers) which are very wide boiling range hydrocarbon mixtures (crude like), and will contain all of the sulfur compounds typically found in the naphtha to VGO range. These cracking feeds typically have a total sulfur contents under one (1.0) weight percent.

Sulfur Addition to Cracking Furnace Feed

The first possible source of sulfur in the pyrolysis furnace effluent has been described. However, in those ethylene plants where very low sulfur containing feedstocks are used, e.g. gas crackers – ethane, propane and E/P mixtures – a deliberate addition of a sulfur compound must be made to the pyrolysis furnace hydrocarbon plus dilution steam feed mix. This would also be true for those cracking furnaces which may be dedicated to the pyrolysis of Recycle Ethane, Recycle Propane, or any other recycle streams from within the cracker/petrochemical complex which have little or no sulfur content because of the processing/treating the stream(s) may have received.

This need to deliberately dose a very low sulfur content hydrocarbon and dilution steam mix with sulfur is tied to the pyrolysis coking mechanisms and coke formation rate on the inside wall surface of the furnace tubes composing the radiant coils.

Coking involves the dehydrogenation of hydrocarbons and the thermal transformation of small aromatic molecules to larger complex poly-aromatic compounds (very low in hydrogen content) when a sufficient temperature is reached. At higher temperatures along the radiant coils the thermal coking mechanism increases. The pyrolysis coke precursors thus formed at the hottest temperatures, which are at the inside tube wall surface, are left as coke deposits on the surface.

In addition, the coking mechanisms – free radical coking and condensation coking – that are driven solely by the hydrocarbon species synthesized in the thermal pyrolysis process, there is also the dehydrogenation and coke precursors that are due to the nickel and iron present in the metallurgy of the stainless steel radiant coils. At the nickel and iron sites of the inside surface of the radiant tubes, dehydrogenation of the hydrocarbons at the high surface temperatures is catalyzed by the nickel leading to accelerated coke formation rates apart from the coke due solely to the pyrolysis mechanisms. This contribution to the radiant coil coking by the nickel and iron and their oxides is known as catalytic coking.

The deposition of pyrolysis coke on the inside surface of the radiant coils is then associated with the formation of the carbon monoxide and carbon dioxide that is yielded in the cracking process as a result of steam-reforming reactions and the water shift reaction. These mechanisms for CO and CO_2 formation will be discussed later. But, the nickel present in the furnace tubes at cracking temperature can also catalyze these reforming and water shift reactions.

Therefore, the nickel and iron sites of a radiant coil inside wall surface need to be passivated to counter the catalytic coking. During the pyrolysis operating mode because some of the nickel from the tube metallurgy is actually moved from the coil into the pyrolysis coke as it is deposited, even the lay down of the coke does not completely block all nickel interaction with the cracking mixture. The interaction with nickel will also depend upon the variable porous nature of the coke being laid down, which will be a function of the feed being cracked.

It is the presence of the naturally occurring or deliberately added sulfur the nickel and iron sites are in essence poisoned. The nickel and iron will mainly be in the oxide form, but can exist as the reduced metals during the cracking process due to the presence of the produced hydrogen. The sulfur reacts with the coil metal as follows:

$$NiO + H_2S \leftrightarrows NiS + H_2O \quad \text{(Equ. 1)}$$
$$Ni + H_2S \leftrightarrows NiS + H_2 \quad \text{(Equ. 2)}$$
$$FeO + H_2S \leftrightarrows FeS + H_2O \quad \text{(Equ. 3)}$$
$$Fe + H_2S \leftrightarrows FeS + H_2 \quad \text{(Equ. 4)}$$

forming the metal sulfide, which is much less reactive. Note that the reactions described here are with hydrogen sulfide, a constituent of the cracked gas due to being part of the hydrocarbon feed or created during the pyrolysis of other sulfur compounds entering the radiant coils. As with the make of CO and CO_2, the production of the H_2S will be described shortly.

Note, that the nickel and iron sites do not remain sulfided during the cracking mode operation if a minimum level of sulfur content is not maintained in the hydrocarbon and dilution steam mix. The metal sulfide is also lost during the decoking operating mode. With a steam-air decoking, the metal sulfide at the radiant coil inside surface will be converted back to the metal oxide form. Even when a "steam only" decoking procedure (very rarely utilized by Operators) is employed the metal sulfide in the equilibrium reaction with the steam will be returned to the oxidized form. Decoking by way of the "steam only" technique also can play a part in the CO_2 flowing to the Caustic Tower. This relationship will be described with the section on CO_2 generation.

For the addition of sulfur to gas cracking feedstocks including ethane, propane, or LPG, it is dimethylsulfide (DMS), dimethyldisulfide (DMDS), or mercaptans which are commonly added to an individual cracking furnace dilution steam feed. The sulfur compounds added should be free of COS and carbon disulfide (CS_2).

Disposition of Sulfur Compounds During Cracking and Hydrogen Sulfide Make

For the naturally occurring sulfur compounds in the fresh cracking feedstocks, the various sulfur containing compounds do not all readily crack and produce H_2S. However, mercaptans (R-SH) and disulfides (RSSR) in liquids cracker feedstocks will readily decompose with H_2S as a result. Mercaptans which crack will yield as follows:

$$R\text{-}SH + H_2 \rightarrow RH + H_2S \qquad \text{(Equ. 5)}$$

Disulfides decompose to RSH and olefins and H_2S as the following example reactions represent.

$$RSSR + H_2 \rightarrow 2RSH \qquad \text{(Equ. 6)}$$

$$CH_3SSCH_3 + H_2 \rightarrow CH_2CH_2 + 2H_2S \qquad \text{(Equ. 7)}$$

COS is not produced in the steam cracking furnace. If the fresh cracking feed contains some COS such as in FCC LPG and propane feedstocks, some of the COS is converted to H_2S in the furnace; the remainder passes on through with the cracking effluent from the furnace.

Thiophenes and higher order sulfur ring compounds are much more resistant to cracking than simple chain sulfur compounds, as sulfur ring compounds are more stable at the steam cracking conditions. During pyrolysis, sulfur ring compounds may react with hydrocarbons to produce higher member ring "poly-aromatic" compounds.

The DMS, DMDS, or mercaptans which can be deliberately injected for very low to nil sulfur content hydrocarbon cracking feeds will rapidly decompose at temperatures above 260°C (500°F), following the conversion pathways already described.

There are several factors which will determine the net resulting concentration of H_2S make. The types and concentrations of the sulfur bearing compounds that are contained in cracking feeds or injected will influence the quantity of H_2S produced. The lighter and straight chain sulfur compounds generate more H_2S and the higher the concentration of sulfur compounds the more H_2S which will be made. Associated with the factor of sulfur concentration is also the cracking severity of the cracking furnace operation. The higher the cracking severity the more of the sulfur compound types which will be converted to H_2S.

Carbon Monoxide and Carbon Dioxide Generation

As has been discussed sulfur's presence is required in the steam-cracking furnace to passivate the radiant coil metallurgy, reducing the rate of pyrolysis coke formation and thus lowering the CO and CO_2 production. The formation of pyrolysis coke at the radiant coil wall surface is represented as a dehydrogenation by the following:

$$C_nH_m \leftrightarrows (m/2)H_2 + nC \qquad \text{(Equ. 8)}$$

where the carbon is deposited at the inside tube surface.

The reforming and water shift reactions during steam cracking produce the CO and CO_2. The steam-reforming reaction (which is also catalyzed by nickel present in the stainless steel radiant tubes which is not passivated) proceeds according to the following equilibrium reaction:

$$C + H_2O \leftrightarrows CO + H_2 \qquad \text{(Equ. 9)}$$

as dilution steam reacts with carbon on the tube surface. Then the water shift reaction will form some carbon dioxide as follows:

$$CO + H_2O \leftrightarrows CO_2 + H_2 \qquad \text{(Equ. 10)}$$

per the operating equilibrium conditions. At the radiant coil inside surface, it is believed that there is some interference with the water shift reaction potential yield of carbon dioxide by the following reaction of carbon dioxide with the pyrolysis coke.

$$C + CO_2 \leftrightarrows 2CO \qquad \text{(Equ. 11)}$$

The amount of CO produced is primarily a function of the hydrocarbon feedstock type, the age of the furnace tube, and the concentration of the sulfiding agent naturally contained in the hydrocarbon feed or deliberately added. Thus, the net CO_2 make will be dependent upon the equilibrium production of CO.

Decoking of the furnace radiant tubes is required to free the radiant coils of the deposited pyrolysis coke, which will have eventually grown too thick, causing high pressure drop and/or insulating the metal tube until the furnace firing is such that the metal reaches the maximum allowable temperature.

Following a steam-air decoke or "steam only" decoke (very rarely utilized by Operators) of a cracking furnace, the inside surface of the radiant cracking coils will be very active again for catalytic coking reactions. This is why it is a typical practice to presulfide and passivate the radiant coils with sulfur addition to the flowing dilution steam, prior to the introduction of the hydrocarbon feed to a freshly decoked cracking furnace. If a freshly decoked furnace is not well passivated before the cracking feed is introduced, upon feeding hydrocarbon the Operator can expect a spike in CO_2 in the cracked gas feed to the Caustic Tower.

If "steam only" decoking of pyrolysis furnaces is being practiced by an Operator, the furnace effluent of the decoking operating mode is still routed directly to the Quench Section, meaning time is saved since the furnace does not have to be swung out of the cracking effluent transfer line header. In the "steam only" decoking method the pyrolysis coke is removed at very high temperature by the steam reforming reaction (Equ. 9) above. Thus, in ethylene plants practicing "steam only" decoking, placing a furnace into the "steam only" decoking mode can result in large spikes of CO and CO_2 in the total cracking effluent and consequently result in acid gas perturbations to the Caustic Tower and CO Methanator.

All CO_2 entering the cracking furnaces with the hydrocarbon feedstocks will end up in the cracked gas feed to the Caustic Tower. Also, if there are additional external light hydrocarbon streams, e.g. FCCU Off Gases, being fed to the ethylene plant for recovery, any CO_2 and H_2S content in those streams must be considered, for that acid gas will contribute to the load on the Caustic Tower.

Acid Gases Distribution in the Olefins Processing Train

The furnace cracking effluent gases will be cooled, and, in the case of a liquids cracker, fractionated in the Quench Section to an overhead cracked gas stream, Raw Pyrolysis Gasoline (RPG) stream, and Pyrolysis Fuel Oil (PFO)/pyrolysis tar stream.

The cracked gas to be treated in the Caustic Tower will contain all the CO_2, H_2S, COS, and a trace amount of any light mercaptans and thiophenes.

For liquid hydrocarbon feed cracking, most of the heavy sulfur compounds remain in the PFO product and the RPG stream. This refers to any heavy mercaptans and heavy organic sulfides which will mostly remain as heavy organically bound cyclic and aromatic ring compounds.

Process Flow Scheme Description for Caustic Tower System

From the overhead of the Quench Water Tower the cracked gas containing all the acid gases flows to the Cracked Gas Compression (CGC) train. Per simple block flow diagram in Figure 1, the cracked gas will be compressed by the first three stages of a four stage CGC or four stages of a five stage system, prior to the cracked gas being treated in the Caustic Tower System.

In the Caustic Tower System, the compressed cracked gas will be contacted with circulating caustic solutions in the tower configured in either two or three acid gas scrubbing sections. Figure 2 presents the simplified process flow scheme of a two section – Weak Caustic Section and Strong Caustic Section – Caustic Scrubber scheme.

From the CG Compressor Stage "X" discharge aftercooler, the stream flows to a Knock-Out Drum to separate any condensed hydrocarbons and water from the cracked gas stream. Before the cracked gas feed enters the bottom of the Caustic Tower, the cracked gas is preheated to sufficient temperature to optimize caustic utilization for the net removal of CO_2 and H_2S which must occur in the tower, while avoiding hydrocarbon condensation and minimizing polymerization. At the bottom of the Caustic Tower, the cracked gas is contacted with the circulating Weak Caustic solution. The cracked gas flows from the Weak Caustic Section through a chimney tray to the circulating Strong Caustic Section, where the acid gas removal is completed.

In the caustic scrubbing, the actual mass of caustic solution required to react with the acid gas constituents is not very large and cannot satisfy the actual liquid loading requirement of a tray (or packing) in the Caustic Tower. Therefore, the recirculation of the caustic solutions from the bottom of a section back to the top of that section is utilized to match the hydraulic criteria as demanded by the particular tray (or packing) type that is being used in the scrubber. The required liquid volume flow flux per active area (gpm/ft^2) of the tray active area must be maintained, developing the required froth height and mass transfer contact. If packing is used, the liquid volume flow rate per volume of installed packing (gpm/ft^3 of packing), and the consistent distribution to maintain this, is even more critical.

The net caustic solution at the bottom of the Weak Caustic Section is pumped from the bottom of the tower to the top of the Weak Caustic Section. The top caustic section, Strong Caustic Section, solution is collected on the chimney tray and then pumped back to the top of the Strong Caustic Section.

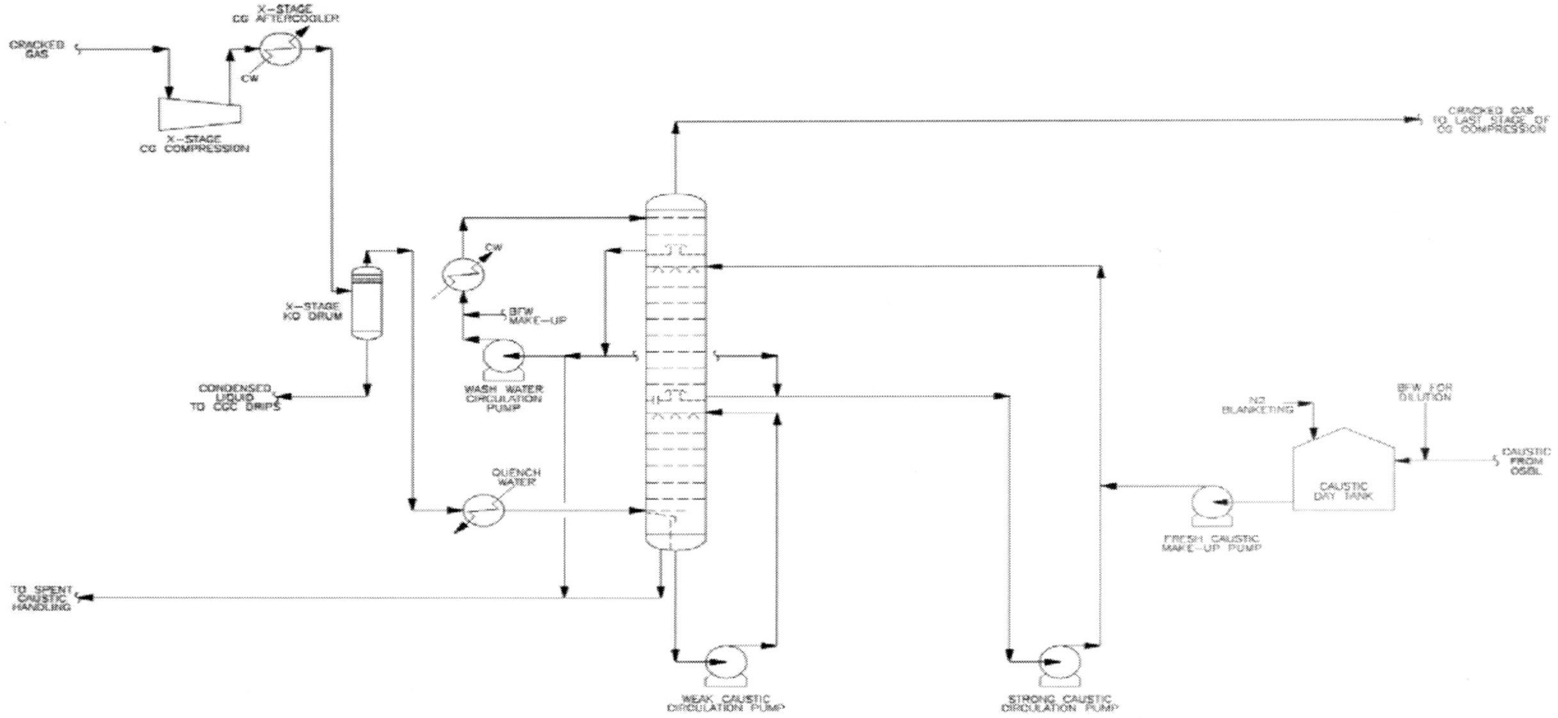

FIGURE 2 – Simplified Caustic Tower Process Flow Scheme for Two Caustic Section Arrangement

On the Strong Caustic circulation loop, the fresh make-up caustic is injected. The excess caustic solution then overflows by internal down pipes from the chimney tray to the top of the Weak Caustic Section. Sufficient excess fresh caustic is fed to the Caustic Tower to assure reliable operation.

Drawn from the bottom of the Caustic Tower is a spent liquid resulting from the caustic make-up feed, water inventory control and acid gas reactions. This bottoms liquid product, the Spent Caustic, contains the remaining free caustic which is unreacted, being in excess to what is stoichiometrically required to react all the acid gas. Spent Caustic flows from the overflow sump on the bottom of the tower to the spent caustic handling system for solvent washing with gasoline, settling, de-gassing and further treatment. With this Spent Caustic any condensed hydrocarbon liquids should be drawn off as well. The condensed hydrocarbon liquid phase should contain any "Red Oil" which may have been created in the tower.

The fresh make-up caustic solution is sourced from a day tank, where a higher concentration caustic solution is diluted with boiler feedwater (BFW) to create the make-up caustic solution. The BFW needs to be oxygen free and the day tank is blanketed with nitrogen to keep the caustic solution from contacting air and being contaminated with oxygen.

The cracked gas scrubbed of acid gas components flows from the Strong Caustic Section through a chimney tray to a Wash Water Section, where the cracked gas is contacted with Wash Water. The Wash Water removes entrained caustic, protecting downstream equipment from possible caustic carry over in cracked gas flow to the last CGC stage. Though some Caustic Tower process flow schemes have Wash Water Section configured as a once through flow arrangement, typically the Wash Water Section will be configured with a Wash Water Circulation Pump. As with the caustic sections, the recirculation of Wash Water in the section loads the contact stages as may be required while lowering the Wash Water make-up water required, but at the expense of pumping power. The circulating Wash Water also cools the treated cracked gas, reducing the compression power requirement.

A slip stream of the Wash Water is blown down to the spent caustic line as required to control build up of any caustic concentration in the circulating Wash Water. Some of the Wash Water will also be used to control the water balance of the circulating caustic solutions by letting down on flow control some of the Wash Water to feed the Strong Caustic Section circulating solution. This helps regulate circulating liquid caustic strengths and salt concentrations in the solutions. Oxygen free BFW is used to make-up and maintain the water level on the Wash Water Chimney Tray.

Adding another circulating caustic section allows for an enhancement of the driving forces for the acid gas removal. The addition of a third section provides the opportunity to optimize total tray count and Strong Caustic Section inlet solution NaOH strength with the overall average caustic concentration profile in the Caustic Tower. Figure 3 presents the simplified process flow scheme where an Intermediate Caustic Section is utilized in the Caustic Tower System.

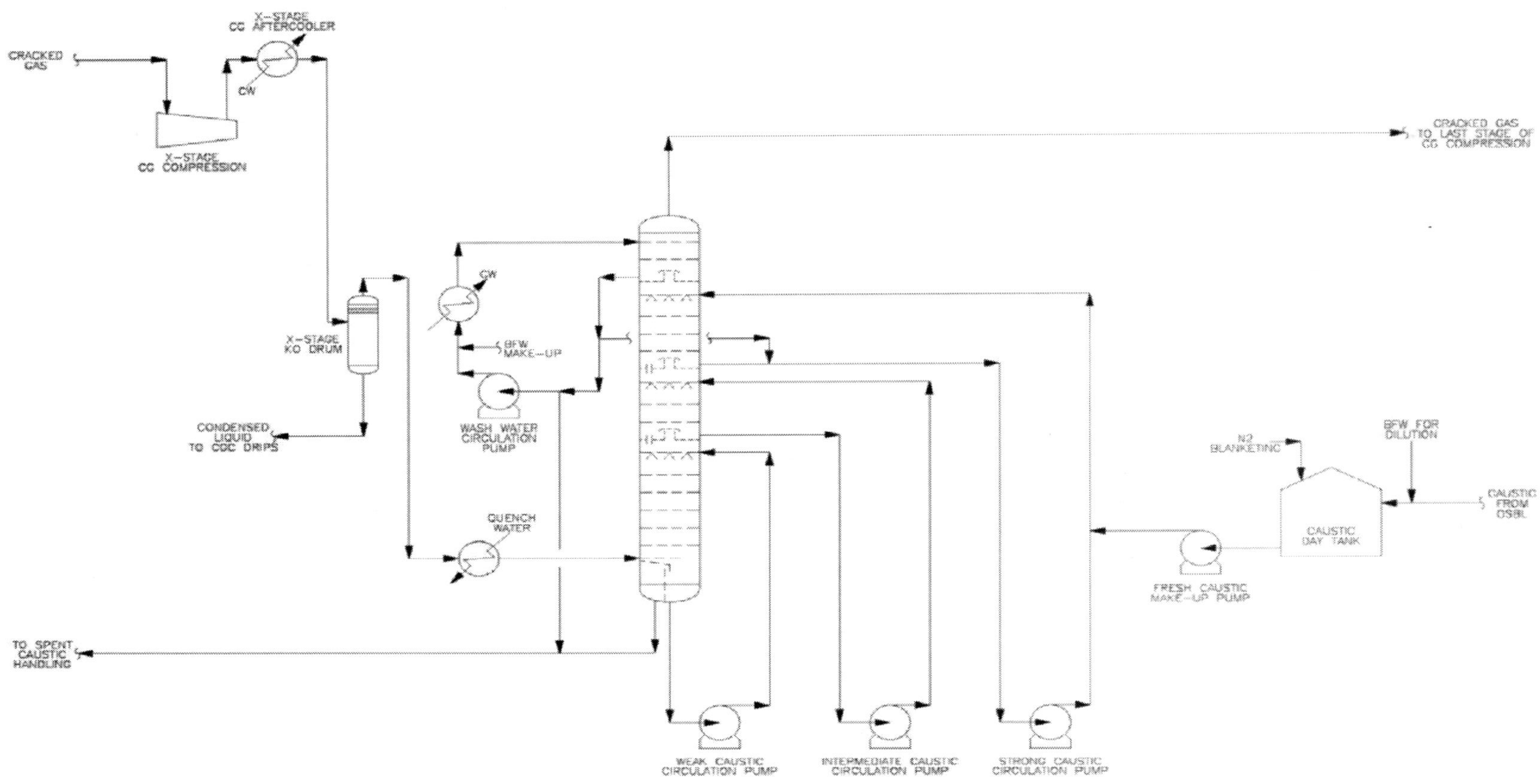

FIGURE 3 - Simplified Caustic Tower Process Flow Scheme for Three Caustic Section Arrangement

Chemistry in Caustic Towers

Within the Caustic Tower of an olefins plant there is a good deal of chemistry possible besides the primary reaction of acid gases with sodium hydroxide. The removal of acid gases from cracked gas in the Caustic Tower is a classic example of absorption with chemical reaction. Thus, the key to the caustic scrubbing process is the solubility of the cracked gas stream H_2S and CO_2 content within the caustic solution, where the H_2S and CO_2 react with the NaOH.

Unfortunately, the circulating caustic solution also absorbs other constituents from the cracked gas stream. Depending upon the solubility limits of the caustic solution for absorbing various cracked gas components, also absorbed is an array of hydrocarbon components such as carbonyls, ketones, organic acids (oxygenates formed during steam cracking) and the unsaturated components that predominate in the cracked gas stream. It is the variety of the chemical mix in the liquid phase of the caustic treating process, which result in a number of polymerization reactions.

The reactions resulting in the formation of polymers are undesired, but inherent to a certain degree. It is these polymers which constitute the essential factor in Caustic Tower fouling (as well as difficulties in Spent Caustic handling equipment). With a number of different initiators possible, these polymers can be produced to levels which can result in serious rates of fouling in the Caustic Tower system, causing serious operating performance problems.

Acid Gas Removal Chemistry

When CO_2 and H_2S are absorbed by the sodium hydroxide solution, the following reactions occur in the liquid phase:

$$CO_2 + 2NaOH \rightarrow Na_2CO_3 + H_2O \qquad \text{(Equ. 12)}$$
$$H_2S + 2NaOH \rightarrow Na_2S + 2H_2O \qquad \text{(Equ. 13)}$$

with these two stoichiometric reactions between the principal acid gases and NaOH being fast and irreversible. It is believed that the reactions actually occur just at the vapor interface with the liquid phase as soon as the CO_2 and H_2S have been absorbed into the liquid.

The H_2S reaction per Equation 13 is much faster than Equation 12. Therefore, for most ethylene plant Caustic Tower operations the H_2S removal will be almost totally accomplished in the Weak Caustic Section.

It is primarily the reaction products of Na_2CO_3, sodium carbonate, and Na_2S, sodium sulfide, which are needed for reliable operation of the caustic scrubbing process.

However, as the salts of Na_2CO_3 and Na_2S are being created, CO_2 and H_2S absorbed into the liquid phase in the presence of water can react with the salts per Equation 14 and 15 to form sodium bicarbonate ($NaHCO_3$) and sodium hydrogen sulfide (NaHS).

$$Na_2CO_3 + H_2O + CO_2 \rightarrow NaHCO_3 \quad \text{(Equ. 14)}$$
$$Na_2S + H_2S \rightarrow 2NaHS \quad \text{(Equ. 15)}$$

These two secondary reactions are much slower than the two desired primary reactions to Na_2CO_3 and Na_2S. In a short residence time, counter-current absorber as the Caustic Tower, the level of CO_2 and H_2S removal via Equations 14 and 15 is low, but solubilities of sodium bicarbonate and sodium bisulfide in the solution are four (4) to six (6) times lower than that of the sodium carbonate and sodium sulfide.

Because of the lower solubility of the $NaHCO_3$ and NaHS, these two salts tend to precipitate from the solution, which can lead to equipment problems. Thus, to maintain the primary acid gas reactions and reliable system operation, the Caustic Tower fresh sodium hydroxide make-up needs to be maintained sufficiently in excess to enhance the potential of the dissolved H_2S and CO_2 to find NaOH with which to react.

Besides the precipitation potential of $NaHCO_3$ and NaHS, when the Caustic Tower is operated under conditions of NaOH deficiency, the following reversible reactions associated with $NaHCO_3$ and NaHS also occur:

$$CO_2 + NaOH \leftrightarrows NaHCO_3 \quad \text{(Equ. 16)}$$
$$H_2S + NaOH \leftrightarrows NaHS + H_2O \quad \text{(Equ. 17)}$$

Because these reactions are reversible, they do not guarantee the acid gas compounds remain converted to a salt. Therefore, again, to assure H_2S and CO_2 molecules, once reacted in the liquid phase, remain removed from the cracked gas stream, the absorber needs to be operated with enough excess NaOH in the Spent Caustic.

Caustic Towers can remove mercaptans (RSH) and COS, as well as hydrogen cyanide (HCN), to a very limited extent. The reactions of mercaptans with NaOH to form mercaptides are reversible, equilibrium reactions, as represented by the following:

$$RSH + NaOH \leftrightarrows NaRS + H_2O \quad \text{(Equ. 18)}$$

which cannot be driven to full conversion to the salt, due to the limited solubility of the mercaptans from the vapor phase into the liquid, and the very slow reaction rates compared to the short contact time provided in the scrubber. But, besides the limited solubility, the caustic solution circulating back to the top of a Caustic Section creates another limitation.

The cracked gas which had been lowered in mercaptans content is now contacted again at the top of the section with the caustic solution now has a concentration of mercaptides. The gas phase lower mercaptans concentration at the top of a section now provides a concentration differential causing a dissociation of mercaptides back to mercaptans, resulting in a net mercaptans removal from the cracked gas of very little on a once through basis.

While there can be a very small disappearance of COS across the Caustic Tower, COS for all practical purposes should be expected to pass through virtually unaffected. The COS which might be able to dissolve into the caustic solution can be converted to some extent by undergoing a possible series of reactions combining hydrolysis (COS + H_2O react to form CO_2 and H_2S) followed by acid gas conversion. However, COS solubility into the caustic solution is very small and any conversion by hydrolysis (be it vapor phase or liquid phase) is also very unfavorable.

Red Oil Formation Reactions

Besides the acid gases, oxygenated compounds, including carbonyl compounds, are formed in the pyrolysis furnaces. The amount of ketones and aldehydes produced can vary widely from 50 to 500 ppm in the cracking effluent depending on the type of hydrocarbon feed being cracked and cracking severity.

Hydrocarbons in the cracked gas feed dissolve in the caustic solution in small amounts proportional to their concentration in the gas phase and solubility in the caustic solution. Highly unsaturated compounds, such as acetylenes and dienes, are appreciably soluble in the circulating caustic. Also, most of the carbonyls are absorbed. Depending upon the operating pressure, temperature, number of Caustic Sections that make up the Caustic Tower, the solution circulation rates and type of tower internals, the amount of aldehydes absorbed will vary. Of these oxygenates, the key one, acetaldehyde, may have anywhere from fifty (50) to ninety-five (95) percent of it removed from the cracked gas in the Caustic Tower.

In the presence of caustic, the absorbed carbonyl compounds at the temperature and pressure conditions will undergo condensation reactions to form heavy materials. This polymerization mechanism is called aldol condensation. The aldol condensation reaction products have a characteristically orange to reddish color. The color arises from the number of carbon-carbon double bonds that have been able to form in the polymerization reaction chain progressions. As the reactions have the conditions to progress, the color will be seen to form and develop as an oily, viscous hydrocarbon liquid phase is produced. If left unchecked, an aldehyde resin material is formed. This is fouling material that composes deposits found in many caustic systems.

The classic aldol condensation reaction is an addition polymerization reaction as shown in the following simplified reaction sequence:

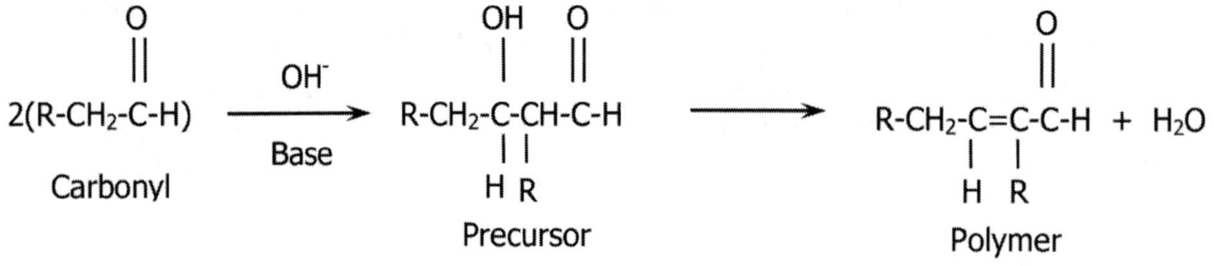

The sodium hydroxide as a base attacks the neutral aldehyde creating an intermediate precursor with another aldehyde molecule. The equilibrium of the reaction is such that the precursor proceeds to give up a molecule of water to become the more stable polymer form.

Caustic Towers, with the large ratio of caustic recirculation flow to Spent Caustic draw off, have in the liquid traffic a long residence time for the polymerization to occur. The anion attack of the caustic on additional aldehyde molecules and carbonyl polymer continues as represented by the following:

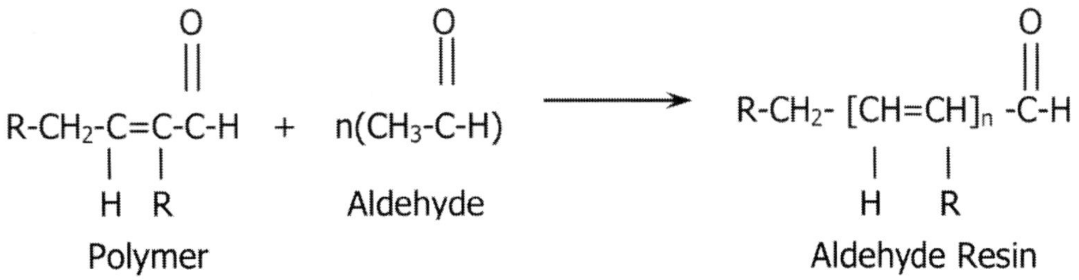

with the formation of the very sticky and viscous resin through multiple aldol condensation reactions.

As mentioned previously, a series of color changes will occur as the reactions can advance. The lower molecular weight carbonyl polymer creates a light straw yellow color. The color changes from yellow to orange and then to red as the polymeric reaction continues and the molecular weight grows. Typically, a deep red hydrocarbon liquid phase is produced, and thus the reason for the term "Red Oil". Should conducive conditions exist, the condensation reactions can eventually form the reddish aldehyde resin sludge in the recirculating caustic loop. Pictures, courtesy of Baker Petrolite, are provided in Figure 4 which present the potential polymerization progression described for carbonyls.

Besides the residence time of the caustic solutions in hold up and recirculation volumes, other key parameters which are factors in the magnitude of Red Oil formation and fouling deposits that an Operator may experience will be operating temperature, cracked gas carbonyl content, and the caustic concentrations in circulating solutions.

Associated with the Red Oil chemistry discussion, it needs to be mentioned that there is chemistry that the Operator can bring to bare to try and counter act the aldol condensation reactions. A carbonyl scavenger, described as an antipolymerant, can be injected into the fresh caustic make-up and/or circulating caustic solutions to attempt to nullify carbonyls entering the tower. Or, an antifoulant dispersant chemical treatment can be applied.

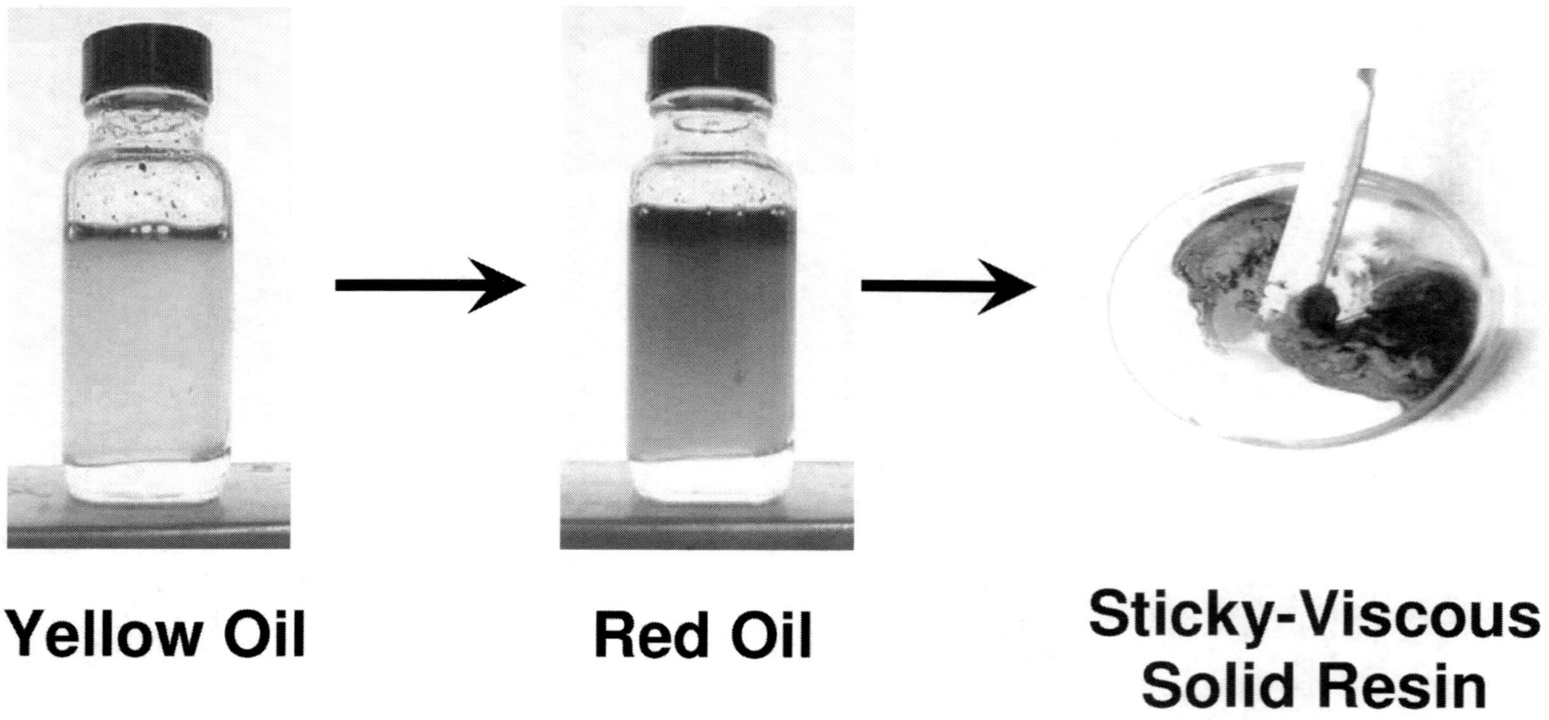

Figure 4 – Aldol Condensation Polymerization Progression
Images Courtesy of Baker Petrolite

An antipolymerant chemical, such as hydroxylamine (NH_2OH), will react with any carbonyl group, e.g. aldehyde, absorbed into the caustic solution that the chemical scavenger molecules comes in contact with to form an inert oxime compound plus water, preventing possible polymerization. However, the antipolymerant will also react with any carbonyl polymer at any point in the possible aldol condensation chain growth. Thus, even if some carbonyl does enter into the aldol condensation reaction pathway, the potential amount of polymer made (and thus the possible fouling rate) can be significantly reduced. The oxime compounds formed are stable and soluble in the caustic solution and would be removed from the tower with the Spent Caustic draw. Therefore, the actual extent of success with an antipolymerant chemical treatment in combating aldol condensation reactions depends greatly on the CG Feed concentration of carbonyls and the amount of chemical injected.

An antifoulant dispersant chemical injection will not eliminate the possibility for aldol condensation reactions from necessarily proceeding, but these antifoulant chemicals, consisting of polymeric compounds which are both polar and non-polar in nature, are designed to keep the carbonyl polymers dissolved within the caustic solutions of the system. The antifoulant molecules are said to attach themselves to the initial, small polymer molecules as they form, blocking the carbonyl polymer from additional anion attack and continued reaction with other carbonyl molecules and further chain growth. But, the dispersant chemicals have also shown the ability to attach to existing polymer deposits on tower internals and piping walls and dissolve these large polymer particles back into the circulating caustic solution. The chemical is injected into the Strong Caustic circulation so that the antifoulant will be made available to all the caustic sections. The dissolved polymer particles with dispersant attached would be removed from the system with the Spent Caustic draw.

One drawback that may be encountered with the use of an antifoulant dispersant chemical treatment program is that an antifoaming agent will need to be added also to guard against foaming inside the tower. Foaming problems have been commonly reported to occur when antifoulant dispersant chemistry has been applied to a Caustic Tower operation, but the extent to which foaming may become an issue and the amount of antifoaming agent that may need to be applied varies widely. The magnitude of any foaming problem in a Caustic Tower System is a function of the amount of polymer formation potential for the operation and the configuration of the tower internals. There appears to be an enhanced potential for foaming in the system when packing is being used in the Caustic Tower as opposed to trays.

Free Radical Polymerization Reactions

The other source for polymer formation in a Caustic Tower is free radical polymerization. The unsaturated components such as acetylenes, dienes, and styrene that dissolve in the caustic solution can polymerize in the liquid phase. Dissolved oxygen, chlorides, chlorates, peroxides, azo compounds and transition metal ions can promote (as initiators) the free radical addition reactions.

Free radical polymerization follows a three step path with first chain initiation, followed by chain propagation, and then termination. Initiation is the step in which a promoter present in the caustic solution generates the free radical. This initiation reaction may be promoted from heat, transition metal reactions or oxygen/peroxide interactions. In the case of most of the possible promoters, the initiation step will be impossible to stop. Chain propagation, the second step, then occurs when the free radical reacts with additional monomer to form a lengthening oligomer or polymer molecule. If and when two free radicals react with each other to form a stable molecule that terminates the potential chain propagation that those free radicals represented.

The three step pathway of polymer formation by free radical additions are shown in the reactions which follow. The possible chain initiation reaction forms are represented by Equations 19 through 21 along with the various reactions shown as Equations 22 through 24 that continue the chain propagation. The reaction forms that represent the terminators of the free radical polymerization are shown as Equations 25 and 26.

Chain initiation reactions are of the following form:

Initiation

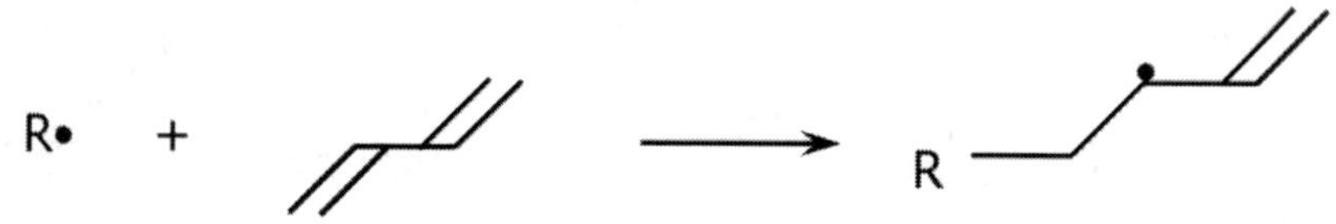

R-H → R• + H• (Equ. 19)

M^{++} + R-H → M^{+}R• + H^{+} (Equ. 20)

M^{++} + R-O-O-H → M^{+}R-O-O• + H^{+} (Equ. 21)

where R• represents reactive monomer, polymer or oligomer, and M^{+} are metal ions.

Chain propagation reactions are of the following form:

Polymer Growth

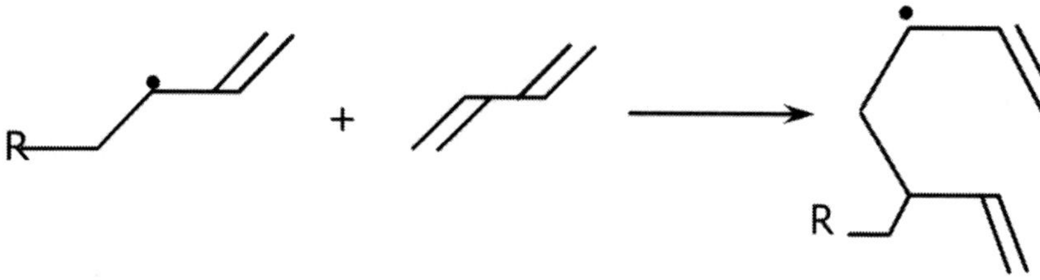

R• + O_2 → R-O-O• (Equ. 22)

R-O-O• + R'-H → R-O-O-H + R'• (Equ. 23)

R'• + C=C → R'-C-C• → Polymer (Equ. 24)

The chain termination reactions are of the following:

R• + R'• → R-R' (Equ. 25)

R• + R'-O-O• + R'-H → R-O-O-R' (Equ. 26)

As with the aldol condensation reactions, there is chemistry that the Operator can bring to bare to try and counter the initiators of free radical polymerization reactions. A free radical inhibitor can be injected into the fresh caustic make-up and/or circulating caustic solutions to attempt to minimize any reactions in the caustic solutions which will generate free radicals.

Within a Caustic Tower System, the Operator may find that fouling is due predominantly to free radical polymerization or aldol condensation. But, it could also be found that both mechanisms are each contributing significantly to polymerization and fouling.

Operating for Effective Performance

It is clear that the critical consideration for the Caustic Tower operation is removal of H_2S and CO_2. Therefore, the most efficient or minimum usage of fresh caustic make-up may not always be the major emphasis at a given moment in time.

Operating the Caustic Tower with virtually no free NaOH in the Spent Caustic allows a very sparing margin for control error or acid gas load changes. If an Operator decides to operate the Caustic Tower System with the Spent Caustic stream at near zero free NaOH content, vigilance and good communication between the operating crew will be needed. Also, care needs to be taken that caustic usage minimization is not practiced such that sodium bicarbonate and sodium hydrogen sulfide concentrations in the circulating caustic exceed a maximum allowable, leading to possible precipitation of the salts within the tower and piping, which eventually could cause restrictions to flow and plugging at points in the system.

Therefore, control of fresh caustic make-up is important. As is paying close attention to the fresh caustic make-up strength. And, considerations for minimizing the rate of acid gas load increases should be a constant consideration as cracking furnace conditions are being changed. As one would expect, it is the possible disturbances of Cracking Furnace Area operations which will account for the majority of the typical sources for swings and drift in the Caustic Tower System performance.

The Operator needs to be mindful of the key control variables of temperature in the Weak Caustic Section and liquid levels. Monitoring of temperatures and levels are crucial to effective control of reactions and maintaining material balances.

Adequate attention also needs to be paid to the factors – of which operating temperature control is one – that can influence the potential formation rate of polymers in the circulating caustic solutions, and thus change the fouling rate of the Caustic Tower System. Quality of purchased caustic is a key factor in combating the formation of polymers which can result in fouling, but also contribute to foaming. Another important factor is the quality of water that is used as make-up into the System, be it for fresh caustic dilution or for control of the basic water balance of the circulating solutions.

As part of the effort to prevent foaming and maintain a minimum rate of fouling, Operations should give consideration to steps to maintain removal of Red Oil from the circulating systems. As a first line of effective operations, an objective should be to eliminate the potential promoters of polymerization and foaming within the Caustic Tower System from entering and/or leading to the formation of a Red Oil liquid phase and possible solids deposition. However, as the formation of Red Oil, other hydrocarbon liquids and solid particles of polymers and salts can occur, it is as well a measure of the Operator's material balance control task to, as may be possible, reject accumulations of such from the systems. It is therefore important to be able to utilize liquid levels, flow meters and stream sampling and analyses to monitor and control the overall mass balance of the systems, concentration changes and any mass accumulations.

Control of Caustic Usage

Control of the caustic usage and the strength of the circulating caustic solutions is typically going to be such that the cracked gas from the overhead of the scrubber will contain less than 1 ppm by mol CO_2.

Frequent analyses of the caustic solutions and Spent Caustic draw are necessary to maintain proper operation of the Caustic Tower. Specifically these analyses will indicate to the Operators:

a. Whether the weak caustic contains sufficient free NaOH to effect a high degree of removal of the acidic components from the entering cracked gas and thereby maintain a sufficiently reduced concentration of acid gases in the scrubbed gas to the upper section of the column. It is desired that caustic usage be minimized, however, a balance must be struck between possible caustic usage efficiency and sodium bicarbonate and sodium hydrogen sulfide build ups in solution.

b. Whether sufficient fresh caustic make-up solution is being added to the tower upper section to ensure complete removal of acidic components from the cracked gas and to limit the bicarbonate concentration in the Spent Caustic. The fresh caustic make-up flow should be varied in response to the analyses results for free caustic and salts concentrations in the circulating caustic loops and Spent Caustic stream.

c. Whether the caustic concentrations are optimum for the fresh caustic make-up and Strong Caustic Section caustic solution feed. A caustic concentration range of 8 to 11 weight percent NaOH in the Strong Caustic Section caustic solution feed has been found to provide optimum absorption rates for CO_2 and H_2S. Acid gas removal in a Caustic Tower is a process controlled by the liquid phase. The absorption rate of the acid gases into the liquid phase is influenced by the caustic solution concentration, and thus of course by the change of that NaOH concentration in the solution as NaOH is reacted. Figure 5 demonstrates the effect of the NaOH concentration on the absorption of CO_2 into caustic solutions as measured by Tepe and Dodge and reported by them in *reference 11*. The reader can see from Figure 5 why it would be anticipated that the NaOH content of the solution feeding the top of the Strong Caustic Section would be targeted for 8 to 11 weight percent. Thus, the fresh caustic make-up NaOH concentration needs to be adequately maintained at a high enough value above the NaOH strength feeding the Strong Caustic Section.

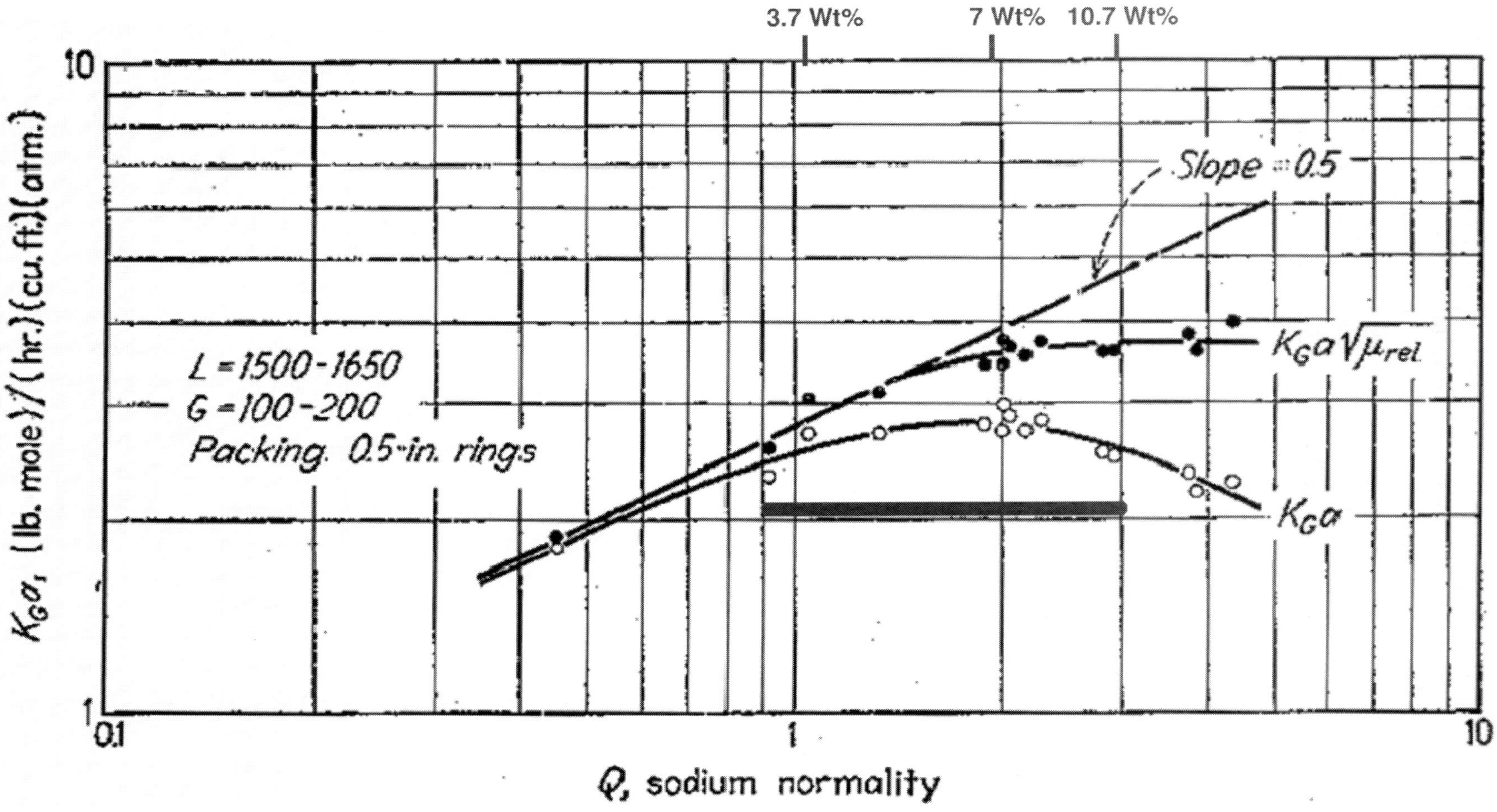

Figure 5 – Effect of Sodium Hydroxide Concentration on Rate of Carbon Dioxide Absorption
(Data of Tepe and Dodge 1943)

d. Whether water is being added to the circulating caustic at the proper rate. As the cracked gas is heated to operating temperature it will saturate with additional water, which will dehydrate the circulating caustic solution. Too low an addition of make-up water to the circulating caustic inventory will cause the concentration of salts to build up. If the concentration any salt becomes too high, it may precipitate within the tower and piping, contributing to system fouling. To be safe, as a rule the recommended limit for the concentration of sodium bicarbonate is about 3 weight percent. Too high a water dilution rate will result in a decrease in efficiency of removal of the acidic components because the average caustic solution strength on trays will be reduced.

Caustic and Water Quality

Caustic make-up quality can cause problems within the Caustic Scrubber when the fresh caustic introduces impurities into the system. There have been three basic grades of caustic which can be purchased:

- Diaphragm (Commercial)
- Membrane
- Rayon

with a fourth option called Mercury Free Cell also available. The typical specifications for these four commercial caustic products are given in Table 1.

Table 1
Typical Specifications of Purchased Caustic Grades

Product	Total Alkalinity as Na_2O wt%	Alkalinity as NaOH wt%	Na_2CO_3 wt% max	NaCl wt% max	$NaClO_3$ wt% max	Na_2SO_4 wt% max	Fe wt ppm max	Cu wt ppm max	Ni wt ppm max	Hg wt ppm max
Diaphragm	38.10-39.60	49.0-51.0	0.2	1.1	0.30	0.05	9	0.2	3	N/A
Membrane	38.40-39.50	49.5-51.0	0.05	0.01	0.0005	0.0025	3	0.3	0.3	N/A
Rayon	38.40-39.50	49.5-51.0	0.09	0.005	0.0003	0.0025	2	0.3	0.3	0.2
Hg Free Cell	38.40-39.50	49.5-51.0	0.1	0.012	0.001	0.005	5	N/A	0.6	0.05

The Diaphragm Grade is not recommended since it has a significant maximum allowable sodium chloride and sodium chlorate content. Both compounds introduce chlorides which are oxidizing reaction agents which therefore contribute to fouling and corrosion problems. Sodium chlorate, with its oxygen, is a very strong oxidizing agent and would be a significant contributor to fouling problems by way of free radical polymerization reactions.

Rayon Grade caustic solution is preferred since the Rayon Grade has the negligible sodium chloride and sodium chlorate impurities. However, it should be noted that Rayon

Grade contains a small amount of mercury, which will show up in the final waste effluent from the plant.

Note as well that if faced with a lack of Rayon Grade availability, the Membrane Grade is much preferred before any consideration of the Mercury Free Cell Grade.

All of the commercially available caustic grades are contaminated with some very small amounts of iron, copper, and nickel. Now the concentrations of these metals will be greatly diluted down as the purchased caustic is diluted down to the final fresh caustic make-up strength, but the metals could still potentially contribute somewhat to free radical polymerization reactions, acting as initiators for the chain reactions. In the case of iron, there can be an indicator of an intrusion of additional iron contamination of the Caustic Tower System beyond that which would be contained in purchased caustic per the product specifications in Table 1. It has been shown that higher concentrations of iron can result in caustic solutions with aldol condensation polymers turning shades of green in color.

Figure 6 shows an example of a Caustic Tower System which was found to have its circulating caustic solutions contaminated with iron. The pictures of caustic samples from the system, courtesy of Baker Petrolite, show the circulating Strong Caustic had a light yellow to greenish tinge color, but the Weak Caustic sample was green with the Spent Caustic after deoiling even a darker green. It was not known if the iron contamination was brought in via the caustic supply system and/or resulted from corrosion of the carbon steel material in the Caustic Tower System. A key point is that Red Oil may not always be red, but a green color.

The example highlights that when a caustic solution and/or Red Oil is green, it is an indicator that there may be an iron contamination issue. High iron contamination can in turn contribute to accelerated polymerization and fouling via free radical reaction initiation, but may also reflect as well a corrosion issue due to perhaps other cracked gas feed and/or caustic make-up contaminates.

Caustic can also absorb oxygen if it is improperly stored, leading to increases in caustic tower fouling. Therefore, it is very important to keep all caustic storage vessels adequately inert blanketed, and an Operator should insist that their caustic supplier assure oxygen free handling.

But, the caustic purchased must be diluted to the fresh caustic make-up strength for the Caustic Tower System; therefore the water used for dilution should be of very good quality. High quality boiler feedwater is preferred. The boiler feedwater should be very low in dissolved oxygen content and other impurities such as calcium and metals and particulate matter (both dissolved and suspended solids). Calcium introduced to a Caustic Tower System will result in the precipitation of calcium carbonate ($CaCO_3$) within the system adding to the fouling of the systems. Of course, as already discussed, the introduction of oxygen and metals are initiators for the free radical polymerization reactions. If particulate matter, e.g. corrosion products, is introduced to the Caustic Tower there is the potential for the particulates to cause foaming of the caustic solution flows in the scrubber.

Strong Section Weak Section De-Oiled Spent

$Fe(OH)_2$ has a green color.

A check for iron contamination is advised.

Figure 6 – Indicator of Iron Being Present In Circulating Caustic Loops
Images Courtesy of Baker Petrolite

In fact, besides caustic dilution and Wash Water Section make-up, the Operator should be mindful of the quality of any water that may be used during operations and maintenance of the piping and ancillary equipment associated in any way with the Caustic Tower Systems. When using water to clean and back wash any part of the caustic supply (outside battery limits or inside battery limits), Caustic Tower and ancillaries, and Spent Caustic handling systems, it is recommended that the quality of that water should be very good in order to avoid potentially any contamination reaching the Caustic Tower.

Operating Temperature

Increasing the operating temperature of a Caustic Tower improves the effectiveness of the acid gas removal by improving the rate of the acid gas reactions relative to the solubility of acid gas constituents in the caustic solution. The rate of diffusion of the acid gases into the caustic solution is improved with increasing temperature because the liquid viscosity is reduced, even if the solubility limit for the acid gas components is lowered.

Controlling the operating temperature is recommended to be done by varying the temperature of the Cracked Gas Feed flowing into the scrubber. Heating the circulating Weak Caustic solution is too prone to acceleration of any polymerization reactions occurring within the circulating liquid. Quench Water is the optimum energy efficient heat source for heating the CG Feed.

However, when attempting to manipulate operating temperature, the Operator must take two factors into consideration. A higher temperature in the Weak Caustic Section will promote a higher rate of polymerization within the tower. A lower temperature, on the other hand, could result in condensation within the tower of less volatile, heavier hydrocarbons contained in the vapor flow. Condensed hydrocarbon liquid could cause foaming of the caustic solution flows in the Caustic Tower. Also, condensation of a liquid hydrocarbon phase with the caustic solutions will contribute to an increase in the rate of polymerization that may be occurring in the circulating caustic sections.

The tower should be operated at the lowest possible temperature consistent with the acid gas removal required and no hydrocarbon condensation. It is suggested that the temperature control should initially be set to about 120°F, and adjusted as may be found to be reasonable for the operating balance between potential Caustic Tower foaming and fouling and acid gas reactions.

As a rule, when the Cracked Gas Feed contains lower amounts of dienes and styrene, such as with ethane and ethane/propane mix cracking, the operating temperature can be higher. For liquids crackers, with higher concentrations of dienes in the CG feed to the Caustic Scrubber, the operating temperature should be lower. However, to prevent condensing hydrocarbons out of the vapor phase, at the typical operating pressure range of 200 to 300 psig, the operating temperature is recommmended to be no lower than 105°F. Operating in a temperature window of 110°F to 130°F should be reasonable with due consideration for the diene content of the CG feed.

From the perspective of the effect that operating temperature has upon the typical Caustic Tower's acid gas removal performance, Figure 7 has been provided. The chart represents CO_2 removal from the gas phase of a Caustic Tower where a total of forty-five (45) caustic contacting trays have been installed. As can be seen by the example basis represented by the chart, the initial vapor phase CO_2 concentration is reduced to well below one (1.0) part per million when the tower operating temperature is 110°F or higher.

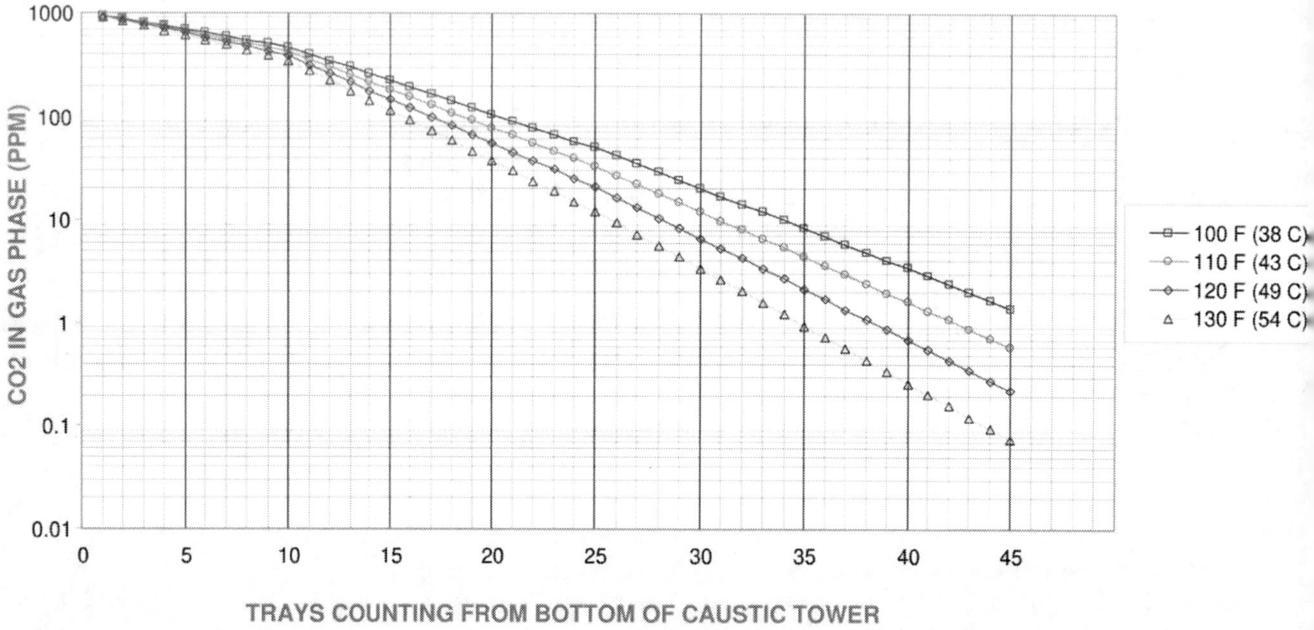

Figure 7 – Effect of Operating Temperature on Carbon Dioxide Removal from Cracked Gas
Courtesy of Shaw Energy & Chemicals Group

Caustic Solution Circulation Rates and Strengths

If the Caustic Tower has sufficient contact stages – trays or total packing height – available, as previously described under caustic usage control, it would be anticipated that the Caustic Tower operating optimum solution strength for the circulating caustic feed to the top of the Strong Caustic Section would be found somewhere in the concentration range of 8 to 11 weight percent NaOH. And, with reasonable flexibility for control error, it would also be expected good operations would have a remaining free caustic content in the Spent Caustic stream of only 1 to 2 weight percent. Running with the free caustic content of the Spent

Caustic at less than 1 weight percent (even at near zero) requires a much higher level of Operator attention for changes to and disturbances of plant operations.

For a Caustic Tower configured with three circulating caustic sections, the Intermediate Caustic Section (middle section) caustic solution strength will be a direct function of the Strong Caustic Section feed solution NaOH content and operating performance. Under good operating conditions, it would be expected for the Operator to find a free caustic strength somewhere in the concentration range of 6 to 10 weight percent NaOH at the top contact stage of the Intermediate Caustic Section.

As has been discussed in the process flow scheme section of this paper, the actual fresh caustic make-up flow required for the amount of acid gases to be reacted is too small to match the requirements of even minimum hydraulic liquid loading for trays or packing in the tower. Therefore, the flow rates for the caustic solution recirculation loops is very important to maintaining good vapor/liquid contacting over the contact stages available in the tower. Circulating the caustic solutions is a pumping power expense that an Operator will likely want to minimize. However, if a sufficient contact stage count is to be achieved, the caustic solution flow setting must always consider not only the minimum load point of the column internals, but the basic flow pattern efficiencies of the tray or packing type installed. For operations there are various considerations which must be made for the various vapor/liquid contacting device types which can be used in a Caustic Tower.

Sieve and Valve Trays In Caustic Tower Service

As cross flow trays, sieve and valve trays have efficiency in liquid contact time with the vapor due to the long flow path of the liquid across the tray. However, there must be an even distribution of liquid over the width of the liquid flow front as the liquid moves across the tray.

Both sieve and valve trays in Caustic Tower service suffer from solid and polymeric deposition on the tray panels and downcomers. As sticky, fouling deposits are made on the tray, the holes of sieve trays and the valves of fix valve trays become blocked. For valve trays with moving valves, the valves can become stuck either closed or open. Deposition of polymeric solids in downcomers and at the outlet of downcomers can disturb the distribution of liquid flowing from the downcomer out onto the tray. All in all over the operation of the tower, with polymeric deposition on sieve and valve trays, the resulting vapor and liquid maldistribution will grow with time. The efficiency of acid gas removal will decrease and caustic consumption will increase with time.

Packing In Caustic Tower Service

Packing offers both the desirable feature of low pressure drop and large interfacial contact area; but the performance of packing in Caustic Tower service is not consistent. The successful wetting of the packing seems to be a problem and depends primarily on the initial liquid distribution and its redistribution along the height of the packing bed. Packing requires high liquid rates to assume proper and effective wetting of the packing surfaces. These high liquid rate, much higher than those used in trayed towers require larger circulation pumps and corresponding higher capital and operating costs.

Trayed caustic towers have a higher tolerance for fouling. Therefore, control of caustic circulation flows is very important to packing in a Caustic Tower. For should there be any spots through the packing volume that have insufficient wetting of the packing surface, deposition of polymers material will be most likely to occur. Within the packing with the growth rate of the polymer deposits, the vapor and liquid maldistribution will, as with the trays, escalate over time resulting in decreasing efficiency and higher caustic usage.

Also, as has been previously mentioned, packing appears to be more susceptible to the possible initiation of and increase of caustic solution foaming tendencies, which may require the use of antifoaming agent injection to control. This increase in the risk of foaming with the use of packing would seem to be a function of the large surface area provided by packing relative to possible dispersed polymer over the surface area. It is speculated that the wide dispersion of polymer particles over such a wide area reduces the surface tension of the flowing caustic solutions, enhancing the risk of foaming.

With packing, there can be more frequent tower clean up required, and of course packing fouled with the polymers does not clean up well. Packing removal and replacement is an expensive maintenance activity.

Ripple Tray™ Technology In Caustic Tower Service

Another contact device which is well proven in Caustic Tower service is the Ripple Tray™. Because of the nature of the Ripple Tray™ design, the tray performs very favorably in Caustic Towers, providing high capacity and self cleaning characteristics. In Caustic Tower service, the self cleaning characteristic of the Ripple Tray™ has been shown to keep the flow path open and maintain the distribution of vapor and liquid throughout the run time between major turnarounds. Thus, the Ripple Tray™ in Caustic Towers provides and maintains excellent contact between vapor and liquid flows.

A Ripple Tray™ has no downcomer, and therefore is a true countercurrent vapor/liquid contacting device. As such, the Ripple Tray™ has a resultant agitation or froth (pulsation), which provides an unusually high heat and mass transfer rate per unit of cross-sectional area. The froth height held on the tray provides for the bubble surface for the efficient exchange of the acid gas mass in the vapor phase with the caustic solution.

The Ripple Tray™ utilizes perforated tray decks, which are corrugated to promote the vapor and liquid contacting. Corrugation geometry and the perforation pattern of the tray are designed specific to the hydraulic capacity requirement at the different locations in the tower. As a tray without downcomers, the entire cross-sectional area of the tower is being used for vapor and liquid contacting and thus tends to counteract the deposition of polymer, which can be formed in the liquid phase. Besides the fact that with no downcomers the entire cross section is available, the corrugation provides more surface area per unit of tower cross sectional area. It is the enhanced available area for vapor/liquid contacting that provides the Ripple Tray™ with its high capacity characteristic. Yet, the turndown of the Ripple Tray™ will be compatible with that of the Cracked Gas Compressor.

With no downcomer, the liquid entering the tray washes the Ripple Tray™ as the liquid flows through the pulsation action to holes of the tray, where the liquid leaves the tray taking polymer with it. Thus, the pulsation action helps the hole area clear. A crossflow tray can only move deposited polymer potentially over into the downcomer, if the polymeric solids get over the outlet weir. Also, as already explained above, the continuous movement of polymeric solids through the flow passages of random and structured packing cannot always be assured. Should liquid flow distribution become skewed such that areas of the packing begin to operate with little liquid flow, the packing surface will be denied the liquid washing action needed to counter polymer formation on the packing.

For a visualization of vapor and liquid contact from Ripple Tray™ to Ripple Tray™, Figure 8 illustrates liquid flow from the bottom of one tray into the froth of the tray below in countercurrent contact with the rising vapor. On the tray, the froth represents the occurrence of a high degree of liquid and vapor mixing.

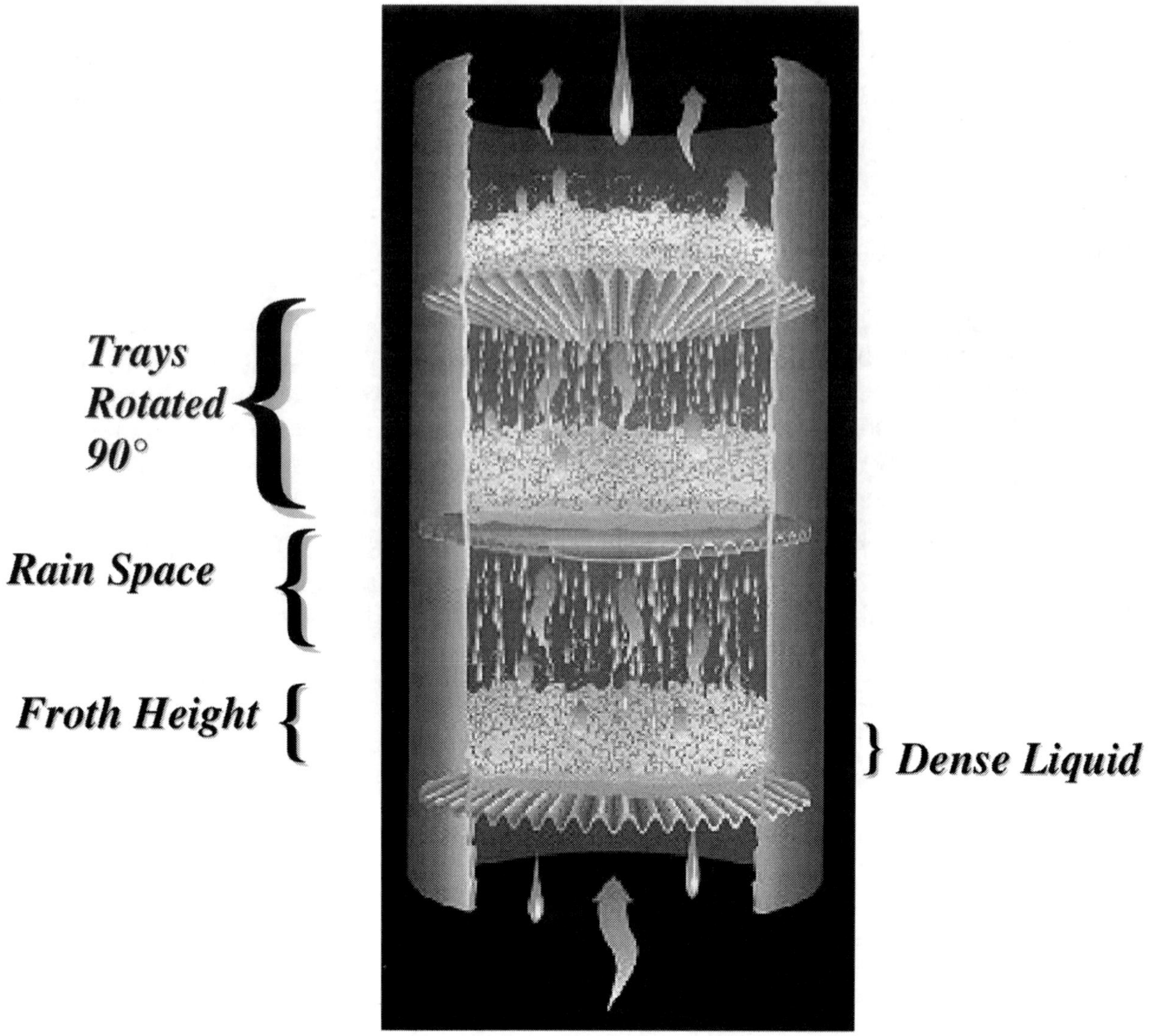

Figure 8 – Ripple Tray™ Vapor and Liquid Flow Schematic
Courtesy of Shaw Energy & Chemicals Group, Houston, Texas

Material Balance Control

For material balance control, after the critical objective of CG leaving the Caustic Tower on specification for acid gases, the next crucial task is minimization of caustic usage and the flow of Spent Caustic within the practical operating constraints posed. Frequent monitoring of the free caustic concentration and salts content of the Spent Caustic draw stream against fresh caustic make-up strength relative to acid gas removal should be done to determine caustic utilization – the actual percentage of NaOH content of the fresh caustic make-up to the

Strong Caustic that is consumed by reactions in the tower. With increasing utilization there will be decreased caustic used (purchased). Within the actual mechanical and process performance limitations of the contact stages installed in the Caustic Tower, this frequent determination of caustic utilization, when compared to the acid gas content allowable which is slipping out with the CG overhead from the tower, will result in minimizing the operating expense of caustic purchased.

With this assessment there should be verification that precipitation of salts, e.g. sodium bicarbonate and sodium bisulfide, are not a problem. The Operator will then know how tight the excess free NaOH in the Spent Caustic can be controlled relative to the constraint of control error philosophy for operating spikes in acid gas load being practiced. Less free NaOH in Spent Caustic again means less caustic purchased, but needs to be weighed against operating safety margin needed/desired.

The material balance on actual sodium hydroxide fed versus that allowed to slip out in the Spent Caustic is one part of minimizing the total Spent Caustic flow and the handling and treating costs of that Spent Caustic. The other part of the control of total Spent Caustic flow is the water balance of the system. The Operator should look to optimize the water balance, using no more than is reasonable to maintain dilution of fresh caustic solution to optimum strength for the system installed, maintain circulating caustic loop concentrations of NaOH and salts with reasonable operating window, and maintain circulating Wash Water quality. Optimizing water use will not only mean less total Spent Caustic stream flow and treating cots, but the operating expense of the water feed itself will be minimized. As it is boiler feedwater that should be used for make-up water to the system, the cost of the water make-up will be an important consideration.

Along with minimization of caustic usage and the flow of Spent Caustic to treatment, the Operator has the important mass balancing task of eliminating the accumulation of Red Oil and polymers and possibility for salt deposits within the tower. As has already been mentioned, the assurance of solubility of the carbonates and sulfides in the caustic solutions is being handled with the caustic and water balancing tasks. The other aspect of accumulation control is to work to minimize as best possible polymerization reactions, but see as well that those Red Oil and polymer solids that may be created are being taken from the system.

As part of the effort to prevent foaming and maintain a minimum rate of fouling, Operations should give consideration to steps to maintain removal of Red Oil from the circulating systems. The tower should have arrangements to allow any possible condensed hydrocarbon liquid phase and/or Red Oil accumulation to be skimmed periodically, as required, from the circulating caustic solution holdup inventories on the chimney trays. At the bottom of the Caustic Tower along with being able to control the Spent Caustic draw off, any hydrocarbon liquid phase with Red Oil formed in the tower should also be drawn off.

Arrangements should be provided, as well as, for the capability to solvent wash on-line, at a reasonable frequency, to help control the build up of polymeric concentrations within the tower. Typically an aromatic gasoline should be injected into the Strong Caustic circulation loop periodically to provide a wash through the caustic compartments of the tower to potentially remove polymer from tower internals.

As part of mass balance control, filters should be provided on the Weak Caustic circulation loop to remove polymeric solids that can build up in the solution inventory. Primarily, the potential promoters of polymerization and foaming within the Caustic Tower System should be eliminated as might be possible. For in those systems where expensive polymer inhibitors and/or dispersant (and perhaps defoaming) chemicals are being used to address system fouling, part of the mass balance control task should be to reduce (if not eliminate) the amounts of these chemicals being injected.

Monitoring of Caustic Tower System

Consistent and regular monitoring of the Caustic Tower System is a must in order to detect changes in the operation of the Caustic Tower and other associated ancillary equipment. The base operating trends and analytical work needed will be crucial to troubleshooting any problem that may arise.

The basic operating conditions of the system must be assessed. These should include:

- Cracked Gas feed rate.
- Temperature of cracked gas feed into Scrubber.
- If cracked gas heater is used, the cracked gas temperature into and out of the heater should be tracked, in order to monitor the performance of the preheat exchanger relative to the temperature of the heat source.
- Temperature of Caustic Tower solutions and temperature profile in the tower. The approach of vapor flowing temperature to caustic solution temperature in Weak Caustic Section is important.
- If Weak Caustic Circulation heating is the flow scheme method for heating and controlling the Caustic Tower temperature, the caustic solution outlet temperature from the Weak Caustic Circulation Heater is very important. The caustic temperature must be high enough to assure good control of the tower temperature profile, but if allowed to be too high the rate of polymerization in the Weak Caustic Section and external circulation system will be high.
- Caustic Tower differential pressure. Tower pressure drop can be a good indicator of possible foaming, which can be an indicator of a hydrocarbon liquid phase accumulation, including Red Oil and other polymeric material, which needs to be skimmed from the system. And/or, differential pressure trends can be an indicator of the fouling condition of the tower internals and the rate of polymerization occurring.
- Caustic Tower overhead temperature. The Wash Water Section not only removes entrained caustic from the cracked gas flow, the Wash Water cools the gas as well. Thus, a higher CG overhead temperature can be an indicator of Wash Water Section problems which may result in caustic carryover from the overhead of the Caustic Tower.

- Wash Water and caustic solution circulation rates, key to adequate vapor/liquid contacting and washing to achieve absorption efficiency.
- Wash Water make-up flow rate, key to adequate inventory control in the Wash Water Section.
- Flow of make-up water to the Strong Caustic Section, in order to assure adequate control of caustic and salts concentrations in the tower liquid solutions.
- Flow of fresh caustic make-up, which is important for sufficient free caustic.
- Liquid levels and inventory control in the various sections of the scrubber, which will be important for solution concentrations and mass balances of salts, polymers, and free caustic in the system. As well as, highlight for the Operator as possible symptom of accumulations of Red Oil, polymers and other condensed hydrocarbons leading to increased fouling and the onset of foaming.
- Pressure differential of filter for removal of polymeric solids from the circulating Weak Caustic loop.

There is stream sampling and analytical testing which must be implemented at a frequency sufficient to alert the Operator to a shift in system condition. The following are the suggested stream sampling and analytical work to be carried out.

- Analyze the Cracked Gas Feed for specific H_2S and CO_2 and carbonyl content.
- Analyze the CG leaving overhead from the Caustic Tower for CO_2 content to assure ethylene product specification and for comparison to CG Feed acid gas content.
- Sample fresh make-up caustic and check for caustic strength and turbidity and contaminants such as dissolved oxygen, metals and chlorides.
- Sample each circulating caustic solution and Spent Caustic stream, making visual observation of samples, measuring for caustic strength, turbidity, free oil phase, and salts content.
- Simple shake testing of bottle samples of circulating caustic solution samples should be done periodically and when foaming is suspected to be occurring to look for foaming tendency and the time it takes for any foam to break.
- Polymer and hydrocarbon and solids content of circulating caustic and Spent Caustic solutions should be measured to help with mass balance tasks and track rates of polymerization.
- For oils that may be skimmed from the system, occasional sampling and measurement of polymer content by precipitation has been suggested.
- Sample circulating Wash Water and measure pH, turbidity and conductivity to assess caustic entrainment and possible carryover in cracked gas overhead.

- Wash Water make-up should be analyzed for oxygen content. If Wash Water make-up is not boiler feedwater quality, then it should be tested for other contaminants and particulate matter.

For performance monitoring, there are a few other key aspects that must not be forgotten. The basic cracking conditions and the operating mode status of all cracking furnaces need to be considered. Operator alerts need to be made for changes which will be made in feedstocks, cracking conditions, and cracking furnace status, so that possible Caustic Tower condition changes can be made also if needed.

Flow rates and compositions of external streams purchased and fed directly to CGC for recovery, such as Refinery Off-Gases, need to be monitored for potential acid gases, oxygen, carbonyls or precursors for such and other contaminants content. Vent streams returning to the CGC from derivate units need to be monitored for possible carbonyl compounds, oxygenates, contributions to the acid gases load and other contaminants. Vent stream that might contain compounds such as vinyl acetate need monitoring because in caustic solutions vinyl acetate is converted to acetaldehyde, which then will contribute to Red Oil formation.

There will also be symptoms of system fouling conditions that should be noted. Just as a higher rate of tower pressure drop increase, liquid level increases, Red Oil and polymers content in samples of circulating caustic solutions are signals, any difficulties with Caustic Circulation Pump services such as cavitation or limitations to flow capacity are possible alerts to fouling issues that indicate a need for gasoline washing and/or skimming of hydrocarbon phase. Of course, symptoms also mean it is necessary to look for and eliminate/minimize those things that may be entering the system which directly contribute to the rate of polymerization. Or, make operating adjustments, such as with operating temperature, to stop hydrocarbon condensation or lower the rate of polymer formation.

Conclusions

No matter what the configuration and installed contact stages of the Caustic Tower System, the basic operational objectives and problems faced by the Operator of the Acid Gas Removal Area of an ethylene plant are the same. All Caustic Tower operations should find the basic guides described in this paper to be useful.

For every Caustic Tower that has sufficient installed contact stages for the acid gas load in the cracked gas feed flow being processed, the minimum usage of caustic should be found with a tower circulating Strong Caustic feed NaOH concentration generally in the range of 8 to 11 weight percent, and about 1 to 2 weight percent of free NaOH left in the Spent Caustic stream leaving the tower. If there are fewer contact stages available than what would allow the free caustic in the Spent Caustic to be as low as 1 weight percent, the utilization of the NaOH provided in the fresh caustic make-up must still be maximized. While meeting the required acid gas removal and providing for the control error flexibility needed, the lowest possible amount of NaOH must be thrown away in the Spent Caustic.

The Operator will want to minimize the total Spent Caustic flow to treatment. Besides throwing away the lowest amount of caustic, the water balance of the system needs to be watched as well. Wash Water Section water make-up flow needs to be watched. Beyond what is needed for adequate performance of the section's contact stages to wash out any caustic entrainment and maintain the Wash Water quality leaving the section, excess water make-up will just add to the Spent Caustic flow to be handled.

As part of watching the water balance, the Operator needs to see that circulating caustic solution strengths are maintained as needed for good acid gas removal, but at the same time with enough dilution water to keep salts in solution.

Controlling the operating temperature of the Caustic Tower within an operating range of 110°F to 130°F should provide good acid gas removal, while minimizing polymerization reaction potential.

Aim to minimize Red Oil and polymer formation in the tower by monitoring and eliminating or minimizing as many as possible of the contaminates that are initiator of aldol condensation and free radical polymerization reactions from entering the systems. The operation should use Rayon grade caustic, and boiler feedwater quality water should be used for make-up to Wash Water and any caustic solution dilution. Oxygen should be eliminated from caustic storage and in caustic handling systems.

The Operator must be mindful of the performance characteristics of the Caustic Tower internals. Trays or packing should be operated at the proper liquid loads needed to achieve vapor/liquid contact. In the case of fouling mitigation, for those towers with packing, an Operator may want to consider switching from packing to trays.

Again, to combat fouling and foaming in the Caustic Tower, the Operator needs to skim as required any condensed hydrocarbon phase and any Red Oil formation. The circulating Weak Caustic solution should have filters to capture polymeric solids which may be formed in the tower. An aromatic gasoline wash should be injected as needed into the caustic circulations of the tower to dissolve polymer deposits in the system.

Chemicals for carbonyl scavenging and polymer dispersion and free radical polymerization inhibiting do work after a fashion. But the objective should be to minimize, if not eliminate, their use if possible because of the high expense. In the case of polymer dispersants there has been found generally that injection of defoaming chemical will be needed as well.

An operations monitoring program with an adequate frequency of sampling and analysis must be put in place for the Cracked Gas feed and overhead product and the fresh Caustic Make-Up stream, Spent Caustic stream and all circulating caustic streams.

References

1. Dankwerts, P. V., Sharma, M. M., "The Absorption of Carbon Dioxide Into Solutions of Alkalis and Amines", The Chemical Engineer, October, 1966.

2. Kohl, A. L., Nielsen, R. B., Gas Purification, 5th Edition, Gulf Publishing Co., pp. 378 – 409, 1997.

3. Perry, R., and others eds., Perry's Chemical Engineers' Handbook, 7th Edition, The McGraw-Hill Companies, Inc., Section 14, "Gas Absorption and Gas-Liquid System Design", 1997.

4. Reyniers, Marie-Francoise S. G., Forment, G. F., (1994), "The Influence of Metal Surface and Sulfur Addition on Coke Deposition in the Thermal Cracking of Hydrocarbons", AIChE 1994 Spring National Meeting, Atlanta, Georgia, April, Paper 26d.

5. Dionne, D., Kurukchi, S., Gondolfe, J. M., (1999), "Caustic Tower Retrofit Using High Capacity Second Generation Ripple Tray™ Technology", AIChE 1999 Spring National Meeting, Houston, Texas, March, Paper 64f.

6. Blaschke, M. W., (2003) "Caustic Tower Fouling: Identifying the Causes", AIChE 2003 Spring National Meeting, New Orleans, Louisiana, March, Paper 56c.

7. Stancato, B., De Haan, S., (2001), "Coke Inhibition", EEPC HSE Conference, The Hague, Netherlands, October, Paper No. 10.

8. De Haan, S., (2006), "An Overview of Self-Inflicted Contaminants", AIChE 2006 Spring National Meeting, Orlando, Florida, April, Paper 31a.

9. Prickett, R., Harper, S., Kurukchi, S., (1995), "Caustic Tower Revamp Using Ripple Trays™", AIChE 1995, Spring National Meeting, Houston, Texas, March, Paper 14f.

10. Mullenix, D., Jordan, M., Bundick, H., Martin, F., Lewis, V., (1996) "Control of Carbonyl Polymer Fouling in Caustic Towers", AIChE 1996 Spring National Meeting, New Orleans, Louisiana, March, Paper 19a.

11. Tepe, J. B., Dodge, B. F., (1943), Transactions of the America Institute of Chemical Engineers, Vol. 39, p. 255.

12. Hallee, L. P.; Zdonik, S. B.; Green, E. J; Manufacturing Ethylene, The Petroleum Publishing Company, The Oil & Gas Journal, Copyrighted 1966, 1967, 1968, 1969, and 1970, pp. 80-84.

Ethylene Plant Operations
Acetylene Hydrogenation Tutorial

Lecture I : Introduction to Subject

2009 Spring National Meeting
AIChE Session 107 - April 29th
Ed Mohundro

Ethylene Plant Operations
Acetylene Hydrogenation Tutorial

AIChE Session 107 - April 29, 2009

This lecture has been prepared for presentation at the AIChE 2009 Spring National Meeting in Tampa, Florida.

Topics to be Covered

- Introduction ---(Ed Mohundro)
  - General Overview of Subject Matter
  - Basic Review of Process Flow Schemes Operator Can Encounter

Topics to be Covered (Cont'd)

- Chemistry & Catalysts ---(Dick Voight)
  - Chemistry - Backend and Front-end
  - Basic Review of Catalysts

Topics to be Covered (Cont'd)

- Front-End Acetylene Hydrogenation – (Rick Callejas)
  - Discussion of Design & Operating Parameters
    - Operating Conditions and Controls Available
    - Basic Practices
    - Condition and Life of Catalyst Considerations
    - Targeting and Monitoring System Performance

Topics to be Covered (Cont'd)

- Start-Up and Shutdown Considerations
- Basic Troubleshooting & Upset Recovery

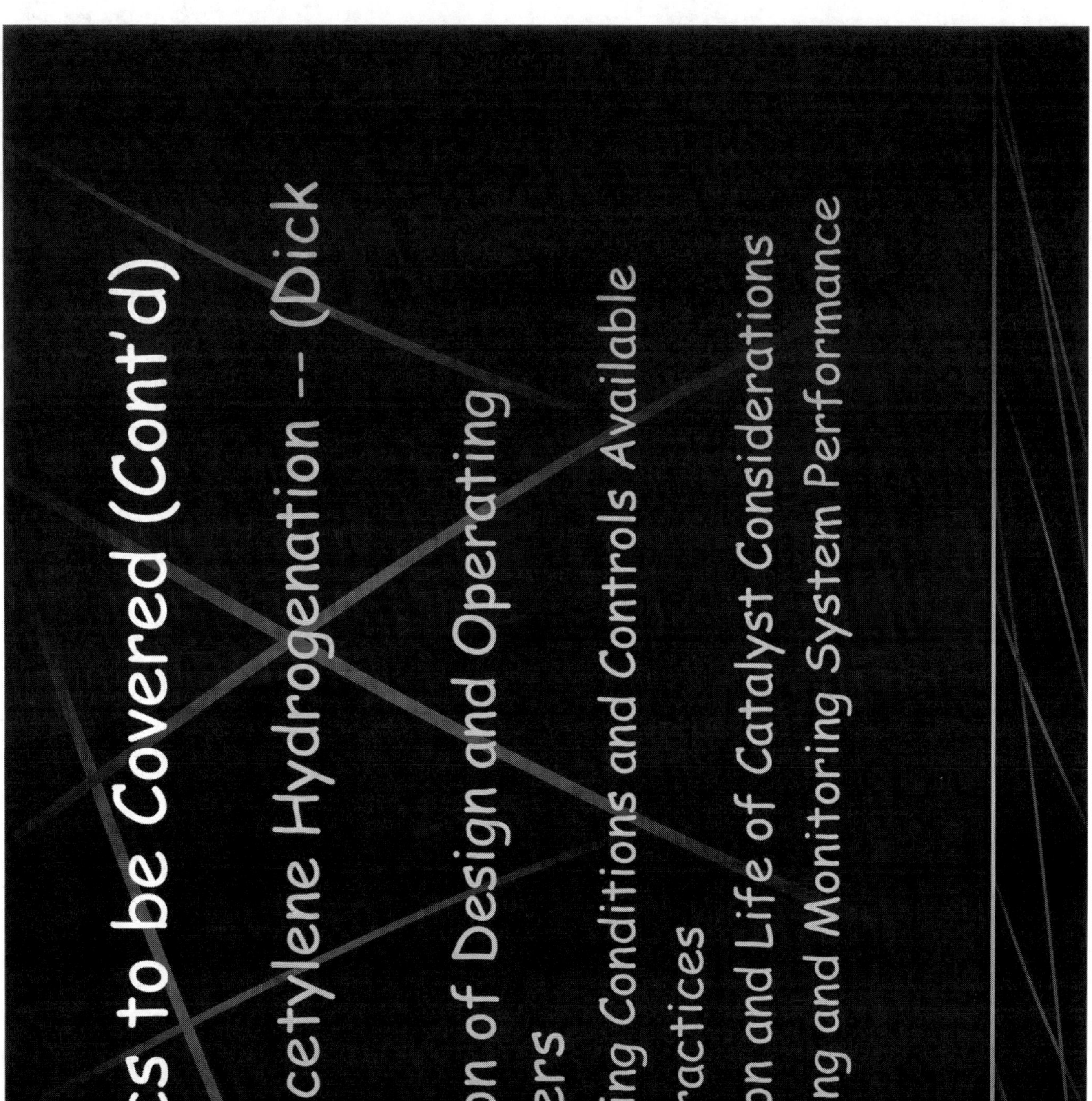
Topics to be Covered (Cont'd)
Backend Acetylene Hydrogenation -- (Dick Voight)
Discussion of Design and Operating Parameters
Operating Conditions and Controls Available
Basic Practices
Condition and Life of Catalyst Considerations
Targeting and Monitoring System Performance

Topics to be Covered (Cont'd)

- Start-Up and Shutdown Considerations
- Basic Troubleshooting & Upset Recovery

Topics to be Covered (Cont'd)

- Catalyst Regeneration and Safety Considerations -- (Ed Mohundro)

Acetylene Hydrogenation Overview Comments

- Trace levels of acetylene, methyl acetylene (MA), and propadiene (PD) are formed in steam cracking furnaces
- Tolerable only at < 1 ppm in polymer grade ethylene and propylene
- Acetylene fractionates with ethylene; MAPD fractionates with propylene
- Dominant method for removal of acetylene and MAPD is selective hydrogenation

Overview Comments

- Our focus will be on selective hydrogenation of acetylene
- Some Comments on selective C3 Hydrogenation are:
  - Predominately carried out via liquid phase process.
  - Details about the design / operation of liquid phase C3 hydrogenation processes are proprietary.
  - However, some MAPD conversion can take place in Front-end acetylene converters.

Introduction

- Basic Requirements for Acetylene Hydrogenation
- Options for Locating Hydrogenation Units Within Ethylene Plant Sequence
  - Pros/Cons for Each Location

Basic Requirements for Acetylene Hydrogenation

- Safe Operation; e.g., No Exothermic Runaway Reactions
- Maintain acetylene in ethylene product below specification or manufacturing target -- this might be below 1 ppm; or perhaps, below 0.5 ppm
- Don't cause interruptions to continuous operation of the ethylene plant

Secondary Requirements - or, Strong Wishes for Acetylene Hydrogenation

- Operating with Ethylene Gain; the Higher, the Better
- "Low Investment Cost"
- Good Catalyst Life and Reasonable Catalyst Cost

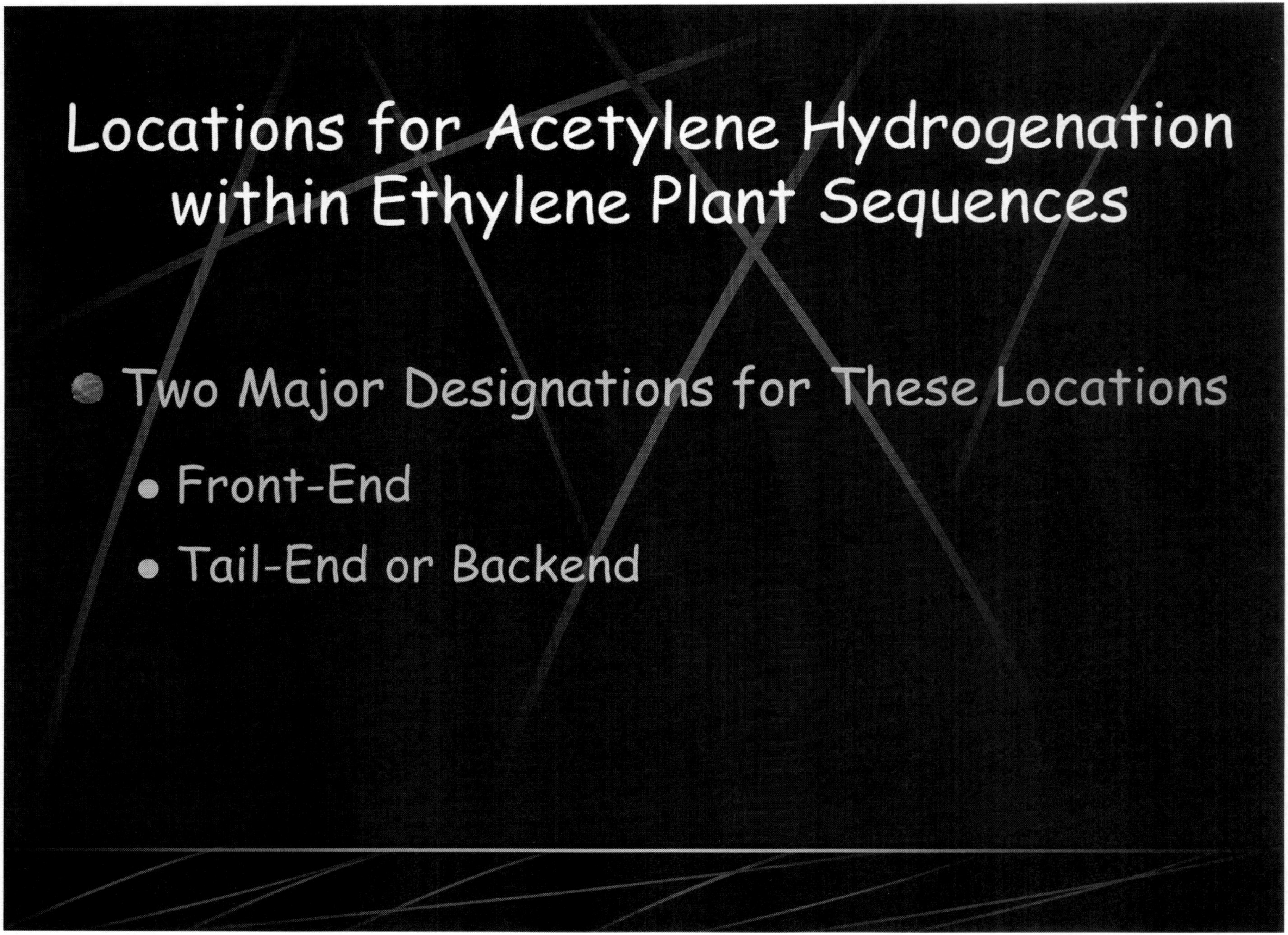
Locations for Acetylene Hydrogenation within Ethylene Plant Sequences
Two Major Designations for These Locations
Front-End
Tail-End or Backend

Front-End Locations

- DeC2 Overhead in DeC2 - 1st Sequence
- DeC3 Overhead in DeC3 - 1st Sequence
- Raw Gas: Within Process Gas Compressor Circuit, Downstream of the Caustic Tower

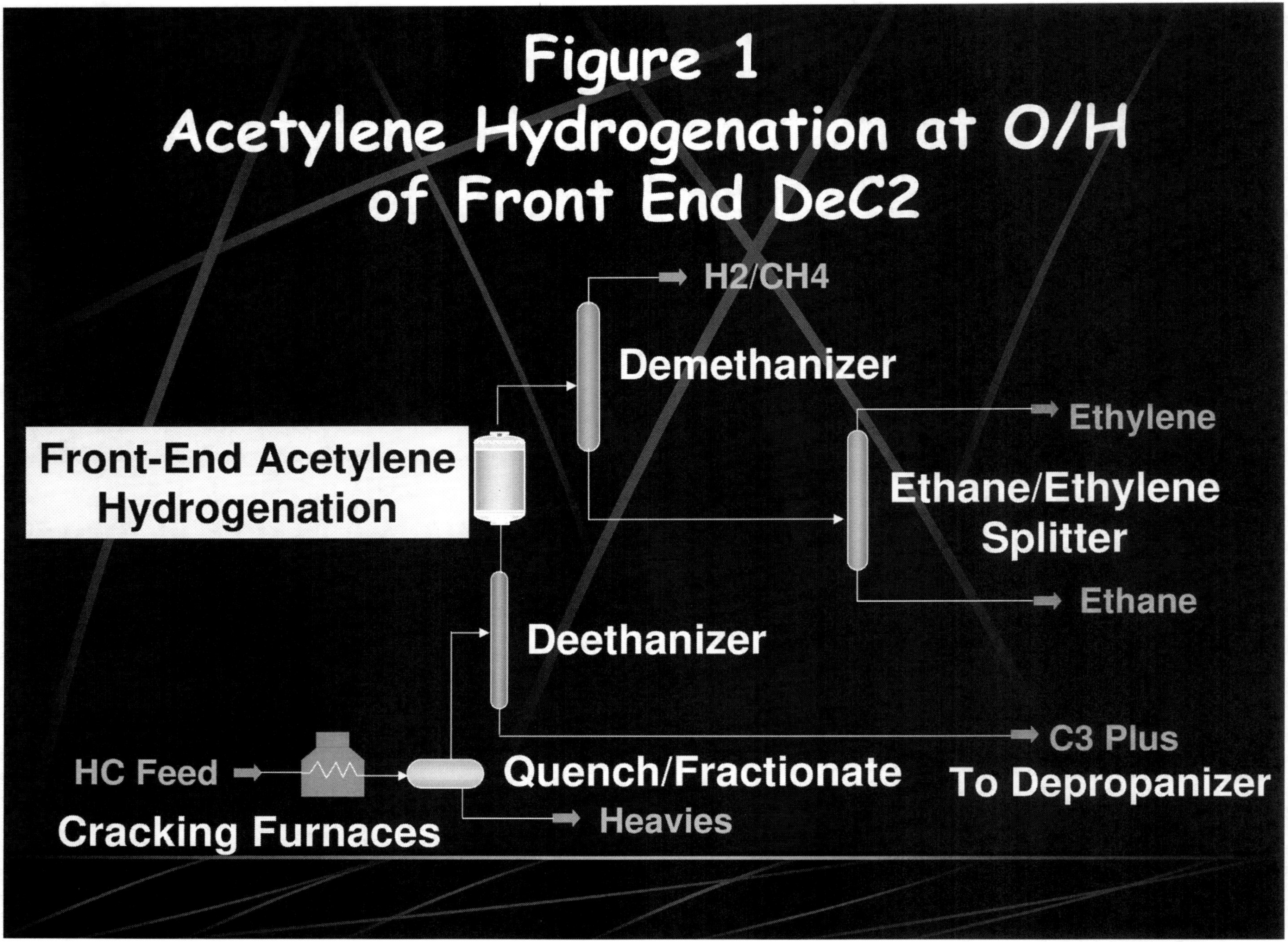
Figure 1
Acetylene Hydrogenation at O/H
of Front End DeC2
H2/CH4
Demethanizer
Ethylene
Front-End Acetylene
Hydrogenation
Ethane/Ethylene
Splitter
Ethane
Deethanizer
C3 Plus
HC Feed
Quench/Fractionate
To Depropanizer
Heavies
Cracking Furnaces

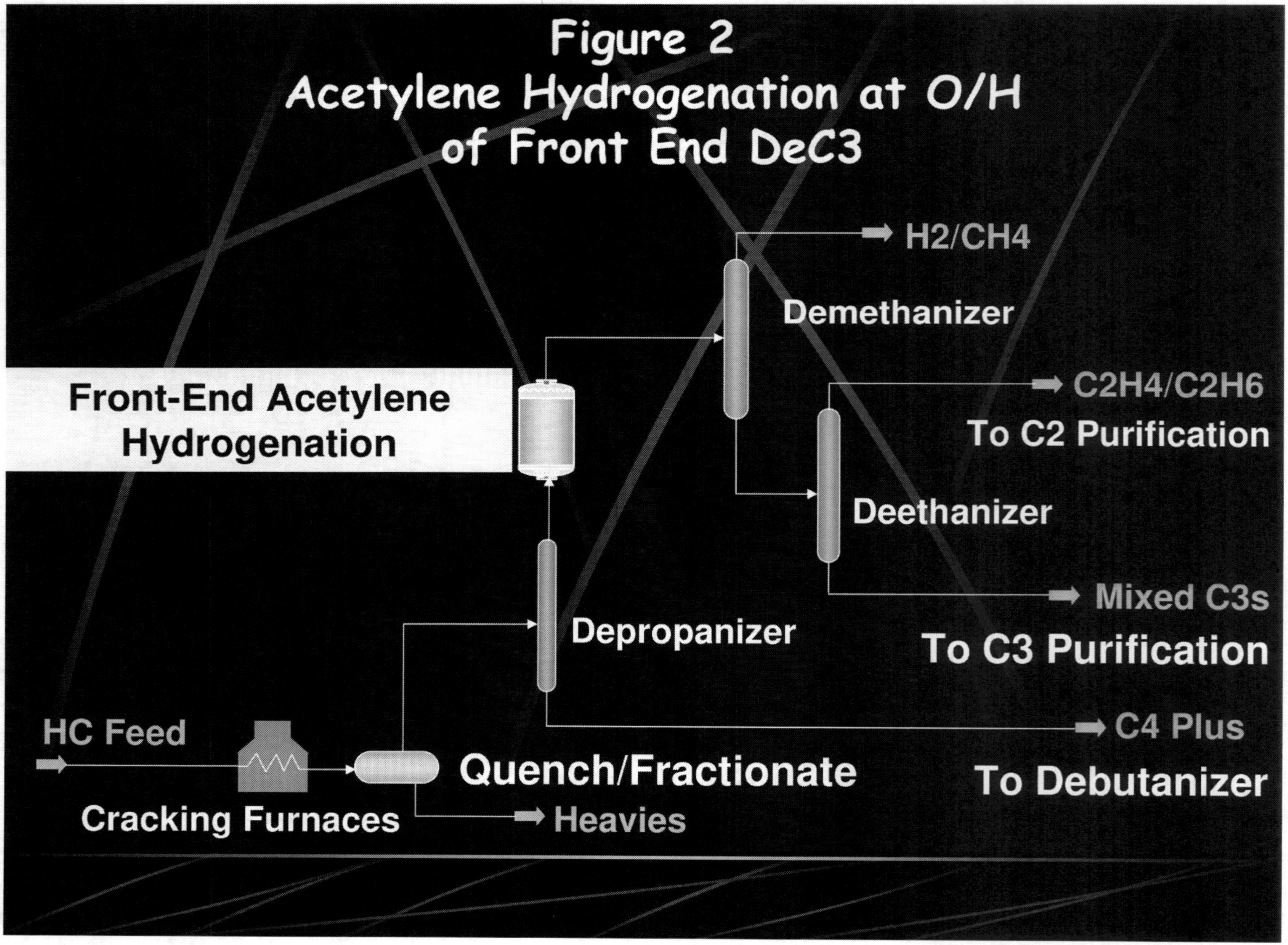
Figure 2
Acetylene Hydrogenation at O/H
of Front End DeC3
H2/CH4
Demethanizer
Front-End Acetylene
Hydrogenation
C2H4/C2H6
To C2 Purification
Deethanizer
Mixed C3s
To C3 Purification
Depropanizer
HC Feed
C4 Plus
Quench/Fractionate
To Debutanizer
Cracking Furnaces
Heavies

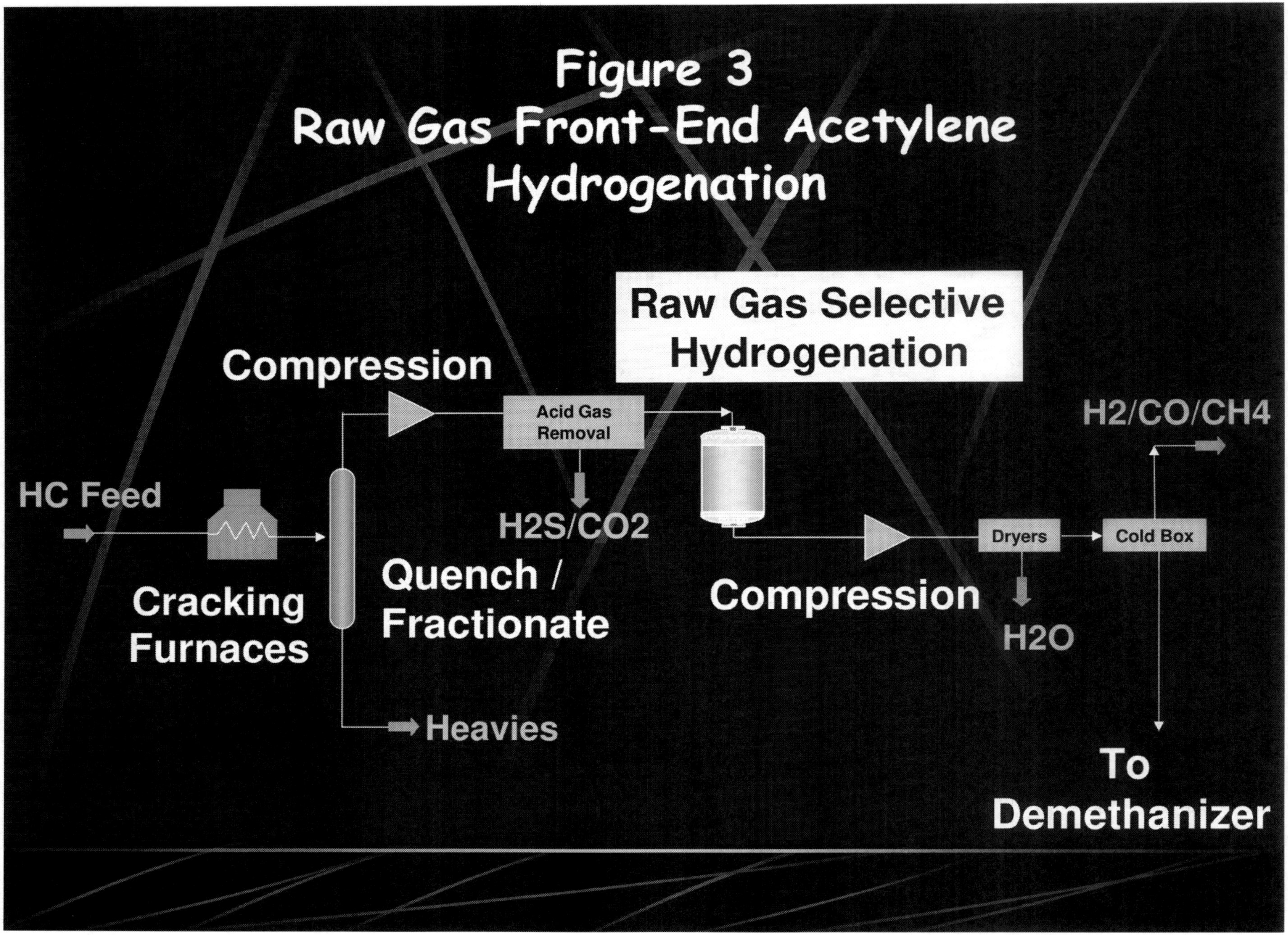
Figure 3
Raw Gas Front-End Acetylene Hydrogenation
Raw Gas Selective Hydrogenation
Compression
Acid Gas Removal
H2/CO/CH4
HC Feed
H2S/CO2
Dryers
Cold Box
Quench / Fractionate
Cracking Furnaces
Compression
H2O
Heavies
To Demethanizer

"Older-Generation" Front-End Hydrogenation Based On Sulfided Nickel Catalyst

- Dominant catalyst is Pd-based, with the "Latest Generation" catalyst being bimetallic or promoted palladium. Typically, silver is the promoter.
- Another type of catalyst is sulfided Ni, *e.g.*, Sud Chemie's C-36
- Introduced about 50 years ago; operates with inlet temperatures that range from 350 F (~175 C) to 450 F (~230 C). Some of these units are still in service
- Must be presulfided and maintained in a sulfided state via continuous injection of sulfur compounds.

Tail-End or Backend Location

- Associated with Demethanizer - First (DeC1 - 1st) Ethylene Recovery Sequence
- Located Upstream of Ethylene/Ethane Splitter

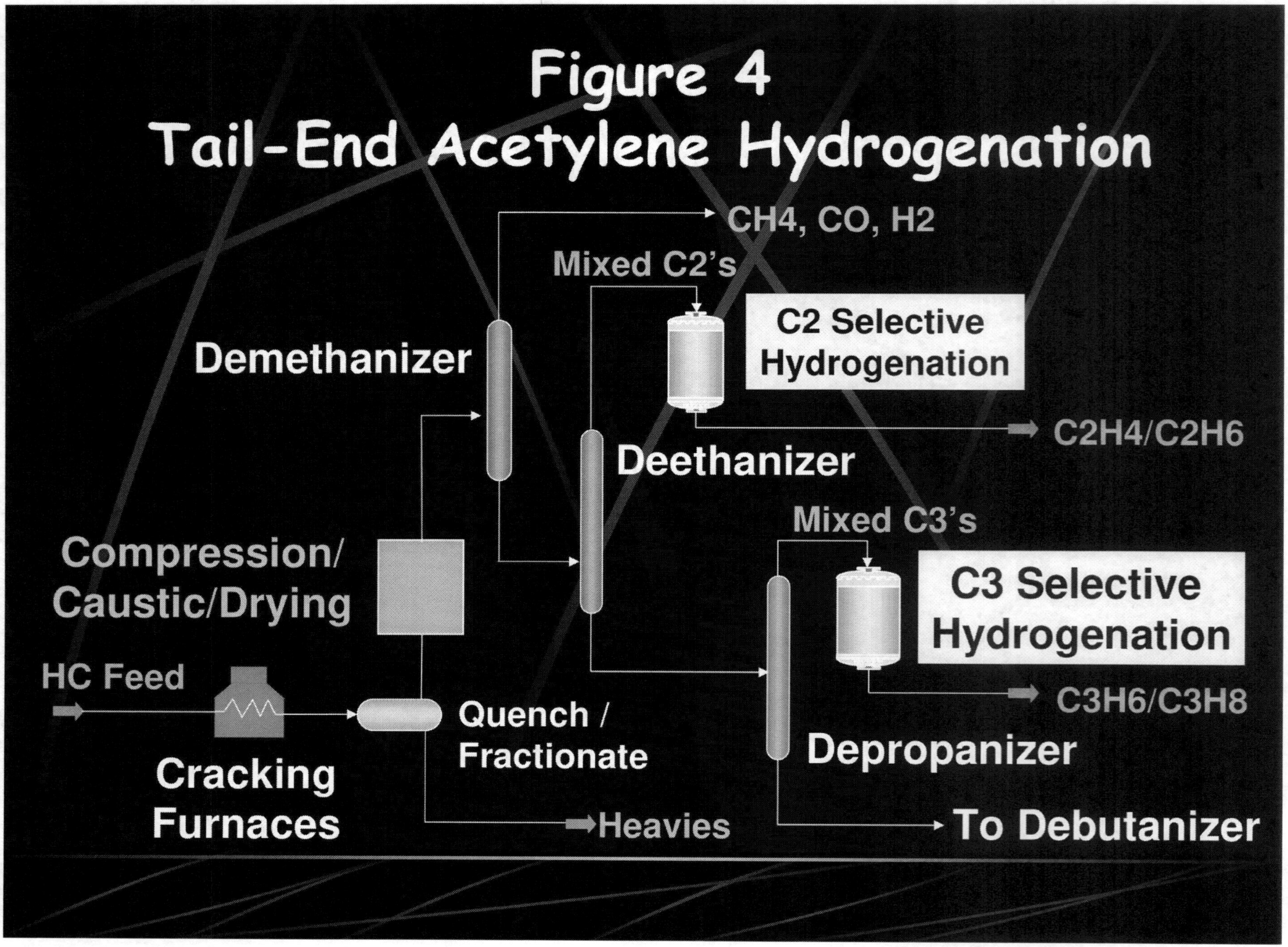
Figure 4
Tail-End Acetylene Hydrogenation
CH4, CO, H2
Mixed C2's
Demethanizer
C2 Selective Hydrogenation
C2H4/C2H6
Deethanizer
Mixed C3's
Compression/ Caustic/Drying
C3 Selective Hydrogenation
HC Feed
C3H6/C3H8
Quench / Fractionate
Depropanizer
Cracking Furnaces
Heavies
To Debutanizer

Pros and Cons for Each Location

Front-End Hydrogenation: DeC2 O/H

- CG Feed Contains C2 and Lighter Material
- Therefore, CG Feed Is:
  - Very Rich In Hydrogen
  - Very Rich In Ethylene
  - Particularly If Furnace Feed Is Ethane

Front-End Hydrogenation: DeC2 O/H

- **Pros:**
  - Clean Stream; i.e., No Traces of BD to Cause Catalyst Fouling
  - Do Not Need Spare Reactor
  - High Concentration of H2 Suppresses Formation of Green Oil
  - Reactors are Smaller, Relative to the Front-End DeC3 O/H System

Front-End Hydrogenation: DeC2 O/H

- **Cons:**
  - Operation Sensitive with High H_2 Concentration:
    - Initiation of Ethylene Hydrogenation
    - May Result in Exothermic Runaway
  - Operations Stability Operation Sensitive to Feed CO Content Fluctuations

Front-End Hydrogenation: DeC3 O/H

- CG Feed Contains C3 and Lighter Material
- CG Feed Somewhat Rich in Hydrogen and Ethylene
- Feed Contains MAPD

Front-End Hydrogenation: DeC3 O/H

- **Pros:**
  - Relatively Clean Stream
  - Moderately High H2 Concentration Suppresses Formation of Green Oil
  - MAPD Inhibits Hydrogenation of Ethylene Mitigating Against Exothermic Runaways
  - 50 % MAPD Hydrogenated
    - Essentially to Propylene
    - Thus, Downstream C3 Unit Smaller

Front-End Hydrogenation: DeC3 O/H

- **Cons:**
  - Acetylene Plus MAPD Impose Higher Reactor "Load"
    - Higher Overall Heat Rise
    - Higher Rate of Catalyst Fouling
  - Traces of BD May be in Feed; Would Cause Additional Catalyst Fouling
  - May be Difficult to Operate Without Spare Reactor for Continuous Operation of 5+ Years
  - Reactors Larger than for Front-End DeC2 O/H Units

Front-End Hydrogenation: Raw Gas

- Located within Process Gas Compressor Circuit, Downstream of Caustic Tower, but Upstream of Main Driers
- Installed in Many Ethane Crackers Designed/Built in the Mid 60's/Early 70's
- Designed Historically for Operation with Sulfided Ni Catalyst
- Plants Now Being Designed and/or Retrofitted with Pd/Promoter Catalysts in Raw Gas Service

Front-End Hydrogenation: Raw Gas

- **Pros:**
  - Hydrogenates C4 Acetylenes/BD
    - Minimizing Fouling Potential in Downstream DeC2 and DeC3
  - Converts Majority of C3/C4 Acetylenes and Diolefins into Olefins
    - Thus Final C3 and C4 Streams More Attractive for Fuel Gas and/or for Recycle Cracking
  - Allows Elimination of Guard Driers

Front-End Hydrogenation: Raw Gas

- **Cons:**
  - Catalyst Deactivation Relatively Rapid
    - Especially for First Bed
    - Due Primarily to C4 and C5+ Acetylene and Diolefins
  - Even with Pd/promoter Catalyst Spare Reactor Neded for Continuous Operation of 5+ Years
  - Difficult to Operate with Ethylene Gain, Especially if Catalyst is Sulfided NiCo
  - Acid Species May Be Formed

Tail-End Hydrogenation: *Characteristics*

- Feed is *C2's*
  - O/H of Backend DeC2
  - Contains 0.5 to 2.5% Acetylene
- Hydrogen Injected Into Each Bed
- Spare Reactors with Facilities for Insitu Catalyst Regeneration/Reduction
  - *Green Oil (Oligomers of Acetylene)*
  - *Causes Rapid Catalyst Deactivation*

Figure 5
2-Bed Tail-End Acetylene Hydrogenation Reactors

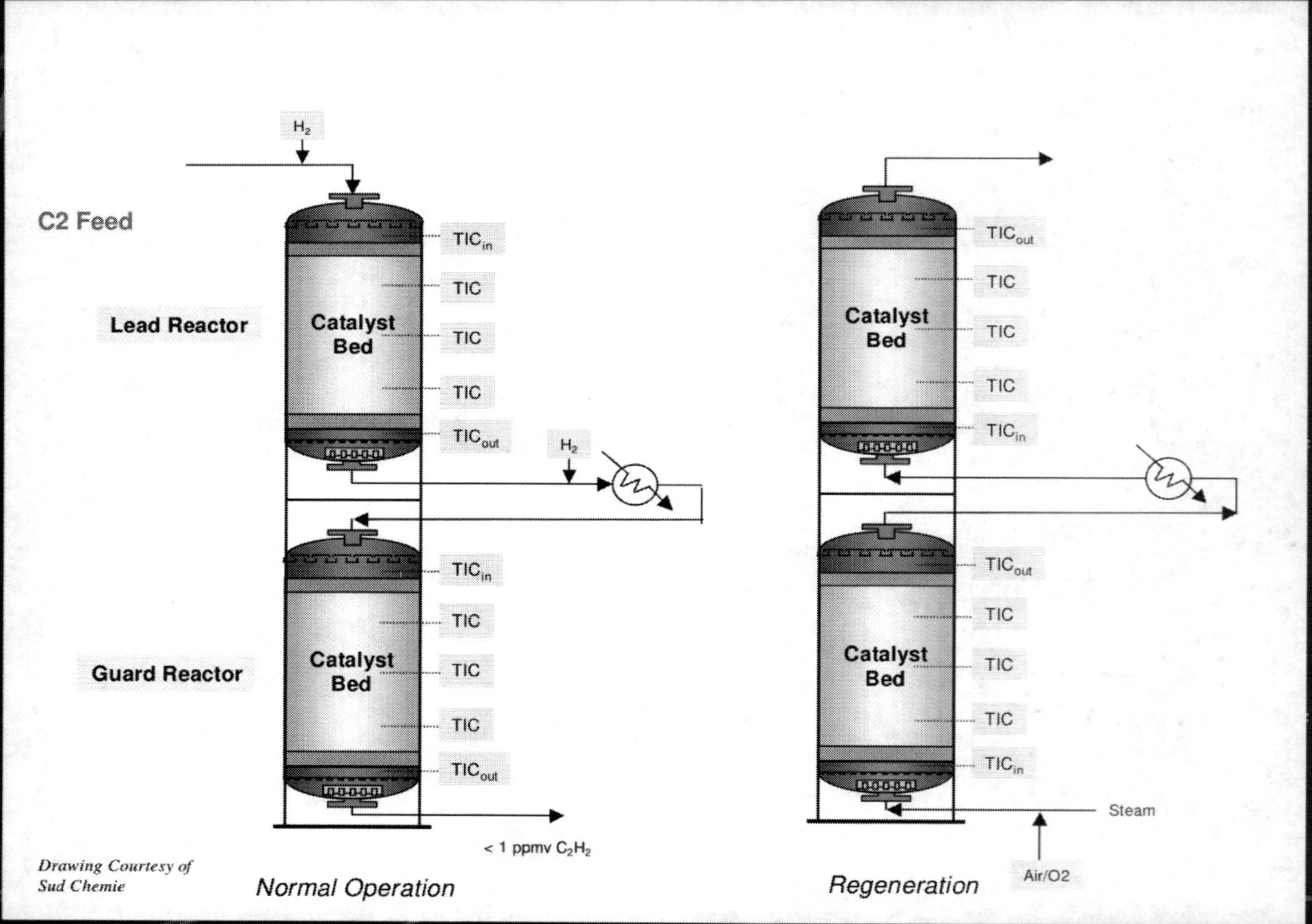

Tail-End Hydrogenation: Concerns/Considerations

- Acetylene Upgraded to Ethylene, Downgraded to Ethane and Green Oil, (Mostly C4's with Small Amounts of C6 to C20+)
- Downstream Green Oil (C6+) Removal Facilities
- H2 in Reactor Effluent Must Be Removed
- Character of Catalyst Changes Throughout Cycle
- Catalyst "Rejuvenated" Via Insitu Steam/Air Regeneration Followed by H2 Reduction

Tail-End Hydrogenation: Concerns/Considerations

- Character of Catalyst Undergoes Slow "Permanent" Changes Through Online Operation and Repeated Regenerations
- Catalyst Life Should be 5+ years; Could Be 10+ years

Tail-End Hydrogenation: Concerns/Considerations

- Economic Considerations Drive Operators Toward Optimizing Units
  - Maximize Ethylene Gain
  - Can Make Operation Vulnerable to Acetylene Breakthrough
  - Off-Spec Incident May Result in Substantial Production Loss

Ethylene Plant Operations
Acetylene Hydrogenation Tutorial

Lecture II : Chemistry & Catalysts

2009 Spring National Meeting

AIChE Session 107 - April 29th

Richard W. Voight

Ethylene Plant Operations
Acetylene Hydrogenation Tutorial

AIChE Session 107 - April 29, 2009

This lecture has been prepared for presentation at the AIChE 2009 Spring National Meeting in Tampa, Florida.

Catalysts

Primary Catalyst Type for Selective Hydrogenation

- Palladium on Alumina
- Low Loading on Catalyst for Ethylene Purification
- Typically 0.02 - 0.03 Wt%
- Typical Surface Area (SA) is 5.0 - 75 m2/g

Catalysts

Higher Pd Concentrations Used:

- For Heavier Hydrocarbon Applications
  - Mixed C3+
  - Typically Liquid Phase Applications

Lower SA Catalysts Typically Used:

- Where Control of Selectivity Using CO Not Desired

Catalysts

- Careful Control of Properties Essential When Produced
- Properties Shaping Overall Catalyst Performance
  - SA
  - Metal Loadings
  - Pore Volumes
  - Acidity

Catalysts

- Newest Generations of Catalysts Are Bi Metallic
- Where Most Common Loading Pd + Ag
- These Bi Metallic Catalysts Tend:
  - To Have Much Higher Selectivity
  - To Have Lower Green Oil Production Rate

Catalysts

Selectivity Is Extremely Important

- To Prevent Ethylene Loss
- But, In Front-End Applications Helps In Runaway Prevention
- High Selectivity Also Promotes Acetylene Removal to Low Levels

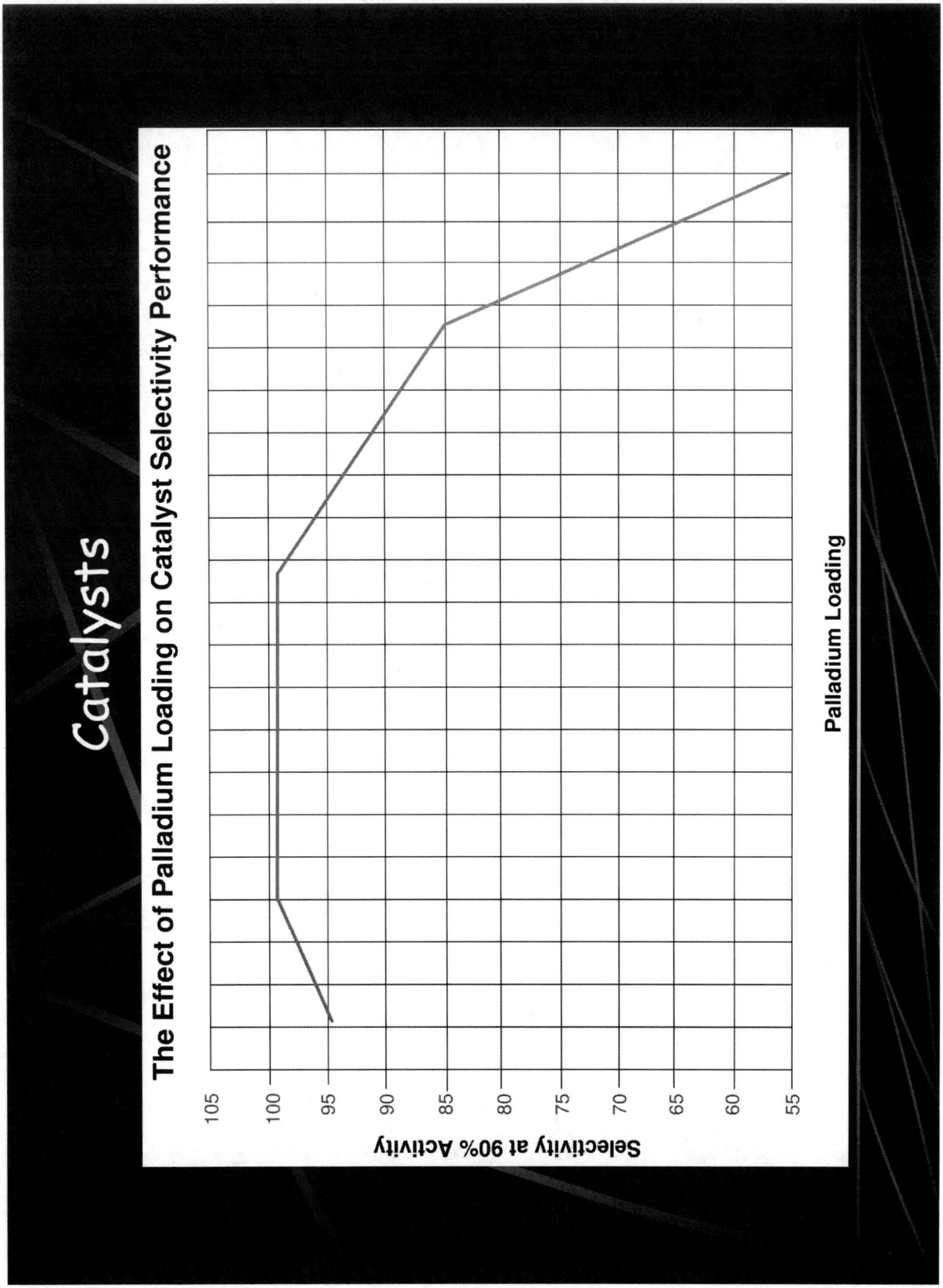
Catalysts
The Effect of Palladium Loading on Catalyst Selectivity Performance
Selectivity at 90% Activity
105
100
95
90
85
80
75
70
65
60
55
Palladium Loading

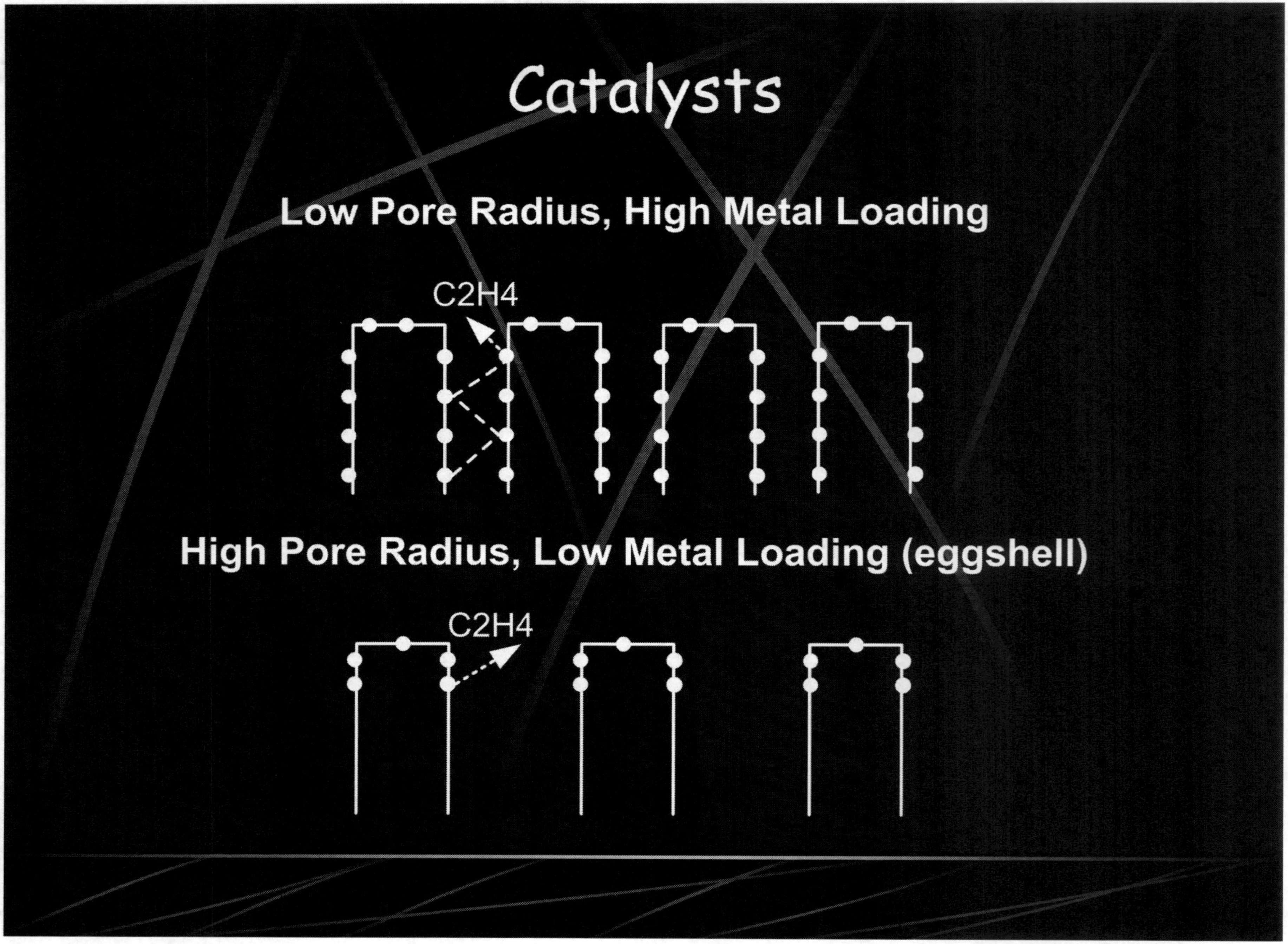

Catalysts
Low Pore Radius, High Metal Loading
C2H4
High Pore Radius, Low Metal Loading (eggshell)
C2H4

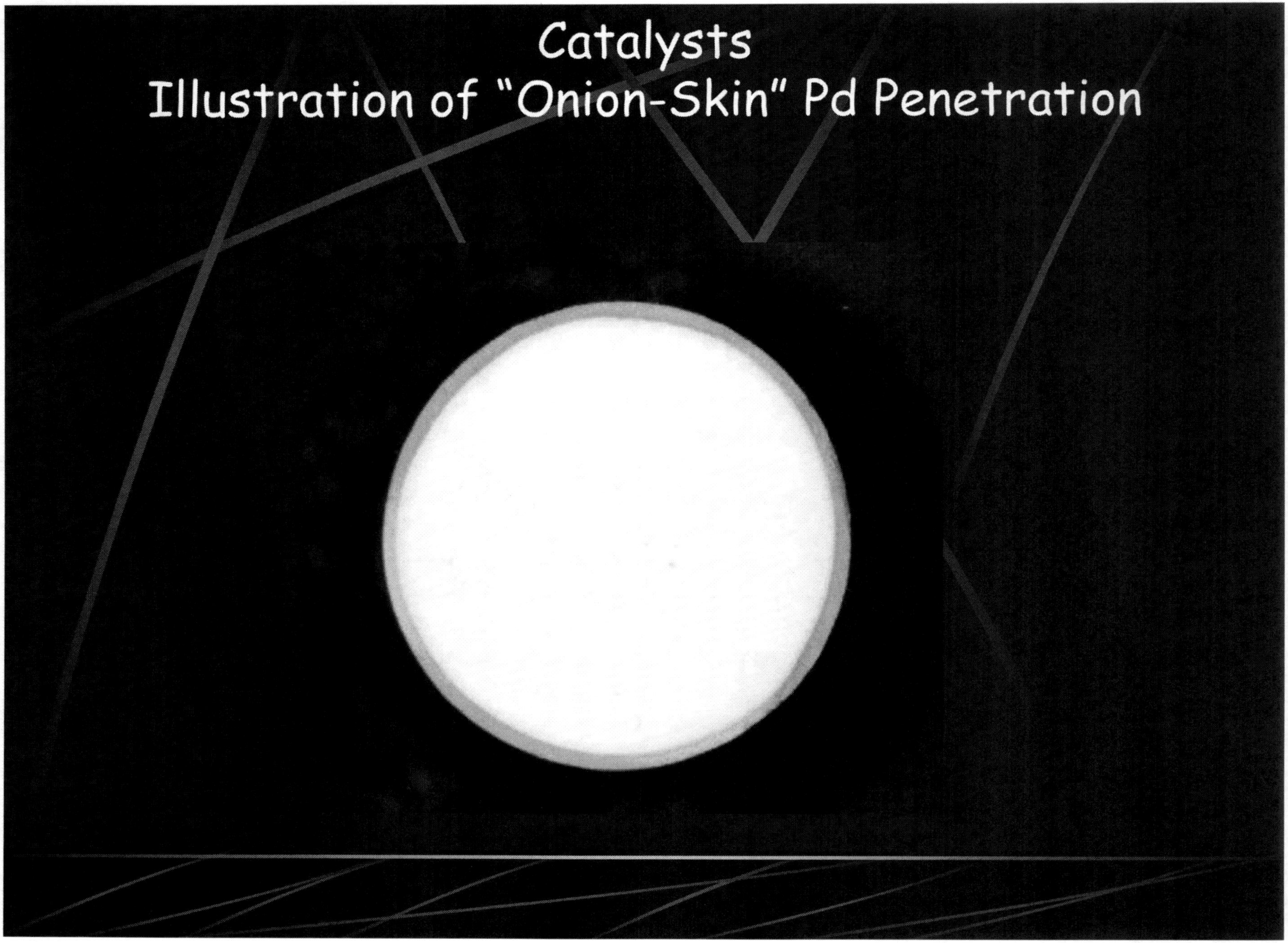
Catalysts
Illustration of "Onion-Skin" Pd Penetration

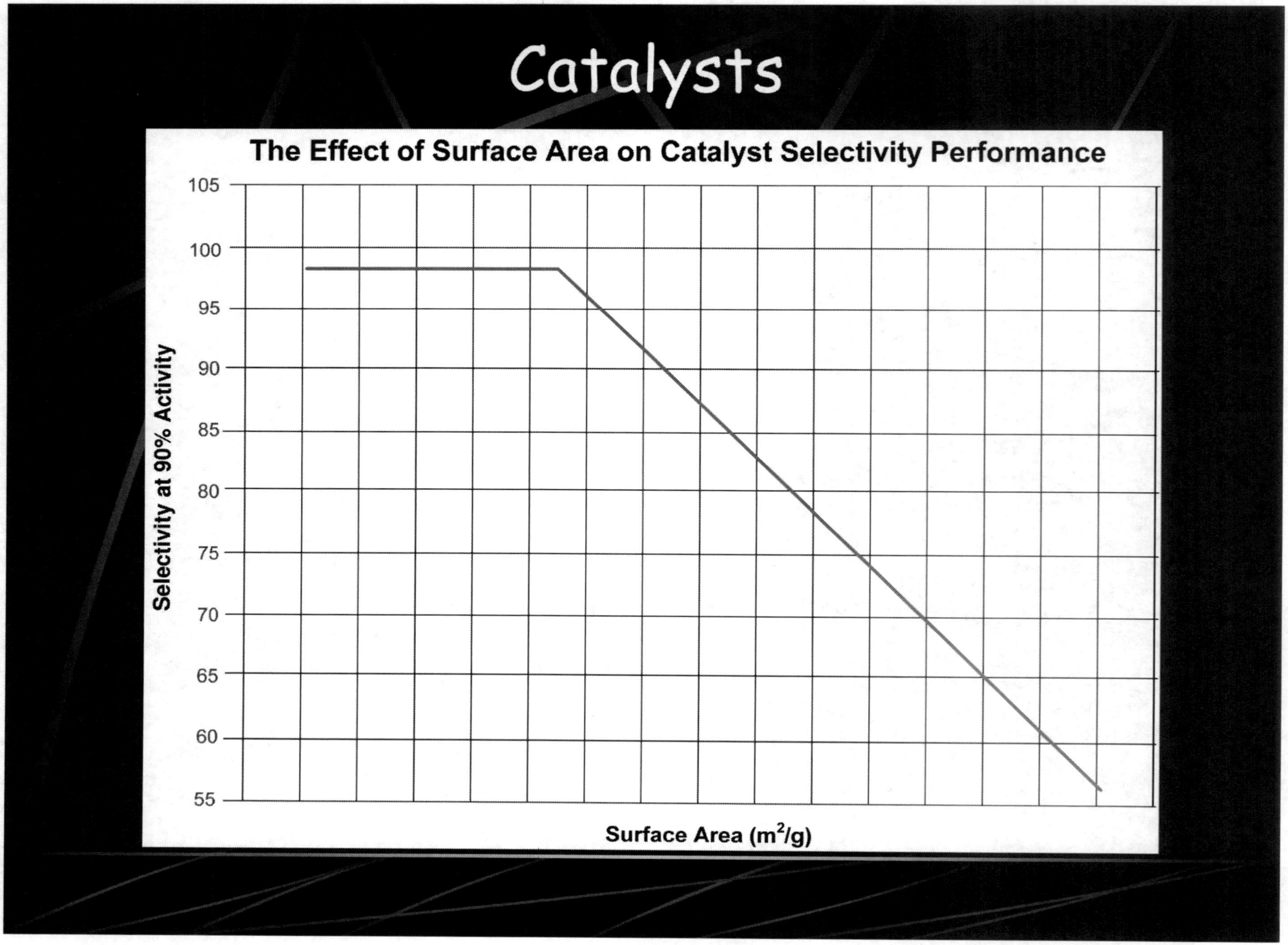
Catalysts
The Effect of Surface Area on Catalyst Selectivity Performance
Selectivity at 90% Activity
105
100
95
90
85
80
75
70
65
60
55
Surface Area (m2/g)

Catalysts

Catalyst Efficiency Affected By:

- Operating Conditions
- Reactor Feed Purity
- Original Design

Catalyst Performance Also Affected By:

- Way Catalyst Placed in Service
- Way Catalyst Reduced Prior to Use

Sulfided Nickel Catalysts

- Also Used for C_2 Selective Hydrogenation
- Normally in Raw Gas Applications

Chemistry for Front-End Hydrogenation

For Deethanized CG Feeds:

Acetylene + H_2 = Ethylene (Desired)

Acetylene + $2H_2$ = Ethane (Undesired)

Ethylene + H_2 = Ethane (Undesired)

Acetylene + Acetylene + H_2 = Oligomer (Undesired) $\rightarrow$ Green Oil

Chemistry for Front-End Hydrogenation

For Depropanized CG Feeds:

Reactions Shown for Deethanized Feed Also Occur

Plus C3 and C4 Reactions as Follows:

MAPD + H_2 = Propylene (Desired)

Propylene + H_2 = Propane (Undesired)

Butadiene + H_2 = Butene (Desired)

Butene + H_2 = Butane (Undesired)

Key Reactions

- All Reactions Highly Exothermic
- Primary Reaction
  $C_2H_2 + H_2 \rightarrow C_2H_4$ Delta H = -174kj/mole
- Secondary
  $C_2H_4 + H_2 \rightarrow C_2H_6$ Delta H = -139kj/mole
- For Front-End hydrogenation These Two Reactions the Most Prevalent
- Can Easily Lead to Reactor Runaways if Selectivity Not Maintained

Chemistry

- Ethylene Rate of Reaction 100X Faster Than Acetylene
- Therefore Once Ethylene Reaction Occurs Excessive Reactor Temperature Excursion Will Occur

Chemistry

All Reactions Are Controlled By:

- Process Conditions
- Catalyst Selection
- Finally By Displacement Reactions, Where:
  - Selectivity A Function of Species Adsorption to Palladium As Follows:
  - CO > Acetylenes > MA > DiOlefins > Olefins

Chemistry

- All of These Factors Play Important Part in Successful Reactor Operation
- Careful Control of CO Levels Also Extremely Important for Stable Operation
- Catalyst Selection Very Important
  - Good Selectivity for Stable Reactor Operations
  - Sensitivity to CO Levels

Green Oil

- Term Used to Describe Heavier Polymers
- Typically Carbon Number Greater Than C10
- Slowly Fouls Catalyst Leading to Loss of Activity

Green Oil

- Green Oil Make Initiated by Acetylene Dimerization Reaction:

  $C_2H_2 + C_2H_2 + H_2 \longrightarrow C_4H_6$

- Once BD Formed, Polymerization Continues:
  - Based on Process Temperature
  - Based on Acidity of Catalyst
  - Based on Amount of BD Formed

Green Oil

- Amount of BD Formation is Based:
  - On the Amount of Acetylene Converted In Reactor
  - Not the Amount of Acetylene in Feed to Reactor

Ethylene Plant Operations
Acetylene Hydrogenation Tutorial

Lecture III: Front-End
Acetylene Hydrogenation

2009 Spring National Meeting
AIChE Session 107 – April 29th
Rick Callejas

Ethylene Plant Operations
Acetylene Hydrogenation Tutorial

AIChE Session 107 - April 29, 2009

This lecture has been prepared for presentation at the AIChE 2009 Spring National Meeting in Tampa, Florida.

Front-End Acetylene Hydrogenation

- Selectivity Factor
  - Function of CG Feed Component Species
  - Adsorption on Palladium Sites Controls Selectivity
  - Initially Acetylene and Ethylene Compete

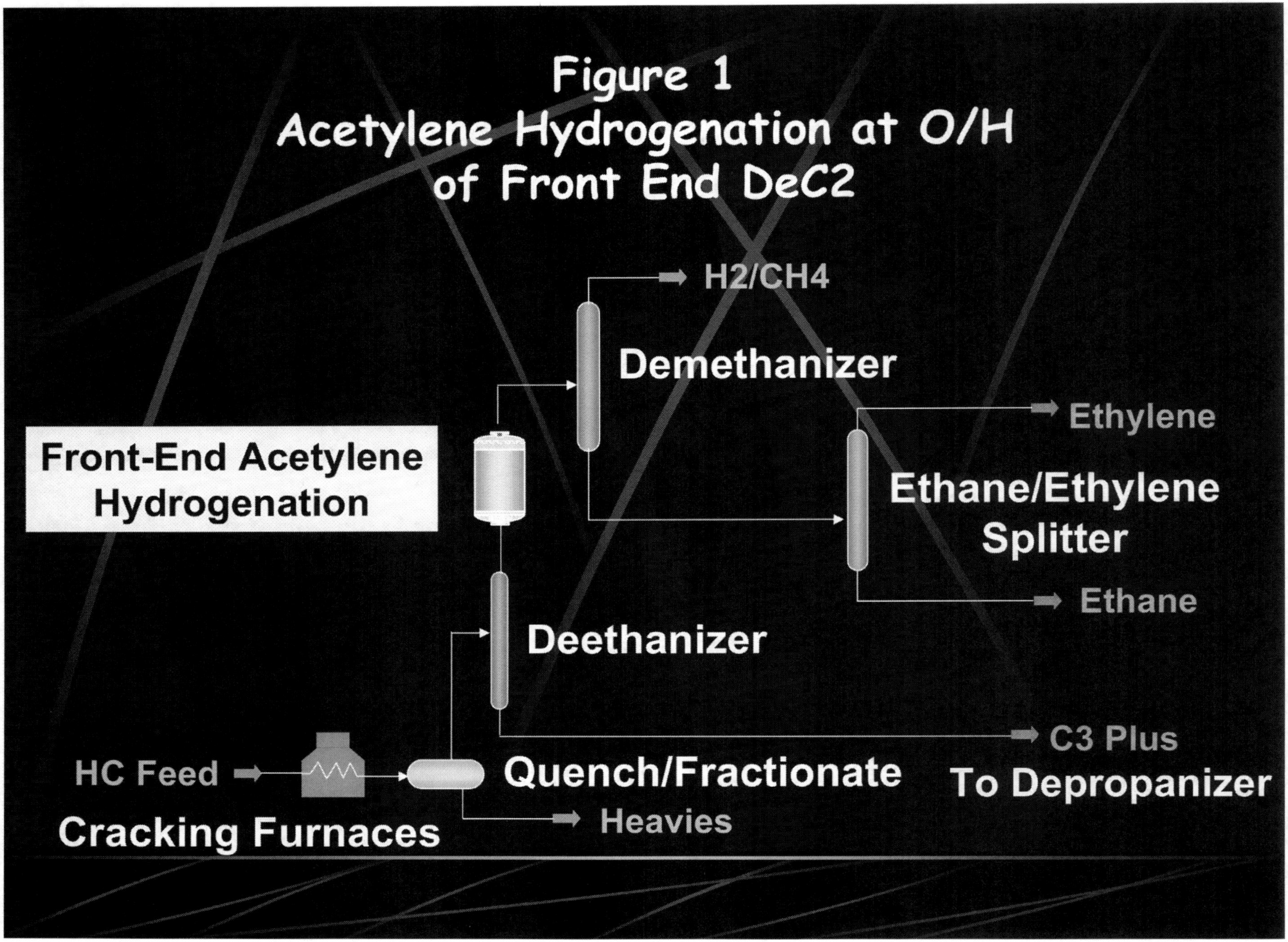
Figure 1
Acetylene Hydrogenation at O/H
of Front End DeC2
H2/CH4
Demethanizer
Ethylene
Front-End Acetylene
Hydrogenation
Ethane/Ethylene
Splitter
Ethane
Deethanizer
C3 Plus
HC Feed
Quench/Fractionate
To Depropanizer
Heavies
Cracking Furnaces

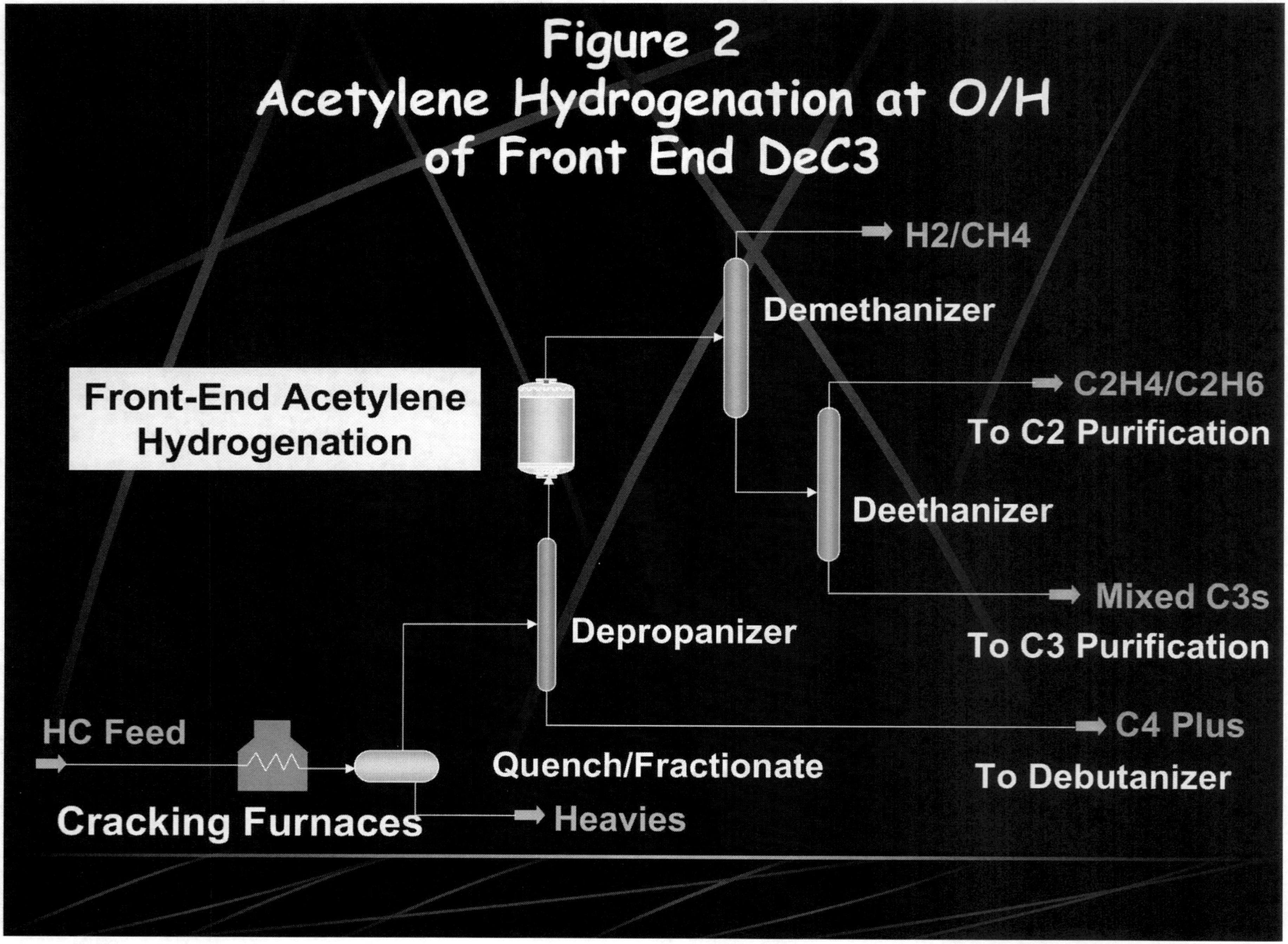
Figure 2
Acetylene Hydrogenation at O/H
of Front End DeC3
H2/CH4
Demethanizer
Front-End Acetylene
Hydrogenation
C2H4/C2H6
To C2 Purification
Deethanizer
Mixed C3s
To C3 Purification
Depropanizer
HC Feed
C4 Plus
Quench/Fractionate
To Debutanizer
Cracking Furnaces
Heavies

Selectivity Factors

- As Acetylene Reduced to <0.2 - 0.5 PPM
  - Acetylene No Longer Competes with Ethylene
  - Ethylene Hydrogenation Begins
- CO Main Inhibitor to Ethylene Adsorption
- CO Stronger to Olefins the Weaker
- MA and Diolefins as Inhibitors in CG

Operation Variables for Good Control

- Steady CO Content in Reactor CG Feed at All Times Recommended
  - Stable Furnace Cracking Effluent CO Output
  - Control CO Make with Sulfur Addition to Cracking Furnance Tubes
- In Case of furnace Area CO Make Upset
  - Adjust Reactor CG Inlet Temperature
  - Maintain Ethylene Product C_2H_2 Spec

Operation Variables for Good Control

- Control Reactor Feed Flow Rate
  - Maintain Constant Reactor Space Velocity
  - Use of Recycle Streams to Achieve
  - Guarantees Good Selectivity
- Loss of Reactor CG Feed
  - Avoid Sudden Temperature Excursion
  - Requires Immediate Reactor Trip
  - Temperature Excursions Decrease Catalyst Life
  - Affect Catalyst Charge Overall Selectivity

Operation Variables for Good Control

BETTER TO GO OFF TEST IN ACETYLENE & AVOID RUNAWAY BECAUSE:

- Recovery Time From this Condition Faster
- Much Production Time Lost
  - Must Depressure Reactors
  - Must Complete Cool Down of All Beds to Min Temperature
  - Reactors Then Pressured Up for Start-Up

Operation Variables for Good Control

- Recovery From Reactor Temperature Excursion
  - Trip to Maximum Cooling
    - Bypassing the Steam Heater
    - Important to Have Way of Cooling CG Feed
- Final Result to Look for:
  - CG Feed Temperatures to Reactors Lower Than Catalyst Clean Up Temperature
  - No Catalyst Bed Hot Spots

Operation Variables for Good Control

- Recovery From Reactor Temperature Excursion
  - 2 to 3 Days of Steps for Safe and No Issues Start-up After a Runaway
  - Recovery Perhaps Only a Few Hours for Trip to Maximum Cooling When No Reactors Depressured

Operation Variables for Good Control

- Catalyst Performance Fastest Analyzer for Reactor System
- Outlet Temperature of Reactors Great Indicator of Upset
  - Outlet Temperature Rate of Change from Norm Requires Immediate Action
  - Change Reactor Inlet Temperature to Stop Outlet Temperature Swing

YOU COULD BE TOO LATE!

Operation Variables for Good Control

- Philosophy of When to Assess Cause of Reactor Temperature Upset
  - Once Outlet Temperature Move Up or Down Stopped Look for Root Cause
  - Do Not Try to Find Cause of Problem Before Stopping Temperature Swing

Operation Variables for Good Control

- If Rate of Change Upward Cannot be Stopped by Normal Control Actions
  - Trip to MAXIMUM Cooling
  - Let Product Go Off Test
- If Rate of Change Down Can't be Stopped By Normal Inlet Temperature Increases
  - Stop and Look for Root Cause
  - Avoid Increasing Temperature Too Much
  - Avoid Increasing Temperature Too Quickly
  - Avoid Action That Results in Runaway

Operation Variables for *Good* Control

- Drastic *Changes* Are Never Recommended
- Use Reactor Outlet On-Line Analyzers As *Guide*
- Remember Analyzer Measurements Are Not Instantaneous
- Remember to *ALWAYS* Leave Some Acetylene in Final Reactor Outlet
  - *Assures Operating in Selective Mode*
  - *Avoids Runaways That Hurt Catalyst*

Acetylene Clean-Up & Temperature Excursion Relationship

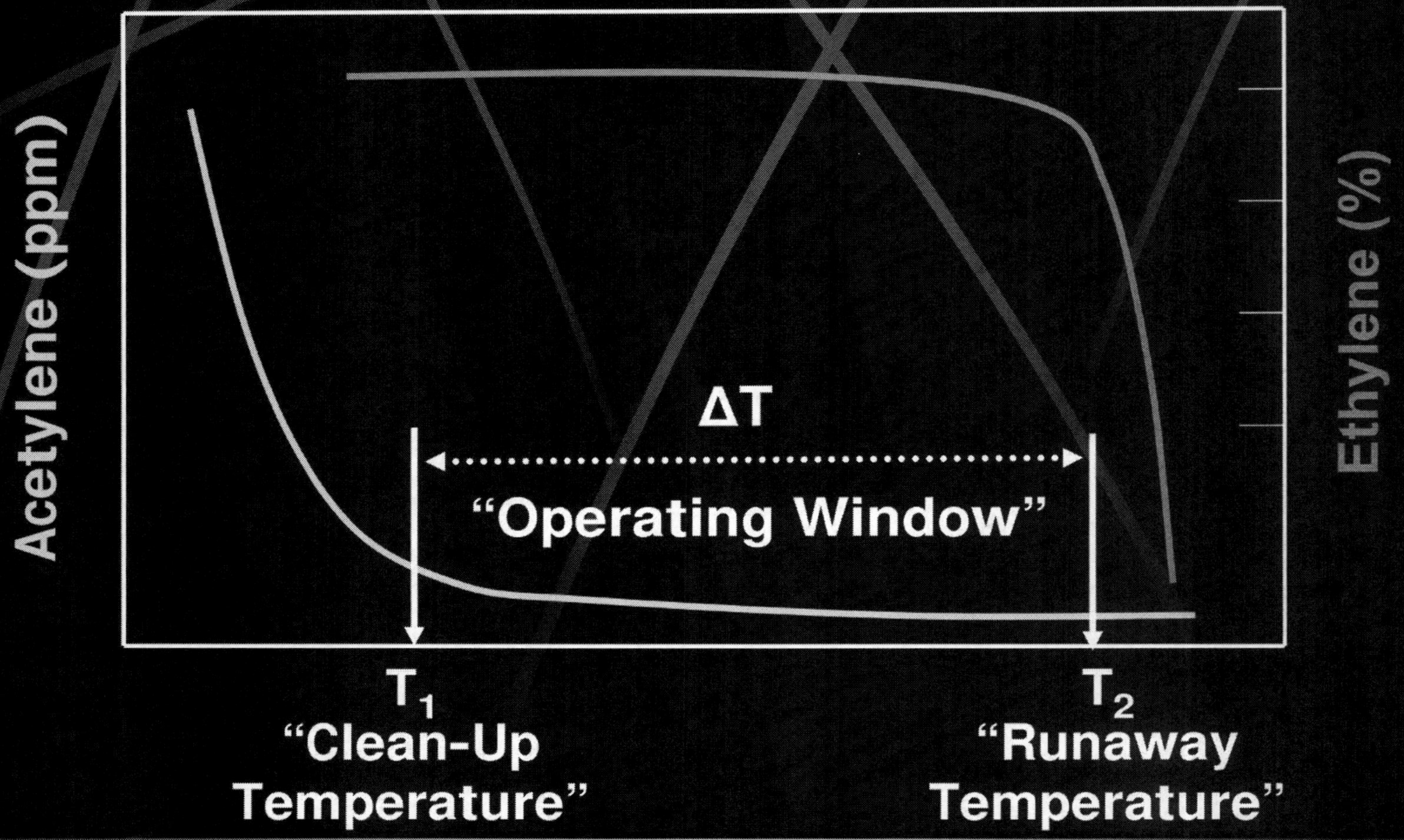

Analytical Equipment Recommended for Good Control

- Cracking Furnace Effluent CO Continuous Analyzer
- Reactor CG Feed Acetylene
  - Short Cycle Times
  - Possible Dual Analyzers
- Reactor Feed Continuous CO Analyzer
- Mid Bed Acetylene Analyzer
- Reactor Outlet Acetylene
  - Dual Analyzers
  - Trying to Minimize Cycle Time

Other Equipment Recommended for Good Control

- Catalyst Bed Thermocouples
  - The More the Better
  - Use Spiral Arrangement
  - Located to Optimize Bed Coverage for Hot Spot Monitoring
  - Temperature In and Out of Each Bed
- Sometimes Heater Between Beds Beneficial to Separate Bed Control
- PROPER SIZING OF INTERBED COOLER A MUST

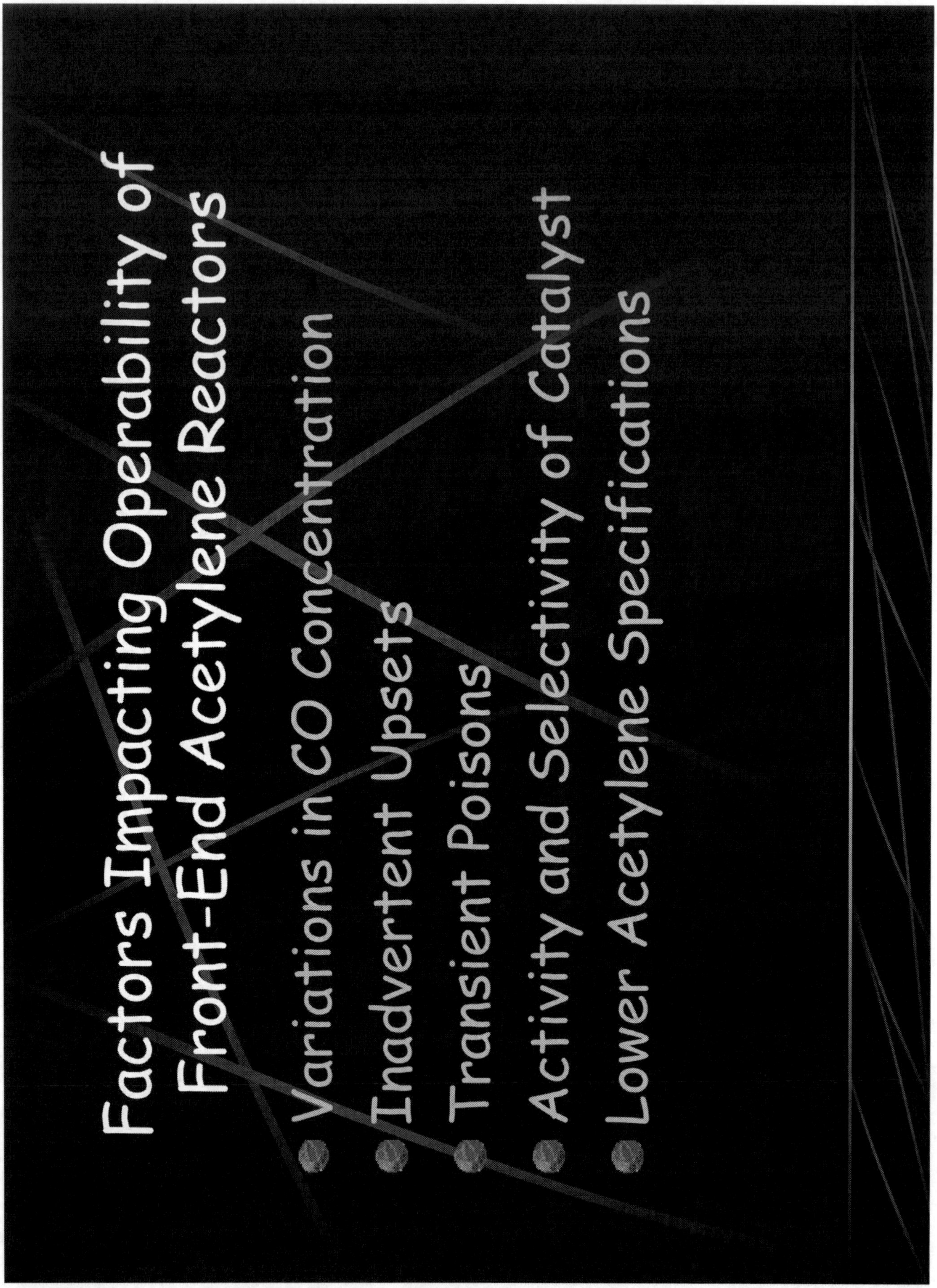
Factors Impacting Operability of Front-End Acetylene Reactors
Variations in CO Concentration
Inadvertent Upsets
Transient Poisons
Activity and Selectivity of Catalyst
Lower Acetylene Specifications

Front-End Acetylene Hydrogenation Getting Started Steps

- Pressurize reactor to pressure near associated CG Compressor Stage suction pressure
- Start CG Compressor and open valves to reactor
- Establish CG flow through the reactor at coldest temperature process allows

Start-Up Considerations #1 for Front-End Acetylene Hydrogenation

Once full flow is established & lowest temperature stable in all beds:

- Start raising CG feed temperature to reactor system very slowly
- Depending on catalyst vendor for catalyst charge there should be experimental data of T1 (Clean Up Temp) & ΔT to excursion

Start-Up Considerations #2 for Front-End Acetylene Hydrogenation

Continue increasing CG feed temperature to reactor system very slowly until:

- Heat wave begins coming through the reactor system
- Stop raising CG feed temperature until reactor system temperatures stabilize
- Check reactor analyzers for Acetylene conversion taking place

Start-Up Considerations #3 for Front-End Acetylene Hydrogenation

Once reactor system temperatures have stabilized after kick off of conversion reaction:

- Continue increasing CG feed temperature to reactor system
- Raising CG feed temperature should be much slower now
  - Want to stay selective
  - Continue to monitor Analyzers and acetylene conversion and reactor ΔT's
  - Never go to ZERO Acetylene in a reactor outlet

Start-Up Considerations #4 for Front-End Acetylene Hydrogenation

Once acetylene clean up is achieved:

- Stop increasing CG feed temperature to reactor system
- Initiate normal operation routine and verify analyzers

Start-Up Considerations #5 for Front-End Acetylene Hydrogenation

- Objective is to stay selective and maintain stability
  - Always remember to leave some trace of acetylene in reactor effluent
  - Continue to monitor Analyzers and acetylene conversion and reactor ΔT's
- Should a reactor bed temperature excursion begin lower the CG feed temperature to that reactor incrementally until excursion is stopped.

Reactor System Excursion Event

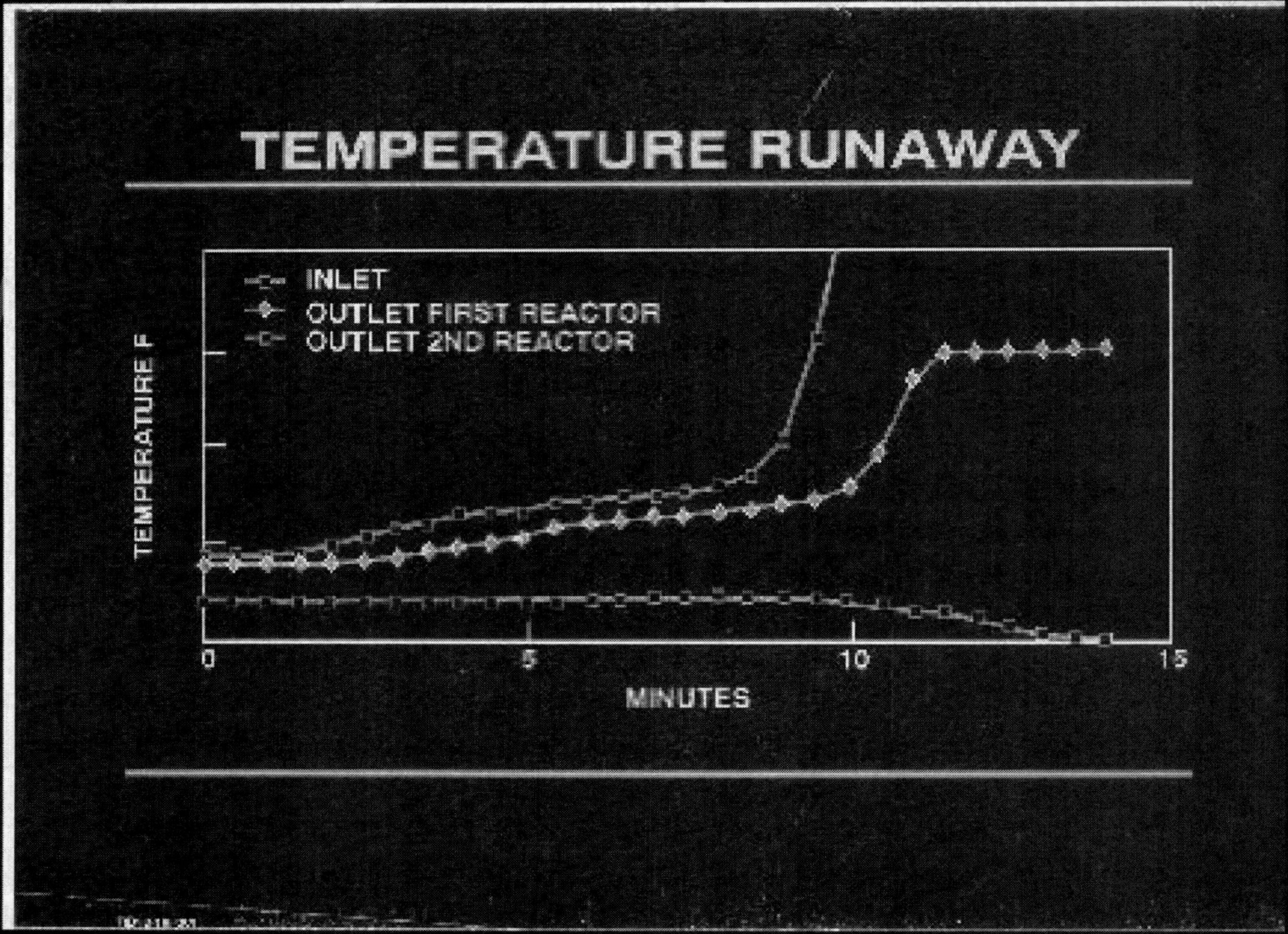

Ethylene Plant Operations
Acetylene Hydrogenation Tutorial

Lecture IV : Backend Hydrogenation

2009 Spring National Meeting
AIChE Session 107 - April 29th

Richard W. Voight

Ethylene Plant Operations
Acetylene Hydrogenation Tutorial

AIChE Session 107 - April 29, 2009

This lecture has been prepared for presentation at the AIChE 2009 Spring National Meeting in Tampa, Florida.

Tail End Selective Hydrogenation

Overview of Process Flow Diagram

Process Control Variables

Temperature

H2/C2H2 Ratio

Tail End Selective Hydrogenation

- Use of Selectivity Modifiers
- Non-Controllable Variables
- Pressure
- Space Velocity
- Catalyst Type

Tail End Selective Hydrogenation

- Effects of Varying Process Variables
  A. Temperature
  B. H2 Ratio
  C. Selectivity Modifier
  D. Effects of C2H2 Concentration
  E. Balancing of C2H2 Conversion Across the System
- Effects of Various Poisons

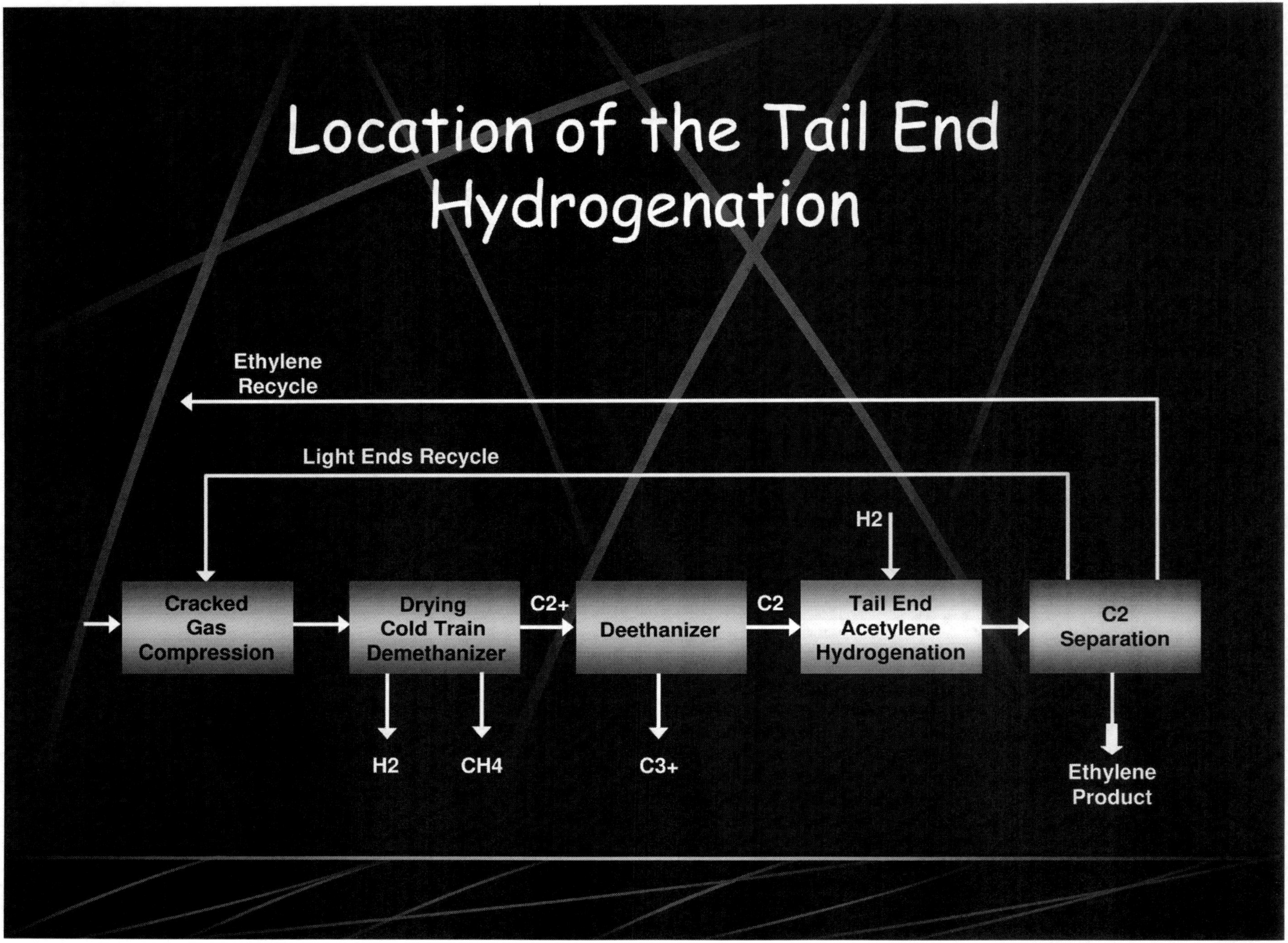
Location of the Tail End Hydrogenation
Ethylene Recycle
Light Ends Recycle
H2
Cracked Gas Compression
Drying Cold Train Demethanizer
C2+
Deethanizer
C2
Tail End Acetylene Hydrogenation
C2 Separation
H2
CH4
C3+
Ethylene Product

Process Flow Scheme

Effects of C2H2 Concentration

Mixed C2 Feed C2H2 Concentration Range	Operating Reactors Required
0.1 to 0.7	One Reactor
0.8 > 1.5	Two in Series
1.6 > 2.5	Three in Series

Tail End Hydrogenation

One of the final steps in the production of polymer grade ethylene.

Tail End Feed Stream may contain 0.03 – 2.5% acetylene.

Acetylene removed down to levels as low as 20 ppb in extreme cases.

Tail End Hydrogenation

Removal must be accomplished in such a way that ethylene loss is minimized.

Green Oil produced during the hydrogenation is removed downstream along with any excess H2 and CO.

Tail End Process Summary

- Uses DeMethanized DeEthanizer Overhead
- Prime Purpose - C2 Acetylene Removal
- Prime Objective - Ethylene Gain to Very Minimum Ethylene Loss
- Designs Typically Limited to C2 Acetylene Concentrations of <2.5%

Tail End Process Summary

Operating Strategy is to stay within the Operating window.

That is where acetylene is removed and ethylene hydrogenation is minimized.

Lower end of the window is minimum temperature and H_2 required for acetylene removal.

Tail End Process Summary

Upper End of the operating window is where the high temperatures and/or excessive H2 injection accelerates hydrogenation of ethylene.

Under these conditions it is almost impossible to achieve acetylene removal to very low levels of <1.0 ppm.

What Effects Operating Window?

Operating Window Varies Due To:

- Acetylene Concentration
- Reactor Feed Rate
- Catalyst Type and Age
- Trace Poison Levels

What Controls Tail End?

- Temperature
- Inlet H2 Content
- CO Content
- Space Velocity
- Pressure
- Non-Controllable Variables

Process Variables

What process variable should be adjusted?

- Temperature
- Hydrogen Flow
- CO Injection Rate

How can process data be used to assess which process parameters to adjust?

Process Variables

TEMPERATURE

What questions are to be asked about temperature control?

- Adjust by how much and at what rate?
- What is to be look for in process response?
- What is maximum temperature limit?

Temperature is primary control for long term operation.

Process Variables

Temperature Adjustments

What data to look at when changes are made?

- Reactor Outlet Acetylene Content Changes
- Rate of change for Acetylene Removal
- Reactor Outlet H2 Changes
- Reactor Temperature Profile Changes
- Rate of Change of Temperatures

Process Variables

HYDROGEN FEED FLOW

During normal operation increase H2 flow only if increasing temperature in small amounts (less than 10 degrees F) fails to improve conversion.

Process Variables

HYDROGEN FEED FLOW

H_2 flow is increased because:

- Upset condition needs to be controlled.
- Because temperature increase will not produce quick enough response to prevent going off test on acetylene spec.

Process Variables

HYDROGEN FEED FLOW

- Rate of ethylene hydrogenation, is first order with respect to H2.
- Too great of an addition of H2 will result in extreme loss of selectivity.
- Cleanup of small amounts of acetylene is almost always achieved with excess H2.

Process Variables

CO ADDITION

When should CO be added and how much?

- Selectivity Poor
- Difficult to Maintain Acetylene Leakage

CO is used to expand the operating window.

Process Variables

USE OF CO INJECTION

- CO flow will depend on the catalyst type and the process conditions at the time.
- As CO injected selectivity will improve.
- As CO injected acetylene removal will become more efficient.
- If excess of CO added activity will be lost and conversion will suffer.

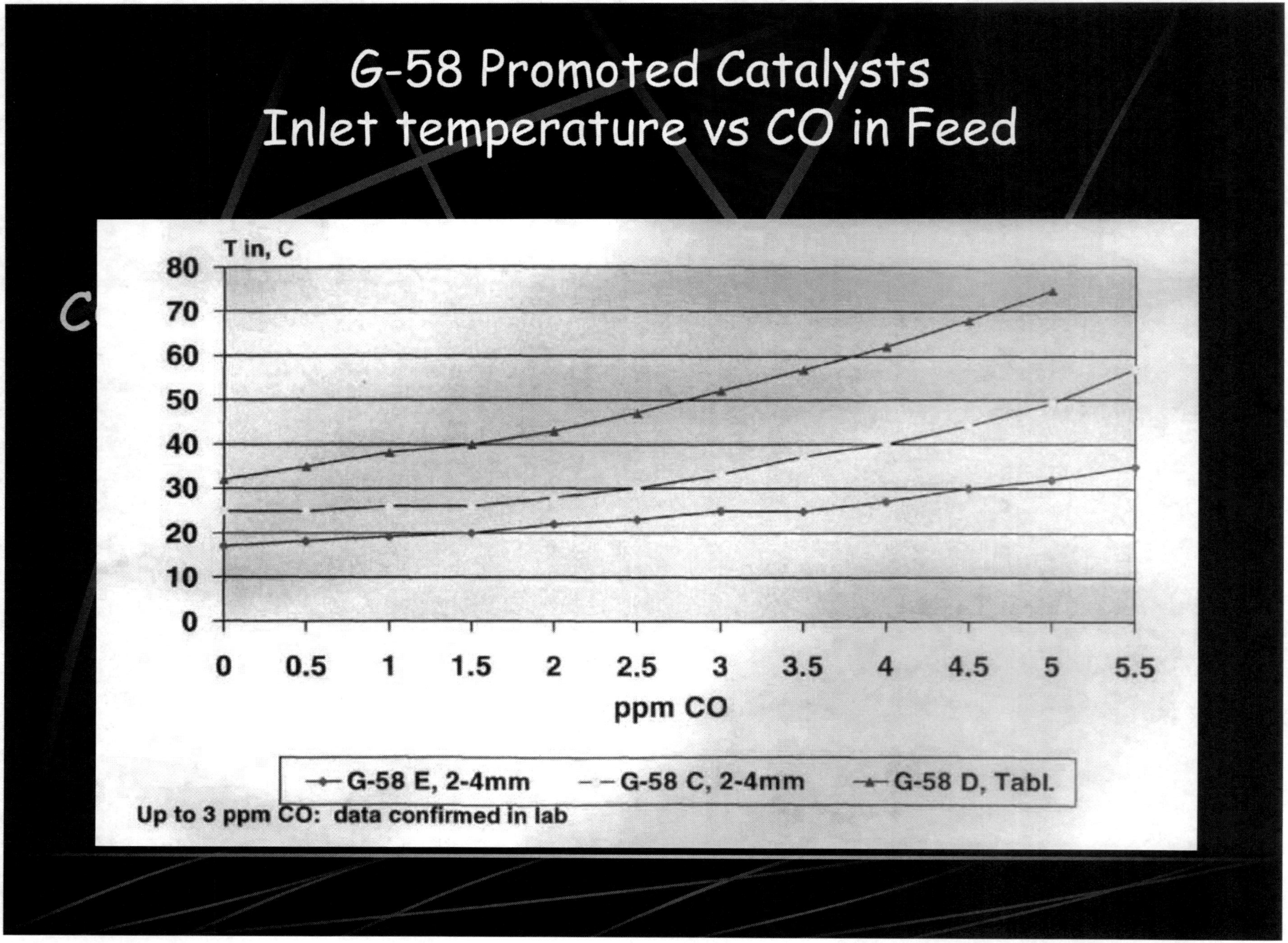
G-58 Promoted Catalysts
Inlet temperature vs CO in Feed
T in, C
80
70
60
50
40
30
20
10
0
0
0.5
1
1.5
2
2.5
3
3.5
4
4.5
5
5.5
ppm CO
G-58 E, 2-4mm
G-58 C, 2-4mm
G-58 D, Tabl.
Up to 3 ppm CO: data confirmed in lab

Process Variables

CO ADDITION

- Add CO in Small Increments
- CO Normal Concentration AT 0.1 PPM
- As Rule Maximum CO Concentration:
  - With Commercial Catalysts About 2.0 PPM
  - With Promoted Catalysts Much Less

Process Variables

EFFECTS OF C2H2 CONCENTRATION

Higher Levels of Acetylene Mean:

- Greater Green Oil Make
- Greater Green Oil Fouling Potential
- Shorter Operating Time Between Catalyst Regenerations

Process Variables

Balancing of Reactor Acetylene Conversion

- Conversion Load Distribution of Reactors
  - Dependent on Catalyst Type
  - Dependent on Condition of the Catalyst
- With A 1 Reactor Operating System:
  - 100% of Required Conversion in Single Stage
  - From SOR to EOR

Process Variables

Balancing of Reactor Acetylene Conversion

- With 2 Reactor Operating System at SOR:
  - Lead Bed Normally 75% of Required Conversion
  - 25% of Conversion in Second Bed
- With 2 Reactor Operating System at EOR:
  - Lead Bed Slips to 65% of Required Conversion
  - Tail Bed Takes 35% of Load

Process Variables

Balancing of Reactor Acetylene Conversion

- With 3 Reactor Operating System at SOR:
  - Lead Bed Normally 60% of Required conversion
  - Second Bed Converts 25% of Required Load
  - Third Bed Converts Remaining 15%
- With 3 Reactor Operating System at EOR:
  - Lead Bed Slips to 50% of Required Conversion
  - Second Bed Takes 35% of Conversion
  - Tail Bed Takes Remaining 15% of Load

Non-Controllable Variables

Non-Controllable Variables Are:

- Not Directly Under Operator Control
- But Impact Performance of Unit
- Mentioned Here for Awareness

Three Non-Controllable Process Variables

- Space Velocity
- Acetylene Concentration
- System Pressure

Non-Controllable Variables

- Space Velocity
  - Value Set at Time of System Process Design
  - Essentially Not Changed
  - Based on Catalyst Kinetics of Catalyst Manufacturer
- Acetylene Concentration
  - Function of Cracking Furnace Operation
  - Cracking Feedstock
  - Cracking Severity
- System Pressure – Not Subject to Much Change

Catalyst Poisons

- Have Critical Effect on Catalyst Performance
- Some Poisons Are Permanent
- Some Poisons Are Temporary

Catalyst Poisons

- CO and Sulfur Are Temporary Poisons
- When Removed from Feed CO and Sulfur Come Off Catalyst with Time
- During Hydrogenation Permanent Poisons
  - Arsine (AsH3)
  - Mercury
  - Other Metals
- Many Plants Now Employ Catalyst Poison Guards

Catalyst Poisons

AsH3 Affect on Catalyst

- AsH3 will give extreme selectivity to the catalyst in levels <5ppb.
- In greater levels AsH3 will cause complete catalyst deactivation.
- AsH3 will essentially be removed during regeneration of the catalyst.

Hg Affect on Catalyst

- Hg causes catalyst deactivation.
- Hg is removed completely by regeneration.

Catalyst Poisons

Metals Considered Permanent Poisons

- Fe, Mn, Sb, Ca, Mg and Si
- Not Removed By Standard Catalyst Regeneration
- Long Term Catalyst Activity Loss

Ethylene Plant Operations
Acetylene Hydrogenation Tutorial

Lecture V: Catalyst Regeneration

2009 Spring National Meeting
AIChE Session 107 - April 29th
Ed Mohundro

Ethylene Plant Operations
Acetylene Hydrogenation Tutorial

AIChE Session 107 - April 29, 2009

This lecture has been prepared for presentation at the AIChE 2009 Spring National Meeting in Tampa, Florida.

Catalyst Regeneration

Front-End Catalysts Regeneration

- Typically Regenerated Exsitu
- Exercise Diligence in Getting Pd/Promoter Catalysts Regenerated.
  - <u>Expectations Are Regenerated Catalyst Will Operate for 3-5+ Years</u>
- Regeneration Company Should Have Experience with Regenerating Pd/promoter Catalysts.
- Confirm Regenerated Catalysts Performance via Laboratory or Bench Scale Activity/Selectivity Test.

Catalyst Regeneration (Cont'd)

"Rejuvenated" Tail-End Pd/Promoter Catalysts

- "Rejuvenated" by Insitu Steam/Air Burn, Followed by H_2 Reduction
- *Well-Executed Regeneration/Reduction Returns Catalyst to Essentially Same SOR Performance Achieved During Previous Cycle.*
- Conversely, A Regeneration/Reduction Poorly Executed Can Return to Service Catalyst that Performs Worse than When Taken Offline.

Catalyst Regeneration (Cont'd)

Comments on Insitu Regeneration:

- *Many Times Regeneration Step Is "Short-Changed".*
- This Could Mean Any or All of the Following:
  - Inadequate Facilities.
  - Inadequate Procedures.
  - Inadequate Training of Operators and Process Engineers.
  - Insufficient Operator Attention.

Ethylene Plant Operations
Acetylene Hydrogenation Tutorial

Lecture VI: Safety Considerations

2009 Spring National Meeting
AIChE Session 107 - April 29th

Ed Mohundro

Ethylene Plant Operations
Acetylene Hydrogenation Tutorial

AIChE Session 107 - April 29, 2009

This lecture has been prepared for presentation at the AIChE 2009 Spring National Meeting in Tampa, Florida.

Safety Considerations

- Following Not Intended to Provide Precise Guidance on Safety Facilities/Procedures Required for Acetylene Hydrogenation Units
  - Provides Only Overview of Safety Considerations
  - May Serve as Background Information for Succeeding Papers
  - Ethylene Producer Responsibile for Safety Facilities/Procedures

Safety Considerations

- Primary Safety Concern with Acetylene/C3 Hydrogenation is to Prevent Situations that get Reactor(s) into Runaway Exothermic Reactions
  - Possible to Reach Conditions Where:
    - Uncontrollable Reactions begin to Occur
    - These Reactions may Include: Hydrocracking, Polymerization, and Ethylene Decomposition
  - Could Lead to Excessive Temperatures/Pressures
  - Raises Possibility for Loss of Containment

Safety Considerations (Cont'd)

Front-End Acetylene Hydrogenation

- Safety Facilities/Procedures -- *Aimed at Preventing and/or Mitigating Against Consequences from Initiation of Ethylene Hydrogenation in Presence of High Concentrations of Hydrogen*
- Maintain Feed above Minimum Rate: Long Contact Times will Initiate Ethylene Hydrogenation
  - Handled by Having Reactors in a Recycle Loop
  - Catalysts Suppliers and Ethylene Contractors Provide Guidelines for Minimum Allowable Feed Rate

Safety Considerations (Cont'd)

Front-End Acetylene Hydrogenation (Cont'd)

- Install Multiple Thermocouples in Catalyst Beds
- Alarm these T/C's, along with Outlet T/C's, When Temperatures Exceed Upper Limit for Normal Operating Temperatures
- Alerts Operator to Take Corrective Action
  - Reducing Inlet Temperature
  - Tripping Reactor Preheat

Safety Considerations (Cont'd)

Front-End Acetylene Hydrogenation Cont'd)

- Safety Facilities/Procedures Should Handle Need to Isolate/Depressure, and Purge Reactor with an Inert Gas

Safety Considerations (Cont'd)

Tail-End Acetylene Hydrogenation

- Primary Consideration is Avoiding High Ratios of Hydrogen to Reactants
  - Normal Control Scheme Addresses by Maintaining Specified H_2/Feed Ratio for Each Bed
  - Abnormal Situations Could Lead to Excess Hydrogen Entering Catalyst Beds. For example:
    - Instrument failure
    - Operator Error
    - and/or Flawed Procedures for Handling Unsteady State Operations

Safety Considerations (Cont'd)

Tail-End Acetylene Hydrogenation (Cont'd)

Safety Facilities and Procedures May Include Following:

- Multiple Thermocouples Within Each Catalyst Bed
  - These T/C's, along with bed outlet T/C's provide high temperature alarms.
  - Lowest level alarm Alerts Operator of temperatures above normal operating range.
  - *Gives Operator Opportunity to take Corrective Action; e.g., Reducing Inlet Temperature.*

Safety Considerations (Cont'd)

Tail-End Acetylene Hydrogenation (Cont'd)

- If Temperatures Continue to Climb, then H_2 Valves Will Be Tripped
  - Prevent Additional H_2 Entering Reactor with High Temperatures at any Point(s) within the Catalyst Beds
- If Temperatures Continue to Increase, Reactor Will Be:
  - Isolated/Depressured, and
  - Purged/Cooled with an Inert Gas

Preliminary Findings on OSHA's Refinery NEP

James Lay, P.E.
Lisa A. Long (Presenter)
Michael Marshall, P.E. (Presenter)

U.S. DOL – OSHA
200 Constitution Ave, NW
Washington, DC 20210
202-693-2409
Long.lisa@dol.gov

Prepared for Presentation at
American Institute of Chemical Engineers
2009 Spring National Meeting
5th Global Congress on Process Safety
43rd Annual Loss Prevention Symposium
21st Ethylene Producers Conference
Tampa, Florida
April 26–30, 2009

UNPUBLISHED

Abstract
On June 7, 2007, U.S. DOL - OSHA (OSHA) initiated a Petroleum Refinery Process Safety Management National Emphasis Program (NEP) in response to a large number of fatal or catastrophic incidents in the petroleum refining industry. The NEP describes OSHA's intent to inspect all refineries (SIC 2911, NAICS 32411) in federal jurisdiction within 2 years, and provides a new protocol for conducting these inspections. OSHA state plan states were also encouraged to inspect refineries within their jurisdiction using the same protocol, and many are doing so.
This paper will briefly describe the inspection protocol, update OSHA's progress in completing these inspections, and review findings of the NEP and other PSM inspections conducted since 1992, including the elements of the PSM standards that were most frequently cited.

Background
Following a major incident at BP's Texas City Refinery in April 2005, OSHA compiled and evaluated data from PSM incidents and found that petroleum refining had more serious PSM incidents than any other industry segment. The data revealed that between 1992 and 2005, fatal (one or more employees killed) and catastrophic (three or more employees hospitalized) PSM incidents in all industries resulted in 138 fatalities and 553 injuries. During this period, 36 of the 240 total serious PSM incidents occurred in petroleum refineries, resulting in 52 employee deaths (38% of total) and 250 employee injuries (45%). As a result of the incident history in petroleum refining, OSHA developed a Petroleum Refinery National Emphasis Program (NEP) that was issued on June 7, 2007. The Refinery NEP outlines policies and procedures for OSHA Compliance Officers to use in conducting inspections of petroleum refineries, and mandates that OSHA inspect all petroleum refineries, excluding those in the Voluntary Protection Program (VPP), within federal jurisdiction over two years. Although VPP sites are excluded from NEP inspections, their PSM programs are evaluated as part of the VPP process. OSHA strongly encouraged State Plan States to adopt the NEP inspection strategy, and all states with refineries have either implemented the Refinery NEP or a similar program.
The Refinery NEP emphasizes *implementation* of PSM programs. Compliance officers use a list of static questions, which is published in the Refinery NEP, and a smaller list of dynamic questions, which is unpublished and periodically changed. The questions focus on issues identified as contributing to previous incidents and contain compliance guidance to assist Compliance Officers in evaluating PSM implementation in a unit selected by the inspection team. Contractors working in or adjacent to the selected unit are also inspected.

Inspection Numbers and Fines
Inspection data[1] demonstrates that OSHA has found a significant number of PSM violations in the refining industry. Between June 7, 2007 and February 12, 2009, OSHA has opened inspections of 42 refineries and 81 contractors. Citations have been issued in 30 of these refinery inspections; the remainder are ongoing. Table 1 illustrates the proposed penalties associated with the violations, broken down by classification.

Table 1: Refinery NEP Penalties (June 7, 2007 to February 12, 2009)

	Refineries		Contractors	
Violation	Number of	Total	Number of	Total

[1] Data presented is from inspections conducted by Federal OSHA. Data does not include inspections conducted by State-Plan States.

Classification	Violations	Penalties	Violations	Penalties
Willful	6	$379,000	0	$0
Repeat	17	$447,500	0	$0
Serious	462	$1,661,705	18	$39,225
Unclassified	9	$429,000	0	$0
Other	29	$31,450	13	$7,487
Total	**523**	**$2,948,655**	**31**	**$46,172**

On average, Refinery NEP inspections at refineries have resulted in 17.4 violations with penalties of $98,300. Refinery NEP inspections of contractors at refineries have resulted in an average of 0.4 citations with penalties of $658.

Violations broken down by Element and Sub-element of the PSM Standard

OSHA inspection data was analyzed to which element and sub-element of the PSM standard were most frequently violated. The following table shows the top 10 elements and sub-elements.

Table 2 – Refinery NEP Most Frequently Cited PSM Elements

Element	Description	Number of Violations
J	Mechanical Integrity (MI)	404
D	Process Safety Information (PSI)	270
F	Operating Procedures (OP)	224
E	Process Hazard Analysis (PHA)	195
M	Incident Investigation (II)	159
L	Management of Change (MOC)	94
O	Compliance Audits (CA)	78
H	Contractors	35
C	Employee Participation	16
G	Training	14

Table 3 - Refinery NEP Most Frequently Cited PSM Sub-elements

Sub-element	Description	Number of Violations
d(3)	PSI – information pertaining to the equipment in the process	246
j(4)	MI – inspection and Testing	198
f(1)	OP – develop and implement written operating procedures	183
e(3)	PHA – requirements of PHA	101
j(5)	MI – equipment deficiencies	95
j(2)	MI – written procedures	77
e(5)	PHA – system to promptly address recommendations	74
m(1)	II – investigate each incident and near miss	63
m(4)	II – prepare an investigation report	48

o(4)	OA – promptly determine and document response to findings	47

As an aid in understanding the nature of the violations, each of the top 10 most frequently cited Sub-elements will be discussed and examples of language from actual citations presented.

Discussion of Most Frequently Cited Sub-elements

d(3) PSI – information pertaining to the equipment in the process	246 Instances

Requirement:

1910.119 d(3) – Information pertaining to the equipment in the process.

(i) Information pertaining to equipment in the process shall include:

(A) Materials of construction;

(B) Piping and instrumentation diagrams (P&ID's);

(C) Electrical classification;

(D) Relief system design and design basis;

(E) Ventilation system design;

(F) Design codes and standards employed;

(G) Material and energy balances for processes built after May 26, 1992; and,

(H) Safety systems (e.g. interlocks, detection or suppression systems).

(ii) The employer shall document that equipment complies with recognized and generally accepted good engineering practices.

(iii) For existing equipment designed and constructed in accordance with codes, standards or practices no longer in general use, the employer shall determine and document that the equipment is designed, maintained, inspected, tested, and operating in a safe manner.

Discussion:

The most frequent instance of violation within the d(3) sub-element was failure to have accurate P&ID's in the PSI, followed by a failure to document that equipment follows RAGAGEP, and failure to document information for relief system basis and design.

When P&ID's contain errors, they typically contain more than one error, which accounts for the large number of instances of P&ID violations. These P&ID errors were for various types of equipment and controls.

There have been numerous citations for failure to follow RAGAGEP. The citation history shows that this violation has been found in many areas including facility siting, relief valve design, safety instrumented systems, fireproofing, control of intervening valves, and electrical classification.

Instances of violations for failure to document relief system design and design basis included many instances of not completely documenting design codes and basis.

Some examples of paraphrased citation language are given below.

Example Citation Language:

- P&ID's are inaccurate: temperature indicator in field is not shown on P&ID, valves on the P&ID were not found in the field, valves in were not on the P&ID, Car sealed valves not shown as such on the P&ID, wrong valve sizes on P&ID.
- The employer failed to ensure that recognized and generally accepted good engineering practices (RAGAGEP) were followed with respect to the set pressures of relief devices. The relief devices on an exchanger were set at 540 and 600 psig while the MAWP for the exchanger was 480 psig.
- The employer did not comply with RAGAGEP when it failed to protect employees inside structures exposed to explosion, fire, and high pressure hazards as a result of a release of highly

flammable materials. A facility siting study found deficiencies that needed correction in the Auxiliary Control Room, Central Control Room, and Laboratory.

- The employer did not document that it complied with RAGAGEP such as:
  - API 753 Section 3 when it failed to protect employees inside a wooden trailer being used as an operator shelter that was placed within the 1 psi overpressure zone;
  - ASME Boiler and Pressure Vessel Code (BPVC), Division 1 Section VIII, UG-135(d) when it failed to provide adequate controls to ensure that the isolation valve for a safety valve was in the open position;
  - API 521 where a number of relief valves were evaluated to be undersized in regards to the external fire case scenario.
  - ISA-S84.01, for calibrating, inspecting, testing, and maintaining Safety Instrumentation Systems (SIS) for emergency shutdown and interlocks.
  - NFPA 496-2003 Section 7.4.7 in that failure of the positive pressure air system inside the control building would not activate an alarm at a constantly attended location.
  - API 752, a knock-out drum was evaluated to not be able to provide adequate retention time to prevent liquid carryover to the flare.
  - NFPA 101 Section 7.8.1 when Illumination of the means of egress was not adequately maintained throughout the unit.
- PSI did not include the relief system design and design basis for: relief valves, removal of fractionator PSV, crude unit blowdown system, coker unit blowdown system, cat unit blowdown system.

j(4) MI – inspection and Testing	198 Instances

Requirement:

1910.119 j(4) – Inspection and testing.

(i) Inspection and tests shall be performed on process equipment.

(ii) Inspection and testing procedures shall follow recognized and generally accepted good engineering practices.

(iii) The frequency of inspections and tests of process equipment shall be consistent with applicable manufacturers' recommendations and good engineering practices, and more frequently if determined to be necessary by prior operating experience.

(iv) The employer shall document each inspection and test that has been performed on process equipment. The documentation shall identify the date of the inspection or test, the serial number or other identifier of the equipment on which the inspection or test was performed, a description of the inspection or test performed, and the results of the inspection or test.

Discussion:

Within the j(4) sub-element, failure to perform inspection and testing was most frequently cited. This was followed by failure to follow RAGAGEP for inspection and testing, inappropriate frequency of inspection and testing, and, finally, failure to properly document inspection and testing. Failures in the specific requirements for the inspection and testing occurred for many different types of equipment and controls. While there were a variety of aspects of inspection and testing failures, there were numerous violations regarding thickness measurements.

Example Citation Language:

- The employer does not ensure that thickness monitoring for process equipment including: a piping circuit, is conducted in accordance with the MI inspection program.

- Inspections and testing procedures did not follow RAGAGEP such as API 510 Section 6.6.2.3 when a bench tested safety valve failed to pop at its set pressure due to fouling. The valve was repaired and returned to service with the same 6 year inspection frequency. RAGAGEP requires that the inspection interval shall be reduced when a pressure relieving device is heavily fouled.
- The frequency of inspections and tests of process equipment was not conducted as specified by the employer's master plan to maintain mechanical integrity; such as, but not limited to; a cathodic protection system, and process analyzers.
- The employer did not follow RAGAGEP, such as API 570 when it failed to resolve anomalous data for pressure vessels and piping, such as increasing thickness measurements over previous readings.
- A critical piping circuit did not have any inspection data established since the line was installed in 1972
- The employer did not ensure that thickness readings were documented for an FCCU.

f(1) OP – develop and implement written operating procedures	183 Instances

Requirement:

1910.119 f(1) – The employer shall develop and implement written operating procedures that provide clear instructions for safely conducting activities involved in each covered process consistent with the process safety information and shall address at least the following elements.

(i) Steps for each operating phase:

(A) Initial startup;
(B) Normal operations;
(C) Temporary operations;
(D) Emergency shutdown including the conditions under which emergency shutdown is required, and the assignment of shutdown responsibility to qualified operators to ensure that emergency shutdown is executed in a safe and timely manner.
(E) Emergency Operations;
(F) Normal shutdown; and,
(G) Startup following a turnaround, or after an emergency shutdown.

(ii) Operation limits:

(A) Consequences of deviation; and
(B) Steps required to correct or avoid deviation.

(iii) Safety and health considerations:

(A) Properties of, and hazards presented by, the chemicals used in the process;
(B) Precautions necessary to prevent exposure, including engineering controls, administrative controls, and personal protective equipment;
(C) Control measures to be taken if physical contact or airborne exposure occurs;
(D) Quality control for raw materials and control of hazardous chemical inventory levels, and,
(E) Any special or unique hazards.

Discussion:

The most frequent citations for developing and written operating procedures involved developing procedures for emergency shutdown. Other commonly cited violations included issues related to safe operating limits, failure to develop operating procedures for all operations, and failure to certify operating procedures.

Example Citation Language:

- The employer did not ensure that the written operation procedures addressed the conditions under which emergency shutdown is required, and the assignment of shutdown responsibilities to qualified operators to ensure that emergency shutdown is executed in a safe and timely manner.
- The employer did not develop and implement written procedures that provide clear instruction consistent with process safety information that address operating limits. The procedure for variable limits was not consistent with information lists on the control board / DCS system. The consequences of deviation list of operating variables did not address the safety and health consequences.
- The employer did not develop or implement written procedures that provided clear instructions for safely conducting activities in the covered process including but not limited to: the use of chains or car seals to ensure that block valves on pressure relief lines were kept in the open position, the Crude Unit blowdown system in all operating procedures of that unit, the Coker Unit blowdown system in all operating procedures of that unit, the Cat Unit blowdown system in all operating procedures of that unit.
- Written operating procedures for Main Flare operations did not contain steps for transition from normal operations to emergency operations when the Main Flare loses flame.
- Procedures were not developed for switching lube oil pumps on a compressor, resulting in the shutdown of the compressor and causing a mechanical integrity incident.
- An emergency shutdown procedure for loss of cooling water was not reviewed to show operating practices were current and that the procedures still included a column which was removed from service over one year ago.

e(3) PHA – requirements of PHA	101 Instances

Requirement:

1910.119 e(3) – The process hazard analysis shall address:

(i) The hazards of the process;

(ii) The identification of any previous incident which had a likely potential for catastrophic consequences in the workplace;

(iii) Engineering and administrative controls applicable to the hazards and their interrelationships such as appropriate application of detection methodologies to provide early warning of releases. (Acceptable detection methods might include process monitoring and control instrumentation with alarms, and detection hardware such as hydrocarbon sensors.);

(iv) Consequences of failure of engineering and administrative controls;

(v) Facility siting;

(vi) Human factors; and

(vii) A qualitative evaluation of a range of the possible safety and health effects of failure of controls on employees in the workplace.

Discussion:

Over half of the citations issued for violations of e(3) have been related to failures to address facility siting or human factors. A substantial number of violations have also been issued regarding failure to address the consequences of failure of the engineering and administrative controls, and failure to identify previous incidents which could have had catastrophic consequences.

Example Citation Language:

- The process hazard analysis did not address facility siting. The PHA did not address the location of a piperack containing hydrogen, benzene, naptha, and butane found to be within 22 ft of the operator shelter. The PHA did not address the discharge of flammable gasses including but not limited to butane, propane, and naptha from the stack of an atmospheric blowdown drum which could impact a furnace and an operator shelter located 43 ft and 91 ft away, respectively.
- The employer did not ensure that the PHA addressed human factors such as: the number of alarms and warning that board operators can handle in an emergency or the hazards associated with process operators isolating critical valves in an emergency.
- The employer did not identify, evaluate, address the consequences of all scenarios that might result in the discharge of hot, heavier-than-air, or liquid hydrocarbons when it conducted the process hazard analysis including but not limited to: crude unit blowdown system, coker unit blowdown system, cat unit blowdown system, or closed valve on a pressure relief line.
- The PHA did not address the lack of fireproofing on pipe rack supports located in close proximity to process equipment such as but not limited to pumps.
- The employer's PHA did not control hazards related to heat tracing and insulation of Condenser outlet piping to reduce the likelihood of low spots freezing, which could result in overpressure of a Solvent Regenerator.
- The PHA revalidation of the Topper unit failed to evaluate and control the hazards identified on the human factor checklist.
- The PHA failed to consider unmarked eqiupment on the unit to include exchangers and safety relief valves which could lead to confusion by operators and/or contractors.
- The process hazard analysis did not identify previous incidents which had a likely potential for catastrophic consequences in the workplace. Sewers beneath a UPS battery room having an explosive atmosphere due to a water seal failure in the sewer system.

j(5) MI – equipment deficiencies	95 instances

Requirement:

1910.119 j(5) – Equipment deficiencies. The employer shall correct deficiencies in equipment that are outside acceptable limits (defined by the process safety information in paragraph (d) of this section) before further use or in a safe and timely manner when necessary means are taken to assure safe operation.

Discussion:

The frequency of violation in the j(5) sub-element is particularly noteworthy because there are no further divisions of this sub-element. All of the violations in this sub-element concerned running equipment outside of acceptable limits. The instances of these violations were for various types of equipment.

Example Citation Language:

- The employer did not correct deficiencies in equipment that were outside acceptable limits, including piping thickness reading below the minimum thickness (3 instances).
- The employer did not correct deficiencies in equipment to ensure safety in operation and maintenance before further use or in a safe and timely manner. Three pressurized enclosures for electrical equipment did not have an alarm or did not have an alarm monitored at a constantly attended location.
- Broken bolts in a flange connection on the splitter reboiler exchanger were not repaired.
- The employer did not correct deficiencies in the equipment that were outside acceptable limits including fireproofing insulation removed from conduits powering EIVs was never replaced, and continued to fail critical instrument inspections.

- A relief valve was evaluated to be undersized for protection against the loss of cooling and power failure relief scenarios
- Gage glasses were not maintained in a clean/readable condition.
- Toxic and flammable gas detectors located in duct work to the control room for H2S, and LEL were not in working order.

j(2) MI – written Procedures	77 instances

Requirement:

1910.119 j(2) - Written procedures. The employer shall establish and implement written procedures to maintain the on-going integrity of the process equipment.

Discussion:

From the data, it appears that most facilities did have procedures for most aspects of mechanical integrity. However, there were many instances of facilities not having procedures to handle unique situations such as on equipment that is insulated, has integrally bonded liners or injection points. There were also a few instances of not having procedures to deal with deficiencies or anomalous data.

Example Citation Language:

- The employer did not develop an inspection procedure to inspect for "corrosion-under-insulation" on pressure vessels.
- The employer did not establish and implement written procedures to maintain the on-going mechanical integrity of equipment, including inspections for non-metallic linings of pressure vessels.
- The employer did not write procedures to maintain the on-going mechanical integrity of process equipment in the refinery such as, but not limited to procedures for anomalous data related to thickness growths recorded in piping with criteria as to what would be considered an increase from a previous thickness measurement.
- The employer did not ensure that mechanical integrity procedures for relief devices were implemented at the facility. Multiple instances of relief valves were found that were past their individual inspection intervals for inspection and testing in the Hydrocracker and Alkylation Units.
- MI procedures did not include Information on alterations of vessels including welding qualifications and certifications for adding a nozzle to a heat exchanger
- MI procedures did not include inspection frequency for pressure vessels
- MI program procedures did not address how the employer determines and establishes a representative number of thickness measurements locations (TML) to satisfy requirements for conducting adequate internal and on-steam inspections.

e(5) PHA – system to promptly address recommendations	74 instances

Requirement:

1910.119 e(5) – The employer shall establish a system to promptly address the team's findings and recommendations; assure that the recommendations are resolved in a timely manner and that the resolution is documented; document what actions are to be taken; complete actions as soon as possible; develop a written schedule of when these actions are to be completed; communicate the actions to operation, maintenance and other employees whose work assignments are in the process and who may be affected by the recommendations or actions.

Discussion:

The majority of violations of e(5) involved instances of failure to resolve recommendation either at all or in a timely manner. A large number of these PHA recommendations could be considered to be "low-hanging–fruit" for employers, and often are relatively easy to resolve, e.g. car sealing isolation valves of relief devices. However, the violation history suggests that many recommendations that do not necessarily require large capital outlays have not been resolved.

Example Citation Language:

- The employer did not establish a system to promptly address the process hazard analysis team's findings and recommendations, as shown, for example, in an audit conducted in 2002 in which twenty three findings from a 1996 PHA were identified as not resolved.
- The most recent PHA listed items that have not been completed since the previous PHA, possibly resulting in loss of containment or the formation of a vapor cloud resulting in an explosion. These included: adding a charge pump suction temperature to the DCS with alarm and local indication to avoid cavitation and pump failure, installing an independent low level alarm on a storage tank to avoid hot vapors reaching a storage tank, converting a local temperature indicator on a propane vaporizer to local indication plus board indication with alarm and control, and addition to the critical process variable list to avoid liquid propane feed to heater burners.
- The employer has not resolved an action item from the 1997 to 2005 facility siting analysis in that occupied facilities, such as but not limited to the unit control room, have not been addressed.
- The employer did not assure that recommendations identified in PHAs for the refinery were resolved in a timely manner and the employer did not document the actions to be taken for resolution of PHA recommendations. Subsequent requirements, such as developing a written schedule for completion of actions and communication of the action to affected employees, were also not completed as required.
- The employer had not completed or resolved actions recommended by the its process hazards analysis teams in an expeditious manner including the recommendation related to the review of the relief valve sizing basis for the Propane Treater to ensure protection during a chemical reaction with HP acid.
- Did not resolve the PHA recommendation on adding a temperature sensor on the external surface of the Iso Treater.
- Actions items developed from a 2002 PHA revalidation were not completed in a timely manner in that an action item to re-rate and/or protect a steam coil had not been completed as of November, 2007.

m(1) - Investigate each incident and near miss	63 instances

Requirement:

1910.119 m(1) – The employer shall investigate each incident which resulted in , or could reasonably have resulted in a catastrophic release of highly hazardous chemical in the workplace.

Discussion:

Citations under this sub-element all addressed failure to investigate potentially catastrophic incidents. Multiple instances were identified in many inspections, with high frequencies for pump problems, process leaks and releases, upsets lifting relief valves, and fires.

Example Citation Language:

- Employers did not ensure that incident investigations were conducted for incidents that could reasonable have resulted in a catastrophic release of highly hazardous chemicals in the workplace. Examples include, but are not limited to:
  - Pump fires
  - Pump seal fires (multiple)
  - Packing blown out of pump
  - Pump tripped
  - Hydrocarbon liquid and vapor release (multiple)
  - Lost level
  - Injection line rupture
  - Pipe rack butane release
  - OVHD receiver leak
  - Acid leaks (multiple)
  - Exchanger leaks
  - Heater leaks
  - Over pressure and release to flare
  - Product leaking from threaded area on level float chamber
  - Corrosion
  - Sulfur fire
  - Slide flange fire
  - FCCU fire
- A near-miss incident was not investigated that could reasonably have resulted in a catastrophic release of HHC when a pressure relief valve and the pipe around it developed a buildup of ice due to propane leaking out of the pressure relief valve rendering the relief valve inoperable when the process was in operation.
- An incident investigation was not conducted when a process tower experienced a rapid pressure rise causing the pressure to be unstable which resulted in propane being vented to the flare for several hours.

m(4) – prepare an investigation report	48 instances

Requirement:

1910.119 m(4) – A report shall be prepared at the conclusion of the investigation which includes at a minimum:

(i) Date of Incident;
(ii) Date investigation began;
(iii) A description of the incident;
(iv) The factors that contributed to the incident; and
(v) Any recommendations resulting from the investigation.

Discussion:

Citations have been issued for all aspects of this sub-element, including failure to prepare any incident investigation reports at all. The most common missing component was failure to identify factors that contributed to an incident. Additionally, when factors were listed, they often had a marginal, if any, relationship to the actual causal factors of the incident.

Example Citation Language:

- Reports prepared at the conclusion of investigations of incidents did not include the date the investigation began (6 instances).
- No determinations of how cross-contamination occurred or how it was corrected.

- Root causes/conditions/factors that resulted in 2 crude heater tube leak fires were not identified in the incident investigation report.
- Reports for incidents did not include the factors that contributed to the incident:
    - Incident report addressing smoke in a substation.
    - Incident report addressing operations exceeding design temperatures on a reactor and associated piping and equipment.
    - Incident report addressing fan shaft breakage.
    - Incident report addressing accidental shutdown of extractor reflux pump.
- The report prepared at the conclusion of an incident investigation did not include the factors contributing to the incident, including, but not limited to, that piping inspections, evaluations, and recommendations were not adequate or effective in detecting internal corrosion on a piping weld.
- The incident investigation report did not include the factors contributing to the incident, including, but not limited to, that the quality control procedure was not adequate or effective in assuring that contractor welding was acceptable.
- Recommendations were not established in a crude heater tube leak fire investigation report.

o(4) - responses to findings	47 instances

<u>*Requirement:*</u>
1910.119 o(4) – The employer shall promptly determine and document an appropriate response to each of the findings of the compliance audit, and document that deficiencies have been corrected.
<u>*Discussion:*</u>
These citations all involved failure to address deficiencies identified in compliance audits, including failure to address management system shortcomings to prevent recurrence of the problems identified.

<u>*Example Citation Language:*</u>

- Pre-start up safety reviews/management of change tracking and approval not completed as identified in the last two compliance audits.
- Incident investigations not promptly addressed as identified in the last two compliance audits.
- Compliance audit finding that operating procedures were not being updated were not adequately addressed.
- More than three years had elapsed since the previous compliance audit and audit findings from previous audits were not completed in a timely manner.
- Appropriate responses to compliance audit findings were not determined and documented:
    - Ventilation system designs for control rooms not available
    - Process safety system information not available
    - Equipment test and inspection frequencies were not consistent with applicable manufacturers' recommendations and good engineering practices.
- Incident investigations were not promptly addressed as identified in the last two compliance audits.
- The employer did not ensure that PSM compliance audit findings were resolved prior to closing the action item as completed. Multiple relief devices were found to be past due for inspection/testing during the 2003 and 2006 compliance audits, however, the action items were closed out without ensuring that all relief devices were properly inspected, or that the management system was adequate to ensure that relief devices would not go past their inspection due date again.

- The employer did not specifically document correction of the deficiencies in the compliance audit related to the management system for the employer's Risk Based Mechanical Integrity (RBMI) program.

Comparison of Refinery NEP Data with prior PSM Inspection Data

The Refinery NEP has been OSHA most significant PSM enforcement effort since the promulgation of the PSM standard in June 1992. Comparison of the Refinery NEP Inspection data with PSM inspection data prior to June 1997 shows several notable differences:

- **The Majority of Refinery NEP inspections have been done as planned or programmed inspections as opposed to unprogrammed inspections resulting from accidents, complaints, or referrals.** Approximately 73% of the inspections with PSM citations conducted between 1992 and June 2007 (prior to the Refinery NEP) resulted from accidents, complaints, or referrals .

- **Refinery NEP inspections are resulting in more violations and higher penalties.** Refinery NEP inspections have resulted in an average of 17.4 violations with penalties of $98,300. Prior PSM inspections (where citations were issued) in all SICs have resulted in an average of 3.6 violations with penalties of $22,865.

- **The inspection strategy used in the Refinery NEP has resulted in more violations than the PSM PQV inspection strategy.** The Refinery NEP focuses on issues identified as contributing to previous incidents and contains Inspection Priority Items (IPI) and compliance guidance to assist Compliance Officers in evaluating PSM implementation. In contrast, the PSM PQV used an audit checklist which mirrored the performance requirements of the PSM standard. In 1993, OSHA conducted 39 planned or programmed inspection under PSM Program-Quality-Verification (PQV). These inspections, conducted in 8 identified high risk SICs, resulted in an average of 2.3 violations with penalties of $2,953. Assuming equivalent experience levels and training for the OSHA compliance personnel conducting these inspections, it appears that the Refinery NEP strategy (17.4 violations per inspection) is a better tool for identifying hazards and violations than the PSM PQV strategy (2.3 violations per inspection).

- **Average PSM violations and penalties have been higher in the refining industry than in the regulated community as a whole.** Inspection data shows that PSM inspections prior to June 2007 in the petroleum refining industry (SIC 2911) resulted in an average of 4.8 violations with penalties of $91,570 compared with inspections in all SIC codes that resulted in an average of 3.6 violations with penalties of $22,865.

- **There are some similarities in the most frequently cited violation; however, the NEP's most frequently cited sub-elements show more violations related to process safety information, incident investigation, and compliance audits, and fewer related to more basic elements such as training and employee participation.** The table below shows the most frequent violations cited during the key periods compared: Refinery NEP Inspections, Inspections from 1992 until the start of the Refinery NEP for the entire covered community (for inspections where a PSM citation was issued), and Inspection from 1992 until the start of the Refinery NEP for the petroleum refining industry (for inspections where a PSM citation was issued).

- **Data for the most frequently cited sub-elements reflects OSHAs focus on PSI and RAGAGEP in the inspection priority items and training of compliance officers.** The most frequently cited sub-element, (d)(3) and more specifically (d)(3)(ii)[2] is subject to numerous IPIs in the Refinery NEP document and is a primary focus for OSHA PSM Refinery NEP training for compliance officers.

- **Focus on Incident investigation during Refinery NEP compliance officer training appears to be beneficial.** OSHA training covered requirements for incident investigation with specific emphasis on determining whether employers were conducting "near-miss" investigations (1910.119 m(1)), and adequately reporting of the factors that contributed to the incident (1910.119(m)(4)(iv)). Data indicates that element (m), Incident Investigation, is the 5th most frequently cited PSM element during the Refinery NEP where it was previously the 9th most cited PSM element. More specifically , (m)(1) and (m)(4) were the 8th and 9th most frequently cited PSM sub-element in the Refinery NEP whereas, neither of these two sub-elements were in the top 10 most frequent during the other two study periods.

- **In general, Refinery employers are not resolving PHA and audit findings and recommendations at a rate expected of large, sophisticated employers.** The Refinery NEP data shows that not resolving PHA and audit findings and recommendations were #7 and #10 most frequently cited sub-elements, respectively. Given that Refineries are typically large employers with both resources for, and knowledge of, PSM application, it is surprising that Refineries would have such a large number of violations related to addressing hazards identified by their PHA and audit teams.

Table 4 - Comparison of PSM Violations

Refinery NEP Violations		*2911 '92 to NEP Violations*		*PSM '92 to NEP Violations*	
Elements	**#**	**Elements**	**#**	**Elements**	**#**
J – MI	404	J - MI	263	F - OP	1189
D – PSI	270	F – OP	177	J – MI	1152
F - OP	224	L – MOC	138	E – PHA	991
E – PHA	195	E – PHA	110	D – PSI	744
M - II	159	D – PSI	51	G – Training	551
L – MOC	94	G – Training	46	L – MOC	518
O – CA	78	H – Contractors	43	H – Contractors	448
H - Contractors	35	M - II	42	C – Employee Participation	386
C – Employee Participation	16	C – Employee Participation	17	M – II	228
G - Training	14	N - ER	13	N – Emergency Planning and Response (ER)	177
Sub-elements		**Sub-elements**		**Sub-elements**	

---

[2] 1910.119(d)(3)(ii) - The employer shall document that equipment complies with recognized and generally accepted good engineering practices.

d(3) PSI – information pertaining to the equipment in the process	246	j(5) MI-equipment deficiencies	113	j(2) MI – written procedures	369
j(4) MI – inspection and Testing	198	l(1) MOC - establish and implement procedures for MOC	57	e(1) II – perform an initial PHA	347
f(1) OP – develop and implement written operating procedures	183	j(4) OP – develop and implement written operating procedures	46	l(1) MOC - establish and implement procedures for MOC	287
e(3) PHA – requirements of PHA	101	l(3) MOC – affected employees must be notified and trained before changes	41	c(1) EP – develop a written plan for employee participation	243
j(5) MI-equipment deficiencies	95	j(2) MI – written Procedures	34	j(5) MI-equipment deficiencies	214
j(2) MI – written procedures	77	e(5) PHA – requirements of PHA	30	g(1)Training – train employees involved in operating a process	205
e(5) PHA – requirements of PHA findings and recommendations	74	f(4) OP – develop and implement safe work practices	30	f(1) OP – develop and implement written operating procedures	202
m(1) II – investigate each incident and near miss	63	f(3) OP – review procedures as often as necessary and certify procedures annually	17	j(4) OP – develop and implement written operating procedures	189
m(4) II – prepare an investigation report	48	g(2) Training – refresher training at least every 3 years with input from employees	17	e(5) PHA – requirements of PHA	188
o(4) OA – promptly determine and document response to findings	47	j(3) MI – Training for process maintenance activities	17	N ER – establish and implement emergency action plan including handling small releases	177

Concluding Remarks

OSHA will continue conducting Refinery NEP inspections until all refineries within federal jurisdiction are complete. It is expected that fewer violations will be found in later inspections as refineries learn from others and improve their PSM program. In 2008, OSHA announced that it is working on a new

National Emphasis Program that will target other PSM covered facilities, such as chemical plants, and use a strategy similar to the Refinery NEP although abbreviated. OSHA will continue to share findings from both of these programs with stakeholders in an effort to continuously improve PSM systems and reduce the number of catastrophic incidents.

References

(1) Occupational Safety and Health Administration, *Directive Number: CPL 03-00-004, Petroleum Refinery National Emphasis Program,* June 7, 2007, http://www.osha.gov/pls/oshaweb/owadisp.show_document?p_table=DIRECTIVES&p_id=3589, Accessed January 2009.

LOSS OF DCS OPERATOR INTERFACE DURING ELECTRICAL DISTURBANCE

Sandra Koblich (Author)
Senior Chemical Engineer
Eastman Chemical Company, Longview, TX
G. Michael McLeod (Co-author)
Assistant Superintendant, HCC-3 Operations
Eastman Chemical Company, Longview, TX

Abstract: During Hurricane Ike in September, 2009, two of Eastman's cracking plants experienced an electrical supply disturbance which caused loss of the cracked gas compressor turbine governors, loss of a refrigeration compressor, loss of the DCS process interface, and loss of several other PLC and process control computers. Power was restored within minutes following the electrical disturbance, and all equipment was brought back on-line in the next several hours.

The root cause of the incident was a UPS failure at one of the cracking plants due to inadequate preventative maintenance practices.

Major corrective actions include:

- Evaluate the UPS source for all critical shutdown items in both cracking plants. Identify any items for which the UPS source is supplied by the alternate cracking plant. Change the UPS source to the same cracking plant's UPS.
- Document the existing preventative maintenance program and evaluate improvements.
- Evaluate options for power supply redundancy for the PLC's, such that one UPS failure does not result in multiple PLC failures.
- Improve DCS operator interface screens for UPS configuration and update operator training to describe actual UPS sequential operations and failure modes. Train operators on appropriate responses to UPS alarms.

Sequence of Events

On September 13, 2008, Hurricane Ike moved north through East Texas, with heavy rain and periods of 70 mph winds. Just after 1 pm, two of Eastman's cracking plants began recording numerous UPS alarms. The alarms indicated power was alternating between primary and alternate sources in one plant, while the other plant's UPS activated the back-up battery. Around 4:35 pm, all UPS alarms cleared, and the UPS appeared to be functioning normally. Just after 5 pm, the cracked gas compressors tripped in both plants. Within seconds, the PLC's for the Acetylene Recovery Unit's (ARU's) for both plants lost power. The ARU reactors were bypassed and depressured. At the same time, the DCS computer monitors for one of the cracking plants lost power resulting in the loss of the control interface for that plant. While the power disruption lasted only seconds, the DCS computer system had to be rebooted at one plant, both plants had to restart cracked gas compressors, the PLC's for ARU's had to be restarted and reprogrammed, and the ARU reactors had to be brought back on-line.

At approximately 5:30 pm, both cracked gas compressors were successfully restarted. Approximately ten minutes later, one of the plant's refrigeration compressors tripped, creating concern there were additional electrical disturbances. It was later determined the compressor trip was related to high liquid levels in the suction drums and not the result of an electrical disturbance. By 7 pm, both plant's ARU's were restarted, following restoration of the PLC's. A few hours later the refrigeration compressor was restarted.

Root Cause

The primary cause of the electrical failures at both cracking plants was a disruption of the site power distribution system that precipitated a UPS failure in one of the cracking plants. This failure was determined to be the result of inadequate preventative maintenance practices. The failure was compounded by a lack of understanding of the UPS configuration by operating personnel, cross-plant UPS configuration of PLC's, and a design flaw with the PLC's allowing the programming to be lost during a loss of power.

UPS Back-up Batteries

On June 24, 2008, the UPS systems for both cracking plants were inspected. Each UPS system consists of 18 batteries. At the time of inspection, one battery was replaced in one of the cracking plants. However, 7 other batteries needed replacing: 3 in one plant and 4 in the other plant. These batteries were ordered, but were not replaced until September 15, 2008, after the incident.

UPS Configuration

A diagram of the UPS configuration is shown as Figure 1.

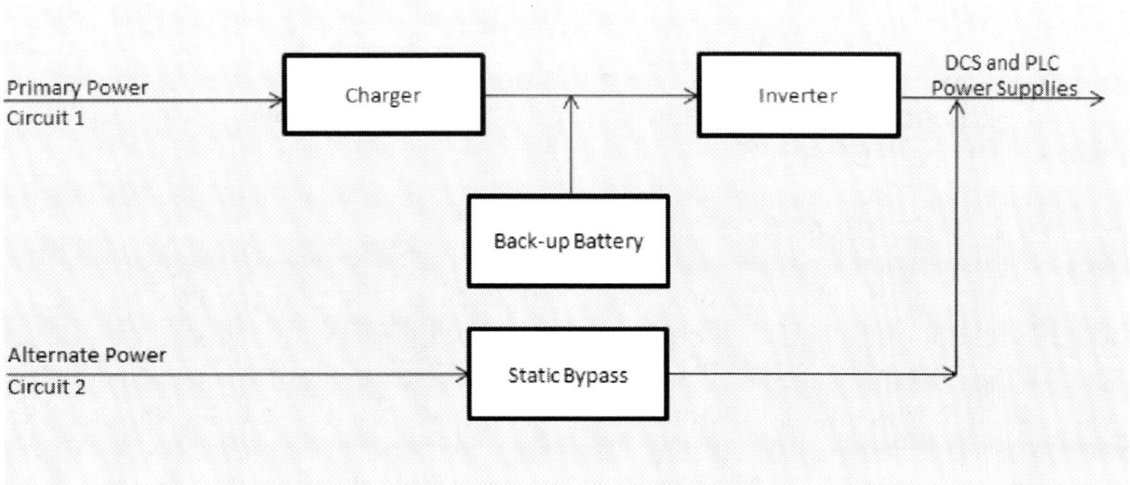

Figure 1 – UPS Configuration for Cracking Plants

The hierarchy of the UPS is as follows:

1. Under normal operation, the UPS primary power supply is routed through the charger and inverter.
2. Should the primary power supply fail, the back-up battery system supplies power.
3. If the back-up battery system fails, then the system utilizes the static bypass and employs alternate power.
4. The system will return to primary power when it becomes stable.

At just after 1 PM on September 13, one of the cracking plant's UPS lost primary power. The system attempted to revert to the back-up battery system; however, the back-up battery system was not functional. The UPS then accessed alternate power. Once primary power became stable, the UPS returned to primary power. This sequence of events repeated itself several times throughout the afternoon during the storm, as recorded by the plant's computer alarm log. Just after 5 PM, this sequence occurred again. However, this time the alternate power also failed. Without primary power and back-up battery, the plant's UPS failed. Shortly after the plant equipment lost power and shutdown, primary power was again restored.

A review of the DCS alarm screens for the UPS revealed that there is no indication alarm for a back-up battery failure. Thus, in the hours before the UPS failure, it was not understood by operating personnel that the system had reverted to alternate power due to a failed back-up battery. In fact, the general understanding has been that if primary and alternate power were to fail, the back-up battery system would be accessed to allow about 30 minutes of operation for a controlled shutdown. However, as detailed above, once alternate power is energized, the back-up battery has already failed.

While the UPS failure resulted in a loss of DCS interface (the consoles had to be rebooted manually after power was restored), control power was restored to the I/O gear when the primary power was restored. The robustness of the DCS system design allowed both crackers to be operated from the remaining DCS consoles without incident until the failed consoles could be rebooted.

Cross-Plant UPS Configuration

As expected, when one cracking plant's UPS failed, the equipment supported by that plant's UPS lost power. This included DCS computer interface (monitors), the cracked gas compressor governor, and the PLC for the ARU. However, the isolated electrical failures at the second cracking plant could not initially be explained; the second plant also lost power for the cracked gas compressor governor and the PLC for the ARU. Further investigation revealed that the first cracking plant provided the UPS for the failed equipment at the second cracking plant. The UPS systems for both plants were not independent, as originally intended.

Initialization Failures of PLC's

Following the shutdown of the cracked gas compressors in both plants the PLC's controlling the ARU's bypassed and depressured the reactors. When the power supplying the PLC's was lost the PLC's design allowed the programming to be erased. When power was restored and the PLC's programming was restored an error was discovered in the programming. This error allowed the PLC program to initialize without verifying the actual unit conditions. By initializing without verifying the status, the DCS system and the PLC indicated different unit status, one indicated "NORMAL" the other indicated "TRIPPED".

Corrective Actions

Corrective actions were taken for each of the causes identified above. The areas of focus of this effort included plant UPS independence, improvement of the preventative maintenance program for UPS systems, PLC programming modifications, and operator training related to UPS configuration.

Plant UPS Independence

Electrical failures would not have occurred at both cracking plants had the two plant UPS systems been completely independent. All critical shutdown items for each plant were investigated, and the source of UPS power identified. A summary was created of all the equipment which receives UPS from the alternate cracking plant. A project was then initiated to move the UPS source to the same cracking plant's UPS.

UPS Preventative Maintenance

The existing preventative maintenance (PM) program has been documented and improvements implemented.

PLC Programming

The PLC programs for the ARU's and cracked gas compressors have been modified to initialize in the "tripped" state. This will also force the DCS to go to the "tripped" state if it is not already. In addition, options for power supply redundancy for the PLC's will be evaluated, such that one UPS failure does not result in multiple PLC failures.

Operator Training

The DCS operator interface screens will be modified to better describe the existing UPS configuration. Alarm points will be added to specifically identify when the back-up battery is activated, and a 30 minute timer will be triggered. Operator training has been updated to describe actual UPS sequential operations and failure modes. Finally, operating personnel will be trained on the appropriate responses to UPS alarms. This is critical since plant staffing outside of day shift does not include PLC and DCS system experts.

Killing Rattlesnakes Before They Bite You!

John B. Prows

Vice President, Manufacturing Excellence

Performance Products Division,

Huntsman Petrochemical

Robert Simon

A3 Olefins Operations Manager,

Port Neches Performance Products Site,

Huntsman Petrochemical

Abstract:

Do you know what lies underneath your insulation? Is the pipe, exchanger or vessel under your insulation in good shape or is it is corroded and/or stress cracked? Do your piping systems and equipment have defects ("rattlesnakes") that you can't see that are ready to bite you at any time?

We learned the answer to these questions the hard way. In April 2006 a jet fire broke out on our LOU (Light Olefins Unit) in Port Arthur, Texas when an eight inch diameter line containing cracked gas suddenly ruptured. Within minutes the entire LOU Cold Train was engulfed in a massive jet fire that totally destroyed the back end of this large scale Olefins plant. Four people working outside near the pipe rupture area were able to outrun the fire. Miraculously no one was injured but the plant reconstruction took 14 months and the rebuild cost exceeded $300 million. Want to know what we learned? We will share with you our views on how to find rattlesnakes and kill them before they bite you!

Rattlesnakes (defects & corrosion) may by lurking dormant underneath your insulation. Once they bite you things can quickly escalate beyond your control.

Huntsman Port Arthur Base Chemical
Light Olefins Unit Fire
April 29, 2006

LOU, Port Arthur, Texas

LOU Fire in Progress, April 29, 2006

Huntsman Port Arthur Base Chemical
Light Olefins Unit Fire
April 29, 2006

Executive Summary

On Saturday, April 29, 2006 at 7:28 a.m., there was a sudden release of hydrocarbons followed by a "blow torch like" fire in the ethylene recovery area of the Huntsman Port Arthur Base Chemicals (PABC) Light Olefins Unit (LOU) in Port Arthur, Texas. There were no fatalities or serious injuries from the fire. There was one case of heat exhaustion associated with fighting the fire.

Two team behaviors and one "act of God" were instrumental in reducing the potential for serious injury during this incident:

- The LOU Operations Team strictly followed procedures for denying work permits and access during unit upsets. The ethylene recovery area was in an upset condition before the "blow torch like" fire started.

- The Huntsman Emergency Response and Sabine Area Mutual Aid teams quickly responded to the fire. They effectively contained the large fire to the initial area of impact and limited softening and collapse of major structures and vessels. While allowing non-isolatable fuel sources in the system to burn, they effectively prevented vapor cloud explosions and/or a Boiling Liquid Expanding Vapor Explosion ("BLEVE").

- The released fuel found an early heat source and ignited. The LOU event was a large fire, not an explosion.

This report documents the events and causal factors of the incident and includes specific recommendations targeted to reduce the risk of a repeat event.

Occupational Health and Safety Administration ("OSHA") Process Safety Management ("PSM") and Environmental Protection Agency ("EPA") Risk Management Program ("RMP") regulations require that an investigation team be assigned within 48 hours after an incident. The Huntsman Incident Investigation Team (IIT list shown in acknowledgements section) was formed and the investigation was initiated the day of the incident. The IIT used the Apollo Why Drill investigation process developed by Apollo Associated Services, Ltd. The IIT placed the highest evidentiary weight on physical evidence and event logging systems such as distributed control system alarm and event logs.

The release of hydrocarbons was caused by a sudden "guillotine" rupture of an 8-inch diameter cracked gas warm-up[1] line. The 27-year old carbon steel line ruptured because the top of the pipe was severely corroded and no longer had sufficient wall thickness to contain even normal operating pressure.

---

[1] A "Warm-up" line was installed as a part of the original 1979 LOU design to prevent potential brittle fracture of the bimetallic Demethanizer Tower. The original LOU design specified a Demethanizer Tower that has a carbon steel lower section and a

The corrosion on the warm-up line was in a localized two-foot long section of a two-hundred foot long horizontal run. The rest of the warm-up line horizontal run was otherwise in good condition. Heavy corrosion in this two-foot area was caused by condensed water dripping intermittently from a pair of cold service uninsulated stainless steel lines crossing at a 90° angle just a few feet above this section of the warm-up line. The most severe corrosion was on top of the ruptured warm-up line and is most likely to have been caused by corrosion under insulation. Insulation was completely removed from the warm-up line and from most other lines in the fire zone by the fire and high pressure fire water.

Although the severely corroded warm-up line could have failed at any time, and would have failed eventually, the line ruptured on April 29, 2006 due to pressure spikes (674 psig maximum vs. 550 psig normal) and pressure cycles that occurred during a unit upset. The original warm up line thickness of 0.332 inches had an 8 fold factor of safety and was designed to contain 4460 psig. However, only 5% of the original pipe wall thickness remained in the thinnest places (0.018 inches vs. the 0.332 inches original nominal thickness). The minimum wall thickness to contain the warm-up line normal operating pressure was calculated to be 0.045 inches. With an actual measured wall thickness of 0.018 inches in the thinnest area, the warm-up line was in immediate danger of rupturing under normal conditions.

The abnormal cracked gas pressures were the result of cracked gas flow being impeded by an ice/hydrate plug that formed in the second Cold Train cold box. Ice/hydrate formation was caused by low ppm levels of water breaking through the cracked gas desiccant dryers over a period of time. The desiccant dryers were overloaded because cooling water exchangers that are designed to remove the water were impaired as a result of two extended unit outages (Hurricane Rita and later a 3rd party steam supplier boiler outage).

When the warm-up line ruptured, cracked gas began to jet out of the south end of the line at an initial release rate of 1.8 million pounds per hour. Cracked gas is a flammable material with a typical composition of ethylene (46%), methane (18%), propylene (14%), ethane (11%), propane (2%), hydrogen (2%) and other (7%). The warm-up line section that ruptured was on a second level pipe rack 24 feet above ground. The south end of this ruptured line pointed at a very dense area of piping and at the ethylene/propylene refrigeration compressor deck. This south end "blow torch" began to rapidly heat the pipes in its path. Nine minutes later, several steam, ethylene and propylene lines began to rupture due to fire impingement. Aside from the 8-inch warm-up line, 20 lines in the path of this "blow torch" ruptured due to flame impingement. Some of these lines were connected to distillation towers and vessels, each of which contained large amounts of fuel (from 10,000 to 700,000 pounds). As the vessels in the cracked gas train got hotter and hotter from the intense radiant heat, the liquid inside them began to vaporize. The vapor flowed backward through the cracked gas system and out the south end of the ruptured warm-up line. This flow path resulted in a self-feeding mechanism for the fire, greatly increasing the intensity and extent of the damage.

---

stainless steel upper section. However, carbon steel can brittle fracture at low temperatures and could cause collapse of the tower. The warm-up line routes 60° F warm cracked gas directly to the tower to warm up the bottom of the tower in case of a temperature upset.

When the fire started, four Huntsman employees were working 150 feet south of the warm-up line rupture area. When the flammable cracked gas mixture ignited, they reported little or no blast wave but saw flames coming toward them, which they were able to outrun.

The large fire was controlled within hours. For safety reasons (e.g., to prevent vapor cloud formation), small residual hydrocarbon fires were allowed to burn until May 3, 2006 when the final small fires were extinguished.

The IIT identified three possible ignition sources in the area of the ruptured pipe:

(1) exposed superheated steam piping;

(2) a superheated steam leak (both the exposed piping and steam leak were about fifteen feet away from the point of rupture); and,

(3) a possible spark created by the 15-inch section of warm-up line that ripped loose from the guillotine rupture area and struck and dented an angle iron beam 3 feet below the rupture point.

The ignition source will never be known: however, the cracked gas ignited almost instantly based on computer data and eyewitness reports. This was a blessing because it prevented an explosive cloud from forming.

Three major flare lines, which would be needed to remove residual hydrocarbons and vent away hydrocarbon vapors from equipment and piping after the incident, were also severely damaged by the fire. For example, the main 42-inch diameter flare line was compromised in many areas including one section that had drooped 8 feet. The once round pipe cross-section became oval, which made it more difficult to plug and to isolate. Repairing these three flare lines so they could be used for unit "safeing" was a very involved process due to the extent of the damage. [2] The flare repair work was completed on May 19, 2006. Once the three flare lines were repaired, the 89 lines traversing from the fire affected area to the non-affected area could be isolated (completed on June 13) and hydrocarbon removal from the fire and non-fire zones could begin. The above work was controlled by written authorization from the IIT so that evidence preservation and incident scene control could be ensured.

Other factors leading to this incident included:

- Failure of the existing Mechanical Integrity (MI) system and inspection processes to identify severe pipe wall thinning caused by corrosion under insulation on a seldom used line;
- Cracked Gas Compressor pressure and surge control system design and operation; and,

---

[2] "Safeing" is a term typically used to describe a sequence of events necessary to reconfigure a system to a lower level of risk.

- Cold Train plugging mechanisms, system design, and procedures and training to remove plugging: [3]
  - Cold Train moisture removal system design and operation,
  - Loss of cracked gas cooling capability after unusual and/or prolonged outages.

Several systemic issues were identified by the IIT:

- Mechanical Integrity (MI)
  - Current MI system did not identify the warm-up line location that corroded, thinned severely, and ruptured in spite of an API 570 based inspection program that inspects 21,000 piping locations and 23,000 equipment locations.
  - Staffing for this inspection work was reduced in 2002 following a recommendation by a third party consultant. Nonetheless, thickness inspections and visual inspections on the warm-up line piping circuit were done on schedule,
  - Documentation of the follow up on warm-up line inspection report repair recommendations was inconsistent.
  - Current MI processes need to establish a greater focus on systemic piping issues. For example, a piping clamp was installed on the top of a 20-inch section of warm-up line near the 8-inch warm-up line take off point. Other corrosion was found on top of the 20-inch warm-up line piping near the clamp location after the incident.

- Training /Education

  While in general the level of training afforded operators is significant and indicative of a strong skill set, the IIT makes the following observations:
  - Hiring and training of console operators to run a unit as complex as the LOU is a lengthy process. Major process upsets occur infrequently, yet normally require an immediate correct response. Retirements and staffing reductions have resulted in loss of experience.
  - Console operators need better tools (e.g., a dynamic simulator, alarm watch dog, etc.) and abnormal situation training to help prepare them to handle upset conditions.
  - Training effectiveness has been impacted by the transfer of training personnel to fill other positions coupled with training staff reductions over the last few years.
  - Incident Key Learning to be shared:
    - Low levels of moisture breakthrough[4] from the cracked gas dryers (1 to 10 ppm moisture) can form ice deep in the Cold Train (-40 to -120 degrees F).

[3] "Plugging" – refers to a condition in Cold Train equipment wherein tubes in heat exchangers or small passages in high efficiency heat exchangers can become flow restricted.

[4] "Breakthrough" describes a condition where water vapor in the cracked gas stream completely saturates the 88,000 pounds of desiccant in a Cracked Gas Dryer bed. When the desiccant is saturated and cannot absorb any more water, the water vapor exits the bottom of the dryer with the cracked gas and enters the Cold Train where it can condense and freeze. On average, a Cracked Gas Dryer desiccant bed can absorb about 7500 lbs or 900 gallons of water before becoming saturated. The IIT calculate that, cracked gas entering the dryers at 60° F and 300,000 pounds per hour (pph) delivers about 110 pph or 13 gallons of water per hour to the dryer beds. Bed saturation at this rate takes about 68 hours. The dryer beds are normally switched every 48 hours to prevent breakthrough. Cracked gas entering the dryers at 70° F and 380,000 pph delivers about 188 pph or 23 gallons of water per hour to the dryer beds and will cause bed saturation in about 39 hours

    - Methanol injection may not remove hard ice from heat exchanger tubes or cold box exchangers since the vapor stream carrying the methanol will take the path of least resistance through the exchanger.
    - Corrosion under insulation occurred on top of the warm-up line and at several other locations.
    - Calcium silicate insulation traps moisture against the pipe. Perlite insulation does not.

- Alarm Management
  - Operators experienced a flood of alarms leading up to the incident, which was a clear distraction while they attempted to address the Cold Train plugging and compressor surging issues.
  - A system is currently in place to prevent operators from changing critical shutdown alarms. Some operating alarms, like the dryer moisture high alarm point, are not included in the current management of change system, and can be changed by operating personnel. Operating discipline for operation alarm management needs to be improved.

- Procedures
  - Standard operating conditions with alarm settings, consequence of deviation, and corrective action for control points need to be provided in some type of summary form for quick reference. Important operating parameters are often times buried in the body of the procedural text and are not easily accessible.
  - Operating procedures in some areas are not being consistently followed. Some procedure elements like moisture breakthrough for the cracked gas dryer were treated by many as "guidelines" while other procedures were strictly followed. Supervisor and management expectations in this area need to be clarified.

- Systemic risk reduction
  - Current operating practice and environmental regulations reinforce the approach of running and working through problems or upset conditions vs. shutting the LOU down. Shutdowns and startups cause flaring, potential impact on the community, stress on the equipment, and introduce a higher level of EHS risk than normal operations. Operations Managers, Operations Team Leaders (OTLs), engineers, and console operators need further training in the risk balance decision-making process for LOU shutdowns.
  - Design modifications to improve Cracked Gas Compressor surge control and improve cracked gas discharge pressure control can lower the risk of pressure cycling and pressure spikes.

Photo 1
LOU Fire Scene Overview

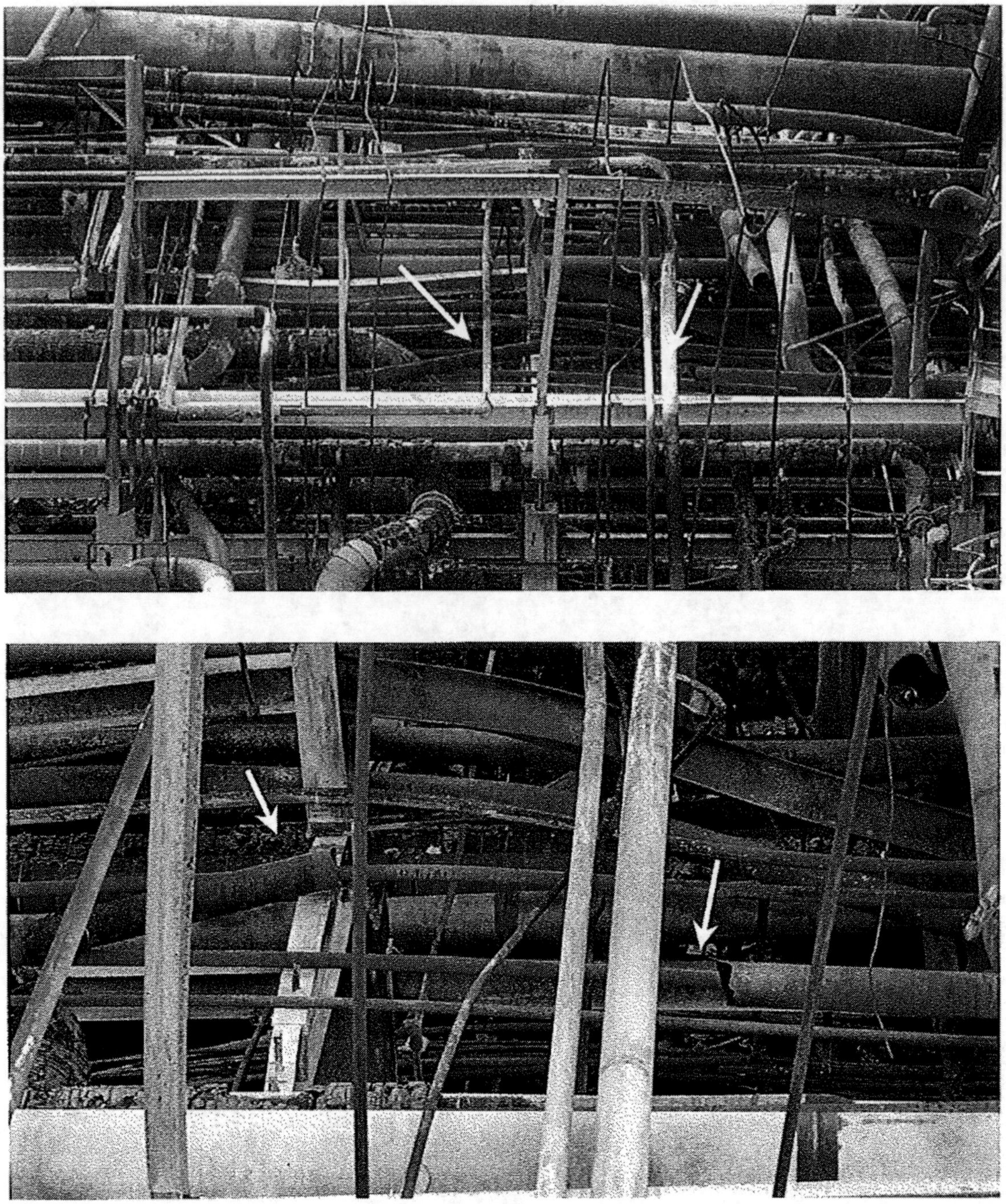

Photo 2
The cracked gas warm-up line separation.

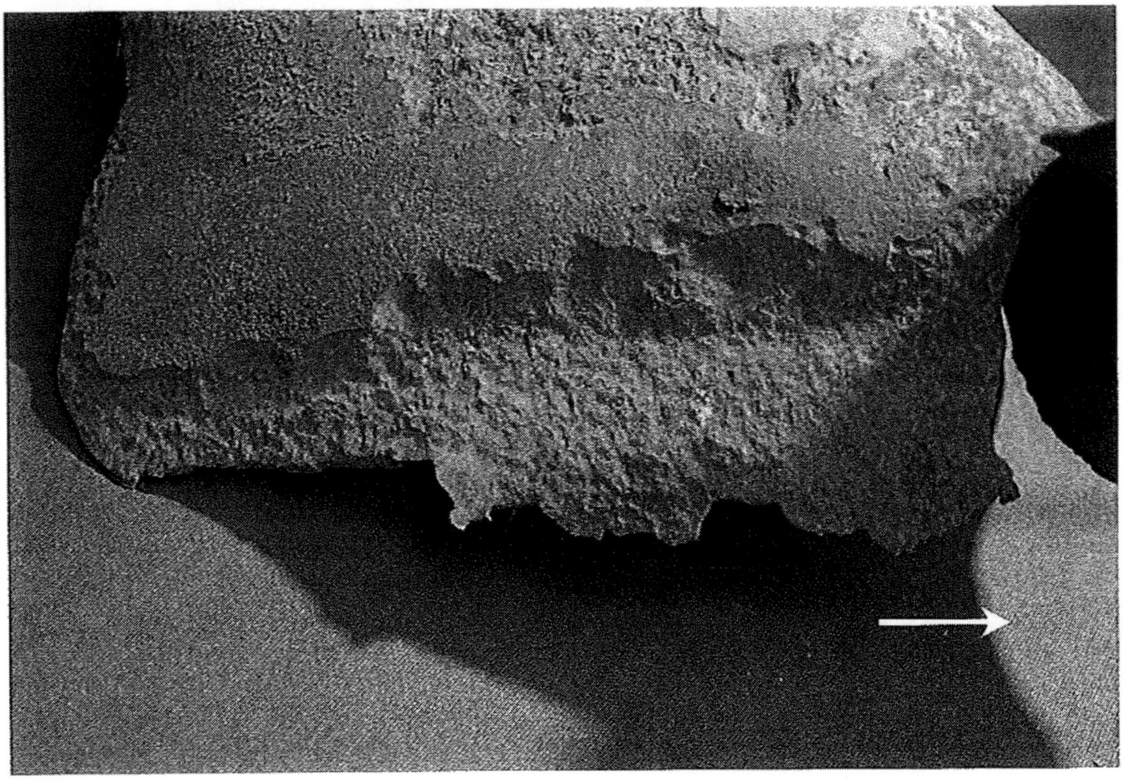

Photo 3
The "flap" from the warm-up line displayed a band of significant external corrosion that ran along what would have been the top (12 o'clock position) of the pipe. The band of corrosion extended between the north and south fracture across the flap.

Independent Analysis: Warm-Up line Corrosion and Rupture

Note : The following independent analysis was prepared by Dr. D. Scott Harding, P.E., of Exponent Failure Analysis Associates, Inc.

When limited access to the unit was finally allowed after the fire, the cracked gas warm-up line, which bypasses the Cold Train and supplies "warm" cracked gas to the Demethanizer Tower, was found to have suffered a circumferential "guillotine" rupture. The circumferential "guillotine" rupture completely severed the warm-up line with the south end of the line being whipped and displaced to the west by the thrust of the escaping nominal 620 psig cracked gas. During the initial entry into the unit, a small flap of metal was found on the ground beneath the "piggy-back" exchangers 10E-223 and 10E-224 .

The flap when recovered and compared to the fractured ends of the warm-up line was found to match the pipe. The flap had been "blown" out of the warm-up line during the rupture. Examination of the pipe rack in the area of the rupture revealed a dent in an angle-beam beneath the warm-up line where the flap struck the beam as it exited the pipe.

The thrust loading and whipping of the south end of the rupture displaced the pipe with sufficient force that it hopped over several other lines to the west and was bent when it struck a vertical line. The thrust loading also displaced the warm-up line to the south, bending the 90-degree easterly run of the line where it enters the pipe rack. The whipping of the south end of the warm-up line caused the open end of the line to point up and directly at the ethylene/propylene refrigeration compressor deck.

The jet of burning cracked gas escaping from the warm-up line impinged directly on 20 other lines in the pipe rack or that were associated with the adjacent Propylene Second Stage Flash Drum (10F-250). Those 20 lines then ruptured as a result of overheating and direct flame impingement. The north end of the fractured warm-up line also experienced some whipping and was displaced slightly upwards and towards the west by approximately one to two feet at the open end of the pipe.

Examination of the fractured ends of the warm-up line revealed that the external surface of the line had experienced significant corrosion at the location of the rupture. The flap, which is the piece of pipe "blown out" between the two fractures, also displayed significant external corrosion particularly along the top (twelve o'clock position) of the pipe.

Measurement of the wall thickness along the edges of the flap and fractured pipe ends with a point micrometer revealed wall thicknesses as thin as 0.018 inch at the top of the pipe near the south circumferential fracture. The rupture of the warm-up line initiated near the south fracture at the top of the pipe. The crack then ran axially through the corrosion groove along the top of the pipe until each end of the propagating crack encountered thicker material. The two ends of the crack then turned and ran circumferentially, extracting the flap from the pipe in a double "guillotine" rupture. Examination of the internal surfaces of the warm-up line revealed no internal corrosion.

At the maximum pressure of 674 psig measured at the Fourth Stage Cracked Gas Compressor Discharge Drum (the pressure would be slightly lower at the location of the rupture) and using the specified minimum tensile strength of the API 5L Grade B Sch-40, 8-inch warm-up line of 60 ksi, the minimum thickness needed to withstand the internal pressure is 0.045 inch (based on uniform wall thickness). The 0.045 inch rupture thickness contains no factor of safety, and does not account for the material potentially having strength above the specified minimum. The specified minimum thickness per ANSI B31.3-1999 (§304.1.2, equation 3b) is 0.137 inch. The ANSI B31.3 minimum thickness of 0.137 inch contains a factor of safety implicit in the Code. Based on the thickness measurements, the warm-up line suffered external corrosion to the point that it was below the minimum thickness specified by Code and was also below the minimum thickness necessary to withstand the applied internal pressure.

In the areas where the insulation is not present on the warm-up line as a result of the fire, the warm-up line appears to have suffered external corrosion at only one location -- the location of the rupture. Insulation was damaged by the fire and by the application of firewater during the fire suppression effort for a significant distance on either side of the rupture. Most of the damaged insulation fell off of the warm-up line. Remaining insulation was subsequently removed from the warm-up line to allow for further checks of its condition.

Visual inspection reports for the warm-up line from April 5, 2004 reported an approximately 6-foot section of insulation missing from the horizontal run of pipe containing the fracture. No records were found that indicate the missing insulation had been repaired. One possible explanation for the unique area of localized external corrosion is that the corrosion occurred at the location of the missing insulation. The nominal 60°F warm-up line would have sweated and stayed wet all but a few days per year and allowed the corrosion to progress in the absence of any coating or paint.

Examination of the corrosion pattern on the warm-up line revealed that the wall loss occurred primarily on the top (12:00 position) and in a circumferential band adjacent to the south fracture. The pattern suggests that water or moist air was entering through seams or gaps in the insulation. The insulation on the warm-up line consisted of individual blocks held in place by aluminum sheathing. The blocks provide a gap that runs axially along the top of the pipe as well as circumferential gaps for the ingress of water/moisture. The seam in the sheathing is also near the top of the pipe.

Besides the warm-up line sweating, other sources of water/moisture include two lines that run perpendicular to and above the warm-up line at the location of the rupture. The lines are the high-pressure ethane/propane line and the ethane/propane dryer purge line. The high-pressure ethane/propane line operates at 60 to 80 °F and the ethane/propane dryer purge line operates well below ambient temperatures when in use. Both lines were not insulated and could intermittently sweat and drip on the warm-up line at the point where the line ruptured.

Summary -- The rupture of the warm-up line was the result of severe localized external corrosion. The severe thinning from external corrosion suffered by the warm-up line left the pipe in a condition where the pipe was in immediate danger of rupturing. If the pipe had not ruptured from the pressure fluctuations of the upset LOU and had survived the impending unit shutdown, the pipe most likely would have ruptured during the next start-up with possibly far greater consequences.

Ice/Hydrate Plugging of Cold Train

After the incident, 5 gallons of water were drained from the second cold box heat exchanger 10E-1308 (-70 degrees F operating temperature). There was no water found in upstream or downstream equipment/piping. There were no breached lines or equipment that led to or exited this exchanger. Therefore, the water in 10E-1308 was pre-incident and was not added as a result of the fire containment effort.

Prior to the incident, the moisture content in the cracked gas stream to the dryers increased due to heat exchanger fouling. As the moisture in the inlet stream increased, there was a need for more frequent dryer switching (roughly every 36 hours). However, the dryers continued to be switched out at the usual frequency (every 48 hours). As a result, there was moisture breakthrough from the dryers at the end of each run. The breakthrough was indicated by the moisture analyzer, AR-203, but the dryers were not switched at the time of breakthrough.

Low levels of moisture (1 to 10 ppm) did enter the Cold Train over a 41 day period of time. The operations high moisture alarm in the DCS for the AR-203 Cracked Gas Moisture Analyzer was found to be set at 10 ppm vs. the 2 ppm maximum stated in the procedure.

The temperature downstream of the dryers is low enough for hydrate formation and ice formation even at these low moisture concentrations. The dew points of the cracked gas are a function of moisture content and were calculated using a simulation. The calculated dew points are tabulated in Table 9 below:

Table 1
Cracked Gas Dew Points as Function of Moisture Content

Temp.	Dew Point Moisture
(°F)	(ppm)
-120	0.8
-100	1.6
-80	2.9
-60	5.6
-50	7.7
-40	10.8
-30	15.0
-20	21.0
-10	29.5
0	41.6

Once the high discharge pressure was noticed and a high Cold Train pressure drop diagnosed, methanol was pumped into heat exchangers upstream of 10E-1307. This seemed to help for a very brief period only. Because the process temperature from the dryer to 10F-217 (downstream of 10E-1307) falls from 60° F to -30° F, the only moisture that would drop out in this section is the moisture from hydrate formation. However, downstream of 10F-217, in heat exchangers 10E-1308 and 10E-148B, the temperature drops from -30° F to -90° F. This is the temperature range of the dew points for the level of moisture breakthrough seen in the moisture data.

The methanol was injected in the right places for previous understanding of how to break up the hydrates. Given that there was some improvement in the pressure drop, it appears to have been effective for that limited purpose. However, the methanol was injected upstream of 10F-217. The methanol remained in the liquid phase and was knocked out in 10F-217.

Methanol was not injected in the feed of 10E-1308, where the temperature range is conducive for water condensation and ice formation but not for methanol to freeze (temperature was too high). This was the most likely spot for ice formation in the Cold Train, which was subsequently confirmed by the post-incident presence of water in 10E-1308.

Corrosion Under Insulation

The IIT requested inspection of non-fire affected areas on the LOU piping and the A3 Olefins unit at Port Neches to see if corrosion under calcium silicate insulation was a problem. Corrosion on the top of pipes was seen in many places. General corrosion under insulation was also seen on A3 piping. Gaps between the bottom of the pipes and the insulation were frequently seen and little to no corrosion was seen in these areas.

Photo 4

The wet insulation on the propylene refrigerant line sagged away from the bottom of the pipe from the additional weight. The direct contact between the top of the pipe and the wet insulation resulted in the observed corrosion.

Photo 5
Propylene refrigerant line from 10E-1401 was found to have under insulation corrosion along the top of the pipe.

Photo 6

The CUI experienced by the top of the propylene refrigerant line from 10E-1401 likely resulted from perforations in the aluminum insulation jacket that allowed water to saturate the underlying insulation. The photographs display different areas on the same line.

Photo 7
CUI on A3 Olefins Unit propylene piping. Hole in pipe plugged with ice.

Photo 8
CUI on A3 Olefins Unit. Deep pits in top of propylene line.

Photo 9
CUI on A3 Olefins Unit. Deep pits in small bore propylene line.

Recommendations

1. Consider improvements to the Mechanical Integrity piping inspection program to better identify Corrosion Under Insulation, especially for equipment and piping that operates in the temperature range of 25° F to 300° F. Also inspect the tops of pipes for thickness and corrosion, and consider higher quality insulation that does not wick water and hold it against the pipe surface.

A) Consider modifying inspection programs for both PABC and other Huntsman facilities to include a more aggressive inspection program for insulated piping that has an operating temperature within 25° to 300° F. The program should include the removal of insulation on suspected areas and coating pipes prior to encasing with insulation.

B) Improve systems to ensure inspection report related work orders are written and completed for all inspection action items.

The cause of the carbon steel Demethanizer warm-up line rupture was external corrosion. The exact mechanism of how water from the overhead dripping lines caused the warm-up line corrosion is not known (this is because it is not known for sure whether the ruptured area was insulated or not). It is interesting that the corrosion was most pronounced near the top of the pipe. There are two possible mechanisms for this phenomenon. The first mechanism is that near constant dripping of water on top of a non-insulated pipe could have accelerated corrosion at the top of the pipe. However, absent a moisture capture mechanism, such as insulation, there is nothing to retain the moisture on top of the pipe. The second mechanism is CUI due to moisture being trapped in or under the insulation and being held against the pipe.

The conventional wisdom had been that the moisture trapped inside breached insulation would pool in the bottom of the insulation and would result in corrosion on the bottom of the piping. Insulation, such as mineral wool or fiberglass, that will absorb water, can hold that water up against the pipe wall for longer periods of time, until it dries out. The IIT learned from post incident inspections that CUI is more likely to happen on the top half of the pipe than it is on the bottom half, because the insulation was observed to lie in contact with the top half of the pipe, whereas on the bottom of the pipe there was an air gap.

Another point of conventional wisdom that is now under question based on IIT post incident inspections is that the moisture will pool down to a low point elbow and that is the best place to check for corrosion rates in a piping circuit. This was not seen in warm-up line case. When the elbow tested in 1999 it still had near original installation thickness.

The IIT learned that LOU piping elbows tend to be insulated with Perlite, a material that does not absorb water. The straight run lines were insulated with calcium silicate which does absorb water. This insulation works well for piping that is not hot enough to drive off any moisture that penetrates the aluminum cladding on the insulation. Therefore, it is necessary to check straight sections of pipe that operate in the 25° F to 300° F temperature range, particularly if the pipe is insulated with a material that will absorb water (i.e., calcium silicate).

2. CUI - Corrosion under insulation

Corrosion under insulation occurred on top of the warm-up line and at several other locations. Looking for water leaks, condensate leaks, cladding damage, insulation damage, bulging and other evidence of integrity failure should be an ongoing job of all Huntsman employees working in the LOU.

3. Review piping geometry to look for possible "Drippers" and ensure that they do not drip onto exposed piping.

The second possible mode of external corrosion of the Demethanizer warm-up line (as noted above) involves two "dripper" pipes located directly above the spot where the worst corrosion occurred. The dripper pipes were two uninsulated stainless steel process pipes that operate in the 0°F to 100°F range, which are temperatures where ambient moisture would condense. The condensed moisture dripped down onto the warm-up line insulation cladding or piping below. This constant exposure to moisture can accelerate the corrosion of both the pipe where the condensation occurs and the pipe onto which the water drips.

Current practice is to repair steam and condensate leaks as maintenance "fill-in jobs." This can lead to delays in getting the leaks fixed when they are small. By taking a more aggressive approach to repairing steam and condensate leaks, potential damage to adjacent piping can be reduced.

4. Investigate anti-surge controls for the Cracked Gas, Propylene Refrigeration and Ethylene Refrigeration Compressors with appropriate training for operations.

The Ethylene Refrigeration Compressor surged for a period of time prior to the incident. The Cracked Gas Compressor surged for a period of time also after the cracked gas feed rate was reduced. The IIT learned that the Huntsman Wilton, UK cracker installed an improved anti-surge system to reduce energy consumption on their Cracked Gas Compressor. This system has been effective in helping to reduce compressor surging.

5. Consider installing a Fourth Stage Cracked Gas Compressor discharge pressure controller.

Because PRV's are not designed for steady control but for sudden relief of pressure, they increased the oscillations of flow and pressure through the Cracked Gas Compressor. A discharge pressure controller -- that can either route cracked gas to the flare or back to the disengaging drum -- would provide for less pressure and flow oscillation and a means to prevent reaching the PRV set pressure. The separate discharge process controllers would provide an additional layer of protection to avoid reaching the MAWP of the compressor or downstream vessels and piping.

6. Formally share incident key learning and provide training for operations, maintenance and technical personnel.

(A) Dryer Moisture Breakthrough

Low levels of moisture breakthrough from the dryers can form ice deep in the Cold Train even at a 10 ppm water level.

Dew point is a better measure of moisture breakthrough from the dryers vs. ppm moisture level because it will let the operations team know where moisture is likely to liquefy and turn to ice/hydrate in the Cold Train system.

The IIT also identified an apparent deficiency in the understanding of the critical interplay between performance data and the dryer operations. Refresher training is needed.

(B) Methanol Injection to Break Hydrate/Ice (and other anti-freeze solvents).

As the Cold Train plugging-related events unfolded, there were attempts to remove the plugging through methanol injection. It is clear that there was confusion over the location of methanol injection points. This confusion created delays in methanol injection. Furthermore, methanol was not injected into 10E-1308, where water was found after the event. An MOC was prepared in anticipation of injecting n-propanol. A permanent procedure is needed that covers methanol, n-propanol, and toluene injections with adequate explanation on where and why those solvents are to be injected.

Ice/hydrate formed over a longer period of time will be frozen more solid and harder to break with methanol.

Methanol injection may not remove hard ice since the vapor stream carrying the methanol will take the path of least resistance through the exchanger.

7. Consider improving moisture measurement to avoid water breakthrough into the Cold Train to limit the potential for Cold Train pluggage.

The current analyzer measures moisture on the outlet of the on-line dryer. The IIT identified five possible improvements to this system.

- Additional moisture analyzer – adding a moisture sensing analyzer to evaluate moisture levels near the bottom of the dryer bed would improve breakthrough protection. It would allow for dryer switches based on potential dryer bed moisture breakthrough in addition to the moisture breakthrough analyzer after the dryer. There are existing sample points within each dryer bed.
- Redundant analyzer - Installing a redundant analyzer would improve confidence in the reading and would improve analyzer reliability.
- Hand held sample point - Providing a sample point for routine operator checks using a hand-held device would give the LOU operators the ability to verify the on-line analyzer readings.
- Larger capacity dryers - Installing larger capacity dryers will increase drying capacity and extend potential dryer run lengths before desiccant must be replaced.
- Install a third dryer - Providing a lead/guard dryer arrangement increase the dryer capacity and provide a back up guard bed to catch any primary bed breakthrough.
- Dew point vs. ppm readout on analyzer – dew point is a better measure for moisture breakthrough to the Cold Train.

Acknowledgements

Huntsman Incident Investigation Team (IIT) Membership

Name	Title	Company	Expertise
John Prows	Vice President, Manufacturing Performance Products	Huntsman	Incident investigation Manufacturing Mgmt, PSM Chemical Engineer
Liz McDaniel	Vice President, EHS Americas	Huntsman	EHS programs and management Chemical Engineer
Keith Moreau	Job Instruction Trainer	Huntsman	Process operations knowledge and operator training
David Nutt	Senior EHS Counsel	Huntsman	EHS regulations and investigations
William Shelden	Six Sigma Black Belt	Huntsman	Process design Process control Chemical Engineer
Sasha Vragolic	Operations Manager, A3/Utilities	Huntsman	Process engineering Process control Operations Management MS Chemical Engineer
Christopher S. Buehler	Senior Manager	Exponent/Failure Analysis Associates	Process engineering PhD Chemical Engineer
D. Scott Harding	Managing Engineer	Exponent/Failure Analysis Associates	Metallurgical engineering/failure analysis, P.E. PhD Mech. Engineering
P. John Palmer	Principal	Vescen Consulting	Incident scene investigation, PSM/RMP Chemical Engineer B.A.S.C./M. Phil. P. Eng.
Larry Matta	Sr. Engineer	Exponent/Failure Analysis Associates	Heat transfer and fluid mechanics modeling Jet and flame propagation PhD Aeronautical Engineering, P.E.

INVESTIGATION OF NO_X RELATED COLD BOX INCIDENT

BASF FINA Petrochemicals Limited Partnership

William Hack, *Operations Manager, BASF Corporation*

Robert Copeland, *Technology Director, BASF Corporation*

Dwight Hines, *Product and Process Technology Manager, BASF Corporation*

Abstract

On the morning of March 13, 2008, the rupture panels on the top of the Port Arthur cold box opened, releasing a mixture of nitrogen, hydrogen and methane to the atmosphere. A root cause investigation team comprised of cross-functional, cross-company experts concluded that the most probable source of the incident was a decomposition-triggered detonation of 'NO_x Gums' in a 1-inch drain line in the coldest portion of the cold box.

The NO_x Gums formed when a conjugated di-olefin, most likely butadiene, carried over from the first two drums in the chilling train to the fifth drum and reacted with NO_x gases that had condensed and accumulated. The heat of reaction was sufficient to initiate a decomposition, which proceeded to a detonation. A 4-foot section of 6-inch piping and a 2-foot length of 1-inch piping were destroyed. Another section of piping and some control cabling adjacent to the detonation source was damaged. Much of the energy of the detonation was absorbed by the pearlite insulation that fills the cold box.

The primary purpose of this paper is to communicate the key lessons learned from this incident to other members of the ethylene industry. The topics to be covered will include:

- Background information
- Cold box process overview
- NO_x chemistry
- History of previous NO_x gum incidents
- What happened?
- How did the incident occur?
- Why did the incident occur?
- Recommendations and learnings

General Plant Information

BASF FINA Petrochemicals Limited Partnership (BFLP) operates a naphtha cracker in Port Arthur, Texas. The cracker is a joint venture between BASF Corporation and TOTAL PETROCHEMICALS USA, INC. At the time of its construction it was the largest single train liquid olefins cracker in the world and BASF's largest investment outside of Europe.

Design capacities for the cracker primary products are: 2.1 Billion Lbs. / yr (950 KTA) of ethylene and 1.3 Billion Lbs. / yr (590 KTA) of propylene.

Background Information

Butadiene and NO_x liquids are found in separate sections of the cold box due to very different vapor pressures. If allowed to come into contact, they react to form a potentially unstable family of compounds referred to as NO_x Gums. The hazards of NO_x Gums in cold box operations have been documented in several incident investigations; most notably, the Shell Berre incident in 1990.

Port Arthur personnel were aware of the potential for NO_x to accumulate in the cold box. Therefore, procedures developed as a result of a 2004 NO_x task force study comprised of BASF, ATOFINA, Lummus and external NO_x consultants were in place and routinely executed when the cold box was removed from service. Areas of the cold box where deposits can form were analyzed for the presence of NO_x. In the history of the plant, NO_x accumulations had not been found; including such recent events as Hurricane Rita (2005) and during plant turnarounds in both 2006 and 2007.

Cold Box Process Overview

The process gas leaving the charge gas dryer(s) is progressively chilled against the process to about -98°F. The condensate is separated in the DeMethanizer Feed Separator No. 1 (D-3001) and fed to the DeMethanizer.

The vapor from the overhead of the DeMethanizer Feed Separator No. 1 (D-3001) is chilled against off gases in the cold box and then ethylene refrigerant to -145°F at E-3005X. The condensate is then separated in separator No. 2 (D-3002) and sent to the DeMethanizer. During normal plant operation the D-3002 bottoms stream contains the remaining butadiene.

Charge gas is then further chilled in the cold box to -204°F by heat exchange with hydrogen and methane off gas via E-3011X and 3012X. The condensate is separated in DeMethanizer Feed Separator No. 3 (D-3003X) and sent to the DeMethanizer. The coldest feed to the DeMethanizer is obtained from this drum and the residual overhead gas is hydrogen of approximately 75 mole percent purity.

Figure 1 – Chilling Train / Cold Box Overview

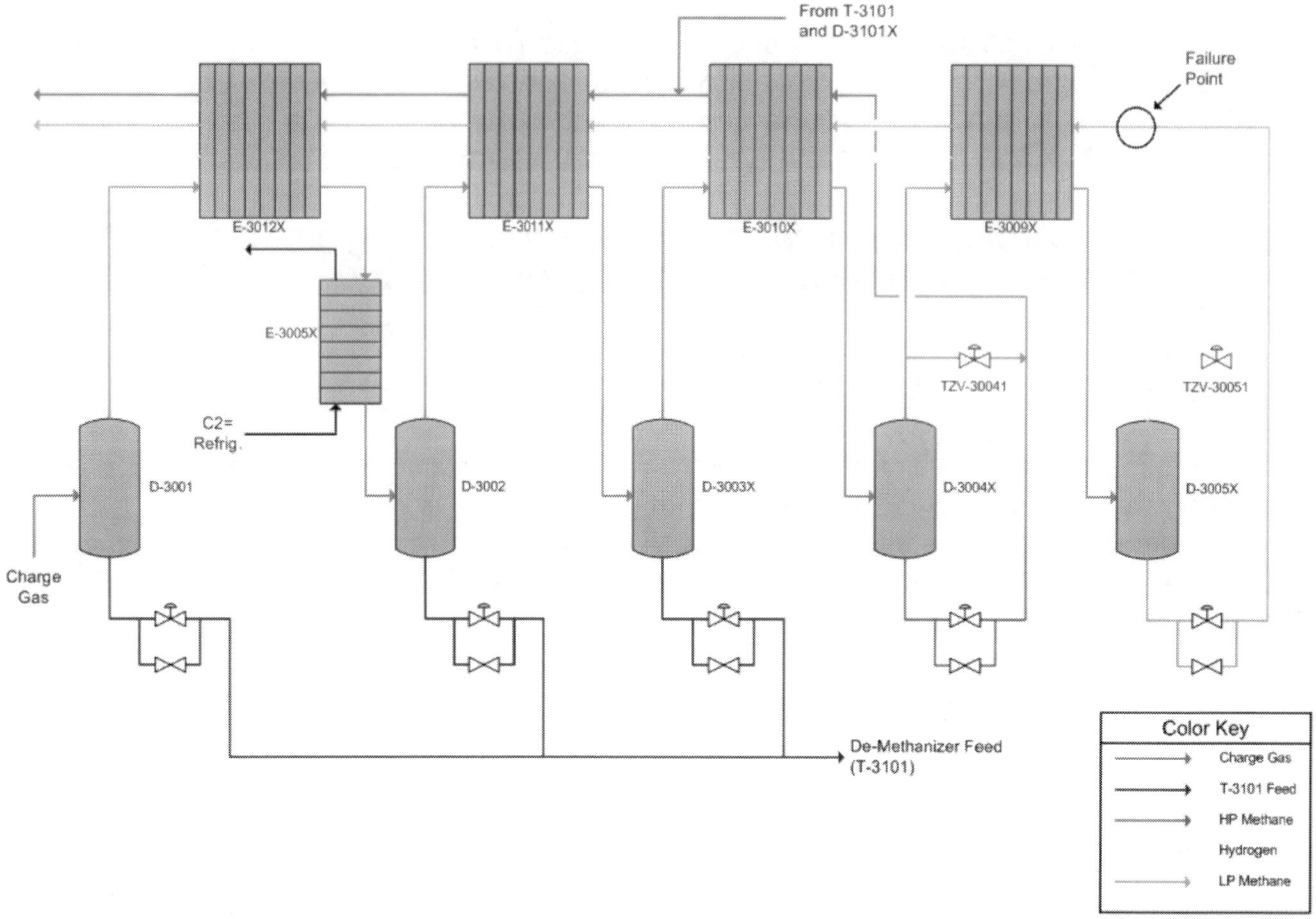

The hydrogen rich stream is further upgraded to 96 mol percent purity in a two stage expansion system. Hydrogen purification takes place in an adiabatic heat exchanger system in which the refrigeration required is derived from expansion of liquefied methane (Joule-Thompson Expansion).

In the first, across TZV-30041, the stream is chilled to -240°F, condensing a portion of the contained methane and essentially all of the C_2's in D-3004X. Refrigeration is provided by expanding the condensed liquid to DeMethanizer overhead pressure via the high pressure methane pass. After reheating (via E-3010X), the methane rich stream is blended with the DeMethanizer overhead, reheated further (via E-3011X through 3013X) and sent to the fuel gas system or methane regeneration system.

In the second, across TZV-30051, the overhead vapor is further chilled to -280°F. The refrigeration for this cooling step is provided by expanding the condensed liquid to fuel gas pressure from D-3005X. The cooling is recovered across core exchangers E-3009X through 3013X. Liquid is expanded, vaporized, reheated and sent to the fuel gas system as low pressure methane.

The second JT Expansion yields approximately 96 mol percent hydrogen. This hydrogen is reheated in the chilling train (via E-3009X through E-3013X) and sent to the Methanator for carbon monoxide removal or to the fuel gas system. The methanated hydrogen is cooled and dried before being used in various hydro-treating services or being sold.

NO_x Chemistry

Accumulation of NO_x compounds in the cold box is due to trace amounts of NO and O_2 contamination in the cold box feed. Other contaminants like NH_3 and NO_2 can also contribute to the accumulation of NO_x compounds. The source of NO has generally been linked to refinery gases especially from FCC Units. However, many plants have found NO_x compounds in cold boxes that do not recover refinery gases or recycle nitrogen compounds known to form NO_x compounds. Usually, NO_2 is not considered to be a large contributor in cold box feed (though it is formed in the cold box) because it is easily caustic scrubbed.

The chemistry of NO_x accumulation is complex and not completely understood in the industry. However, the major reactions are outlined below.

$$2NO \text{ (nitric oxide)} + O_2 \rightarrow 2NO_2$$

Nitric oxide reacts with trace amounts of oxygen to form NO_2 which can then react with itself or another NO group.

$NO_2 + NO \leftrightarrow N_2O_3$ (dinitrogen trioxide or blue liquid). This reaction is favored at low temperature because the N_2O_3 drops out as a liquid (or solid at very cold temperatures). Since this reaction is reversible, removing the N_2O_3 from the vapor keeps the reaction going toward the product.

N_2O_3 is not explosive. However, since the reaction is reversible, at room temperature it will revert back to NO_2 and NO. The NO in air will form NO_2 quickly and the resulting NO_2 is extremely toxic.

The liquid N_2O_3 (or possibly the dimer of NO_2 which is N_2O_4) can react with ammonia to form NO_x salts or with a diolefin such as butadiene to form NO_x gums which can be very unstable and even decompose explosively at temperatures well below ambient. The NO_x gums, however, can also be quite stable. The stability depends on the degree of nitration, accumulation at a given point (or agglomeration), amount of diolefin and possibly other unknown factors.

NO_x salts are formed by the reaction below:

$N_2O_3 + NH_3 \rightarrow NO_x$ salts. NH_4NO_2 (ammonium nitrite is explosive around 140°F) NH_4NO_3 (ammonium nitrate) is more stable and has a melting point of around 338°F.

Table 1 – Physical Data on Nitrogen / Oxygen Compounds [1]

	NO	NO_2	N_2O_3	N_2O_4	N_2O_5
Melting Point C	-163.7	Freezing Point -11.2	-102	10.2	30
Boiling Point C	-151	21.2	3	22.4	47
Density (Vapor) Kg/m3	1.3	3.9	1400	1450	2050
Specific Gravity (Air=1)					2.62
Appearance at Room Temp	Colorless	Reddish Brown	Pale Blue	Colorless	White Crystals
Phase at Room Temp	Gas	Gas	Unstable	Unstable	Solid

Previous NO_x Gum Incidents and History

The potential for a NO_x compound detonation in ethylene plant cold boxes has been recognized since the mid 1980's. The classic case discussed in many safety reviews is the Shell Berre cold box rupture following a NO_x gums decomposition which occurred in 1990.[2] This incident sparked several industry group investigations which are documented in the EPC (Ethylene Producers Committee) and EEPC (European Ethylene Producers Committee) meeting minutes.

There have been several reported incidents of NO_x decompositions that resulted in loss of containment in steam crackers. Including the Shell Berre case, three other incidents resulting in loss of containment have been reported in the literature. The BFLP incident which occurred on March 13, 2008 will bring that total to five documented cases.

In addition to these five incidents that involved actual equipment/piping damage, there have been over a dozen near misses where either NO_x gums or NO_x salts have been discovered in the process but did not detonate or decompose to the point that equipment failure resulted.[3,4]

Both the EPC and EEPC have conducted surveys of their members to document NO_x issues.[5] In about one third of the reported NO_x findings, the source of the NO contamination has not been positively identified. In these cases, the Nitrous Oxide cannot be traced back to either the recovery of FCC gases or the breakdown of nitrogen containing additives, which are usually identified as prime suspects in the NO ingress.

Plant Conditions Prior to Incident

At around 4:30 PM on Tuesday, March 11, as a result of an upset in the hot section of the cracker, the heaters were placed in hot steam standby and feed was cut to the charge gas compressor. The cold box refrigeration loops were kept in operation, and the cold box was 'bottled-up' to allow for an easier restart of the unit when the hot section issues were resolved.

At approximately 3:30 AM on Wednesday, March 12, the cold section was placed in standby operation. During this time the three large process compressors were left in operation. The Charge Gas Dryer discharge valve was closed and the Cold Box was isolated. The back-end towers were either isolated or placed in total reflux. Refrigeration was available to all users. The cold section was in this mode to allow the cracker to return to normal operations quickly once the problem in the hot section was resolved.

During this time, a cold box temperature control valve leaked allowing the low pressure methane system to slowly increase in pressure. As a result, the pressure control valve on the low pressure methane system opened to maintain the system at approximately 60 psig. The leaking temperature control valve allowed the rest of the cold box, normally operating at 480 psig, to slowly depressurize to approximately 100 psig. The DeMethanizer Feed Separator No. 2 (D-3002) overhead temperature rose from approximately -143 ^{O}F to -39 ^{O}F.

Just prior to the incident, at approximately 7:50 AM on Thursday, March 13 the following process measurements were recorded at the cold box:

- The Cold Box was isolated
    - The "A" discharge valves on both Charge Gas Dryers were in the closed position.
    - All control valves between the Cold Box and the DeMethanizer were closed.
    - The temperature and flow control valves on D-3004X were closed.
    - The temperature and flow control valves on D-3005X were closed.
    - The automated block valve at the outlet of E-3013X in the low pressure methane system was closed.
    - The automated block valve at the outlet of E-3013X in the hydrogen system was closed.
    - The pressure control valve at the outlet of E-3013X in the high pressure methane system was closed.
- Pressure at the overhead of D-3003X had fallen from 480.6 to 106.3 PSIG.
- Pressure in the low pressure methane system had risen to 85.2 PSIG with the pressure control valve open 2.5%. The set point had been raised to 85 PSIG.
- All Cold Box Drums had liquid levels.

What Happened?

At approximately 8 AM, on the morning of March 13, 2008, the C4 operations group notified the control room that a loud bang had been heard from the Cracker and that a white powder material was on the ground between the Cracker and C4 Complex.

Cold section personnel immediately responded and found the cold box insulation material (Pearlite) covering the ground in the area surrounding the cold box and in the area to the north. Upon further investigation, it was discovered that the rupture doors on the top of the cold box were open and relieving pressure. An LEL meter was used to determine that the gasses escaping from the box were high in LEL.

The Cracker and C4 Complex unit emergency alarms were activated and the incident command system was initiated. All non-essential personnel were cleared from both units and a hot zone was established. Cold section operations verified the isolation of the Cold Box. The chilling train was then purged with nitrogen to remove all process gasses.

Upon further investigation we believe that NO_x gums, which had formed in a 1-inch drain line, decomposed explosively. The detonation destroyed a 4-foot section of 6-inch piping and a 2-foot length of 1-inch piping in the low pressure methane system upstream of the hydrogen methane chiller number 2 (E-3009X).

Picture 1 – Damaged 6-inch Vapor Line Upstream of E-3009X

How did the incident occur?

The incident was most probably the result of butadiene coming into contact with NO_x liquids in an isolated drain line on the low pressure methane pass of the cold box.

Due to the failure of a Joule-Thompson valve to provide a tight shutoff, the isolation of the cold box failed allowing the pressure to fall. This caused the box to warm allowing butadiene to be carried over to the back end of the system. It should be noted that the operators isolated the cold box correctly per standard operating procedures. A curve showing the change in butadiene concentration in the overhead of D-3002 as the box warmed is shown below.

Graph 1 – BD Concentration in D-3002 overhead versus temperature

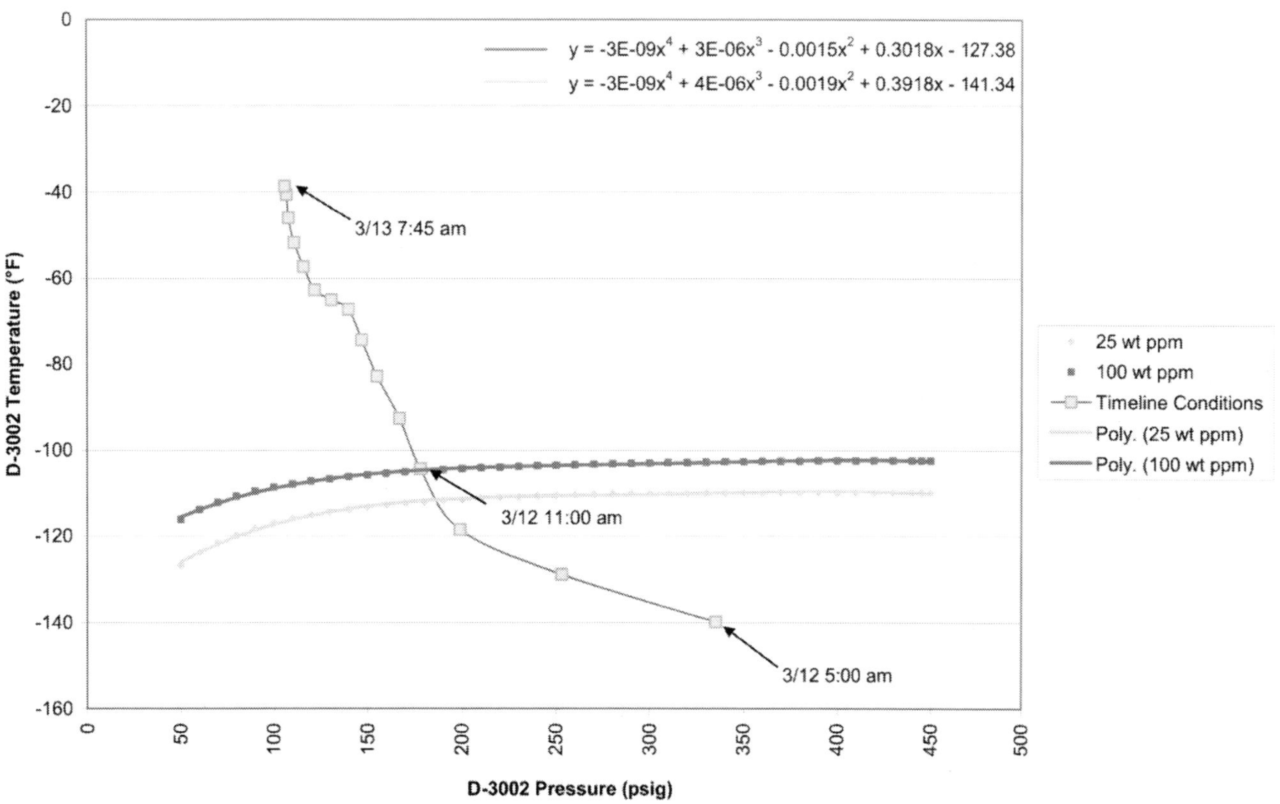

When the butadiene vapors were carried over from the second drum, we believe some portion of the material contacted the NO_x liquid that had collected in the drain line. The reaction that formed the NO_x Gums proceeded rapidly even at cryogenic temperatures. It is believed that the well-insulated drain line allowed the system to behave in a near adiabatic manner. This allowed the energy released by the NO_x gum formation to raise the local temperature to the decomposition temperature.

The decomposition proceeded to a detonation. The gas in the system was rapidly vented through the damaged piping into the cold box. A flammable mixture was avoided due to the nitrogen inerting of the cold box. The material escaping into the cold box exceeded the relief pressure of the rupture panels installed to prevent an overpressure failure. The panels opened, relieving a mixture of nitrogen, hydrogen and methane to the atmosphere. The energy from the detonation was largely absorbed by the Pearlite insulation which limited the damage.

Why did the incident occur?

The reduction in pressure and the accompanying warming of the cold box allowed butadiene to be carried over into the coldest drum. The operating procedures designed to prevent the carry-over of dienes were based on temperature measurements assuming normal cold box pressures. The liquid from the fifth drum is sent to a separator prior to passing through a cross exchanger to cool the incoming process streams (See Figure 2). The vapor line from this separator is designed with a drain at its lowest point, providing a location where NO_x liquids (NO, NO_2, N_2O_3, and N_2O_4) could accumulate. The plant was not aware of this low point drain.

Figure 2 – Sketch of Piping and Equipment related to E-3009X and E-3010X

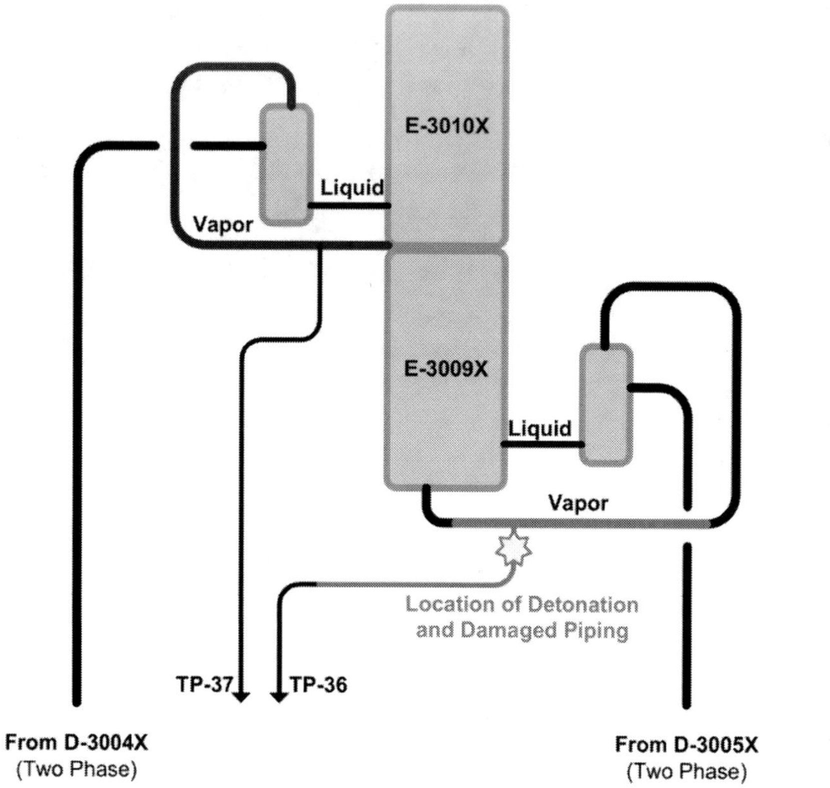

An Incident Event Model and a Root Cause Analysis were created to identify the various contributing and root causes which lead to the incident.

Listed below are the identified root causes. Some have been combined due to the similarity of their contributing causes.

Review LTA (Less than Adequate) or Not Performed

- The HAZOP performed during initial engineering and construction of the facility, did not identify the specific hazards associated with diolefins reaching the low pressure and high pressure methane passes of the cold box due to leaking temperature control valves. As a result, a safety function was not identified to address this issue.
- While the Joule-Thompson valves (TZV-30041 / TZV-30051) were specified with tight shutoff; they were part of a different safety function designed to protect the downstream piping from brittle fracture failure due to excessively cold temperatures after Off Gas Exchanger No 3 (E-3013X). Secondary Isolation was not available between the control valves and the heat exchangers.

Recommendations

- Consider reviewing the HAZOP for your facility, based on the information presented in this paper, to determine if the loss of containment due to leaking valves has been integrated into your cold box's safety concept. In the case of this incident, the location of the leaking valve made the situation worse. The leaking control valve allowed the cold box gases, which had been bottled up, to escape into the low pressure methane pass. As a result, butadiene was pulled across the area of the cold box where NOx liquids tend to concentrate and collect. (See Graph 1)
- Consider draining all cold box drums that contain butadiene whenever forward flow to the cold box is lost. This will minimize the amount of butadiene that can be carried over to the back of the cold box.

Problem Analysis LTA and Facts Wrong / Requirements Not Covered

- It was not anticipated that the loss in pressure in the drums of the cold box, due to a leaking control valve at D-3004X or D-3005X, would allow carry over of diolefins into the sections of the cold box that could potentially contain NO_x liquids.
- It was not anticipated that NO_x gums would form in the 1-inch drain on the vapor line to E-3009X in the low pressure methane pass. The drain line acted as an insulated 'dead leg' allowing a near adiabatic condition to exist. As a result, it is believed, that the heat of reaction warmed the drain line to a temperature that allowed the reaction to proceed to decomposition while surrounding equipment was still at temperatures below the expected decomposition temperature.
- The initial cold box design does not provide a reliable means of preventing butadiene from being carried over into the colder portions of the system during startup and shutdown operations.

- Previous draegering, at points around the cold box, during scheduled and unscheduled outages did not indicate NO_x gases were present. The belief that NOx was not present in the cold box may have given a false sense of security. Subsequently, we have determined that draegering is not an effective means of determining the need for a cold box alcohol wash.
- An alcohol wash may have prevented the immediate incident. However, had this been done, the lack of knowledge of the 1-inch drain line could have resulted in the concentration of NO_x gums in the drain lines with the potential to create a more serious event.

Recommendations

- Verify that personnel have been adequately trained in the identification and response to NO_x Gum and NO_x Salt formation risks associated with the operation of a cold box.
- Consider installing secondary isolation for all valves where tight shut off is required.
- Consider adding a vent to flare on the overhead of the last drum that condenses butadiene. This allows the overhead of the drum to be flared until the pressure and temperature at the drum are such that butadiene is condensed and will not carryover to the backend of the cold box.
- Consider specifying all drain locations in your operating procedures to assure that all alcohol is removed from your cold box after washing. This should prevent the possible concentration of NO_x Gums in a drain line due to the evaporation of the wash alcohol.
- Identify the critical pressure / temperature relations in your cold box to assure that your operation will remain in the safe zone (see Graph 1). Consider adding this information to you operator control graphics.
- Consider developing plans and contingencies based on the assumption that NO_x will always be present. Consider doing an alcohol wash whenever a cold box is warmed.
- Consider reviewing your current draegering procedures for effectiveness. Are sampling methods adequate? Are detection limits and cold box temperatures during sampling taken into account?

Documentation Content LTA and Product Inspection LTA

- The P&IDs misrepresented the 1-inch drain lines near the liquid / vapor separators upstream of E-3009X and E-3010X. The drains were shown on the liquid and not the vapor lines where they are located. In addition, the P&IDs indicated that the drains were internal to the cold box, while they actually penetrate the wall of the cold box near grade. The cold box was delivered to the site under a nitrogen pad, and was immediately filled with pearlite insulation.

- The interior of the cold box was a "black box" to operations. The equipment and piping systems were not available for review prior to placing them in service.
- The contracting strategy used for "packaged units" was different than the remainder of the cracker process and equipment.

Recommendations

- Consider requiring a member of the operations team to attend all final equipment shop acceptances for packaged equipment that is to be received on site in a condition that prevents a complete inspection or review by operations after it is received.
- Verify that P&IDs represent the actual plant configuration. Flag any piping or equipment system that cannot be easily verified for special consideration.

No Procedure for Task

- The Operations group was not aware of the drain lines located on the bottom of the vapor passes to E-3009X and E-3010X. As a result they were not included in the cold box procedures. Had the drain line been identified the accumulation of NOx liquids could have been prevented or minimized by the routine purging of the lines to flare or the process.
- The purpose of the 1-inch drain lines was unknown at the time of the incident.

Recommendations

€ Consider installing cold service piping to allow the blow down of cold box low point drains to flare on a monthly basis to minimize accumulation of NOx liquids.

Lesson Content LTA

- The training programs provided all relevant information known by the BFLP prior to the incident. The training programs need to be updated based on the new information resulting from this incident and investigation.

Recommendations

- Review the content of your cold box training program, based on the information provided in industry literature, to determine if new information is available and should be added.

References

1. Linde, D.R., CRC Handbook of Chemistry and Physics, 72nd Edition, CRC Press, 1991-1992.
2. Kohler, J., "Cold Box Explosion at Shell Steam Cracker in Berre, France". Paper presented at the 3rd Ethylene Producers Conference at the 1991 National Meeting of the AIChE.
3. Halle, R.T., "Hazards of NOx Accumulations in Cryogenics Ethylene Recovery Facilities", Paper presented at the 5th Ethylene Producers Conference at the 1993 National Meeting of the AIChE.
4. Henstock, W.H. "NOx in the Cryogenic Hydrogen Recovery Section of an Olefins Production Unit", Plant/Operations Progress. Vol 5, No.4. , Page 232 (October, 1986).
5. Grenoble, D.C., "Report on Activities of the NOx in Ethylene Recovery Task Group", Paper presented at the Ethylene Producers Conference at the 2004 National Meeting of the AIChE.

Hazards of NOx in Cryogenic Recovery Systems - Roundtable Discussion

Richard Rodriguez

Shell Oil Company

Abstract:

The EPC Safety Sub-committee will sponsor a roundtable discussion of issues associated with the Hazards of NOx in Cryogenic Recovery Systems. The discussion will provide participants the opportunity to review in more detail the paper presented by BASF (NOx Platefin Exchanger Failure). The NOx Safety Group Members will provide updates on the status of any new issues regarding the Hazards of NOx in Cryogenic Recovery Systems.

ROUNDTABLE PRESENTATION - PAPER NOT AVAILABLE

Energy Incentives & Emission Reductions

Colin P. Bowen
Vice President, Olefins Technology Business Development
Shaw Energy & Chemicals Group

ABSTRACT

The olefin industry has responded over the last 30 years to increasing costs of fuels and feedstocks. Specific energy consumption (SEC) in today's new units is less than 50 percent of earlier units. The capacities of many older units have been expanded, and, in conjunction, efficiencies have been improved.

In North America, due to the relatively low costs of natural gas feedstocks and fuels, until about 2005, energy efficiency revamps have been minimal. To contrast, W. Europe and Japan, with plants of equivalent age, have responded more extensively to efficiency upgrades. Today, their enhanced efficiencies are being further directed by semi-global CO_2 emission reduction requirements.

Now, North America is also recognizing the financial and legal directives to improve efficiencies and emission reductions (CO_2, NOx, et al.) and could take similar action to those which have been implemented by our global competitors. Related revamp incentives include alternative lower cost feedstocks and capacity expansions. Our current challenge is to improve both energy demand and associated emissions. Related improvements will be cited.

INTRODUCTION

Energy efficiency papers and presentations have been recently included in the 2007 AIChE Ethylene Producers Conference and the 2008 European Ethylene Producers Conference. Additional related data has been included in other recent conferences. This paper will, therefore, summarize key statements, recognize future directions, and include relevant efficiency and emission reduction objectives.

ENERGY COSTS

The table below compares energy costs for associated feedstocks. We are all aware of the past/present energy costs; refer to the related data and associated feedstock and product costs:

		June 2008	January 2009
Crude Oil	($/BBL)	140	~40
	($/MMbtu)	25	~7
	($/lb)	45	~13
Natural Gas	($/MMbtu)	13	~6
Light Naphtha	(¢/lb)	76	29
Gas Oil	(¢/lb)	93	55
Ethane	(¢/lb)	60	21
LPG	(¢/lb)	54	28
Ethylene	(¢/lb)	70	21
Propylene	(¢/lb)	82	18

When global economics stabilize, it is expected that energy prices will approach 2008 levels. Although today's energy costs are significantly lower than earlier last year, sensitivities to energy efficiency requirements remain.

US POLITICAL ENERGY POLICY

A further recent change in our energy attitude is linked to the election of President Barack Obama. His key statements following his election are summarized below:

IMPORTED CRUDE & VEHICLE IMPROVEMENTS US import of crude oil and related products (~13 MMBPD) must be reduced; this is particularly linked to our gasoline consumption. Vehicle fuel consumption standards will be increased 4 percent per annum. Associated government financial inputs will be provided to manufacturers to re-tool their units. Tax credits (~$7 thousand) will be allocated for purchasing advanced technology vehicles. By 2015, the target is to deploy 1 million plug-in hybrid cars that achieve 150 miles/gallon.

Crude Tax and Concessions US crude oil import companies will be additionally taxed for the next 5 years. This fund will be allocated to low-to-middle income families to offset energy costs.

Activate Leased US Oil and Gas Locations Encourage companies to proceed with further domestic oil and gas recovery developments (68 million acres available). Further offshore allocations will be made.

Biofuels Target biofuel domestic production incentives will be assigned. Additional cellulose ethanol of 2 billion gallons to be produced per year during his first term (2013). Objective is to produce 60 billion gallons per year by 2030.

Nuclear Power Future units have been confirmed/supported. For example, The Shaw Group has recently been awarded six units in the US and four in China. Nuclear waste disposal is still an issue.

Carbon Dioxide Emissions By 2020, our emissions should be no higher than those of 1990. Within that time frame, fuels will be upgraded to achieve 10 percent less carbon content. By 2050, emissions should be reduced to only 80 percent of 1990 output.

Renewable Electricity By 2012, we should generate 10 percent of our electricity from renewable sources (wind, solar, geothermal). Tax concessions will be allocated accordingly. The future target is 25 percent renewable by 2025.

More Efficient Homes Financial incentive is planned to improve both existing and future home energy consumptions.

Impact on our Industry Instructions regarding emissions and indirect incentives, along with feedstock and energy costs/consumptions, are anticipated.

Current Global Industry Objectives

The global economy has affected the olefin industry with respect to feedstock, fuel, and product prices. Consequently, global operating percentages have recently dropped from ~95 percent to ~80 percent. Several construction projects have been delayed and future projects deferred or cancelled.

As a consequence, several production companies have temporarily or even permanently shut down certain units. In preparation for future crude oil and fuel cost increases, these companies are asking contractors to recommend lower cost feedstock substitutions, minimum cost energy efficiency improvements, and associated emissions reductions. Previous revamp projects always focused on higher production capacities; that is not today's objective for almost all US and EU plants. Their Mid-East and Far-East joint ventures are still focusing on that aspect.

Current Specific Energy Consumption (SEC)

North America has many (~20) ethylene plants currently operating which were originally built over 30 years ago. At that time typical SEC values were ~12,000 btu/lb C_2H_4 for ethane crackers and ~18,000 btu/lb C_2H_4 for naphtha crackers.

North American operating companies have made certain SEC improvements for these older plants, but only to a limited extent. Many natural draft furnaces (~86 percent firing efficiency) have been upgraded with induced draft fans (~93 percent firing efficiency). Several recovery units have been expanded and converted (~30 percent energy savings). Operation process control advances have also reduced actual SEC to its designed guarantee level. Such improvements have reduced SEC by ~25 percent max. However, current SEC values in latest designs are ~50 percent lower than the earlier units. The key design improvements were listed in my 2007 EPC paper; refer reference number one at the end of this paper.

Most North American producers now recognize the SEC disadvantage of their plants versus those which have been constructed elsewhere more recently. We also recognize that older plants of similar age in Europe and Japan have been much further upgraded. The North American industry is actively reviewing improved SEC options. Capacity expansion initiatives are secondary to those that provide improved efficiency. Alternative feedstocks, improved selectivity, extended operations, and future CO_2 emission savings are also linked to the SEC improvements and associated costs. Improvements will be recommended in the following sections.

SEC Improvements

In reviewing the SEC situation, it is logical to compare existing equipment efficiencies against current available efficiencies. Several such improvements could be accomplished within a plant revamp.

Table 1
Current Equipment Efficiencies

Cracking Furnaces:	**Thermal η ~94%**
Main Frame Compressors:	**Polytropic η ~85%**
Main Frame Steam Turbine Drivers:	**Isentropic η ~80%**
Main Fractionator Towers:	**R/D ~1.1 - 1.3 x R/D min**
Principal Heat Exchangers:	**Ambient ~10°F ΔT**
	Cryogenic ~ 5°F ΔT
Major Pumps:	**Efficiency ~70%**
Turbo-Expanders:	**Isentropic η ~90%**

In this section, recent design improvements will be restated, which will enhance energy efficiencies. This is a brief description of multiple options; further details can be subsequently supplied.

Cracking Furnace Improvements

Many existing US cracking furnaces were installed over 30 years ago. Many have been metallurgically and mechanically upgraded, but further improvements could be realized. Refer to the improvements listed below, many of which have been incorporated into overseas furnace revamps:

- ID fans to allow higher convection section ΔP and heat recovery
- Gas turbine integration (GTI) retrofits
- Combustion air pre-heat systems – reduced fuel firing
- Lower temperature flue gas; corresponding enhanced metallurgy to avoid contaminant condensation and corrosion (~85°C)
- Improve radiant coil metallurgy, geometry, uniform firing
- Conversion from expensive to lower cost feedstocks
  - either naphtha/atmospheric gas oil (AGO) to hydro-cracker residue (HCR) or LPG condensate
  - or naphtha/AGO to ethane/LPG
- Older furnaces completely replaced, or convection sections replaced, or converted from two separate units to a single larger dual firebox unit
- Selective catalytic reduction (SCR) with convection installation/replacement to reduce NOx
- Improve burners to reduce NOx
- Further quench exchanger heat recovery; flexible TLX-1, TLX-2 arrangements to maximize recovery for alternative feedstocks

Quench Oil Tower (QOT)/Quench Water Tower (QWT) Improvements

- When pyrolysis fuel content in the furnace effluent is above 2 percent by weight, the QOT bottoms temperature is sufficiently high to generate MP steam.
- If flowrate has increased significantly, the vapor-liquid contacts and fractionation results often deteriorate significantly. Upgrade the vapor-liquid inlet distributor and improve uniform non-fouling trays. Optimize operating pressure with respect to upstream furnaces and downstream cracked gas compressor (CGC).
- For ethane cracking QWT, the major problem is to effectively remove the pyrolysis tar from the QW to avoid tower and exchanger fouling. Key success is linked to the pre-saturator section of the tower and its associated tar-water-gasoline separator.
- Dilution steam feedwater must also be free of any remaining tar and gasoline. Success can be achieved via our proprietary pre-treat contactor.
- Current ethane crackers now employ feedstock dilution steam saturators, a reduced energy route.
- Improved QW exchangers incorporating twisted tubes to improve heat transfer coefficient (HTC) and avoid fouling.

Cracked Gas Compressor (CGC)

- Current compressors employ aerodynamically improved flow paths, impellers, dry gas seals, anti-fouling internal coatings, and enhanced control facilities, resulting in higher operating efficiency and availability than previous designs.
- To minimize polymerization of cracked gas on the compressor internals, individual compressor stage gas temperatures are suppressed by continuously injecting cool boiler feedwater (BFW). In the event polymer build up occurs, wash/solvent oil is injected on an intermittent basis for cleaning.
- Several 30-plus-year plants have replaced their cracked gas compressors and steam turbines to improve operational capacity, availability, and efficiency.
- The latest units have installed high capacity, efficient intercoolers; often a single unit for each service. The corresponding shell side/tube side allocation will depend on the cooling water quality. CG inter-stage ΔP can thereby be reduced.
- Minimum internal vapor-liquid recycles are specifically achieved. Condensates can be pumped forward rather than flashed backward.

Acid Gas Removal

- Earlier designs based on heavy feedstocks anticipated relatively high NH_3, CO_2 and oxygenates. The acid gas removal system often incorporated a pre-treater (amine) and a follow-up treater (NaOH). These towers were sometimes equipped with a spare unit.
- Today, Shaw has improved the caustic tower design significantly and employed a higher quality NaOH contact fluid. Accordingly, most units can achieve the necessary acid gas removal with a single tower incorporating multi-stage circulations.

- Spent caustic from the tower is then treated with a patented pre-treat system to produce a hydrocarbon-free caustic sulphide. This treated caustic stream can either be sold as an organic-free product for the pulp/paper industry and noble metal manufacturers, or be more efficiently treated in a wet air oxidation (WAO) unit. That treatment converts sulphides to sulphates prior to environmental emissions.

Front-End Hydrogenation System

- Virtually every new unit employs such a system. Very few older units with back-end hydrogenation have converted to front-end systems.
- The hydrogenation unit typically reacts the C_3 lighter or the C_2 lighter systems. Certain recent units employ coordinated fractionation and hydrogenation for C_4 lighter units. Of course, such a system also converts butadiene to butenes.
- Another recent announcement involved pre-condensing the C_2 or C_3 lighter streams to effectively eliminate the excess hydrogen prior to the reaction.
- Catalyst manufacturers have also improved catalyst quality. Certain schemes employ adiabatic hydrogenation as many ethylene licensors did earlier on back-end units. Selectivity is slightly better than adiabatic systems.
- Ethane crackers sometimes route the entire cracked gas to the hydrogenation unit prior to related fractionation facilities. Circuit pressure drop is somewhat lower.
- The major advantages of front-end hydrogenation systems are conversion selectivity, much higher operating cycles and constant higher quality, energy efficiency, installed cost, and feedstock flexibility.
- Another associated aspect is the renewed interest in extracting acetylene rather than hydrogenating it. In turn, this will reduce overall energy consumption by eliminating the hydrogenation facilities, reducing system pressure drop, and increasing overall product cost benefits.

Cryogenic Fractionation

- Improved energy efficiencies in cryogenic systems have been incorporated into recent grassroots units. Several plants were consequently expanded in capacity and reduced in their recovery SEC. However, if today's primary objective is to improve efficiency at essentially constant capacity, it is also necessary to understand the associated need to replace existing older equipment. Such replacements have included plate-fin exchangers or cryogenic towers which corroded, cryogenic pumps with poor reliability, and associated reaction facilities. In certain cases, Shaw has? assisted long-term clients with redesign in conjunction with such equipment replacements.
- In certain cases, we increased the recovery of their hydrogen. In other cases, we actually reduced cryogenic duty by producing dilute ethylene product for alternative derivative facilities. Several current steam cracker and catalyst FCC units produce only dilute ethylene and their refrigeration duty is significantly lower.

- Various refrigeration systems also can be improved by revamping or replacing existing refrigeration compressors and drivers with current technology. Similar improved designs for hydrogen/methane expander-recompressor systems are also available.

Improved Operational Routes

- Most major petrochemical companies have recently focused on improving actual operations of their units. This involved analyzing critical equipment performance, instructing and encouraging operators to minimize energy margins, and making the necessary operational adjustments as alternative feedstocks are substituted and/or cracked under alternative conversions.
- Extensive details of such operational improvements have been previously presented (Texas Technology Showcase, Dec. 2006). A similar outline related to operational analysis and investigation was presented by LyondellBasell (AIChE STS, Feb. 2009). Several companies claimed typical energy efficiency improvements of up to 15 percent for such operational enhancements. Typical operating enhancements include:
  - Optimized steam system arrangement to minimize condensation and let-downs
  - Optimized cooling water (CW) pump operation to avoid unnecessary duties
  - Tighter operating inputs similar to performance test conditions
  - Advanced/optimized process control schemes
  - Upgraded inside battery limit (ISBL)/outside battery limit (OSBL) cogeneration facilities
  - Minimize steam condensate rejection
  - Minimize deaerator steam input and vents
  - Flare system recovery as supplementary fuel gas

Future Efficiency Opportunities

Most overall design improvements have been previously listed in earlier AIChE energy sessions. As most recent plants are relatively high capacity, one can justify incorporating high efficiency designs and equipment.

A critical energy decision is linked to the upstream/downstream OSBL elements. In the case of several recent ethane crackers, the SEC can be significantly reduced if the feedstock is delivered as liquid. The ethylene product delivery condition also affects the SEC. Another element is linked to the import and export of various steam streams. If condensation of various steam elements can be conveniently carried out in external units, overall efficiency is improved. A similar link is the electricity generation, which is typically carried out with an associated cogeneration facility in the OSBL power unit. A few ethylene units have integrated the gas turbine cogen system with cracking furnaces. Such schemes can reduce the SEC by up to 10 percent.

Several current designs focus extensively on linked ISBL/OSBL energy generation. Key decisions are related to respective steam generation pressures, associated condensate rejection and treatment, and standby supplies during startup/shutdown.

Potential energy efficiency concepts, which were announced in preceding sessions, include the following:

- Pre-fractionation via ceramic zeolite separation devices
- Pressure swing adsorption (PSA) for product recovery from waste streams
- Inorganic membrane separation to remove hydrogen without organic condensate fouling, which could be applied to off-gas streams and cracked gas streams. Latest systems operate at convenient temperature/pressure conditions at high flux rates. Single phase ~95% mol H_2; double phase ~99% mol H_2
- Further membrane applications to pre-separate olefins from paraffins
- Integrated catalytic hydrogentation – fractionation schemes (where catalyst performance is adequate)

NON-ETHYLENE ENERGY IMPROVEMENTS

It is important to be aware of recent developments in the power generation industry. A future intent is to generate steam at much higher pressures and higher temperatures (~275 barg, ~600°C, or ~4000 psig, ~1100°F) in order to upgrade the electrical power production efficiency (~43 percent) versus standard ethylene unit efficiency (~30 percent). This potential super-critical power generator scheme incorporates multiple series steam turbines with the initial exhaust steam re-superheated. The final steam exhaust pressure is also low (~1.4 in. Hg) versus typical exhaust (~4 in. Hg). Power generation steam turbines are designed to operate at constant rotational speed, and will require design modifications to operate as variable speed drivers in ethylene units.

The typical olefin unit highest steam pressure/temperature is ~130 barg, ~520°C (~1900 psig, ~970°F). The industry has employed such schemes for almost 40 years and they are well proven. However, when power generation partners have proven their latest scheme, related specialists should evaluate and recommend what similar schemes might be incorporated in future olefins units.

Another efficiency improvement in future power units is the extended recovery of flue gas heat content. LP steam recovery schemes have been developed which incorporate the BFW pre-heat where flue gas exhaust temp is ~ 50°C. Olefin unit stream cracking furnace flue gas exhaust temperature is rarely much lower than ~ 120°C, although certain overseas units have been constructed down to 85°C. This requires the related convection section elements to be of either higher quality metallurgy or ceramics. As we know, if the exhaust temperature could be reduced from 120°C to 50°C per latest power plant designs, it would increase overall furnace efficiency from ~ 93percent to ~ 97 percent.

We are all aware of the previous and potential links between nuclear electrical technology and the olefins industry. When France installed 80-plus nuclear units, their conventional ethylene units were converted from steam turbine compressor drivers to variable speed electric motor drivers. Operating reports are very positive and such facilities are now operating at a lower cost. As additional nuclear units are now being installed in several global regions (The Shaw Group is engaged in many such units), this may well be an appropriate power source for new olefins units.

I personally moved from our European office to our US office exactly on the fateful day of 28 March 1979. Unfortunately, that was when the Three Mile Island nuclear power plant failure occurred. This event and others worldwide, essentially blocked further nuclear units for this continent; only 30 years later have we proceeded with new capacity.

Additional alternative energy sources are now of increasing interest because of their non-hydrocarbon fuel sources and resulting lack of CO_2 exhausts (or non-taxed CO_2 exhausts). Current energy capacities are relatively small; therefore, their direct contribution to our olefin units is unlikely. However, we should recognize these options which may, in the future, contribute to our imported energy.

Wind power generation is increasing. In 2005, total US wind power was 7,000 MW, which represented only 3.5 percent of renewable energy. Today at ~30,000 MW (installed and under construction), wind power represents ~15 percent of renewable energy, which is almost 1percent of total energy generation. Key state wind power capacities (MW) are:

State	**Existing MW**	**MW Under Construction**
Texas	7,116	1,651
Iowa	2,790	20
California	2,517	275
Minnesota	1,752	0
Washington	1,375	70
Colorado	1,068	0
Oregon	1,067	250

Current unit power cost based on capital cost is ~ \$0.03/kWh.

Larger capacity wind units (up to 5 MW) cost relatively more than current conventional units (~\$0.05/kWh) because of the much higher component costs of the large diameter blades (~60 M radius) and associated facilities. The standard wind power generators produce DC current which is then fed to an inverter system that delivers AC at the specific system frequency. Of course, power generation is directly linked to the atmospheric wind conditions, so for such a system to be associated with an olefin unit would be difficult.

To avoid the electricity production variations, a different wind turbine scheme is under development. Instead of DC generation, the wind turbine nacelle is fitted with a lightweight four-stage compressor/intercooler system discharging at ~100 barg. The high-pressure air is then delivered, stored, and expanded through a regional turbine to produce electricity as needed. Such a scheme, based on compressed air energy storage, also could possibly be integrated with olefin units in the future where compressors and cooling systems could be driven by such units.

There are solar energy systems now being added to many industrial and domestic facilities. Current cost is relatively attractive and tax incentives are a key element. Major solar schemes have been installed in several US states to generate electricity. Earlier solar systems generated steam, but today the proven photovoltaic system based on silicon semiconductors to process the solar-generated photons is the accepted scheme. Based on limited size, such systems are only of minor potential value to olefins units today.

A further scheme which could be linked indirectly to olefin units is the low impact hydro-power system. Earlier conventional hydro-power units produced high energy but today, various countries are incorporating lower impact systems into their power generation. A major potential system is the tidal energy scheme which has already been installed in several European areas.

Geothermal heat generation might also be relevant to several units. Underground heat sources typically heat existing water to ~100-200°C. Certain units actually circulate such water. A few geothermal facilities produce high pressure water at up to 700°C. However, it should be recognized that, in most cases, the water contains various contaminants (H_2S, NH_3, CH_4, CO_2, et al.).

PLANNED ENERGY EMISSION REDUCTIONS

A related aspect to energy efficiency is emission reduction. We are all aware of current policies regarding NOx emissions. Many companies today employ selective catalytic reduction (SCR) systems to essentially eliminate NOx for flue gas. We are all also aware of improved steam cracker firebox/burner improvements which have reduced NOx from ~120mg/NM^3 to ~40mg/NM^3. Related details have been provided in several previous EPC presentations.

The key future emission challenge, however, is CO_2 removal – carbon capture and sequestration (CCS). Several imminent European electric power generators have incorporated such a system. European olefin units are examining related options. Unfortunately, the related energy cost to add CCS to power generators is ~20 percent of total power production. In order to extract CO_2 from the boiler flue gas, it is necessary to treat it with amine absorption system, compress it, remove major contaminants, condense it, and deliver liquefied CO_2.

Given that the anticipated CO_2 emission tax cost could be up to 60EU/ton CO_2 in Europe by 2012, it is necessary to reduce emissions directly or indirectly.

An alternative scheme to achieve CO_2 removal, which is being installed in a few power generation boilers, is the oxy-combustion approach. This process provides high oxygen combustion vapor; hence the flue gas consists essentially of CO_2 and H_20. Correspondingly, the CO_2 can be more conveniently condensed and sequestered. The radiant firebox conditions will be a much higher temperature; therefore, it may be necessary to recycle flue gas to avoid overheating. Radiant heat transfer coefficients also need to be checked and adjusted if such schemes are revamped rather than newly constructed.

Similar issues apply to standard olefin units. One interesting aspect of such an oxy-combustion scheme is linked to future designs where radiant coils are upgraded to higher temperature operation. This is done to further reduce thermal cracking residence time and selectivity. A related element of such a system is the change in radiant versus convection heat duties, enhanced refractory linings, modified burners, and adjusted heat transfer coefficients.

FUTURE ETHYLENE OPTIONS

As industry seeks to improve ethylene production energy and emissions, certain potential future routes must be recognized. This is a brief concluding section. More extensive details have been presented, and future operational results will be subsequently presented.

Last year, I presented the latest fluid catalytic cracking routes to the production of propylene and ethylene. I described our latest catalytic pyrolysis process (CPP) which can achieve olefin yields similar to heavy liquids steam cracking. Energy consumption and cost is lower since coke combustion in regenerators is essentially zero cost, and associated regenerator steam boilers produce additional energy sources. FCC catalytic units consume heavier, lower-cost feedstocks, but associated cracked gas contaminants are more extensive. Such contaminant removal is of minimum energy cost and ultimate contaminants and emissions are essentially zero.

The industry is aware of the latest methanol-to-olefins and methanol-to-propylene (MTO/MTP) schemes. This is a potential lower cost feedstock source (low-cost natural gas and/or coal gasification) but is a higher cost installation system and energy consumption route. Similar schemes based on alternative low-cost feedstock gasification are under review but not imminent. Gasification projects are becoming increasingly attractive to coal-based countries (e.g., USA, China), with principle objectives being fuel production rather than MTO/MTP.

Bio-routes to produce ethanol to subsequently produce ethylene have been installed in a few countries. The latest versions are underway at some sites. This is a potentially lower energy cost route, but ethylene capacity is minimal. Most ethanol is still directed towards gasoline fuel injection.

Associated cracking/conversion of heavier olefins is also a related route. Several such processes are in service and under development.

Presented in this paper are the imminent alternative routes to lower olefins. The industry needs to update and upgrade existing units and install current and/or future technologies to achieve energy and emission objectives. Professional details are available to clients to our mutual benefit, and, as mentioned, potential project performance information will be presented at future EPC sessions.

References:

1. Bowen, CP, (2007), "Olefins Energy Management Review," *AIChE EPC*, Houston, Paper No. 124A.

2. Bowen, CP, (2008), "Energy Consumption of Crackers," *Tutorial Session EEPC*, Madrid.

3. President Obama, (2008), "Energy Policy," *Hart Energy Solutions Conference*, Houston.

4. Bowen, CP, Chris Santner and Dilip Dharia, (2008), "Advances in Catalytic Routes to Olefins," *AIChE EPC*, New Orleans, Paper No. 219D.

Energy Management System (EMS) development and implementation in Ethylene Plants

Camila H. Wakano

Consultant

Shell Global Solutions (US) Inc.

Jack Buehler

Consultant

Shell Global Solutions (US) Inc.

Abstract: Rising energy costs and environmental concerns increasingly require companies to improve their energy efficiency in order to reduce operating costs, increase profits and reduce greenhouse gas emissions. This paper describes the development and implementation of an Energy Management System (EMS), a tool that helps promote energy efficient operation. The paper describes work done for two Shell sites, each having 2 ethylene plants. The average benefit of an effective EMS is estimated to be 3% to 5% reduction in the site's energy cost. Consequently, an EMS can be an effective tool to help sites minimize CO_2 emissions.

An EMS is a combination of structured management processes and monitoring tools, linked to real-time plant data systems, that aims to continuously reduce energy costs, bottlenecks and emissions, by human intervention.

Monitoring tool elements include items like measurement requirements, identification of key performance indicators (KPI) and drivers, (i.e. any process variable that influences the changes and variability in the selected KPI), software and hardware specifications and graphical user interface requirements. As an example, a KPI for an olefin plant would be the plant's total energy consumption per unit high-valued chemical (HVC) production, while for the ethylene cracker furnace it could be the total energy loss through the furnace stack caused by deviation of drivers like excess O_2 and/ or high stack temperatures away from their targets.

Management process elements, which include the definition of responsibilities and accountabilities for energy utilization and encouragement of participation and involvement of personnel across all levels of the site, ensure energy optimal operation of the plant.

Energy Management System (EMS) development and implementation in Ethylene Plants

Rising energy costs and environmental concerns increasingly require companies to improve their energy efficiency in order to reduce operating costs, increase profits and reduce greenhouse gas emissions.

This paper describes the development and implementation of an Energy Management System (EMS), a tool that helps promote energy efficient operation by sharing and analyzing data in an orderly manner that identifies poor performance and provides means to improve it. The paper describes work done for two Ethylene Plants.

The average benefit of an effective EMS is estimated to be 3% to 5% reduction in the site's operating energy cost, which is the typical benefit of maintenance of energy operational excellence against erosion and neglect. Consequently, an EMS can be an effective tool to help sites minimize CO_2 emissions.

1. What is an Energy Management System (EMS)?

An Energy Management System (EMS) is a combination of structured management processes and monitoring tools, linked to real-time plant data systems, aiming to continuously reduce energy costs, bottlenecks and emissions, by human intervention.

Monitoring tool elements include

- measurement requirements
- identification of key performance indicators (KPI)
- identification of drivers, i.e. any process variable, external influence or operating constraint that influences the changes and variability in the selected KPI
- software and hardware specifications
- graphical user interface requirements

As an example, a KPI for an olefin plant would be the plant's total energy consumption per unit high-valued chemical (HVC) production, while for the ethylene cracker furnace it could be the total energy loss through the furnace stack caused by deviation of drivers like excess O_2 and/ or stack temperatures away from their targets.

Management process elements ensure energy optimal operation of the plant through energy centric work practices, and it includes

- definition of responsibilities and accountabilities for energy utilization
- encouragement of participation and involvement of personnel across all levels of the site

Ultimately, an EMS provides information to appropriate personnel within an organization to help them manage energy use and costs.

2. What does an EMS deliver?

The key deliverables of an EMS program are as follows:

- *Early detection of poor performance*

Comparison of actual performance against target combined with frequent performance monitoring allows early detection of poor performance.

- *Support for investigation of poor performance*

An effective drill-down mechanism helps in the diagnosis of possible causes of poor performance. Information is structured and organized in order to facilitate the energy analysis, guiding the user from the observed deviation in energy usage (e.g. overall poor performance of the plant) to the root cause (e.g. incorrect control set-points on the equipment).

- *Identification and justification of energy projects*

A proper EMS fosters discussions and investigations on energy usage across all levels of the site and those can easily lead to identification of energy projects. Those projects can either involve purely operational excellence or low to high capital expenditure. Energy data from EMS helps in the justification of capital projects.

- *Energy reporting*

Regular and periodic reporting to management, engineers, operators and other relevant personnel ensures that those responsible for performance are taking effective action. In addition, easy to understand energy reports tailored for different user groups at the site demonstrate the importance of proper energy management.

- *Evidence of success*

By providing energy information in an organized and structured way, EMS is a means to quantitatively demonstrate reduction in energy use and/or cost.

- *Maintenance of energy performance against erosion*

Once energy goals have been achieved, the EMS helps sustain the energy performance achieved, thereby preventing its erosion over time.

The average benefit of an effective EMS is estimated to be 3% to 5% reduction in the site's energy cost, which is the typical benefit of the maintenance of energy operational excellence against erosion and neglect.

3. EMS Development in Ethylene Plants

The Energy Management System (EMS) described in this paper was recently implemented in two Shell Chemical sites, each having two ethylene plants.

The EMS was developed by Shell Global Solutions[1] consultants in conjunction with the energy focal points and process engineers at each site. This initiative served as a pilot of the development of a standard EMS to be implemented globally in all Shell Ethylene Plants.

In this paper, the EMS development consists of the system design, concept and functionality phases. The implementation phase of the EMS is the system building. Although the system design can be changed during the implementation phase, the bulk of the design decisions were made during the *"EMS Design Workshop"*, which is a two-day meeting with the personnel involved in the EMS development to determine the EMS characteristics.

This section summarizes the EMS design decisions made during the workshop.

3.1 Management Process Elements

The EMS design phase focuses heavily on the monitoring elements, e.g. KPI and driver identification or view structure design; however, some aspects of the Management Process Elements should not be overlooked during this phase.

3.1.1 Management Commitment

Before the EMS development phase starts, at least the management team should be fully engaged and committed to the program. This is critical because only with management buy-in is it possible to keep energy efficiency and, consequently, the EMS development, implementation, maintenance and application as a priority to the appropriate personnel within the organization.

A plan to engage the lower levels of the organization can be discussed during the development and implementation phases of the EMS.

3.1.2 Expectations

Since EMS is a people oriented program, it is vital to manage the site's expectations. The deliverables expected for the system should be clear to all involved.

In this particular case, the sites' management made two specific requests that significantly impacted the decisions regarding the EMS development:

- EMS should be kept as simple as possible;
- Development and Implementation should be completed in 3 months.

3.1.3 Manpower

The EMS development mainly occurs during the *"EMS Design Workshop"*, a two-day meeting between the personnel involved in the EMS development to determine the specific site's EMS characteristics.

For this particular EMS development, each site assigned an energy focal point. Usually, the energy focal point is a process engineer familiar with most areas of the site and can provide information and contacts within the site. The energy focal point is also responsible for making all the decisions regarding the EMS development and implementation, or escalating them to the proper personnel. The energy focal point is the ultimate owner of EMS.

Besides the energy focal point, two Shell Global Solutions consultants were assigned to the program, one for each site. Basic functions of the Shell Global Solutions consultants were

---

1 Shell Global Solutions is a trading style used by a network of technology companies of the Shell Group

to facilitate the sites' decisions regarding EMS by providing information and knowledge based on past experiences with the program and to build the system.

Additionally, another Shell Global Solutions consultant, an olefins specialist, participated in the EMS design.

3.2 Monitoring Tool Elements

Decisions regarding the monitoring tool elements for ethylene plants are summarized in this section.

3.2.1 *Hardware and Software Specifications*

An EMS requires sensors and instruments to collect process information and monitoring software to analyze the data.

Sensors and instruments include utility meters (electricity, fuel gas, steam, etc.) and other meters associated with the energy use such as temperature, pressure, flow and composition meters. The site usually already has most of the necessary sensors and instruments installed. However, often times extra measurements and re-calibrations are required.

The monitoring software requirements for an EMS include:

- Calculation requirements: EMS calculations require general algebraic calculations of multiplication, addition, etc. In addition, steam table calculations such as enthalpies of steam at determined temperature, pressure and/or quality conditions are recommended for accuracy.
- Display requirements: The effectiveness of an EMS hinges significantly upon the ability to display real time current and historical trends of several key calculated variables in an organized and structured manner. Capability of trend plots at user selected frequencies and over a user selected time span is absolutely essential. The existence of hot buttons to facilitate navigation within the system is preferable.

Both Ethylene Plants discussed in this paper have been using RADDICAL INSIGHT®, a user friendly interface to the process data historian developed by Shell Global Solutions, to monitor and manage their process. The same software was used as the platform for the EMS application.

3.2.2 *Hierarchical Structure Definition*

In EMS, information should be available and displayed in a structured and organized manner. Since the functionality and the purpose of the information vary according to the level of the organization, a hierarchical structure is recommended.

A chemical site can be divided into 4 different levels:

1. Site level (e.g. chemical site)
2. Unit level (e.g. ethylene plant A, ethylene plant B, butadiene plant, etc.)
3. Sub-unit level (e.g. group of furnaces, group of compressors, part of a unit, etc.)
4. Equipment level (e.g. column, furnace, compressor, etc.)

An illustration of the 4 levels of a chemical site is presented in Figure 1.

This paper limits the EMS discussion from the unit level of an ethylene plant down to its equipment level only.

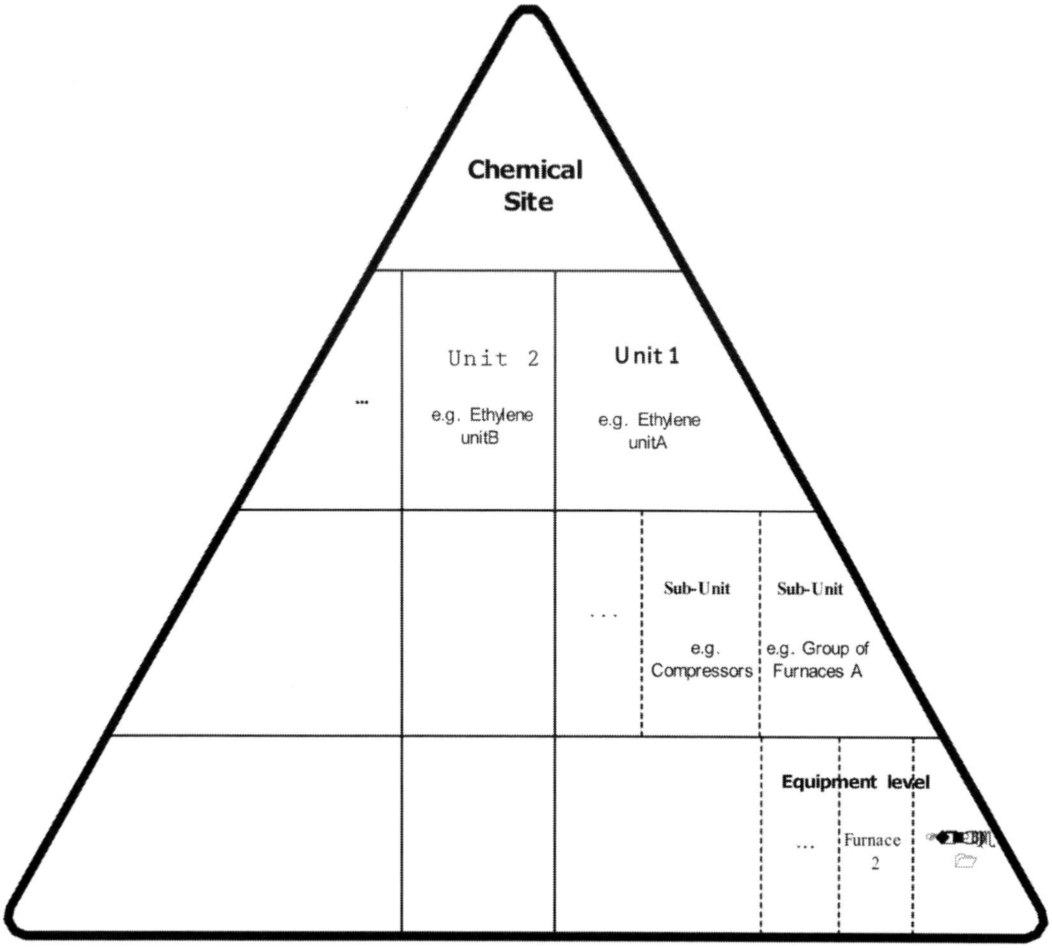

Figure 1 – EMS hierarchical structure

3.2.3 *System Boundaries Definition*

In this paper, the unit system boundaries are already determined: ethylene plant. Next step is to determine the sub-unit system boundaries.

In order to properly define the sub-unit system boundaries for the EMS the following criteria should be weighed

- Existing system boundaries already defined by common process and/or common management;
- Metering availability;
- Time availability for EMS implementation.

In the ethylene plants discussed in this paper, generally the following sub-units were selected to be included in the EMS:

1. Group of Furnaces A
2. Group of Furnaces B
3. Group of Furnaces C
4. Compressors
5. Pyrofrac, or Pyrolysis Quench Tower

Due to the large number of furnaces in the ethylene plant, a group of furnaces was selected. Its selection could be determined by common functionality (e.g. ethane furnaces), physical proximity or common management. In this case, all compressors were combined in the "Compressors" sub-unit, but more than one group could have been identified. Similarly for the pyrofrac, if more than one tower exists, it could be another sub-unit (pyrofrac A, pyrofrac B) or they could be combined in one sub-unit.

It was decided not to include the remainder of equipments, because of time constraints. Besides, the furnaces and compressors alone represent most of the total energy consumption of the ethylene plants. Pyrofrac was included due to its steam generation capability.

The equipment level system boundaries are straightforward for the group of furnaces and compressors. Each furnace and each compressor train is considered as one equipment in the equipment level. For the pyrofrac, each steam producer of the pyrofrac system was considered as one "equipment" at the equipment level.

3.2.4 *Key Performance Indicator Identification*

A Key Performance Indicator, or KPI, is an overall performance factor used to measure or compare progress towards a target. Performance deviations from the target are used as the initiation of corrective and improvement actions.

For EMS purposes, KPIs have to be established at all levels of the organization. In this case, the KPIs identified for the unit, sub-unit and equipment levels in the Ethylene Plants are presented in Table 1.

Table 1 – Key performance indicators identified for an ethylene plant

Level	Area	KPI	KPI description	Metric unit
Unit	Ethylene Plant	Specific Energy Consumption	Total energy consumption of the Ethylene Plant per unit high-valued chemical (HVC) production	BTU/lb of HVC
Unit	Ethylene Plant	Economic Gap	Economic deviation of Unit KPI and its target	M$/day*
Sub-unit	Group of Furnaces A/B/C	Specific Energy Consumption	Total energy consumption of group of furnaces per unit of total group of furnaces feed	MMBTU/Mlb of feed **
Sub-unit	Compressors	Specific Energy Consumption	Total energy consumption of compressors per unit of total discharge	BTU/lb of Discharge

Sub-unit	Pyrofrac	Specific Energy Generation	Total generated energy production per unit high-valued chemical (HVC) production	Generated MMBTU/Mlb of HVC
Equipment	Furnace	Specific Energy Consumption	Energy consumption of the furnace per unit of feed	MMBTU/Mlb
Equipment	Furnace	Stack O2 Energy Loss	Energy loss due to stack O2 deviation from its target	MMBTU/h
Equipment	Furnace	Stack Temperature Energy Loss	Energy loss due to stack temperature deviation from its target	MMBTU/h
Equipment	Furnace	Economic Gap	Economic deviation of Furnace KPI from its target	M$/day
Equipment	Compressor	Specific Turbine Steam Mass Consumption	steam consumption per unit discharge	lb of steam/lb of discharge
Equipment	Compressor	Specific Turbine Steam Energy Consumption	Energy consumption per unit discharge	BTU/lb of discharge
Equipment	Compressor	Economic Gap	Economic deviation of the compressor KPI from its target	M$/day
Equipment	Part of the Pyrofrac	Specific Steam generation	Steam generation per unit high-valued chemical (HVC) production	lb of generated steam/lb of HVC
Equipment	Part of the Pyrofrac	Specific Steam generation	Low-pressure steam generation per unit total dilution steam generation	lb of generated low-pressure steam/lb of dilution steam generated

* The prefix "M" before the metric unit refers to thousands.
** The prefix "MM" before the metric unit refers to millions.

The unit level has two KPIs: specific energy consumption and the economic gap. The specific energy consumption in BTU per unit high-valued chemical production was the metric already used in the sites and remained the same. The economic gap helps quantify the gap between actual and target KPIs in dollars (of lost opportunity due to inefficient operation), which can be an effective tool to drive awareness of energy efficiency.

For the group of furnaces and compressors, the KPI chosen was specific energy consumption. Since the pyrofrac is an energy recovery unit (i.e. it generates energy), its KPI is energy generation per unit high-valued chemical production.

At the equipment level, each furnace has four KPIs: specific energy consumption, stack O_2 energy loss, stack temperature energy loss and economic gap. Understanding where energy is lost and what process parameters drive that loss, are means to ensure that these losses are minimized. Due to time constraints, energy losses were calculated only for furnaces.

Specific steam consumption for the compressors were calculated in two different ways: mass of steam per unit discharge flow and energy of steam per unit discharge flow. This was done because the process engineers for the area were accustomed to the units of steam in pounds (lb). The specific energy in BTU per discharge is more accurate because it takes into account the temperature and pressure of the steam. This KPI will help pick up any drop off in superheat temperature or inlet pressure of the steam at the inlet to the turbine and vacuum deterioration on condensing turbines. Therefore, both KPIs were included.

The pyrofrac was separated into different parts, each containing a steam generation system. Specific steam generation was calculated in two different ways: mass of steam generated per unit high-valued chemical production and mass of low-pressure steam generated per unit high-pressure steam produced (dilution steam). The latter was calculated because in the case of the sites studied in this paper, one of the goals at the pyrofrac system was to maximize the amount of the dilution steam produced from heat recovered in the pyrofrac.

3.2.5 *Driver Identification*

A driver is any process variable that influences the changes and variability in the selected KPI. Hence, drivers form the focus of attention in the attempts to optimize the KPIs relative to their target values.

Driver variables can be classified broadly into two categories: non-actionable parameters versus actionable variables that can be manipulated.

Non-actionable parameters are those influential parameters that are beyond operator control. Examples include ambient temperature, product prices, or operating constraints due to safety.

Actionable variables, on the other hand, are variables that operators can change to affect the operation. Examples are reflux ratios, reboil rates, recycle loops flow and furnace air-fuel ratio.

The term "driver" is used in this paper for any causal and controllable variable that directly or indirectly influences the KPIs.

Table 2 presents the drivers selected for each piece of equipment in the ethylene plants discussed in this paper. They were identified by means of engineering judgment and plant experience. It's important to realize that all plants are different in their detailed design and there may be some specific equipment, or an unusual utility configuration, or an extra recycle etc. that is unique to this plant. Therefore, site personnel involvement in the identification of drivers is required in order to identify such specific drivers.

Table 2 – Drivers identified for an ethylene plant

Equipment	Drivers
Furnace	- Stack O_2 - Stack temperature - Draft - Steam generation per fuel gas consumption
Compressor	- Kickback flow - Discharge flow - Suction pressure - Suction temperature - Discharge pressure - Surface condenser vacuum - Interstage temperatures
Pyrofrac	- Tower pressure - Tower overhead temperature - Reflux ratio - Quench oil gravity - Dilution steam generation flow

3.2.6 *Target Setting*

Target setting is an important element of the EMS because the visual identification of the gap between the trend of the actual KPI and the trend of its target enables the user to detect poor performance.

Targets can be defined using one or more considerations such as safety limits, economic margin optimization, energy efficiency, industry benchmark, etc.

For the ethylene plants under consideration in this paper, targets were defined for the unit-wide KPI, i.e. specific energy consumption, and for most of the controllable drivers. Initially, the target of the specific energy consumption was set as the best sustainable performance achieved in the past. Driver targets were set according to existing guidelines of safety and historical best practice.

The key to success is that targets are sound. This can be challenging since there are many influencing factors outside staff control that dictate the process performance and, hence, its best achievable performance, such as different feed flow and composition, different product specifications, or ambient conditions among other factors.

None of those influencing factors were taken into account in the target setting process of the ethylene plants due to time constraints. A more comprehensive analysis using process simulators, data mining and statistical tools could have been used in order to better understand the process variability. However, it is expected that by initially having the best achievable performance as a guideline, more experience and information are obtained to make better target setting decisions in the future.

It is vital that the energy target setting is fully harmonious with the overall production target setting, the latter usually being closely monitored and sought. Minimizing energy consumption can at times conflict with the optimum production targets. Therefore, energy targets should be reconciled with the production goals and updated as production plans change.

Therefore, it's important to constantly review the targets and challenge them, making sure that changes in the process are considered in the target setting process.

3.2.7 *Display Structure*

An important feature of an EMS is a drill-down mechanism designed to effectively help in the diagnosis of possible causes of poor performance. Information has to be structured and organized in order to facilitate the energy analysis, guiding the user from the observed deviation in energy usage (e.g. overall performance degradation of the plant) to the root cause (e.g. incorrect control set-points in an equipment).

The rationale of the drill-down mechanism utilized in the ethylene plants studied in this paper is described below:

Unit level: At the unit display, all the sub-units KPI trends are available. The sub-unit with the biggest gap performance is determined and it's display selected.

Sub-unit level: Each equipment in the sub-unit has a KPI trend that is displayed and the equipment with the biggest gap performance can be determined. The equipment view is selected.

Equipment level: Trends of all drivers of a particular equipment are depicted and the driver causing the inefficient operation can be determined.

This root cause analysis is a fundamental feature of an EMS, and is executed effectively when a proper drill-down mechanism is in place.

There are two different ways of determining the performance gap of a unit, sub-unit or equipment: qualitatively or quantitatively. In a qualitative analysis, engineering judgment and experience, coupled with offline gap analysis, lead to the next level "culprit" causing the inefficiency observed in the higher-level KPI. In a quantitative analysis, however, all KPIs are being calculated on the same basis (e.g. economic gap or energy losses) and the "culprit" is readily determined. In the ethylene plants discussed in this paper, the method with the quickest and simplest development was used, i.e. the qualitative method.

There are some features that can facilitate the drill-down mechanism and they are listed below:

Trends vs. Point Values: a trend offers much more information than a point value. Consideration of usage of trends over point values should always be made.

Links/Hot buttons to/from other EMS views: links offer an easy and quick way to navigate within EMS, facilitating the drill-down analysis.

Links/Hot buttons to/from site process flow views (non-EMS views): more details on a determined equipment can be easily and quickly achieved by usage of links to the site process flow diagrams, if already existent in the same platform used for EMS. Similarly the other way around: links to EMS views from the site existent process flow diagrams, are a quick way of obtaining more energy information.

Figures 1 to 5 show examples of EMS views applied to ethylene plants. The examples presented are close replicas of the real views developed for the plants, however the data numbers are omitted.

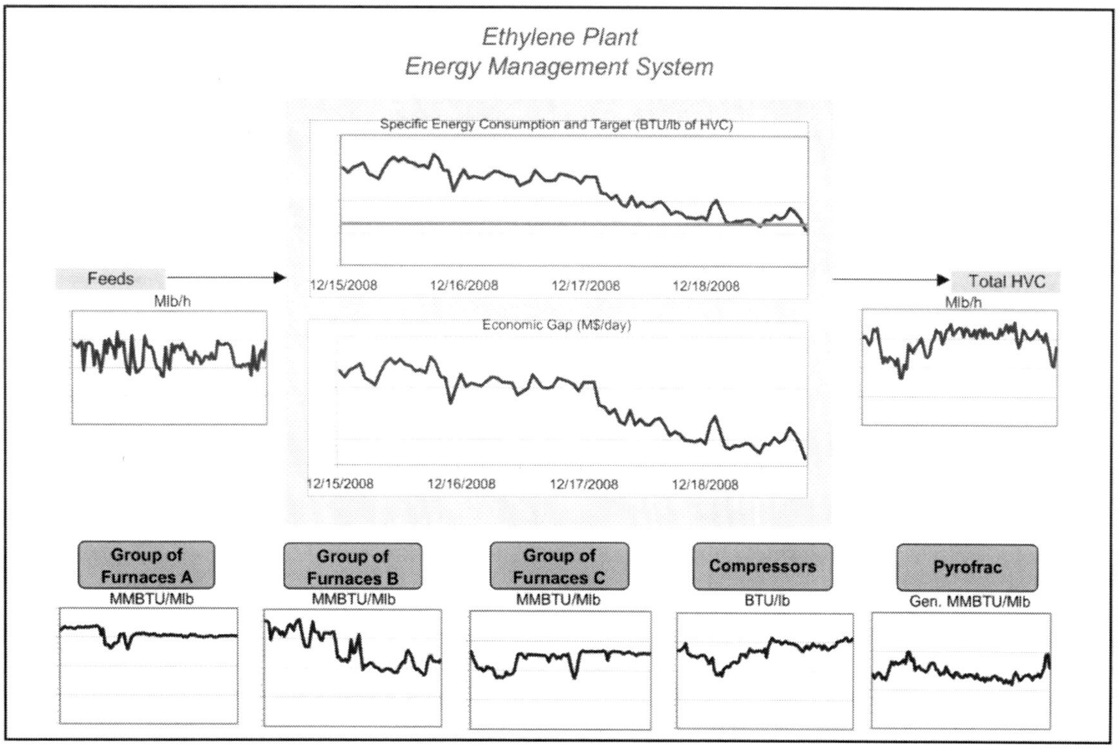

Figure 1 – Unit level view – Ethylene Unit

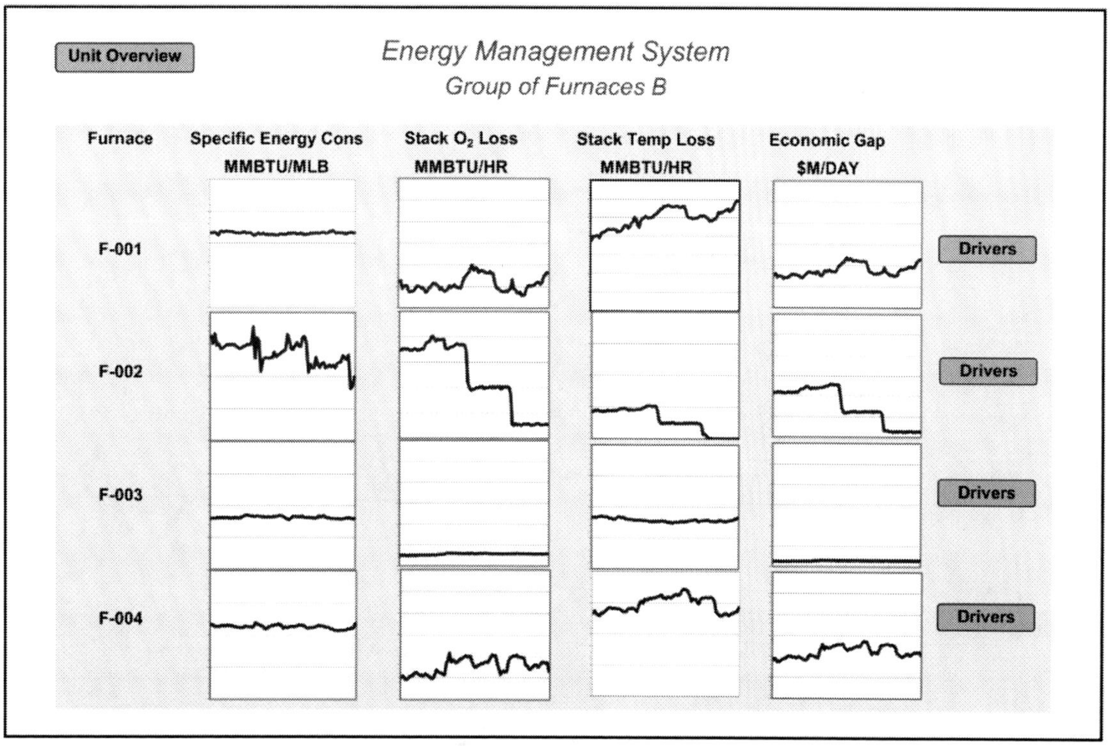

Figure 2 – Sub-unit level view – Group of Furnaces

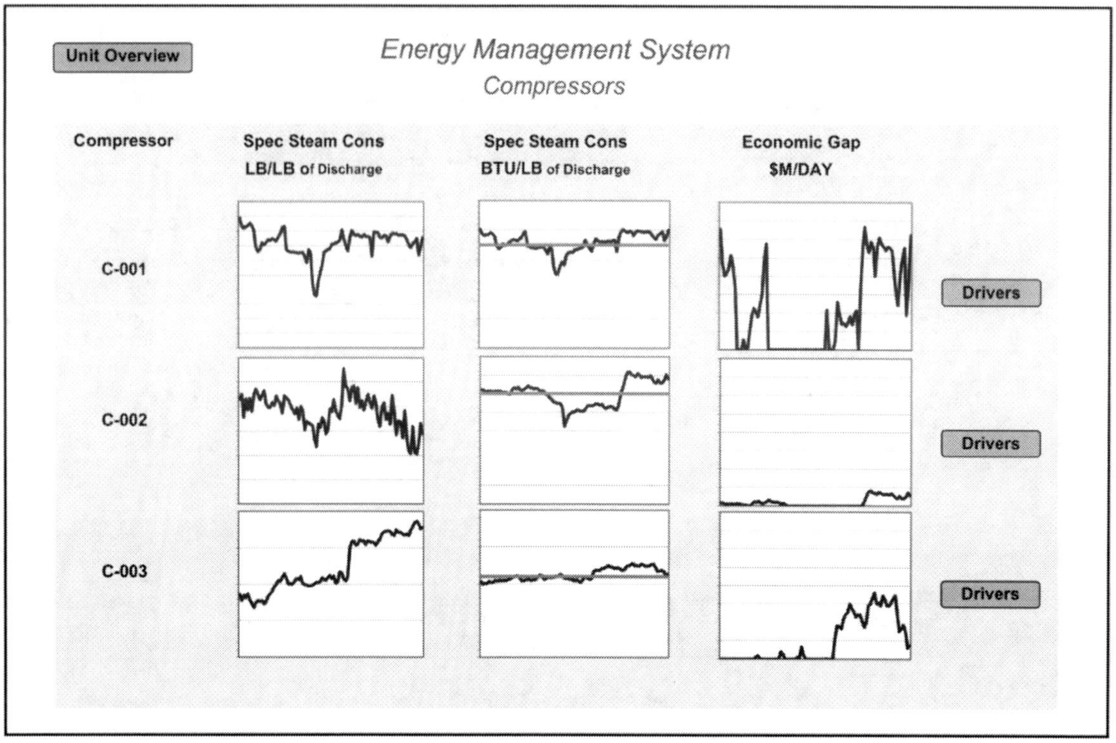

Figure 3 – Sub-unit level view – Compressors

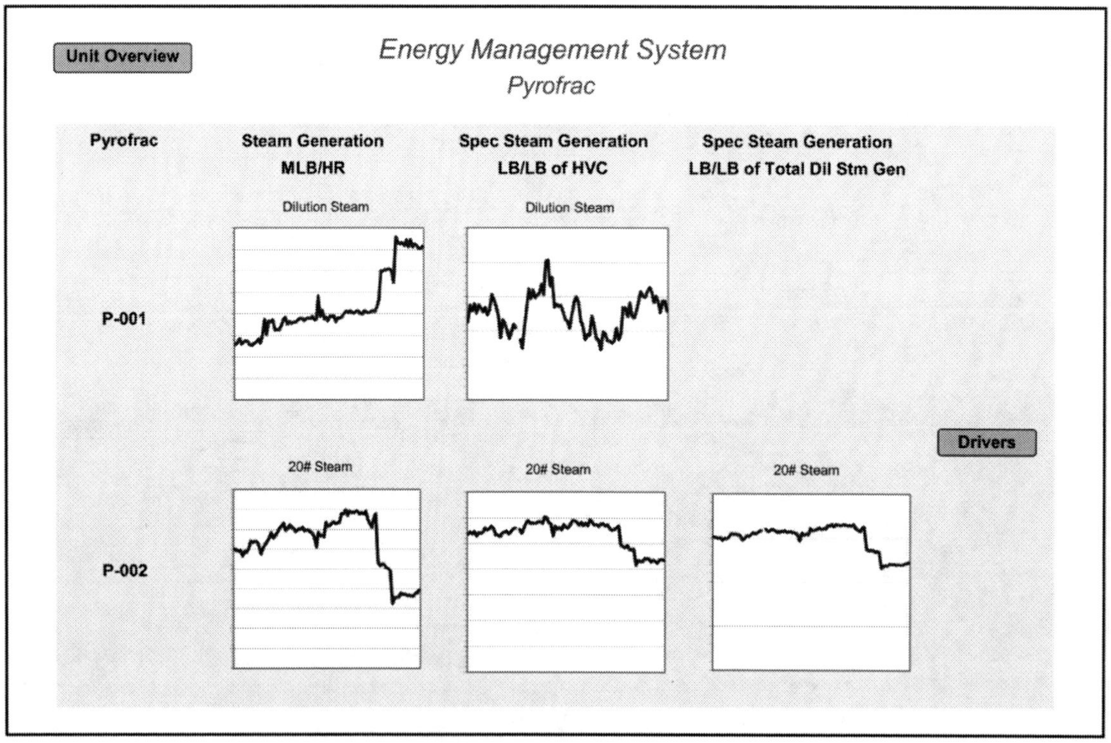

Figure 4 – Sub-unit level view – Pyrofrac

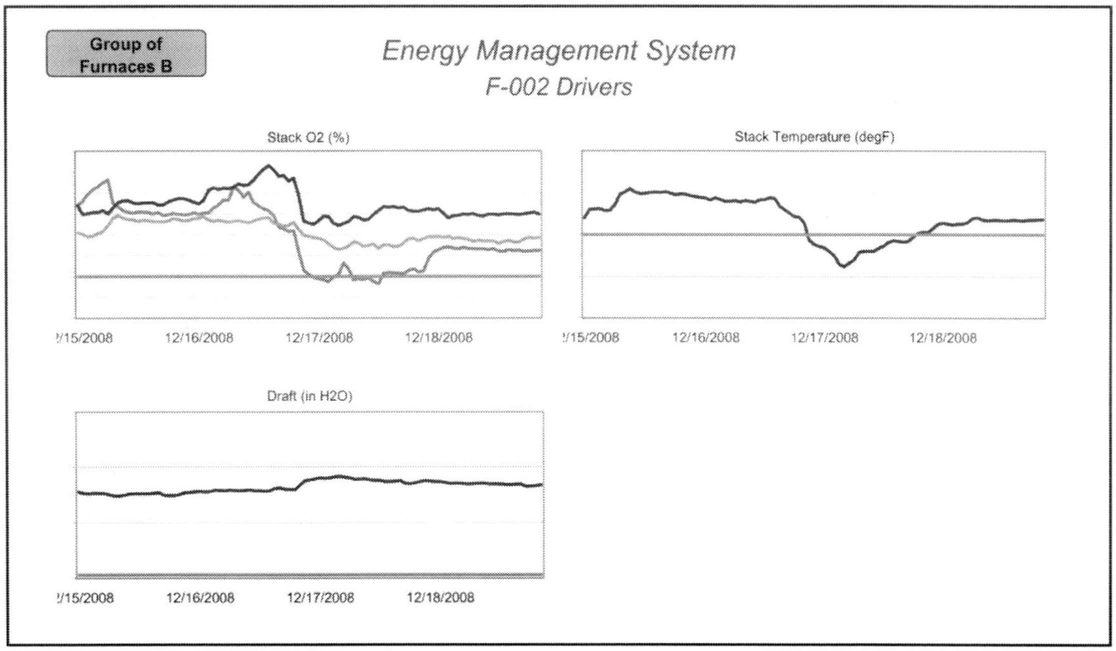

Figure 5 – Equipment Level – Furnace drivers

4. EMS Implementation in Ethylene Plants

Once the EMS design is defined, the EMS implementation can begin. In the implementation phase, calculations and views are programmed and EMS starts to take shape.

Several decisions regarding main features and characteristics of the EMS were made during the development phase. The detailed and more time consuming decisions, such as calculation methods or identification of certain KPIs, can be made during the implementation phase.

Details of calculation methods are outside the scope of this paper because of their variability from site to site. This is due to differences in meter availability, need to ensure consistency with existing site calculations, expected accuracy, among other site-specific differences that affect calculation methods.

For this program, it was estimated that half of the site focal point time would be required within a period of 3 months to finalize the EMS implementation. The energy focal point is responsible for providing necessary information (e.g access to the IT system, process data, energy cost information, etc.) and contacts within the site. The energy focal point is also responsible for making all the decisions regarding the EMS development and implementation, or escalating them to the proper personnel.

Besides the energy focal point, two Shell Global Solutions consultants were assigned to the program, one for each site. They worked full time for 3 months to finalize both sites' EMS. Basic functions of the Shell Global Solutions consultants were to facilitate the sites' decisions regarding EMS by providing information and knowledge based on past experiences with the program and to build the system.

For this particular EMS development, less than half of the time of an intern for three months was used to help build the system. Additionally, an olefins specialist was consulted.

It should be expected that site personnel would be actively involved during the EMS implementation. Frequent meetings are scheduled with appropriate personnel to make sure that the EMS will suit its purposes. Identification of unique KPIs and drivers is often obtained from site personnel. Management should highly encourage participation of all the levels of the organization.

Three months of implementation were not enough to properly set all the management process elements, i.e. define responsibilities and accountabilities for energy utilization across all levels of the site, an essential element for EMS success. Therefore, when the Shell Global Solutions consultants left the sites, the management process elements weren't yet completely defined.

One of the sites decided that their weekly production meeting would include energy reporting in their agenda. Every week, the energy performance is reported and, most importantly, discussed by relevant staff (process engineers, production managers, operators) who then escalate and address issues and solutions when necessary.

The other site decided not to have frequent official meetings on energy yet. However staff is already aware of the existence of the system. EMS has been used mainly by the energy focal point, who now is able to satisfactorily quantify energy savings obtained by ongoing energy projects using the EMS and by process engineers who are looking for improvement in their areas.

Although the average benefit of an effective EMS is estimated to be 3% to 5% reduction in the site's energy cost, by the time this paper was written, there was not enough time to properly quantify the savings obtained with either EMS. Nevertheless, both sites rapidly reported that the EMS provides essential information regarding energy, which leads to identification and justification of energy projects, and substantially simplified the process of energy reporting.

It must be clear that the end of the EMS implementation does not finish EMS work. EMS, when used in its full capacity, cannot be seen as a regular project with a beginning and an end, because there is no end to it. EMS work is ongoing, requiring continuous attention from all levels of the organization. And that's why it's so important to motivate staff and be able to sustain it. Effective communication, which encourages participation and involvement of staff and the development of methods of rewarding good performance are part of the needed organizational culture that can drive and sustain energy cost reduction initiatives.

5. Concluding remarks

This paper describes the EMS development and implementation in Ethylene Plants. Although the EMS design, i.e. KPIs and drivers identified and display structure, was presented in detail, it should only be used as reference. Not even the two ethylene plants discussed in this paper ended up having the same set of KPIs, drivers and display structure as presented. Therefore, the specific characteristics and needs of the site should always be considered. The described EMS concepts and ideas can be applied to most ethylene plants and other industrial processes as well.

The EMS described in this paper is very streamlined and contains only the basic main features of an EMS. Improvements to the tool are strongly encouraged. Amongst the more advanced EMS techniques are the intelligent target setting (i.e. dynamic targets that varies according production targets or other influencing factor outside staff control) and advanced energy gap analysis (i.e. automated identification of the different causes of bad performance).

Despite the fact that most of this paper discussed and presented the monitoring tool elements of an EMS, key to an effective and successful EMS program is to have personnel across all levels of the organization actively pursuing energy improvements. This can be ensured by proper definition of responsibilities and accountabilities combined with an organizational culture that encourages and rewards good performance.

EMS is an ongoing initiative. It's very important to have everybody involved motivated when they first start using the tool and it's equally important to sustain their motivation in order to sustain and keep the achievement of improvements in site. Besides promoting achieved successes with the tool, credibility in the tool is another very important means to sustain motivation. Therefore, the tool should be checked for necessary updates frequently, such as target updates or adjustments due to process or equipment modifications. Proper maintenance of the tool guarantees reliability and credibility.

Finally, assurance that help is available from the management team to correct poor performance is another means to sustain and enhance motivation.

THE IMPACT OF FOULING AND CORROSION ON ETHYLENE PLANT EFFICIENCY

Authors: Stephen De Haan, Barbara Stancato, Brian Sullivan
(Lummus Technology, Bloomfield, New Jersey)

Introduction

Fouling and corrosion can impact almost every area of the plant. The result can be reduced capacity, reduced efficiency and even safety issues.

Some fouling and corrosion result from factors inherent to the process. Other problems are caused by impurities introduced in the feeds or utilities supplied. In essence, the plant begins to corrode and foul the day that it is placed in operation. While fouling and corrosion of piping and equipment can be controlled to acceptable levels, it can never be eliminated. The key measures of success are efficiency, capacity, and time between turnarounds.

Efficiency

Since nearly every area of the plant can be affected by fouling and corrosion, many of the fouling and corrosion problems will impact efficiency. The largest impacts will be in the areas of largest energy consumption, including:

- Cracking furnaces
- Charge gas compressor
- Major compressor drivers
- Cooling water system

Lesser but important impacts can occur in the:

- Gasoline fractionator
- Water quench tower
- Dilution steam system
- Acid gas removal systems
- Dryers
- Coldbox
- Deethanizer, depropanizer, and debutanizer
- Catalyst beds

Furnaces

The furnaces consume about 60% (net) of the total energy required to run the plant. The energy that is not consumed is converted to steam for use elsewhere. Fouling of some parts of the furnace is routine and accepted as part of normal operation. Routine fouling occurs in the radiant coil as a result of coke formation and in the transferline exchangers (TLEs) by coke formation reactions and condensation mechanisms. While both radiant coil and TLE fouling occur during the furnace run, TLE fouling is more significant in terms of efficiency loss.

Radiant Coil Coking

The resistance to heat transfer from the burner flames to the process gas increases as the radiant coils coke. The firebox temperature must increase to maintain the same feed rate and cracking severity, and may increase by as much as 50°C over the course of the run, depending on many factors including the coil design and run length. As the firebox temperature increases, the radiant efficiency decreases and more fuel is consumed, sending more and hotter flue gas to the convection section. Additional steam is generated but since the convection section has a fixed heat transfer surface, the flue gas exits the furnace somewhat hotter thus reducing efficiency.

In modern furnaces with efficiencies of 92 to 94%, the decrease in efficiency is relatively minor. A furnace with an efficiency of 93% at start of run (SOR) will lose about 0.2 percentage points in efficiency over the run (the average efficiency during the run is therefore 0.1% less than the SOR efficiency). For a 1000 kta plant, this equals about 0.7 mmkcal/h or US$150,000 per year (at $25 per mmkcal). The penalty is far greater for furnaces with much lower efficiency.

Radiant coil corrosion does not directly play a significant role in efficiency loss.

Transferline Exchangers

The efficiency penalty for fouling of TLEs can be far greater for certain feedstocks and cracking conditions.

The TLEs cool the furnace effluent, generating steam in the process. There are two mechanisms for TLE fouling. The first type is a continuation of coil coking with carbon depositing on the inlet tube sheet of the TLE. The cracking reaction continues and long residence times in the inlet cone transition to the tubes can result in substantial coke deposition. The accumulated coke can plug tubes, thereby increasing pressure drop and reducing the heat transferred. The second type of fouling is as a result of condensation of heavy material in the TLE tubes. If the furnace effluent contains enough heavy material, it can condense on the relatively cold TLE tubes. The temperature of the inside tube surface increases as heavy tars accumulate, thus reducing heat transferred. When the surface temperature reaches the dew point of the effluent, fouling slows dramatically.

Gas-cracking TLEs foul as a result of the first mechanism (coking). Modern TLE designs minimize or avoid this fouling issue by using:

- Conventional tubesheet type exchangers with aerodynamic inlet cones and large tubes that are resistant to plugging
- Double pipe TLEs (no tubesheet)
- Conventional exchangers with aerodynamic inlet chambers (no tubesheet)

Lummus has used all three types with the selection dependent on the type of radiant coil, feedstock, cracking conditions, and economics.

Liquid-cracking TLEs foul as a result of both coking reactions and condensation. The relative importance of each mechanism is a function of the feedstock type. The fouling rate is not affected significantly by the TLE type but rather is a function of the feedstock, cracking conditions, steam pressure, and effluent outlet temperature. Feedstocks and cracking conditions that favor tar production, lower super-high pressure steam (SHP) temperature, and a low TLE outlet temperature at SOR also favor rapid and significant fouling.

Even in a properly designed system, cracking naphtha at maximum severity might result in the fouling curve shown in Figure 1 where fouling is represented by an increase in TLE outlet temperature.

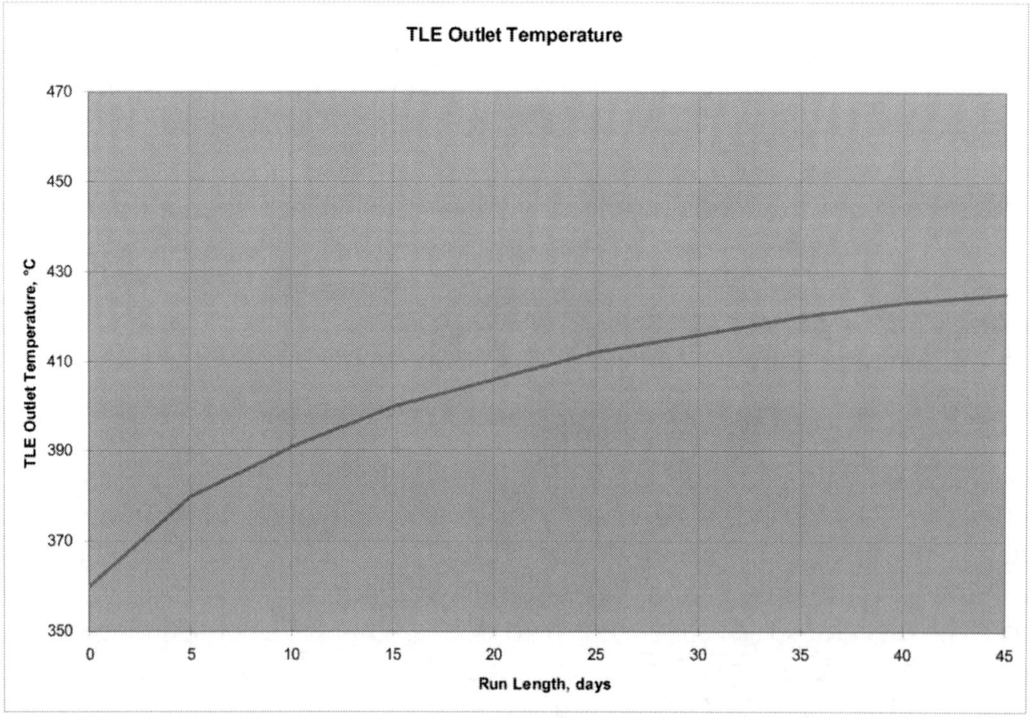

Figure 1 – Fouling curve when cracking naphtha at maximum severity

The TLE outlet temperature initially increases rapidly but levels off when the inside of the TLE tube reaches the dew point of the effluent at all points along the tube. Subsequently, the temperature rises very slowly as a result of reaction mechanism fouling.

Steam generation in the TLE decreases as the tubes foul and the outlet temperature rises, thus lowering plant efficiency. The heat that was recovered as SHP when the TLE was clean is now lost downstream to quench oil and recovered at lower levels.

A lighter feedstock or a feed cracked at moderate severity would foul more slowly and less severely or possibly not at all.

Examining the fouling curve provokes the question: how much efficiency is lost and should the TLE cleaning schedule be altered to improve energy recovery? Considering the naphtha fouling curve shown above, we can compare two ways of operating the plant. In the first method (Figure 2), the TLEs are cleaned every third run (at the end of every third radiant decoke). In the alternate approach shown in Figure 3, the TLEs are cleaned every run.

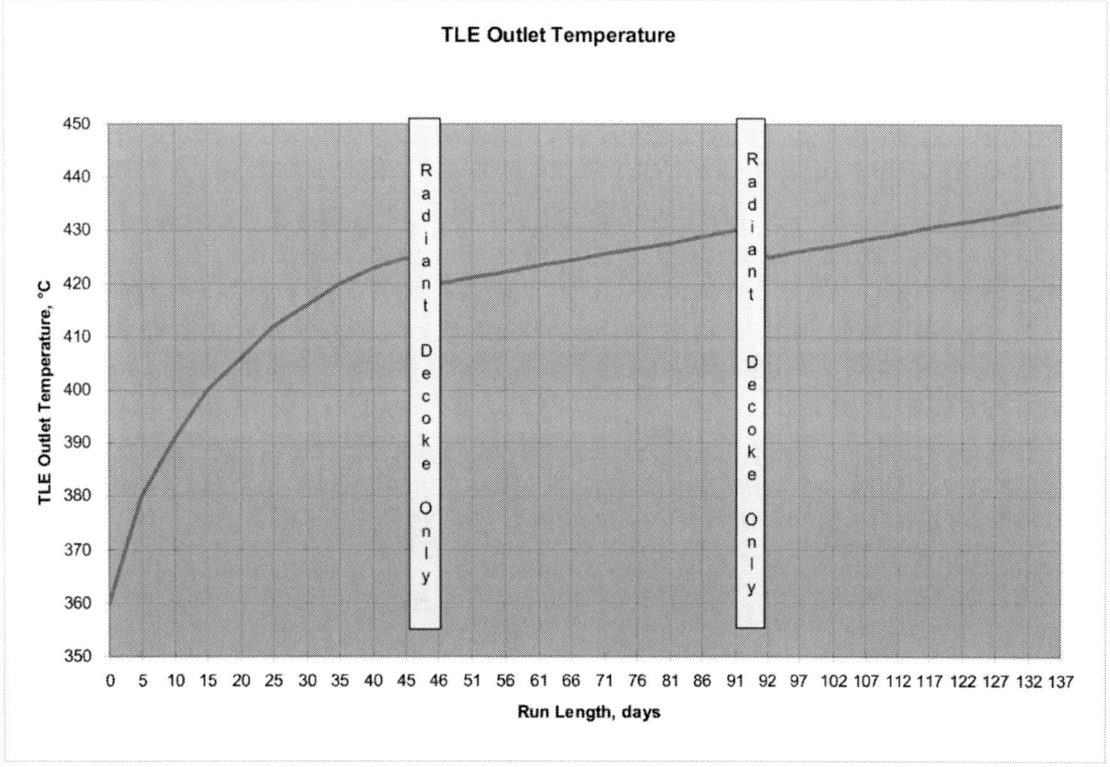

Figure 2 – TLEs cleaned every third run

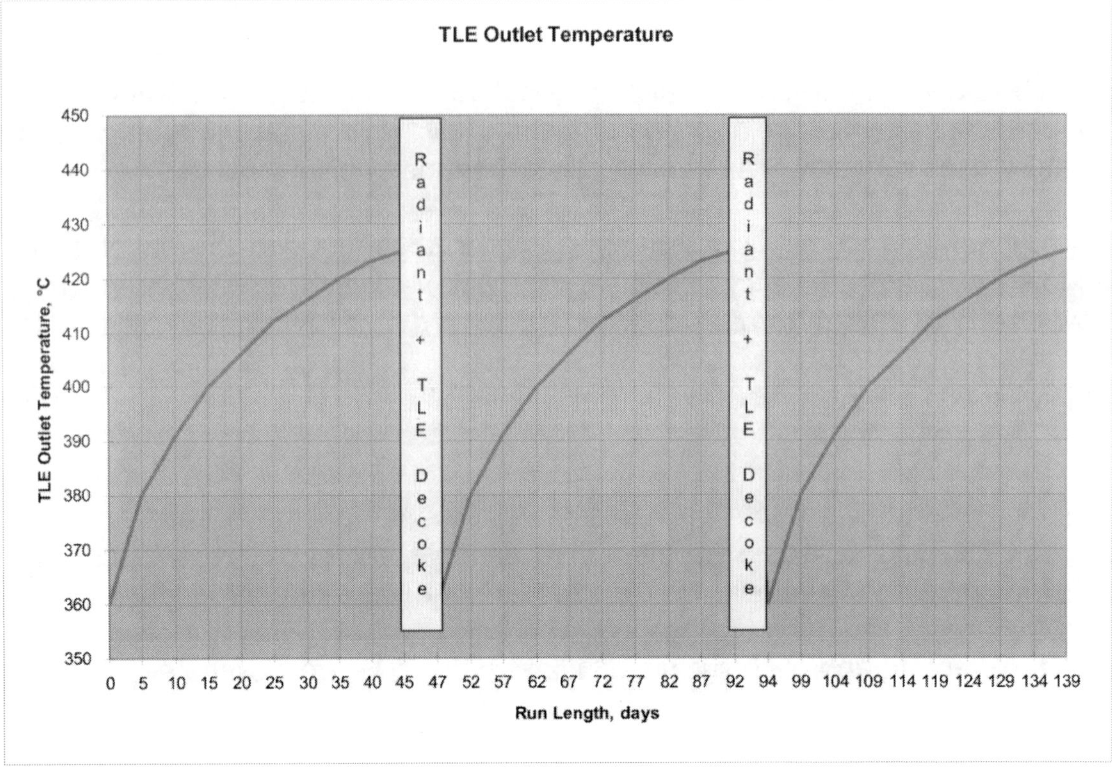

Figure 3 – TLEs cleaned every run

Based on a high-efficiency furnace of 167,000 MTA ethylene capacity, the second method (Figure 3) would save about $250,000 per year in energy costs (at $25 per mmkcal). The optimal strategy would be to clean each time if the plant has a full spare furnace and production is not affected by extending decoking time to clean the TLE. If production is affected, not cleaning each time would be a better strategy. The actual result would of course depend on many variables, including feedstock/product pricing, the time required to clean the TLE, and the plant steam balance. Available methods to clean TLEs on line make cleaning each cycle much more attractive.

Corrosion does not usually play a significant role in reducing TLE efficiency. However, if boiler feedwater (BFW) quality is not maintained, the TLEs can be damaged, requiring repair or intentionally plugging off the tubes. Each tube that is plugged reduces the available surface and results in an efficiency loss. As shown above, the actual penalty will depend on the feed and cracking conditions.

Convection Section

The flue gas side of the convection section fouls as a result of contaminants in the combustion air, contaminants in the fuel, refractory dust, and decoking effluent (if the furnace is decoked into the firebox). Most furnaces that are more than 10 years old show some signs of fouling on the finned surfaces. Contaminants collect between the fins, reducing efficiency.

Corrosion, in the form of oxidation, is expected and can also result in a loss of efficiency by damaging fins. Oxidation is accelerated by operating at higher temperature. Plants often push the furnace capacity to measureable limits such as burner capacity. However, since accurate measures of flue gas and convection metal temperatures are often not available, the convection section tubes can be pushed beyond their design temperature limits, accelerating fin oxidation and tube life in general. A furnace rating is advisable before pushing beyond design ethylene capacity.

Approximately each 18°C rise in flue gas stack temperature represents a one percentage point loss in efficiency. For a 1000 kta plant , a loss of this magnitude is worth about $1.5 million per year (at $25 per mmkcal). Higher flue gas temperatures can of course be caused by other changes, including feedstock properties, excess air, and capacity increases. An analysis is therefore required before concluding the cause of a higher-than-expected stack temperature and to ascertain the potential gain from cleaning.

If flue-gas-side fouling is the issue, several cleaning methods have been utilized:

- Water washing
- Air lancing
- Steam lancing
- CO_2 blasting
- Chemical methods

Water washing is easiest in the sense that physical access is not an issue. Water can be introduced at the convection section top. Care must be taken to avoid wetting the refractory more than necessary and the subsequent start up should include a dry-out hold. Clients, however, have not generally reported great success with water washing. Some have concluded that a lot of dirty water is generated without much gain in efficiency. The other methods have therefore become more popular.

The lancing and blasting methods require access and while labor intensive, do improve efficiency. There are claims that the CO_2 blasting method is most effective because solid CO_2 penetrates the foulant and then vaporizes, breaking the material off.

We are aware of applications of a chemical cleaning program that makes the deposits on the convection coils soft and powdery, enabling soot blowers to remove the deposits easily.

It is also possible to foul the inside of the convection section tubes, particularly when cracking heavy feedstocks. Proper superheating of dilution steam and design of injection points generally prevents this type of fouling. Heavy feeds may require special dilution steam injection nozzles. However, contaminated feedstock can foul the best of systems. Feeds delivered by barge are prone to contamination from sea water and/or the previous cargo.

Pipelines that alternate materials can lead to similar problems. Finally, storage tanks can cause additional problems (see below).

If corrosion of fins is the cause of the efficiency loss, replacement of tubes is an option, though it is usually difficult and time consuming. Downtime and production losses impact the potential benefits of any project of this type. Careful analysis and planning is therefore a must.

Many furnace convection sections are designed with spare rows so new surface can be added to compensate for lost efficiency. Adding tubes in these open spaces is easier than replacing existing tubes but can still require substantial downtime.

Burners

Ethylene furnaces utilize wall burners, hearth burners, or a combination of the two. Proper maintenance and operation of the burners is a key to the efficient operation of the furnace. Poor burner performance can be caused by fouling, physical damage, and improper adjustments, as well as other causes.

Modern low-NOx hearth burners have very small fuel ports that can clog with debris, such as rust and scale from the fuel piping, and distort the flame profile.

Fouling of the burners, in particular, can occur as the result of polymerization of fuel components or from dirt or scale in the fuel. Unsaturated compounds in the fuel can polymerize, particularly if they reach the burners as a liquid. The material discharged into the fuel system as a result of reactor and dryer regenerations, purges and excess production in other plants should be monitored. If fuel with unsaturated compounds is used, some level of fouling will occur and must be tolerated.

Fouling most commonly occurs at the burner tips, which are hot; however, fouling can occur in the fuel gas piping and burner riser to the tips as well. Utilization of gas turbine exhaust can make problems worse because of the increased air temperature.

Fouled burners can result in a less than optimal heat flux profile and, in more extreme cases, after-burning. A poor heat flux profile produces shorter than expected run lengths and reduced efficiency. After-burning can result in shorter radiant coil tube life, shorter than expected run lengths, damage to convection tubes, and problems in meeting environmental targets.

Fouling on the tip of wall burners can result in overheating. The polymer can ignite, heating the burner tip and providing a source of ignition for flashback into the throat of the burner. To address this issue, an appropriate cleaning cycle must be adopted to maintain burner performance. According to our database, the median frequency for performing burner cleaning in ethylene plants is every five months.

Plants should also have a filter coalescer on the fuel to the furnaces to remove particles of rust and debris. The coalescer will also help remove fine particles of entrained heavy liquids.

Fouling of burners in a modern high efficiency furnace generally has more effect on furnace run length and radiant coil life than on efficiency.

Feed Preheating

The feeds to ethylene furnaces are preheated in exchangers using waste heat (such as quench water) before they enter the convection section. The temperature chosen during the design is optimized for the particular plant and feedstock, taking into account the plant energy balance, the furnace steam production, and the vaporization profile of the feed in the convection section.

Several plants have reported fouling on the process side of feed preheaters for naphtha crackers but not for gas crackers. Fouling in the process side of feed preheaters is almost always caused by sludge or rust in the feed. Fouling on the quench water side can also contribute to poor performance.

Process-side fouling is usually the result of sludge in the naphtha. As noted earlier, naphtha is often delivered in tanker ships or barges. Small amounts of sea water left in the ships' tanks can contaminate the naphtha. In a storage tank, the water will settle out, causing corrosion at the bottom plate. The rust can catalyze the formation of larger chain molecules that result in sludge. Likewise, the previous load on the barge or ship may have been heavier material, such as crude oil, and residual material can contaminate the naphtha.

The feed outlet temperature decreases as the exchanger fouls, which actually increases the efficiency of the furnace. However, the crossover temperature drops, causing the firing to increase. The net result is a slight loss in overall plant efficiency.

Extreme cases of fouling increase pressure drop, limiting furnace capacity due to feed and dilution steam valve openings.

Our furnace model estimates that a decrease of 10°C in preheater outlet temperature has a negligible effect on the heat balance of the furnace. It results in a minor change in crossover temperature, a more significant change in stack temperature, but negligible changes in fuel usage and steam production. However, there are other consequences of feed preheater fouling:

- Reducing the feed temperature could make the convection tubes cold enough to cause water condensation and potential corrosion.

- In addition for liquid feeds, the vaporization profile will be shifted. The colder feed will be preheated to a lower temperature at the dilution steam injection point. If the feed is not 100% vaporized after the point of dilution steam injection, then convection section fouling will occur.

Preheater fouling can be reduced by filtering out the sludge or rust from the feed before it enters the preheater. Most plants use an 80 to 100 mesh filter for their naphtha feed. Some clients have used filters with mesh size smaller than 120, but this requires more frequent cleaning of the filters.

Gasoline Fractionator (Oil Quench Tower)

The upper sections of the gasoline fractionator can foul resulting in increased pressure drop. The fouling tendency of the fractionator varies significantly from plant to plant. Trace impurities in plant feedstock, including oxygenates, are believed to accelerate fouling. For a 1000 kta plant, an increase of 0.07 kg/cm^2 (1 psi) in the fractionator pressure drop results in an efficiency penalty of 1600 kW in increased charge gas compressor (CGC) power or $300,000 per year (at $25 per mmkcal).

The fouling is the result of polymerization of compounds such as indene and styrene catalyzed by oxygenated compounds. Iron oxide is also believed to be a catalyst. Chemical treatments have been useful in greatly slowing or even stopping fouling. The treatments consist of oxygen scavengers and chain terminators. This material of course can be a significant cost.

Both packing and trays have been used in the upper sections of the gasoline fractionator. Packing should be made of stainless steel, which is much less reactive. Packing requires much more careful irrigation than trays since fouling increases when surfaces are not completely wet. Even with proper irrigation, packing has been prone to fouling in some plants. Fixed valve trays or other fixed orifice type trays are generally preferred in areas of low liquid flow.

Fouling can also occur in the quench oil heat recovery circuit. The heaviest components of the quench oil (poly-nuclear aromatics PNAs) and coke particles tend to form sticky, tarry material that fouls exchangers. Current plant designs use cyclone filters to remove coke from the quench oil, while viscosity control systems are used to minimize the concentration of PNAs in the quench oil.

Corrosion does not generally play a significant role in efficiency issues in the gasoline fractionator.

Quench Water Tower

In liquid-feed plants, or gas feed plants that have a gasoline fractionator, quench water towers (QWT) are not prone to fouling. The QWT is protected by proper operation of the gasoline fractionator. In gas-feedstock plants, however, QWTs can foul. The risk of significant fouling is related to the feedstock, cracking selectivity, and cracking conversion.

High conversion ethane crackers (70% plus) have a much higher tendency to foul than plants operated at lower conversion. This is particularly true if the furnaces are older, less selective types or cracking conditions are less selective (low steam-to-oil ratios and/or high back pressures). The fouling is related to tar production and the lack of aromatic gasoline compounds in the furnace effluent. The problem is compounded by coke fines circulating in the quench water (QW). Ethane cracking at high conversion produces tar and substantial naphthalenes, both of which can deposit on the QWT internals. Unfortunately, ethane cracking produces very little aromatic gasoline that could help dissolve the tar and naphthalene, carrying them away in the gasoline product.

Similarly, gas cracking plants that use too much C_4 feedstock can encounter fouling issues as the result of heavy tar production. C_4 feed that contains substantial C_5s or other heavy compounds will further increase fouling. Once again, less selective cracking at high conversions will compound the problem.

A loss in efficiency occurs when the QWT pressure drop increases, resulting in increased power consumption in the CGC. As noted earlier, for a 1000 kta plant, an increase of 0.07 kg/cm^2 results in an efficiency penalty of 1600 kW or $300,000 per year (at $25 per mmkcal).

In all cases, the proper design of tower internals helps. In many cases, packing has been successfully used. For high conversion high-risk cases, coke catchers, on-line gasoline washing, and systems such as cyclones for QW clean up should be considered.

The cost of on-line gasoline washing can be significant since the aromatic wash oil that seems to work best – containing little or no dienes – is expensive. Some clients are now looking at recirculating systems.

Corrosion does not generally play a significant role in efficiency issues in the QWT.

Dilution Steam System

The dilution steam (DS) system can be prone to both fouling and corrosion. The system re-uses poor quality water while trying to minimize losses (blowdown). As a result, water containing organic acids and other impurities is vaporized, sometimes at relatively high temperatures.

In the traditional DS system, recycled process water is vaporized with heat supplied by quench oil and/or medium pressure steam. The water is taken from the QWT, sometimes coalesced and steam stripped to remove lighter acidic compounds. The water in a gas cracking plant should be treated in a DOX or Pall coalescer system to minimize free oil. Vaporization takes place at about 170°C and blowdown is limited to 7 to 10%.

Data from Lummus Technology's technical service database shows that the inlet water to the vaporizers can contain up to 300 ppm free oil, 200 ppm organic acids, and 10 ppm styrene along with significant dissolved and suspended solids. The water often contains significant levels of chlorides, sodium, and phenols as well.

The heat transfer surfaces foul by forming a scale. As heat transfer is degraded, DS may need to be supplemented by direct injection of medium pressure steam, which results in additional water consumption and waste treatment costs. If the scaling is not removed periodically, pitting can occur under the scale. Many plants therefore clean individual exchanger bundles on a one year or shorter cycle.

Various levels of treatment are applied to minimize corrosion and fouling. The most fundamental treatment is to control pH in the QW (6 to 7.5 at 25°C) and dilution steam drum (8.5 to 9.5 at 25°C). Both caustic and amines have been used to control pH. Caustic is cheapest but carryover to the furnaces can cause radiant tube damage. Amine is more expensive and requires proper control to protect the exchangers. There is also a chance that amine carryover to the furnaces can ultimately cause ammonia contamination of the ethylene product. The Lummus database indicates that the plants are approximately evenly split between the use of caustic and the use of amines to control the pH of the process water/dilution steam system.

In gas cracking plants, DS is often supplied by direct saturation. The feed is heated and brought into direct contact with the process water in a saturator tower. By controlling the temperature, the feed is humidified to the correct steam-to-oil ratio. Since this operation takes place at low temperatures (around 130°C), corrosion is not significant; however, fouling is a concern. Proper preparation of the water and design of the saturator system have proven successful in minimizing fouling and its impact.

The formation of emulsions in the quench water settler causes poor oil-water separation that allows hydrocarbons to circulate through the quench water system. Hydrocarbons can separate in stagnant areas and form polymers and gums, which increases the risk of fouling in the quench water coolers and users. Any heavy hydrocarbons carried forward into the coalescer system or the saturator will result in fouling of these units as well.

Normally, oil and water, having different specific gravities, will separate in a properly designed system. However, if the oil and water form an emulsion, separation can be greatly inhibited. Emulsions in the quench water system can be caused by a number of factors, including:

- Poor separation in the gasoline fractionator or changes in feedstock or cracking conditions can result in a heavier gasoline with a specific gravity close to that of water, thus making oil-water separation more difficult. While this in itself does not cause an emulsion, it will inhibit separation and increase the amount of gasoline in the circulating quench water.
- A higher quench water pH will cause lower surface tension, which will tend to reduce the efficiency of separation and favor the formation of emulsions.
- The stability of an emulsion is determined by the properties of the interfacial film between the oil and water. The presence of emulsifying agents (surfactants) and solids can substantially impact the properties of the film and the stability of the emulsion. The lower the interfacial tension, the more stable the emulsion. Surfactants can lower the interfacial tension and also charge the individual droplets, causing them to repel one another rather than to coalesce. Surfactants can enter the quench water from the feedstock (phenols, sulfur compounds and organic acids), in streams recycled from other areas of the plant (caustic tower and wet flare drum), from other plants, or in chemical additives injected into the process. Streams recycled from within the plant or from other units are often the major sources of surfactants and the cause of emulsion problems.

Charge Gas Compressor (CGC)

Ethylene plants' charge gas compression systems are prone to fouling. Nearly half of all ethylene plants report substantial fouling in the charge gas compressor, aftercoolers, or both. Moreover, nearly all of the gas cracking plants report problems with fouling in their charge gas compressor system. In many cases, fouling of the CGC has been cited as a prime reason for the scheduling of plant turnarounds.

Fouling of the charge gas compressor is most often the result of polymerization and condensation reactions. The predominant reactions involve materials present in the charge gas that polymerize and deposit on the internal surfaces of the compressor. Condensation reactions do not normally play as important a role. These fouling reactions are sensitive to temperature, pressure, concentrations of reactants, and the presence of precursors and catalysts. Temperature is a key variable, with the reaction rates more than doubling for every 10°C increase in temperature.

As polymers form, they roughen the internal surfaces of the compressor, alter the cross-sectional area and angles in the flow path, and impact the rotor characteristics. These changes lead to increased frictional losses and altered aerodynamics. The result is a reduction in the efficiency and head produced. In practice, this loss manifests itself as an increase in operating speed and power consumption. To maintain the operating pressures, the compressor speeds up and more power must be delivered by the steam turbine or variable speed motor.

A fouled compressor will use more energy to process the same gas at the same conditions. For a 1000 kta plant, a loss of one percentage point in efficiency translates to about $180,000 in additional energy costs. A loss of one percentage point per year over a four-year plant run therefore costs the owner about $1.8 million in additional energy consumption.

If the compressor reaches the maximum speed or the turbine cannot produce the required additional power, the flow must be reduced or the suction pressure increased. As suction pressure is increased, the yield of products becomes less selective, favoring the increased production of fuels and aromatics while producing less ethylene and propylene per ton of feed. The loss in selectivity thereby increases energy consumption by making downstream separation more difficult.

Vibration analysis, bearing temperatures, and other measurements allow continuous monitoring of the compressor operation; however, none of these provides a direct unambiguous measurement of fouling. Increased vibration can result from fouling deposits that imbalance a rotor, but could also result from other mechanical problems unrelated to fouling. In addition, vibration is seldom an early warning sign of fouling.

Many plants calculate the efficiency of the individual compressor stages in an attempt to monitor the machine performance. The fundamental equations show that the polytropic efficiency of a centrifugal compressor depends upon the temperature and pressure in and out of each stage, the properties of the charge gas, and the performance characteristics of the compressor and turbine. Production rates, feedstock, and other operating conditions change, sometimes impacting efficiency more than increases in fouling. Accurate online calculation and tracking of compressor efficiency is therefore difficult, requiring consideration of the compressor performance curve.

CGC fouling is inhibited by changing the operating environment. From an operational perspective, the methods for combating CGC fouling fall into four basic categories, and more than one method is often employed:

1. Wash Water

   Direct water injection has become a popular method of reducing operating temperatures to control fouling and has been practiced for over 15 years. Water is injected into the suction or separately at each rotor. The water is vaporized, thus reducing the gas temperature. Temperature reductions in the range of 8 to 14°C are possible by injecting water at a rate equal to 1 to 2% of the process gas per stage. While incidents of water-induced erosion have occurred, water can be injected at relatively high rates without causing damage. However, feedback from plants that use water injection as their primary method of fouling mitigation indicates it is not nearly as successful as wash oil. In addition, some plants have found that fouling in the intercoolers has increased as the fouling in the compressor has decreased.

2. Wash Oil

A solvent that is capable of removing the deposit is added directly to the compressor. The most common approach is to use what is generically called "wash oil." The basic characteristics of a suitable wash oil are:

- High aromatic content – The minimum recommended aromatics content of a suitable wash oil is 50 to 60 wt%. The wash oil needs to have high solvency to dissolve the polymer deposits.
- "Free" of fouling precursors and suspended solids – Wash oil should be essentially free of styrene and diene compounds. In addition, since the wash oil may at least partially evaporate in the compressor, it should be low in suspended solids.
- High boiling point – A minimum end boiling point (EBP) of 250°C is typically recommended. It is necessary to maintain a liquid state in the later wheels of each stage, where the temperature is higher.

3. Additives

The following classes of additives are potentially useful in combatting charge gas compressor fouling:

- Free radical inhibitors that react with free radicals before they can initiate polymer chains.
- Chain stoppers that react with polymer chains to stop any further growth.
- Metal deactivators that deactivate Fe_2O_3 and prevent the catalysis of polymer reactions.
- Corrosion inhibitors that prevent the formation of Fe_2O_3.
- Dispersants that dissolve polymers that have already formed.

4. Compressor Coatings

Plants have coated rotors and other internals with non-stick coatings. Coatings have helped but not resolved the problem. Coatings also seem to move the problem downstream to the aftercoolers.

Caustic Tower

The caustic tower can foul by free radical polymerization and aldehyde condensation, with the latter being more important. Aldehydes in the furnace effluent undergo condensation polymerization reactions in either an acid or base environment. The caustic tower, therefore, is a good environment to form the polymers commonly known as red or yellow oil.

Properly designed towers and chemical treatment, as required, can keep fouling to a minimum. Packed towers are acceptable for this service and provide minimum pressure drop. The keys to proper design are sufficient liquid flow, proper distribution, and minimizing residence time of any red oil that is formed.

Deethanizer and Depropanizer

The bottom sections and reboilers of both the deethanizer and depropanizer are prone to free radical polymerization. The polymer is largely a cross-linked polybutadiene that can coat the trays, tower walls, and reboiler tubes.

In the early years of the ethylene industry, these columns were typically operated at relatively high bottom temperatures. For example, early depropanizer designs used cooling water as a condensing medium and operated with bottom temperatures higher than 95°C. The result was substantial fouling and the need for a spare column bottom and reboiler.

Current depropanizer designs typically use a maximum bottom temperature of 80°C and modern treatments to minimize fouling. Spare reboilers are provided and desuperheated steam is often used as a reboil medium.

Generally, the treatment chemicals are added in the deethanizer feed, exit the bottom, and treat the depropanizer bottom as well. Deethanizers are also provided with spare reboilers. Since the deethanizer is usually reboiled by quench water, the spare reboiler typically is based on desuperheated steam. Once again, the 80°C maximum bottom temperature should be observed.

Fouling in modern deethanizers and depropanizers has relatively little impact on plant efficiency. The fouling is limited to the reboilers, which have relatively long run lengths and can be switched as required. The Lummus technical service database indicates the switching frequencies and cleaning times shown in Table 1.

Table 1.
Switching Frequency and Cleaning Duration for Towers with Spare Reboilers

	Median Switching Frequency (Months)	Median Time to Clean (Hours)
Gasoline Stripper	24	48
Condensate Stripper	14	30
Deethanizer (QW reboiler)	15	48
Deethanizer (LPS reboiler)	12	44
LP Depropanizer	18	30
HP Depropanizer	12	48
Debutanizer	12	27

Debutanizer

The bottom of the debutanizer has less fouling issues than the previous two towers. The treatment chemicals added to the deethanizer reach the debutanizer reboiler and generally slow any fouling to an acceptable rate.

The key issue with the debutanizer can be overhead fouling: the butadiene concentration is sufficient to cause polymerization. While the clients in our technical service database have not reported this as a serious issue, others have, with fouling severe enough to block pressure safety valve inlets.

Steam System

BFW is generally processed to an ultra-clean standard prior to use in the TLEs, where the SHP steam is generated. However, depending upon the degree of maintenance and operations monitoring performed on the BFW purification system, some level of unacceptable impurities may enter with the water.

Impurities in the SHP BFW will result in fouling and corrosion issues for the TLEs. Deposits reduce heat transfer and cause concentration cells, which increase corrosion potentials. Contaminants – such as iron, calcium, magnesium and copper – transported into a TLE will form deposits on the high-heat-flux area of the tubes.

Non-volatile BFW impurities can carry over into the saturated steam by entrained water droplets. Contaminated saturated steam causes fouling in the superheaters and the downstream turbines.

Another pathway for non-volatile contaminants to get into the steam is attemperation (desuperheating) water. Systems that use BFW must have a BFW that has AVT (All Volatile Treatment) chemistry at the point of takeoff and a cation conductivity less than 0.3 micromhos. BFW is not a preferred source of attemperation water: the preferred sources are condensed steam from a "sweet water" condenser or polished condensate.

Finally there are selective volatile contaminants that need to be regulated in the steam. These are generally vapor-phase SiO_2 and organics. SiO_2 must be maintained below specified maximum levels in the boiler water so the vapor-phase silica stays below 20 ppb. The maximum boiler water SiO_2 levels are based on boiler water pressure. Organic limits are set by the turbine manufacturer. Dissolved oxygen can be considered a volatile contaminant and should be maintained below 7 ppb in all parts of the system.

Minerals will cause deposits on the turbine blades and the steam side of the surface condenser tubes, thereby reducing efficiency or even causing failure. Caustic and chlorides will cause stress corrosion cracking and corrosion. Organics will cause acidic corrosion.

To prevent corrosion and fouling, it is important to monitor the quality of the BFW and steam.

Cooling Water Exchangers

Cooling water (CW) heat exchangers are used throughout the ethylene plant. Fouling of the CW exchangers can impact the plant's performance by reducing efficiency, capacity, and the interval between plant turnarounds.

The areas where fouling of CW exchangers has the greatest impact on plant efficiency are:

- The turbine surface condensers,
- Charge gas compressor intercoolers, and
- The propylene refrigeration compressor discharge condenser.

To a lesser extent, fouled CW exchangers in the QWT coolers also impact plant efficiency.

The raw CW make-up supply, even after filtering, often contains impurities such as dirt (sediment), minerals, and microbiological organisms. In addition, dust is extracted from the air in the cooling tower and hydrocarbons can leak into the water. There are several mechanisms by which these contaminants can foul the CW circuit.

- Low velocity in an exchanger or in piping can allow the solids that are suspended in the CW to settle out and deposit on the surfaces of the exchanger/piping. In general, a minimum velocity of 1 m/s is recommended to minimize the settling of deposits. Velocity can become a problem if CW throttling is used for control.

- Dissolved minerals in the water will precipitate out under certain conditions to create a scale deposit. The rate of deposition is a function of the water's pH, temperature, concentration of minerals, and other factors. The most common salts to deposit are calcium carbonate and calcium phosphate, although there are others. The solubility of the salts decreases as the temperature increases. Therefore, as the exchanger fouls and its outlet temperature rises, the problem gets worse. Iron from corrosion of carbon steel exchangers can also foul heat exchangers.
- Leaks in CW exchangers can result in hydrocarbon entering the CW supply. Depending upon the nature of the hydrocarbon, they can foul by different mechanisms (i.e., polymerization, etc.)
- Microbes can grow rapidly enough under CW system conditions (which typically have a pH range of 7.0-9.0 and temperature levels (supply and return) of 21-48°C) to plug exchanger tubes and distributors in the cooling tower or to initiate Microbially Influenced Corrosion (MIC). Once the distribution of water through the cooling tower is disrupted, there is a loss of cooling efficiency and the CW temperature rises, accelerating the problem. MIC can cause severe corrosion that produces leaks that allow hydrocarbon into the CW system.

Initially, fouling deposits limit the heat transfer and cause a rise in outlet temperature. This correlates to a reduction in efficiency of the related equipment. Subsequently, if the deposits build up sufficiently, the pressure drop through the CW circuit will increase and limit the water circulation flow, which in turn will limit heat transfer more dramatically and often result in a plant capacity limitation. Ultimately, corrosion can occur that limits the life cycle of the equipment and potentially could cause an unplanned shutdown. Even without corrosion failure, excessive fouling can limit the interval between plant turnarounds.

Below are some examples that illustrate the economic sensitivity of a typical 1000 kta ethylene plant to fouling in the CW exchangers (at $25/mmkcal fuel value):

- An increase of 2°C in the outlet of all the charge gas compressor intercoolers will increase the CGC kW by 480, costing about $90,000/yr.
- An increase of 2°C in the outlet of the propylene refrigeration compressor outlet cooler will increase the kW by 1100, costing about $210,000/yr.
- An increase of 2°C in the charge gas compressor turbine surface condenser will increase the SHP steam consumption by 1.2 t/h, costing about $270,000/yr.

If fouling does occur in the CW circuit, it may be possible to clean exchangers while the plant is in operation. To clean the scale, it is necessary to acidify the water and flush the equipment to remove the salts. Acidifying the entire system could lead to increased corrosion and difficulty in purging the salts, so it is highly recommended that the fouled exchanger shell be isolated from the rest of the system prior to cleaning in this manner.

The best methods to prevent the degradation of performance due to CW fouling and premature equipment aging due to corrosion include:

- Improve/maintain inlet strainers/filters.
- Maintain acceptable velocities throughout the system. This tends to be a problem during plant turndown or right after a turnaround when the exchangers are fully clean and operators are tempted to throttle the CW supply to the exchangers. In this case, maintaining full or acceptable velocities may result in excessive heat transfer (i.e., reducing tower pressure greatly in a CW condenser). When possible, it is preferred to isolate a single shell in a multiple shell arrangement rather than throttling the CW.
- Avoid control systems that require CW throttling.
- Where needed and practical, put in back-flush connections to allow for blowing down of debris that may enter the CW side of exchangers.
- Properly monitor and use chemical treatment of the system, including adequate on-line monitoring equipment and chemical feed automation.

Monitoring and control of all cooling water parameters is key, including:

- pH
- Cycles of concentration
- Free halogen levels
- Treatment program concentrations such as phosphate, polymeric dispersants, and azoles

Of particular importance in controlling fouling are scale inhibitor dosages. Monitoring the amount of scale inhibitor polymer (by methods such as the use of a fluorescent molecule bonded to the dispersant/scale inhibiting polymer) has been effective in responding to daily operational and water chemistry fluctuations.

A higher variation in the water quality can be tolerated at a pH range of 6.8-7.8 as compared to operating at a pH range of 8.4-9.0 since most inorganic scales are less soluble as pH increases. Table 2 gives some examples.

Table 2.
Water Quality Variation

pH Range	**7.0 – 7.6**	**7.6 – 8.2**	**8.2 – 9.0**
Allowable pH Control Variability	0.6	0.4	0.2
Allowable Cycle Variability	2 cycles	1 cycle	0.5 cycle
Allowable PO_4 Variability	4 ppm	2 ppm	1 ppm

Conclusion

Fouling and corrosion start the day the plant is commissioned. Their impact on plant efficiency and economics can be managed by the actions taken to mitigate their progress and effects. As noted in this paper, the major impacts occur in the cracking furnaces, charge gas compressor, compressor drivers, and cooling water system. The economic consequences are quite significant as can be seen by the examples cited. Other potential costs, such as lost production, more frequent turnaround intervals, and shortened equipment life add to the economic impact. With proper attention to operation, maintenance, and treatment programs, the impact of fouling and corrosion can be controlled but never eliminated.

Acknowledgements

- We would like to thank Daniel Meier, David Cody, and Robert Presenti of the Nalco Company for providing their valuable input into this paper.
- We would like to thank Ulf Walden of Polarchem International Ltd. for his important contribution to this paper.

Continuous Energy Improvement at EU-1544

Jennifer L. Romig

Operations Assistant

Chevron Phillips Chemical Compan LP

Charles T. Polito

Ethylene Technical Staff Engineer

Chevron Phillips Chemical Company LP

Eric Niemeyer

Process Engineer

Chevron Phillips Chemical Company LP

Chris Bealing

Senior Staff Consultant

KBC Advanced Technologies plc

Allan Rudman

Vice President

KBC Advanced Technologies plc

Abstract: The ethylene unit (EU-1544) at the Port Arthur Plant of Chevron Phillips Chemical Company LP (Chevron Phillips Chemical) was originally commissioned in 1970 with a major unit expansion in 1997. A cyclohexane unit and a third-party cumene unit are also located at the site. Due to rising energy costs, significant focus has been placed on reducing energy consumption. In 2005, Chevron Phillips Chemical committed to performing process audits as part of each ethylene unit's turnaround cycle. For the 2008 turnaround, the first process audit was performed for EU-1544 at Port Arthur in 2006. The goal of the pre-turnaround process audit was to identify operational changes and projects to improve the energy performance of EU-1544. The work from the 2006 process audit became the starting point and the data set for KBC's Strategic Energy Review (SER). KBC reviews the entire site when benchmarking, performing operational reviews, and Pinch Analysis for the SER. Results from the SER include a ProSteam model of the steam system and a Road Map of identified projects for implementation. Following the turnaround, a process audit was done to accomplish three major goals: to identify and value the remaining gaps at "clean unit conditions," to establish baseline performance for a clean unit, and to confirm benefits from projects and from maintenance work done during the turnaround.

This paper will discuss the methods, implementation, and benefits of performing process audits to achieve greater energy efficiency at EU-1544.

Introduction

Process audits are a concerted effort to gain an understanding of the performance of the unit. Chevron Phillips Chemical Company (Chevron Phillips Chemical) implemented process audits as a part of the turnaround schedule for the ethylene units. EU-1544 at the Port Arthur Plant was the first unit to have a turnaround with the new process. Audits were done before and after the 2008 turnaround to evaluate current performance and changes between the audits. The data was used to evaluate opportunities for capital projects, expense projects, repair work, fouling issues, and improvements. KBC performed an Energy Audit using their techniques to confirm and supplement the process audit work. These detailed evaluations and examinations have led to a better understanding of energy usage in EU-1544 and a formal approach for gap closure.

Process Audit Overview

Purpose

Process audits evaluate the performance of the ethylene unit and compare it to previous audits and to design. A pre-turnaround audit is completed two to three years before a turnaround to allow for project development and lead time for any new equipment. Following the turnaround, another process audit is done to establish a new baseline for the unit's operation. From this post-turnaround audit, we can quantify the benefit of the work and projects done during the turnaround and quantify the performance reduction over time. A RoadMap can be developed to identify a path of operational changes, maintenance work, and capital projects to improve energy use.

Methods

For each process audit, data from field surveys and the process historian are used as inputs for the simulations of the unit clarify and include reconciliation. The field survey includes the steam and other utility systems. This set of data provides a snapshot of the unit's performance by matching the simulations to the data.

During the field survey, samples, pressures, and temperatures are taken at predetermined locations. In addition, the planned locations include areas that are suspected to have operational issues or hydraulic limitations. Prior to the survey, a set of P&IDs is marked-up indicating the location of each measurement and the type of measurement. Some examples of pressure measurements are listed below:

- Suction and discharge pressures on compressor stages
- Feed, overhead, and bottoms pressures on towers
- Inlet and outlet of exchangers

Temperatures were measured on the suctions and discharges of the compressor stages and at locations between exchangers. Samples were analyzed to verify feed compositions and to verify tower cutpoints. Any unusual situations in the field were documented (significant elevation changes, unexpected liquid in bleeders, etc).

To augment the field work, process data from the data historian is collected for the duration of the field survey. The data is used in furnace and unit simulations. In the simulations, cutpoints, utility usage, flow rates, temperatures, and pressure are matched as closely as possible to the data collected during the audit.

With the match case completed, equipment performance is compared to design and other process audits. From the comparisons, the gaps in performance are identified. Also, gap closures are confirmed. Further data analysis is done to investigate the reasons for the gap (flawed design, fouling, a change in operation, etc). With this information, appropriate solutions are developed such as capital projects, maintenance work orders, and turnaround scope recommendations.

2006 Process Audit

Chevron Phillips Chemical proprietary unit simulation, a unit simulation model, was first used to match overall unit material balance with unit limits. This match included furnace configuration, feed type and conversion. Raw data was collected around the unit in January 2006. Where appropriate specific single gauge pressure survey data was collected. If some critical flows were suspect, then control valve outputs were used and with their data sheet and pressure drop data across the valve, flows were estimated, e.g. deethanizer side reboiler quench water flow and C2 splitter pasteurization section vent condenser refrigerant flow.

Furnace effluent compositions and flows were taken from the unit simulation model as a starting basis for introduction into the unit process models. The process models for the rigorous heat and material balances were several discrete sections of the process.

The streams were run through each discrete model with several iterations among the connected models. After process side models were finalized, refrigeration loads and operating data from the process were put in utilities refrigeration model to better match plant operation. The following match methodology was done for major equipment types:

Furnaces:
Furnaces were modeled and compared to operating data using the full detailed furnace model which included firebox, convection and TLE. Results of simulation match were tabulated and compared to design data sheets. Stand-alone boiler feed water economizer were modeled and compared to operating data using the rigorous heat and material balance simulator.

Compressors/Turbines/Surface Condensers:
Cracked gas and refinery fuel gas compressors were matched with operating data on inlet, outlet conditions. Water injection and wash oil was accounted for in crack gas compressor, CGC, modeling. Speed was also matched to operating conditions. The resulting polytropic efficiencies were then tabulated. Modeled power from compressor side was then used to calculate turbine adiabatic efficiency. The turbine was modeled as an expander with actual steam inlet and outlet conditions, and efficiency was calculated based on power requirements from the compressor side. Surface condenser performance was rated based on inlet conditions and vacuum. Cooling water rates from the performance audit were used to model and rate surface condensers.

Ethylene and propylene compressors were matched to the design curves only and no fouled polytropic efficiencies were calculated since these compressors are considered to be in clean service. Refrigerant loads from process models were used to determine compressor operating points. Speed, flows and inlet conditions for each stage were matched and agreed with the operating data. The modeled power from compressor side was then used to calculate turbine adiabatic efficiency. Turbine was modeled as an expander with actual steam inlet and outlet conditions, and efficiency was calculated based on power requirements from the compressor side. Surface condenser performance was rated based on inlet conditions and vacuum. Cooling water rates from the performance audits were then used to model and rate surface condensers.

Expanders/Recompressors:
Actual inlet and outlet conditions were set in the model based on operating data, which included single gauge pressure survey around expanders and recompressors. Expander by-pass rates and recompressor by-pass rates were set in model based on operating data. The expander and recompressor efficiencies were then calculated to match inlet and outlet conditions. For expanders, inlet and outlet pressures were the main metrics used to calculate efficiency. Shaft work efficiency between expander and recompressor was varied to help match operating data.

Distillation Towers:

Towers were modeled to match operating data. Tower pressures and pressure profiles were fixed from the operating data. Depending on accuracy of on-line analyzers, either component specifications or overhead/bottoms temperatures were used to specify the tower operation. Reflux rate meter, if good, was either used as a specification or as a check. Condenser and reboiler duties were calculated and used as a check for the match case. Total number of ideal stages in the tower were varied to match the operating conditions.

Heat Exchangers:

A rigorous heat exchanger model was used to match operating data (used simple heat exchangers across board with FT factors). FT factors were inputted in model for each exchanger's shell and tube configuration. For cooling water exchangers, if cooling water rates were known from audit, they were inputted in model. For quench exchangers, if quench water rates were known, they were inputted in model. Model then calculated actual heat transfer coefficient and LMTD. For refrigeration exchangers, refrigerant flow was calculated based on the process requirements, however actual pressure survey data was used to set actual refrigerant temperature or pressure level for that exchanger.

Pre-Chill exchangers:

With actual temperature and pressure data, heat exchanger LMTD's were calculated and compared to design to determine approach to refrigeration.

LNG exchangers:

For LNG exchangers, actual inlet and outlet conditions on multipass exchangers were set. Process flows and compositions were based on process modeling match for demethanizer ARS area. Model then calculated MITA (Minimum Internal Temperature Approach) and compared to design MITA. The team also met with Shaw to review operating data around ARS and compared temperature approaches to refrigerations, and compared differences between dephlegmator passes.

Cooling Water System:

A detailed hydraulic model was developed for the current operation, using calculated process duties to determine and verify cooling water rates. Survey pressure and pump data were used to tune model. Heat exchanger K values were used to adjust model flow rates and pressure drops to match operating data. Clean heat exchanger pressure drops were checked with HTRI. Model was then used to predict cooling water circulation rate with clean heat exchangers.

Steam System Model:

Deaerator steam was calculated via an overall heat balance. Slow roll steam, flare steam and miscellaneous estimated from correlations. Import steam from nearby refinery and Cyclohexane plant based on actual steam conditions and rates taken from manufacturing cost statement. Steam generation based on furnace model from the OMB match case. Compressor turbines based on process matches, measured surface condenser vacuums and cooling water flow rates. Dilution steam rates based on metering and furnace rates from OMB. Process heat exchanger duties used from process model matches, whenever operating data available, actual steam chest pressures were used in exchanger model to better steam usage and heat exchanger performance. Pump driver turbines (topping turbines) used either design efficiency or calculated turbine efficiency if data was available for match. Heat losses were added for high pressure level.

Miscellaneous Pressure Profiles:
Single gauge pressure survey data was compared to theoretical in critical CGC compression train, chilling train and refrigeration trains. Areas in CGC were identified for horsepower savings. Areas in refrigeration were identified for both compressor power savings and improved demethanizers and ARS performance.

2006 Process Audit Findings and Recommendations

Furnace Area

- Excess air, elevated stack temperature, and high TLE inlet pressures
- New secondary TLE opportunities
- Optimize Lummus stack dampers, rerate combined effluent economizer
- Debottle remote desuperheater to minimize higher pressure steam letdown at high rates
- Insulation of effluent line benefits
- Clean convection sections

Steam System

- An imbalance of 50 mlbs/hr on the LP steam header. Part of the discrepancy was considered to be metering.
- Ethylene and propylene refrigeration compressor surface condenser leaks identified
- Cracked gas, ethylene and propylene refrigeration compressor surface condenser fouling – clean during turnaround and implement continuous surface condenser cleaning project
- Refrigeration system energy inefficiencies
- Cracked gas compressor degraded polytropic efficiency
- Fully utilize cyclohexane vent steam in ethylene plant
- Eliminate slow roll on spare turbines for major pumps

Cooling tower performance and water circulation rate

- Low flow through surface condensers and cracked gas compressor interstage coolers

- Reduced propylene production
- New more efficient cooling tower extraction turbines
- Propylene refrigeration horsepower savings
- Cooling tower high efficiency packing fill

Quench system improvements

- New quench tower replacement internals for better approach temperatures
- New quench water clean-up facilities for improved circulating quench water heat transfer, minimizing LP steam

Refrigeration system energy gaps

- High pressure drop between second and third stage ethylene refrigeration
- Expander/recompressor performance gap from design
- Opportunities to optimize propylene refrigeration discharge pressure to the available cooling water temperature
- Better optimum use of reheat in the core to save ethylene refrigeration
- Economic incentive to derime cold box
- Use of e/p feed vaporization to supplement propylene refrigeration (long term project)

Propylene capacity

- Process control improvements needed to maintain stable quench water reboiler heat during furnace swaps and stabilize tower pressure at varying cooling water temperature to permit operation closer to flood
- Reduced cooling water fouling and higher cooling water rates to condensers
- Install new feed distributer and feed partial vaporizer for incremental production

Ethylene capacity

- Better heat integration to lower ethylene fractionator top feed temperature back to design by modifying heat exchanger train
- Derime C2 splitter vent condenser to minimize recycle and debottleneck C2 splitter

Feedstock flexibility

- Condensate stripper debottleneck
- Increased segregated propane capacity capabilities
- Purity ethane feedstock opportunities

Integration of the KBC Energy Audit

Following the completion of the simulation work for the pre-turnaround process audit, KBC used the audit data to perform a Strategic Energy Review (SER). This work complimented the work of the process audit by identifying additional opportunities to optimize energy usage in the plant, and validated our energy findings. Findings and projects identified in the SER were re-evaluated and re-quantified on subsequent process audits.

The SER also benchmarked the performance of the Port Arthur site using the KBC Best Technology (BT) index. The BT index is defined as the ratio of the total energy consumption (including fuels, power and steam import) to the sum of BT energy consumption allowances (standards) for each process unit and associated infrastructure.

KBC BT standards have been developed by fundamental analysis and design studies on individual process units. They take into consideration the key process parameters affecting energy consumption along with the specific site economics and a first quartile BT index is achievable in a grass roots unit built today.

The calculated BT for the base case operation at Port Arthur site including both Ethylene and Cyclohexane units placed the site on the border of the third and fourth quartiles in the extensive KBC petrochemical database.

A total of 22 projects were included in the full SER RoadMap. These projects were split into stages as follows:

- Group 1 – Projects planned for implementation in the forthcoming shutdown
- Group 2 – Other low cost or quick return projects recommended for rapid implementation
- Group 3 – Medium/longer payback projects or projects requiring further investigation. Not proposed for rapid implementation
- Group 4 – Long-term strategic investment

A major issue in developing the RoadMap was the overall site steam balance. Steam was being vented from the 50# level in one area of the site while simultaneously being let down from 220# to 50# in another area. The use of the ProSteam model was critical in defining compatible groups of projects that eliminated venting while at the same time minimizing letdown flows.

The Group 1 and 2 projects combined achieved savings of several million dollars with an overall payback of just one year. The Group 3 projects then doubled the savings potential with an incremental payback of less then two years.

The Group 3 projects are medium-term opportunities while the Group 4 projects encompass longer-term options such as major revamps of the cracking furnaces.

Chevron Phillips Chemical has implemented the Group 1 and 2 projects. Groups 3 projects are scoped out for implementation using the KBC energy road map in the next several years.

The proposed projects move the Port Arthur site to a position at the top end of the second quartile in the KBC petrochemical BT database. In order to improve further, the site would need to consider projects with significantly higher levels of capital investment that are not normally justifiable in a retrofit situation.

Post-Turnaround Audit

The post-turnaround audit is similar to the pre-turnaround audit with some significant differences. The pre-turnaround audit was used to form the road map for several years worth of projects. The post-turnaround audit's scope was narrower. While it some of the relative merits of these projects' economics changed due to energy pricing there were no major changes to the RoadMap. The post-turnaround audit was preformed using the same methodology outlined above, with emphasis on implemented capital projects and results from other turnaround work.

The field survey, process models, and data historian information from the pre-turnaround audit serves as a starting point because it has enough data to complete the full unit match simulation and full material balance. Additional sample points are added strategically to evaluate capital project projects. The same data historian information is gathered along with any new readings as necessary.

After the simulation match work was completed, the process audit team worked to identify gaps between each operating process system and the unit design model. These are compared to the 2006 audit to learn how some gaps were closed and why others remain. Direct comparison of some systems was difficult since some of the operating conditions and economics were significantly different from 2006.

The closed gaps were from capital improvements, better operation or "unknown". For instance, there was a major improvement in steam usage from 2006 to 2008. Some of the improvement was classified as turnaround cleaning improvements in major compressors, improvements in distillation tower operation, and energy efficiency capital improvements.

The post-turnaround findings also were compared to the KBC RoadMap. Some items were completed while others were on hold. The impacts of these differences were evaluated to highlight key items to accomplish on the long-term RoadMap.

In the short term progress, high impact operational focus areas were identified. For instance, a recurring opportunity in the distillation section was cut point changes in distillation towers. Other towers could benefit in energy and production with additional focus. The cut point changes in two towers had improved following the 2006 audit but had since regressed to their original values. The team recommended developing a system to provide more consistency for cut point operation in the distillation section.

The gap closures, open gaps, and the KBC RoadMap audit were then used to recommend actions, highlight opportunities and show results to plant management.

Conclusion

At Chevron Phillips Chemical, process audits are done around each turnaround to evaluate the performance of the ethylene units and identify opportunities. The Port Arthur Plant performed process audits in 2006 and 2008. A KBC SER Energy Audit was done in 2006 following the audit to augment the evaluation by looking at energy usage in EU-1544 including heat recovery and integration. The post-turnaround audit allows for the evaluation of a "clean unit" and improvement resulting from the turnaround. The pre-turnaround audit in 2006 resulted in identified maintenance activities, capital projects, and operation issues could be addressed both immediately and during the planned outage. Ultimately, this process has lead to making continual improvements in the energy usage of the ethylene unit and a better understanding of the opportunities.

OLEFIN PLANT ENERGY SAVINGS THROUGH ENHANCED AUTOMATION

Douglas C White
Emerson Process Management
Houston, Texas
doug.white@emerson.com

Abstract: Energy is the single largest controllable cost for olefin plants and the rise in prices has caused most plants to examine even more closely their energy usage. Automation enhancements can significantly reduce energy consumption across all areas of the plant. Some of these savings can be achieved with no investment, only changes in normal operating procedures. In other cases improvements to on-line analyses, measurements and control action are justified but generally require relatively modest investments. The management of the utilities at a major olefin site can be difficult with many daily operating decisions that must balance competing economic and production issues. Real time modeling of process and utility equipment and monitoring of the energy usage in plants permits allocation decisions to be made much more frequently and accurately, often resulting in substantial savings. In this paper, results from several installations are summarized to provide guidance to olefin plant staff on likely areas for savings. Olefin plants are complex and highly integrated and analysis of potential energy savings must recognize this complexity and integration. A systematic evaluation methodology is presented to insure that projected savings are both realistic and attainable with the proposed investment. Examples from several different plants are included.

Introduction and Background:

Olefin plants are large energy consumers with energy the largest variable operating cost after feedstocks. Using energy efficiently has been and remains a primary goal for olefin producers. For most plants the marginal fuel is natural gas and the change in average US natural gas costs (1) over the last few years, as illustrated in Figure 1, illustrates both the general rise in energy prices and the volatility in these prices. At a price of $6 per million BTU's (mBTU), a 5% energy saving is worth approximately $4.5 million per year for a typical North American 500,000 T/ yr ethylene plant with naphtha feed.

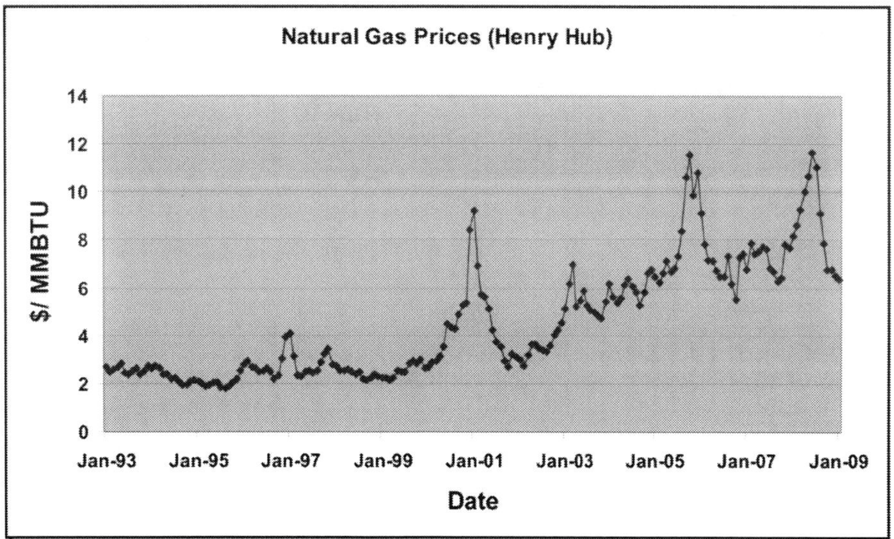

Figure 1: Natural Gas Prices

In addition to the direct price of energy, it seems likely that the US will adopt some regulations regarding greenhouse gas emissions in the next few years. If the regulations in other countries are a guide, these may take the form of a "Cap and Trade" on CO2 emissions or a carbon tax. This will place an increased value on energy reductions since these reductions can be used to offset increases in other areas or can be sold under a "Cap and Trade" system. If these CO2 reductions are valued at $20 / ton, which is at the low range of recent prices in Europe, then reductions in natural gas or equivalent fuel usage would have an additional value of approximately $1.3 / mBTU. This would add a value of approximately $1 million per year to the 5% energy savings above.

Although many olefin plants have been investing to reduce energy consumption, there is still a wide variation in energy usage, even after correction to standard conditions for feed quality, product composition and process configuration as illustrated in Figure 2. Surveys, such as reference 2, repeatedly show wide gaps between the most efficient plants and the least efficient. There is at least a 40% spread in energy usage between them, even after correction to standard conditions for process configurations, product grades and feed types. This variation is primarily due to the age of the equipment with older plants often having less efficient equipment and less heat integration than newer plants. The rise in average energy costs has increased the incentive for additional investments and there are typically still additional opportunities to be found even in better performing sites, generally estimated to be of the order of 20%. Note that the theoretical minimum energy for these plants is still well below current levels at roughly 25% of the current average value.

There are a large number of possible investments that can be made to reduce energy with different costs and different impacts as illustrated in Table 1. These range from low cost/ relatively low impact programs such as those to reduce steam trap and other steam leaks to installation of integrated gas turbines with very high costs and long implementation times and correspondingly high impact. In this paper the focus will be on automation investments. These generally fall in the low to medium cost range with savings that are typically mid range. As a result, the expected return on investment for these programs can be quite high.

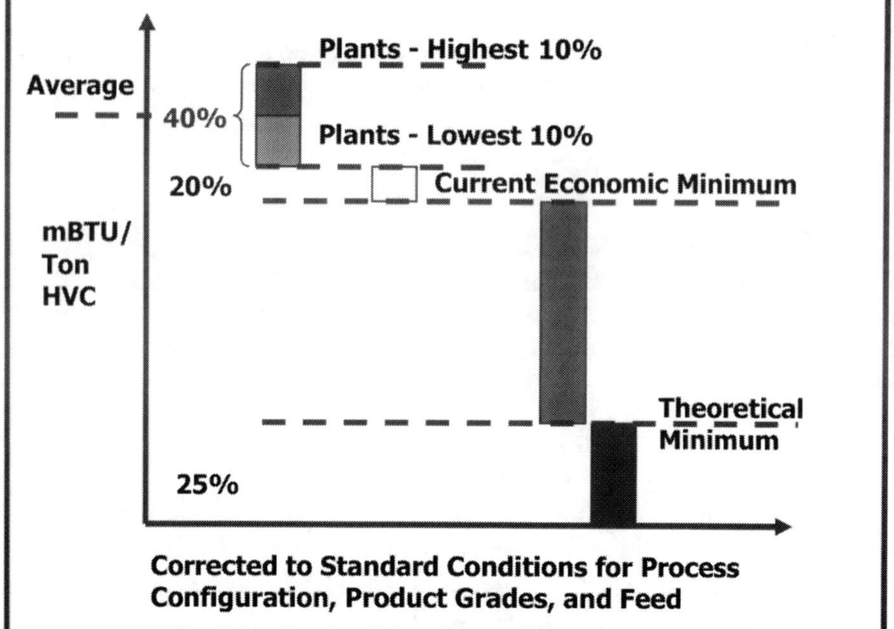

Figure 2: Olefin Plant Energy Usage

Table 1 Potential Olefin Energy Investments			
	Capital Cost (Time To Implement)		
Potential Energy Savings	**Low**	**Medium**	**High**
High			Integrated Gas Turbine: Cogeneration; Furnace / Separation Process Upgrade; Replacement Low Efficiency Process Equipment
Medium	Improved Automation; Operating Procedure Changes; Energy KPI Monitoring	Advanced Control/ Optimization; Site Energy Management Systems	
Low	Increased Insulation; Steam Trap/ Leak Management; Exchanger Maintenance Condensate Recovery		

Automation Impact on Energy

Typical automation and related systems in olefin plants are shown in Figure 3. The ISA 95 specification groups them in Levels 1 to 4 as shown. Almost all of these systems will have some impact on energy usage but the focus of this paper will be on those in Levels 1 to 3. Economic values calculated in the text following for potential improvements will be based on a typical North America (NA) Naphtha feed plant producing 500,000 T/ Year ethylene with the assumed properties listed in the Appendix. No credit is taken for the value of greenhouse gas reductions.

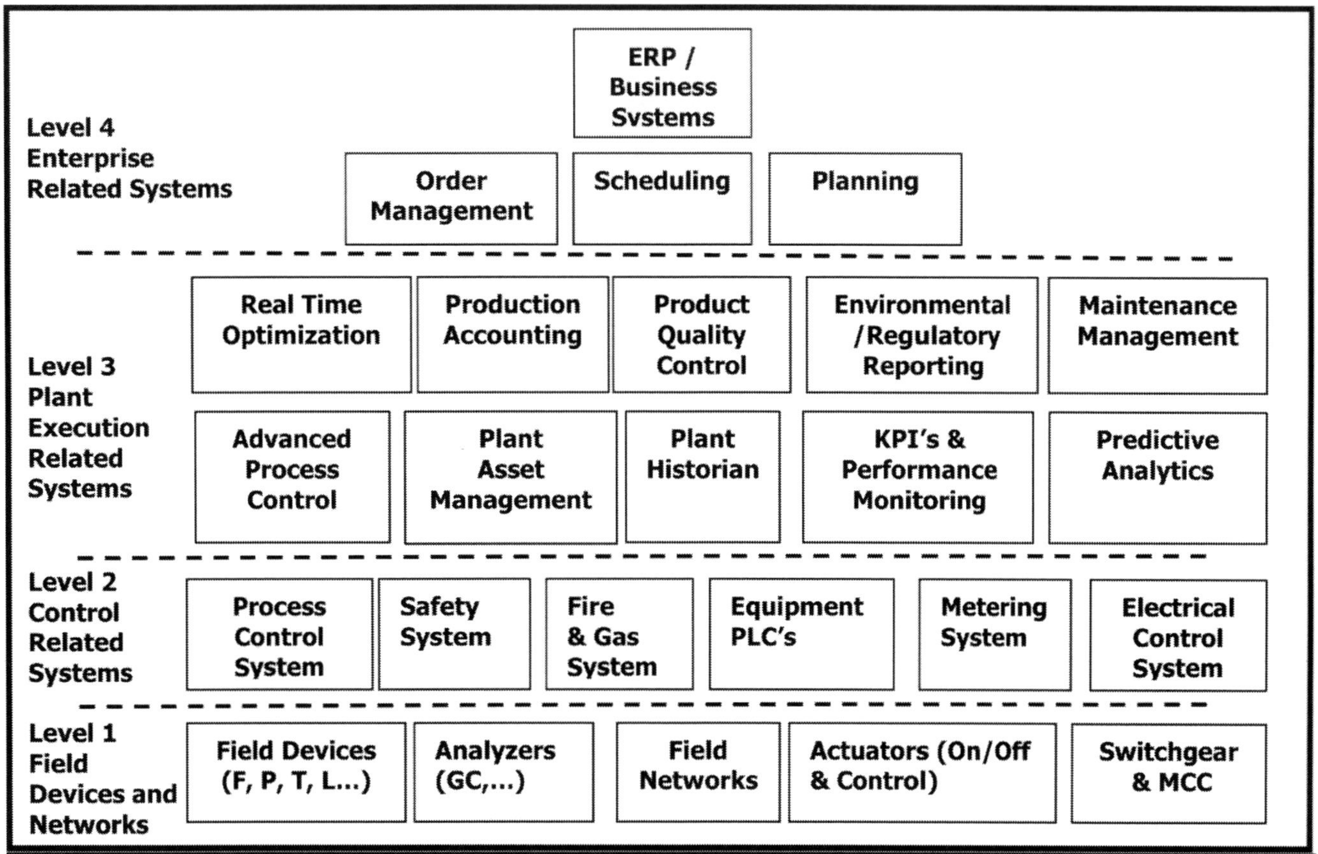

Figure 3: Olefin Plant Automation Systems

There are two primary mechanisms by which new investments can reduce energy usage at specified production levels - either by reducing the cost of supply or by reducing process energy demand. Reducing supply costs can be subdivided further into investments that reduce external purchase costs or those that increase internal energy production efficiency. Table 2 lists some of the specific ways that automation and related systems can reduce energy within this framework. There are a large number of impact areas, too many to be covered in any single paper. Olefin plant advanced control and optimization systems are one of these and can obviously have a large effect on energy usage. However, their general functionality and energy impact has been covered previously in many papers, for example reference 3, and will not be discussed further here. In this paper a few specific automation impact areas will be covered that, in the author's experience, are not as widely recognized as others.

Table 2
Potential Automation Energy Saving Strategies

Reduce Process Energy Demand	**Reduce Energy Supply Costs**	
	Increase Internal Utility Production Efficiency	**Reduce External Purchase Costs**
• Advanced Control/ Optimization • Furnaces • Quench/ Frac • Compressors • Distillation • Improved Control Loop Performance • Maximize Process Heat Recovery /Minimize Losses • Maximize Recovered Steam • Minimize Pressure Drop • Minimize standby turbines and boilers • Better control at low rates	• Improve Combustion Efficiencies • Steam System Management • Steam Header Management (minimize venting, letdown, pressure) • Maximize Recovered Steam • Blowdown Control • Steam vs. Electric Turbine Optimization	• Energy Management System • Power, Fuel, Steam purchase

Control Loop Performance Improvements

Efficient and effective execution of the basic control loops at the plant is obviously essential to successful operation of the plant and to other functions such as advanced control and real time optimization. Control loops are composed of a measurement element; an actuator, most commonly a valve; and an executed control algorithm. As suggested in Figure 4, improvements in each of these elements can lead to reduced energy usage. In the sections below, some specific improvement areas applicable to olefin plants are discussed.

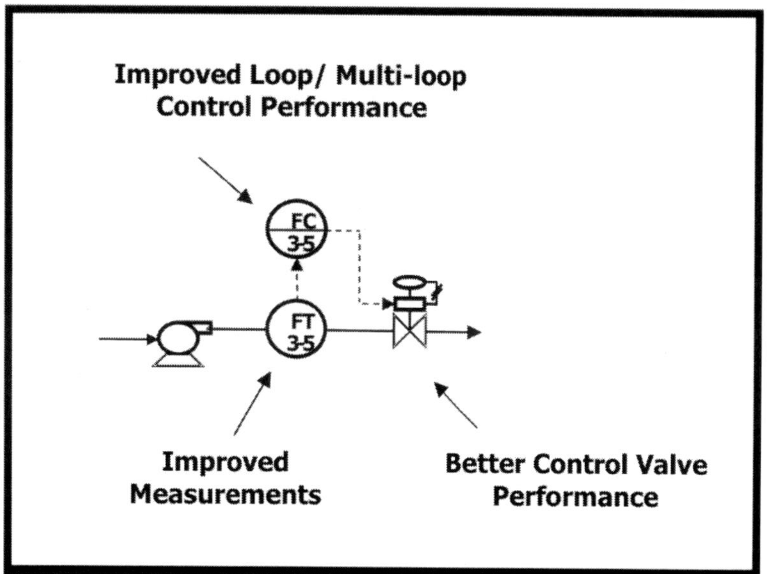

Figure 4: Energy Saving Through Control Loop Improvements

Measurement

One of the first areas to evaluate for potential automation changes is improving the measurements of key plant variables in terms of accuracy, location, frequency and number. Physically relocating a measurement to reduce dead time in a loop can often do more to improve performance than any tuning or algorithm. Increasing the frequency of measurement, for example for a gas chromatograph, can be similarly beneficial for loop performance. Increasing the accuracy of the measurement can sometimes be used to reduce operating constraint margins and change operating conditions to save energy. In the past it has typically been very expensive to consider adding measurements after a plant has been built because of the permitting and wiring requirements. Wireless transmitters, which are now more widely accepted in plant use, significantly reduce the cost of adding additional measurements.

There are some specific measurement improvements that can have a significant impact. Many olefin plants experience wide variation in their plant fuel gas composition and the corresponding heating value. This variability induces variability in the cracking furnace combustion control and increases the required operating constraint margin above the optimum air/ fuel ratio.

In Table 3 below volumetric heats of combustion for standard fuel gas components are compared with their mass equivalents. Note that there is much less variability on the later basis. Even hydrogen, which is one of the major causes of volumetric heating value variability, has a ratio on a mass basis much closer to the other components than its volumetric equivalent.

Table 3
Fuel Gas Component
Heating Value

Component	Heat of Combustion BTU/ cu ft (gross)	Heat of Combustion BTU/ lb (gross)
Hydrogen	320	60957
Methane	1011	23861
Ethane	1783	22304
Propane	2563	21646
Butane	3374	21293

This suggests that fuel gas control on a mass rather than volume basis will eliminate much of the variability and that is the experience of plants that have adopted such control. With modern flow measuring devices that will directly measure mass and also provide a gas density measurement it is easy to implement such controls. The value of this improvement will depend on the normal variability of the fuel gas composition and its effect of the furnace excess air operating margin. However, for a typical NA plant this improvement might be 0.2% in furnace efficiency which would be worth approximately $200,000 per year, more than enough to give a one year simple payback on fuel gas mass meters.

Another key measurement improvement for cracking furnace combustion is the use of stack gas CO measurements. While some sites have installed these analyzers previously, their use has been limited by the perception that the sampling systems were high maintenance items and the analyzers themselves were somewhat unreliable. However, CO analyzers are now available for in situ installation with no sampling system and no calibration gas requirements and have proven to be quite reliable.

The recent research by The John Zink Company, as shown in Figure 5, and discussed further in reference 4, provides further incentive for improved online combustion gas analysis. The research shows that ambient air temperature and relative humidity changes have a surprisingly strong effect on combustion conditions, even with constant fuel composition. The first graph shows that a heater tuned to satisfactory combustion conditions at 80 deg F and 0% relative humidity (14.7 psia) with constant air/ fuel ratio control will experience incomplete combustion if the ambient temperature changes to 100 deg F and 80% relative humidity with an ambient pressure of 14.1 psia. Such changes are well within the range of ambient condition changes experienced on the US Gulf Coast. The second graph shows the specific effect of these changes on CO composition.

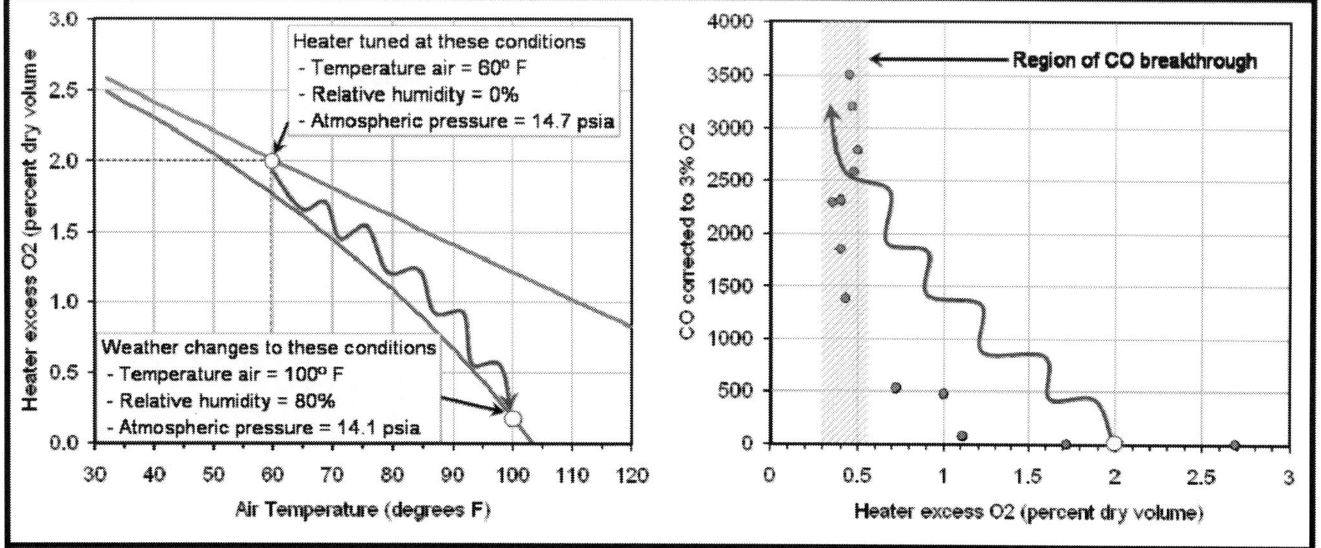

Figure 5: Ambient Effects On Combustion:
From Reference (4): ***Bussman and Baukal***

With varying fuel composition, direct CO measurement is the best way to monitor combustion efficiency and enable close control of the furnaces.

Valve Performance

Olefin plants often operate at varying load conditions, due to market conditions, feedstock availability, furnace decokings, and process equipment availability. Control strategies need to be designed and implemented that reflect efficient performance over the full range of operating conditions and field devices need to be chosen and sized appropriately. The cracked gas and refrigeration compressors are major energy consumers in olefin plants and proper anti-surge control, as shown in Figure 6, is an important factor in minimizing energy usage at lower load conditions. The stage recycle or anti-surge valve is controlled by the surge control system which typically includes compensation for changes in speed, suction temperature and pressure, and gas composition, if measured. The recycle valve has to open very quickly and accurately to re-circulate gas from the stage discharge to the suction. The ability to stably and

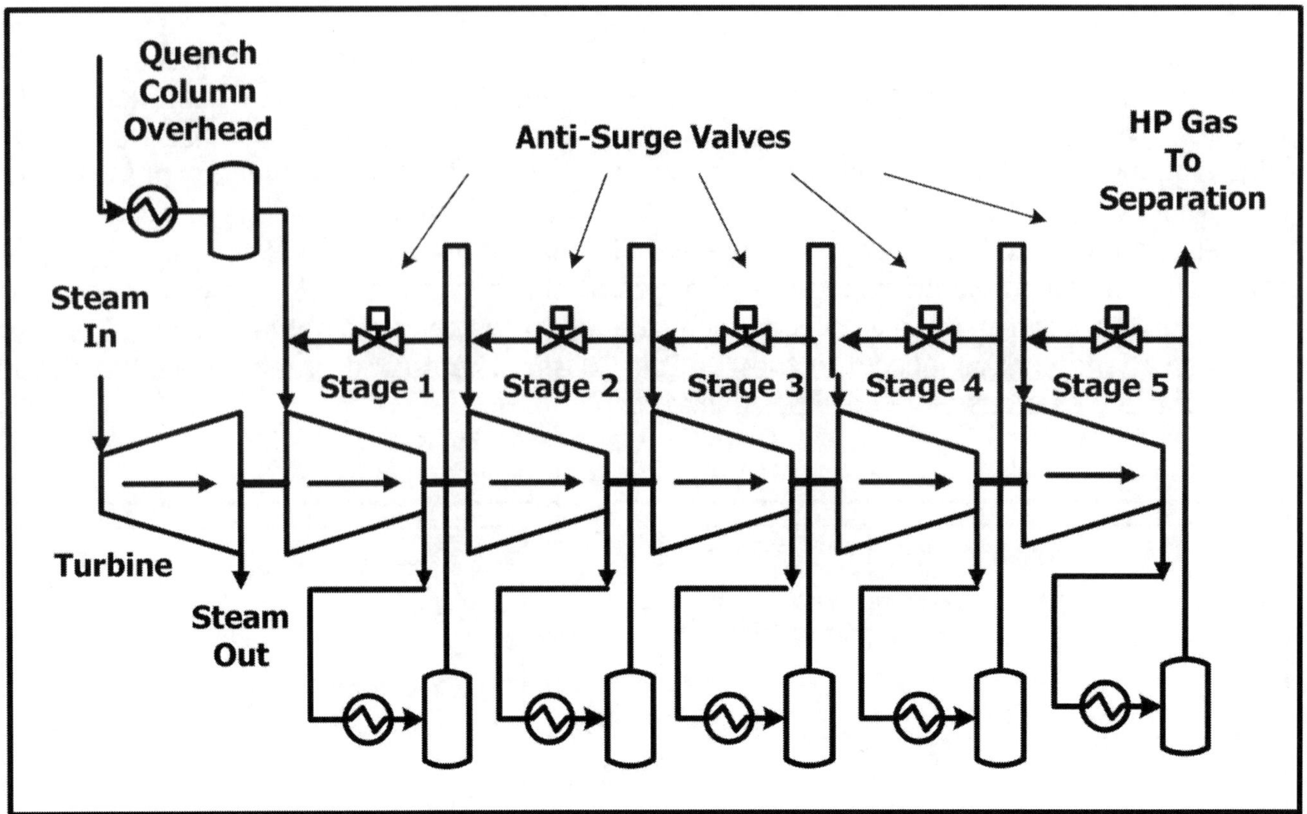

Figure 6: Compressor Surge Control

safely operate the compressor closer to the surge limit saves energy and the size of the required operating constraint margin is dependent on the response characteristics of the anti-surge valve. Developments in valve technology now permit even very large recycle valves to go from fully closed to fully open and vice versa in under two seconds.

In figure 7 a typical surge line for a compressor is shown. The actual normal operating surge limit will be set at some margin from the actual limit. By installing a higher

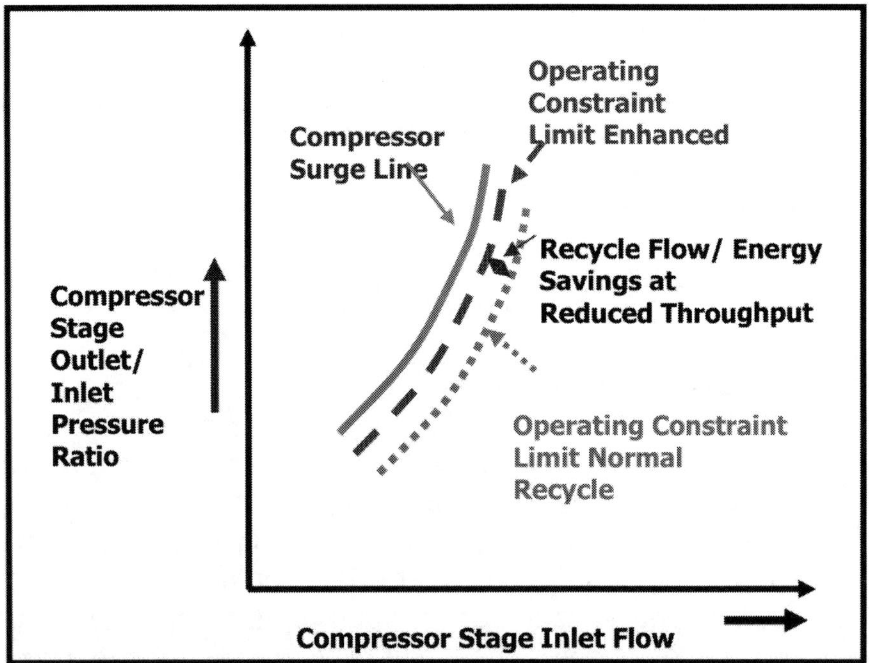

Figure 7 – Compressor Energy Savings With Improved Valve Performance

performing valve it will be possible to reduce the operating margin. The value of such improved control depends on the current operating margin and the percentage of time that the plant runs at lower load conditions. If the plant is in low throughput conditions one-quarter of the time, the value of the reduced recycle would be approximately $120,000 per year for the reference plant. In addition, at normal operating flow conditions, a higher performing valve would permit the cracked gas compressor suction pressure to be reduced, resulting in increased ethylene yield, if economically justified.

Loop Dynamic Analysis and Tuning

One of the most cost effective measures that can be taken to improve energy control performance is to analyze the dynamic response of key basic control loops, identify problems and correct them. If poor performance exists, the problem(s) could be in the measurement, the control valve or other actuator, or in the loop dynamic tuning. It is also important to identify when the problems are external to the loop either in terms of process disturbances or due to multi-loop interaction. The analysis can be facilitated by use of a hardware/ software system that permits high speed loop data capture and automates more sophisticated analyses. Assuming that the field automation equipment is performing adequately, the next step is to tune the loops with a rigorous approach that clearly recognizes the required tradeoffs among stability, robustness, performance and total system dynamics. The Lambda tuning method (5) is a well proven, mathematically sound methodology that deals explicitly with the tradeoffs above. The top set of graphs in Figure 8, from reference 6, shows the performance of the

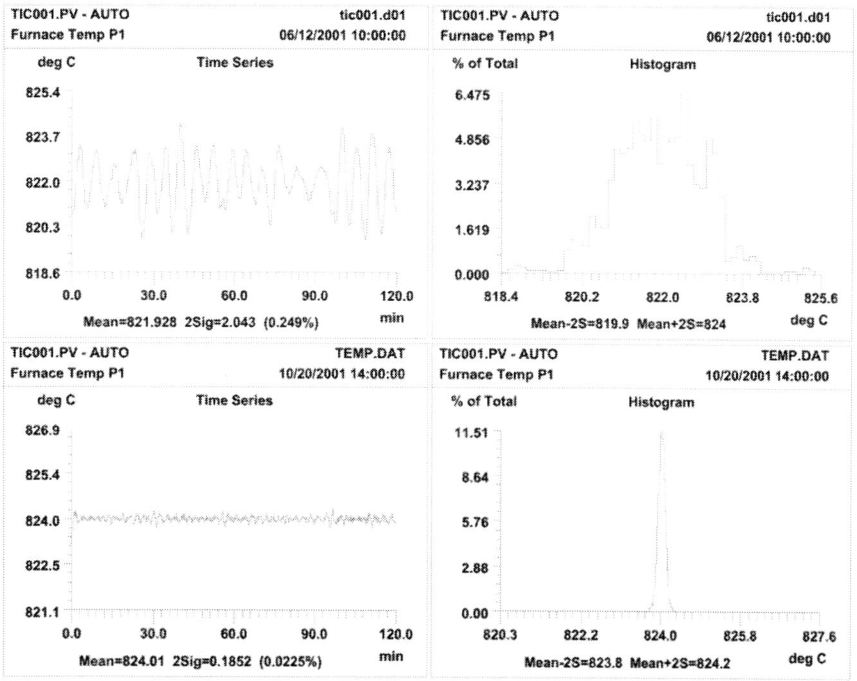

Figure 8 – Cracking Furnace Outlet Temperature Control Tuning Effect; From Reference 5; *Filho*

outlet temperature controller on a cracking furnace when tuned using older methodologies, in this case Ziegler-Nichols Quarter Amplitude Damping. The right hand graph is the histogram of the temperatures. The lower graph shows the performance of the same control loop, with the same valve and measurement, after tuning using the Lambda methodology. The reduced variability leads to reduced energy consumption at the same target setpoint. Since furnace fuel consumption is a non-linear concave function of outlet temperature, excursions above the setpoint consume more fuel than is saved by excursions below the setpoint, at the same mean. In addition, tube coking rates are similar non-linear functions of temperature and reduced variability leads to reduced coking rates, again at the same mean temperature. As the furnace tubes coke, the heat transfer efficiency is reduced, resulting in increased fuel usage. Alternately, the reduced variability can be used to increase the target outlet temperature setpoint at a constant coking rate equal to original value. Such an increase will increase once through conversion on the cracking furnaces. For ethane and propane feed furnaces this will decrease overall recycle rates at constant net conversion with a potential net energy savings. The proper balance between these two possibilities for energy saving can be calculated from an overall plant economic analysis.

Steam System Management and Control

Control of the overall steam system in an olefin plant is one of the most challenging automation areas. Figure 9 shows a typical system though every site tends to have a unique configuration, particularly sites that are older and have experienced several debottlenecking projects and equipment upgrades. The current control systems tend to have evolved over time with new controls added to handle a particular issue, sometimes without full consideration of their effects over the complete range of expected operation.

There will be normally be multiple steam headers at different pressure levels, each with multiple users and suppliers. Since most of the VHP steam is supplied by heat recovery on the cracking furnaces, this supply varies with furnace changes in rate, feed type, and severity. Demand at each level is similarly changing with changing plant conditions and changes in one header can induce disturbances in the others. There is often varying steam import or export to other plants at the site further adding to the disturbances to the system. The large steam turbines for the cracked gas and refrigeration compressors are the largest utility consumers on the site after the cracking furnaces but there are typically many other users. In some plants there will be on site power generation as well.

A further complexity with regard to the control of steams system is the hierarchy of objectives as shown in Table 4, which lead to different control responses depending on the state of the system. The lowest priority is utility optimization with the objective of determining the most economic operating conditions for the current plant demand. They are normally calculated by a steady state optimization model. Typical control actions at this level might be to specify targets, such as the turbines' extraction steam flows, which would minimize venting and steam letdown from one level to the next lower level and maximize power production. More sophisticated optimizer models might actually calculate new header pressure targets, within a limited range, which would increase unit capacity or increase efficiency. The next higher control level is the normal dynamic operating region control which responds to typical disturbances in supply and demand and adjusts conditions to compensate and maintain stable operation. There are many constraints on operation, many manipulated variables and lots of interaction among the variables leading to use of multivariable control for this region. Feedforward action on measured disturbances is used to further improve dynamic performance. The adjustments required for dynamic compensation can be in short term conflict with the optimization objectives. For example, assume there are two manipulated variables with roughly equal steady state gains with regard to header pressure. Assume further that one has a small coefficient in the optimizer objective functions but a large time constant with associated slow dynamic effect while the other has the opposite characteristics. From an optimization point of view it would be desirable to primarily manipulate the first to control header pressure since changes would have a small effect on the objective function while dynamic compensation priorities would favor the second. This leads to control in priority levels 2 and 3 with short term dynamic considerations governing while the optimizer is setting targets for these variables that are only able to be satisfied in relatively steady operation. .

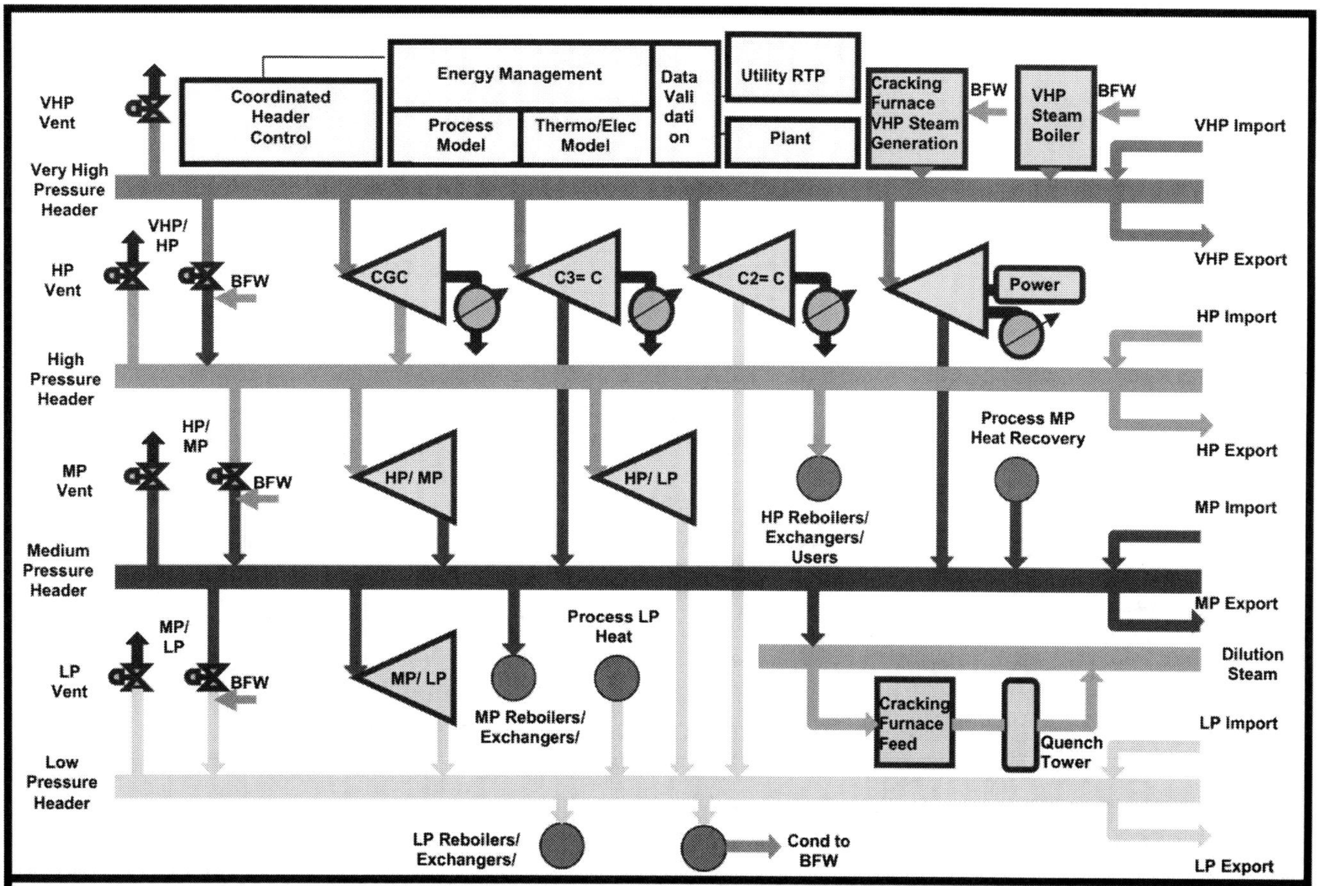

Figure 9: Typical Olefin Plant Steam System

Table 4 Steam System Control Hierarchy				
Priority	**Objective**	**Typical Issue**	**Typical Response**	**Control Function**
1.	Insure Safe Operation	Boiler Trip Compressor Trip Furnace Trip	Load Shedding; Maintain dilution steam flow; Provide steam to flare Maintain flows to main turbines	Pre-programmed Sequence Control
2.	Correct Abnormal Operation	High / low header pressure outside range Furnace decoking	Steam Letdown Control; Venting	Overrides
3.	Respond to Normal Disturbances	Meet varying turbines steam Requirements Meet varying user steam requirements Compensate for varying furnace VHP steam supply	Control Turbines' Extraction Steam Steam Letdown Control Supplemental Boilers Load Controls Steam Import/ Export Vary power generation	Cascaded Multivariable Control
4.	Optimize Utility Supply	Economics	Maximize internal power generation Minimize letdown Manage Import/ Export Steam Steam/ Electric Turbine Selection	Steady State Optimization Model

However, there are relatively common situations where larger disturbances move the controlled variables outside their normal operating region into what might be termed an abnormal state. These disturbances can be either on the demand side or the supply side, for example a decoking of a furnace where there is an increased demand for steam for decoking simultaneous with a loss of the furnace's VHP steam production. The objectives in this control region are to smoothly return the steam system back to the normal control range. One of the characteristics of this state is that the control actions are non-symmetric, i.e. there are different desired control actions if the variable is below its operating range than above. For example, if the pressure in the HP header was above the control range the control action might be to open the HP/ MP letdown as long as the MP header pressure was not above its target range. If the pressure went higher then the action would be to open the HP vent to atmosphere. Conversely, if the pressure was below its range the action would be to open the VHP/ HP letdown, assuming the VHP header was in range. If the pressure continued to drop the action might be to shut the HP/ MP letdown, if open. If not open, then the action might be to reduce feed to the plant. Again this assumes that normal control actions were at their limits. There are further complications because one of the headers can be in an abnormal state while the others are in a normal control situation. Generally this control action is handled via override functionalities with some obvious complexities.

Priority one control is assuring safe operation and activation of this level typically is a response to a furnace or other equipment trip. The header controls have to be prioritized in order of importance in the event that a severe upset occurs. This means that the headers with the highest priority will be kept as close to the targets as possible while other header targets will be relaxed as required in the order of their priority. The priorities might be to keep enough MP steam in the dilution steam headers to prevent decoking of the furnace tubes, provide dispersion steam to the flare, and also enough VHP steam to allow the main turbines to be safely dropped to minimum rates or shut down. This is clearly a challenging control issue requiring careful dynamic coordination and is often handled via pre-programmed sequences along with manual intervention.

Successful control design and implementation then involves integration among these four priority levels with smooth transition from one level to the next as shown in Figure 9. The number of manipulated variables, constraint variables, their interactions, and the frequency and magnitude of disturbances provide significant potential benefits for improved automation. Improved control loop performance via improved measurements, better actuator response, and better tuning are all beneficial.

Modeling the overall steam and utility system and using this model for realtime economic decisions can have a large benefit. Figures 10 and 11 from reference 7 show the results from implementation of such a modeling system on a large petrochemical site. Figure 10 shows the calculation of the optimum operation and 11 the difference in the objective function if the optimum is implemented. It is common to find that the benefits from such an installation can range from one to three percent of the ongoing utility costs. This

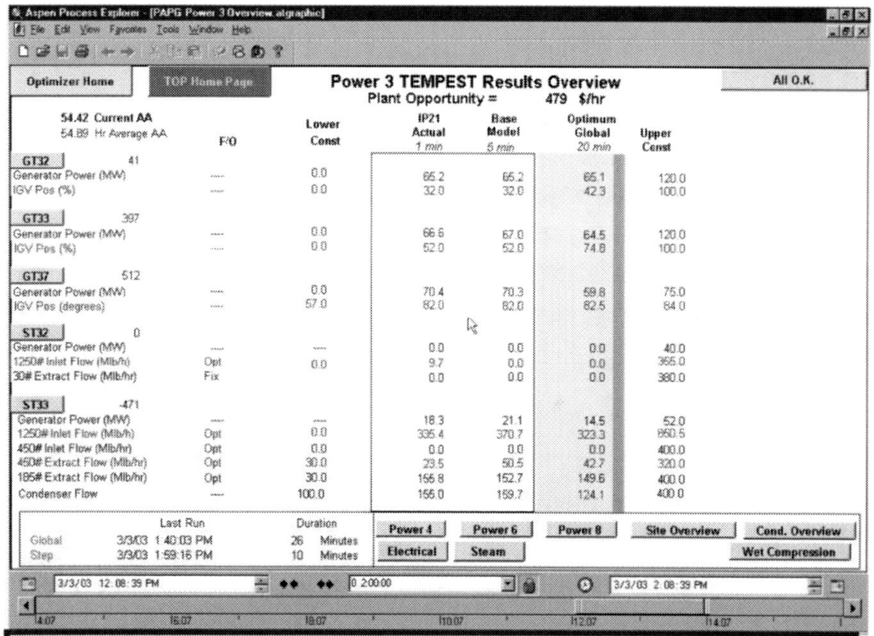

Figure 10: Utility Modeling and Optimization; From Reference (7); *Bearden and McNally*

would yield a potential value of approximately 400,000 $/ Yr to 1,200,000 $/ year for our reference plant.

In addition, the use of the model for off line "What if" analysis permits evaluation of investment options including improved measurements and other operating equipment configurations. Calculating incremental energy costs for each level of steam at current conditions can be helpful for setting priorities.

Ongoing equipment energy use monitoring is provided by the system and is a key to sustaining reductions in energy usage. Having a centralized single point of calculations reduces internal disagreements over actual usage figures.

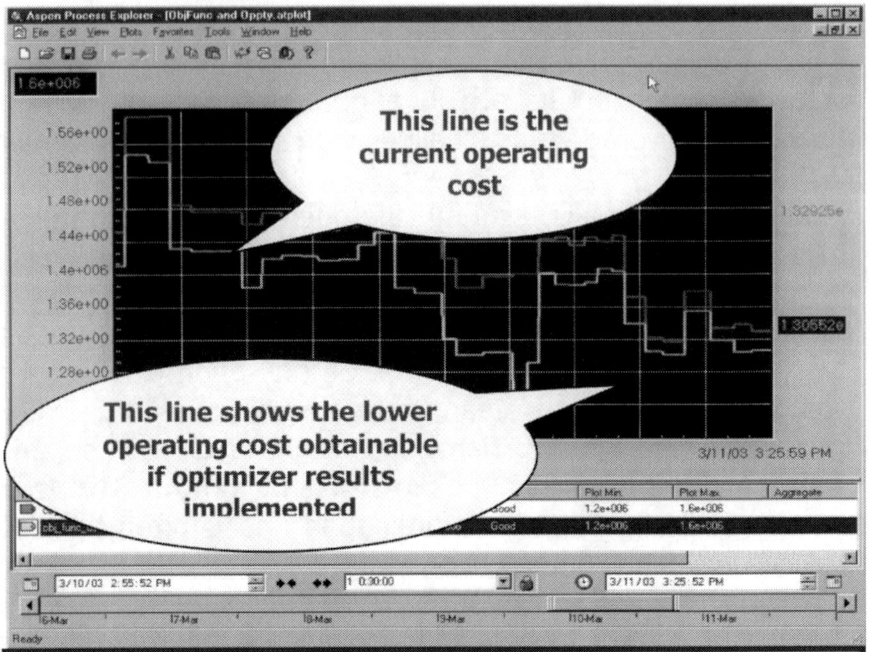

Figure 11: Optimization Results From Reference (7): *Bearden and McNally*

Developing An Automation Energy Saving Program

Most North American olefin plants have plans to reduce energy use. However, a structured program to assess, implement and sustain automation energy savings can still have high value.

Assessment

A typical automation assessment would have the steps shown in Figure 12. The objective of the assessment is to develop a detailed action plan with a timeline for

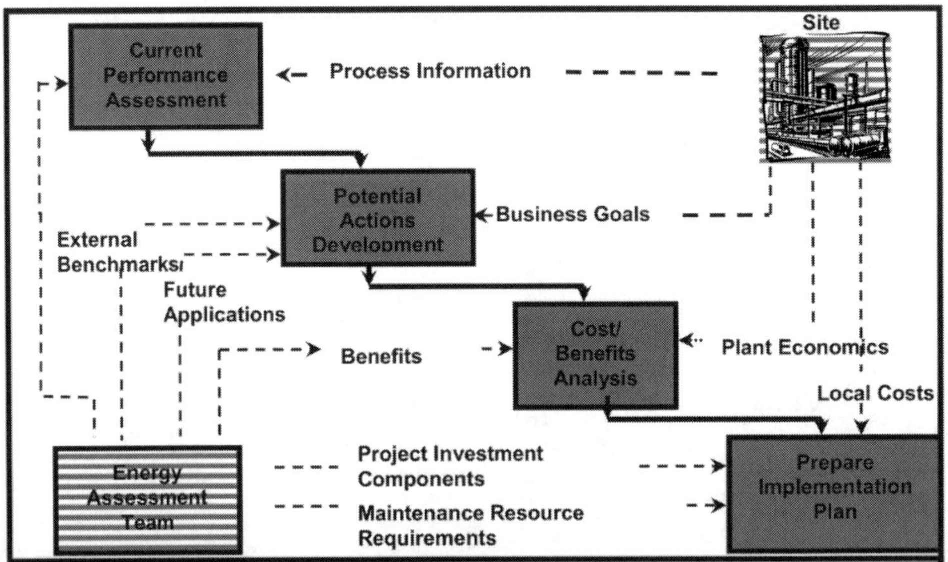

Figure 12 – Assessment Phase

execution. The first step in the program is then to identify where the plant is today with respect to functionality, use, and integration of existing energy monitoring and control automation and information infrastructure.

Part of this assessment is an identification of control loops, including advanced control loops, which have a major impact on energy usage. These loops will be the initial target for an ongoing performance analysis program. To identify these loops it is helpful to categorize them as shown in Table 4. First the control loops are ranked relative to the impact of significantly degraded performance of the loop on incremental energy consumption, generally expressed in monetary terms. This ranking can be done either using engineering knowledge of the energy relationships or by a correlation analysis of plant data or some combination of both approaches. Next the historical frequency of significantly degraded performance is determined, generally from the site data historian and maintenance records. From this analysis, loops are prioritized based on the product of their impact and problem frequency. A possible result of the analysis is also shown in Table 4.

Table 4
Critical Energy Loop Analysis
Priority For Monitoring/ Maintenance

	Historical Frequency Significantly Degraded Performance		
Energy Loss Consequences Significantly Degraded Performance	**≤ Once A Month**	**From Once A Month to Once A Year**	**≥ Once A Year**
≥ $500k per year	**Priority 1** - Fuel Gas Pressure Control - Combustion Air Control - Steam Header Letdown Control	**Priority 2** - Coil Outlet Temperature Control - Fuel Gas Flow Control - Steam Header Pressure - Turbine Extraction Steam Control	**Priority 3** - Dilution Steam Ratio Control - CGC Suction Pressure Control
$50k to $500k per year	**Priority 2** - Steam Drum Level Control	**Priority 3** - Coil Differential Temperature Control - Quench Column Overhead Temperature - Separation Column Product Quality Control	**Priority 4** - HC Feed flow controllers - Separation Column Pressure Control - Reboiler Flow Control

Table 4 Critical Energy Loop Analysis Priority For Monitoring/ Maintenance			
	Historical Frequency Significantly Degraded Performance		
Energy Loss Consequences Significantly Degraded Performance	**≤ Once A Month**	**From Once A Month to Once A Year**	**≥ Once A Year**
≤ $50k per year	**Priority 3** - BFW Flow Control	**Priority 4** - Acetylene Converter Outlet Temperature Control	**Priority 5** - Fuel Gas Pressure Control

For the higher priority loops, further analysis is done to identify the source of the problems, i.e. measurement, actuator, tuning or external factors such as disturbances or interactions with other loops. From this analysis, a set of potential actions to correct the problems is produced. At this stage it is also important to include analysis of possible new improvements to the automation such as increased measurements of energy critical variables including additional analyzers, advanced control implementation, infrastructure improvements such as networking to the motor control center, and resizing of pump rotors and associated flow valves for current operating conditions. It is common to identify savings that can be obtained with just a change in operating procedures, for example relaxing an operating constraint that was set based only on historical experience rather than real process limits. In addition, it is useful to benchmark the plant against industry leaders to help identify areas for potential improvements.

Next is the cost / benefit analysis to provide a screening for potential project evaluation. It is important is to match the uncertainty in both the cost and benefits. For screening purposes, it is not necessary to have a cost estimate +/- 10% when the benefit estimate is +/- 50%. Recommended projects are evaluated on the basis of return on investment. The recommended sequence is then set by:

- Size of investment and schedule duration to achieve benefits
- Required infrastructure changes needed
- Cash flow
- Available resources
- Leveraging existing applications
- Project dependencies

With this information, a funding program request is prepared and submitted for internal approval. At this point it is necessary to refine cost estimates to the +/ - 10% accuracy level. It is generally advisable to choose some relatively quick low investment projects for initial implementation. Even if there are larger investment items with higher ROI, quick success and actual benefits help build internal support for these larger projects. After these initial activities, it is typical to choose some projects for re-assessment in more detail. When funding is authorized, implementation proceeds.

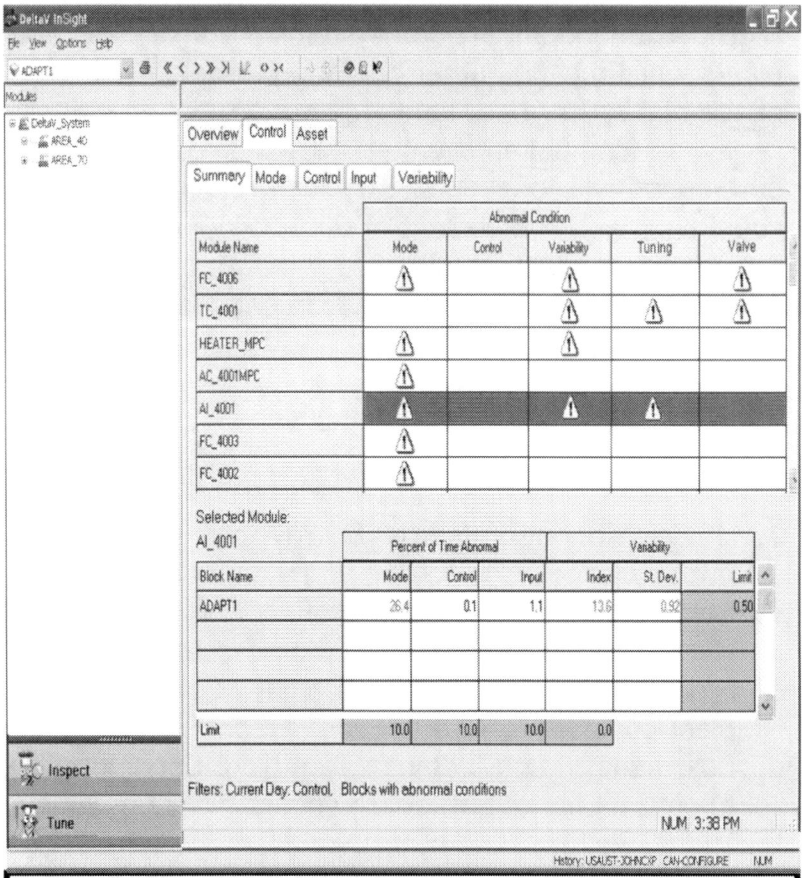

Figure 13 – Loop Monitoring

Implementing and Sustaining the Program

After the program is approved, initial implementation proceeds. However, it is important to build into the program the tools and procedures to maintain the savings. There have, unfortunately, been too many programs that were initially successful but the savings degraded over time with changes in personnel and management attention. For success, there needs to be management "buy-in" and assignment of specific accountability to individuals with regular review. Most companies find that this is aided by routine calculation, tracking and reporting of automation metrics. Figure 13 shows a modern control performance monitoring package that identifies when key loops are operating outside their desired range, provides an alert to operations and maintenances of the degraded loop performance and provides suggested areas for further investigation with regard to the cause of the conditions. Providing visibility of results to operators and supervisors helps to keep the performance at the desired levels.

Conclusions

High performing automation systems are a key to maintaining minimum energy usage in olefin plants while meeting other operating targets. Enhancements to automation can provide some of the highest payback items to consider for energy saving process upgrades. In this paper a review of some improvement areas is presented. The potential benefits from improved automation will, of course, vary with the current state of the plant. However, these benefits could approach five percent of the energy usage with a value of six million dollars per year for the reference plant considered, comparing a plant with a low utilization of automation with one with current best in class. Greenhouse gas savings would be additional. This certainly provides sufficient incentive to consider investments in this area as plants strive to continually reduce their usage.

References

1. US Department of Energy; Energy Information Agency web site; http://www.eia.doe.gov/oiaf/forecasting.html
2. Ren, T.; M. Patel, K. Blok; "Olefins from conventional and heavy feedstocks: Energy use in steam cracking and alternative processes;" *Energy*; Vol. 31; pp. 425 – 451 (2006)
3. Nasi, M.; M. Sourander, M. Tuomala, D. C. White, M. L. Darby; "Experience with ethylene plant computer control"; *Hydrocarbon Processing*; June, 1983; pp. 74- 82
4. Bussman, W. and C. Baukal; "Ambient Condition Effects on Heater Efficiency;" Texas Chemical Council EHS Conference; Galveston; June, 2008; also Bussman, W.; C. Baukal and R. Sexton; "An Automatic Advantage"; *Hydrocarbon Engineering*; September, 2008
5. Olsen, T. and Bill Bialkowski; "Lambda Tuning as a Promising Controller Tuning Method for the Refinery," *Proceedings 2002 AIChE Spring National Meeting*; New Orleans; Paper 42a
6. Filho, A.G.; "Benefiting From the Use of Smart Instruments and Asset Management Software in Naphtha Cracking;" *Proceedings 2002 NPRA Computer Conference*; Austin, Texas; Paper CC-02-156
7. Bearden, T. and J. McNally; "Case Study: Dow Chemical, Freeport, Texas;" *Proceedings Texas Technology Showcase; Session C3: Plant Energy Optimization*; March, 2003; Houston, Texas

Appendix

Reference Plant - Basis For Economic Calculations

The following assumptions were used for economic calculations in the paper:

- Olefin Plant with Naphtha Feed
- 500,000 Tons/ Year Ethylene Production
- Total Energy Usage - 31 mBTU/ T Ethylene (16 mBTU/ T HVC)
- Incremental Fuel - Natural Gas at $6/ mBTU
- No credit taken for value of CO2 reductions
- Cracking Furnace Base Case Thermal Efficiency – 88%
- Compressors; CGC – 60000 HP; Propylene – 40000 HP; Ethylene – 15,000 HP.
- Total Steam Usage in plant – 11 mBTU/ T Ethylene

AUTHOR INDEX

AUTHOR INDEX